Genomics in the Clinic

Genomics in the Clinic

A Practical Guide to Genetic Testing, Evaluation, and Counseling

Edited by

Antonie D. Kline, MD
Director of Clinical Genetics, Harvey Institute for Human Genetics, Greater Baltimore Medical Center, Baltimore, MD, United States

Ethylin Wang Jabs, MD
Chair of the Department of Clinical Genomics, Mayo Clinic, Rochester, MN, United States; Adjunct Professor of Genetics and Genomic Sciences, Icahn School of Medicine at Mount Sinai, New York, NY, United States; Adjunct Professor of Genetic Medicine, Johns Hopkins University School of Medicine, Baltimore, MD, United States

ELSEVIER

ACADEMIC PRESS
An imprint of Elsevier

Academic Press is an imprint of Elsevier
125 London Wall, London EC2Y 5AS, United Kingdom
525 B Street, Suite 1650, San Diego, CA 92101, United States
50 Hampshire Street, 5th Floor, Cambridge, MA 02139, United States
The Boulevard, Langford Lane, Kidlington, Oxford OX5 1GB, United Kingdom

ISBN: 978-0-12-816478-5

For information on all Academic Press publications visit our website at
https://www.elsevier.com/books-and-journals

Publisher: Stacy Masucci
Acquisitions Editor: Peter B. Linsley
Editorial Project Manager: Susan E. Ikeda
Production Project Manager: Swapna Srinivasan
Cover Designer: Matthew Limbert

Typeset by TNQ Technologies

Dedication

Together because of our love for the discipline of medical genetics, we dedicate this book to our patients and their families for the lessons they have taught us. We thank all our colleagues who have enhanced our clinical and research experiences.

Individually, we dedicate this book to:

Laird G. Jackson, M.D., who inspired me to enter the field of genetics and who was my mentor for many years; Tilde S. Kline and Irwin K. Kline, who were my role models as parents and in medicine; Julia, Eli and Conrad Clemens for their support always; and Douglas K. Clemens, who always provides encouragement and promotes resilience.

-Antonie D. Kline

Barbara Migeon, M.D., my mentor who helped me realize my goals, and my family of Shih Yi Wang, Chun Lien Wang, Robert T. Wang, Douglas A. Jabs, and Alexandra W. Jabs, who have always provided encouragement and support.

-Ethylin Wang Jabs

Contents

Contributors

Daniah Albokhari, MBBS, Assistant Professor, Department of Pediatrics, Taibah University College of Medicine, Medina, Saudi Arabia

Houriya Ayoubieh, MD, Assistant Professor, Department of Medical Education, Paul L. Foster School of Medicine, Texas Tech University Health Sciences Center, El Paso, TX, United States

Manisha Balwani, MD, MS, Professor and Chief, Division of Medical Genetics and Genomics, Department of Genetics and Genomic Sciences, Icahn School of Medicine at Mount Sinai, New York, NY, United States

Jessica C. Barry, MS, CGC, Section of Genetic Counseling, Clinical Genetics Center, 22q and You Center, The Children's Hospital of Philadelphia, Department of Pediatrics, Perelman School of Medicine of the University of Pennsylvania, Philadelphia, PA, United States; LVPG Maternal Fetal Medicine, Lehigh Valley Health Network, Allentown, PA, United States

Natalie Blagowidow, MD, Director, Harvey Institute for Human Genetics, Greater Baltimore Medical Center, Baltimore, MD, United States

Joann Bodurtha, MD, MPH, Professor of Genetic Medicine, Pediatrics and Oncology, McKusick-Nathans Department of Genetic Medicine, Johns Hopkins University School of Medicine, Baltimore, MD, United States

Alexander Bottini, MD, Vitreoretinal Fellow, The Retina Institute, St. Louis, MO, United States

Takae M. Brewer, MD, Genomic Medicine Institute, Cleveland Clinic, Cleveland, OH, United States

Scott Brodie, MD, PhD, Professor of Ophthalmology, New York University School of Medicine, New York University Langone Medical Center, New York University Eye Center, New York, NY, United States

Emily E. Brown, MGC, CGC, Genetic Counselor, Division of Cardiology, Johns Hopkins University School of Medicine, Baltimore, MD, United States

Lynn Wein Bush, PhD, MS, MA, Faculty, Scientist, Bioethicist, Division of Genetics and Genomics, Department of Pediatrics, Boston Children's Hospital, Boston, MA, United States; Lecturer on Pediatrics, Harvard Medical School, Boston, MA, United States; Member of Faculty, HMS Center for Bioethics, Harvard Medical School, Boston, MA, United States

Russell J. Butterfield, MD, Associate Professor, Division of Pediatric Neurology, Department of Pediatrics, University of Utah Health, Shriners Hospital for Children, Salt Lake City, UT, United States

Kirk Campbell, MD, Professor of Medicine, Division of Nephrology, Icahn School of Medicine at Mount Sinai, New York, NY, United States

Douglas K. Clemens, DMD, Cross Keys Dental Associates, Baltimore, MD, United States

Virginia L. Corson, MS, CGC, Genetic Counselor, Associate Professor, Gynecology & Obstetrics and Pediatrics (retired), Johns Hopkins University School of Medicine, Baltimore, MD, United States

Cheryl Cytrynbaum, MS, CGC, Genetic Counsellor, Department of Genetic Counselling, Project Investigator, Genetics and Genome Biology, The Hospital for Sick Children, University of Toronto, Toronto, ON, Canada; Lecturer, Department of Molecular Genetics, University of Toronto, Toronto, ON, Canada

Vaidehi Dedania, MD, Assistant Professor of Ophthalmology, New York University School of Medicine, New York University Langone Medical Center, New York University Eye Center, New York, NY, United States

George A. Diaz, MD, PhD, Professor, Department of Genetics and Genomic Sciences, Department of Pediatrics, Icahn School of Medicine at Mount Sinai, New York, NY, United States

Harry C. Dietz, III, MD, Investigator Howard Hughes Medical Institute, William S. Smilow Center for Marfan Syndrome Research, Victor A. McKusick Professor of Medicine and Genetics, Assistant Professor of Neurosurgery, Associate Professor of Medicine, Professor of Pediatrics, McKusick-Nathans Department of Genetic Medicine, Johns Hopkins University School of Medicine, Baltimore, MD, United States

Mary Beth Palko Dinulos, MD, Associate Professor of Pediatrics and Pathology, The Geisel School of Medicine at Dartmouth, Section Chief of Genetics and Child Development, Department of Pediatrics, Dartmouth-Hitchcock Medical Center, Lebanon, NH, United States

Christine M. Eng, MD, Professor, Department of Molecular and Human Genetics, Baylor College of Medicine, Houston, TX, United States

Charis Eng, MD, PhD, Sondra J. and Stephen R. Hardis Endowed Chair of Cancer Genomic Medicine, Chair and Director, Genomic Medicine Institute, Chair and Director, Center for Personalized Genetic Healthcare, American Cancer Society Clinical Research Professor, Cleveland Clinic Lerner Research Institute, and Taussig Cancer Institute, Cleveland Clinic, Cleveland, OH, United States; Department of Genetics and Genome Sciences, Germline High Risk Focus Group, Case Comprehensive Cancer Center, Case Western Reserve University School of Medicine, Cleveland, OH, United States

Audrey L. Fan, MS, CGC, Cancer Genetic Counselor, Cancer Genetics and Genomics, Stanford Healthcare, Palo Alto, CA, United States

Clair A. Francomano, MD, Professor of Medical and Molecular Genetics, Indiana University School of Medicine, Indianapolis, IN, United States

Steven J. Frucht, MD, Professor, Department of Neurology, The Marlene and Paolo Fresco Institute for Parkinson's and Movement Disorders, New York University Grossman School of Medicine, New York, NY, United States

Jaya Ganesh, MD, Associate Professor of Pediatrics, Department of Genetics and Genomic Sciences, The Icahn School of Medicine at Mount Sinai, New York, NY, United States

Bruce D. Gelb, MD, Director and Gogel Family Professor, Mindich Child Health and Development Institute, Professor of Pediatrics and Genetics and Genomic Sciences, Departments of Pediatrics and Genetics and Genomic Sciences, Icahn School of Medicine at Mount Sinai, New York, NY, United States

Lediana Goduni, MD, Retina Fellow, Bascom Palmer Eye Institute, Miami, FL, United States

Shen Gu, PhD, Department of Molecular and Human Genetics, Baylor College of Medicine, Houston, TX, United States; Assistant Professor, School of Biomedical Sciences, Faculty of Medicine, The Chinese University of Hong Kong, Hong Kong S.A.R., China

Isha Gupta, MS, LCGC, HBSc, Genetic Counselor, Igentify, New York, NY, United States

Randi J. Hagerman, MD, Medical Director of the MIND Institute, Distinguished Professor, Endowed Chair in Fragile X Research, UC Davis Health System, Sacramento, CA, United States

Judith G. Hall, MD, Professor Emerita, Medical Genetics and Pediatrics, University of British Columbia, Vancouver, BC, Canada

Julie Hoover-Fong, MD, PhD, Professor of Genetic Medicine and Pediatrics, McKusick-Nathans Department of Genetic Medicine, Director, Greenberg Center for Skeletal Dysplasias, Johns Hopkins University School of Medicine, Baltimore, MD, United States

Louanne Hudgins, MD, Professor of Pediatrics, Division of Medical Genetics, Stanford University, Director, Perinatal Genetics, Lucile Packard Children's Hospital, Stanford, CA, United States

Ayuko Iverson, MD, Assistant Professor, Department of Genetics and Genomic Sciences and Medicine, Icahn School of Medicine at Mount Sinai, New York, NY, United States

Ethylin Wang Jabs, MD, Chair of the Department of Clinical Genomics, Mayo Clinic, Rochester, MN, United States; Adjunct Professor of Genetics and Genomic Sciences, Icahn School of Medicine at Mount Sinai, New York, NY, United States; Adjunct Professor of Genetic Medicine, Johns Hopkins University School of Medicine, Baltimore, MD, United States

Cynthia A. James, PhD, CGC, Assistant Professor, Division of Cardiology, Johns Hopkins University School of Medicine, Baltimore, MD, United States

Shama Jari, MD, Prenatal Geneticist, Harvey Institute for Human Genetics, Greater Baltimore Medical Center, Baltimore, MD, United States

Kim M. Keppler-Noreuil, MD, Professor of Pediatrics, Division Chief of Genetics and Metabolism, Department of Pediatrics, University of Wisconsin School of Medicine and Public Health, Madison, WI, United States

Lynne M. Kerr, MD, PhD, Professor, Division of Pediatric Neurology, Department of Pediatrics, University of Utah Health, Shriners Hospital for Children, Salt Lake City, UT, United States

Amy Kimball, MS, CGC, Genetic Counselor, Harvey Institute for Human Genetics, Greater Baltimore Medical Center, Baltimore, MD, United States

Antonie D. Kline, MD, Director of Clinical Genetics, Harvey Institute for Human Genetics, Greater Baltimore Medical Center, Baltimore, MD, United States

Joel N. Kline, MD, Professor and Vice-Chair, Division of Pulmonary, Critical Care and Occupational Medicine, Department of Medicine, University of Iowa School of Medicine, Iowa City, IA, United States

Amy Kritzer, MD, Attending Physician, Division of Genetics and Genomics, Department of Pediatrics, Boston Children's Hospital, Boston, MA, United States; Assistant Professor, Harvard Medical School, Boston, MA, United States

Michele P. Lambert, MD, MSTR, Associate Professor of Pediatrics, Perelman School of Medicine at University of Pennsylvania, and Medical Director, Special Coagulation Laboratory, Director, Pediatric Platelet Disorder Program, Children's Hospital of Philadelphia, Philadelphia, PA, United States

Cheryl D. Lew, MD, MSEd, MS, HEC-C, Director, Pediatric Rare Lung Disease Program and Director, Home Mechanical Ventilation Program, Division of Pediatric Pulmonology and Sleep Medicine, Children's Hospital Los Angeles, Los Angeles, CA, United States; Clinical Professor of Pediatrics, Keck School of Medicine at University, Southern California, Los Angeles, CA, United States; Clinical Professor of Pediatrics, Emeritus Keck School of Medicine at University Southern California, Los Angeles, CA, United States

Shao-Tzu Li, MS, CGC, Cancer Genetics Service, Division of Medical Oncology, National Cancer Centre Singapore, Singapore

Gretchen MacCarrick, MS, CGC, Genetic Counselor, McKusick-Nathans Department of Genetic Medicine, Johns Hopkins University School of Medicine, Baltimore, MD, United States

Dena R. Matalon, MD, Clinical Assistant Professor, Department of Pediatrics, Division of Medical Genetics, Associate Program Director, Medical Genetics Residency, Stanford University, Lucile Packard Children's Hospital, Stanford, CA, United States

Donna M. McDonald-McGinn, MS, CGC, Chief, Section of Genetic Counseling, Clinical Genetics Center, 22q and You Center, The Children's Hospital of Philadelphia, Department of Pediatrics, Perelman School of Medicine of the University of Pennsylvania, Philadelphia, PA, United States; Department of Human Biology and Medical Genetics, Sapienza University, Rome, Italy

Francis J. McMahon, MD, Chief, Human Genetics Branch, National Institute of Mental Health Intramural Research Program, National Institutes of Health, Bethesda, MD, United States

Kristin Meliambro, MD, Assistant Professor, Icahn School of Medicine at Mount Sinai, Department of Medicine, Icahn School of Medicine at Mount Sinai, New York, NY, United States

Rebekah A. Moore, MS, LGC, Genomic Medicine Institute and Taussig Cancer Institute, Cleveland Clinic, Cleveland, OH, United States; DDR Precision Medicine, AstraZeneca Pharmaceuticals, Gaithersburg, MD, United States

Brittney Murray, MS, CGC, Genetic Counselor, Division of Cardiology, Johns Hopkins University, Baltimore, MD, United States

Tara Newcomb, MS, CGC, Neuromuscular Genetic Counselor, Division of Pediatric Neurology, University of Utah Health, School of Medicine, Papillion, NE, United States

Joanne Ngeow, MBBS, MPH, Head, Cancer Genetics Service, Division of Medical Oncology, National Cancer Centre Singapore, Singapore; Lee Kong Chian School of Medicine, Nanyang Technological University Singapore, Singapore

Jessica Ogawa, MD, Assistant Professor, Department of Genetics and Genome Sciences, Case Western Reserve University School of Medicine, Cleveland, OH, United States; Professor of Genetic Medicine, Pediatrics and Oncology, McKusick-Nathans Department of Genetic Medicine, Johns Hopkins University School of Medicine, Baltimore, MD, United States

Dhruv K. Patel, BSc, Philadelphia College of Osteopathic Medicine, Philadelphia, PA, United States

Toni I. Pollin, MS, PhD, CGC, Associate Professor, Medicine and Epidemiology and Public Health, Track Leader, Human Genetics, University of Maryland School of Medicine, Baltimore, MD, United States

Pankaj Prasun, MD, Assistant Professor of Genetics and Genomics, Assistant Professor of Pediatrics, Department of Genetics and Genomic Sciences, Icahn School of Medicine at Mount Sinai, New York, NY, United States

Maksym Puliaiev, MD, Fellow, Pulmonary and Critical Care Medicine, University of Iowa School of Medicine, University of Iowa Health Care, Iowa City, IA, United States; Pulmonary and Critical Care Medicine, MercyOne Healthcare System, Des Moines, IA, United States

Reed E. Pyeritz, MD, PhD, William Smilow Professor of Medicine and Professor of Genetics, Emeritus, Perelman School of Medicine at the University of Pennsylvania, Philadelphia, PA, United States

Sonja A. Rasmussen, MD, MS, Professor, McKusick-Nathans Department of Genetic Medicine, Johns Hopkins University School of Medicine, Baltimore, MD, United States

Giulietta Maria Riboldi, MD, Instructor, Department of Neurology, The Marlene and Paolo Fresco Institute for Parkinson's and Movement Disorders, New York University Grossman School of Medicine, New York, NY, United States

Scott M. Schecter, MD, Assistant Professor, Division of Pulmonary and Critical Care Medicine, University of Virginia Health, Charlottesville, VA, United States

Angela E. Scheuerle, MD, Professor, Department of Pediatrics, Division of Genetics and Metabolism, Professor, Department of Pathology, Division of Genetic Diagnostics, Director UTSW Rare Disease Center of Excellence, UT Southwestern Medical Center, Dallas, TX, United States

Stuart A. Scott, PhD, Professor, Department of Pathology, Stanford University, Stanford, CA, United States; Clinical Genomics Laboratory, Stanford Medicine, Palo Alto, CA, United States

Suma Shankar, MD, PhD, Professor, Chief, Division of Genomic Medicine, Pediatrics & Ophthalmology, University of California Davis, Sacramento, United States

Anne Slavotinek, MBBS, PhD, Professor of Pediatrics, Medical Genetics, Division Director, Department of Pediatrics, Cincinnati Children's Hospital Medical Center, Cincinnati, OH, United States

Constance L. Smith-Hicks, MD, PhD, Associate Professor of Neurology, Neurology and Developmental Medicine, Kennedy Krieger Institute, Johns Hopkins University School of Medicine, Baltimore, MD, United States

Rosalyn W. Stewart, MD, MS, MBA, Professor, Internal Medicine and Pediatrics, Johns Hopkins School of Medicine, Baltimore, MD, United States

Cristina Trandafir, MD, Assistant Professor, Division of Pediatric Neurology, Department of Pediatrics, University of Utah Health, Salt Lake City, UT, United States; Division of Pediatric Neurology, Department of Pediatrics, Baylor College of Medicine, Texas Children's Hospital, Houston, TX, United States

Vivian Narcisa Triano, MS, LCGC, Genetic Counselor, Kaiser Permanente Northern California, North Valley Genetics, Sacramento, CA, United States

Hilary J. Vernon, MD, PhD, Associate Professor of Pediatrics, McKusick-Nathans Department of Genetic Medicine, Johns Hopkins University School of Medicine, Baltimore, MD, United States

Melissa P. Wasserstein, MD, Chief, Division of Pediatric Genetic Medicine, Professor of Pediatrics and Genetics, The Children's Hospital at Montefiore, The University Hospital for Albert Einstein College of Medicine, Bronx, NY, United States

Bryn D. Webb, MD, Assistant Professor of Genetics and Genomic Sciences, Assistant Professor of Pediatrics, Department of Genetics and Genomic Sciences, Icahn School of Medicine at Mount Sinai, New York, NY, United States; Associate Professor, Department of Pediatrics, University of Wisconsin School of Medicine and Public Health, Madison, WI, United States

Rosanna Weksberg, MD, PhD, Clinical Geneticist, Division of Clinical and Metabolic Genetics, Senior Associate Scientist, Genetics and Genome Biology Program, The Hospital for Sick Children, University of Toronto, Toronto, ON, Canada; Professor of Pediatrics and Molecular Genetics, Institute of Medical Science, University of Toronto, Toronto, ON, Canada

Bo Yuan, PhD, Assistant Professor, Department of Molecular and Human Genetics, Baylor College of Medicine, Houston, TX, United States; Clinical Assistant Professor, Department of Laboratory Medicine and Pathology, University of Washington, Seattle, WA, United States; Laboratory Director, Department of Laboratories, Seattle Children's Hospital, Seattle, WA, United States

Foreword

At the start of my career, I was planning on becoming a geneticist. I was in graduate school earning my master's degree in genetics, and the cystic fibrosis gene had just been discovered. The polymerase chain reaction (PCR) was new and took hours to complete. The human genome project had begun. It seemed both portentous and full of potential. Soon thereafter, I was an eager medicine-pediatrics resident whose favorite book was Smith's *Recognizable Patterns of Human Malformation*. However, my career path changed and I became a primary care physician. Now, I am a practicing general internist and general pediatrician. In my daily life, understanding the effects of genetics on disease is undeniable, from the obvious genetic diseases, such as sickle cell anemia and cystic fibrosis, to the more complex disorders such as cardiovascular disease, asthma, diabetes, and cancer susceptibility. Genetics is everywhere and includes classic Mendelian inheritance as well as other mechanisms of inheritance and effects of phenotype (e.g., mitochondrial, sex-linked expression, mosaicism, incomplete penetrance, epigenetics, etc.). Patients are now solicited by private companies to "learn their ancestry breakdown, genetic health risks" with increasing frequency. There are recommendations for diagnosis of many disorders by genetic testing and determination of potential risks with genetic counseling. The field of genetics has evolved tremendously since my master's degree. My once facile knowledge is now passé.

Genomics in the Clinic is a must read to help understand the myriad genetic usages in today's practice and to help clinicians improve their communication with patients. This book reviews basic terminology, and applicability of genetics in prenatal diagnosis, cancer care, and prevention. Practitioners will find this book useful to understand the effects of genetics in clinical practice. The authors, Drs. Kline and Jabs, have been studying and practicing genetics for over 30 years and provide a practical and easy-to-understand guide through complex genetic patterns of inheritance and their applicability in practice.

Dr. Antonie Debra Kline is the Director of Clinical Genetics at The Harvey Institute for Human Genetics, Greater Baltimore Medical Center. She is a clinically active medical geneticist, board certified in Pediatrics, and Clinical Genetics and Genomics. She is the Chair of the Baltimore-Washington Genetics Group. Dr. Kline has studied the clinical characterization, growth and development, natural history, and aging of Cornelia de Lange syndrome. She has been a strong proponent of patient advocacy and is the Medical Director of the Cornelia de Lange Syndrome Foundation and nationally recognized for her exceptional work with this condition. She also has conducted research on the quality of life of individuals with Ehlers–Danlos syndrome and the natural history of the Alstrom and Au–Kline syndromes.

Dr. Ethylin Wang (Mimi) Jabs* is Mount Sinai Professor of Developmental Genetics in the Departments of Genetics and Genomic Sciences, Cell Developmental and Regenerative Biology, and Pediatrics at The Icahn School of Medicine at Mount Sinai. Dr. Jabs is a clinical geneticist who is board certified in Clinical Genetics, Clinical Cytogenetics, and Molecular Genetics. She is an educator and previously directed the Mount Sinai's ACMGE accredited Medical Genetics Residency and Fellowship Training Programs and a Primary Care Provider Education on Common Disease Genetics program funded by the National Institutes of Health. Her major research interests are in craniofacial disorders. She has authored more than 300 original articles, reviews, and chapters.

Their broad expertise combines with the clinical experience of multiple clinical geneticists and specialists to create an easy-to-read, highly relevant book. They review genetic terminology, the many heritable patterns, and explain techniques and methods of genetic testing and result interpretation. With this book, clinicians will understand everything from genetic screening in pregnancy and taking a family history of cancer, to the interpretation of genetic and genomic results and their management. They review the implications of direct-to-consumer marketing for genetic testing and ethical and psychosocial implications associated with genetic testing results. Their approach of using case-based examples to explain basic genetic concepts and complex genomics adds to the everyday relevance and enjoyment of this text. Readers will gain a clearer understanding of genetics and its influence on nearly all aspects of healthcare. A read of this book is essential for understanding, preventing, treating, and curing disease.

Rosalyn W. Stewart, MD, MS, MBA
Professor of Clinical Medicine
Departments of Internal Medicine and Pediatrics
Johns Hopkins University School of Medicine

---

* Author EWJ was employed by Icahn School of Medicine at Mount Sinai at the time of writing the manuscript, but is currently employed by Mayo Clinic.

Acknowledgments

We would like to acknowledge those who significantly helped with this project. We greatly appreciate Erin Brittain for her editing and formatting of text, figures, tables, and references for the entire book. We also thank the following individuals for their review and suggestions on concepts, chapters, and cases for the book: Jennifer Billiet, Karen Brown, Emalyn Cork, Audrey Fan, Lisa Karger, and Lakshmi Mehta.

Chapter 1

Introduction

Ethylin Wang Jabs[1,*] and Antonie D. Kline[2]

[1]*Department of Genetics and Genomic Sciences, Department of Pediatrics, Department of Cell, Developmental and Regenerative Biology, Icahn School of Medicine at Mount Sinai, New York, NY, United States;* [2]*Harvey Institute for Human Genetics, Greater Baltimore Medical Center, Baltimore, MD, United States*

Abstract

The field of genetics and genomics is expanding rapidly, and clinical skills in genetics are needed in all aspects of medicine. The genetic community agrees that an understanding of the basic tenets and evolving concepts of genetics and genomics will improve healthcare throughout all disciplines. The overall goal of this book is to be used not only as a reference when managing patients with potential genetic conditions but also as instruction on the significant role that genetics and genomics are playing and will continue to play in understanding, preventing, treating, and curing disease in the future.

Genetic concepts are ancient, mentioned in the hieroglyphics of ancient Egypt (2500–200 BCE) and in early bibles (4th century) (Rosner, 1969; Litwins, 1972; Scaiewicz and Levitt, 2015). The migration of the earliest humans before 2300 BCE from Africa to the rest of the world has been elucidated through genetic studies (Bergström et al., 2021). Most significantly, the work of Gregor Mendel (1822–1884) (Schacherer, 2016) points to the importance of genetic concepts in the outcome of not only an individual but also future generations. Nobel prize–winning research, with the discovery of the DNA structure, Sanger sequencing, polymerase chain reaction, and CRISPR, created tools instrumental in the development of genetics and genomics. Moreover, the Human Genome Project jumpstarted the genetic revolution with the completion of the first draft of the human genome sequence in 2001 (Venter et al., 2001; Lander et al., 2001). In the last decade, there has been an explosion of data with the use of diagnostic interventions of digital and imaging health tools and artificial intelligence, as well as innovative molecular and sequencing technologies.

The impact of these major historic milestones on gene discovery and genetic testing ultimately led to the recognition of clinical genetics and genomics as a specialty globally (Rimoin 2011; Zhang et al., 2014; Ferreira et al., 2017; Regier et al., 2017). Norway was the first country in recognizing medical genetics as a specialty in 1971, followed by other European countries, but not until 2011 was clinical genetics officially recognized as a European Union–wide medical specialty. In Brazil, the first medical residency program in medical genetics at the University of Sao Paulo was recognized in 1977, and medical genetics was recognized as a medical specialty by the Federal Council of Medicine in 1983. In South Africa, medical genetics was recognized as a subspecialty in medicine in 2007. The African Genomic Medicine Training Initiative was conceived in 2016 to develop knowledge and skills and improve competency in genomic medicine in nursing (Nembaware et al., 2019). In China, medical genetics was formally recognized as a clinical specialty in 2014 (Sun et al., 2019). There are, however, relatively few training programs for residents in clinical genetics and allied health professionals in genetic counseling in most of Latin America, South America, Asia, and Africa to date.

In the United States, both the American Board of Medical Specialties (ABMS) (http://www.abms.org) and the Council on Medical Education of the American Medical Association independently approved acceptance of Clinical Genetics as a

* Author EWJ was employed by Icahn School of Medicine at Mount Sinai at the time of writing the manuscript, but is currently employed by Mayo Clinic

Genomics in the Clinic. https://doi.org/10.1016/B978-0-12-816478-5.00001-7

new board within the ABMS in 1991. The ABMS now approves the certification of an individual with an MD and/or a PhD in the areas of clinical genetics and genomics, medical biochemical genetics, clinical biochemical genetics, and laboratory genetics and genomics (combining the former areas of clinical cytogenetics and genomics, and clinical molecular genetics and genomics) (http://www.abmgg.org). The American Board of Genetic Counselors was incorporated in 1993 (https://www.abgc.net), establishing their own board for certification and the accreditation of graduate level training programs for a Master's degree as well as a PhD in Genetic Counseling. With the use of genetic technologies in diagnostics, molecular training certification also can be obtained through other boards at the doctorate level, such as molecular genetic pathology from the American Board of Pathology.

Although the discipline of genetics in the clinical setting is relatively young compared to other major clinical disciplines, its principles are applicable in all aspects of medicine. The cornerstone of genetics is the concept of Mendelian inheritance, named after Mendel's work (Schacherer, 2016). He showed by cross-breeding peas of two distinct color traits (yellow and green) that the next generation would have one of these colors in a ratio of one green and three yellow peas or all only in the yellow color. He termed the yellow peas as having "factors" that were dominant and the green peas as having recessive "factors." We now know these "factors" to be "genetic variants." Mendelian conditions, including sickle cell disease and cystic fibrosis, long have been classic examples for the study of clinical genetics. Although much of genetics is represented by rare syndromes, collectively these conditions are abundant. In turn, non-Mendelian inheritance encompasses complex diseases secondary to disruptions in the regulatory networks of multiple genes and environmental factors. Examples of complex diseases include heart disease, cancer, and psychiatric disorders.

Clinical medicine essentially comprises all conditions along a spectrum, with those at one end caused by a variation in a single gene of strong effect (Mendelian) and at the other end caused by the collective contribution of small effects of multiple genes (non-Mendelian). Most pathogenicity in medicine can be, at least in part, due to genetic and genomic variation. All conditions, even those due to a single gene variant, are modified by the effects from all of the other genetic information in our cells as well as environmental exposures, leading to variability in phenotypic outcome. Our collective DNA varies by about one-hundredth of one percent. No two individuals have exactly the identical genomic make-up. Thus, no two individuals have the same phenotype, severity, and outcome of a disease. Even monozygotic twins do not have the same genomes, transcriptomes, or epigenomes (Jonsson et al., 2021; Koch, 2021). The precision in quantitating phenotypes and associating genotypes in patients has ushered in the era of personalized medicine (Topol, 2014).

Genetics and genomics have moved to the forefront of practicing medicine and are being integrated into all clinical disciplines. The implementation of novel technologies of next-generation sequencing now enables sequencing an individual's genomic DNA from a very small amount of sample, a variety of sources, and more rapidly at lower and lower cost. Thus, the combination of detailed phenotyping and "omic" sequencing (i.e., global analysis such as genomics, proteomics, and metabolomics) in large data sets has fostered the growth of genomic medicine, especially in the areas of genetic syndromes, prenatal genetics, pharmacogenomic testing, and cancer genomics.

Newer research discoveries of non-Mendelian concepts of mosaicism, mitochondrial genetics, epigenetics, transcriptomics, and systems biology highlight that the interpretation of genomics as well as other "omic" data are rapidly evolving and not always straightforward. Careful interpretation of DNA sequences is challenging, and ethical use of this information is important. Because of this, those empowered to care for patients using "omic" data must be educated and understand their benefits as well as their limitations. A negative result on a genetic test does not necessarily imply lack of genetic disease; it merely means that a positive result may not have been found yet due to the limitations of the test and/or our current inability to conclusively interpret the sequence results. A variant of unknown clinical significance may be a common result on genetic testing but also may be resolved by reanalysis when new published information becomes available from standardized cohort or population studies (https://www.ncbi.nlm.nih.gov/clinvar). Lastly, a positive result does not necessarily have clinical utility, especially if a genetic diagnosis is not preventable or treatable, as in most complex disorders such as psychiatric conditions. Providers need to weigh whether giving a presymptomatic positive result to an individual truly does "no harm", as in Huntington disease with no current treatment but with potential of clinical trials, or a positive predictive result for some breast cancers, with surgical treatment and increased risks for additional malignancies, but no preventative measures.

Furthermore, there are far fewer clinical geneticists than needed, even when combined with PhD geneticists. Results from United States surveys estimate that there are approximately two clinical geneticists for every million people (Maiese et al., 2019; Jenkins et al., 2021; https://www.gao.gov/assets/710/708545.pdf). There are estimated 1.3−1.5 genetic counselors for every 100,000 individuals in Canada and United States (Costa et al., 2021). In addition, genetic counseling is a profession that is projected to grow 21% from 2019 to 2029, much faster than the average for all occupations of 7% (https://www.bls.gov/ooh/healthcare/genetic-counselors.htm). Nongenetic healthcare providers have begun to order genetic and genomic tests and deliver care based on these results. Patients are ordering their own genetic kits and are asking for help in interpretation from their primary care doctors. Thus, there is a need to educate more healthcare professionals, not just clinical or laboratory genetic trainees, to understand the principles of genetics and genomics, how they can be applied in the clinic, and when to refer to clinical genetics.

The nuances of a genetic evaluation and the various genetic technologies are beyond the scope of what is taught in many healthcare training programs (Campion et al., 2019). Knowledge of criteria for an appropriate genetics referral and having the ability to set expectations, to understand basic testing and molecular nomenclature, to interpret test results, to communicate implications of the results for the individual and other family members, to provide management and treatment, and to grasp the ethical and psychosocial implications of genetics and genomics in medicine are becoming essential clinical skills for the average healthcare provider. The new technologies can be incorporated into primary care. Some centers are testing all comers for genome variation, "building evidence for next-generation sequencing in the context of routine primary care" (Peterson et al., 2019). Other areas of medicine will consider some form of genetic testing, evaluation or treatment likely in all patients in the not-so-distant future, both for preventative and for pathologic reasons. Standardization for such testing is developed in some cases but not available for many clinical scenarios. Nongeneticists may find aspects of this book a useful reference prior to explaining the topic to their patients.

This book translates the salient concepts of clinical genetic care and testing into basic principles for nongenetic physicians, geneticists in training, and other healthcare professionals. It should be relevant within all practices of medicine. Chapters explain genetic concepts for all categories of care, from preconception through aging, of clinical genomics relevant to both Mendelian and complex disorders. Genetic issues in psychiatry are described in detail as an example of the latter. Direct-to-consumer testing and relevant concepts are reviewed, as this is a newer area of inquiry targeted at consumers. Likewise, the topic of ethical issues is introduced, as the potential for these increases when discussing genetic basis for disease and testing. More than 100 cases are presented that are representative of common clinic visits with typical presentations of concern to the nongenetic practitioner. When to consider a referral to a geneticist, and the recommended evaluations, workup, caveats, outcome, and diagnosis of each case, as well as implications for family members are described as guidelines. The lay language and accompanying figures of the final chapter can be used by healthcare providers to effectively communicate complicated concepts and types of tests and results to patients and their families. The hope is that appropriate genetic testing will be accessible to diverse populations and implemented to avoid disparity, stigmata, and loss of confidentiality, while being provided equitably and with inclusivity. With recent technological advances in genetic and genomic testing, a book for this purpose is essential to prepare clinicians for the new age of precision medicine with application of individualized management and treatment, with gene therapy and new molecular therapeutics on the horizon. The editors and authors intend for the readers to utilize Genomics in the Clinic as a reference guide, and also to enjoy the stories behind the concepts and appreciate the complex challenges and the nuances in providing the best clinical genetics and genomics care.

References

Bergström, A., Stringer, C., Hajdinjak, M., Scerri, E.M.L., Skoglund, P., 2021. Origins of modern human ancestry. Nature 590 (7845), 229–237. PMID 33568824.

Campion, M.A., Goldgar, C., Hopkin, R.J., Prows, C.A., Dasgupta, S., 2019. Genomic education for the next generation of health-care providers. Genet. Med. 21 (11), 2422–2430. PMID 31110330.

Costa, T., Gillies, B., Oh, T., Scott, J., 2021. The Canadian genetic counseling workforce: perspectives from employers and recent graduates. J. Genet. Counsel. 30 (2), 406–417. PMID 33131103.

Ferreira, C.R., Regier, D.S., Hadley, D.W., Hart, P.S., Muenke, M., 2017. Medical genetics and genomic medicine in the United States of America. Part 1: history, demographics, legislation, and burden of disease. Mol. Genet. Genomic. Med. 5, 307–316. PMID 28717657.

Jenkins, B.D., Fischer, C.G., Polito, C.A., Maiese, D.R., Keehn, A.S., Lyon, M., et al., 2021. The 2019 US medical genetics workforce: a focus on clinical genetics. Genet. Med. 23 (8), 1458–1464. PMID 33941882.

Jonsson, H., Magnusdottir, E., Eggertsson, H.P., Stefansson, O.A., Arnadottir, G.A., Eiriksson, O., et al., 2021. Differences between germline genomes of monozygotic twins. Nat. Genet. 53 (1), 27–34. PMID 33414551.

Koch, L., 2021. Same genomes, but different. Nat. Rev. Genet. 22 (3), 135. PMID 33442001.

Lander, E.S., Linton, L.M., Birren, B., Nusbaum, C., Zody, M.C., Baldwin, J., et al., 2001. Initial sequencing and analysis of the human genome. Nature 409, 860–921. PMID 11237011.

Litwins, J., 1972. Genetics in the Bible. N. Y. State J. Med. 72 (8), 972. PMID: 4502339.

Maiese, D.R., Keehn, A., Lyon, M., 2019. Current conditions in medical genetics practice. Genet. Med. 21 (8), 1874–1877. PMID 30686822.

Nembaware, V., Genomic Medicine Training Initiative, A., Mulder, N., 2019. The African Genomic Medicine Training Initiative (AGMT): showcasing a community and framework driven genomic medicine training for nurses in Africa. Front. Genet. 10, 1209. PMID 31921282.

Peterson, J.F., Roden, D.M., Orlando, L.A., Ramirez, A.H., Mensah, G.A., Williams, M.S., 2019. Building evidence and measuring clinical outcomes for genomic medicine. Lancet 394, 604–610. PMID 31395443.

Regier, D.S., Ferreira, C.R., Hart, S., Hadley, D.W., Muenke, M., 2017. Medical genetics and genomic medicine in the United States. Part 2: reproductive genetics, newborn screening, genetic counseling, training, and registries. Mol. Genet. Genomic. Med. 5, 621–630. PMID 29178644.

Rimoin, D.L., 2011. The American College of Medical Genetics, the first 20 years. Genet. Med. 13 (3), 179–187. PMID: 21311339.

Rosner, F., 1969. Hemophilia in the Talmud and rabbinic writings. Ann. Intern. Med. 70 (4), 833–837. PMID 4890512.

Scaiewicz, A., Levitt, M., 2015. The language of the protein universe. Curr. Opin. Genet. Dev. 35, 50–56. PMID 26451980.

Schacherer, J., 2016. Beyond the simplicity of Mendelian inheritance. C R Biol. 339 (7–8), 284–288. PMID 27344551.

Sun, L., Liang, B., Zhu, L., Shen, Y., He, L., 2019. The rise of genetics counseling profession in China. Am. J. Med. Genet. C Semin. Med. Genet. 181 (2), 170–176. PMID 30860676.

Topol, E.J., 2014. Individualized medicine from prewomb to tomb. Cell 157 (1), 241–253. PMID 24679539.

US Government Accountability Office, 2020. Genetic services: information on genetic counselor and medical geneticist workforces. (GAO-20–593). Report prepared for Congressional Committees.

Venter, J.C., Adams, M.D., Myers, E.W., Li, P.W., Mural, R.J., Sutton, G.G., et al., 2001. The sequence of the human genome. Science 291, 1304–1351. PMID 11181995.

Zhang, H., Yu, J., Ming, Q., Bao, L., Wu, B.L., Li, P., 2014. On the globalization and standardization of medical genetics and genomics as clinical and laboratory specialties. N A J. Med. Sci. 7 (4), 194–198.

Chapter 2

Basic Principles of Genetics and Genomics

Shen Gu[1,2], Bo Yuan[1,3], Ethylin Wang Jabs[4,*] and Christine M. Eng[1]

[1]Department of Molecular and Human Genetics, Baylor College of Medicine, Houston, TX, United States; [2]School of Biomedical Sciences, Faculty of Medicine, The Chinese University of Hong Kong, Hong Kong S.A.R., China; [3]Human Genome Sequencing Center, Baylor College of Medicine, Houston, TX, United States; [4]Department of Genetics and Genomics Sciences, Department of Pediatrics, Department of Cell, Development and Regenerative Biology, Icahn School of Medicine at Mount Sinai, New York, NY, United States

Abstract

The basis of genetics and genomics are the units of heredity including all chromosomes and genes that instruct cells of our body how to function. Many different kinds of changes in chromosomes, DNA, genes, and proteins lead to diversity and disease and can be inherited in an autosomal, sex-linked, mitochondrial, or multifactorial fashion. This chapter will define the key principles of genetic and genomics as a foundation to understanding how variation in our DNA and its inheritance can cause disease.

The human genome is composed of all the chromosomes or DNA in each cell of an individual. Human DNA is present in the cell nucleus and cytoplasm organelles, the mitochondria. Our understanding of the complex nature of DNA, genetics, and genomics has evolved over time through molecular discoveries driven by sequencing technologies. Healthcare workers must have knowledge of the basic principles of genetics and genomics in order to understand how to evaluate and diagnose genetic disorders, order appropriate tests, and interpret results. More importantly, comprehension of these principles enhances the ability of healthcare teams to communicate effectively genetic and genomic information to patients and their families. This chapter defines normal and common abnormalities in chromosomes, DNA, RNA, genes, coding and noncoding regulatory sequences, repeat sequences, mitochondria, and the epigenome found in individuals and populations. These mechanisms serve to explain various modes of inheritance of genetic disorders including Mendelian, autosomal, X-linked, mitochondrial, and multifactorial inheritance, as well as concepts of mosaicism, genetic heterogeneity, phenotypic variability, penetrance, and modifying factors.

2.1 Chromosome Structure and Function

Chromosomes in the nucleus are basic units of inheritance in the form of threadlike complexes that are primarily composed of deoxyribonucleic acid (DNA) and proteins including histones (Fig. 2.1) (Therman and Susman, 1993). The DNA–protein complex is referred to as chromatin. Chromosomes are present in all nucleated cells of the human body, so they are not present in mature red blood cells and cornified cells of the skin, hair, and nails that do not have a nucleus, as well as in mature hair cells that do not contain nuclear DNA.

[*] Author EWJ was employed by Icahn School of Medicine at Mount Sinai at the time of writing the manuscript, but is currently employed by Mayo Clinic

Genomics in the Clinic. https://doi.org/10.1016/B978-0-12-816478-5.00002-9

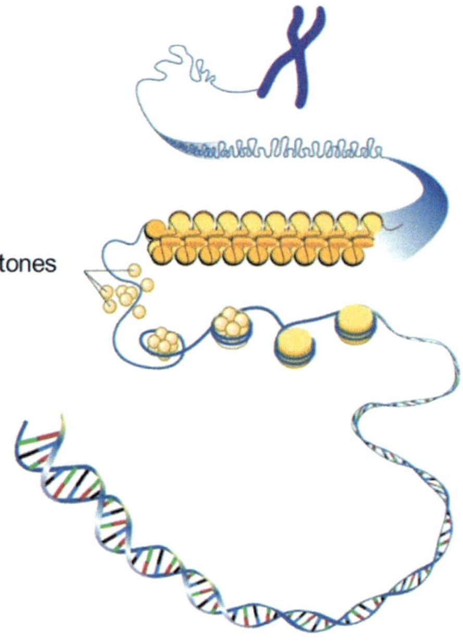

FIGURE 2.1 Chromosome structure. Chromosomes are composed of the DNA double helix of ~2 nm in diameter that is wound around nucleosomes, a protein complex formed from histone octamers, each with ~140 base pairs (bp) of DNA and of ~10 nm in size. The structure is condensed further into chromatin fibers with each loop containing ~100,000 bp and of ~30 nm in diameter that are looped to form the compact chromosome. *Courtesy: National Human Genome Research Institute, https://www.genome.gov/genetics-glossary/histone.*

Normally there are 46 individual chromosomes or 23 pairs of chromosomes (diploid set of chromosomes, 2n with n = 23 chromosomes) in the nucleus of human somatic cells. Twenty-two pairs are autosomal chromosomes and one pair consists of the sex chromosomes, either two X chromosomes in females or one X and one Y chromosome in males. In human gametes, there are 23 individual chromosomes (haploid set of chromosomes, 1n) and with fertilization, the chromosomes from the egg and sperm join to form the first cell of a zygote with 46 chromosomes or 23 pairs of chromosomes (2n). These are homologous pairs of chromosomes or the same chromosomes, one from the mother and one from the father. Chromosomes change in shape and function during the cell cycle as they undergo replication, segregation, and division when a single cell generates two daughter somatic cells during mitosis or four gametes during meiosis.

2.1.1 Mitosis

During mitosis or the one cell division process of somatic cells, 46 chromosomes are in their least condensed state during the synthesis phase (S phase) of interphase when they unwind in order to undergo replication (Therman and Susman, 1993). Each of the 46 chromosomes with one chromatid is duplicated to consist of two chromatids (2n 4c, with c = 23 chromatids). Chromosomes then become more condensed from the prophase and prometaphase to metaphase of the mitotic phase (M phase) when the duplicated homologous chromosomes are aligned along the metaphase plate to prepare for cell division. Chromosomes are held together at the centromere, a specialized structure where spindle fibers attach to mediate their proper movement during cell division. During anaphase and telophase the chromosomes separate so that when the parent cell divides into two daughter cells, each receives one chromatid of the 46 chromosomes (2n 2c). Thus, the process of mitosis creates two somatic cells with 46 chromosomes, each with 46 chromatids, the same number that was present in the original parent cell (Fig. 2.2).

2.1.2 Meiosis

During meiosis or the two cell division process of gametogenesis, oogenesis, or spermatogenesis, 46 chromosomes undergo replication (2n 4c). During prophase I, the chromatids from homologous chromosomes undergo synapsis, when their DNA can undergo crossing over (Therman and Susman, 1993). The chromatids of homologous chromosomes can exchange chromosomal DNA segments, or undergo recombination. These recombined chromatids are not identical to their parental chromatids. With the first cell division, or reduction division, the homologous chromosomes separate so that there

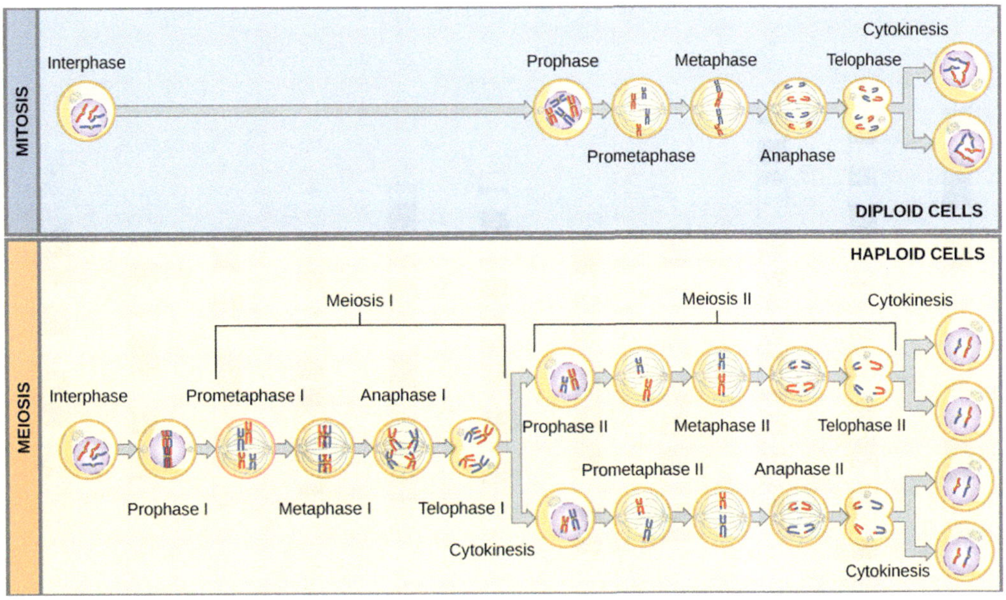

FIGURE 2.2 Mitotic and meiotic cell cycles. At the top in mitosis, diploid cells replicate chromosomes and segregate sister chromatids so that diploid daughter cells are produced. At the bottom in meiosis, two chromosome-segregation phases, meiosis I and meiosis II, follow a single round of DNA. During meiosis I, homologous chromosomes (shown in red and blue) undergo crossing-over and then segregate to opposite poles. During meiosis II, sister chromatids then segregate to opposite poles that result in the formation of nonidentical haploid gametes. *Source: MHCC Biology 112: Biology for Health Professions, Section 80, Comparing Meiosis and Mitosis, OpenStax, Licensed under CC BY 4.0, https://mhccbiology112.pressbooks.com/chapter/comparing-meiosis-and-mitosis/. Modified from the original.*

are now 23 chromosomes each with two chromatids in two new daughter cells (1n 2c). With the second division, the two chromatids separate so that there are 23 chromosomes each with one chromatid in four new daughter cells (1n 1c). Thus, the process of meiosis creates four gametes, sperms, or eggs, each with 23 chromosomes with 23 chromatids that may not be identical in DNA content because of recombination, allowing for variation among offsprings (Fig. 2.2).

2.1.3 Karyotype

Karyotype is a microscopic image of an individual's chromosomes (Gersen and Keagle, 2011; Barch et al., 1997). During prometaphase and metaphase of the M phase, chromosomes are in a condensed state when it is easiest to visualize them under the microscope. The normal male karyotype is designated 46,XY and normal female karyotype is 46,XX, indicating the total number of chromosomes and the sex chromosomes. Chromosomes are different in size and shape (Fig. 2.3). The largest chromosome is numbered 1 and the smaller chromosomes are numbered 21 and 22. The chromatids constrict at the centromere giving a waistlike appearance at the middle of metacentric chromosomes, at an asymmetric position of submetacentric chromosomes, and near one end of acrocentric chromosomes. Generally, the large chromosomes are metacentric in shape such as chromosome 1, while the middle-sized chromosomes are submetacentric and smaller chromosomes 13, 14, 15, 21, and 22 are acrocentric. The centromere location defines two arms of the chromosome, referred to as the "p" or short and "q" or long arms. Thus, the acrocentric chromosomes have the smallest P arms. The two ends of the linear chromosomes are referred to as telomeres. Of note, the X is the largest submetacentric and the Y is a small acrocentric chromosome.

For karyotype analysis, chromosomes are stained to produce a banding pattern, with dark and light bands to define subregions of the chromosome. The number of banded regions of a chromosome depends on how condensed the chromosomes are. High-resolution banding used in clinical cytogenetic laboratories captures the chromosomes at the prophase to prometaphase stage and displays a total of 550−800 bands (Fig. 6.1). Bands are numbered on the P or q arms with the lower numbers near the centromere and the higher numbers near the telomeres (Simons et al., 2013). For example, band 1p36.2 is on the chromosome 1 short arm within band 36 at subband 2. Band 8q21.1 is on the chromosome 8 long arm within band 21 at subband 1. Band 8q21.1 is closer to the centromere and farther from the telomere of the chromosome 8 long arm than band 8q24.2.

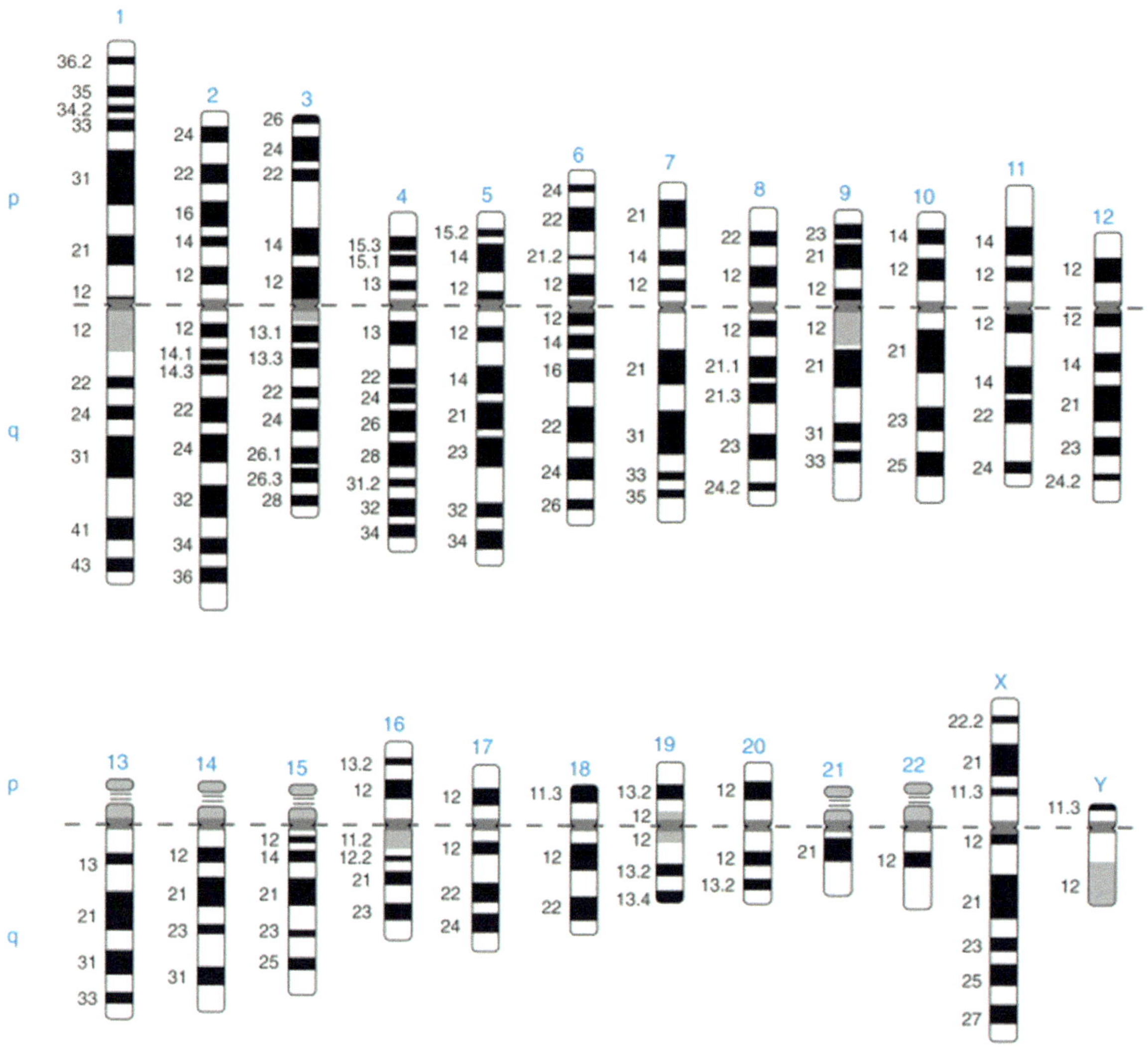

FIGURE 2.3 Karyotype of normal human chromosomes. Ideogram showing G-banding (Giemsa staining) patterns of human metaphase chromosomes of a normal male, 46,XY. Chromosomes are noted by numbers 1–22, X and Y in *blue*. Centromeres are indicated by the *dark gray* regions with a dash separating the short (*P*) arms from the long (q) arms. Bands and subbands are numbered to the left in *black*. *Reprinted from Human Genes and Genomes, Leon E. Rosenberg and Diane Drobnis Rosenberg, Chromosome Abnormalities, Fig. 11.2, Copyright 2012, with permission from Elsevier.*

2.2 Chromosome Variation

The intricate processes of cell division and chromosome replication and segregation into daughter cells are usually precise and tightly regulated, yet errors can occur. Specifically, dysregulated meiosis producing a gamete with a chromosomal abnormality results in an abnormal conceptus, generating a constitutional abnormality that is present from conception and involves cells of the entire body (Gardner et al., 2012).

2.2.1 Chromosomal Aberrations

Variations in chromosomes can be in number or structure. Change in number can occur by the loss of an entire set of chromosomes, the gain of one or more complete sets of chromosomes, or the addition or loss of all or part of a chromosome. Each of these conditions is a variation on the normal diploid number of chromosomal material. The structure of chromosomes may change in shape and rearrangement that is unbalanced with loss or gain of DNA material or balanced with no apparent deletion, duplication, or insertion.

Abnormal Ploidy

A chromosome complement with an abnormal chromosome number that deviates from the normal diploid 46 is defined as heteroploidy. Aneuploidy and polyploidy have been observed in recognized conceptions and/or affected live births.

Aneuploidy is the presence of an abnormal number of chromosomes other than the normal cell number of 46, not including a difference of complete set(s) of chromosome homologs as seen in polyploidy cases. It is the most common type of human chromosome disorder, observed in at least 5% of clinically recognized conceptions. It could be trisomic ($2n + 1 = 47$) or monosomic ($2n - 1 = 45$), with monosomies more deleterious than trisomies. Except for monosomy X (Section 11.7.5), complete monosomies are not viable, while complete trisomies are viable for chromosome 13, 18, 21, X, and Y (Table 2.1, Sections 11.1.5, 11.1.3, 11.13.2).

Full aneuploidy has been attributed to meiotic nondisjunction in the great majority of situations (Fig. 2.4). Nondisjunction is defined as the failure of homologous chromosomes to segregate properly at cell division, and it could happen during either meiosis I or II division. Following the meiotic nondisjunction errors, the conceptus upon fertilization would be either trisomic or monosomic, assuming the other gamete to be normal.

Down syndrome (DS) is the most common and well-known chromosome disorder, characterized by a constellation of clinical findings including distinct facial features, moderate intellectual disability, congenital heart defects, and other recognizable phenotypes. Approximately 95% of DS is due to full trisomy 21, with maternal nondisjunction at the first meiotic division being the predominant cause, and advanced maternal age is a known risk factor (Sections 4.2.2, 11.1.3, 11.13.2).

Polyploidy is the state of more than two sets of paired homologous chromosomes seen in normal human cells (2n chromosomes = 46). Triploidy ($3n = 69$) and tetraploidy ($4n = 92$) have been observed in fetuses (Table 2.1). Triploidy occurs in approximately 1% of conceptions, which results from an extra haploid chromosome set that is of either maternal (digynic) or paternal (diandric) origin. Diandric triploidy is usually the consequence of dispermy, while occasionally it is caused by fertilization with a diploid sperm due to complete nondisjunction during spermatogenesis, whereas digynic triploidy generally results from complete nondisjunction during oogenesis (Gardner et al., 2012). Most triploid conceptions

TABLE 2.1 Outcome of 10,000 pregnancies[a].

Outcome	Pregnancies	Spontaneous abortions	Live births
Total	10,000	1500 (15%)	8500
Normal chromosomes	9200	750 (8%)	8450
Abnormal chromosomes	800	750 (94%)	50
Specific abnormalities			
Triploid or tetraploid	170	170 (100%)	0
45,X	140	139 (99%)	1
Trisomy 16	112	112 (100%)	0
Trisomy 18	20	19 (95%)	1
Trisomy 21	45	35 (78%)	10
Other trisomies	209	208 (99.5%)	1
47, XXX/47, XXY/47, XYY	19	4 (21%)	15
Unbalanced rearrangements	27	23 (85%)	4
Balanced rearrangements	19	3 (16%)	16
Other	39	37 (95%)	2

[a]Estimated frequencies based on observed chromosome abnormalities in spontaneous abortuses and liveborn infants. The actual frequencies may be higher since many spontaneously abort before they are recognized clinically.

Reprinted from Thompson and Thompson Genetics in Medicine, Eighth Edition, Nussbaum, R. L., McInnes, R. R., Willard, H. F., and Hamosh, A. Chapter 5, Principles of Clinical Cytogenetics and Genomic Analysis, Page 73, Table 5.2, 2016, with permission from Elsevier. Modified from original.

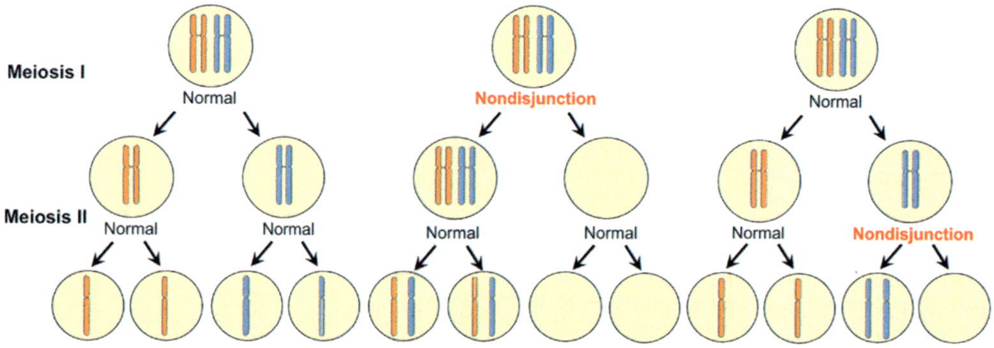

FIGURE 2.4 Consequences of nondisjunction at meiosis I (*middle*) or meiosis II (*right*), compared with normal disjunction (*left*). Reprinted from Thompson and Thompson Genetics in Medicine, Eighth Edition, Nussbaum, R.L., McInnes, R.R., Willard, H.F., and Hamosh, A. Chapter 5, Principles of Clinical Cytogenetics and Genomic Analysis, Page 68, Fig. 5.10, 2016, with permission from Elsevier. *Modified from original.*

lead to spontaneous abortions during the first trimester, and survival beyond the first trimester is rare. Diandric triploids mostly abort between 10 and 20 weeks gestation (mean at 12 weeks) while digynic triploids usually abort around 10 weeks (Gardner et al., 2012). There are no obligate clinical features in triploidy; a wide spectrum of major malformations, growth restriction, and/or distinctive placenta changes has been observed by prenatal ultrasound (Massalska et al., 2017; Toufaily et al., 2016).

Structural Rearrangements

Structural rearrangements are events of chromosome breakage, recombination, or exchange, followed by reunion of chromosomal segments in an abnormal combination. Classical structural rearrangements include translocations, insertions, inversions, isochromosomes, deletions, duplications, marker and ring chromosomes, and complex rearrangements (Figs. 2.5 and 2.6). They are less common events than aneuploidy (Table 2.1). Commonly, structural rearrangements are microscopically visible changes (Section 6.1.1). Smaller events with a size discernible on high-resolution banding, or of submicroscopic size such as microdeletions or microduplications, may be referred to as "genomic rearrangements" or "genomic disorders" (Section 6.1.3) (Gu and Lupski, 2008), although these terminologies are often used interchangeably.

Structural rearrangements can be balanced, with a normal amount of chromosomal material, or unbalanced, with gain or loss of genetic material (Table 2.1). All structural rearrangements arose de novo at one point, either initially developed in the index patient or proband with the abnormality, commonly an unbalanced event with phenotypic consequences that drew medical attention, or inherited from a parent or a more distant ancestor, usually with a balanced rearrangement, but with the potential of generating unbalanced gametes and abnormal progeny.

Translocations

Translocations are generated by the exchange of chromosomal segments between two chromosomes, or sometimes more than two chromosomes. Two major types are reciprocal and nonreciprocal translocations. Reciprocal translocations involve the breakage or recombination between parts of two nonhomologous chromosomes (Fig. 2.5A) (Nussbaum et al., 2016). The total chromosome number remains unchanged, but the components are just rearranged. Individuals carrying balanced reciprocal translocations usually are phenotypically normal other than in their reproductive health, in that they may have higher frequencies of spontaneous abortions, or be infertile (Section 11.1.7). They have a high risk of having a phenotypically abnormal child with an unbalanced translocation with gain or loss of chromosomal material. Generally, the involved chromosomal segments and breakpoints are unique for each family; therefore, the risk and phenotypic consequences for a child with an unbalanced translocation are also unique and require case-by-case evaluation (Gardner et al., 2012).

The most common type of nonreciprocal translocation is Robertsonian translocation, involving breaks at or near the centromeres of two acrocentric chromosomes. The resultant karyotype may have only 45 chromosomes, with the fusion of the long arms of the two acrocentric chromosomes involved and the loss of their short arms, which has hundreds of repeated copies of ribosomal genes, so that loss of some copies does not affect their function (Fig. 2.5B). Robertsonian translocations between chromosomes 13 and 14 and between chromosomes 14 and 21 are the most common combinations (Gardner et al., 2012). Carriers of Robertsonian translocations in a balanced form are phenotypically normal, with one other normal copy of each of the acrocentric chromosomes involved in the translocation, so they have only two copies of

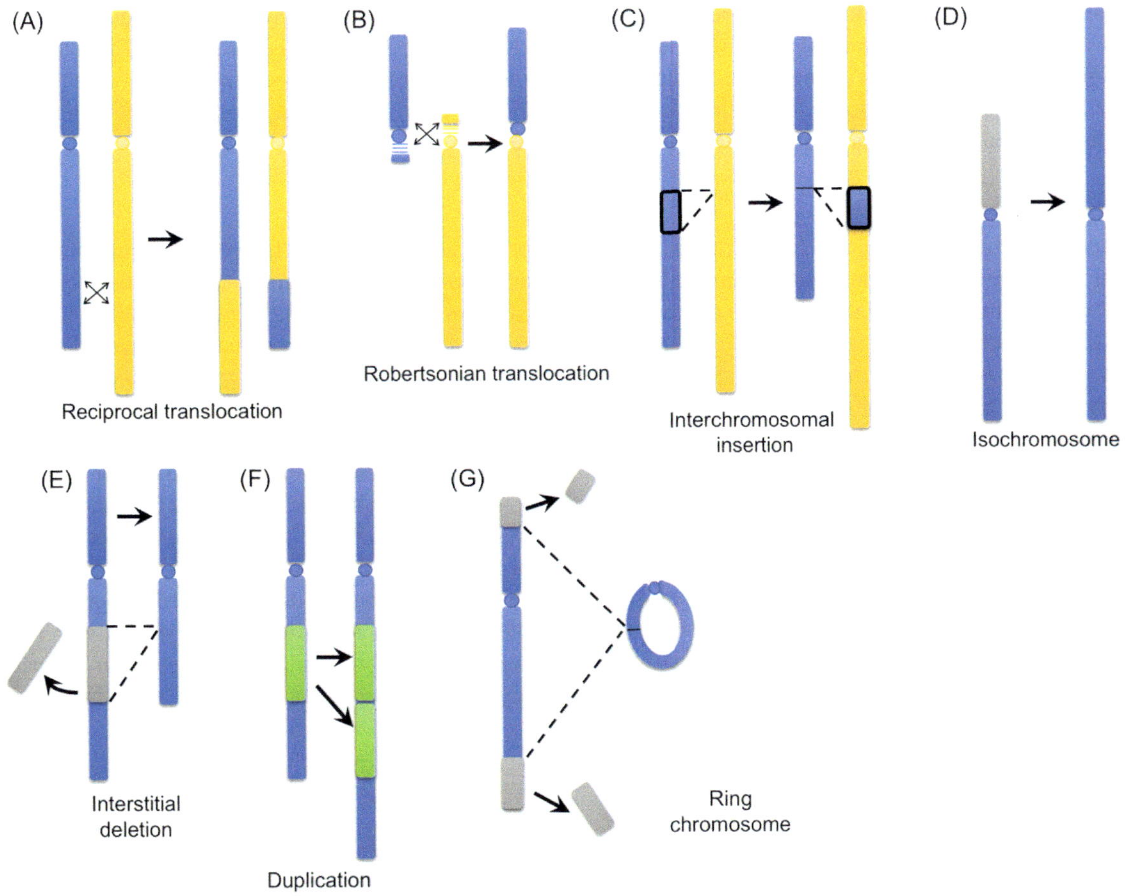

FIGURE 2.5 Structural rearrangements of chromosomes. (A) Reciprocal translocations. (B) Robertsonian translocations. (C) Insertions. (D) Isochromosome generated with a mirror image of one of the arms of a chromosome. (E) Deletion, interstitial within the arm rather than terminal at the end of the arm. (F) Duplication. (G) Ring chromosome, generated with the loss of two acentric fragments from both terminal deletions resulting in marker chromosomes. *Reprinted from Thompson and Thompson Genetics in Medicine, Eighth Edition, Nussbaum, R.L., McInnes, R.R., Willard, H.F., and Hamosh, A. Chapter 5, Principles of Clinical Cytogenetics and Genomic Analysis, Page 68, Fig. 5.11, 2016, with permission from Elsevier. Modified from original.*

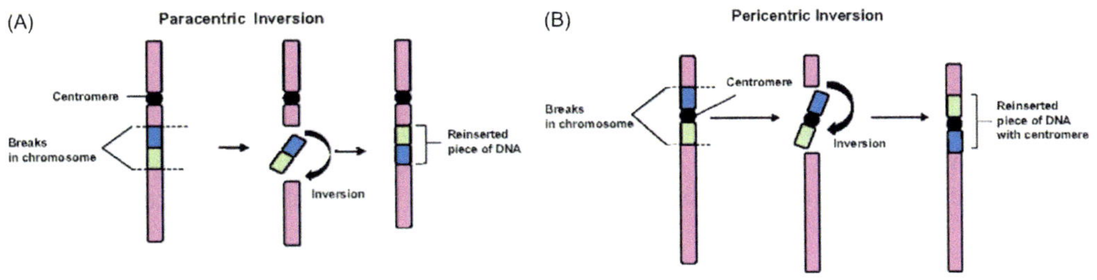

FIGURE 2.6 Structural inversion of chromosomes: (A) Paracentric inversion and (B) Pericentric inversion. *Reprinted from Animal Biotechnology, Madhumita Roy Chowdhury, Sudhisha Dubey, Role of Cytogenetics and Molecular Genetics in Human Health and Medicine, Fig. 24.4E and F, 2014, with permission from Elsevier.*

each chromosome. Like reciprocal translocations, they can be associated with an increased risk for unbalanced offspring by passing on the translocated chromosome. For example, approximately 4% of patients with DS have 46 chromosomes, one of which is a Robertsonian translocation between a chromosome 21 long arm and the long arm of another acrocentric chromosome, causing three copies or doses of chromosome 21q.

Another type of nonreciprocal translocation is an insertion (previously referred as insertional translocation). A chromosomal insertion occurs when a segment of one chromosome is translocated and inserted into a nonhomologous chromosome or a nonadjacent location on the same chromosome or the other homologue, constituting approximately 2% of nonrecurrent copy-number gains (Fig. 2.5C) (Gu et al., 2016; Kang et al., 2010; Neill et al., 2011). These are considered to be complex chromosomal rearrangements or complex genomic rearrangements since they require at least three chromosome breakage events (Section 6.1.3.c). Insertion carriers also have the risk to produce abnormal offspring with gain or loss of the inserted segment that require case-by-case evaluation.

Inversions

An inversion is generated when two breaks occur in one chromosome and the segment between the breaks flips 180° (i.e., the centromeric portion becomes more telomeric and the telomeric portion becomes more centromeric) before rejoining with the two end fragments (Fig. 2.6). There are two types of inversions: paracentric and pericentric. When the two breaks occur within one chromosome arm with no centromere involvement, it is defined as a paracentric inversion (Fig. 2.6A), whereas when the two breaks occur in each arm and the inverted segment contains the centromere, it is called a pericentric inversion (Fig. 2.6B). Similar to carriers with balanced translocations, carriers with balanced inversions usually are phenotypically normal, however are at risk for producing abnormal offspring with unbalanced changes. When the inversion is paracentric, the unbalanced recombinant chromosomes in the gametes of the inversion carrier typically do not lead to viable offspring because of instability of chromosome segregation when spindle fibers are not anchored properly in the absence of a centromere (acentric recombinant) or in the presence of two rather than one centromere (dicentric). In contrast, unbalanced recombinant chromosomes in the gametes of a pericentric inversion carrier contain both gain and loss of segments distal to the inversion. Depending on the size and genetic content of the unbalanced fragments involved, offspring of a pericentric inversion carrier may be viable, and the overall risk for producing a liveborn with unbalanced changes is estimated to be 5%−10% (Gardner et al., 2012). Again, the viability and phenotypic consequences in the offspring depend on the size and genetic content of the chromosomes segments involved, requiring case-by-case evaluation.

2.2.2 Chromosomal Mosaicism

Occasionally during mitosis when the replication and segregation process is abnormal, a chromosome(s) or chromatid(s) becomes abnormal. After conception, this causes an additional cell line(s) with a different chromosomal complement to arise during development and in diseases such as cancer. Such events result in only a proportion of the organism having chromosomally abnormal cells, a situation defined as somatic mosaicism. When nondisjunction happens postzygotically during mitosis, mosaic aneuploidy occurs. This is commonly seen in about half of the cases of monosomy X, also called Turner syndrome (Section 11.7.5), and in 3%−5% of cases of DS (Section 11.13.2). Live births with mosaic trisomies for each autosomal and sex chromosome have been documented. Similarly, balanced and unbalanced structural rearrangements could be present in mosaic form in affected individuals but are relatively rare (Gardner et al., 2012).

2.3 Molecular Structure and Function

The complete set of chromosomes or genetic material present in a cell or organism is referred to as the genome. This molecular structure codes for the inherited basis of how an individual functions as well as ongoing spontaneous changes influenced by the environment. Each individual is unique, yet with 99.9% conservation in the genomes among the human population. It is important to have an understanding of the building blocks of the molecular structure of the human genome.

2.3.1 Deoxyribonucleic Acid

DNA of chromosomes is composed of four different nitrogenous bases including purines, adenine (A) and guanine (G), and pyrimidines, cytosine (C) and thymine (T) (Fig. 2.7) (Strachan and Read, 2018; Krebs et al., 2018). The bases form pairs with A and T forming hydrogen double bonds and G and C forming triple bonds. Each base is modified with a five carbon sugar molecule (deoxyribose) and is referred to as a nucleoside. This is joined with a group of one to three phosphates to form a nucleotide. Nucleotides are strung together like "beads on a string" to form two polynucleotide strands that form the double helical structure of chromatids. The two strands run in opposite directions. At the 5′ end of a strand there is an exposed phosphate group and at the 3′ end is an exposed hydroxyl group on the deoxyribose. The order of the nucleotide sequence on each strand is complementary. A sequence of one strand is read from the 5′ end to the 3′ end,

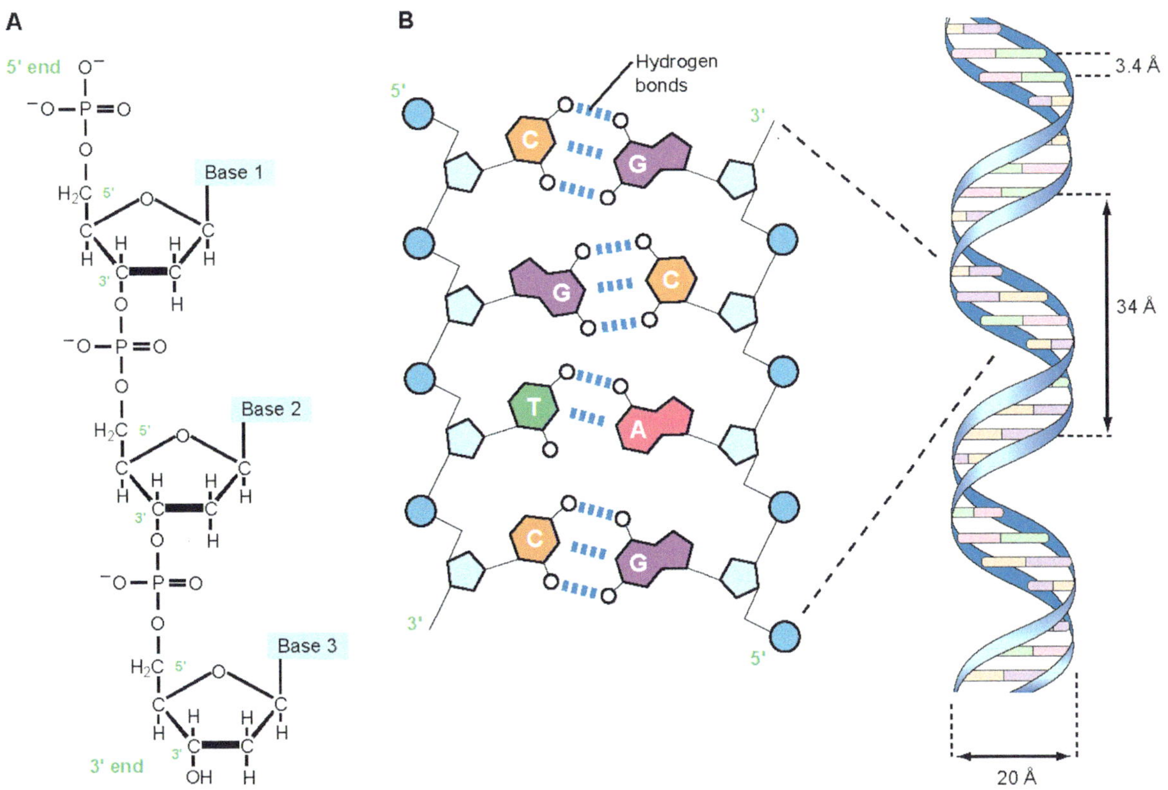

FIGURE 2.7 The structure of DNA. (A) Portion of a DNA polynucleotide chain, showing the 3′–5′ phosphodiester bonds that link the adjacent nucleotides. (B) Schematic of complementary base pairing in DNA. Dotted lines represent hydrogen bonds between bases. A, adenine; T, thymine; C, cytosine; G, guanine. Simplified model of the DNA double helix. The sticks represent base pairs, and the ribbons represent the deoxyribose phosphate backbones of the antiparallel strands. *Reprinted from Thompson and Thompson Genetics in Medicine, Eighth Edition, Nussbaum, R.L., McInnes, R.R., Willard, H.F., and Hamosh, A. Chapter 2, Introduction to the Human Genome, Page 6, Fig. 2.3, 2016, with permission from Elsevier.*

such as 5′-ATTCGTAGT-3′, and is referred to as the forward or sense strand. This strand is paired with the sequence 3′-ACTACGAAT-5′ of the opposite strand or the reverse or antisense strand.

There are approximately 3.1 or 6.2 billion base pairs (bp) of DNA sequence in a haploid or diploid set of chromosomes in a cell, respectively. Chromosome 1 being the largest chromosome has 249 million base pairs (Mb or Mbp), representing 8% of the genome, while chromosome 21 being the smallest consists of 48 Mb, or only 1.5%–2% of the genome. A chromosomal band has approximately 3–5 Mb of DNA.

2.3.2 Ribonucleic Acid

RNA is synthesized from DNA and instructs the synthesis of proteins (Fig. 2.8) (Krebs et al., 2018; Strachan and Read, 2018). Ribonucleic acid (RNA) is a chain of nucleotides composed of four nitrogenous bases like DNA but differs in several ways. Three of the nucleotides are the same in RNA and DNA, adenine, guanine, and cytosine, but the fourth found in RNA is uracil, which replaces thymine found in DNA. In RNA, the nucleotides are denoted by the letters A, C, G, and U. The sugar molecule in RNA is ribose, with a hydroxyl group attached to the 2′ position of the pentose ring that is not present in the deoxyribose found in DNA. RNA is a single strand of nucleotides rather than the double strand found in DNA.

2.3.3 DNA to RNA to Protein

The interpretation of the DNA code to form RNA, which then produces protein, is considered the central dogma of life (Krebs et al., 2018; Strachan and Read, 2018). This tenet requires three key sequential steps of transcription, processing,

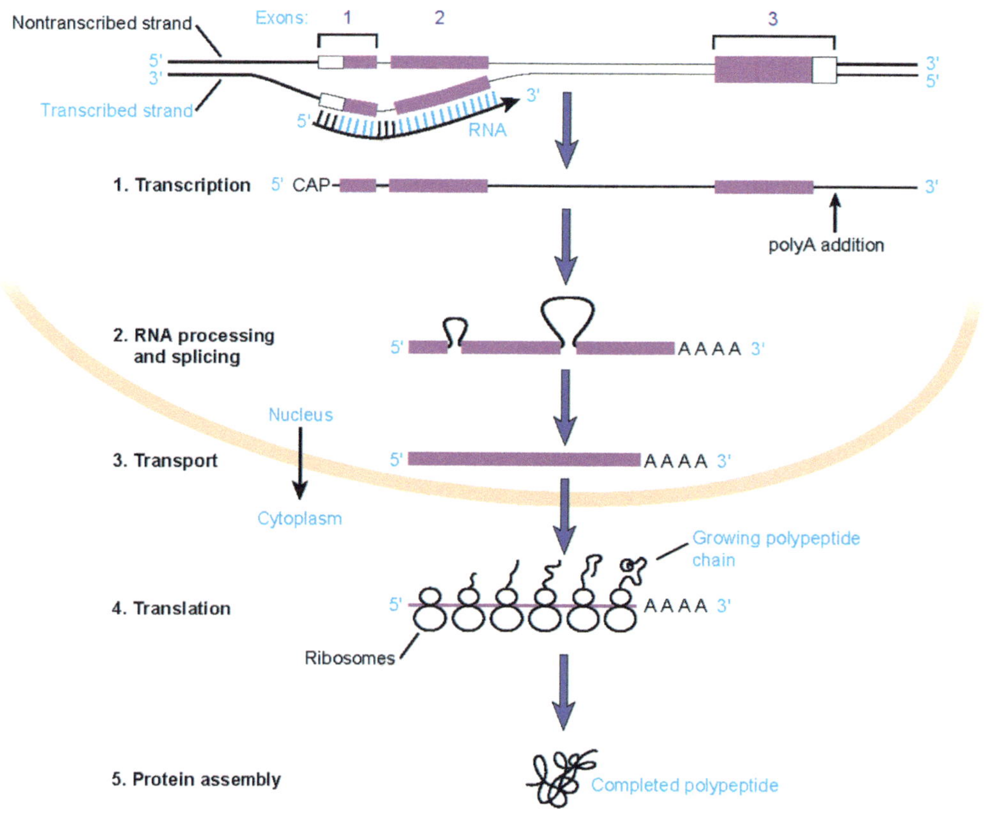

FIGURE 2.8 Transcription from DNA to RNA, RNA processing and splicing, RNA transport from nucleus to the cytoplasm, and translation. A representative gene with three exons and two introns is shown here. The exons are coding sequences (*purple*) and the introns are noncoding sequences (lines between the exons). *Reprinted from Thompson and Thompson Genetics in Medicine, Eighth Edition, Nussbaum, R.L., McInnes, R.R., Willard, H.F., and Hamosh, A. Chapter 3, The Human Genome, Gene Structure and Function, Page 27, Fig. 3.5, 2016, with permission from Elsevier.*

and translation. The goal of the DNA is to generate RNA and proteins to perform structural and regulatory roles in human development, growth, and homeostasis.

Transcription

Transcription is the process by which DNA produces RNA. RNA is synthesized by using DNA as a template and RNA polymerase, a multisubunit enzyme. Transcription of DNA to RNA begins with the binding of the enzyme to a promoter sequence in the DNA, usually found before or "upstream" of a 5′ end of a gene sequence. The helicase activity of the enzyme unwinds the DNA double helix in the 3′ to 5′ direction and synthesizes a complementary RNA strand by an elongation process in the 5′ to 3′ direction (Fig. 2.8). This RNA is referred to as cDNA, a copy of complementary DNA.

The order of a three-nucleotide unit or codon, along the DNA strand in the 5′ to 3′ direction or the RNA strand, codes for an amino acid (e.g., AUU codes for isoleucine, CGU for arginine, and AGU for serine) (Fig. 2.9), and a sequence of multiple codons together is translated to produce a protein sequence. Stretches of DNA or RNA sequence that code for amino acids are referred to as coding regions. The order of other groups of nucleotides gives instructions for regulatory functions and is known as noncoding regions. The RNA sequence has a STOP codon (UAA, UGA, and UAG) that determines the location where the amino acid synthesis is terminated.

Processing

The primary or precursor transcript RNA (pre-mRNA) is modified by enzymes after transcription (Fig. 2.8). A 5' cap (a specialized guanine nucleotide) and a 3′ poly(A) tail (150 or more adenine nucleotides) are added to the pre-mRNA to guide its movement out of the nucleus and stabilize it against degradation. Intronic sequences, which do not code for the final amino acid sequence or protein, are removed by RNA splicing using the splicesome, a large RNA–protein complex. Exons, regions that code for amino acids, are spliced together to form the mature RNA (mRNA) to instruct the final amino acid sequence of a protein.

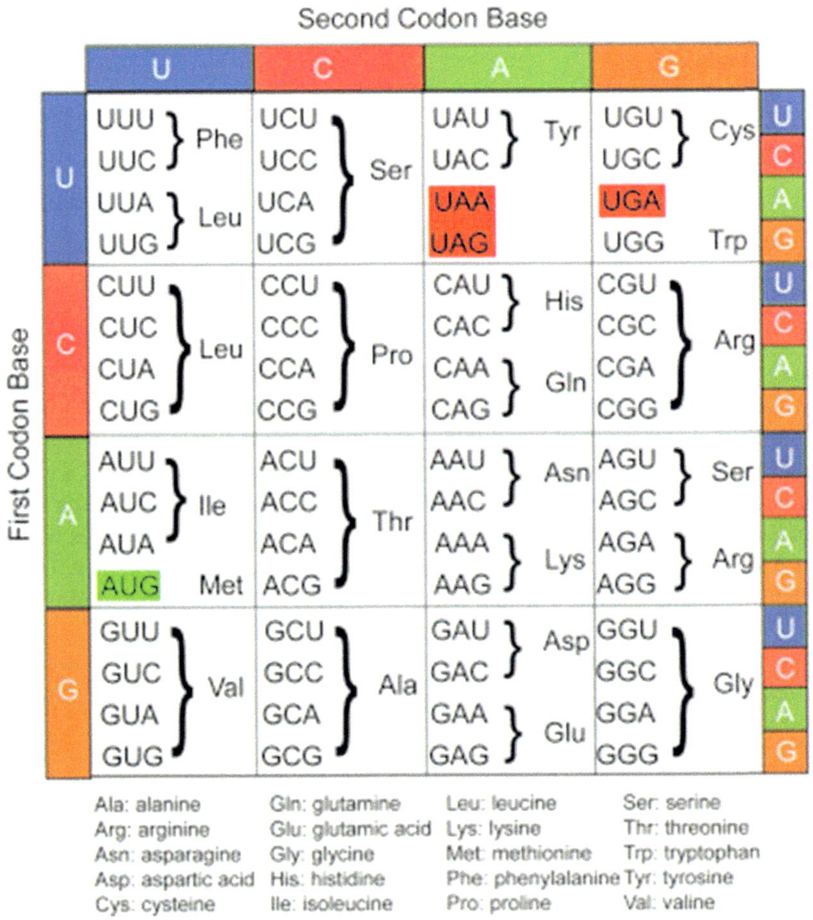

FIGURE 2.9 The genetic code. The sequence of amino acids within a protein is determined by the nucleotide sequence of the mature RNA. To use the table, the first base of the codon is in the leftmost column, the second base of the codon is on the top row, and the intersection of the column and row is the target box in which the codon is located. The third base of the codon is in the rightmost column to identify on which line of the target box the codon is located. Next to the codon sequence in the target box is the amino acid that corresponds to the codon. More than one codon sequence code for a given amino acid. The list of amino acid abbreviations is located below the table. AUG, the start codon, is highlighted in green and codes for methionine. The three stop codons, UAA, UAG, and UGA, are highlighted in red. Stop codons encode a release factor, rather than an amino acid, that causes translation to cease. *Reprinted from Essential Human Virology, Jennifer Louten, Chapter 3, Features of Host Cells: Cellular and Molecular Biology Review, Page 41, Fig. 3.19 and 2016, with permission from Elsevier.*

Translation

Translation is the process by which mRNA produces proteins. mRNA serves as the template that directs the synthesis of an amino acid chain or polypeptide (Fig. 2.8). Transfer RNA (tRNA) molecules link the mRNA sequence code that directs specific amino acids to ribosomes, and ribosomal RNA (rRNA) links the amino acids together to form the coded protein. There are other types of RNA molecules that do not code for proteins but play an active role within cells in catalyzing biological reactions and controlling gene expression (*i.e.*, transcription and translation).

Approximately 360,000 mRNA molecules or transcripts are present in a cell, consisting of approximately 12,000 different transcripts with a typical length of around 2 kilobases (kb). Some highly expressed mRNAs comprise 3% of the mRNA pool whereas others of low expression account for less than 0.1%. RNA transcripts are variable in size. Long RNAs greater than 200 bases include large RNAs such as long noncoding RNA (lncRNA) and mRNA. Small RNAs include ribosomal RNAs (rRNAs), tRNA, and those involved in the RNA interference pathway (RNAi) such as microRNA and small interfering RNA (siRNA), which may inhibit gene expression of targeted mRNA molecules. These RNAs are shorter than 200 bases.

2.3.4 Gene

A gene is a unit of sequence of the DNA that codes for the synthesis of a specific RNA. At the DNA level, a gene consists of many elements (Fig. 2.10) (Krebs et al., 2018; Strachan and Read, 2018). The exons, or protein coding sequences, can be a small part of the gene. The genes also contain noncoding sequences such as introns that are between exons. There are other untranslated regions (UTRs) that can be at the 5′ UTR or 3′ UTR of the open reading frame or coding region that contain regulatory sequences needed to form RNA transcripts or express the gene. A promoter region at the 5′ UTR of the gene is where transcription factors, which are proteins that interact with specific DNA sequences like a TATA consensus

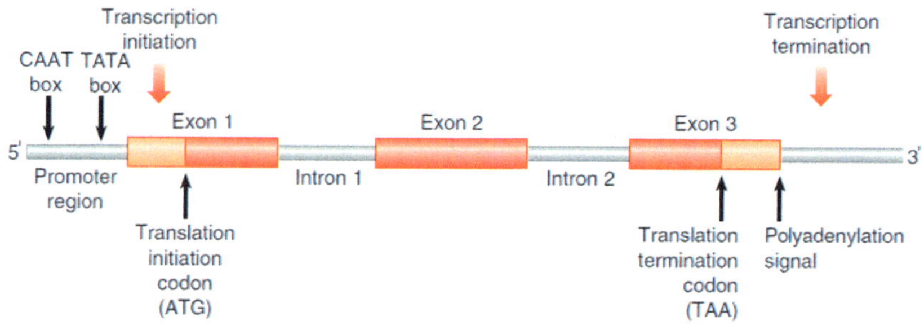

FIGURE 2.10 Gene structure. *Reprinted from Emery's Elements of Medical Genetics, Peter Turnpenny and Sian Ellard, Chapter 2, The Cellular and Molecular Basis of Inheritance, Page 17, Fig. 2.6, 2012, with permission from Elsevier.*

box (rich in thymines and adenines), CAT box (CCAAT), or CG rich CpG islands (regions rich with 5′-C-phosphate-G-3' sequence) to help RNA polymerase bind to an initiation codon ATG to start DNA transcription to RNA. Usually the promoter region is about 25−30 bp 5′ of the start site, but there can be multiple promoter sites of a certain gene to regulate which and/or when transcription factors could interact, thus allowing a diversity of RNA or protein expression in a given cell or cell types. Genes can have other regulatory sequences that are many kilobases, or for some genes a million base pairs away, either upstream or downstream of the open reading frame. They contain enhancer or silencer sequences where activator or repressor proteins could interact, which affect the ability of RNA polymerase to bind to the promoter. These properties allow gene expression to be further modulated within a given cell or cell type at various times during the development and lifetime of an individual.

Pseudogenes are nonfunctional DNA sequences that resemble the sequence of functional genes, which evolved over time by DNA duplication or indirectly by reverse transcription of mRNA transcripts. Their accumulated sequence variants affected the open reading frame, making their sequences nonfunctional and not transcribed (Section 2.4.1).

There are sequences in each intron that direct splicing of exons together or removal of introns. Adjacent to the 3′ end of an exon in the intron is the splice donor site with almost invariant sequence GU, and adjacent to the 5′ end of an exon in the intron is the splice acceptor with another almost invariant sequence AG. In addition, 5′ of the acceptor site in the intron is a polypyrimidine sequence (rich with cytosines and uracils) and further 5′ in the intron is a branchpoint including an adenine nucleotide that forms a lariat shape of the RNA. This recruits proteins to splice at the junction of intron and exons and remove introns to form the mRNA, or open reading frame (Fig. 2.11). The splicing process can create different proteins by alternating the exon composition of the same mRNA. Alternative splicing can occur in many ways. Exons can be extended or skipped, or introns can be retained. Alternative splicing allows one gene to be transcribed into alternative or different mRNAs coding for different proteins.

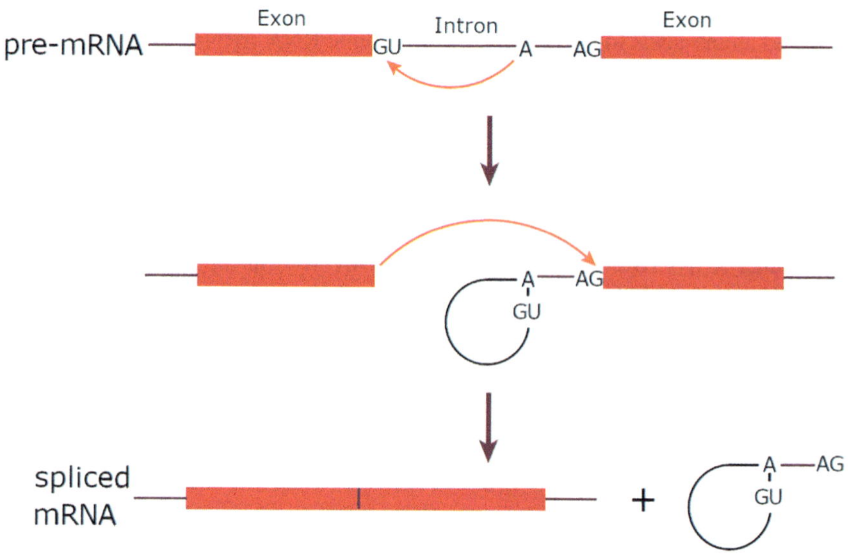

FIGURE 2.11 RNA splicing. The specific branchpoint adenine nucleotide within the intron finds the first nucleotide of the intron at the 5′ splice site with consensus sequence GU, forming the lariat. The released exon then finds the first nucleotide following the last nucleotide of the intron at the 3′ splice site with the consensus sequence AG, thus joining the exons and releasing the intron lariat. *Source: BCSteve—Own work, CC BY-SA 3.0, https://commons.wikimedia.org/w/index.php?curid=30096313.*

It is estimated that there are about 20,000–25,000 genes that code for proteins and 20,000–25,000 noncoding genes producing regulatory RNAs in each cell of the body. In somatic cells, there are homologous pairs of genes, one copy being maternally derived and the other copy being paternally derived. These genes are considered homozygous if the sequence is the same on both copies or heterozygous if the sequences vary from each other. All genes on the single X chromosome in males present as one copy and are considered hemizygous. Genes vary in size from a few hundred to 2 million base pairs. Chromosome 1 being the largest chromosome has 2000–2100 genes, while chromosome 21 being the smallest consists of 200–300 genes. A locus refers to the position or location on a chromosome for a specific gene or DNA sequence.

2.4 Molecular Variation

The human genomes of two unrelated individuals are 99.9% identical, but with at least 5 million base pairs variation in DNA sequences among any unrelated individuals.

2.4.1 DNA Sequence Variation

An allele refers to one of two alternative forms or variants of a gene. When variants are on separate genes or chromosomes, they are "in trans" (Section 11.5.2), as opposed to "in cis" when located on the same gene or chromosome. A genetic variant may be a polymorphism that is common in the general population, occurring in 1% or more individuals, or a "mutation" (now defined as pathogenic variant) that may be rare and disease-causing. Whether a benign polymorphism or a deleterious mutation leading to a pathogenic phenotype, variants have been categorized based on size. Variants larger than 5 million base pairs (Mb) are defined as microscopic copy-number variants (CNVs) or structural variants (SVs) as they are identifiable under the microscope, whereas variants from approximately 1 kb to 3 Mb in size are called submicroscopic SVs (e.g., microdeletions and microduplications) (Feuk et al., 2006) (as discussed in Section 2.2.1). This section will focus on genetic changes shorter than 1 kb that are readily detected by DNA sequencing, including nucleotide substitutions or single-nucleotide variations (SNVs), small deletions, insertions and rearrangements, and dynamic mutations. Percentages observed for each disease-causing variant type are listed in Table 2.2. Notably, with the increasingly use of genome sequencing, the operational range of CNVs or SVs has been widened to include smaller events (e.g., greater than 50 bp in size) (Alkan et al., 2011). Therefore, definition of variants' type based on size should be considered an evolving rule.

TABLE 2.2 Types of mutations in human genetic disease.

Type of mutation	Percentage of disease-causing mutations
SNV	
Missense	50%
Nonsense	10%
RNA processing	10%
Splice site	10%
Long-range regulatory	Rare
Indel	
Small indel (a few base pairs)	25%
Larger gene deletions/inversions/fusions/duplications (may be mediated by repetitive DNA sequence)	5%
Retrotransposition ("jumping" sequences or transposons via RNA intermediates that copy RNA back to DNA by reverse transcription and integrate into different genome locations)	Rare
Dynamic mutation	Rare

Reprinted from Thompson and Thompson Genetics in Medicine, Eighth Edition, Nussbaum, R. L., McInnes, R. R., Willard, H. F., and Hamosh, A. Chapter 4, Human Genetic Diversity: Mutation and Polymorphism, Page 52, Table 4.4, 2016, with permission from Elsevier. Modified from original.

Single-nucleotide Variation (SNV)

The DNA sequence of a gene can be altered in numerous ways. SNVs can refer to nucleotide substitutions that are silent changes or synonymous changes, meaning that such base substitution(s) at the DNA level do not alter the encoded amino acid sequence, as the majority of the amino acids have more than one corresponding codons (Section 2.3.3, Fig. 2.9). SNVs also can be nonsynonymous changes or missense changes that alter the codon to a different amino acid (Fig. 2.12). When an amino acid coding codon is replaced by a termination codon or a stop codon, it is defined as a nonsense change, and the resultant stop codon is called a premature termination codon. Mature RNA carrying a premature termination mutation may be rapidly degraded through the process of nonsense-mediated decay without translation. Alternatively, if there is a nonsense change on one copy of the gene, a truncated (shortened) or an abnormal protein may be generated, which sometimes causes more severe consequences with a dominant negative effect affecting the normal function of the protein formed by the second unaffected copy of the gene. Regarding RNA transcription, processing, and translation, SNVs located at or near the donor or acceptor site could alter normal RNA splicing, create alternative or "cryptic" splicing sites (located in intronic or exonic regions that do not typically code for splice sites), or change mRNA stability or alter translation efficiency when located in 5′ and 3′ UTR. As a result, abnormal RNA splicing or disturbed mRNA expression levels may be created due to these SNVs (Sections 11.3.2, 11.3.5, 11.5.5).

Deletions, Insertions, and Rearrangements

Variants with one or more base pairs being deleted (deletions) or being added (insertions) are small deletions or insertions, which are referred to as indels (Fig. 2.12, Table 2.2). When an indel occurs in the coding sequence and the number of bases altered is not a multiple of three (integral number of codons), the triplet reading frame will be disrupted starting from the

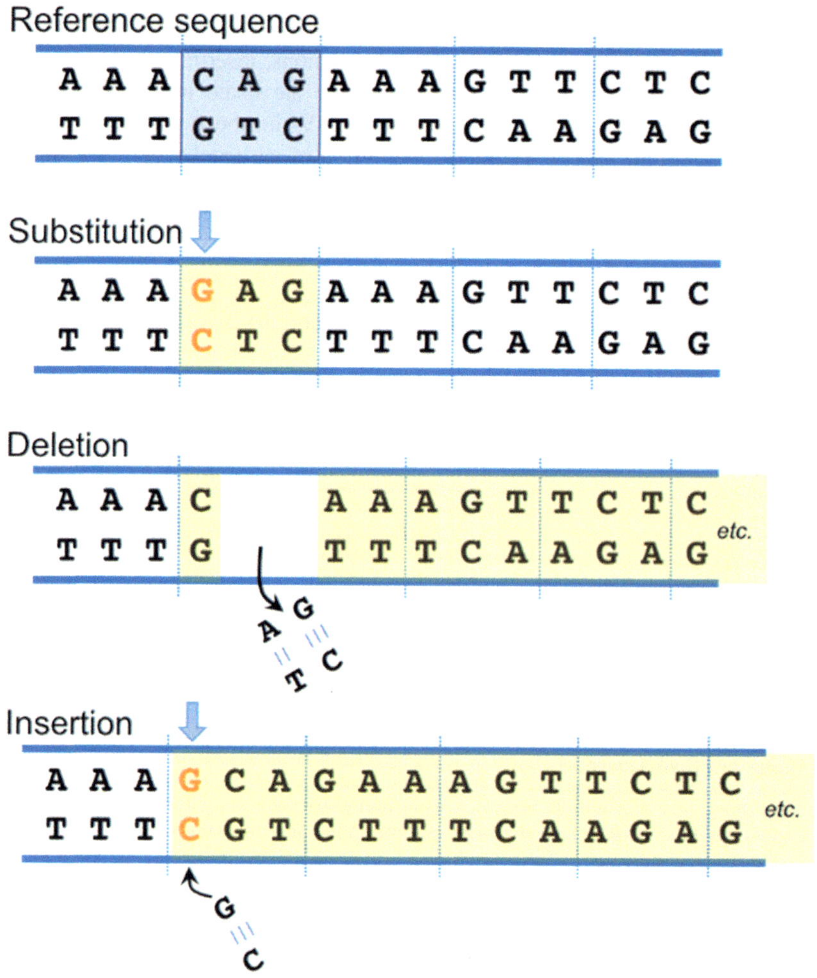

FIGURE 2.12 Examples of variants in a portion of the *UBAP1* gene with five codons shown. The base pairs of the second codon in the reference sequence shaded in (*blue*) are mutated by a base substitution, deletion, or insertion. The base substitution or missense change of a C into G at this position leads to a codon change shaded in (*orange*) and a predicted nonsynonymous change from a glutamine to a glutamic acid in the encoded protein (see genetic code in Fig. 2.9); all other codons remain unchanged. Both the two base pair deletion and the single base pair insertion lead to a frameshift mutation in which the translational reading frame is altered for all subsequent codons shaded in (*orange*), until a termination codon is reached. The two base pair deletion has been reported in affected patients, while the missense change and the one base pair insertion are hypothetical variants. *Modified from Thompson and Thompson Genetics in Medicine, Eighth Edition, Nussbaum, R.L., McInnes, R.R., Willard, H.F., and Hamosh, A. Chapter 4, Human Genetic Diversity: Mutation and Polymorphism, Page 50, Fig. 4.4, 2016, with permission from Elsevier. Modified from original.*

locus of the indel and result in a frameshift mutation. A sequence is considered "in-frame" when the order of nucleotides and codons produce the desired protein product. If there is a shift of the nucleotide sequence by an indel, resulting in different codons, the altered sequence is "out-of frame." Frameshift mutations typically generate functionally altered protein products or create premature termination codons that lead to mRNA degradation with reduced or no protein product. On the other hand, an in-frame indel, in which the number of base pair changes is equal to the integral of number of three for codons, results in gain or loss of codon(s) leading to lengthening (elongation) or shortening of the protein product.

Dynamic Mutations

Dynamic mutations are caused by the expansion of existing polymorphic DNA repeat sequences beyond a copy number threshold. Disorders caused by dynamic mutations are referred to as repeat expansion diseases that primarily affect the nervous system (Paulson, 2018). The involved DNA repeat sequences typically are simple repeats commonly with three to six nucleotides in length, such as (CCG)n, (CAG)n, (CCTG)n, or (GGGGCC)n. These simple repeats may be located in a coding exon, in an intronic region, or within the 5′ and 3′ UTR region that leads to the generation of abnormal protein products or interferes with normal gene expression. Repeat expansion diseases generally show a correlation between repeat length and disease severity, with patients having longer repeats presenting with a more severe phenotype and/or an earlier disease onset age. Such dynamic mutations are also unstable with a changing size when transmitted to the next generation, a phenomenon known as anticipation.

Triplet Repeat Expansions

Triplet repeat (a.k.a. trinucleotide repeat) disorders are a group of genetic diseases caused by the expansion of adjacent triplet repeats at the corresponding gene locus. Triplet repeats are a subset of microsatellites that are frequently seen across the human genome. They are typically repeated 5 to 50 times. These loci are featured by tandemly arrayed units of three nucleotides, which may be unstable and undergo dynamic expansion of the copy number of the tandem units due to DNA replication slippage or DNA repair–induced recombination. Such expansion may lead to genetic defects by various mechanisms including loss-of-function, gain-of-function (new or enhanced expression or function of a gene), or RNA toxicity. Some examples of this include fragile X syndrome (Sections 11.12.3, 11.13.11), Friedreich ataxia, spinocerebellar ataxia (Section 11.13.10), and Huntington chorea (Section 11.13.12).

2.4.2 Genetic Mosaicism

Mosaicism refers to having more than one genetic complement, either germline or somatic in origin (Biesecker and Spinner, 2013). DNA variants such as CNVs, SNPs, deletions, insertions, and triplet repeats can all present in a mosaic form. Germline mosaicism is often present in an unaffected parent, but then can be passed on so that an offspring has the variant in all cells. With increasing paternal age, men have germline mosaicism for mutations, especially de novo single-nucleotide variants that accumulate in their sperm (Section 11.3.2). If mosaicism occurs in an embryo after conception and after at least the second cell division, the offspring has somatic mosaicism with a variant in only a proportion of cells. In these cases, the phenotypic presentation is usually milder than when the variant is present in all cells (Section 11.10.2). Somatic mosaicism is often associated with autosomal dominant conditions of variable presentation within an individual such as McCune–Albright and Proteus syndromes. McCune–Albright syndrome presents with polyostotic fibrous dysplasia, café au lait spots, and precocious puberty. It is caused by pathogenic variants in a gene called *GNAS1*. The variants occur after fertilization of the embryo as a somatic mutation and is therefore not inherited, nor will affected individuals pass the mutation onto their children. Proteus, which is characterized by disproportionate, asymmetric overgrowth, is caused by somatic mutation in a growth regulatory gene called *AKT1*. Affected individuals may experience a wide variety of complications that may include progressive skeletal malformations, benign and malignant tumors, vascular malformations, bullous pulmonary disease, and certain skin lesions. Cancers represent somatic mosaicism from clonal expansion (Chapter 8).

2.5 Mitochondrial Structure and Function

Mitochondria are cytoplasmic organelles in the cell for energy production (Scheffler, 2008; Wallace, 2018). They metabolize or break down carbohydrates and fatty acids in order to generate energy. Although nuclear genes encode hundreds of proteins involved in mitochondrial biogenesis and oxidative phosphorylation, the mitochondria have their own DNA (mtDNA) or "chromosome", which are circular in shape and require the nuclear DNA to function.

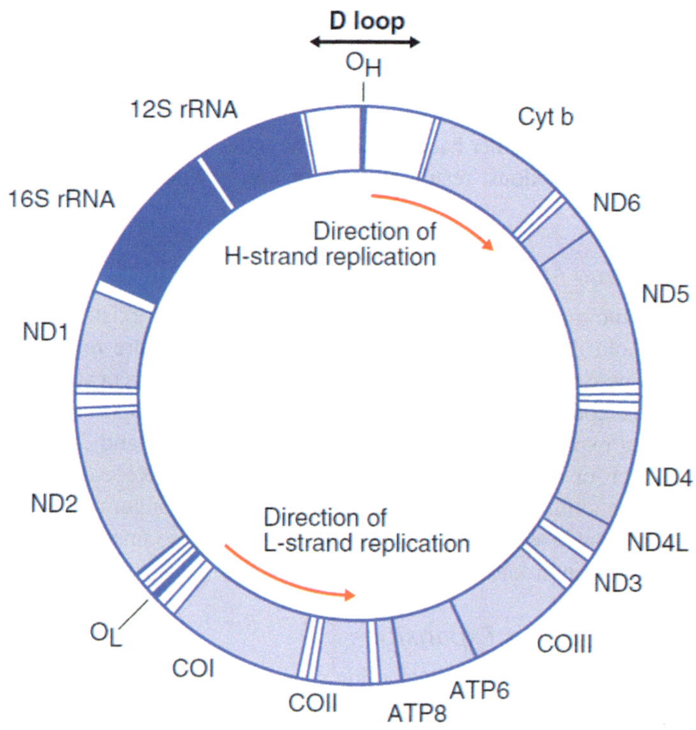

FIGURE 2.13 Human mitochondrial genome. *Reprinted from Emery's Elements of Medical Genetics, Peter Turnpenny and Sian Ellard, Chapter 2, The Cellular and Molecular Basis of Inheritance, Page 18, Fig. 2.7, 2012, with permission from Elsevier.*

The mitochondrial DNA is distinctive in having two strands: a heavy strand and a light strand. The heavy strand (H strand) is rich in guanine and encodes 12 subunits of the oxidative phosphorylation system, 2 ribosomal RNAs (12S and 16S), and 14 tRNAs. The light strand (L strand) encodes one subunit and eight tRNAs. Altogether, mtDNA encodes for 37 genes, 13 proteins subunits, 2 rRNAs, and 22 tRNAs, all of which are involved in the oxidative phosphorylation process. mtDNA sequence consists of 16,569 base pairs (Fig. 2.13).

Mitochondrial DNA is passed essentially from mother to offspring through the egg cell. An egg cell contains on average 200,000 mtDNA molecules, whereas a healthy sperm cell has been reported to contain less than 75 mtDNA molecules. The typical cell has 1000–2000 mitochondria. Among various tissues, the highest level of mtDNA expression is observed in heart, followed by brain and steroidogenic tissues. Mitochondrial DNA is replicated by the DNA polymerase gamma complex. All the polypeptides of this complex are encoded by the nuclear DNA.

During embryogenesis, replication of mtDNA is strictly downregulated from the fertilized oocyte through the preimplantation embryo. The resulting reduction in per-cell copy number of mtDNA plays a role in the mitochondrial "bottleneck" effect (Fig. 2.14). The mtDNA has a 100-fold higher mutation rate than the nuclear DNA, resulting in mtDNA cell-to-cell variability, a concept referred to as heteroplasmy in contrast to having identical mtDNA or homoplasmy. The bottleneck effect can cause different variants and proportions of variant mitochondrial DNA to be passed from one generation to the other. Thus, the mitochondrial DNA in an embryo can be very different from that of its mother.

2.6 Epigenome

The epigenome includes modifications of DNA and histone proteins of the genome (Byrne and Drake, 2019; Monk et al., 2019). Changes of the epigenome can result in changes to the DNA–protein complex or chromatin structure and to the function of gene(s). The most common epigenetic changes are DNA methylation and histone modifications, which can remain as cells divide, and in some cases are inherited. DNA methylation involves attaching methyl groups to segments of DNA. When methyl groups are added to a gene, particularly in the promoter region, the gene is usually turned off or silenced, and less protein is produced from that gene. In other instances, methylation and other modifications can enhance expression of genes. Environmental influences, such as a person's diet and exposure to pollutants, can also impact the epigenome.

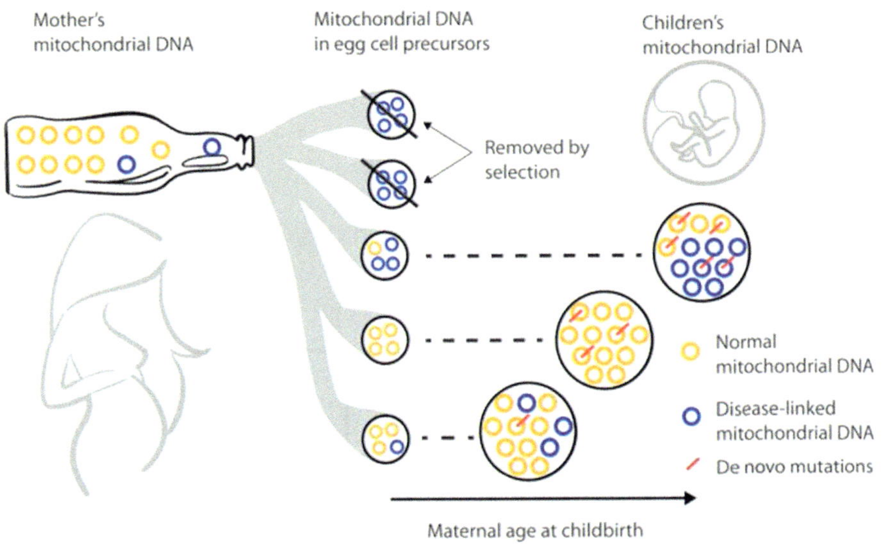

FIGURE 2.14 Mitochondrial DNA inheritance from mother to child. The bottleneck phenomenon causes different copies to be passed to different children, and copies with disease-causing mutations (*blue*) could be passed on to only some children. A process called natural selection helps purge the precursors of egg cells with too many disease-causing copies. Children born to older mothers have more copies of mitochondrial DNA with mutations, which may have important implications for genetic counseling of women planning pregnancies. *Credit: Arsian Zaidi and Kateryna Makova, Pennsylvania State University.*

Chromosome X is inactivated through an epigenetic mechanism (Migeon, 2017). As females have two X chromosomes, X-inactivation prevents them from having twice as many X chromosome gene products as males, who only possess a single copy of the X chromosome. The choice of which X chromosome will be inactivated is random in each cell, but once an X chromosome is inactivated during the late blastocyst stage after implantation, it will remain inactive throughout the lifetime of the cell and its daughter cells and descendants. Thus females are mosaic, with some of their somatic cells having one X chromosome inactive and other cells with the homologous X chromosome inactive. The inactive X chromosome in primordial germ cells becomes reactivated before meiosis. The *XIST* gene encodes a large noncoding RNA (lncRNA) that is responsible for mediating the specific silencing of the X chromosome by coating it. *XIST* is expressed from the inactive X chromosome. However, at least 15%–25% of genes on the inactive X chromosome escape inactivation and are expressed. The expression levels of these escaped genes contribute to sex-dependent, phenotypic variability in health and disease (Posynick and Brown, 2019).

2.7 Modes of Inheritance

Mendelian inheritance is defined simply as inheritance of a condition caused by variant(s) in a single locus or gene. Autosomal and X-linked dominant and recessive inheritance are forms of Mendelian inheritance. There are other non-Mendelian modes of inheritance and mechanisms that modify the expression of a gene, or the presentation of a condition, that add to the complexity of how and if a disease is inherited or occurs in an individual (Nussbaum et al., 2016). Not all genetic variants or mutations are inherited. If a mutation is present only in one member of a family, so that only that individual has a condition and the parents do not have either the mutation or the condition, the mutation is considered de novo. On the other hand, due to modifying genetic, epigenetic, or environmental factors, it is possible that a variant or mutation is inherited from a parent(s), who does not express the condition.

The inheritance of phenotypic feature(s) or condition(s) in a family is illustrated using a pedigree, a tool to present family history in the form of a diagram. Construction of a pedigree uses a standardized set of symbols: squares represent males and circles represent females. If the sex of the person is unknown, a diamond is used. Someone with the phenotype is represented by a filled-in or dark symbol. Heterozygotes, when identifiable, are indicated by a dot inside a symbol or a half-filled symbol for recessive conditions. Genotype or additional specific information can be added under or near a given symbol. Fig. 2.15 represents an index of symbols.

2.7.1 Autosomal Inheritance

Autosomal inheritance of a genetic condition is mediated by variants of an autosomal chromosome or gene. The variants occur on one of the autosomal chromosomes 1–22 and not on a sex chromosome. There are two forms of autosomal inheritance: dominant and recessive.

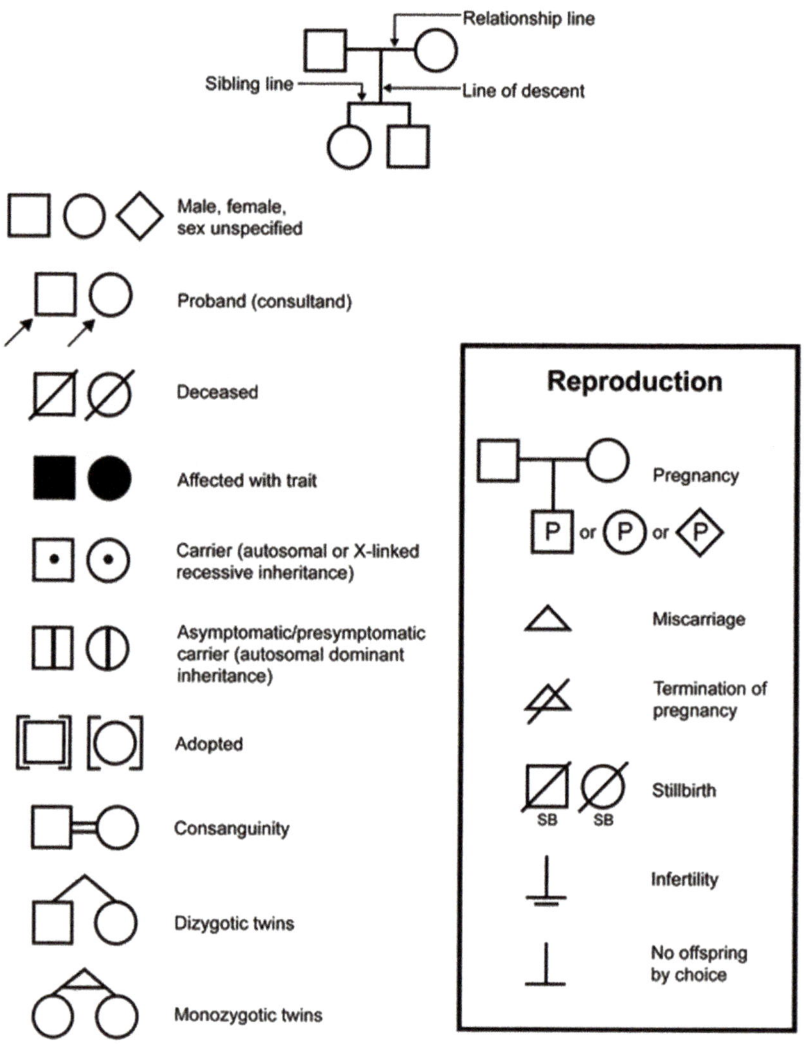

FIGURE 2.15 Standard pedigree nomenclature. *Source: https://www.cancer.gov/publications/dictionaries/genetics-dictionary/def/pedigree, Dictionary of Genetics Terms, originally published by the National Cancer Institute.*

Autosomal Dominant Inheritance

Autosomal dominant inheritance means that only one copy of a variant or mutation on an autosomal chromosome or gene causes the condition. The mutation on one gene can be inherited from an affected parent. During meiosis, it is random whether the mutant gene or the unaffected gene is present in any given germ cell of an affected parent, so that 50% of them will have the mutation and 50% will have an unaffected gene. Thus, there is a 50% recurrence risk that the mutation will be passed on to an offspring. The condition can occur in either males or females and can be passed from one generation to the next (Fig. 2.16).

Autosomal dominantly inherited conditions include Marfan syndrome, a connective tissue disorder (Section 11.10.4), and neurofibromatosis, typically a noncancerous condition with growth of the nervous system and skin, as well as café au lait macules (Section 11.11.2). As an example, the autosomal dominant condition tuberous sclerosis has major findings of numerous benign tumors of multiple organs including the skin (hypomelanotic macules, angiofibromas or fibrous cephalic plaques, shagreen patches, and ungual fibromas), brain (subependymal nodules, cortical dysplasias, and subependymal giant cell astrocytomas), kidney (angiomyolipomas), heart (rhabdomyomas), lungs (lymphangioleiomyomatosis), and eye (multiple retinal nodular hamartomas). Central nervous system tumors are its leading cause of morbidity and mortality, with seizures, intellectual disability/developmental delay, and psychiatric illness. Renal disease is the second leading cause of early death. Two-thirds are de novo cases due to *TSC1* or *TSC2* gene mutations, and the other third is inherited and can be passed from one affected individual with a 50% risk in either males or females from generation to generation. Most autosomal dominant conditions exhibit phenotypic variability or variable phenotypic expression, meaning not all

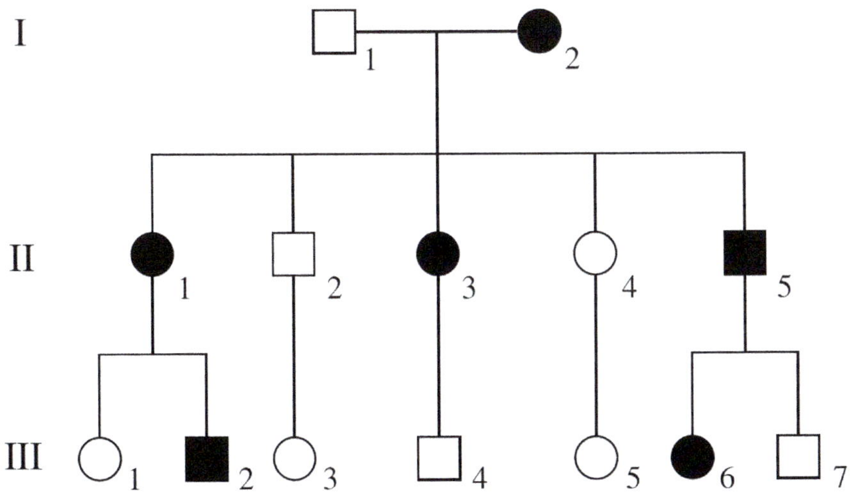

FIGURE 2.16 Autosomal dominant inheritance. *Reprinted from Genetics in Practice: A Clinical Approach for Healthcare Practitioners, Jo Haydon, Chapter 2, The Family History, Page 38, Fig. 2.1, 2007, with permission from John Wiley and Sons, Ltd.*

individuals within the family have all features or the same features. The differences in phenotypic expression even with the same gene mutation in members of the same family are because of the contribution of modifying genes and environmental factors.

Codominant Inheritance

Codominant inheritance refers to having one copy of a variant on one chromosome or one gene and a different variant on the homologous chromosome or gene, each expressing different phenotypes so that the same individual has a combination of both phenotypes. An example of this is the ABO blood groups where the allele A and allele B are both expressed. If a child inherits an allele A from their mother and an allele B from their father, the individual has blood type AB.

Autosomal Recessive Inheritance

Autosomal recessive inheritance means that two copies of a gene each have a variant or mutation, and together cause the condition. Usually, one copy of a variant is present on the gene inherited from the mother and the other copy of a variant is present on the gene inherited from the father. Each parent also has a normal gene. Such parents are designated as carriers or heterozygotes. Unlike autosomal dominant inheritance, one mutation on one copy of a gene is not sufficient to result in an autosomal recessive condition. There is a 25% chance that both mutant copies can be passed on through the egg and the sperm, producing an affected offspring, who has two copies of the mutation or is homozygous for the mutation. A recessive condition can occur in either males or females (Fig. 2.17).

Autosomal recessive conditions are usually not passed from one generation to the next unless the affected individual produces offspring with an individual who is also a heterozygous carrier of the same gene mutation. With the latter mating, there is a 50% risk that they would have an affected offspring. This pattern of inheritance is termed pseudodominance (Section 11.5.6). Should an affected homozygous individual produce offspring with another affected homozygous individual, each with two copies of a variant, then with their mating there is a 100% risk of having an affected offspring.

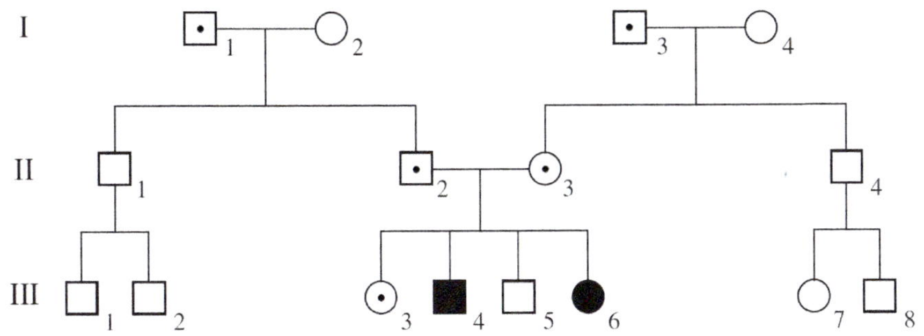

FIGURE 2.17 Autosomal recessive inheritance. *Reprinted from Genetics in Practice: A Clinical Approach for Healthcare Practitioners, Jo Haydon, Chapter 2, The Family History, Page 39, Fig. 2.1, 2007, with permission from John Wiley and Sons, Ltd. Modified from original.*

Homozygous individuals with two copies of the identical variant are more likely to occur with consanguineous (closely related) matings or in an inbred community (Sections 3.2.4, 11.5.1). At a population level, a founder effect resulting in reduced genetic diversity from a small number of ancestors also can increase the risk of homozygosity. More commonly, if the two variants are different but affect the same gene, an individual is affected as a compound heterozygote.

Autosomal recessively inherited conditions include hemochromatosis (Section 11.5.6), metabolic disorders such as urea cycle defects (Section 11.6.4) and phenylketonuria (Section 11.6.6), and sickle cell disease (Section 11.1.1). As an example, sickle cell disease presents with intermittent vaso-occlusive events and chronic hemolytic anemia. Vaso-occlusive events arise from red blood cells that take on a sickle shape, leading to blockage and damage to vital organs and tissue. The tissue ischemia results in acute and chronic pain. The disease-causing mutation (glutamic acid at position 6 on the protein is changed to a valine, p.Glu6Val or p.E6V) in the hemoglobin beta gene, *HBB*, results in the hemoglobin S allele, and two S alleles are present in 60%–70% of sickle cell disease cases in the United States. Other mutations lead to compound heterozygotes with one S allele and a different *HBB* variant such as a C, beta^{+}-thalassemia, beta^{0}-thalassemia, D-Punjab, O-Arab, or E allele. Sickle cell disease is more prevalent among African American and Hispanic American populations for whom 1 in 12 or 1 in 100, respectively, are carriers, or have sickle cell trait. In Africa the prevalence of the S allele is as high as 25%–30%. Parents who are both carriers have a 25% risk of having an affected child, 50% of having children who are carriers, and 25% of having an unaffected child. See Section 4.1 for further discussion of hemoglobinopathies and thalassemias.

2.7.2 Sex-linked Inheritance

Sex-linked inheritance of a genetic condition is mediated by variants or mutations on either the X or Y chromosome or gene. There are two forms of X-linked inheritance, dominant and recessive, and the pattern of inheritance varies between females and males because females have two X chromosomes and males have only one X chromosome and one Y chromosome. For dominant X-linked inheritance, females are more frequently affected because no sons of an affected man will be affected. For recessive X-linked inheritance, males are more frequently and often more severely affected than females because of the process of X-inactivation.

X-linked Dominant Inheritance

X-linked dominant inheritance means that one copy of the mutation on an X chromosome or gene causes the condition. The mutation can act in a dominant manner in both hemizygous males and heterozygous females. All females but no male offspring from an affected male inherit the condition, while 50% of female and male offspring inherit a mutation from an affected mother. The condition can occur in either males or females and can be passed from one generation to the next (Fig. 2.18). An example is X-linked hypophosphatemia, an X-linked dominant form of rickets due to a *PHEX* mutation. Patients have low serum phosphate concentrations, reduced tubular resorption of phosphate corrected for glomerular filtration rate, and absence of the normal physiologic response to hypophosphatemia of an elevation of 1,25(OH)$_2$vitamin D. They may also present with signs of rickets including decreased height velocity and severe lower-extremity bowing.

FIGURE 2.18 X-linked dominant inheritance. *Reprinted from Genetics in Practice: A Clinical Approach for Healthcare Practitioners, Jo Haydon, Chapter 2, The Family History, Page 40, Fig. 2.4, 2007, with permission from John Wiley and Sons, Ltd.*

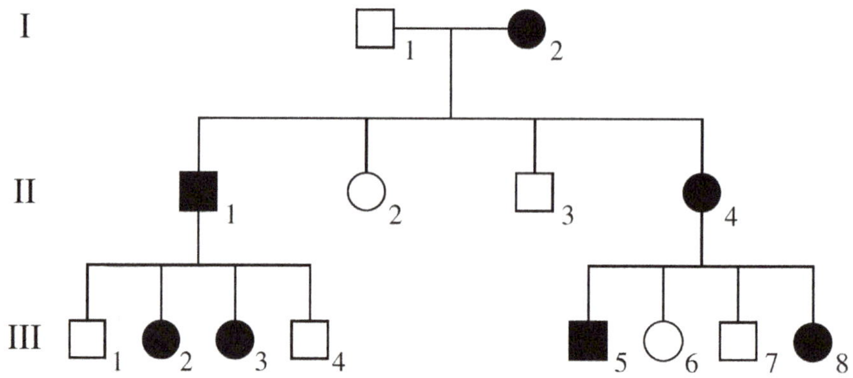

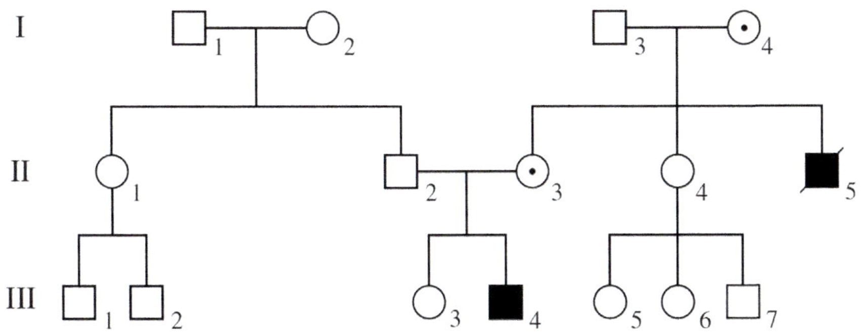

FIGURE 2.19 X-linked recessive inheritance. *Reprinted from Genetics in Practice: A Clinical Approach for Healthcare Practitioners, Jo Haydon, Chapter 2, The Family History, Page 40, Fig. 2.3, 2007, with permission from John Wiley and Sons, Ltd. Modified from original.*

X-linked Recessive Inheritance

X-linked recessive inheritance means that in females, two copies of a variant or mutation, one on each of the X chromosomes, cause the condition. Females with one copy of the mutation are heterozygous carriers who are usually unaffected. In males, one copy of the mutation on their only X chromosome is sufficient to cause the condition (Fig. 2.19). X-linked conditions can generally be more severe in males than females because every cell expresses the X chromosome or gene mutation in males, while females have some cells that do not carry or express the mutation due to X-inactivation or lyonization. As inactivation of the normal X chromosome occurs, female can be symptomatic, especially with skewed lyonization, when inactivation of the normal X has a selected advantage because there is a genetic preference in keeping other gene allele(s) on the chromosome X with the mutation active (Section 2.6).

If both parents are carriers of an X-linked recessive condition, there is a 25% risk that both copies of the variant can be passed through the gametes producing an affected female offspring with two copies of the mutation, or homozygous female. There is a 25% risk of an affected male offspring with one copy of the X chromosome mutation. There is a 25% of a female heterozygous carrier as well as a 25% of an unaffected male. Thus, the condition can occur in either hemizygous males or homozygous or compound heterozygous females. It usually is not passed from one generation to the next unless the affected female with two mutant copies produces offspring with an affected male with one copy of the mutation. With such a mating, there is a 100% risk they would have an affected male or female offspring who are variably affected. X-linked recessive conditions include Duchenne muscular dystrophy (Section 11.13.5) and adrenoleukodystrophy (Section 11.13.9). As an example, glucose-6-phosphate dehydrogenase (G6PD) deficiency presents with hemolysis and hemolytic anemia. Patients with G6PD deficiency can have symptoms of a hemolytic crisis (jaundice, dark urine, shortness of breath, and weakness) triggered by infections, stressors, fava beans, aspirin, and other medications such as sulfa drugs. This condition is particularly common in areas of Africa, Asia, and Mediterranean. Males are affected, with a mutation on their single copy of the *G6PD* gene. There is more variability in the severity of symptoms in females with skewed lyonization. They are generally symptomatic if they are homozygous or in some cases if heterozygous with lyonization of more of the normal alleles. Genetic testing for diagnosis and avoidance of these triggers is recommended.

Y-linked Inheritance

Y-linked inheritance means that one copy of a variant or mutation on the Y chromosome causes the condition. With males, the Y chromosomes would be passed on to each son so that there is 100% risk that the sons would be affected. No females would be affected. An example is Swyer syndrome, 46,XY genetic constitution with female external genitalia. The uterus and fallopian tubes are formed, but the gonads are not functional and underdeveloped as "streak gonads." Mutations in the *SRY* (sex-determining region Y) gene on the Y chromosome are identified in approximately 15% of individuals with this condition.

2.7.3 Multifactorial Inheritance

Digenic inheritance is the simplest form of multigenic inheritance for a genetic disorder when variants or mutations on two different genes are required to cause the condition. Having a mutation on one gene is necessary but not sufficient to cause a condition. An example of digenic inheritance is a rare form of retinitis pigmentosa type 7 due to the simultaneous presence of the PERIPHERIN 2 (*PRPH2* p.Leu185Pro or p.L185P) and a variable rod outer segment protein 1 (*ROM1*) mutations. Individuals with this condition are heterozygotes at the two different loci.

Many other conditions are associated with familial clustering but do not follow traditional Mendelian inheritance patterns. These are due to oligogenic or polygenic inheritance (influenced by variants on several or many genes, respectively) and multifactorial inheritance that involve variants on several or thousands of genes and environmental factors causing the condition. These complex patterns of quantitative inheritance explain the occurrence of many common traits in families and populations such as height, athletic ability, and sociability. The increased risk of having a condition is due to the accumulative effect of many different variants susceptible to environmental effects. There are many examples of multifactorial inheritance for common disorders, for example, congenital malformation such as cleft lip and palate (Section 11.3.10) and adult onset and chronic diseases such as obesity, diabetes, and psychiatric conditions (Sections 5.5.2, 5.7).

2.7.4 Mitochondrial Inheritance

Mitochondrial inheritance is different from dominant or recessive inheritance. A healthy individual's mtDNA molecules are mostly identical. However, in individuals with diseases caused by mtDNA variants or mutations, normal and mutant mtDNA molecules typically coexist in a single cell, which is referred to as heteroplasmy. Disease severity is often associated with the amount of mutant mtDNA in cells, which is determined by events that occurred when the individual's mother was still an embryo. The developing eggs in the female embryo go through an mtDNA "bottleneck," in which the number of mtDNA copies is first reduced and then amplified to more than 100,000 copies (Section 2.5). Accordingly, variable amounts of mutant and unaffected mtDNA are present in the mature eggs of an individual woman and, therefore, in the cells of her offspring. This phenomenon influences the severity of diseases caused by mtDNA mutations and can lead to very different clinical presentations among individuals from the same family or different families (intrafamilial or interfamilial variation, respectively). Thus, the condition can be passed from mother to either male or female offspring, but the severity and clinical features of the condition can be variable.

Individuals can present with variable symptoms of almost any organ including the brain, nerves, muscles, kidneys, heart, liver, endocrine, eyes, ears, and/or pancreas. Some diseases are degenerative with neuromuscular symptoms. Examples of mitochondrial diseases include mitochondrial myopathy, diabetes mellitus, and deafness; neuropathy, ataxia, retinitis pigmentosa, and ptosis; myoneurogenic gastrointestinal encephalopathy, myoclonic epilepsy with ragged red fibers (MERFF); mitochondrial DNA depletion syndrome; Leber hereditary optic neuropathy, Leigh syndrome (Section 11.6.2), and mitochondrial encephalopathy with lactic acidosis and strokelike episodes (MELAS) syndrome (Section 11.6.5). It should be noted that some mitochondrial diseases are caused by mutations in nuclear genes that code for mitochondrial components and they are inherited as other nuclear chromosomal or gene disorders, typically in a recessive manner.

2.7.5 Imprinting Inheritance

Uniparental disomy (UPD) occurs when both copies of a chromosome (complete UPD), or both copies of part of a chromosome (segmental UPD), are derived from the same parent, rather than one copy being inherited from each parent as in the normal situation. If the two chromosomes are derived from identical sister chromatids, it is defined as isodisomy, whereas when both homologues from one parent are present, it is defined as heterodisomy. There are several mechanisms for the generation of complete UPD; among these, the most common mechanism is trisomic rescue, which is the process of attempted restoration of a disomic state after the formation of a trisomic conceptus due to chromosome nondisjunction. In extreme situations, UPD of all chromosomes, also known as uniparental diploidy or genome-wide UPD, could be generated, which has been reported to be viable only in mosaic status (Kotzot, 2008). UPD has been observed for every chromosome. For most chromosomes, there is no apparent phenotypic consequences.

Genomic imprinting is a phenomenon when the ability of a gene to be expressed depends upon the parent who passed on the gene. In some cases, imprinted genes are expressed only when they are inherited from the mother. In other cases, they are expressed only when inherited from the father. When chromosomal regions are subject to a parent-of-origin effect, UPD could result in the inheritance of either two active or two silenced copies of a gene and cause an abnormal phenotype. The nature of imprinting is epigenetic rather than DNA sequence dependent. A mechanism for imprinting is differentially methylated regions regulating genes or clusters of gene expression (Section 2.6). Pathogenicity of several imprinting disorders has been attributed (or partly attributed) to UPD (Table 6.4). Besides causing imprinting disorders, another phenotypic consequence of UPD is that isodisomy could unmask recessive diseases by transmitting two copies with deleterious change from one heterozygous parent carrier (Spence et al., 1988).

Classic examples of imprinting disorders are Prader–Willi and Angelman syndromes, which occur as a result of having both copies of chromosome 15 derived from one parent. The loss of the maternal contribution is linked to Angelman syndrome and the loss of the paternal contribution linked to Prader–Willi syndrome (Section 11.13.1). Prader–Willi syndrome is characterized by hypotonia and feeding difficulties in early infancy, followed by excessive eating with morbid obesity and cognitive impairment; while Angelman syndrome has severe developmental delay, speech impairment, gait ataxia, and inappropriate happy demeanor. Three main molecular mechanisms that result in Prader–Willi syndrome include paternal deletion, maternal UPD of chromosome 15, and rarely an imprinting defect (ID) at chromosome 15q11.2-q13. Angelman syndrome results from ID, paternal UPD of the same chromosome 15 region, and mutations in the *UBE3A* gene. Beckwith–Wiedemann syndrome, a growth disorder with variable neonatal hypoglycemia, macrosomia, macroglossia, hemihyperplasia, omphalocele, embryonal tumors, renal abnormalities, and ear creases, is caused by changes in maternal imprinting, paternal UPD, and genomic alterations at chromosome 11p15 (Sections 11.7.1, 11.7.4, 11.13.1).

2.8 Phenotypic Variation

Phenotype is a term used to describe the amalgamation of characteristics or traits of an individual (Nussbaum et al., 2016). The term can encompass a number of traits such as an individual's physical appearance, biochemical measurements, intelligence, or behavior. The phenotype can result from genetic and environmental factors. A single genetic variant with a strong effect can be associated with a specific phenotype (e.g., a syndrome or specific dysmorphic feature) and is often rare. Multiple genetic variants and environmental factors, each singularly weak, can collectively have an effect(s) that causes a multifactorial condition that is often common (e.g., diabetes, obesity, and high blood pressure) or leads to normal variation in humans (e.g., height, athletic ability, and creativity).

The multitude of variants in the genome can act to modify the severity of a phenotype; such genetic variants can be referred to as modifying factors. Although the terms Mendelian or single gene disorders are used, no phenotype is truly a result of only one gene mutation. Even the large phenotypic effect of one mutant gene can be modulated or modified by other factors so that members in the same family that have inherited the same single gene mutation may have differing severity of a specific feature or a different spectrum of features as in a syndrome. This phenomenon is termed phenotypic variability or variable phenotypic expression, respectively, and is more common in dominant conditions (Section 2.7.1). Phenotypic variability is a term used to denote variation in type or severity of clinical features of a condition within members of a family or among unrelated families such as in neurofibromatosis and ectodermal dysplasia (Sections 11.11.2, 11.11.3). Several different genetic mechanisms may lead to phenotypic variability such as multifactorial factors and somatic mosaicism (Sections 2.2.2, 2.7.3).

The term penetrance is used to describe whether an individual with a variant expresses the phenotype or condition. In some cases, the effect of a variant is modified by other factors so that an individual with the variant does not manifest the condition at all and this phenomenon is considered an example of nonpenetrance (Sections 11.11.2, 11.11.3). Some cases that are thought to be sporadic, meaning an individual is the only member of a family to have the condition, may actually be examples of nonpenetrance in other family members, particularly in dominant conditions (Section 2.7.1).

Genetic heterogeneity refers to variants or mutations at two or more genes that produce similar or the same phenotypes. For example, mutations in different genes of the same pathway and processes can cause a similar phenotype as in the biochemical disorders of the urea cycle (Section 11.6.4) or cohesinopathies (Section 11.5.3).

The many complex concepts of inheritance and variation defined in this chapter lay the foundation for understanding how our genome plays an instrumental role in our health and our ability to make diagnoses and effectively treat and manage a multitude of diseases. This information is the knowledgebase needed to understand genetics and genomic practice presented in following chapters.

References

Alkan, C., Coe, B.P., Eichler, E.E., 2011. Genome structural variation discovery and genotyping. Nat. Rev. Genet. 12, 363–376. PMID 21358748.

Barch, M.J., Knutsen, T., Spurbeck, J.L., Association of Genetic Technologists., 1997. The AGT Cytogenics Laboratory Manual. Lippincott-Raven Publishers, Philadelphia.

Biesecker, L.G., Spinner, N.B., 2013. A genomic view of mosaicism and human disease. Nat. Rev. Genet. 14, 307–320. PMID 23594909.

Byrne, L., Drake, A.J., 2019. Paediatrician's guide to epigenetics. Arch. Dis. Child. 104, 297–301. PMID 30655261.

Feuk, L., Carson, A.R., Scherer, S.W., 2006. Structural variation in the human genome. Nat. Rev. Genet. 7, 85–97. PMID 16418744.

Gardner, R.J.M., Sutherland, G.R., Shaffer, L.G., 2011, fourth ed. Chromosome Abnormalities and Genetic Counseling. Oxford Monographs on Medical Genetics No 61. Oxford University Press, Oxford.

Gersen, S.L., Keagle, M.B., 2011. The Principles of Clinical Cytogenetics, third ed. Humana Press.

Gu, S., Szafranski, P., Akdemir, Z.C., Yuan, B., Cooper, M.L., Magrina, M.A., Bacino, C.A., Lalani, S.R., Breman, A.M., Smith, J.L., Patel, A., Song, R.H., Bi, W., Cheung, S.W., Carvalho, C.M., Stankiewicz, P., Lupski, J.R., 2016. Mechanisms for complex chromosomal insertions. PLoS Genet. 12, e1006446.

Gu, W., Lupski, J.R., 2008. CNV and nervous system diseases—what's new? Cytogenet. Genome Res. 123, 54–64. PMID 19287139.

Kang, S.H., Shaw, C., Ou, Z., Eng, P.A., Cooper, M.L., Pursley, A.N., Sahoo, T., Bacino, C.A., Chinault, A.C., Stankiewicz, P., Patel, A., Lupski, J.R., Cheung, S.W., 2010. Insertional translocation detected using FISH confirmation of array-comparative genomic hybridization (aCGH) results. Am. J. Med. Genet.A. 152A, 1111–1126. PMID 20340098.

Kotzot, D., 2008. Complex and segmental uniparental disomy updated. J. Med. Genet. 45, 545–556. PMID 18524837.

Krebs, J.E., Goldstein, E.S., Kilpatrick, S.T., 2018. Lewin's Genes XII. Jones & Bartlett Learning, Burlington, MA.

Massalska, D., Bijok, J., Ilnicka, A., Jakiel, G., Roszkowski, T., 2017. Triploidy—variability of sonographic phenotypes. Prenat. Diagn. 37, 774–780. PMID 28573747.

Migeon, B.R., 2017. Choosing the Active X: the human version of X inactivation. Trends Genet. 33, 899–909. PMID 28988701.

Monk, D., Mackay, D.J.G., Eggermann, T., Maher, E.R., Riccio, A., 2019. Genomic imprinting disorders: lessons on how genome, epigenome and environment interact. Nat. Rev. Genet. 20, 235–248. PMID 30647469.

Neill, N.J., Ballif, B.C., Lamb, A.N., Parikh, S., Ravnan, J.B., Schultz, R.A., Torchia, B.S., Rosenfeld, J.A., Shaffer, L.G., 2011. Recurrence, submicroscopic complexity, and potential clinical relevance of copy gains detected by array CGH that are shown to be unbalanced insertions by FISH. Genome Res. 21, 535–544. PMID 21383316.

Nussbaum, R.L., Mcinnes, R.R., Willard, H.F., Hamosh, A., 2016. Thompson & Thompson Genetics in Medicine, eighth ed. Elsevier/Saunders, Philadelphia, PA.

Paulson, H., 2018. Repeat expansion diseases. Handb. Clin. Neurol. 147, 105–123. PMID 29325606.

Posynick, B.J., Brown, C.J., 2019. Escape from X-chromosome inactivation: an evolutionary perspective. Front. Cell Dev. Biol. 7, 241. PMID 31696116.

Simons, A., Shaffer, L.G., Hastings, R.J., 2013. Cytogenetic nomenclature: changes in the ISCN 2013 compared to the 2009 Edition. Cytogenet. Genome Res. 141, 1–6.

Scheffler, I.E., 2008. Mitochondria, second ed. John Wiley & Sons, Hoboken, N.J.

Spence, J.E., Perciaccante, R.G., Greig, G.M., Willard, H.F., Ledbetter, D.H., Hejtmancik, J.F., Pollack, M.S., O'Brien, W.E., Beaudet, A.L., 1988. Uniparental disomy as a mechanism for human genetic disease. Am. J. Hum. Genet. 42, 217–226. PMID 2893543.

Strachan, T., Read, A., 2018. Human Molecular Genetics, fifth ed. Taylor & Francis Group, Milton.

Therman, E., Susman, M., 1993. Human Chromosomes : Structure, Behavior, and Effects. Springer-Verlag, New York.

Toufaily, M.H., Roberts, D.J., Westgate, M.N., Holmes, L.B., 2016. Triploidy: variation of phenotype. Am. J. Clin. Pathol. 145, 86–95. PMID 26712875.

Wallace, D.C., 2018. Mitochondrial genetic medicine. Nat. Genet. 50, 1642–1649. PMID 30374071.

Chapter 3

Genetic Counseling and Referrals

Antonie D. Kline[1] and Virginia L. Corson[2]

[1]*Harvey Institute for Human Genetics, Greater Baltimore Medical Center, Baltimore, MD, United States;* [2]*Johns Hopkins University School of Medicine, The Academy at Johns Hopkins, East Baltimore Campus, Baltimore, MD, United States*

Abstract

Clinical genetics is usually a referral-based practice. There may be a single entity or constellation of findings that can prompt the referral to determine if there is an underlying genetic cause. Referrals can occur in the preconceptual or prenatal arena, in the neonatal or pediatric period, or in adulthood, including oncology and internal medicine, and with positive family history of genetic entities. Figuring out inherited versus sporadic causation and genetic risks are objectives of a visit. Recommendations for a counseling visit are described, including the process of obtaining a three-generation pedigree, as well as addressing cost and insurance issues. The role of genetic counseling is one of the most important aspects of any medical genetics visit.

Genetic counseling is a critical part of a genetic evaluation. Whether preconceptually, prenatally, during childhood, or in adulthood, referrals occur for personal or family history reasons. The detailed family history, typically compiled by the genetic counselor, is key to assessing trends and risks. Certain populations have higher risks for genetic variants that may lead to specific clinical recommendations, as do offspring of parents who are closely related genetically. The genetic counselor investigates cost and insurance coverage for the individual undergoing genetic testing and helps coordinate the studies. Additionally, identification of genetic variants in an individual can have implications for other family members, and the genetic counselor can assist in sharing or distributing this information.

3.1 Preparation for Genetics Appointment

When concern is raised about the possibility of a genetic diagnosis in a child or an adult because of clinical symptoms or a positive family history, or when prospective parents are at increased risk for having a child with a genetic condition, a referral to genetics clinic is appropriate. Depending on the type of clinic visit, various medical records and information provided by the family will be necessary for the appointment. For the purpose of discussion, "parents" below may also refer to guardians or caregivers.

3.1.1 Medical Records and Questionnaires

Each clinical setting will request a variety of records for review, often in advance of the visit. For example, a medical genetics clinic will typically request growth parameters and charts, laboratory results, office notes, and imaging reports from the primary care provider for the child or adult who is being evaluated. Surgical records and pathology reports may also be helpful. For a prenatal clinic, it would be important to have the obstetric laboratory results, ultrasound reports, and office notes from the current pregnancy. Subspecialty clinics, such as oncology, will request pathology reports from any relative known to have cancer. In any of these settings, if a family member has had genetic testing, a copy of those results may provide valuable information in considering a particular diagnosis for the individual being evaluated.

Genomics in the Clinic. https://doi.org/10.1016/B978-0-12-816478-5.00003-0

Many genetics clinics will provide an intake questionnaire in advance for completion by the patient or parent. Questions related to family history, pregnancy history, demographics, and general health parameters may be included. In some cases, it may be helpful for patients to contact relatives to gather information about the family history to facilitate answering the questions. Sometimes, photographs of other potentially affected family members or baby pictures of adult patients or parents may be helpful.

3.1.2 Clinical Settings and Referrals

Based on the indication for referral, the patient will be seen by various personnel during the appointment. Children or adults seen in a pediatric or medical genetics clinic often are examined by a medical geneticist, who may be assisted by a genetic counselor, a clinical genetics resident, a physician assistant, or a genetic nurse. Prospective parents seen in a prenatal setting will likely be discussing their concerns with a genetic counselor. Adults being evaluated for a family history of cancer or heart disease are often seen by a genetic counselor as well as a geneticist, an oncologist, or a cardiologist.

When a patient is referred for an appointment, all parties involved should ascertain whether an electronic or written referral is necessary and whether an insurance pre-authorization is required. For some patients, insurance coverage will be jeopardized without following the prescribed process. Likewise, any laboratory testing or radiologic studies recommended during the evaluation may require insurance pre-authorization, and the laboratory chosen may need to be a contracted laboratory. For patients seen in the medical genetics clinic, referral to other subspecialists, such as ophthalmology or cardiology, or for specific testing, such as X-rays or echocardiogram, may be recommended at the conclusion of the counseling session. Again, insurance approval may be necessary prior to scheduling these appointments.

The structure of the genetics appointment will depend on the indication for referral. A detailed family history will be obtained, or reviewed if supplied in advance. The physician will review with the patient or the parents the history, based on the medical records supplied and their recollections. The individual being evaluated will have a pertinent and/or detailed physical examination. At the conclusion of the session, impressions of the team will be shared with the patient or the parents, and genetic testing or additional evaluations will be discussed. In many cases, the patient or family will return for a follow-up session, once these test results are available. In some cases, periodic follow-up, perhaps annually, may be recommended for coordination of subspecialty care and reevaluation for any new manifestations of the condition. For children being treated for a metabolic disorder, such as phenylketonuria (Section 11.6.6), more frequent appointments and monitoring will be advised to monitor the special diet. Virtual visits may be preferable in many of these situations.

In the prenatal setting, a family and pregnancy history will be obtained, the indication for referral will be discussed, and appropriate testing options will be outlined. The genetic counselor will arrange for testing elected by the patient and/or partner and contact them, usually by telephone, with the results. If the results of fetal or parental testing are abnormal or indicate an increased risk for a fetal condition, a follow-up genetic counseling session will be available to discuss results in detail, additional testing, and pregnancy management options.

3.2 Family History and Genetic Background

In each genetic clinic setting described, it is typical to incorporate the family history obtained into a pedigree. These pedigrees will have much in common regardless of the situation but may vary in emphasis.

3.2.1 Pedigree

In all settings, information will usually be requested on three generations of the family (Section 2.7). In instances with a positive family history, information about more distant relatives may be helpful. Information about intellectual disability, sudden death before age 50 years, birth defects, or health concerns may be significant to the indication for referral (Bennett et al., 2008). In the prenatal clinic, special attention will be paid to any reproductive loss or infertility (Section 11.1.7). For oncology patients, detailed information, such as the type of cancer and age at diagnosis about any individual with cancer, will be sought (Sections 11.16.2, 11.16.3). Often, information gained from the family history will contribute to the differential diagnosis or help guide further testing options.

3.2.2 Symptom or Known Syndrome Inherited in Family

Family members will be asked about specific familial conditions. For example, polydactyly, one of the most common malformations in the world, is a dominant trait that can be familial and isolated. It can also be part of a broader underlying genetic syndrome that can be familial or sporadic. Likewise, sensorineural hearing loss can be familial, isolated, and/or syndromic (Sections 11.2.3, 11.3.7, 11.3.8, 11.3.9), as can hyperlipidemia (Section 11.4.6) or colon cancer (Section 11.16.3). Accompanying findings in the family can be very important in helping identify the cause. For any of these, a referral for a genetic evaluation can provide the family with specific information that could help with pregnancy decisions, future precautions or reassurances, and further recommendations to help manage medical complications.

3.2.3 Ethnicity

In most cases, the family members will be asked about their ethnic background. This information may be useful in discussing carrier screening options in the prenatal setting or in considering the differential diagnosis in the medical genetics clinic. There is a known increased chance to be a carrier of a pathogenic variant in certain populations. For example, individuals of Ashkenazi Jewish ancestry have a 1 in 30 chance of being carriers of Tay–Sachs disease, non-Hispanic white individuals have a 1 in 35 chance of spinal muscular atrophy carrier status (Sections 11.1.1, 11.2.4), and African Americans have a 1 in 10 chance of being sickle trait carriers (Section 11.11.1). In addition, individuals have different risks for carrying a pathogenic variant for a specific condition depending on ethnicity (Hoskovec et al., 2018). For example, the risk for carrying a mutation leading to cystic fibrosis is highest in Ashkenazi Jewish individuals with a rate of 1 in 24 and non-Hispanic white individuals with a rate of 1 in 25, next highest in Hispanic individuals with a rate of 1 in 58, then in African American individuals with a rate of 1 in 64, and rarest in Asians with a rate of 1 in 94 (Section 11.5.1). Tables are available for risks and recommendations (King and Klugman, 2018). For carrier screening options based on ethnicity, please see Section 4.1.

3.2.4 Consanguinity

When the biological parents of a child are related to one another, there is an increased risk for autosomal recessive disorders (Section 2.7.1) as they share more genes in common than usual for a couple in the general population. First cousins, for example, share one-eighth, on average, of their genetic information. For a child presenting with failure to thrive, multiple anomalies or developmental delays, an autosomal recessive condition would be highly suspected if the parents are first cousins. Some populations, such as the Amish, Hutterite, Middle Eastern, and Southeast Asian, marry more frequently within the family for religious or traditional reasons and are more likely to both carry the same recessive pathogenic variant (Section 11.5.1). In the prenatal setting, universal or expanded carrier screening may be especially appropriate for consanguineous couples concerned about the increased risk of having a child with an autosomal recessive disorder. For these couples, unless there is a family history of concern, the risk of a having a child with a recessive condition or anomalies varies with the closeness of the relationship and is slightly increased over the general population risk of 2%–3% for anomalies alone (Becker et al., 2015). There may also be ethical issues related to consanguinity (Section 10.3.3).

3.3 Genetic Counseling

As described by the National Society of Genetic Counselors, genetic counseling is the process of helping people understand and adapt to the medical, psychological, and familial implications of genetic contributions to disease (https://www.nsgc.org). It is a specialized field that addresses rare clinical situations, both for diagnosis and testing, as well as for recommendations and management. Whether due to personal or family histories, fetal anomalies, maternal complications, childhood onset issues, adult disorders, or a subset of cancer, these situations set the family and/or patient apart from the general population. The genetic counselor obtains the family history; guides genetic discussions; discusses risk, outcomes, and choices; and helps coordinate the genetic evaluation and plan for further diagnostic workup. Discussions can include descriptions of the gene(s) involved, possible mechanisms for occurrence, and also future possibilities of offspring, as well as what types of future genetic screening and/or testing to pursue. This role has evolved as newer testing (e.g., next-generation sequencing) has produced previously unsuspected information (Yang and Kim, 2018). Risk figures are addressed by genetic counselors, especially carrier risks, recurrence risks, and risks for developing certain conditions (McGlynn and Langfelder-Schwind, 2020). There is also a deep understanding of ethical issues and ramifications (Chapter 10 and Sections 11.18.1, 11.18.2). Direct-to-consumer testing is a newer area that overlaps ancestry, general medicine, and pharmacogenomics with clinical genetics (Chapter 9 and Sections 11.17.1, 11.17.2, 11.17.3, 11.17.4, 11.17.5, 11.17.6). Finally, the genetic counselor provides overall support for the family and/or patient and connects the family to other similarly affected families or groups (Section 3.5).

3.4 Genetic Risks for Family Members

When a genetic disorder or carrier status has been identified in an individual, genetic counseling and testing may be indicated for relatives. In some cases, a clinical examination may provide sufficient additional information, while in other instances genetic testing may be available for risk assessment. Risks for recurrence can then be conveyed to families (Mendes et al., 2018).

In the case of single gene disorders (Section 2.7), the risk of being affected or a carrier of a disorder diagnosed in a family member will depend on the inheritance pattern of the condition and the closeness of the relationship. For autosomal recessive disorders (Section 2.7.1), such as cystic fibrosis or Tay–Sachs disease, siblings of a carrier will have a 50%, or 1 in 2, chance of being a carrier. Unaffected full siblings of an affected individual will have a 66%, or 2 in 3, chance of carrier status. If molecular testing of the proband, or affected individual, has been performed, targeted testing for the variant(s) identified can be performed on relatives to evaluate for carrier status. If molecular testing has not or cannot be performed on the affected individual or carrier, mutation analysis for the relative can still be performed; however, the interpretation of the result may be less conclusive.

For autosomal dominant disorders (Section 2.7.1), such as neurofibromatosis (NF) and Huntington disease (HD), an offspring of an affected individual will have a 50% chance of being affected. That risk can be modified by other factors, such as the results of a clinical examination in the case of NF, or the age of an individual at risk for HD. A 9-year-old girl whose younger brother has NF is unlikely to be affected if she has normal development and her examination is normal, including no neurofibromas, freckling, café au lait spots, and Lisch nodules (Section 11.11.2). A 65-year-old with no signs of involuntary jerking or twitching, difficulty speaking or swallowing, or a decline in mental faculties will have some reduction in a 50% risk of HD based on the family history, compared to a 25-year-old with a similar family history (Section 11.13.12). For both of these conditions, molecular testing would be available to clarify or further evaluate the status of the at-risk relative.

For sex-linked disorders, such as Duchenne muscular dystrophy (DMD) (Section 11.13.5), female relatives of the mother of an affected male will be at increased risk of being carriers and should be offered genetic counseling. The specific carrier risk for females in the family will depend on the relationship to the affected boy and other pedigree information, such as whether there are other affected and/or unaffected males. The identification of a pathogenic variant in the affected boy will allow for targeted testing of these females.

The determination of the chance of other family members to carry a familial mutation for some conditions is complicated by the possibilities of a new mutation in the affected individual, variable expressivity, reduced penetrance, or germline mosaicism (Sections 2.4.2 and 2.8). For example, some boys diagnosed with DMD (Section 11.13.5) who have a negative family history may represent a new mutation. In this case, female relatives such as maternal aunts would not be at increased risk to be carriers of this sex-linked condition. For some autosomal dominant conditions such as NF, the expression of the condition could be so mild in one individual that a diagnosis will not be suspected until a relative with more classic features of NF is diagnosed. Some women who carry a pathogenic variant in a gene that increases the risk of breast and ovarian cancer may never get cancer (Sections 11.16.1, 11.16.2, 11.17.1, 11.18.1). Although those women remain cancer-free, they may pass the variant to their daughters who could go on to develop cancer. Finally, even when individuals may clinically be unaffected with features of a condition, they may be at increased risk to have another affected offspring due to germline mosaicism. The birth of a child with achondroplasia, an autosomal dominant form of short stature, may result from a new mutation in the egg or the sperm, or rarely, could be secondary to germline mosaicism in either parent (Section 11.10.2). Recurrences of achondroplasia in the offspring of individuals with normal stature have occurred secondary to this phenomenon.

If a mitochondrial disorder (Section 6.4.2) is identified in an individual, genetic testing of family members may be indicated. For example, in the case of a disorder caused by a known autosomal recessive (Section 2.7.1) nuclear DNA mutation, testing of unaffected relatives would clarify carrier status. For autosomal dominant (Section 2.7.1) nuclear DNA disorders, one parent of a proband may have the same mutation as the affected individual or the proband may represent a de novo mutation. Results of parental testing will inform whether their other children are at risk and should be offered testing.

In contrast, if the gene variant is not in the nuclear DNA but is in the mitochondrial DNA, the condition is maternally inherited (Sections 2.5, 2.7.4). The mother of an affected individual usually has the pathogenic variant (Section 7.1.3) and may or may not have symptoms. Her children and relatives in the maternal line may also have a variable amount of mutated mitochondrial DNA in their cells. Genetic testing of these relatives may help quantify the level of heteroplasmy, the presence of more than one type of genome within a cell, which may correlate with any clinical symptoms (Section 11.6.5).

For conditions caused by imprinting (Section 2.7.5), in which gene expression depends on parental origin, testing of family members may be available based on the mechanism established for the diagnosis. For example, if uniparental disomy is determined to be the cause of the condition, no further testing is necessary unless a chromosome rearrangement is identified in the proband. If, however, a pathogenic variant in an imprinting gene is detected, testing of relatives who could be carriers of the variant will be recommended.

In some cases, a family member will be diagnosed with a chromosome rearrangement (Section 2.2.1), such as a translocation. An unbalanced translocation will usually result in multiple anomalies and/or developmental delay. A balanced rearrangement might be identified in the evaluation for recurrent pregnancy loss (Section 11.1.7) or incidentally during prenatal diagnosis by amniocentesis or chorionic villus sampling. In any of these situations, close relatives would be at increased risk to carry the same rearrangement, and a blood karyotype can be performed to evaluate the individual's status, as well as the risk to more distant relatives.

3.5 Psychosocial Issues and Support Organizations

The diagnosis of a genetic disorder in a child or an adult is usually accompanied by a variety of emotional reactions. These reactions may include denial, guilt, depression, and anger before the stage of acceptance is reached. Dealing with the new burdens of multiple medical appointments, financial strains and decision conflicts can lead to stress for all involved. In the pediatric setting, parents may have difficulty finding enough time to spend with their unaffected children. In some families, individual, family, or couples therapy may be beneficial during this time of anxiety and adjustment.

When a diagnosis is genetic in origin, there may be implications for other family members, and many questions can arise. Who else in the family might be a carrier for the condition? What relatives should be informed about the diagnosis? Is it possible that other family members are affected with the condition? What is the best way to share this information with relatives? Who can answer the questions of relatives who live far away? All of these issues are typically addressed by the medical genetics providers. Identifying genetic counselors in other areas of the country can be facilitated through the National Society of Genetic Counselors (www.nsgc.org) directory.

Often the diagnosis is unfamiliar to the family, and they will benefit from information found through the websites of national or local parent support organizations. In addition, the Alliance of Genetic Support Groups (www.geneticalliance.org) has many resources that can be helpful to families. Social media may lead parents to posts by families with similar experiences. Parents should be cautioned that some information they read may be misleading or inaccurate and be encouraged to verify any questions or concerns with their medical providers. For professionals involved with the families, informative summaries of the condition may be available online at GeneReviews (www.GeneReviews.org) or the Online Mendelian Inheritance in Man (OMIM).

Parents who are affected with, or are carriers for, a genetic condition often have some degree of guilt if they have an affected child. Sometimes an at-risk individual does not inherit the familial mutation and feels guilty that, by chance, he or she has escaped the burden of the genetic disorder affecting the relatives. Identifying and talking about parental guilt or "survivor guilt" can help mitigate this discomfort over time. Some at-risk individuals will have great difficulty deciding whether to have presymptomatic testing for a late-onset condition, such as HD (Section 11.13.12). Relatives at similar risks may make different decisions, sometimes leading to tension within a family.

How one copes with these situations is often influenced by an individual's history of dealing with other difficult life experiences. For example, if a child were to have a genetic condition that is terminal, a parent's prior experience (or lack thereof) with death in the family may provide a context for his/her adjustment to such an unexpected and sad event. Likewise, one's understanding of the medical implications of a diagnosis may be impacted by educational background or cultural belief systems.

In the prenatal setting, couples who learn of an abnormal prenatal diagnosis may experience similar emotional stages of denial, guilt, depression, and anger. In addition, they may be faced with the difficult decision of whether to continue or terminate a pregnancy. This decision, which often must be made quickly, can lead to a period of crisis. Written materials, short-term therapy, and specialty consultations may be helpful to the decision-making process. For couples who terminate a pregnancy, obtaining photographs or footprints and holding the baby (when possible) can provide tangible memories of the experience, which provide comfort to some. Attending support groups for individuals with similar experiences and follow-up genetic or therapeutic counseling may be beneficial during the period of recovery.

References

Becker, R., Keller, T., Wegner, R.D., Neitzel, H., Stumm, M., Knoll, U., et al., 2015. Consanguinity and pregnancy outcomes in a multi-ethnic, metropolitan European population. Prenat. Diagn. 35, 81–89. PMID 25174624.

Bennett, R.L., French, K.S., Resta, R.G., Doyle, D.L., 2008. Standardized human pedigree nomenclature: update and assessment of the recommendations of the National Society of Genetic Counselors. J. Genet. Counsel. 17, 424–433. PMID 18792771.

Hoskovec, J.M., Stevens, B.K., 2018. Genetic counseling overview for the obstetrician-gynecologist. Obstet. Gynecol. Clin. N. Am. 45, 1–12. PMID 29428277.

King, J.R., Klugman, S., 2018. Ethnicity-based carrier screening. Obstet. Gynecol. Clin. N. Am. 45, 83–101. PMID 29428288.

Mcglynn, J.A., Langfelder-Schwind, E., 2020. Bridging the gap between scientific advancement and real-world application: pediatric genetic counseling for common syndromes and single-gene disorders. Cold Spring Harb. Perspect. Med. 10, a036640. PMID 31570386.

Mendes, A., Metcalfe, A., Paneque, M., Sousa, L., Clarke, A.J., Sequeiros, J., 2018. Communication of information about genetic risks: putting families at the center. Fam. Process 57, 836–846. PMID 28714147.

Yang, M., Kim, J.W., 2018. Principles of genetic counseling in the era of next-generation sequencing. Ann. Lab Med. 38, 291–295. PMID 29611378.

Chapter 4

Prenatal Genetics and Referrals

Virginia L. Corson[1] and Antonie D. Kline[2]

[1]*Johns Hopkins University School of Medicine, The Academy at Johns Hopkins, East Baltimore Campus, Baltimore, MD, United States;*
[2]*Harvey Institute for Human Genetics, Greater Baltimore Medical Center, Baltimore, MD, United States*

Abstract

Prenatal genetics encompasses aspects of each member of the couple as well as the fetus. Parents may wish to know about potential risks that can be assessed before and during pregnancy for their future children. Biological parents considering pregnancy, may be aware of increased risk to the fetus, depending on genetic and ethnic background, family history, parental age, screening test results, or exposures during pregnancy. Seeking genetic counseling for any of these situations can provide knowledge, options for further investigation, and reassurance for future parents. Fetuses found to have a physical anomaly on ultrasound, or abnormality on prenatal screening or testing, will prompt referral for genetic counseling. A discussion of reproductive options, including assisted reproductive technology, may be indicated as well.

Discussions of fetal issues and genetic counseling began in the late 1960s (Fraser, 1968) and have continued through to the present. These included both congenital malformations from exposures to organisms and medications, genetic etiologies, either inherited or spontaneous, and multifactorial causes. As knowledge has progressed, offering carrier testing for parents and reviewing prenatal screening have become standard of care, with prenatal diagnosis expanding in scope and accuracy. Reproductive options also have increased, giving potential parents a broad possibility for achieving a healthy fetal outcome.

4.1 Carrier Screening

Carrier testing for mutations in known genes that could cause specific conditions in future offspring is available through various companies and should be offered to all women contemplating pregnancy. Ideally, carrier screening is done prior to the pregnancy through the primary care provider or obstetrician, or a genetic counselor, to be able to identify couples at risk. Most of the conditions on the panels are autosomal recessive (Section 2.7.1), including sickle cell disease, cystic fibrosis, and spinal muscular atrophy (Sections 11.11.1, 11.2.4). Some are X-linked (e.g., Duchenne muscular dystrophy, Section 11.13.5) with less effects in the carrier woman, yet with the potential to lead to a more severe phenotype in the future children. If an autosomal recessive mutation is found in the one member of the couple, the partner can be offered carrier testing for the same condition. If the test is positive, this will merit a discussion with genetic counseling about the couple's likely 25% chance of having an affected child and potential management options (Section 11.1.1). If not preconceptual, this screening should be done early enough in pregnancy to allow for additional testing if indicated, and to plan for prenatal diagnostic testing (Section 4.4.2) if desired. The couple can decline testing as well.

Ethnic-specific, panethnic, and expanded carrier screening are all possible strategies, as recommended by the American College of Obstetricians and Gynecologists (ACOG Comm 690, 2017; King and Klugman, 2018). With ethnic-based screening, populations known to be at increased risk for certain disorders are offered targeted testing, such as Tay—Sachs disease in the Ashkenazi Jewish population, although the panels vary depending on the testing laboratory (Section 3.2.3) (Hoffman et al., 2014). In light of the increasing ethnic admixture of all individuals, the alternative of panethnic

Genomics in the Clinic. https://doi.org/10.1016/B978-0-12-816478-5.00004-2

screening could be considered, in which a panel of 10−20 disorders is offered to all individuals regardless of ethnicity, and includes cystic fibrosis and spinal muscular atrophy (Section 11.1.1), with a more recent recommendation to include fragile X syndrome (Sections 5.2, 11.12.3, and 11.13.11) (Owens et al., 2018). In addition, testing is now available for less common conditions, leading to the availability of expanded carrier screening panels for over five hundred conditions, with some of these conditions not having clear guidelines for clinical course and/or management if detected (Edwards et al., 2015).

More recently, carriers of hemoglobinopathies are being identified through carrier screening gene panels. Mutations in the recessive genes (Sections 2.7.1) coding for the globin chains of hemoglobin are among the most prevalent in the world. They occur in 5% of humans (Aguilar Martinez et al., 2014) and produce the hemoglobinopathies, such as sickle cell disease (SS) (Section 11.1.1), SC disease, and the thalassemias. Each hemoglobin, the protein complex within red blood cells that binds, transports, and delivers oxygen, is made up of two alpha-globin chains and two other globin chains. In all adults, alpha- and beta-globin chains form hemoglobin A. There are four copies of the alpha-globin gene, two in tandem on each chromosome, inherited in pairs from each parent. There are two copies of the beta-globin gene, one on each chromosome, each from a parent. Mutations causing abnormal structure of the beta-globin chains will lead to abnormally formed hemoglobin, such as hemoglobin S or hemoglobin C. Mutations or deletions leading to an abnormal number or function of globin chains produce the thalassemias, such as alpha-thalassemia or beta-thalassemia. Although all of these disorders are more commonly found in individuals of African, Asian, and Mediterranean background, carrier screening for the hemoglobinopathies is offered to all pregnant women/pregnant people (subsequently referred to as pregnant women). Prenatal diagnosis through CVS or amniocentesis (Section 4.4.2) is available if both parents are found to be carriers of mutations in these genes, which, when combined, cause clinical symptoms.

Although hemoglobin electrophoresis is still the gold standard for detecting different types of hemoglobins, such as hemoglobin A, S, or C, sequencing of the beta-globin genes will detect the mutations causing beta-thalassemias to determine carrier status for all of the beta-globinopathies (Section 11.1.1). Homozygous (e.g., SS, CC) or compound heterozygous (e.g., SC) mutations causing abnormal structure of beta-globin chains do not bind or transport oxygen effectively. Symptoms of sickle cell disease (SS) or SC disease include chronic anemia and vaso-occlusion in organ systems leading to complications (e.g., pain crises, chest syndrome, hyposplenism, aplastic crises). Treatment includes specific organ management, blood transfusion, hydroxyurea and hematopoietic stem cell transplantation (Kohne, 2011). Other mutations or deletions in the beta-globin genes lead to the beta-thalassemias. A beta-thalassemia carrier has one abnormal (thalassemia minor) or one absent beta-globin chain (thalassemia intermedia). Individuals with beta-thalassemia minor are generally asymptomatic or present with mild anemia. Individuals with beta-thalassemia intermedia can have moderate to severe anemia and need blood transfusions. Individuals with both beta-globin chains functionally absent have beta-thalassemia major (also called Cooley anemia), which presents with severe anemia needing blood transfusions for life, and without treatment would lead to hepatosplenomegaly, cardiomegaly and brittle bones, and early death (Kohne, 2011). In the beta-thalassemias, a decreased MCV (below 80 fL) and an increased A2 on hemoglobin electrophoresis are often indications of carrier status, but sequencing of the hemoglobin genes is needed to confirm the diagnosis. Finally, if one beta-globin gene has a mutation causing hemoglobin S or C and the other gene has a mutation or deletion for beta-thalassemia, then the individual has sickle-beta-thalassemia, which clinically can be as severe as sickle cell disease.

An abnormal number of functional alpha-globin chains will lead to the various alpha-thalassemias (Aguilar Martinez et al., 2014). Individuals with one abnormal alpha-globin gene (likely with three instead of four globin chains on each hemoglobin) are considered silent carriers with no clinical significance. Those with two abnormal genes (and two functional alpha-globin chains) have alpha-thalassemia trait, also called a carrier, and will have mild anemia. Carriers of alpha-thalassemia should be suspected in individuals presenting with mild hypochromic (low MCH) and microcytic (low MCV below 80 fL) anemia with a normal hemoglobin electrophoresis, especially if they are of Asian or African ancestry. Molecular testing is needed to confirm carrier status. Three abnormal genes (thus a single functional alpha-globin chain) lead to hemoglobin H disease, with hemolytic anemia, and splenomegaly. When there are zero functional alpha-globin genes (no alpha-globin chains), hemoglobin Bart occurs, with severe anemia and hydrops fetalis and likely death if not treated (Kohne, 2011). Treatment includes blood transfusion and stem cell transplantation.

When assessing reproductive risks for couples in which both partners are hemoglobinopathy carriers, the specific genes must be assessed. For alpha-thalassemia carriers, it is important to determine if carriers have two abnormal alpha-globin genes in cis (on the same chromosome) or in trans (on opposite chromosomes). Individuals of Asian ancestry are more likely to have the abnormal genes in cis and thus are at increased risk for hemoglobin Bart if their partner is also a carrier. Carriers of African ancestry are more likely to have the abnormal genes in trans and thus would not be at risk for a fetus with hemoglobin Bart disease. Finally, regarding the beta-globin genes, for a partner who has sickle cell trait or is a beta-thalassemia carrier, it is important to assess the other partner for both hemoglobin S or C and beta-thalassemia, since sickle-beta-thalassemia disease clinically can be severe. However, for partners in whom one is an alpha-thalassemia carrier and the other has sickle cell trait or is a beta-thalassemia carrier, the offspring should not be at risk for one of these hemoglobinopathies because this involves two separate genes, which function independently.

4.2 High-risk Pregnancy

A high-risk pregnancy is one that involves significant morbidity or even mortality to the mother and/or the fetus(es). It can be due to inherent or acquired medical conditions of the pregnant woman or the fetus, to which genetic etiologies can contribute. Often these are indications for referral to genetic counseling.

4.2.1 Maternal Exposures

Various maternal conditions and medications taken during pregnancy can have teratogenic effects on the growth, organ development, and/or well-being of the fetus. These include fetal exposures to medications taken for conditions such as epilepsy (e.g., phenytoin, valproic acid) or psychiatric disorders (e.g., lithium), addictions (e.g., alcohol), and use of vitamin A derivatives (e.g., tretinoin). Exposure to alcohol during pregnancy can lead to fetal alcohol spectrum disorders, with a broad number and severity of effects, particularly in growth (intrauterine growth retardation, small size, microcephaly), development (cognitive, memory), and behavior (attention-deficit disorder with hyperactivity). Structural malformations occur as well, including congenital heart disease, renal malformations, and cleft palate or other craniofacial anomalies including characteristic facies with short palpebral fissures, smooth philtrum, and thin upper lip (Dejong et al., 2019) (Section 11.3.1). Diabetic embryopathy can be seen in fetuses of women with type 1 diabetes mellitus (Section 11.1.2), and less typically from women with gestational diabetes, with cause attributed to hyperglycemia, oxidative stress, abnormal mitochondrial function, and other factors (Stevenson and Hunter, 2013). Malformations associated with diabetic exposure include congenital heart defects, neural tube defects, holoprosencephaly, and caudal dysplasia (e.g., sirenomelia) (Castori, 2013). Associated conditions include femoral hypoplasia-unusual facies, oculo-auriculo-vertebral spectrum, and VACTERL association. Lastly, illnesses or infections in pregnant women can lead to other embryopathies, such as those attributed to cytomegalovirus and the more recently described Zika virus (e.g., microcephaly brain anomalies) (Section 11.3.1). If there is an increased risk, testing options would include maternal screening, detailed high-resolution ultrasound at midpregnancy, and amniocentesis for further viral or other testing (see below).

There are substances recommended for pregnant women for maintenance or preventative measures. Medications needed to control maternal pathology, such as hypertension (e.g., atenolol), asthma (e.g., albuterol), or hypothyroidism (e.g., levothyroxine), are continued during pregnancy. Preventative medications against entities such as blood clots (e.g., Lovenox) or hyperglycemia (e.g., insulin) are endorsed as well. In general, for prescribed medications, the health benefits of using a medication during pregnancy should be weighed against possible fetal risks. Most medications do not increase the baseline 2%–3% risk for a fetal anomaly that all pregnancies have. With folic acid supplementation now recommended for all pregnant women, the incidence of neural tube defects, particularly anencephaly and spina bifida, has decreased (CDC MMWR Recomm Rep, 1992). The dose of folic acid should be increased 10-fold prior to subsequent conceptions if there has been a prior neural tube defect.

4.2.2 Advanced Maternal Age and Parental Factors

Increasing parental age can impose risks for the fetus. Advanced maternal age (Section 11.1.3) and its association with aneuploidy is well-established and known to be due to nondisjunction during meiosis (Section 2.2.1). An increasing number of meiotic errors in the oocytes occur with increasing maternal age, likely due to reduced recombination, although the exact mechanisms remain unclear (Duncan et al., 2012). The most common pathologic aneuploidies able to survive are 45,X, or Turner syndrome (although most of these do not make it to term), and trisomy 21, or Down syndrome, then trisomies 18 and 13 and other sex chromosome aneuploidies. The most common aneuploidy found in early pregnancies, trisomy 16, is incompatible with life and results in spontaneous abortion. Charts are available with risk estimates for fetus or newborn with Down syndrome or chromosome abnormality (Table 4.1).

Down syndrome (Sections 2.2.1, 6.1.1, 11.1.3, 11.13.2) is mostly due to nondisjunction leading to trisomy 21, with 5% from a Robertsonian translocation involving chromosome 21. The sex chromosome abnormalities due to age-related nondisjunction are 47,XXY or Klinefelter syndrome (Section 11.12.5) and 47,XXX. Other common sex chromosome disorders are 45,X, or Turner syndrome (Section 11.7.5), mosaic in over 90% of cases, and XYY, neither associated with

TABLE 4.1 Estimate of chromosomal abnormalities at midtrimester and term according to maternal age[a]

	Midtrimester		Term	
Maternal age in years	*Down syndrome*	*All chromosomal abnormalities*	*Down syndrome*	*All chromosomal abnormalities*
33	416	185	625	285
34	322	153	500	238
35	250	132	384	204
36	192	105	303	167
37	149	83	227	130
38	115	65	175	103
39	89	53	137	81
40	69	40	106	63
41	53	31	81	50
42	41	25	64	39
43	31	19	50	30
44	25	15	38	24
45	19	12	30	19

[a]Expressed as 1/X.
Source: From Timms, D., Campbell, W.A. 2010. Modifying risk for aneuploidy with second-trimester ultrasound after a positive serum screen. Clin Lab Med 30, 677–92. PMID 20638581; original data from Hook, E.B., Cross, P.K., Schreinemachers, D.M. 1983. Chromosomal abnormality rates at amniocentesis and in live-born infants. JAMA, 249, 2034–2038. PMID 622016.

advanced maternal age. The incidence of sex chromosome disorders is approximately 1 in 400 pregnancies as detected prenatally (Linden et al., 1996).

There is a continuum of aneuploidy with increased age (Table 4.1), and the original cut-off of 35 years for offering any prenatal diagnostic procedure has loosened with improved testing. In 2007 and confirmed in 2020, the American College of Obstetricians and Gynecologists stated that all women should be eligible for invasive testing if interested (ACOG Comm 77, 2007; ACOG Comm 226, 2020). A number of screening and diagnostic tools have been developed to assess risk and provide information to expectant parents. These include high-resolution ultrasound, first trimester and noninvasive screening, and prenatal diagnostic testing (Sections 4.3, 4.4).

Advanced paternal age above 40 years is associated with a higher number of sperm with chromosome abnormalities, leading to increased infertility, although likely no increased aneuploidy in offspring. There is also an increased risk for single gene defects from faulty DNA replication and/or mismatch repair, as well as for autism and affective disorders (Yatsanko and Turek, 2018).

For couples with two or more spontaneous abortions or stillbirths with unknown etiology, chromosomal analysis is offered on both parents to detect a balanced rearrangement (e.g., translocation, inversion) carrier (Section 2.2.1). If found, prenatal diagnostic testing can be offered to test the developing fetus (Section 11.1.7). Significant conditions or birth defects in the family or consanguinity may also prompt genetic counseling (Section 3.2). For couples who screen positive on carrier screening (Section 11.1.1), again, prenatal diagnostic testing can be offered.

4.3 Fetal Imaging

4.3.1 Imaging Techniques

Ultrasound has been used for many years in conjunction with pregnancy. High-resolution ultrasound should be performed in a center with expertise, and with genetic counselors available to provide information about the fetus, specific risks and management options. Fetal ultrasound can detect half of all major organ malformations present during the first trimester, and many during the second trimester, but less well the smaller anomalies or functional defects (Edward and Hui, 2018).

In the first trimester, several specific anatomical areas are assessed to screen for potential genetic conditions. Nuchal translucency is measured between 11 weeks through 13 weeks gestation (Section 11.1.3) and the fetal nasal bone is also evaluated in the first trimester (Section 4.3.2). The ideal anatomy scan is performed at 18–20 weeks gestation. The organ systems that are most amenable to evaluation for anomalies include the craniofacial, cardiovascular, gastrointestinal, pulmonary, genitourinary, and musculoskeletal, as well as the brain. If a congenital heart defect is suspected, or the fetus is thought to have an increased risk for this (e.g., a prior affected pregnancy to the couple), a fetal echocardiogram will be recommended in addition to the high-resolution ultrasound; midtrimester is also the best time for a detailed view of the whole heart. Normal ultrasound surveillance can be very reassuring.

A fetal MRI may be recommended if there is a question of less precise characterization by the ultrasound. This has the highest sensitivity for further assessment of suspected fetal brain anomalies (Manganaro et al., 2017), but also is used to assess other anomalies, such as abdominal findings, lung volume with diaphragmatic hernia, and myelomeningocele in a presurgical setting (Edwards and Hui, 2018).

4.3.2 Abnormal Fetal Imaging

Abnormalities found in the fetus by prenatal ultrasound typically prompt genetic counseling and specific recommendations. Increasing nuchal translucency in the first trimester is one of the risk factors for chromosomal disorders. Above 3.5 mm, there is concern for other entities including congenital heart defects and single gene disorders, particularly in conjunction with other ultrasound findings. The chromosomal disorders include trisomies 21, 13, and 18 as well as Turner syndrome. Fetal edema can be due to hydrops fetalis, which when nonimmune can be due to maternal factors (e.g., parvovirus exposure) or fetal disorders (e.g., cardiovascular or urinary tract anomalies, lymphatic dysplasia, alpha-thalassemia). Single gene disorders also can present with nuchal edema, including Noonan syndrome as well as other rasopathy disorders, those due to variants in the RAS-MAPK pathway (Sections 11.4.3, 11.12.2), as can Turner syndrome (Section 11.7.5). Screening (Section 4.4.1) can be offered for detecting chromosomal disorders, but prenatal diagnostic testing (Section 4.4.2) would be needed for the single gene disorders.

Although a common variant in some ethnic groups (e.g., Afro-Caribbean), an absent or hypoplastic nasal bone also will increase the risk of fetal Down syndrome in the first trimester or may indicate certain single gene disorders especially when other findings are present (e.g., cleidocranial dysostosis) (Cicero et al., 2004; Soto et al., 2006) (Section 11.1.3).

In the second trimester, ultrasound can detect other "soft markers" for aneuploidy, especially Down syndrome (Sections 2.2.1, 6.1.1, 11.1.3, 11.13.2), that on their own usually have no clinical significance. These include echogenic focus of the heart and mild renal pyelectasis associated with Down syndrome, choroid plexus cysts seen in trisomy 18 (Section 11.1.5), and single umbilical artery associated with an increased risk for aneuploidy if there are additional findings (Gurram et al., 2018; Ebrashy et al., 2016). Findings that may have more significance include shortened long bones, associated with Down syndrome but also with skeletal dysplasias, and echogenic bowel, which can be seen with Down syndrome but could indicate bowel obstruction, cystic fibrosis, or congenital infection (e.g., cytomegalovirus) (Norton, 2013). Additional minor ultrasound findings can indicate aneuploidy, such as facial features (e.g., flat facial profile in Down syndrome), finger abnormalities (short or absent second phalanx of the fifth finger in Down syndrome, overlapping fingers in trisomy 18), and toe or foot findings (widened space between first and second toes in Down syndrome, rocker-bottom feet in trisomies 13 and 18). Major findings can also indicate aneuploidy, such as cleft lip and palate in trisomies 13 and 18, or duodenal atresia in Down syndrome. Please see Sections 4.4, 11.1.3, 11.1.5 for screening and testing options when abnormal findings are present.

The second trimester, at 18–20 weeks gestation, is the ideal time to assess organ system anomalies (see also Section 5.1 for specific congenital anomalies). Malformations can be isolated or can indicate an underlying condition, either chromosomal, single gene disorder, or teratogenic exposure. For the oral cavity, cleft lip with or without cleft palate is much more easily detected than cleft palate alone (Section 11.3.10). Numerous single gene disorders can present with cleft lip and/or palate (Section 11.3.10). Limb anomalies may be visible in the first trimester, but should be detected during the second. These can include clubfoot (Section 11.13.4), absent digits (e.g., oligodactyly or absent long bone), or extra digits (polydactyly). Many conditions may also present with congenital heart defects. Some of the more common associations include atrioventricular canal or ventricular septal defect in Down syndrome, conotruncal heart defect in 22q11 deletion syndrome (Section 11.9.2), ventricular or atrial septal defect in trisomy 18 (Section 11.1.5), atrial septal defect with limb reduction anomaly in Holt–Oram syndrome, and pulmonic stenosis in Noonan syndrome (Sections 11.4.3, 11.12.2). Decreased fetal movement can also be noted, which could be due to a number of causes including primary muscular or neurologic and which could lead to joint contractures (Section 11.1.6). Decreased or no fetal movement (fetal akinesia) also will lead to no fetal breathing, which will result in pulmonary hypoplasia. Prenatal diagnostic testing for chromosomes

(Sections 6.1.1, 6.1.3) or specific gene panels (Section 6.2.3) will be suggested, with the family able to discuss the situation and choose the subsequent course of action best for them.

Open fetal defects, such as neural tube defects or abdominal wall defects, are also easily detected, especially in conjunction with prenatal biochemical testing (e.g., elevated fetal alpha-fetoprotein (AFP)). Abdominal wall defects have specific ultrasound characteristics. Gastroschisis, often isolated and sporadic, has the contents of the abdomen floating free in the amniotic cavity through a nonumbilical opening, whereas omphalocele, typically due to a trisomy or an underlying single gene disorder (e.g., Beckwith–Wiedemann syndrome—Section 11.7.1), has a membrane extruding through the umbilicus, enclosing the abdominal contents. Anencephaly is the most severe open neural tube defect and is usually seen in isolation or with bilateral adrenal hypoplasia, but can be due to an underlying condition. Myelomeningocele (spina bifida) can occur at any vertebral level, but is most commonly lumbar and may present with hydrocephalus. Encephaloceles are much rarer (Section 11.1.8). When diagnosed prenatally, isolated neural tube defects have a low risk for an associated aneuploidy or single gene disorder, especially if no additional anomalies are present.

4.4 Prenatal Evaluation

Some couples at increased risk to have a child with a genetic condition or a birth defect may elect screening or diagnostic testing during pregnancy (Section 4.4.2), natural conception without prenatal diagnosis, or consideration of adoption, while others may choose some form of assisted reproductive technology (Section 4.5). Couples will seek the best plan of action for themselves, based on their own beliefs and understanding of the genetic risk and emotional, psychosocial, as well as financial burden of the condition of concern. Based on the most recent American College of Obstetricians and Gynecologists (ACOG) clinical management guidelines (ACOG Comm 226, 2020), all women regardless of age or risk should be offered prenatal screening for chromosomal disorders (Section 6.1.1) and prenatal diagnostic testing for chromosomal disorders as well as for microdeletions/microduplications via microarray (Section 6.1.3), in addition to standard current management. Through genetic counseling, the detection rates and risks of the different screening and diagnostic testing modalities can be explored, and decisions made with which they feel most comfortable.

4.4.1 Serum Screening

Prenatal screening is a noninvasive form of assessment typically coordinated by the obstetrician. Screening test results provide risk factors which can be shared with prospective parents to decide if diagnostic testing could be indicated. Maternal serum screening was initially performed in the second trimester by measuring fetal AFP levels in the maternal blood to assess for Down syndrome if low and open fetal defects if elevated. This evolved into the triple and then the quadruple ("quad") screen, performed at 15–22 weeks gestation, to provide improved aneuploidy screening. Down syndrome is indicated by low levels of AFP and unconjugated estriol (an estrogen steroid) and high levels of inhibin A (a pituitary suppressing hormone) and free beta–human chorionic gonadotropin (hCG) (Driscoll and Gross, 2009). Specific patterns can detect other aneuploidies.

Further research demonstrated ways to detect fetal outcomes using maternal serum in the first trimester, offering more and earlier options for diagnostic testing if indicated. In the highest risk groups for fetal abnormalities, such as mothers of advanced maternal age, the detection rates are higher. First trimester combined screening is typically performed on maternal serum from 11 through 13 weeks gestation and combines maternal age and fetal ultrasound with biochemical testing. The most sensitive ultrasound findings in predicting Down syndrome are the absence of the nasal bone and an increased nuchal translucency measurement (Section 4.3.2). Pregnancy-associated plasma protein A is produced by the placenta; a low level indicates a potential problem with the placenta or fetus and can indicate Down syndrome, and also an increased risk for pregnancy complications, including stillbirth, preterm birth, and intrauterine growth restriction. A high level of free beta-hCG also is associated with Down syndrome with this early testing. Specific patterns can detect other aneuploidies as well. First and second trimester testing, including ultrasound markers, can detect at least 90% of cases of Down syndrome and slightly less for trisomies 18 and 13, with a false-positive rate of about 1%–3%. AFP, inhibin A, and placental growth factor have been included in first trimester screening at some laboratories and can also be utilized as screening markers for preeclampsia (Carmichael et al., 2017). There can be variations of these levels based on ethnicity, maternal weight, and multiparity.

The most recent mode of aneuploidy screening on maternal serum is known as noninvasive prenatal screening, or NIPS, also called noninvasive prenatal testing (NIPT) and cell-free fetal DNA (cffDNA). Fetal DNA fragments from the placenta (fetal fraction) circulating in the maternal bloodstream can be measured after 9 weeks gestation. An abnormal amount of DNA may suggest an aneuploidy such as Down syndrome (Section 11.1.3), trisomy 18 (Section 11.1.5), and trisomy 13. Results have both clinical and statistical validity, with higher detection rates than by first trimester screening, and in a high-risk population. Over 95% of cases of Down syndrome can be identified by NIPS, with a false-positive rate below 1% (Kagen et al., 2017). The positive predictive value varies depending on the laboratory used and is the highest for Down syndrome and often the lowest for sex chromosome abnormalities such as Klinefelter syndrome, Turner syndrome, or triple X (Section 11.1.4). These tests may, however, have nonreportable results, which can be more frequent if the woman is obese, if there are twins, or for any other reason that the fetal fraction could be low. With abnormal results, couples should be referred for genetic counseling, since informed decisions may need to be made based on these results, including deciding about more invasive testing.

Initially, NIPS was offered for pregnancies at higher risk for aneuploidies due to advanced maternal age, a previous pregnancy with trisomy, abnormal first trimester screening, or with markers for Down syndrome on ultrasound (Gregg et al., 2016). Its purpose was to screen for fetal trisomies and sex chromosome disorders (Section 11.1.4). More recently, however, it has been used to identify other aneuploidies, partial aneuploidies, and copy number variants using microarray, exome sequencing, and genome sequencing. Some centers in the United States and other countries are now investigating or offering NIPS as a first-line test for all pregnancies (van der Meij et al., 2019).

Because maternal DNA also is measured on NIPS, direct information about the mother, such as an undetected malignancy or a previously unknown maternal sex chromosome aneuploidy, can also be obtained (Bianchi et al., 2018). There are also legal and ethical implications for this testing that have recently been debated (Dukhovny and Norton, 2018). For more details on ethical and psychosocial implications of prenatal testing and potential outcomes and management, please see Chapter 10.

4.4.2 Prenatal Diagnostic Testing

Prenatal diagnostic testing is a highly accurate invasive form of fetal assessment. This is offered for specific indications, including advanced maternal age, ultrasound abnormalities, parental carriers of chromosome rearrangements (Section 11.1.7) or genetic disorders, and abnormal first trimester or noninvasive screening results (Section 11.1.4). Chorionic villos sampling (CVS) is performed at 10−14 weeks gestation either transabdominally or transcervically, under ultrasound guidance. Villi are cultured and can be tested for chromosomes, specific gene changes, panels, or microarrays. There is a 1 in 350 chance of miscarriage following the procedure, often far lower than the risk for an affected fetus (Beta et al., 2018). There can be a 1% risk for a mosaic result, which could be fetal or, more likely, placental (Sections 2.2.2, 2.4.2). Amniocentesis is performed transabdominally starting at 15−17 weeks gestation, under ultrasound guidance. The amniotic fluid is cultured for cells from the fetal kidney and skin. There is a 1 in 350 or lower chance of miscarriage following this procedure (Beta et al., 2018). Results from CVS and amniocentesis are considered to be diagnostic. With the advancement of NIPS, these diagnostic procedures are performed much less in recent years (Khalifeh et al., 2016). Amniocentesis can also be performed later in gestation to obtain samples for lung maturity in near-term infants.

4.4.3 Prenatal Risk and Family Planning

For pregnant couples at increased risk for a fetal chromosome abnormality based on maternal age (Section 11.1.3), previously affected offspring, or a parental chromosome rearrangement (Section 11.1.7), several testing options are available, including first trimester screening, NIPS/circulating cell-free DNA, ultrasound surveillance, or the diagnostic chorionic villus sampling or amniocentesis. A discussion with the genetic counselor will help clarify the option best suited for the situation.

Pregnant women at increased risk to have a child with a single gene disorder, such as Tay−Sachs disease or spinal muscular atrophy (Section 11.2.4), should be offered chorionic villus sampling or amniocentesis for molecular testing. In most cases, previous testing of the parents or an affected proband will have identified a gene mutation(s) that can be detected in a future fetal sample. For some enzyme deficiencies or metabolic disorders, measurement of a biochemical product in the villus sample or amniotic fluid cells can be performed to provide the diagnosis.

For many couples who have had a child with a birth defect or a fetal anomaly diagnosed during pregnancy, no specific genetic etiology will be determined. In this situation, ultrasound surveillance can be offered for subsequent pregnancies (Section 4.3.1).

Based on results from the prenatal testing, couples will either be reassured about the reduced risk for an affected offspring or will be informed of an abnormal fetal diagnosis. In the latter case, couples will have options for continuing the pregnancy with appropriate surveillance and specialty consultations or for terminating the pregnancy. Placing a baby for adoption is another option considered by some couples who feel that they cannot care for a child with particular special needs, yet do not make the decision to terminate the pregnancy. While these complicated decisions can only be made by the patient and her partner, consultations with genetic counselors, pediatric specialists, maternal-fetal medicine obstetricians, therapists, or clergy can provide valuable insight and support. Please see also section 3.5 for further discussion.

4.5 Other Reproductive Options

Assisted reproductive technology can be used to help with conception and reduce a couple's chance of having offspring with a particular genetic condition and/or birth defect(s).

4.5.1 Preimplantation Genetic Testing

As some couples consider their options for a future pregnancy, they may wish to decrease their risk for having a child with a particular genetic condition by utilizing a combination of in vitro *fertilization* (IVF) and testing of their embryos prior to transfer. If the testing performed is for a single gene disorder (Section 2.4.1), the testing is described as preimplantation genetic testing—monogenic (PGT-M), formerly known as preimplantation genetic diagnosis (PGD). If the testing is to evaluate for chromosomal trisomy and monosomy (Section 2.2.1), the testing is called preimplantation genetic testing—aneuploidy (PGT-A), formerly known as preimplantation genetic screening (PGS). In some cases, the IVF procedure and the PGT will be performed in the same center, while in other settings the IVF group will send the embryo biopsies out to a selected laboratory for the genetic testing. The specific biopsy and laboratory testing techniques have changed over time and continue to evolve (Simpson et al., 2019; Treff and Zimmerman, 2017).

The process usually begins with a consultation with a reproductive endocrinologist, who will take a family and medical history and order some blood work. For couples having PGT-M, samples from family members will be sent to the genetic laboratory, so that a test for the condition of concern can be designed in advance. Testing of the embryos may utilize markers linked to the familial mutation(s) and/or testing for the actual mutation(s) itself (Section 11.1.8). Family samples will also need to be studied if one parent is a balanced translocation carrier (Section 2.2.1) and preimplantation genetic testing—structural rearrangement (PGT-SR) is sought to reduce the risk of having an offspring with an unbalanced rearrangement. Some couples with a history of infertility or recurrent pregnancy loss may seek PGT-A to select an embryo with a normal chromosome complement for transfer.

Once the genetic laboratory has reviewed the patient's records and performed relevant family studies, as needed, the women undergoes ovarian stimulation and egg retrieval. Fertilization of the eggs is carried out with a sperm sample, and typically, the viable embryos are biopsied after 5 days. Sometimes, especially in the case of PGT-M, sperm are injected directly into the eggs, a process known as intracytoplasmic sperm injection (ICSI), to minimize the chance of contamination with a second sperm or to increase the chance of successful fertilization. The biopsies retrieved from each embryo are then transported to the genetic laboratory for analysis. The sampled embryos are cryopreserved while the testing proceeds. When PGT-M is performed for the risk of a single gene disorder, aneuploidy screening is also carried out on each embryo. Once testing is complete, the results are reported back to the IVF center and shared with the couple. The patient can then use the results of this testing to decide which embryo(s) will be transferred to her uterus.

The successful use of IVF with PGT allows for an embryo at low risk for the condition of concern to be transferred to the patient. The costs of the process are significant, however, and may not be covered by insurance. In addition, there is some evidence to suggest that the process of IVF may increase the risk slightly for certain genetic conditions (e.g., imprinting disorders) and birth defects (Okun and Sierra, 2014). Results of aneuploidy testing for a particular embryo may be mosaic (Section 2.2.2) raising the question of whether the embryo has two cell lines or whether the abnormal cells are confined to the portion of the embryo analyzed. Finally, although the results of testing are highly accurate, there remains some chance of error secondary to technical laboratory issues. Follow-up testing through prenatal screening or diagnosis (Section 4.4) is typically offered to confirm the PGT prediction.

4.5.2 Donor Sperm/Egg

Couples at increased risk for a single gene disorder may elect to use donor sperm or an egg donor to increase their chance of a healthy offspring. In the case of a carrier couple at 25% risk to have a child with an autosomal recessive disorder (Section 2.7.1), use of a sperm sample from a donor who has had carrier testing with normal results for the condition of concern would significantly reduce the risk of an affected offspring. Using an egg donor would accomplish the same risk reduction, but is medically more complicated and expensive.

For men who carry a gene for an autosomal dominant condition (Section 2.7.1), conception for his partner using a donor sperm sample would prevent the genetic mutation from being transmitted to his offspring. If a woman is the carrier of a similar autosomal dominant or sex-linked gene mutation, use of a donor egg would have the same effect.

Some couples may find this option more acceptable than natural conception with prenatal diagnosis or PGT-M. Artificial insemination by donor and IVF with a donor egg can be arranged through a reproductive endocrinologist.

References

Aguilar Martinez, P., Angastiniotis, M., Eleftheriou, A., Gulbis, B., Manu Pereira Mdel, M., Petrova-Benedict, R., et al., 2014. Haemoglobinopathies in Europe: health & migration policy perspectives. Orphanet. J. Rare Dis. 9, 97. PMID 24980780.

American College of Obstetricians and Gynecologists (ACOG), 2007. ACOG Practice Bulletin No. 77: screening for fetal chromosomal abnormalities. Obstet. Gynecol. 109, 217–227. PMID 17197615.

American College of Obstetricians and Gynecologists (ACOG), 2017. Committee Opinion No. 690 Summary: carrier screening in the age of genomic medicine. Obstet. Gynecol. 129, 595–596. PMID 28225420.

American College of Obstetricians and Gynecologists (ACOG), 2020. ACOG Practice Bulletin No. 226: screening for fetal chromosomal abnormalities. Obstet. Gynecol. 136, 859–867. PMID 32976375.

Beta, J., Lesmes-Heredia, C., Bedetti, C., Akolekar, R., 2018. Risk of miscarriage following amniocentesis and chorionic villus sampling: a systematic review of the literature. Minerva Ginecol. 70, 215–219. PMID 29161799.

Bianchi, D.W., 2018. Cherchez la femme: maternal incidental findings can explain discordant prenatal cell-free DNA sequencing results. Genet. Med. 20, 910–917. PMID 29215645.

Carmichael, J.B., Liu, H.P., Janik, D., Hallahan, T.W., Nicolaides, K.H., Krantz, D.A., 2017. Expanded conventional first trimester screening. Prenat. Diagn. 37, 802–807. PMID 28613385.

Castori, M., 2013. Diabetic embryopathy: a developmental perspective from fertilization to adulthood. Mol. Syndromol. 4, 74–86. PMID 23653578.

Centers for Disease Control and Prevention, 1992. Recommendations for the use of folic acid to reduce the number of cases of spina bifida and other neural tube defects. MMWR Recomm. Rep. (Morb. Mortal. Wkly. Rep.) 41 (RR-14), 1–7. PMID 1522835.

Cicero, S., Rembouskos, G., Vandecruys, H., Hogg, M., Nicolaides, K.H., 2004. Likelihood ratio for trisomy 21 in fetuses with absent nasal bone at the 11–14-week scan. Ultrasound Obstet. Gynecol. 23, 218–223. PMID 15027007.

Dejong, K., Olyaei, A., Lo, J.O., 2019. Alcohol use in pregnancy. Clin. Obstet. Gynecol. 62, 142–155. PMID 30575614.

Driscoll, D.A., Gross, S., 2009. Clinical practice. Prenatal screening for aneuploidy. N. Engl. J. Med. 360, 2556–2562. PMID 19516035.

Dukhovny, S., Norton, M.E., 2018. What are the goals of prenatal genetic testing? Semin. Perinatol. 42, 270–274. PMID 30195989.

Duncan, F.E., Hornick, J.E., Lampson, M.A., Schultz, R.M., Shea, L.D., Woodruff, T.K., 2012. Chromosome cohesion decreases in human eggs with advanced maternal age. Aging Cell 11, 1121–1124. PMID 22823533.

Ebrashy, A., Kurjak, A., Adra, A., Aliyu, L.D., Wataganara, T., De Sa, R.A., et al., 2016. Controversial ultrasound findings in mid trimester pregnancy. Evidence based approach. J. Perinat. Med. 44, 131–137. PMID 26506099.

Edwards, J.G., Feldman, G., Goldberg, J., Gregg, A.R., Norton, M.E., Rose, N.C., et al., 2015. Expanded carrier screening in reproductive medicine-points to consider: a joint statement of the American College of medical genetics and genomics, American College of obstetricians and Gynecologists, national society of genetic counselors, perinatal quality foundation, and society for maternal-fetal medicine. Obstet. Gynecol. 125, 653–662. PMID 25730230.

Edwards, L., Hui, L., 2018. First and second trimester screening for fetal structural anomalies. Semin. Fetal Neonatal Med. 23, 102–111. PMID 29233624.

Fraser, F.C., 1968. Genetic counselling and the physician. Can. Med. Assoc. J. 99, 927–934. PMID 4235633.

Gregg, A.R., Skotko, B.G., Benkendorf, J.L., Monaghan, K.G., Bajaj, K., Best, R.G., et al., 2016. Noninvasive prenatal screening for fetal aneuploidy, 2016 update: a position statement of the American College of Medical Genetics and Genomics. Genet. Med. 18, 1056–1065. PMID 27467454.

Gurram, P., Figueroa, R., Sipusic, E., Kuhnly, N., Clark, S., Janicki, M.B., 2018. Isolated single umbilical artery and fetal echocardiography: a 25-year experience at a tertiary care city hospital. J. Ultrasound Med. 37, 463–468. PMID 28850695.

Hoffman, J.D., Park, J.J., Schreiber-Agus, N., Kornreich, R., Tanner, A.K., Keiles, S., et al., 2014. The Ashkenazi Jewish carrier screening panel: evolution, status quo, and disparities. Prenat. Diagn. 34, 1161–1167. PMID 24996053.

Hook, E.B., Cross, P.K., Schreinemachers, D.M., 1983. Chromosomal abnormality rates at amniocentesis and in live-born infants. JAMA 249, 2034–2038. PMID 6220164.

Kagan, K.O., Sonek, J., Wagner, P., Hoopmann, M., 2017. Principles of first trimester screening in the age of non-invasive prenatal diagnosis: screening for chromosomal abnormalities. Arch. Gynecol. Obstet. 296, 645–651. PMID 28702698.

Khalifeh, A., Weinter, S., Berghella, V., Donnenfeld, A., 2016. Trends in invasive prenatal diagnosis: effect of sequential screening and noninvasive prenatal testing. Fetal Diagn. Ther. 39, 292–296. PMID 26898118.

King, J.R., Klugman, S., 2018. Ethnicity-based carrier screening. Obstet. Gynecol. Clin. N. Am. 45, 83–101. PMID 29428288.

Kohne, E., 2011. Hemoglobinopathies: clinical manifestations, diagnosis, and treatment. Dtsch. Arztebl. Int. 108, 532–540. PMID 21886666.

Linden, M.G., Bender, B.G., Robinson, A., 1996. Intrauterine diagnosis of sex chromosome aneuploidy. Obstet. Gynecol. 87, 468–475. PMID 8598978.

Manganaro, L., Bernardo, S., Antonelli, A., Vinci, V., Saldari, M., Catalano, C., 2017. Fetal MRI of the central nervous system: state-of-the-art. Eur. J. Radiol. 93, 273–283. PMID 28668426.

Norton, M.E., 2013. Follow-up of sonographically detected soft markers for fetal aneuploidy. Semin. Perinatol. 37, 365–369. PMID 24176161.

Okun, N., Sierra, S., Genetics, C., Special, C., 2014. Pregnancy outcomes after assisted human reproduction. J. Obstet. Gynaecol. Can. 36, 64–83. PMID 24444289.

Owens, K.M., Hohany, L., Holland, C., DaRe, J., Mann, T., Settler, C., et al., 2018. FMR1 premutation frequency in a large, ethnically diverse population referred for carrier testing. Am. J. Med. Genet. 176, 1304–1308. PMID 29603880.

Simpson, J.L., Kuliev, A., Rechitsky, S., 2019. Overview of preimplantation genetic diagnosis (PGD): historical perspective and future direction. Methods Mol. Biol. 1885, 23–43. PMID 30506188.

Soto, E., Richani, K., Goncalves, L.F., Devers, P., Espinoza, J., Lee, W., et al., 2006. Three-dimensional ultrasound in the prenatal diagnosis of cleidocranial dysplasia associated with B-cell immunodeficiency. Ultrasound Obstet. Gynecol. 27, 574–579. PMID 16619383.

Stevenson, R.E., Hunter, A.G., 2013. Considering the embryopathogenesis of VACTERL association. Mol. Syndromol. 4, 7–15. PMID 23653571.

Timms, D., Campbell, W.A., 2010. Modifying risk for aneuploidy with second-trimester ultrasound after a positive serum screen. Clin. Lab. Med. 30, 677–692. PMID 20638581.

Treff, N.R., Zimmerman, R.S., 2017. Advances in preimplantation genetic testing for monogenic disease and aneuploidy. Annu. Rev. Genom. Hum. Genet. 18, 189–200. PMID 28498723.

Van Der Meij, K.R.M., Sistermans, E.A., Macville, M.V.E., Stevens, S.J.C., Bax, C.J., Bekker, M.N., et al., 2019. TRIDENT-2: national implementation of genome-wide non-invasive prenatal testing as a first-tier screening test in The Netherlands. Am. J. Hum. Genet. 105, 1091–1101. PMID 31708118.

Yatsenko, A.N., Turek, P.J., 2018. Reproductive genetics and the aging male. J. Assist. Reprod. Genet. 35, 933–941. PMID 29524155.

Chapter 5

Clinical Genetics and Referrals

George A. Diaz[1], Francis J. McMahon[2], Stuart A. Scott[1,3], Randi J. Hagerman[4] and Antonie D. Kline[5]

[1]*Department of Genetics and Genomic Sciences, Icahn School of Medicine at Mount Sinai, New York, NY, United States;* [2]*Human Genetics Branch, National Institute of Mental Health Intramural Research Program, National Institutes of Health, Bethesda, MD, United States;* [3]*Department of Pathology, Stanford University, Stanford, CA, United States;* [4]*Mind Institute, Department of Pediatrics, University of California, Davis, Sacramento, CA, United States;* [5]*Harvey Institute for Human Genetics, Greater Baltimore Medical Center, Baltimore, MD, United States*

Abstract

A genetic evaluation should be considered for any unusual constellation of findings that perplexes the primary care provider or other subspecialist, known entity associated with genetic change(s), significant family history that could be relevant to care, and/or atypical age of presentation. This includes outpatient referrals for children or adults with malformations, learning and/or behavioral issues and metabolic abnormalities, as well as inpatient evaluations for more acute or complex situations. From albinism to Zellweger spectrum disorder and from fetal life through adulthood, knowing when to refer to clinical genetics is important. Specific body organ systems can each have their own anomalies and malformations that are either isolated or indicate an underlying condition. Genetic testing also has a limited but growing role in adult common and complex disorders including psychiatry. The judicious use of genetic testing in select individuals may provide information that is of personal utility to patients and families, can detect rare genetic syndromes with common manifestations, and may help avert adverse drug events. Pharmacogenomics is playing an increasing role in the practice of medicine, in which genetics itself is a burgeoning field.

Clinical genetics historically included the study and management of chromosomal disorders, and inborn errors of metabolism (IEMs), and later broadened to include the field of dysmorphology, or the study of abnormally formed tissues with evaluation of features, anomalies, and malformations leading to syndrome identification. Although DNA and genes were identified early, the association of genes with specific conditions arrived much later. Prenatal genetics with diagnostic capabilities (Chapter 4) and cancer genetics with current multigene associations (Chapter 8) followed. More recently, adult genetics has expanded, and common and complex disorders, previously known solely as multifactorial, are being identified as having associations with genetic variants. The newer fields of pharmacogenomics and infectious disease genetics illustrate the current and ever-expanding nature of clinical genetics. Because clinical genetics is evolving to embrace almost all of medicine, this chapter covers many areas. It begins by describing findings on the medical history and physical examination that should prompt consideration of an underlying genetic condition, and the latter part relates to more specialized and newer, more complex areas of interest in clinical genetics.

5.1 Congenital Anomalies and Dysmorphic Features

Congenital anomalies are abnormalities present at birth, differing from the normal standard and occurring in about 3% of all live births. Major congenital anomalies involve an organ system anatomically or functionally, leading to some degree of morbidity, and most require medical or surgical intervention, such as severe congenital heart disease. Minor anomalies usually do not affect function, and typically need no correction; single minor anomalies, such as minor ear differences (Fig. 5.1A) or unusual hair growth patterns (Fig. 5.1B), are often familial and benign. Anomalies incorrectly formed during

Genomics in the Clinic. https://doi.org/10.1016/B978-0-12-816478-5.00005-4

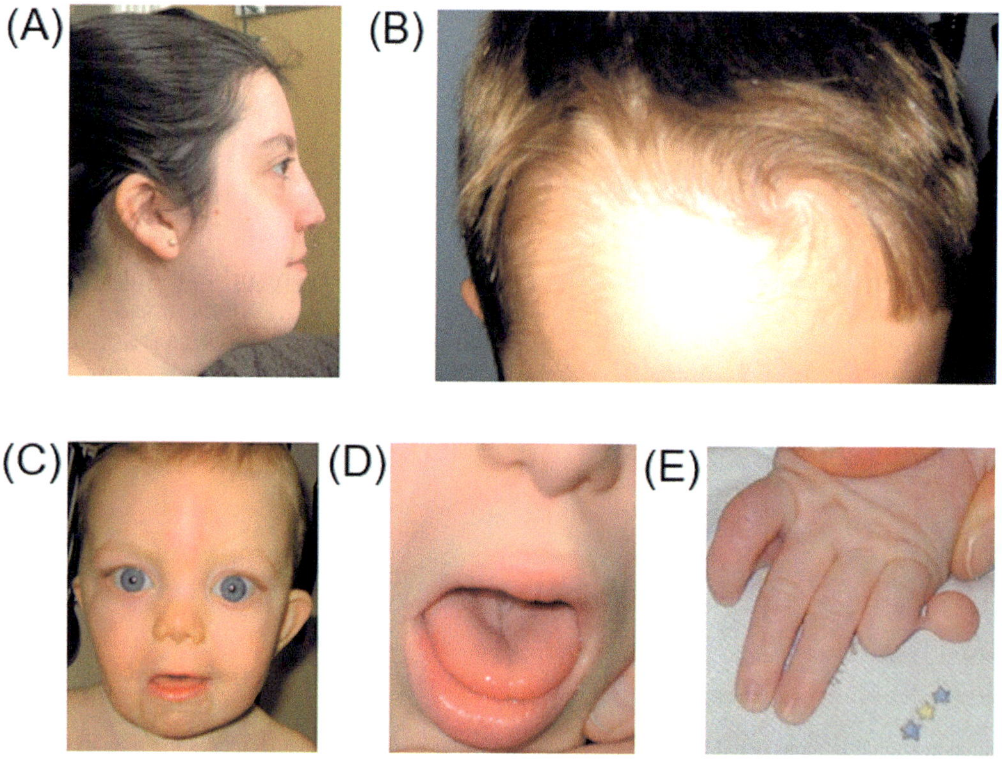

FIGURE 5.1 Anomalies can be familial or syndromic. (A) Posteriorly rotated ear, often a dominant trait [*publishing permission obtained*]; (B) frontal hair upsweep and whorl, found as isolated finding in normal individuals; (C–E) individual with Au-Kline syndrome due to a pathogenic variant in the gene *HNRNPK*. Note long palpebral fissures, prominent ears, metopic suture prominence in (C), hypoplastic alae nasi and grooved tongue in (D), and postaxial polydactyly in (E) [*publishing permission obtained*].

embryogenesis can be due to different mechanistic causes, whether genetic (up to 30% and including chromosomal or single gene changes), teratogenic (up to 10%, an external agent affecting the embryo, such as a medication taken by the mother or a viral infection), or multifactorial (up to 35% with a combination of environmental, genetic, and unknown) causes, and of unknown etiology for the remainder (Shawky and Sadik, 2011). Environmental causes include maternal infections, diabetes (Section 11.1.2), and other health-related entities, as well as maternal medications, and exposure to drugs and alcohol (Section 11.3.1). A single congenital anomaly is often isolated, sporadic, and possibly repairable by surgery but still can be genetic in origin. When two or more major anomalies are present or a set of features occurring together suspected to be due to a single cause, there is a stronger indication of an underlying genetic syndrome, and a referral to clinical genetics should be considered.

A detailed physical exam is paramount in clinical genetics. There are standard published recommendations for descriptions of parts ("elements of morphology") of the face (Allanson et al., 2009; Carey et al., 2009; Hall et al., 2009; Hennekam et al., 2009; Hunter et al., 2009), and there are standard graphs for measurements (Jones et al., 2021). There are also published descriptions of abnormally formed "elements of morphology" for hands and feet (Biesecker et al., 2009) and genitalia (Hennekam et al., 2013). Variants in genes may lead to embryological differences that may produce changes in bone structure of the face, size, shape, and position of the eyes, ears, and nose, and differences in the mouth including lip, tongue, palate, and chin. They can also produce inner organ system malformations, extremity differences, and/or skin anomalies. One or two differences could be familial and inconsequential (Fig. 5.1A, B), but three or more could indicate an underlying genetic syndrome (Fig. 5.1C–E). It is helpful to assess the individual from head to toe, taking into account the various body systems, and to have a low threshold for a referral to genetics.

Dysmorphic features refer to patterns of features that together are abnormally formed, usually facial, externally seen. These can be due either to embryologic factors, whether resulting from an underlying syndrome or condition, or to familial features. Trained clinical geneticists focus on patterns of dysmorphic features that evolve over time to help identify the underlying syndrome. There are often cardinal minor but distinctive facial features that help differentiate certain syndromes, such as those seen in Fig. 5.2.

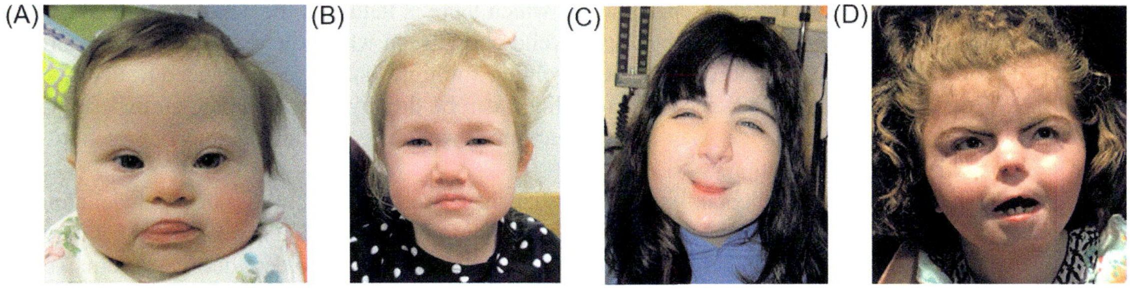

FIGURE 5.2 Facial features can be characteristic of specific genetic conditions. (A) Down syndrome, due to trisomy 21. Note flat midfacial region, upslanting palpebral fissures, epicanthal folds, prominent tongue [*publishing permission obtained*]; (B) 22q11 deletion syndrome. There is hooding of eyes, broad nasal root in most cases, small mouth and small chin, more obvious with increasing age [*courtesy of McDonald-McGinn, D., M.S., Children's Hospital of Philadelphia; publishing permission obtained*]; (C) Rubinstein-Taybi syndrome, a disorder of transcription. Note downslanting palpebral fissures, long lashes, prominent nose with beaked nasal tip and small chin [*publishing permission obtained*]; (D) Cornelia de Lange syndrome, also a disorder of transcription. Note synophrys, arched brows, long lashes, small upturned nose with anteverted nares, small mouth and chin [*publishing permission obtained*].

5.1.1 Head and Skull

The size and shape of the head, size of the fontanelles, and presence and status of the cranial bone sutures are important. Whether the individual has microcephaly (head circumference lower than two standard deviations below the mean) or macrocephaly (head circumference greater than two standard deviations above the mean) can be due to isolated or syndromic brain malformations, underlying genetic syndromes, teratogens, or a single gene change. For example, polymicrogyria, very small, smooth, disrupted brain gyri due to numerous gene changes, can be associated with microcephaly or macrocephaly, as well as intellectual disability (ID) and other findings. Exposure to teratogens in utero can produce brain changes; for example, the TORCH infections can produce microcephaly with brain abnormalities (Section 11.3.1). There is also both benign familial microcephaly and benign familial macrocephaly. Anencephaly, due to incomplete closure superiorly of the neural tube, is typically seen in isolation or with adrenal absence, but occasionally as part of syndromes. Encephaloceles are also neural tube defects and can occur in conjunction with other anomalies in a syndrome (Section 11.1.8).

The skull is made up of multiple bones separated by sutures, which come together at the fontanelles. Typically, the posterior fontanel is closed at or shortly after birth, and the majority of anterior fontanelles will close by 24 months (Pindrik et al., 2014). Early fusion of the sutures, craniosynostosis, can lead to brachycephaly and dolichocephaly (short or long anterior-posterior length, respectively), trigonocephaly (prominent forehead), or plagiocephaly (unilateral fusion and asymmetry), which can be isolated or accompany an underlying syndrome (Section 11.3.2), often with hand and foot findings. Ridging over the suture will be palpable and may cause abnormal growth around the site. Fig. 5.1C shows an example of ridging of the metopic suture of the forehead. The head shape also may be determined by external forces in utero, such as multiple births or uterine fibroids, and can produce abnormal shapes such as dolichocephaly (elongated head, often seen in premature infants), brachycephaly (flat back of head, as seen in Down syndrome, due to an extra chromosome 21—Fig. 5.2A, Section 11.13.2), or plagiocephaly (often due to positioning, either in utero or acquired after birth).

5.1.2 Facies

Facial features are noted almost instantly when encountering any individual. They can be broken down by site, as in the physical exam. Anomalies may affect form as well as function. One major malformation with two or more minor anomalies should prompt consideration of an underlying syndrome.

Eyes

For the eyes, there can be anomalies of the external and inner eye but also isolated involvement or those associated with underlying conditions. Microphthalmia, or smaller globe size, is on the spectrum with anophthalmia, absent eye. Accompanying microphthalmia may be colobomas, or deficiencies, of different parts of the inner eye, such as the iris, retina, or optic nerve, or of the eyelids, and often indicate an underlying syndrome. Macrophthalmia is not typically present at birth but could arise from inner eye pressure due to glaucoma or a mass behind the eye, such as hemorrhage or optic glioma (Section 11.11.2). The shape and position of the eye can be widely variable. The eyes may be hyperteloric (widely spaced) or hypoteloric (close set) (e.g., holoprosencephaly in which there is failure of the developing brain to split apart). The orbits can be shallow or deep-set. There can be upper eyelid hooding, in which the skin below the eyebrows droops over the upper eyelids (e.g., 22q11 deletion syndrome involving multiple body systems and due to missing part of chromosome 22—Fig. 5.2B, Section 11.9.2). The upper eyelid(s) can be droopy (ptosis) due to an underlying syndrome (e.g., Noonan syndrome, a rasopathy disorder with short stature, neck webbing, and cardiac defects—Sections 11.4.3 and 11.12.2) or due to cranial nerve dysfunction.

The majority of eyes have neutral palpebral fissures. Unusual shapes to the palpebral fissures include: almond-shaped (e.g., Prader–Willi syndrome, due to deletions or abnormal methylation of imprinted genes—Section 6.4.3—on the long arm of chromosome 15—Section 11.13.1), short (e.g., fetal alcohol spectrum disorder, due to maternal exposure of alcohol), long (e.g., Kabuki syndrome, a disorder involving low muscle tone, skeletal anomalies, and specific facial features due to variants in the *KMT2D* and *KDM6A* genes), upslanting (e.g., Down syndrome, Fig. 5.2A), or downslanting (e.g., Sotos syndrome, an overgrowth syndrome, due to mutations in the *NSD1* gene—Section 11.17.2; or Rubinstein–Taybi syndrome, a disorder of transcription regulation, due to mutations in the *CREBBP* gene—Fig. 5.2C).

Eyebrows and eyelashes may be involved in genetic syndromes. Synophrys is the term for brows joined together above the nose (e.g., Cornelia de Lange syndrome, a small stature syndrome with distinctive facial features—Fig. 5.2D, Section 11.5.3, due to mutations in genes associated with cohesin). Sparse brows can have less hairs medially or laterally. Flaring of the brows indicates abnormal direction of hair growth medially or laterally. Lashes can be sparse, absent, long, thick, curly, and ingrown. Epicanthal folds, folds over the inner canthi, are common in certain ethnic backgrounds, such as Asian and Pacific Islander, but also can accompany syndromes, such as Down syndrome (Fig. 5.2A). Congenital cranial nerve abnormality can also lead to eye movement abnormalities and facial paralysis, such as in Moebius syndrome, a syndrome with both cranial nerves VI (abducens nerve) and VII (facial nerve) involvement.

The inner eye can contain many different types of anomalies that usually imply an underlying condition. Inherited disorders of the eyes are an important cause of visual impairment in children and adults. Nearly every tissue in the eye is susceptible to inherited disease. Diseases of the retina (Sections 11.3.3–11.3.6, 11.3.9, and 11.5.4) are of considerable interest as possible targets for gene replacement therapy. One example of a cone-rod retinal dystrophy is Alström syndrome, a disorder of ciliary function with hepatic fibrosis, insulin resistance, and deafness due to mutations in the *ALMS1* gene. Congenital cataracts can accompany genetic syndromes such as in forms of muscular dystrophy (Section 11.13.5) but also can be due to inborn errors of metabolism (IEMs) such as galactosemia, due to a disorder of galactose metabolism. There can be heterochromia of the irides (e.g., Waardenburg syndrome that also includes hearing loss—Section 11.3.7). Optic nerve involvement (e.g., hypoplasia or anomaly) usually indicates an underlying condition.

Periodic vision screening is recommended throughout childhood. Visual impairment can be present from birth, such as in albinism, a disorder of decreased pigmentation (Section 11.11.1), usually accompanied by congenital nystagmus (Section 11.3.3), or with severe myopia and detached retinas, as seen in some cases of Stickler syndrome, a hereditary disorder of connective tissue due to mutations in collagen genes (Section 11.3.5). Childhood vision problems have an array of causes. Cortical blindness can occur due to brain or nervous system deficits. Acquired ocular findings could indicate storage of an underlying IEM (e.g., cherry red spot in Tay–Sachs disease or Niemann–Pick syndrome, both lysosomal storage disorders due to enzyme deficiencies).

Ears

Likewise, for the ears, the size, shape, and position, as well as anomalies of the external and inner ear, may be involved. Microtia, or small ear (length less than two standard deviations below the mean), usually is accompanied by hypoplastic or absent external ear canal and has differing degrees of involvement and conductive hearing loss. Microtia can be isolated, or due to an underlying syndrome, such as the oculo-auriculo-vertebral spectrum or craniofacial microsomia, due to defects of first and second branchial arches, often accompanied by hemifacial microsomia (hypoplasia of the ipsilateral mandible and smaller lower side of the face). Anotia, absent ears, are very rare. Longer larger ears are seen in males with the fragile X syndrome (FXS), an ID condition (Sections 5.2, 11.12.3, and 11.13.11). The overall shape of the external ear can help identify a condition: ears can be abnormally folded and positioned in CHARGE syndrome, a syndrome with choanal atresia and other malformations due to variants in the *CHD7* gene; squared with unusual tragus in 22q11 deletion syndrome (Section 11.9.2); cup-shaped; or narrow and flat. The ears may also be positioned differently, such as low-set (e.g., Noonan syndrome—Sections 11.4.3 and 11.12.2) or posteriorly rotated (e.g., chromosomal disorders or isolated; see Fig. 5.1A). There are many familial versions of minor ear findings as well.

Preauricular ear pits typically occur anterior to the ear on the cheek, but rarely can be along the helix, and can be isolated or due to an underlying syndrome (e.g., cat eye syndrome). Preauricular ear tags can be variable in size and occur anterior to the ear or along the cheek toward the mouth and also can be isolated or due to an underlying syndrome (e.g., Treacher Collins syndrome, a disorder of symmetric craniofacial anomalies). Dysplasia of the helix or antihelix can be due to a specific condition (e.g., Down syndrome with small dysplastic ears—Section 11.13.2). Inner ear anomalies can involve the ear bones (absent or dysplastic), the cochlea (malformed), vestibular aqueduct (enlarged), and the semicircular canals (dysplastic or absent).

Hearing loss is considered a major malformation, especially if sensorineural in nature. It is recommended by the American Academy of Pediatrics that all newborns be screened for hearing by one month of age; all states have implemented newborn screening for hearing (American Academy of Pediatrics, 2007). In most screening programs, otoacoustic emissions testing is used for the screening, with referral for audiology testing (e.g., auditory brainstem response) if the newborn does not pass after two attempts (Section 11.2.3). In addition, regular screening throughout childhood is also recommended. Hearing loss accompanies many genetic conditions along with other findings (Sections 11.3.7, 11.3.8, and 11.3.9), but most instances of congenital hearing loss are genetic, but nonsyndromic. Of the gene changes causing hearing loss, the majority are autosomal recessive in nature. Gene panel testing (Section 6.2.3) is available to delineate the causal mutation to see if the individual is at any increased risk for other findings. Conductive hearing loss can accompany certain genetic conditions, particularly if due to a malformation involving the ear (e.g., oculo-auriculo-vertebral spectrum with microtia and atretic ear canals), or secondarily, as in cleft palate (Section 11.3.10).

Nose

Nose shape and size also is often familial, however can be characteristic in certain conditions. The shape can be short, broad, bifid, tubular, or pear-shaped. There may be a displaced nose, or proboscis, in alobar holoprosencephaly. Anomalies may occur in the nasal bridge, which can be broad and flat, high, or nearly absent as in Stickler syndrome (Section 11.3.5). There can also be a broad nasal root as in the 22q11 deletion syndrome (Section 11.9.2), or abnormal nasal tip (e.g., flat in Potter sequence, due to oligohydramnios). There can be a nasal pit below the nares.

The philtrum (area between nose and mouth) has typical characteristics in certain syndromes. It can be long, short, prominent, or flat without pillars (e.g., fetal alcohol spectrum disorder). The philtrum can be involved in cleft lip (Section 5.1.2.4).

A sense of smell can be low (hypoosmic) or absent (anosmic), often due to a cranial nerve I abnormality. A classic genetic condition with anosmia is Kallmann syndrome; one of the hypogonadotropic hypogonadism conditions leading to early puberty can be due to mutations in the *PROK2* gene (Section 11.17.6).

Mouth and Oral Cavity

The mouth may be small or wide, often in a familial pattern, but when small and accompanied by small chin (micrognathia) and/or recessed jaw (retrognathia), might indicate an underlying syndrome (e.g., Stickler syndrome, Section 11.3.5). The mouth can also be extended laterally (e.g., Treacher Collins syndrome). The corners of the mouth may be downturned (e.g., chromosomal disorders, such as in 22q11 deletion syndrome; Fig. 5.2B). The upper lip may be thin without a clear Cupid's bow (e.g., fetal alcohol spectrum disorder), normal with an elongated M-shaped Cupid's bow (e.g., Au-Kline syndrome; Fig. 5.1C), everted, or cleft. Cleft lip can be unilateral, bilateral, incomplete (not crossing the vermilion border), or complete, into the philtrum and reaching the nose with resultant nasal deformity. Cleft lip also may be associated with a cleft palate (Section 11.3.10), which can occur in the hard (anterior) or soft (posterior) palate or both. Both cleft lip and cleft palate can be isolated, usually multifactorial in origin, or secondary to an underlying syndrome (e.g., trisomy 13, which often includes both cleft lip and cleft palate). Lower lip may be full, everted, or contain pits or mounds (e.g., Van der Woude syndrome, also seen with cleft lip and palate, due to mutations in the *IRF6* gene). The tongue can protrude when generalized hypotonia is present (e.g., Down syndrome; Fig. 5.2A). It can be enlarged, due to hyperplasia either symmetrically (e.g., Beckwith−Wiedemann syndrome, an overgrowth syndrome due to imprinted genes on the short arm of chromosome 11, Section 11.17.1) or of half (e.g., hemihyperplasia, partial overgrowth, Section 11.10.1), or due to storage of abnormal material in storage disorders along with enlarged gums, as in the mucopolysaccharidoses, IEMs due to faulty breakdown, and storage of mucopolysaccharides in lysosomes. Glossoptosis, or tongue tethered posteriorly, can be a part of the Pierre Robin sequence, which also includes cleft palate, micrognathia, and potential airway obstruction, and which can accompany a number of syndromes (e.g., 22q11 deletion syndrome, Section 11.9.2; or Stickler syndrome, Section 11.3.5).

Dental anomalies are variable and accompany underlying genetic conditions. These include missing or absent teeth, pointed teeth, peg teeth (all three found in types of ectodermal dysplasia, Section 11.11.3), fused teeth, duplicated teeth, or giant teeth. There can be delayed eruption of primary teeth and premature or delayed loss of primary teeth. Natal teeth, those present at birth, are usually isolated.

Facial features evolve with aging, and clinical geneticists are familiar with these changes. With enumerable genetic conditions, the exact pattern of features can provide clues for etiology, for overlapping features among different syndromes caused by mutations in genes of the same pathway or gene family. There are also a number of genetic catalogues (e.g., Online Mendelian Inheritance in Man, www.omim.org) and facial recognition software available on the internet to help with this.

5.1.3 Cardiovascular System

The heart is one of the most commonly involved systems in human disease, and although heart disease is often caused by lifestyle decisions, a certain portion can be genetic. Most pathogenic gene variants that lead to cardiac or vascular complications are autosomal dominant, and thus making the correct diagnosis can impact not only the individual, but also affect future generations. It is important to think about an underlying genetic condition when certain cardiovascular conditions are present.

Congenital heart disease, or heart anomalies, present in approximately 1 in 100 newborns on average, with highest incidence in Asia, then Europe, and then the Americas (van der Linde et al., 2011). Ventricular septal defect (VSD) is the most common cardiac anomaly worldwide, occurring in 40%, followed by atrial septal defect, aortic stenosis, and pulmonic stenosis (Samanek and Voriskova, 1999). Conotruncal cardiac malformations are the next most common, especially tetralogy of Fallot. Congenital heart disease can be isolated or part of an underlying syndrome, most commonly including Down syndrome (VSD, atrioventricular canal defect) (Section 11.13.2) and 22q11 deletion syndrome (conotruncal cardiac malformation, especially tetralogy of Fallot) (Section 11.9.2) (Hoang et al., 2018). Interruptions of the aorta are commonly associated with underlying syndromes such as coarctation in Turner syndrome, having an absent sex chromosome such as a 45,X karyotype (Section 11.7.5), or supravalvular aortic stenosis in Williams syndrome, a disorder of small stature due to mutations in the elastin gene. Less severe common congenital heart anomalies include bicuspid aortic valve, which can be syndromic (e.g., Marfan syndrome, a hereditary disorder of connective tissue, due to variants in *fibrillin 1*—Section 11.10.4, or Turner syndrome) and mitral valve prolapse (often associated with hereditary disorders of connective tissue). Malformations of the coronary arteries are relatively common and most are benign (Fuller and Ravishankar, 2016).

Laterality malformations usually include the heart (e.g., situs inversus, dextrocardia) and may be associated with other conditions, most commonly polyasplenia or Ivemark syndrome, and the primary ciliary dyskinesias, which also involve recurrent sinopulmonary infections and the GI tract (Section 5.1.4).

Pathogenic mutations also can lead to more specific abnormalities of the cardiovascular system. These include functional abnormalities of the heart muscle due to single gene variants, causing dilated or hypertrophic cardiomyopathy or arrhythmogenic right ventricular cardiomyopathy (Section 11.4.5) (Girolami et al., 2018). Acquired cardiomyopathy can also develop in genetic disorders (e.g., Noonan syndrome—Sections 11.4.3 and 11.12.2); in some metabolic disorders (e.g., long-chain acyl-coA dehydrogenase deficiency, a fatty acid oxidation defect, and Fabry disease, a lysosomal storage disorder); and in primary neurologic disorders (e.g., Duchenne muscular dystrophy, due to the dystrophin gene—Section 11.13.5; or Friedreich ataxia, a neurodegenerative disorder due to variants in the frataxin gene). Cardiomyopathy can also be one mode of presentation in hereditary transthyretin amyloidosis (Section 11.4.8), with deposition and accumulation of amyloid in the heart. Conduction defects can also be seen in amyloidosis. There can be inherited heart rhythm disorders, classified as channelopathy disorders, such as prolonged QT syndrome due to mutations in the *KCNQ1* gene (Section 11.4.4) or ST elevation in the Brugada syndromes (Girolami et al., 2018; Kim 2014). There can be genetic disorders of the blood vessels, including the great vessels. These would include the aortopathies as isolated conditions or associated with underlying syndromes leading to aortic aneurysms (e.g., Marfan syndrome—Section 11.10.4; Loeys–Dietz syndrome, or vascular Ehlers–Danlos syndromes—Section 11.10.5, all hereditary disorders of connective tissue), as well as arteriopathies (Section 11.4.7).

There are a number of genetic conditions related to the hematologic system, particularly the cells that contribute to blood. Many conditions can present with a bleeding tendency, either at birth (e.g., hemophilia, Section 11.9.1) or acquired, such as in bone marrow failure (Section 11.9.5). Each of the cell types in blood can have genetic causes of dysfunction. The white cells are involved in leukemias (Sections 11.16.6 and 11.16.7), lymphocytes are involved in conditions with immune dysfunction and lymphomas, and a number of genetic conditions can present with platelet dysfunction (Sections 11.9.3 and 11.9.4). Sickle cell disease produces a tendency for the red blood cells to form a sickle shape in low oxygen settings, which can prevent blood flow to vital organs, not allow full oxygenation, and lead to crisis. See also Section 4.1 for a discussion of hemoglobinopathies and thalassemias.

5.1.4 Gastrointestinal System

Gastrointestinal (GI) malformations can involve abdominal wall defects, situs anomalies, and more specific organ anomalies, such as stenosis or atresias. Omphalocele, or extrusion of abdominal contents through the umbilicus, is highly associated with genetic conditions (e.g., trisomy 13; trisomy 18—Section 11.1.5; or Beckwith-Wiedemann syndrome—Section 11.17.1). Gastroschisis, or extrusion of abdominal contents through the abdominal wall lateral to the umbilicus, is generally sporadic and can be minimal or major and associated with demise. Situs anomalies include asplenia or polysplenia, and complete situs inversus, abdominal organs in reversed or mirrored from their normal positions. These can be isolated or associated with an underlying syndrome such as the primary ciliary dyskinesias (Section 5.1.3).

Esophageal atresia is often associated with tracheoesophageal fistula and is usually isolated. Pyloric stenosis usually is a multifactorial condition due to thickening of pyloric muscle presenting at 2–4 weeks of age but can also be associated with an underlying syndrome. Duodenal atresia is the most common intestinal atresia and is associated with Down syndrome in about 25% of cases (Dalla Veccia et al., 1998). Intestinal atresias can be seen with other anomalies, including genitourinary. Imperforate anus (also called anal atresia) can be isolated or part of an underlying syndrome (e.g., Pallister–Hall syndrome, a condition with hamartomas and pituitary dysfunction, due to variants in the *GLI3* gene).

Congenital biliary atresia is a rare malformation that can be due to a number of genetic disorders (e.g., Alagille syndrome, due to mutations in the *JAG1* gene, and also with eye and cardiac anomalies). There are genetic disorders leading to hyperbilirubinemia, both direct (e.g., Alagille syndrome; Zellweger spectrum, a peroxisomal disorder; bile acid synthesis disorders; and other IEMs) and indirect (e.g., several IEMs). There are also many genetic hepatic disorders, particularly secondary (e.g., metabolic disorders leading to storage in the liver, such as the glycogen storage disorders, Section 11.5.2).

Pancreatic anomalies seen in some syndromes include an annular pancreas, which also can accompany duodenal atresia. Pancreatic involvement is one of the major findings in cystic fibrosis (Section 11.4.1), but also in diabetes mellitus type 1 with decreased production of insulin, and type 2 with decreased function of the insulin receptors. There are also single gene disorders leading to rarer types of diabetes (Section 11.7.3).

Functional bowel disorders can be genetic in etiology. Acute abdominal pain can be a presenting feature in the porphyrias, genetic disorders with a build-up of porphyrin, which aids hemoglobin (Section 11.5.5), or hemochromatosis (Section 11.5.6), a recessive disorder of the hematologic system leading to absorption of excessive iron. Emesis can be due to hyperammonemia, such as in the urea cycle disorders (Section 11.6.4). Constipation can be due to hypotonia or could indicate Hirschsprung disease, or aganglionic intestine, which can be seen in several genetic conditions, such as Down syndrome (Section 11.13.2). Hereditary transthyretin amyloidosis (Section 11.4.8) can affect the intestines and liver. Liver cirrhosis can also be seen with alpha-1-antitrypsin deficiency (Section 11.4.2). Liver fibrosis can be due to genetic causes, such as in Alström syndrome. Autoimmune conditions, such as celiac disease, have a multifactorial (Section 2.7.3) genetic component with inherited genetic susceptibility (Section 11.17.2). Please see Section 5.5 for additional cases with adult-onset presentations.

5.1.5 Genitourinary System

Genitourinary malformations comprise the second most common malformations after cardiac, comprising 35%–40% of all congenital malformations. These can involve the kidney and urinary system as well as the genitalia. Absence of both kidneys leads to the oligohydramnios (Potter) sequence, incompatible with life; unilateral renal agenesis can be benign and be detected incidentally, but also can be syndromic (e.g., branchio-oto-renal syndrome, which also includes branchial cleft remnants and hearing loss—Section 11.3.8). Prenatal exposure to maternal diabetes (Section 11.1.2) can also lead to renal and genital anomalies. There are multiple potential additional kidney anomalies, including hydronephrosis, which is usually sporadic and often due to ureteral obstruction, with the most common cause being posterior urethral valves. Autosomal dominant polycystic kidney disease is the most common of the cystic kidney disorders and is known to be associated with an increasing incidence of renal cysts with age, as well as hepatic cysts and other locations (Section 11.8.3). Polycystic kidneys can also be autosomal recessive, often lethal, as well as syndromic (e.g., ciliopathy disorders—Sections 11.1.8 and 11.5.4).

Functional disorders of the kidney are often genetic in origin either isolated or as part of an underlying condition. These can include types of glomerulonephritis (Section 11.8.2); nephropathy, as can be seen in hereditary transthyretin amyloidosis (Section 11.4.8), Fabry disease, and Alport syndrome, a hereditary disorder of connective tissue with renal involvement and sensorineural deafness; and even kidney stones. Progressively poor renal function leading to renal failure can also develop in Alström syndrome due to fibrosis.

Disorders leading to abnormal genitalia can be chromosomal, genetic, hormonal, teratogenic, or multifactorial. The most common is cryptorchidism, found in 20% of premature infants assigned male but also associated with underlying conditions (e.g., Prader–Willi syndrome, Section 11.13.1). Hypospadias has increased in incidence and occurs in 1 in 2–300 male births (van der Horst and de Wall, 2017). It can be isolated most commonly, or associated with other genital anomalies in those assigned male, and also associated with underlying syndromes (e.g., trisomy 13; trisomy 18—Section 11.1.5). Sex chromosome disorders (Section 2.7.2) can lead to smaller size of genitalia (e.g., Klinefelter syndrome with small testes, Section 11.12.5). Large testicles are noted in fragile X syndrome (FXS) after puberty (Sections 5.2, 11.12.3, and 11.13.11). Ambiguous genitalia are rare and typically associated with disorders of sex development; they can be due to hormonal dysfunction (e.g., 21-hydroxylase deficiency or other interruptions of the cholesterol metabolism pathway), chromosomal defects (e.g., trisomy 13), or single gene disorders (e.g., partial androgen insensitivity disorder—Section 11.8.1). Abnormal genitalia in those assigned female are much rarer, and functionally more benign.

5.1.6 Skeletal System

The musculoskeletal system includes bone, ligaments and tendons, and muscles, providing the framework of the body. The extremities, chest, pelvis, and the back all comprise part of the skeletal system, closely aligned with the nervous system in terms of function and thus deficits. See also Section 5.1.1 for skull.

There are primary genetic conditions producing skeletal dysplasias in which formation of the developing skeleton is compromised, resulting in abnormal shape and/or size of the bones, often with abnormal proportions. There are also genetic conditions that can lead to abnormally formed bones (e.g., hypophosphatasia with decreased bone density), "brittle" bones as in osteogenesis imperfecta (Section 11.10.3), increased bone density, or increased bone formation (e.g., multiple exostoses with random bony projections, due to variants in the *EXT1* gene). Significant underlying components of the musculoskeletal system are collagen and other fibers, precisely formed and providing stability; when genetic variants disturb these alignments, the hereditary disorders of the connective tissue occur, such as the Ehlers–Danlos syndromes (Section 11.10.5), Marfan syndrome (Section 11.10.4), and Stickler syndrome (Section 11.3.5).

Neck

The neck has few congenital anomalies. Increased nuchal translucency in a fetus can indicate a chromosomal disorder (Section 4.3.2). There can be broad or webbed neck noted at birth, particularly with remnants of lymphedema in utero, such as in Noonan syndrome (Sections 11.4.3 and 11.12.2) or Turner syndrome (Section 11.7.5). Both of these conditions can have low posterior hairline. Torticollis, or shortening of the sternocleidomastoid muscle causing a tendency to turn the neck to that side, may be present at birth if there is abnormal positioning in utero, including with crowding as in multiple births, but also can arise from neurologic abnormalities such as hypotonia. This may lead to positional plagiocephaly as well. Branchial cleft sinus or pit laterally on either side can indicate an underlying syndrome (e.g., branchio-oto-renal syndrome—Section 11.3.8).

Chest

Anomalies of the chest include the pectus deformities of the sternum and costochondral junction, with pectus excavatum protruding inward (e.g., Marfan syndrome, Section 11.10.4) and pectus carinatum protruding outward (e.g., Noonan syndrome, Sections 11.4.3 and 11.12.2). There also can be absence of the pectoralis muscle(s), as in Poland syndrome, which also includes breast and hand anomalies on the same side. Hypoplastic nipples can be seen isolated or as part of a genetic condition, and extra nipples (polythelia) are common in the general population. The most significant anomaly involving the diaphragm is the diaphragmatic hernia, which can be small but also very large, potentially involving lobes of the liver and the GI tract. Diaphragmatic hernias can be isolated but also can be associated with specific conditions (e.g., Cornelia de Lange syndrome—Section 11.5.3; Fryns syndrome, which also includes abnormal facies and digits, and is due to variants in the *PIGN* gene).

Pulmonary malformations are uncommon but can include absent lobes of the lung and cystic adenomatoid malformation, also called congenital pulmonary airway malformation, which is abnormal overgrowth of specific lung tissue with cyst formation. There can be acquired pulmonary disease changes such as in cystic fibrosis (Section 11.4.1). Functionally, alpha-1-antitrypsin deficiency can present with early emphysema, chronic obstructive pulmonary disease, and liver disease, with jaundice in childhood and cirrhosis when older (Section 11.4.2). Also see Cardiovascular Section 5.1.3 above for heart and vascular system.

Extremities

Hypoplasia or absence of the long bones can occur unilaterally or bilaterally, symmetrically or asymmetrically. These can be isolated, nonsyndromic, familial, or can occur as part of an underlying condition, particularly skeletal dysplasias, especially when symmetric. The skeletal dysplasias are a heterogeneous group of over 450 genetic conditions due to abnormal bone and/or cartilage growth typically resulting in short stature (Krakow, 2015), most with normal intelligence and few with internal malformations. They present symmetrically, involving the trunk and/or limbs, and often with disproportionate extremities; the location of the shortening helps classify the condition. There can be proximal limb or rhizomelic shortening (e.g., achondroplasia, the most common disproportionate skeletal dysplasia due to mutations in the *FGFR3* gene—Section 11.10.2), distal limb or mesomelic shortening, or acromelic shortening of the hands and feet.

Other limb anomalies may be present, not in the setting of a skeletal dysplasia. Upper limb deficiencies of all bones may occur, such as absent or hypoplastic radius and/or thumbs (e.g., in the Fanconi anemias, a heterogeneous group also with bone marrow failure and genomic instability due to variants in genes involved in DNA damage repair—Section 11.9.5; or in Holt–Oram syndrome, also with congenital heart disease, due to mutations in the *TBX5* gene). Lower limb deficiencies occur as well, but less commonly. There can be characteristic limb size differences in syndromes, such as proportionately long arms and legs (e.g., Marfan syndrome—Section 11.10.4), small hands and feet (e.g., Prader–Willi syndrome postpubertally—Section 11.13.1), or unilaterally large hands and/or feet (e.g., in hemihyperplasia—Section 11.10.1). There also can be overgrowth of part of an extremity (e.g., as in Proteus syndrome, a disorder of sporadic overgrowth due to somatic mutations in the *AKT1* gene).

Single digit anomalies may occur sporadically or as part of an underlying syndrome. Thumb findings include short broad proximally placed thumbs, digitalized thumbs, or duplicated thumbs (e.g., several craniosynostosis syndromes—Section 11.3.2). Polydactyly can be preaxial, on the radial/thumb or tibial side, or postaxial, on the ulnar or fibular side. As previously mentioned, it can be familial or sporadic, isolated or syndromic (Fig. 5.1E). Syndactyly may occur between all or some digits, including complete syndactyly (e.g., craniosynostosis disorders—Section 11.3.2) and partial syndactyly (e.g., Smith–Lemli–Opitz syndrome, a disorder of cholesterol metabolism, with the second and third toes in a Y shape). Abnormal palmar creases can indicate abnormal movement (e.g., fisting of the hand or no movement) in utero, such as in neurologic conditions, and also can accompany underlying syndromes (e.g., single palmar crease in Down syndrome—Section 11.13.2). Most genetic conditions do not have major distal extremity anomalies. See Fig. 5.3 for other examples of minor hand and foot anomalies.

Both congenital talipes equinovarus deformity (clubfoot) and congenital hip dysplasia can accompany underlying genetic syndromes (e.g., clubfoot in vascular Ehlers–Danlos syndrome—Section 11.10.5; hip dysplasia in many skeletal dysplasias). They also can have a neurologic origin, such as lack of movement in utero leading to multiple contractures, or arthrogryposis (e.g., amyoplasia—Section 11.1.6). Congenital joint contractures of the extremities can include pterygia or webbing.

Back

Anomalies of the back generally refer to the spine, although rarely scapular or muscular anomalies may occur. Vertebral anomalies include bifid vertebrae, butterfly vertebrae, and absent or duplicated vertebrae and are often associated with underlying syndromes (e.g., VACTERL association, a well-known composite of findings—Vertebral defects, Anal atresia, Cardiac defects, Tracheo-Esophageal fistula, Renal anomalies, Limb anomalies—without known gene etiology). A meningomyelocele, due to neural tube and bony defects at any level of the spine, can be open or partially closed. Scoliosis, kyphosis, and lordosis can occur in many syndromes, particularly in skeletal dysplasias or hereditary disorders of connective tissue. They are also often acquired.

5.1.7 Integument

The skin is a huge organ and can be involved in a number of underlying syndromes. It derives from ectoderm embryologically, and thus the ectodermal dysplasias encompass findings in the skin, hair, teeth, nails, and sweat glands (Section 11.11.3). Specific anomalies of the skin include vascular lesions, disorders of abnormal pigment, or abnormal texture such as in the ichthyosis (scaling) conditions, or very soft stretchy skin (as seen in the Ehlers–Danlos syndromes—Section 11.10.5).

Abnormal pigmentation can include hypo- or hyperpigmentation. The most common anomalous pigmentation seen in genetic conditions is the café au lait spot, which can range in size and shape. Multiple café au lait spots are one of the most frequent reasons for referral in clinical genetics. They can indicate a neurocutaneous disorder (e.g., neurofibromatosis type 1,

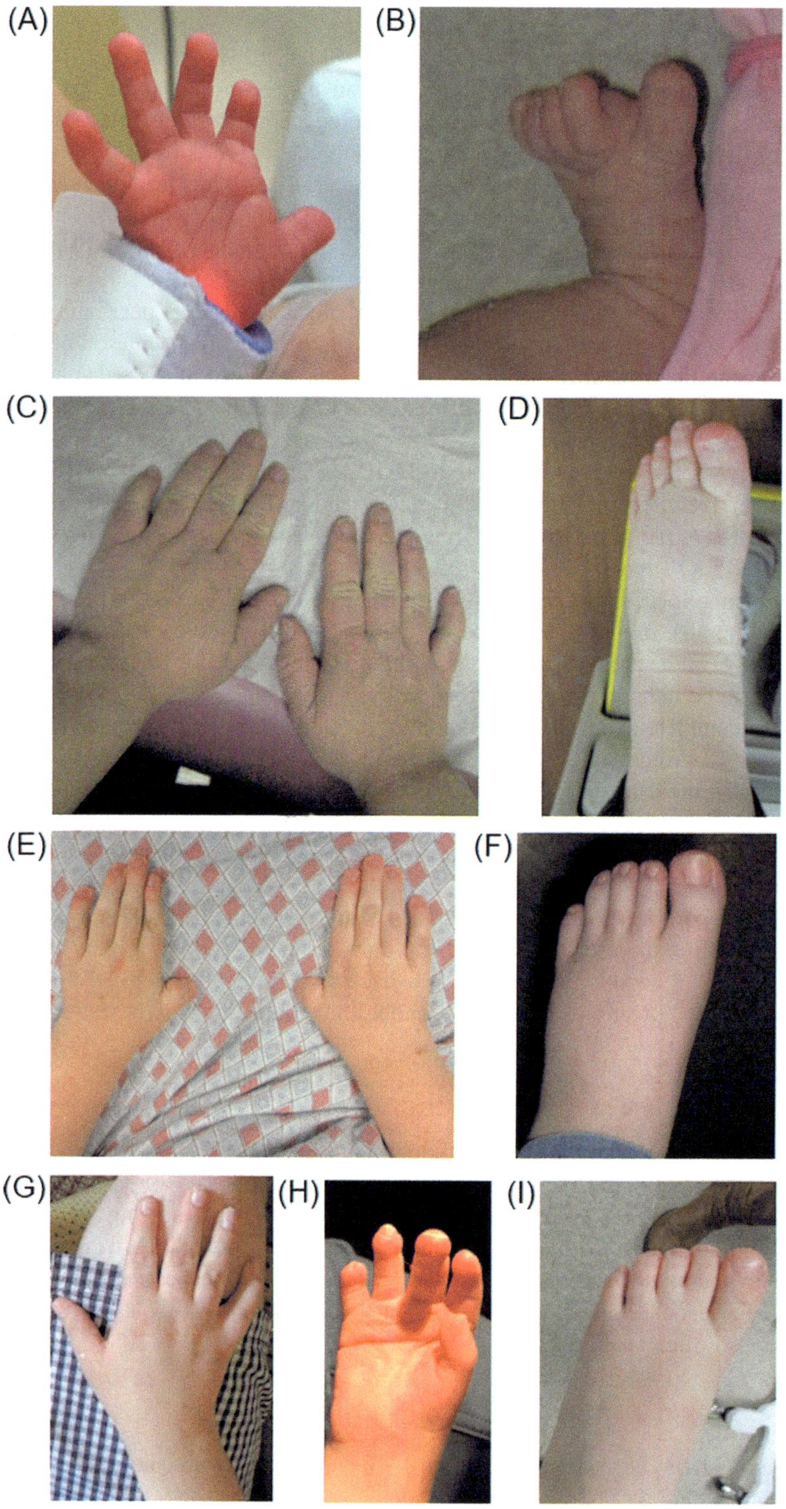

FIGURE 5.3 Minor extremity findings characteristic of specific genetic conditions. (A, B) Down syndrome. There is brachydactyly (short fingers), clinodactyly especially of the 5th finger, single palmar crease, as well as a wide space between first and second toes; (C, D) Alström syndrome, a disorder of cilia. There are broad hands, thick proximal phalanges, distally tapered fingers, and broad great toes with convex toenails; (E, F) Rubinstein–Taybi syndrome. Note very short broad thumbs and broad great toes; (G–I) Cornelia de Lange syndrome. Note brachydactyly, single palmar crease, and partial syndactyly of toes 2, 3.

which includes neurofibromas, due to mutations in the *NF1* gene—Section 11.11.2; or tuberous sclerosis, which includes brain, retina, lung, and renal lesions, due to mutations in several *TSC* genes) but can also be present in the Fanconi anemias (Section 11.9.5), occasionally in Russell–Silver syndrome, an imprinting disorder syndrome of small stature (Section 11.7.4), and larger and more irregularly shaped in other conditions (e.g., McCune–Albright syndrome, a somatic disorder that includes polyostotic fibrous dysplasia, due to somatic mutations in the *GNAS1* gene). There can also be multiple unusual nevi (e.g., Noonan syndrome with multiple lentigines—Section 11.4.3; Proteus syndrome, with epidermal nevi or cerebriform connective tissue nevi). Examples of other increased pigmentation findings include freckling (e.g., axillary or inguinal freckling seen in neurofibromatosis type 1—Section 11.11.2) and streaking (e.g., incontinentia pigmenti, a neurocutaneous syndrome with other skin findings, due to mutations in the *IKBKG* gene). Hypopigmentation may also occur either in multiple lesions (e.g., "ash leaf spots" seen in tuberous sclerosis), or globally as in albinism and its myriad causes (Section 11.11.1). Acquired hyperpigmentation can be seen with the iron deposition in hemochromatosis (Section 11.5.6), "bronze diabetes."

Lipomas may arise either sporadically, or due to an underlying condition (e.g., Proteus syndrome; or Cowden syndrome, a cancer syndrome that also includes macrocephaly and autism, due to mutation in the *PTEN* gene—Section 11.17.1). Other acquired skin findings from genetic disorders can include thickened skin from storage disorders after deposition of substrate (e.g., mucopolysaccharidoses). Most of these are not evident immediately after birth. Localized absence of skin, cutis aplasia, is typically located on the scalp (e.g., trisomy 13).

Nail anomalies occur often in ectodermal dysplasias (thickened) (Section 11.11.3), but also can be duplicated, "spoon-shaped," convex (e.g., Turner syndrome, Section 11.7.5), or absent (e.g., 5th finger absent nails in Coffin–Siris syndrome, a disorder of chromatin function due to mutations in multiple genes). Many of these may not be evident immediately at birth.

Abnormalities of the hair, in addition to sparseness related to ectodermal dysplasias, and sparseness or brittleness in ichthyosis, may include excessive hair (e.g., hypertrichosis in Rubinstein–Taybi syndrome, Fig. 5.2C), localized excessive hair (e.g., Wiedemann–Steiner syndrome with hairy elbows), kinky hair, or lack of hair pigment (e.g., Waardenberg syndrome, Section 11.3.7).

5.1.8 Neuromuscular System

The central nervous system too may have congenital abnormalities, which can be isolated or syndromic. The brain can have many potential anomalies, in all areas, detectable by brain scan. Cortical changes include abnormalities of the gyri (e.g., lissencephaly, smoothened gyri, or polymicrogyri), fusion abnormalities (e.g., holoprosencephaly), ventricular abnormalities (e.g., hydrocephalus, excessive cerebrospinal fluid in the ventricles), and acquired defects (e.g., schizencephaly, cleft in the brain). Midline defects include hypoplasia or agenesis of the corpus callosum, which can be seen incidentally on autopsy, or as part of an underlying syndrome (e.g., 22q11 deletion syndrome—Section 11.9.2), and absent septum pellucidum, which can be seen with septo-optic dysplasia. There can be degrees of hypoplasia in nearly all areas of the brain such as the cerebellum (e.g., Dandy–Walker malformation), temporal lobe, parietal lobe, occipital lobe, and brainstem. Neural tube defects include anencephaly and encephalocele, which may occur anteriorly or occipitally.

Spinal cord anomalies are rare, but include myelomeningoceles, which also present with vertebral defects. Most common are lower spinal cord defects, also called spina bifida. The peripheral nervous system can also be involved in genetic conditions, although much less commonly. Neuropathies can be genetic in cause (e.g., the hereditary sensory and autonomic neuropathies, and peripheral neuropathy seen in hereditary transthyretin amyloidosis—Section 11.4.8) or acquired (e.g., as a result of diabetes).

Muscle function is directly related to innervation. Muscular dystrophy can present at all ages and leads to muscular degeneration (e.g., Duchenne muscular dystrophy becomes symptomatic in early childhood, due to mutations in the *DMD* gene—Section 11.13.5). Myotonic dystrophy causes loss of muscle function (Section 11.13.4). Myopathies are disorders of abnormally formed muscle tissue (e.g., Nemaline myopathy presenting with hypotonia and weakness).

Functional problems of the nervous system are many, and abnormal findings on the neurologic exam can have a genetic cause. This could include abnormal gait (e.g., ataxias—Section 11.13.10), abnormal movements (e.g., dystonias—Sections 11.13.6 and 11.13.7), or tremor (e.g., fragile X tremor/ataxia syndrome, Sections 5.2 and 11.13.11). Parkinson disease can be acquired or have a genetic cause (Section 11.13.8). Seizures and epilepsy can accompany a host of genetic disorders, especially in infancy (Section 11.13.3). A decline of mental function due to genetic cause can occur in the pediatric realm (e.g., Rett syndrome, a regressive encephalopathy, due to methylation defect—Section 11.12.1), or later in life such as dementia (some forms due to genetic causes, either early—Section 11.14.1, or late—Section 11.17.5). Causes of autism spectrum disorder (ASD) are largely unknown, but there are some specific genetic causes (Sections 5.2 and 11.12.4) and a standard recommended genetic workup.

5.2 Developmental Disabilities and Behavioral Abnormalities

Intellectual disability (ID) is one of the main reasons for referral in the pediatric population. Approximately 3% of children have a developmental disability (DD), with an intelligence quotient (IQ) below 70 recognized in early childhood (Reichenberg et al., 2016). The majority of these individuals have mild ID, and approximately 0.5% of children have a severe ID with an IQ of 35 or less. There is a strong familial component in mild ID, whereas no familial factors typically are found in severe ID, with point mutations, deletions, or duplications usually being de novo (Reichenberg et al., 2016). The recommended workup for ID thus includes initially an SNP microarray (Section 6.1.3), followed by or concurrent with a gene panel (Section 6.2.3) or exome sequencing (Section 6.3.1).

The rising prevalence of autism spectrum disorder (ASD) meeting the Diagnostic and Statistical Manual of Mental Disorders, 5th edition (DSM5) behavioral criteria, includes a heterogeneous group of patients from a genetic standpoint. The prevalence of ASD is assessed regularly through two different survey mechanisms. One is through parent report of DD or ASD in children 3−17 years of age, with the latest assessment demonstrating that 5.3% of children meet criteria for one of these (Zablotsky et al., 2015). A second mechanism is the Autism and Developmental Disability Monitoring Network, which found the latest prevalence of ASD to be 16.8 per 1000 children evaluated at age 8 years at 11 sites throughout the United States (Baio et al., 2018). There are both genetic and nongenetic causes of autism (Sections 11.12.4 and 11.17.1), which in the majority of cases cannot be determined.

The most common heritable cause of ID and the most common single gene cause of ASD is fragile X syndrome (FXS) (Hagerman et al., 2017) (Sections 11.12.3 and 11.17.3). This disorder is caused by a triplet expansion (Sections 2.4.1 and 6.4.1) of CGG repeats at the 5′ end of the *FMR1* gene on the distal long arm of the X chromosome. The use of "male" and "female" below refers to assignment at birth. Expansion occurs primarily through maternal meiosis, and having an increase in the number of repeats above the normal, a premutation, confers the risk for having an affected male with a full expansion that is typically methylated. Affected males have ID, and, in 60%, ASD. In females with FXS, only one X chromosome has the full mutation, so that approximately one-third of girls with FXS have an average IQ (100 or above), although learning problems are common, one-third have ID but usually on the mild end, and one-third have a borderline IQ (between 70 and 100). Clinicians should order a fragile X DNA test whenever they identify a child with ID or ASD, and this testing is usually covered by insurance.

Although most individuals with a fragile X premutation are normal intellectually, symptoms such as anxiety, ADHD, or social deficits are common, and approximately 10% of males have ASD (Farzin et al., 2006). The premutation is also the most common genetic cause of early menopause, and approximately 20% of women with the premutation experience fragile X−associated primary ovarian insufficiency (FXPOI) (Sullivan et al., 2005). In adulthood, the premutation also commonly causes a variety of neuropsychiatric conditions, including depression, anxiety, chronic fatigue, chronic pain, fibromyalgia, and hypothyroidism; these conditions fall under the umbrella term of fragile X−associated neuropsychiatric disorders (FXAND). The most serious problem associated with premutation for FXS, however, is the neurodegenerative fragile X−associated tremor/ataxia syndrome (FXTAS), in which tremor or ataxia develop first, followed by neuropathy and cognitive decline (Hagerman et al., 2018) (Section 11.13.11).

Other causes of ID include most chromosomal disorders (e.g., Down syndrome—Section 11.13.2; trisomy 13; trisomy 18—Section 11.1.5) and partial deletion syndromes (e.g., 22q11 deletion syndrome—Sections 11.9.2 and 11.14.2), both of which should be detected on the microarray, as well as many single gene disorders (e.g., Cornelia de Lange syndrome—Section 11.5.3), certain metabolic disorders (e.g., Zellweger spectrum disorder, a peroxisomal disorder), and teratogen exposures (e.g., fetal alcohol spectrum disorder—Section 11.3.1). Regression, or decline in previously acquired milestones, can also be due to single gene disorders (e.g., Rett syndrome—Section 11.12.1), metabolic disorders with storage in the central nervous system (e.g., Tay−Sachs disease), or entities leading to Alzheimer disease affecting memory and behavior (Section 11.14.1). A tiered approach to diagnosis is recommended, with metabolic and/or gene testing following the microarray and fragile X testing (Section 11.12.3). Management typically includes referral for early intervention therapy, mandated in each state for all at-risk children (see also Section 11.3.7), and close monitoring of development with aging.

Behavioral abnormalities often accompany many conditions with ID. These can include self-injury (e.g., Lesch−Nyhan syndrome, a metabolic disorder of purines), aggression (e.g., Smith−Magenis syndrome due to microdeletions on the long arm of chromosome 17) (Powis and Oliver, 2014), anxiety (e.g., Williams syndrome), depression, and, very commonly, ADHD (e.g., neurofibromatosis—Section 11.11.2). Management typically includes referral to behavioral psychology and/or child psychiatry, as well as counseling, not only for the individual affected but also the family/caregivers.

5.3 Growth Disorders

Disruption of growth can occur prenatally (intrauterine growth restriction, IUGR), during infancy, early or late childhood, or in puberty with lack of growth spurt. Many of these will have an underlying genetic cause. Small stature or small in all growth parameters (Section 11.7.4), with or without microcephaly, should be differentiated from short stature which can be proportionate, often due to hormone imbalance, or disproportionate, usually due to an underlying skeletal dysplasia (Section 11.10.2). Uniformly small stature can be due to chromosomal disorders (e.g., trisomy 13; trisomy 18—Section 11.1.5; untreated Turner syndrome—Section 11.7.5), single gene disorders (e.g., primordial dwarfism), or hormonal causes such as growth hormone deficiency, which itself can be genetic in origin or sporadic. Failure to thrive usually refers to low weight parameters and can be genetic (Section 11.5.3), gastrointestinal, nutritional, psychiatric, or social (e.g., neglect). Genetic causes often include chromosomal or single gene etiology.

There can be generalized overgrowth, with all growth parameters involved including macrocephaly (Sections 11.7.1 and 11.7.2), or regional overgrowth, with hemihyperplasia of all or part of the limbs, digits, face, tongue, and/or genitalia (Section 11.10.1). Generalized overgrowth can be teratogenic (e.g., diabetic embryopathy), syndromic (e.g., Beckwith–Wiedemann syndrome—Section 11.7.1; or Klinefelter syndrome—Section 11.12.5), or familial (Section 11.7.1). Regional overgrowth can be due to single gene disorders (e.g., Klippel–Trenaunay syndrome, an autosomal dominant condition with hemangiomata and asymmetric hypertrophy of limb), somatic changes (e.g., Proteus syndrome), or chromosomal (e.g., WAGR complex, or Wilms tumor, aniridia, genitourinary malformations, and growth retardation or ID, due to deletion on chromosome 11). Growth hormone excess is rare and acromegaly affects primarily the head and the hands and feet.

Obesity can be dietary in cause most commonly, but also can be familial and/or genetic. Single gene syndromes may include obesity, as seen in both Alström and Bardet–Biedl (Section 11.5.4) syndromes, both with insulin resistance and diabetes type 2. Prader–Willi syndrome classically included obesity after the age of 2 years, but with growth hormone treatment, this has been largely alleviated (Section 11.13.1). There is an association between leptin resistance and obesity. Also, monoamine oxidase A deficiency can present with obesity and behavioral problems, but minimal other findings.

5.4 Metabolic Disorders

Inborn errors of metabolism (IEMs) are a broad class of disorders encompassing defects in the synthesis or degradation of carbohydrates, amino acids, or lipids (Table 5.1). The accumulation of metabolites that results from reduced flux through a synthetic or catabolic pathway provides an effective means to diagnose individuals affected with inborn errors. Importantly, significant accumulation of metabolites can occur prior to the development of clinical symptoms. The recognition that early treatment with a specialized medical formula could prevent severe ID in the case of phenylketonuria (PKU) motivated the initiation of newborn screening for this disorder starting in the 1960s (Guthrie, 1962). The development of tandem mass spectrometry as a diagnostic tool opened the door to more widespread screening for inborn errors starting in the late 1990s (Wilcken et al., 2003). Currently all states screen for a core set of 26 IEMs (e.g., PKU, maple syrup urine disease) and most screen for many more than these (Section 11.2.1). In addition to inborn errors, the core conditions on the Recommended Uniform Screening Panel also include cystic fibrosis, hearing loss, severe combined immunodeficiency,

TABLE 5.1 Examples of different types of inborn errors of metabolism.

Inborn errors of metabolism	Examples
Urea cycle disorders	Ornithine transcarbamylase deficiency, citrullinemia, argininosuccinic aciduria, argininemia, citrin deficiency
Organic acidemias	Propionic acidemia, methylmalonic aciduria, isovaleric acidemia, glutaric acidemia, maple syrup urine disease
Fatty acid oxidation defects	Medium-chain acyl-CoA dehydrogenase deficiency, carnitine palmitoyl transferase 1 deficiency, long-chain hydroxyacyl-CoA dehydrogenase deficiency
Amino acidopathies	Tyrosinemia, phenylketonuria, homocystinuria
Carbohydrate disorders	Glycogen storage disorders, galactosemia, fructosemia

Adapted from: https://www.genome.gov/Genetic-Disorders/Inborn-Errors-of-Metabolism.

hemoglobinopathies, congenital cardiac disease, hypothyroidism, congenital adrenal hyperplasia, and spinal muscular atrophy (Section 11.2.4).

Although many treatable metabolic disorders are included on newborn screening panels (Section 11.2), it is important to note that not all treatable metabolic disorders are currently screening targets. Importantly, severe forms of some disorders commonly present before screening results are available or, in the case of milder variant forms, may not be flagged by the screening assay. Severe cases of organic acidemias, urea cycle, or fatty acid oxidation disorders will usually present in the first few days of life and require management in an intensive care environment. Milder forms of screened disorders may not manifest until a situation of significant catabolic stress occurs and may be recurrent with no symptoms between episodes (Section 11.6.1). Such a metabolic decompensation can happen at any point during a patient's life, including into adulthood. It is thus important for primary care and other nongeneticist providers to keep in mind clinical scenarios for which referral to a metabolic specialist may be appropriate.

5.4.1 Clinical Symptoms

Common reasons for referral to a metabolic specialist can be grouped into several general categories. These include the evaluation of patients with neurological signs/symptoms (developmental delay or regression, neurobehavioral changes, neuroimaging findings), gastrointestinal symptoms (cyclic or recurrent vomiting, failure to thrive), laboratory findings (unexplained hypoglycemia—Sections 11.5.2 and 11.6.3, or metabolic acidosis), or physical signs (organomegaly, unusual body odor—Section 11.6.1). Some examples of features that may be observed in different types of inborn errors are provided in Table 5.2. These features may overlap in suggestive patterns for different disorders such as the co-occurrence of unusual odor and encephalopathy in maple syrup urine disease (Section 11.6.1), developmental regression and organomegaly in mucopolysaccharidoses, and behavioral changes, vomiting, and encephalopathy in urea cycle disorders (Section 11.6.4). The most favorable outcome of a metabolic evaluation is the identification of a treatable disorder in order to maximize the outcome for the patient. However, even in the case of disorders without an effective therapy, establishing a diagnosis for genetic counseling purposes is important as most metabolic disorders have an autosomal recessive inheritance pattern (Section 2.7.1) and a recurrence risk of 25% with each pregnancy.

TABLE 5.2 Signs/symptoms associated with different types of inborn errors.

Sign/Symptom	Inborn error examples
Symptoms manifest during catabolic stressor (fever, surgery, prolonged fasting, corticosteroids)	Organic acidemias, urea cycle disorders, amino acidopathies, fatty acid oxidation disorders, acute intermittent porphyria
Cyclic or recurrent vomiting	Organic acidemias, urea cycle disorders, amino acidopathies, mitochondrial disorders
Hepatosplenomegaly	Gaucher disease, Niemann–Pick disease types A, B, C, mucopolysaccharidoses
Cardiomyopathy	Pompe disease, Danon disease, Barth syndrome, mitochondrial disorders, carnitine uptake deficiency
Recurrent rhabdomyolysis	Fatty acid oxidation disorders, mitochondrial disorders
Myopathy	Glycogen storage disorders, mitochondrial disorders, fatty acid oxidation disorders
Unusual body odor	Glutaric aciduria type I or II, isovaleric acidemia (sweaty feet), phenylketonuria (musty or mousy), maple syrup urine disease (maple syrup), tyrosinemia (boiled cabbage)
Marfanoid habitus	Homocystinuria
Abnormal brain MRI findings	Leukodystrophies, Leigh syndrome
Abnormal eye findings (cataract, clouding, retinal findings)	Galactosemia, lysosomal storage disorders
Failure to thrive or short stature	Mucopolysaccharidoses, mucolipidoses, urea cycle disorders, glycogen storage diseases

Adapted from: https://www.genome.gov/Genetic-Disorders/Inborn-Errors-of-Metabolism.

Neurobehavioral Concerns

A history of developmental delay or ASD is a common reason for referral to genetics. Whereas chromosomal microarray (CMA) (Section 6.1.3) is diagnostic in 15%–20% of these patients (Miller et al., 2010), the diagnostic yield of metabolic evaluations in patients with nonsyndromic autism has been reported to be as low as 0.5% (Schiff et al., 2011). These statistics support an approach in which metabolic evaluation is part of a tiered approach in patients with nonsyndromic ID or global developmental delay who have had a negative first-line genetic screening (CMA, fragile X testing) (Section 5.2).

Recent referral and evaluation guidelines for nongeneticist providers (Michelson et al., 2011; Moeschler et al., 2014; Press et al., 2016; Belanger and Caron, 2018) suggest consideration of a metabolic evaluation in patients in whom some aspect of the clinical history raises suspicion for an underlying inborn error. Important historical elements and physical findings for which to assess in patients with developmental delay or unexplained ID include a history of regression of milestones (Sections 11.12.1 and 11.13.9), hypotonia (Sections 11.6.2 and 11.13.1) or hypertonia, or findings suggestive of myopathy.

Episodic Symptoms

IEMs that result in the accumulation of toxic metabolites may present as recurrent episodes of unexplained encephalopathy, a history of recurrent vomiting (Section 11.6.4) or failure to thrive or complaint of an unusual body odor (Section 11.6.1). The underlying inborn error will dictate the clinical presentation but in general, amino acidopathies will present with isolated neurological symptoms (e.g., late-diagnosed PKU, maple syrup urine disease) whereas organic acidemias (e.g., methylmalonic acidemia, propionic acidemia) may present with encephalopathy, metabolic acidosis, and hyperammonemia. Patients with tyrosinemia and galactosemia who were not picked up on newborn screen, such as infants adopted from other countries or whose parents opted out of screening, may present with liver and kidney dysfunction as well as encephalopathy.

Ophthalmologic findings suspicious for an underlying IEM include the presence of a cherry red spot, cataracts, or maculopathy. Extra-neurological findings of particular interest include organomegaly, coarse facial features, and joint contractures.

Important historical elements to assess in the case of a patient with repeated episodes of emesis, acidosis, encephalopathy, or neurological symptoms (dystonia, ataxia, hallucination) include dietary habits, such as fasting intervals, preferred food types, and aversions. In milder cases of intoxication disorders, parents may note a history of selective eating including protein aversion in the case of urea cycle disorders or organic acidemias, or carbohydrate aversion in the case of hereditary fructose intolerance or citrin deficiency.

Multisystemic Presentation

Hepatomegaly and/or splenomegaly noted on a physical examination is concerning for malignancy or primary liver or hematological disease but may also signal an underlying inborn error (Section 11.5.2). Assessments of patients with organomegaly should include careful evaluation for associated signs of storage in other body compartments (Section 11.5.2). The most common lipid storage disorder, Gaucher disease, will not result in facial findings but will generally have readily detectable skeletal involvement in the femurs (Section 11.9.4). The less common acid sphingomyelinase deficiency (Niemann–Pick type A or B) may have a similar degree of organomegaly but will have more prominent pulmonary pathology with increased interstitial markings on chest radiographs. In contrast, patients with glycogen storage disease type 1 have a characteristic facies with prominent cheeks and a doll-like face resulting from tissue storage in addition to characteristic biochemical features including hypertriglyceridemia, hyperuricemia, hypoglycemia, and lactic acidosis (Section 11.5.2). Even more striking tissue accumulations are observed in the mucopolysaccharides with variable degrees of neurological involvement but generally significant storage in skin and joints leading to coarse facies and joint and skeletal pathology evolving over time.

Mitochondrial disorders can present in myriad forms and can be difficult to diagnose. Mitochondrial disorders are usually multisystemic but in some cases clinical symptoms may be restricted to a single organ system, further complicating their diagnosis. Features that should raise suspicion for a mitochondrial condition include unexplained lactic acidosis, apparent life-threatening event, strokelike episodes, seizures, myopathy, neuropathy, or developmental delay, especially if clinical signs or symptoms occur in three or more organ systems (Sections 11.6.2 and 11.6.5). In cases of mitochondrial disease caused by mutations in mitochondrial DNA, symptoms may be transmitted through the maternal lineage. Maternal family history of early-onset diabetes, vision and/or hearing loss, or other neurological disorders should trigger suspicion for a potential mitochondrial disorder. It is important to remember that most mitochondrial components are encoded by nuclear, not mitochondrial, DNA so autosomal recessive and dominant modes of inheritance are more commonly observed in this class of disease than pure maternal transmission.

5.4.2 Abnormal Laboratory Findings

In patients with developmental delay, laboratory abnormalities may provide a clue to the presence of an undiagnosed inborn error. Recurrent hypoglycemia and metabolic acidosis are among the more common laboratory findings leading to a metabolic referral (Section 11.6.3). In the course of working up patients with unexplained delay or other neurological symptoms, some providers may check ammonia and/or lactate levels. It is important to note that hyperlactatemia/hyperammonemia may be a laboratory artifact if samples are not collected free-flowing (without a tourniquet) and placed immediately on ice. Truly elevated lactic acid (Section 11.6.5) and ammonia levels, as well as severe hypoglycemia and metabolic acidosis, should be considered emergencies and a cause to refer to the metabolic specialist for immediate management. Respiratory alkalosis in the setting of an elevated ammonia is also concerning for the presence of a urea cycle disorder. If an amino acid profile is sent for suspicion of an amino acidopathy, the sample should be sent after fasting as postprandial samples will generally have nonspecific elevations which may cause concern.

Molecular carrier screening for genetic disorders (Section 4.4.1) has long been known to on occasion incidentally identify putatively asymptomatic individuals carrying pathogenic variants on each copy of a screened gene (Verlingue et al., 1993; Balwani et al., 2010). Increasingly, as prenatal screening panels are including an expanding number of target genes, apparently healthy individuals who carry two pathogenic variants in a particular gene associated with metabolic diseases may be identified. Assessment for the presence of associated symptoms is warranted but care should be taken to not unduly increase anxiety in these patients as they may require no additional medical intervention. Examples of such referrals include individuals who carry mild hyperphenylalaninemia-associated variants in the phenylalanine hydroxylase (*PAH*) gene or partial biotinidase deficiency—associated alleles in the *BTD* gene. Generally, such individuals will not require medical management but may benefit from referral for purposes of reassurance, particularly if these test results are returned during a pregnancy.

5.4.3 Management of Metabolic Disorders

Patients who have been diagnosed with an IEM are generally managed with substrate restriction, amino acid supplementation and diversion, replacement or chaperone therapy as appropriate to the underlying disorder. For organic acidurias, urea cycle disorders, and amino acidopathies, substrate restriction involves limitation of natural protein intake and supplementation with specialized medical formulas that lack the specific offending amino acids relevant to each disorder (e.g., phenylalanine for PKU—Sections 11.2.2 and 11.6.6; branched chain amino acids for maple syrup urine disease, or nonessential amino acids for urea cycle disorders). Similarly, amino acids that are downstream of the metabolic block will need to be provided in excess (e.g., tyrosine for PKU, citrulline or arginine for specific urea cycle disorders). In addition to dietary therapy, new pharmaceutical therapeutic options have become increasingly available for IEMs. In Section 11.6.6, sapropterin dihydrochloride is used as a chaperone therapy to increase the residual phenylalanine hydroxylase activity in a patient with PKU. As an example of diversion therapy, reduction of total body nitrogen to lower ammonia levels in patients with urea cycle disorders (Section 11.6.4) is accomplished with phenylbutyrate or phenylacetate-based medications. New therapies, including enzyme replacement (Section 11.9.4) and gene therapies, are currently in development for a variety of different IEMs. See Section 7.4.2 for information about practice guidelines for abnormal newborn screening results.

5.5 Adult Onset Conditions

Many genetic conditions have obvious findings at birth, especially those affecting growth, facial features, organ system development, or neurologic status, but might continue to have morbidity through adulthood. Developmental delays are often unclear until over 2 years of age when language skills should be solidifying, and ID is usually diagnosed somewhat later. Other conditions, especially with a spectrum of involvement and those on the milder end, take longer to manifest. There are a number of genetic conditions that do not arise until adulthood. These can be divided by inheritance and body system.

5.5.1 Mendelian Disorders

Most malformations and syndromes involving the craniofacial system arise at birth, but morbidity from these can persist long into adulthood. For example, genetic eye conditions also can have both childhood and adult onset, such as retinitis pigmentosa, which can be isolated but also can be associated with deafness, as in Usher syndrome (Section 11.3.9). Types of sensorineural deafness, genetic in etiology, can be adult onset, such as that due to a schwannoma in neurofibromatosis type 2. Tooth abnormalities in ectodermal dysplasia often need life-long management with prosthodontics, orthodontics, and endodontists (Section 11.11.3).

Genetic cardiac disorders often occur in adulthood, particularly channelopathy disorders (Sections 5.1.3 and 11.4.4). Cardiomyopathy in adults can be genetic in origin, such as in hereditary transthyretin amyloidosis (Section 11.4.8). There are lipid metabolism disorders that give rise to hyperlipidemias (Section 11.4.6), and there are many genetic factors predisposing to hypertension, atherosclerosis, and stroke. Adult-onset pulmonary genetic conditions could include some forms of cystic fibrosis that present more mildly (Section 11.4.1). Genetic causes of gastrointestinal conditions in adulthood primarily involve the liver (e.g., porphyrias—Section 11.5.5) and could also involve the lung (e.g., alpha-1-antitrypsin deficiency—Section 11.4.2—that can present with emphysema). Hemochromatosis (Section 11.5.6) can affect multiple organs in adulthood; it can lead to cirrhosis of the liver, congestive heart failure, and diabetes. Diabetes itself can be familial and genetic (Section 11.7.3) in a certain subset of the population, more likely when two or more generations are involved.

Renal conditions due to genetic causes arise much more commonly in adulthood (Section 5.1.5), especially autosomal dominant polycystic kidney disease, that also can affect the liver (Section 11.8.3). Infertility in those assigned both male and female at birth can have genetic causes, including sex chromosome defects (e.g., Klinefelter syndrome—Section 11.12.5); in cystic fibrosis (Section 11.4.1), those assigned male at birth nearly always have infertility due to congenital bilateral absence of the vas deferens, and those assigned female at birth may also have some decreased fertility (Ahmad et al., 2013).

There are also many genetic neuromuscular conditions that have delayed age of onset into adulthood (e.g., Charcot–Marie–Tooth disease, a degenerative nerve condition, often due to variants in the *PMP22* gene; some forms of limb-girdle muscular dystrophy, and myotonic dystrophy—Section 11.13.4). There are genetic forms of Parkinson disease (Section 11.13.8). Alzheimer disease can be genetic (Section 11.14.1), including with autosomal dominant inheritance, or associated with genetic risk factors, especially if both alleles have apolipoprotein E4 on chromosome 21 (Section 11.17.5). In Down syndrome (Section 11.13.2), although all affected individuals will show pathologic brain findings characteristic of Alzheimer disease, dementia will present in only slightly over half of these individuals.

5.5.2 Multifactorial Disorders

There are many adult-onset conditions with multifactorial (Section 2.7.3) influence. For many complex disorders, such as in mental illness, there are certain principles for which clinical genetic testing does not yet apply in the overwhelming majority of cases. Hypertension can have a genetic cause, such as when present with neurofibromatosis type 1 (Section 11.11.2), but much of the time is multifactorial and has a number of influences. Obesity can have many causes, typically multifactorial, although some forms can have a genetic influence, such as with variants in the melanocortin 4 receptor gene, which typically functions in the brain and adrenal gland. Some cases of type 2 diabetes, very common in the general population, are directly due to genetic conditions (e.g., Alström syndrome, or epigenetic factors) (Zanuso et al., 2017), but most are multifactorial and environmental. HLA class 1 antigens can influence some multifactorial conditions, such as type 1 diabetes (Saberzadeh-Ardestani et al., 2018). Immunologic conditions can also be related to HLA-B27 specifically, which has highest association with ankylosing spondylitis and other autoimmune conditions. Potential reactions to multiple drugs can be also influenced by presence of HLA antigens (Tables 5.2 and 5.4; Section 11.14.3). Finally, cancers too can be multifactorial, as well as sporadic, inherited, or secondary to a new mutation in the individual (Chapter 8). See Section 5.7 below for psychiatric disorders, as an example of a group of heterogeneous and multifactorial conditions with onset primarily in adulthood.

5.6 Pharmacogenomic Testing

Pharmacogenomics is the study of the genetic determinants of interindividual drug response variability, and increasing enthusiasm for implementing clinical pharmacogenomic testing is evidenced by the growing number of personalized medicine programs that preemptively genotype germline pharmacogenomic variants and the recent availability of clinical practice guidelines for pharmacogenomic-guided therapy (Scott, 2011; Shuldiner et al., 2013; Relling and Klein, 2011; Relling et al., 2020). Additional support for the implementation of pharmacogenomics for the use of healthcare providers includes the increasing use of clinical decision support (CDS) with pharmacogenomic-guided recommendations deployed through the electronic health record (Hicks et al., 2016; Goldspiel et al., 2014). Practice guidelines and CDS are critical for interpreting pharmacogenomic test results, as provider education on pharmacogenomics has historically been very limited. Other considerations for providers when incorporating pharmacogenomics into their practice include the level of evidence supporting the clinical validity of a gene/drug response association, the evolving regulatory and reimbursement landscape as it relates to pharmacogenomics, the availability of pharmacogenomic tests from certified clinical laboratories, and the appropriate translation of genotype results into inferred phenotypes and their clinical significance.

5.6.1 Ordering Pharmacogenomic Testing

CLIA-certified laboratories can offer pharmacogenomic tests using either validated laboratory-developed tests or commercially available products approved by the US FDA. However, given that there currently are no professional standards that explicitly govern which genes should be included in clinical pharmacogenomic tests, providers typically are expected to pursue the appropriate gene or gene panel based on clinical indication and/or medication(s) of interest. A small number of medications have FDA requirements for preemptive pharmacogenomic testing in their drug labels and/or black box warnings (e.g., *HLA-B*5701* for abacavir hypersensitivity, Mallal et al., 2008); however, the majority of available pharmacogenomic tests are not currently required prior to prescribing an associated medication. As such, providers should always consider pharmacogenomic test results in the context of all available clinical information, as drug response variability typically is influenced by multiple factors. Moreover, given the recent availability of direct-to-consumer pharmacogenomic testing (Lu et al., 2017), patients should always be counseled to never independently modify their treatment based on personal pharmacogenomic test results and always consult with their care providers (Section 9.3).

Single gene pharmacogenomic tests are available for providers and typically include common drug-metabolizing enzymes with variable activity (e.g., CYP2D6, CYP2C19, CYP2C9) and well-characterized variant alleles that encode abnormal enzyme activity (e.g., *CYP2D6*4*, *CYP2C19*2*, *CYP2C9*3*) (Cavallari et al., 2017). These single gene tests are practical when a provider is interested in a potential predisposition to a drug response phenotype for selected medications with well-known pharmacogenomic effects (e.g., codeine, clopidogrel, warfarin, see Sections 11.15.1 and 11.15.2); however, multiplexed pharmacogenomic gene panels (Section 6.2.3) based on clinical specialty are increasingly available for providers, which can provide results that inform on multiple drugs and/or drug classes. Examples of clinical specialties with available pharmacogenomic panel tests include, but are not limited to, cardiology, psychiatry, pain management, and oncology. Although the gene content within these pharmacogenomic panels currently may vary between clinical laboratories, many professional organizations affiliated with clinical genetic testing (e.g., ACMG, AMP, CAP, CDC, FDA) are increasingly producing practice guidelines and policy statements to harmonize pharmacogenomic testing across laboratories and facilitate its appropriate use.

5.6.2 Pharmacogenomic Results

Clinical pharmacogenomic reports typically include an interpretation that categorizes drug-metabolizing enzymes (e.g., CYP2D6, CYP2C19, CYP2C9) into specific phenotypes for the tested patient: ultrarapid (UM), rapid (RM), normal (NM), intermediate (IM), or poor (PM) metabolizer. The phenotypes of other pharmacogenomic genes can include an inference of function (e.g., SLCO1B1), drug sensitivity (e.g., the warfarin target VKORC1—Section 11.15.1), increasing intolerance to a drug (e.g., mercaptopurine in acute lymphoblastic leukemia—Section 11.15.3), or simply for the presence or absence of a risk allele (e.g., HLA-A and HLA-B—Section 11.14.3). Nomenclature for pharmacogenomic alleles is governed by the Pharmacogene Variation (PharmVar) Consortium (Gaedigk et al., 2019) and standardized nomenclature for allele function and phenotypes have been defined and reported by the Clinical Pharmacogenetics Implementation Consortium (CPIC) (Caudle et al., 2017).

An ongoing challenge for clinical providers is translating pharmacogenomic test results into actionable therapeutic prescribing decisions, as several sources exist, and the available evidence is continually evolving. Although the drug labels of a select few medications do provide some pharmacogenomic and/or metabolizer phenotype-based dosing information, publicly available evidence-based clinical practice guidelines are also provided by the NIH-funded CPIC (Table 5.3). Importantly, CPIC guidelines do not determine or recommend if pharmacogenomic testing should be performed or not; rather, they specifically are intended to provide clinical recommendations to providers when a pharmacogenomic test result is available or already known (Sections 11.15.1, 11.15.2, and 11.14.3). The CPIC guidelines are structured in a consistent

TABLE 5.3 Clinical Pharmacogenetics Implementation Consortium (CPIC) guidelines.

Drug or drug class	Gene(s)	CPIC references in the journal, Clinical Pharmacology and Therapeutics
Abacavir	*HLA-B*	Martin et al. (2012, 2014)
Allopurinol	*HLA-B*	Hershfield et al. (2013), Saito (2016)
Atazanavir	*UGT1A1*	Gammal et al. (2016)
Atomoxetine	*CYP2D6*	Brown et al. (2019)
Carbamazepine and Oxcarbazepine	*HLA-B* and *HLA-A*	Leckband et al. (2013), Phillips (2018)
Clopidogrel	*CYP2C19*	Scott et al. (2011, 2013)
Codeine	*CYP2D6*	Crews et al. (2012, 2014)
Efavirenz	*CYP2B6*	Desta et al. (2019)
Fluoropyrimidines	*DPYD*	Caudle et al. (2013), Amstutz (2018)
Interferon-alpha	*IFNL3*	Muir et al. (2014)
Ivacaftor	*CFTR*	Clancy et al. (2014)
Ondansetron and Tropisetron	*CYP2D6*	Bell et al. (2017)
Phenytoin	*CYP2C9* and *HLA-B*	Caudle et al. (2014)
Rasburicase	*G6PD*	Relling et al. (2014)
Simvastatin	*SLCO1B1*	Wilke et al. (2012), Ramsey (2014)
Selective serotonin reuptake inhibitors (SSRIs)	*CYP2D6* and *CYP2C19*	Hicks et al. (2015)
Succinylcholine	*RYR1* and *CACNA1S*	Gonsalves et al. (2019)
Tacrolimus	*CYP3A5*	Birdwell et al. (2015)
Tamoxifen	*CYP2D6*	Goetz et al. (2018)
Tricyclic antidepressants (TCAs)	*CYP2D6* and *CYP2C19*	Hicks et al. (2013, 2017)
Thiopurines	*TPMT* and *NUDT15*	Relling et al. (2011, 2013, 2018)
Voriconazole	*CYP2C19*	Moriyama et al. (2016)
Warfarin	*CYP2C9* and *VKORC1*	Johnson et al. (2011, 2017)

For current list of available CPIC guidelines and citations, refer to https://cpicpgx.org/guidelines/.

format that is designed to enable practitioners to efficiently identify the translation from reported genotypes and diplotypes to drug response phenotypes and clinical recommendations, including background information on the strength of evidence supporting the genotype/phenotype association and recommended therapeutic actions.

Although the clinical utility of broadly implementing pharmacogenomic testing for many medications is still being debated and evaluated in ongoing clinical trials, the emerging evidence supporting clinical validity coupled with increasing enthusiasm from precision medicine programs and selected medical centers has resulted in an escalating demand on current and future clinical providers to have appropriate education, CDS, and competencies in the field of pharmacogenomics.

5.7 Genetics of Mental Illness

As previously mentioned, most recently there has been enormous growth in studying the genetics of complex medical conditions. This latest expansion in the all-encompassing field of clinical genetics has been fast-moving and far-reaching into all areas of medicine. Comprehension of the integration of both familial heredity and the many aspects of disease risk is a goal for both patients and families seeking genetic evaluation and the medical professionals evaluating numerous common medical conditions. As an example, psychiatric disorders that are indeed multifactorial (Section 2.7.3) with a strong polygenic component have inspired countless, collaborative studies to date, and contributed greatly to our further understanding of the brain and its function, in particular. Learning about the genetics of psychiatric disorders is an excellent way to understand the application of clinical genetic principles.

Psychiatric disorders comprise a broad range of illnesses whose primary presentation involves distressing or impairing disturbances in mood, perception, and behavior. Major classes include mood and anxiety disorders, such as depression, bipolar disorder, and panic disorder; psychotic disorders, such as schizophrenia; and behavioral disorders, including eating disorders, addictions, and phobias. A range of neurodevelopmental disorders, such as autism and attention deficit hyperactivity disorders, may present in the psychiatric clinic as well, either as the primary focus or as conditions comorbid with other psychiatric disturbances. See Section 5.2 for genetic evaluation of patients with neurodevelopmental disorders.

Taken together, psychiatric disorders constitute one of the largest contributors to death and disability around the world. This is because psychiatric disorders typically begin early in life, pursue a chronic course, and lead to substantial risk for personal suffering, occupational failure, and suicide. Available treatments are often effective in relieving symptoms but are not a cure. There are no proven forms of primary prevention, and relapse is common. Despite this rather bleak picture, there is increasing optimism that substantial clinical advances are imminent. This optimism rests largely on recent progress in genetic research.

5.7.1 Psychiatric Disorders and Heritability

It has long been known that many psychiatric disorders are substantially heritable, with genetic contributions to risk as high as 90%. Among the most heritable conditions are autism, schizophrenia, and bipolar disorder. Genetics appear to play a significant role even in common disorders such as depression, anxiety, and addiction. In the absence of established etiologies or biomarkers, genes have come to be viewed as key risk factors, the elucidation of which could shed light on causes, treatment, and prevention.

5.7.2 Genetic Epidemiology

Genetic understanding of the origins of psychiatric disorders predates molecular approaches by many decades. Long before it was possible to perform a genetic association study in humans, researchers used the methods of genetic epidemiology to investigate mental illness. Genetic epidemiology exploits naturally occurring patterns of illness in relatives to draw inferences about the sources and magnitude of illness risk. While observational in nature, genetic epidemiology has established a solid basis from which molecular and experimental studies of mental illnesses have been able to go forward.

Family and Twin Studies

Families are a natural unit of genetic investigation. They provide a constrained genetic and environmental background against which individual risk alleles for psychiatric disorders can be evaluated. On the other hand, in families it is difficult to disentangle purely genetic effects from those attributable to shared environment. This is especially problematic when gene–environment correlations play a big role. For example, families at high genetic risk for alcoholism share risk alleles, but also may share an environment where drinking behavior is modeled for offspring and where alcohol is more readily accessible.

Twin studies seek to disentangle genetic from environmental influences by comparing rates of mental illness in monozygotic twins, who share all inherited genetic risk factors, to dizygotic twins, who share, on average, only half. Environmental risk factors are assumed to be equally shared by both types of twins. The difference in illness rates between monozygotic and dizygotic twins thus provides a measure of heritability, or how large a role genetic risk factors play in the illness. This basic design can be applied to other classes of relatives, such as half-siblings and first cousins.

Twin studies have consistently demonstrated that most psychiatric disorders are heritable. The magnitude of heritability ranges from 80% to 90% in autism, schizophrenia, and bipolar disorder, to about 10% in anxiety or alcohol use disorders, while most psychiatric conditions show intermediate values. Twin studies also demonstrate significant sharing of genetic risk factors across different psychiatric disorders, known as coheritability (Kendler et al., 1995). The greatest coheritability is observed between depression and anxiety disorders, between various substance use disorders, and between various psychotic disorders.

Adoption Studies

Adoption studies provide an additional strategy for dissociating genetic from environmental influences. In the typical design, adoptees with a given diagnosis are compared to their adoptive and biological parents. This approach is limited by the difficulties that often attend ascertainment of adoptees whose adoptive and biological parents are all available for study, so has not been as widely applied to psychiatric conditions. Landmark adoption studies of schizophrenia and of bipolar disorder clearly demonstrated elevated rates of illness only in the biological parents of ill adoptees (Kety et al., 1971; Mendlewicz and Rainer, 1977). More recent studies based on large, population-based registries have produced similar results in much larger samples and have also demonstrated elevated rates of illness in adoptive parents of adoptees with other psychiatric disorders (Lichtenstein et al., 2009).

5.7.3 Molecular Genetics of Psychiatric Illnesses

The clear familiality and heritability of most psychiatric disorders contributes to major efforts to identify specific genes. These efforts initially took the form of genetic linkage studies, but the complex genetic architecture of mental illness led to slow progress and uncertain results. Candidate gene association studies were also largely unsuccessful, partly due to the inherent limitations of the method and partly due to our inability to really make educated guesses about the genes to study. Genome-wide association studies (GWAS), which allow an unbiased, genome-wide search for common genetic variants in unrelated individuals, finally have provided a study design better suited to the complex genetic architecture of psychiatric illnesses. These studies have now successfully identified hundreds of genetic markers that are reliably associated with major psychiatric diagnoses. These include over 250 loci associated with schizophrenia, one of the largest case–control samples studied so far (Pardiñas et al., 2018). Fewer significant associations have been identified for other major disorders, such as bipolar disorder (Stahl et al., 2019) and depression (Howard et al., 2018), but both theoretical models and empirical evidence predict that many additional risk loci will be identified as sample sizes grow.

Some broad conclusions can already be drawn with confidence. Individual risk alleles have small effect sizes, with genotype relative risk values on the order of 1.1 and none above 1.4 in large samples. Common variants appear to account for 5%–30% of the individual variation in risk (heritability), leaving a substantial heritability gap in comparison to estimates from twin studies (Manolio et al., 2009). While the great success of the GWAS approach has provided important research opportunities, from the standpoint of clinical applications, however, GWAS findings so far offer little utility.

Copy Number Variants

One class of genetic variation that appears to exert a more substantial impact on risk in psychiatric disorders is copy number variations (CNVs) (Section 2.4.1). While individually rare (below 1% frequency), taken together, CNVs comprise an important class of genetic risk factors for neuropsychiatric disorders, especially neurodevelopmental disorders (Cook and Scherer, 2008). To date, several dozen recurrent CNVs have been described in association with neurodevelopmental disorders, schizophrenia, or bipolar disorder (Table 5.4). Importantly, CNVs typically confer much higher risk for disease than common SNPs (Section 2.4.1), with odds ratios on the order of 5–50 (Kirov, 2015). However, these CNVs have little diagnostic specificity within the broad range of neuropsychopathology. Many occur within the context of contiguous gene syndromes, with significant and sometimes life-threatening pathology in cardiovascular, renal, or other organ systems outside the central nervous system. For example, as shown in Table 5.4, in the 22q11.2 deletion syndrome (Sections 11.9.2 and 11.14.2) with a known risk for neurodevelopmental issues, there is about a 12% chance of developing schizophrenia and a 3% chance of developing bipolar disorder.

Penetrances are estimated with CalPen (Addepalli et al., 2020) using carrier frequencies and sample sizes reported in Ref. (Kirov, 2015) and prevalence values of 1% schizophrenia (Scz) or 2% bipolar disorder (BD).

Sequencing

Most of the substantial heritability of mental illness has not been explained by either common risk alleles identified by GWAS or rare CNVs. This has prompted a search for other kinds of genetic variation that may be responsible. Next-generation sequencing technologies now allow for routine and cost-effective sequencing of most of the coding regions of the genome (exome), and whole-genome sequencing that includes most of the rest of the genome is now widely used in research (Sections 6.2.2, 6.3.1, and 6.3.2). While sample sizes remain underpowered to detect most rare sequence variants, early results suggest that rare single-nucleotide variants in coding regions (Section 2.4.1) play an important role at least in schizophrenia (Genovese et al., 2016).

TABLE 5.4 Penetrance estimates for schizophrenia (SZ) and bipolar disorder (BD) for some of the most frequently observed copy number variants.

CNV	Fraction of group carrying the CNV (%)			Estimated Penetrance (%)	
	Controls	Scz	BD	Scz	BD
1q21.1 del	0.021	0.170	0.030	7	3
1q21.1 dup	0.037	0.130	0.099	3	5
NRXN1 del	0.020	0.180	0	8	<1
3q29 del	0.001	0.082	0.025	20	17
7q11.2 dup	0.006	0.066	0	9	2
15q11.2 del	0.280	0.590	0.170	2	1
15q12-13 dup	0.006	0.083	0	10	2
15q13.3 del	0.019	0.140	0.043	6	4
16p13.11 dup	0.130	0.310	0.011	2	<1
16p11.2 dup	0.030	0.350	0.130	10	8
22q11.2 del	0.000	0.290	0.012	12	3

De novo Studies

Studies of de novo DNA sequence variation, which typically compare exome sequences (Section 6.3.1) in two parents and their affected offspring, show that 2–3 new mutations or variants (Section 2.4.1) typically occur de novo with each meiosis (Gratten et al., 2013). If one or more of these mutations happens to exert a deleterious effect on a gene important in neurodevelopment, the offspring may exhibit a range of neuropsychiatric symptoms. Early studies detected deleterious de novo variants in neurodevelopmental disorders, such as ID, ASD, and childhood-onset epilepsy (Iossifov et al., 2014). Later studies have found similar results in schizophrenia, albeit in a smaller fraction of cases (Fromer et al., 2014). Since each de novo event is unique, de novo studies typically implicate a large variety of genes involved in neurodevelopment. Statistical evidence in support of a pathogenic role for an individual gene depends upon the detection of recurrent de novo events in that gene, typically requiring very large sample sizes.

Case–Control Studies

Parental DNA may be difficult or impossible to collect, especially in adult-onset disorders. For this reason, exome (Section 6.3.1) and whole-genome sequencing (Section 6.3.2) is often performed in large samples of cases and controls. Study designs must carefully correct for ancestry mismatching between cases and controls (population stratification), which can be particularly problematic with rare alleles.

Unlike GWAS, rare variant studies must usually exploit statistical methods that improve statistical power by estimating the burden of risk alleles within a gene or set of genes (Liu and Leal, 2012). Genetically isolated populations, in which certain otherwise rare alleles have drifted to higher frequencies, may provide the advantage of increased statistical power when multiple cases share the same risk allele (Zuk et al., 2014).

5.7.4 Pharmacogenomics in Psychiatry

Genetic variation influences the pharmacokinetics (absorption, distribution, and metabolism) or pharmacodynamics (biologic target) of many drugs. Genetic variation may also play an important role in serious adverse events, which are often influenced by immune pathways encoded by complement and major histocompatibility genes.

Among genes involved in pharmacokinetics, those encoding the drug-metabolizing cytochrome P450 enzymes have been especially well studied. Common genetic variants in these genes are known to influence the metabolism of several medications commonly used for treatment of psychiatric disorders (Hodgson et al., 2014). The best characterized variants are summarized in Table 5.5.

TABLE 5.5 Some pharmacogenetic assays relevant for psychiatric medications.

Gene	Risk alleles	Medications	Outcome
HLA-A	*31:01	carbamazepine, oxcarbazepine	Severe adverse cutaneous events
HLA-B	*15:02	carbamazepine, oxcarbazepine	Severe adverse cutaneous events
CYP2D6	slow metabolizer	antidepressants	Elevated blood levels
CYP2C19	slow metabolizer	antidepressants	Elevated blood levels

While many studies have sought to identify pharmacodynamic variants, replicated findings are few and remain controversial (McMahon, 2014). The lack of robust findings probably reflects the difficulties inherent in measuring drug response, limited sample sizes, and genetic heterogeneity of the targeted illnesses. Polygenic analyses based on large numbers of markers spanning the genome have detected significant effects, but the proportion of variance explained remains too small for clinical relevance (GENDEP Investigators et al., 2013).

Serious adverse events are rare with drugs commonly used in psychiatry but have occurred in people taking some less commonly used drugs. The anticonvulsant, carbamazepine, for example, which is occasionally used as an adjunctive treatment for bipolar disorder, can provoke severe adverse cutaneous reactions, such as Stevens–Johnson syndrome (Section 11.14.3), especially in individuals with certain alleles in HLA genes.

The second major group of serious adverse events of relevance for psychiatry comprises agranulocytosis and other blood dyscrasias in patients exposed to the antipsychotic drug clozapine. The risk of agranulocytosis and consequent need for regular monitoring of blood counts severely limits the use of this highly effective medication. Two rare HLA haplotypes are each independently associated with high risk of agranulocytosis in patients exposed to clozapine (Goldstein et al., 2014). However, more than half of clozapine-associated agranulocytosis cases do not carry either risk haplotype, so genetic testing can only be used to identify high-risk patients.

5.7.5 Clinical Genetic Testing in Psychiatry

Currently genetic testing is not widely used in clinical psychiatry. While the evidence base is still incomplete, findings from ongoing studies of common and rare risk alleles, as well as pharmacogenomics, suggest that future clinical practice will be increasingly informed by the use of genetic information (Nurnberger et al., 2018). This information will likely come from SNP microarray (Section 6.1.3), exome sequencing (Section 6.3.1), or whole-genome sequencing (Section 6.3.2) data collected as part of the medical record. For the near future, however, psychiatrists will need to select genetic tests that are most appropriate for specific clinical questions. Several professional and regulatory organizations provide guidelines that may help with decisions about when to use which genetic tests (e.g., Genetic Testing Statement, 2019) but with the exception of neurodevelopmental disorders presenting in children (Miller et al., 2010), no clinical practice standards have yet been established.

Family risk assessment, based on a thorough review of the mental health history of first-degree relatives, provides the most information and is unlikely to be supplanted by molecular genetic testing in the foreseeable future. This underscores the importance of professional training aimed at how to take a good family history, knowledge of recurrence risks, and interpretation of this information for patients and their families (Besterman et al., 2019). In some situations, such as families that are heavily loaded with psychiatric and other illnesses, referral to a genetic counselor may be indicated.

CNV testing can play a role in clinical psychiatric evaluation. Some rare genetic syndromes, such as 22q11 deletion syndrome, often present with a psychiatric illness complicated by additional medical problems that require specialized evaluation and treatment (Section 11.14.2). Aside from clinical utility, the establishment of a "genetic diagnosis" may have considerable "personal utility" for patients and their families (Kohler et al., 2017), allaying some of the guilt and stigma that often accompany a psychiatric diagnosis and potentially improving medication adherence.

Polygenic Risk Scores

Large numbers of common genetic variants seem to account for the greatest share of genetic risk for common mental illnesses. This fact has raised interest in the use of a "polygenic risk score" (PRS) in clinical psychiatry in differential diagnosis or treatment selection (Section 11.17.5). Current evidence suggests that psychiatric PRSs lack both sensitivity and specificity, complicating the interpretation of these results in clinical settings. The predictive power of PRSs depends on heritability and the size of informative reference samples (Middeldorp and Wray, 2018). Thus, better reference samples may lead to more informative PRS results, especially in those patients who fall into the extremes of the score distribution. Empirical evidence from other complex phenotypes suggests that people in the extreme upper tail of the PRS distribution may be at substantially increased risk of illness. If this also proves true for psychiatric disorders, then it may be possible to use PRSs to identify individuals who could benefit from primary prevention. This information could also conceivably inform the selection of appropriate treatments early in the course of illness. For example, individuals who present with a depressive episode and are found to have a very high PRS for bipolar disorder might do better if treated with medications other than antidepressants, since these are known to provoke manic episodes in people with bipolar disorder.

Pharmacogenomic Tests

While pharmacogenetic tests are already widely used to inform treatment selection in clinical psychiatry, the evidence base supporting clinical utility remains quite limited. As a result, there is considerable controversy as to when and if pharmacogenomic testing should be employed. There is also little consistency in the genetic composition of the most widely marketed test panels. Most published recommendations focus on specific drug–gene pairs, such as HLA and carbamazepine (Amstutz et al., 2014). It seems clear that pharmacogenomic testing (Section 5.6), when employed in clinical psychiatry, should be considered as just one part of the complex set of information that clinicians must consider in treatment selection (McMahon, 2022).

References

Addepalli, A., Kalyani, S., Singh, M., Bandyopadhyay, D., Naga Mohan, K., 2020. CalPen (Calculator of Penetrance), a web-based tool to estimate penetrance in complex genetic disorders. PLoS One 15, e0228156. PMID 31995602.

Ahmad, A., Ahmed, A., Patrizio, P., 2013. Cystic fibrosis and fertility. Curr. Opin. Obstet. Gynecol. 25, 167–172. PMID 23429570.

Allanson, J.E., Cunniff, C., Hoyme, H.E., McGaughran, J., Muenke, M., Neri, G., 2009. Elements of morphology: standard terminology for the head and face. Am. J. Med. Genet. 149A, 6–28. PMID 19125436.

American Academy of Pediatrics, Joint Committee on Infant Hearing, 2007. Year 2007 position statement: principles and guidelines for early hearing detection and intervention programs. Pediatrics 120, 898–921. PMID 17908777.

Amstutz, U., Shear, N.H., Rieder, M.J., Hwang, S., Fung, V., Nakamura, H., et al., 2014. Recommendations for HLA-B*15:02 and HLA-A*31:01 genetic testing to reduce the risk of carbamazepine-induced hypersensitivity reactions. Epilepsia 55, 496–506. PMID 24597466.

Amstutz, U., Henricks, L.M., Offer, S.M., Barbarino, J., Schellens, J.H.M., Swen, J.J., et al., 2018. Clinical Pharmacogenetics Implementation Consortium (CPIC) guideline for dihydropyrimidine dehydrogenase genotype and fluoropyrimidine dosing: 2017 Update. Clin. Pharmacol. Ther. 103, 210–216. PMID 29152729.

Baio, J., Wiggins, L., Christensen, D.L., Maenner, M.J., Daniels, J., Warren, Z., et al., 2018. Prevalence of autism spectrum disorder among children aged 8 Years - autism and developmental disabilities monitoring network, 11 sites, United States, 2014. MMWR Surveill. Summ. 67, 1–23. PMID 29701730.

Balwani, M., Fuerstman, L., Kornreich, R., Edelmann, L., Desnick, R.J., 2010. Type 1 Gaucher disease: significant disease manifestations in "asymptomatic" homozygotes. Arch. Intern. Med. 170, 1463–1469. PMID 20837833.

Belanger, S.A., Caron, J., 2018. Evaluation of the child with global developmental delay and intellectual disability. Paediatr. Child Health 23, 403–419. PMID 30919832.

Bell, G.C., Caudle, K.E., Whirl-Carrillo, M., Gordon, R.J., Hikino, K., et al., 2017. Clinical Pharmacogenetics Implementation Consortium (CPIC) guideline for CYP2D6 genotype and use of ondansetron and tropisetron. Clin. Pharmacol. Ther. 102, 213–218. PMID 28002639.

Besterman, A.D., Moreno-De-Luca, D., Nurnberger Jr., J.I., 2019. 21st-Century genetics in psychiatric residency training: how do we get there? JAMA Psychiatr. 76, 231–232. PMID 30601877.

Biesecker, L.G., Aase, J.M., Clericuzio, C., Gurrieri, F., Temple, I.K., Toriello, H., 2009. Elements of morphology: standard terminology for the hands and feet. Am. J. Med. Genet. 149A, 93–127. PMID 19125433.

Birdwell, K.A., Decker, B., Barbarino, J.M., Peterson, J.F., Stein, C.M., Sadee, W., et al., 2015. Clinical Pharmacogenetics Implementation Consortium (CPIC) guidelines for CYP3A5 genotype and tacrolimus dosing. Clin. Pharmacol. Ther. 98, 19–24. PMID 25801146.

Brown, J.T., Bishop, J.R., Sangkuhl, K., Nurmi, E.L., Mueller, D.J., Dinh, J.C., et al., 2019. Clinical Pharmacogenetics Implementation Consortium guideline for cytochrome P450 (CYP)2D6 genotype and atomoxetine therapy. Clin. Pharmacol. Ther. 106, 94–102. PMID 30801677.

Carey, J.C., Cohen Jr., M.M., Curry, C.J., Devriendt, K., Holmes, L.B., Verloes, A., 2009. Elements of morphology: standard terminology for the lips, mouth, and oral region. Am. J. Med. Genet. 149A, 77–92. PMID 19125428.

Caudle, K.E., Thorn, C.F., Klein, T.E., Swen, J.J., McLeod, H.L., Diasio, R.B., et al., 2013. Clinical Pharmacogenetics Implementation Consortium guidelines for dihydropyrimidine dehydrogenase genotype and fluoropyrimidine dosing. Clin. Pharmacol. Ther. 94, 640–645. PMID 23988873.

Caudle, K.E., Rettie, A.E., Whirl-Carrillo, M., Smith, L.H., Mintzer, S., Lee, M.T.M., et al., 2014. Clinical Pharmacogenetics Implementation Consortium guidelines for CYP2C9 and HLA-B genotypes and phenytoin dosing. Clin. Pharmacol. Ther. 96, 542–548. PMID 25099164.

Caudle, K.E., Dunnenberger, H.M., Freimuth, R.R., Peterson, J.F., Burlison, J.D., Whirl-Carrillo, M., et al., 2017. Standardizing terms for clinical pharmacogenetic test results: consensus terms from the Clinical Pharmacogenetics Implementation Consortium (CPIC). Genet. Med. 19, 215–223. PMID 27441996.

Cavallari, L.H., Beitelshees, A.L., Blake, K.V., Dressler, L.G., Duarte, J.D., Elsey, A., et al., 2017. The IGNITE Pharmacogenetics Working Group: an opportunity for building evidence with pharmacogenetic implementation in a real-world setting. Clin. Transl. Sci. 10, 143–146. PMID 28294551.

Clancy, J.P., Johnson, S.G., Yee, S.W., McDonagh, E.M., Caudle, K.E., Klein, T.E., et al., Clinical Pharmacogenetics Implementation Consortium, 2014. Clinical Pharmacogenetics Implementation Consortium (CPIC) guidelines for ivacaftor therapy in the context of CFTR genotype. Clin. Pharmacol. Ther. 95, 592–597. PMID 24598717.

Cook Jr., E.H., Scherer, S.W., 2008. Copy-number variations associated with neuropsychiatric conditions. Nature 455, 919–923. PMID 18923514.

Crews, K.R., Hicks, J.K., Pui, C.H., Relling, M.V., Evans, W.E., 2012. Pharmacogenomics and individualized medicine: translating science into practice. Clin. Pharmacol. Ther. 92, 467–475. PMID 22948889.

Crews, K.R., Gaedigk, A., Dunnenberger, H.M., Leeder, J.S., Klein, T.E., Caudle, K.E., et al., Clinical Pharmacogenetics Implementation Consortium, 2014. Clinical Pharmacogenetics Implementation Consortium guidelines for cytochrome P450 2D6 genotype and codeine therapy: 2014 update. Clin. Pharmacol. Ther. 95, 376–382. PMID 24458010.

Dalla Vecchia, L.K., Grosfeld, J.L., West, K.W., Rescorla, F.J., Scherer, L.R., Engum, S.A., 1998. Intestinal atresia and stenosis: a 25-year experience with 277 cases. Arch. Surg. 133, 490–496. ; discussion 496-7. PMID 9605910.

Desta, Z., Gammal, R.S., Gong, L., Whirl-Carrillo, M., Gaur, A.H., Sukasem, C., et al., 2019. Clinical Pharmacogenetics Implementation Consortium (CPIC) guideline for CYP2B6 and efavirenz-containing antiretroviral therapy. Clin. Pharmacol. Ther. 106, 726–733. PMID 31006110.

Farzin, F., Perry, H., Hessl, D., Loesch, D., Cohen, J., Bacalman, S., et al., 2006. Autism spectrum disorders and attention-deficit/hyperactivity disorder in boys with the fragile X premutation. J. Dev. Behav. Pediatr. 27, S137–S144. PMID 16685180.

Fromer, M., Pocklington, A.J., Kavanagh, D.H., Williams, H.J., Dwyer, S., Gormley, P., et al., 2014. De novo mutations in schizophrenia implicate synaptic networks. Nature 506, 179–184. PMID 24463507.

Fuller, S., Ravishankar, C., 2016. Acquired and congenital coronary artery diseases. Pediatr. Crit. Care Med. 17, S356–S361. PMID 27490623.

Gaedigk, A., Sangkuhl, K., Whirl-Carrillo, M., Twist, G.P., Klein, T.E., Miller, N.A., PharmVar Steering Committee, 2019. The evolution of PharmVar. Clin. Pharmacol. Ther. 105, 29–32. PMID 30536702.

Gammal, R.S., Court, M.H., Haidar, C.E., Iwuchukwu, O.F., Gaur, A.H., Alvarellos, M., et al., 2016. Clinical Pharmacogenetics Implementation Consortium (CPIC) guideline for UGT1A1 and atazanavir prescribing. Clin. Pharmacol. Ther. 99, 363–369. PMID 26417955.

GENDEP Investigators, Mars Investigators, Star D. Investigators, 2013. Common genetic variation and antidepressant efficacy in major depressive disorder: a meta-analysis of three genome-wide pharmacogenetic studies. Am. J. Psychiatr. 170, 207–217. PMID 23377640.

Genetic Testing Statement, 2019. Genetic Testing and Psychiatric Disorders; A Statement from the International Society of Psychiatric Genetics [Online]. Brentwood, TN: International Society of Psychiatric Genetics (ISPG). Available: https://ispg.net/genetic-testing-statement/. Accessed December 2019.

Genovese, G., Fromer, M., Stahl, E.A., Ruderfer, D.M., Chambert, K., Landen, M., et al., 2016. Increased burden of ultra-rare protein-altering variants among 4,877 individuals with schizophrenia. Nat. Neurosci. 19, 1433–1441. PMID 27694994.

Girolami, F., Frisso, G., Benelli, M., Crotti, L., Iascone, M., Mango, R., et al., 2018. Contemporary genetic testing in inherited cardiac disease: tools, ethical issues, and clinical applications. J. Cardiovasc. Med. 19, 1–11. PMID 29176389.

Goetz, M.P., Sangkuhl, K., Guchelaar, H.J., Schwab, M., Province, M., Whirl-Carrillo, M., et al., 2018. Clinical Pharmacogenetics Implementation Consortium (CPIC) guideline for CYP2D6 and tamoxifen therapy. Clin. Pharmacol. Ther. 103, 770–777. PMID 29385237.

Goldspiel, B.R., Flegel, W.A., Dipatrizio, G., Sissung, T., Adams, S.D., Penzak, S.R., et al., 2014. Integrating pharmacogenetic information and clinical decision support into the electronic health record. J. Am. Med. Inf. Assoc. 21, 522–528. PMID 24302286.

Goldstein, J.I., Jarskog, L.F., Hilliard, C., Alfirevic, A., Duncan, L., Fourches, D., et al., 2014. Clozapine-induced agranulocytosis is associated with rare HLA-DQB1 and HLA-B alleles. Nat. Commun. 5, 4757. PMID 25187353.

Gonsalves, S.G., Dirksen, R.T., Sangkuhl, K., Pulk, R., Alvarellos, M., Vo, T., et al., 2019. Clinical Pharmacogenetics Implementation Consortium (CPIC) guideline for the use of potent volatile anesthetic agents and succinylcholine in the context of RYR1 or CACNA1S genotypes. Clin. Pharmacol. Ther. 105, 1338–1344. PMID 30499100.

Gratten, J., Visscher, P.M., Mowry, B.J., Wray, N.R., 2013. Interpreting the role of de novo protein-coding mutations in neuropsychiatric disease. Nat. Genet. 45, 234–238. PMID 23438595.

Guthrie, R., 1962. Screening for "inborn errors of metabolism" in the newborn infant: a multiple test program. Birth Defects Original Article Series IV 92–98.

Hagerman, R.J., Berry-Kravis, E., Hazlett, H.C., Bailey Jr., D.B., Moine, H., Kooy, R.F., et al., 2017. Fragile X syndrome. Nat. Rev. Dis. Prim. 3, 17065. PMID 28960184.

Hagerman, R.J., Protic, D., Rajaratnam, A., Salcedo-Arellano, M.J., Aydin, E.Y., Schneider, A., 2018. Fragile X-associated neuropsychiatric disorders (FXAND). Front. Psychiatr. 9, 564. PMID 30483160.

Hall, B.D., Graham Jr., J.M., Cassidy, S.B., Opitz, J.M., 2009. Elements of morphology: standard terminology for the periorbital region. Am. J. Med. Genet. 149A, 29–39. PMID 19125427.

Hennekam, R.C., Cormier-Daire, V., Hall, J.G., Mehes, K., Patton, M., Stevenson, R.E., 2009. Elements of morphology: standard terminology for the nose and philtrum. Am. J. Med. Genet. 149A, 61–76. PMID 19152422.

Hennekam, R.C., Allanson, J.E., Biesecker, L.G., Carey, J.C., Opitz, J.M., Vilain, E., 2013. Elements of morphology: standard terminology for the external genitalia. Am. J. Med. Genet. 161A, 1238–1263. PMID 23650202.

Hershfield, M.S., Callaghan, J.T., Tassaneeyakul, W., Mushiroda, T., Thorn, C.F., Klein, T.E., et al., 2013. Clinical Pharmacogenetics Implementation Consortium guidelines for human leukocyte antigen-B genotype and allopurinol dosing. Clin. Pharmacol. Ther. 93, 153–158. PMID 23232549.

Hicks, J.K., Swen, J.J., Thorn, C.F., Sangkuhl, K., Kharasch, E.D., Ellingrod, V.L., et al., 2013. Clinical Pharmacogenetics Implementation Consortium guideline for CYP2D6 and CYP2C19 genotypes and dosing of tricyclic antidepressants. Clin. Pharmacol. Ther. 93, 402–408. PMID 23486447.

Hicks, J.K., Bishop, J.R., Sangkuhl, K., Muller, D.J., Ji, Y., Leckband, S.G., et al., 2015. Clinical Pharmacogenetics Implementation Consortium (CPIC) guideline for CYP2D6 and CYP2C19 genotypes and dosing of selective serotonin reuptake inhibitors. Clin. Pharmacol. Ther. 98, 127–134. PMID 25974703.

Hicks, J.K., Dunnenberger, H.M., Gumpper, K.F., Haidar, C.E., Hoffman, J.M., 2016. Integrating pharmacogenomics into electronic health records with clinical decision support. Am. J. Health Syst. Pharm. 73, 1967–1976. PMID 27864204.

Hicks, J.K., Sangkuhl, K., Swen, J.J., Ellingrod, V.L., Muller, D.J., Shimoda, K., et al., 2017. Clinical Pharmacogenetics Implementation Consortium (CPIC) guideline for CYP2D6 and CYP2C19 genotypes and dosing of tricyclic antidepressants: 2016 update. Clin. Pharmacol. Ther. 102, 37–44. PMID 27997040.

Hoang, T.T., Goldmuntz, E., Roberts, A.E., Chung, W.K., Kline, J.K., Deanfield, J.E., et al., 2018. The Congenital Heart Disease Genetic Network Study: cohort description. PLoS One 13, e0191319. PMID 29351346.

Hodgson, K.J., Shelton, K.H., van den Bree, M.B., 2014. Mental health problems in young people with experiences of homelessness and the relationship with health service use: a follow-up study. Evid. Base Ment. Health 17, 76–80. PMID 25043432.

Howard, D.M., Adams, M.J., Shirali, M., Clarke, T.K., Marioni, R.E., Davies, G., et al., 2018. Genome-wide association study of depression phenotypes in UK Biobank identifies variants in excitatory synaptic pathways. Nat. Commun. 9, 1470. PMID 29662059.

Hunter, A., Frias, J.L., Gillessen-Kaesbach, G., Hughes, H., Jones, K.L., Wilson, L., 2009. Elements of morphology: standard terminology for the ear. Am. J. Med. Genet. 149A, 40–60. PMID 19152421.

Iossifov, I., O'roak, B.J., Sanders, S.J., Ronemus, M., Krumm, N., Levy, D., et al., 2014. The contribution of de novo coding mutations to autism spectrum disorder. Nature 515, 216–221. PMID 25363768.

Johnson, J.A., Gong, L., Whirl-Carrillo, M., Gage, B.F., Scott, S.A., Stein, C.M., et al., 2011. Clinical Pharmacogenetics Implementation Consortium guidelines for CYP2C9 and VKORC1 genotypes and warfarin dosing. Clin. Pharmacol. Ther. 90, 625–629. PMID: 21900891.

Johnson, J.A., Caudle, K.E., Gong, L., Whirl-Carrillo, M., Stein, C.M., Scott, S.A., et al., 2017. Clinical pharmacogenetics implementation Consortium (CPIC) guideline for pharmacogenetics-guided warfarin dosing: 2017 update. Clin. Pharmacol. Ther. 102 (3), 397–404. PMID 28198005.

Jones, K.L., Jones, M.C., Campo, M.D., 2021. "Normal Standards" in Smith's Recognizable Patterns of Human Malformation, eighth ed. Elsevier Saunders, Philadelphia.

Kendler, K.S., Walters, E.E., Neale, M.C., Kessler, R.C., Heath, A.C., Eaves, L.J., 1995. The structure of the genetic and environmental risk factors for six major psychiatric disorders in women. Phobia, generalized anxiety disorder, panic disorder, bulimia, major depression, and alcoholism. Arch. Gen. Psychiatr. 52, 374–383. PMID 7726718.

Kety, S.S., Rosenthal, D., Wender, P.H., Schulsinger, F., 1971. Mental illness in the biological and adoptive families of adopted schizophrenics. Am. J. Psychiatr. 128, 302–306. PMID 5570994.

Kim, J.B., 2014. Channelopathies. Korean J. Pediatr. 57, 1–18. PMID 24578711.

Kirov, G., 2015. CNVs in neuropsychiatric disorders. Hum. Mol. Genet. 24, R45–R49. PMID 26130694.

Kohler, J.N., Turbitt, E., Biesecker, B.B., 2017. Personal utility in genomic testing: a systematic literature review. Eur. J. Hum. Genet. 25, 662–668. PMID 28295040.

Krakow, D., 2015. Skeletal dysplasias. Clin. Perinatol. 42, 301–319 viii. PMID 26042906.

Leckband, S.G., Kelsoe, J.R., Dunnenberger, H.M., George Jr., A.L., Tran, E., Berger, R., et al., 2013. Clinical Pharmacogenetics Implementation Consortium guidelines for HLA-B genotype and carbamazepine dosing. Clin. Pharmacol. Ther. 94, 324–328. PMID 23695185.

Lichtenstein, P., Yip, B.H., Bjork, C., Pawitan, Y., Cannon, T.D., Sullivan, P.F., et al., 2009. Common genetic determinants of schizophrenia and bipolar disorder in Swedish families: a population-based study. Lancet 373, 234–239. PMID 19150704.

Liu, D.J., Leal, S.M., 2012. A unified framework for detecting rare variant quantitative trait associations in pedigree and unrelated individuals via sequence data. Hum. Hered. 73, 105–122. PMID 22555759.

Lu, M., Lewis, C.M., Traylor, M., 2017. Pharmacogenetic testing through the direct-to-consumer genetic testing company 23andMe. BMC Med. Genom. 10, 47. PMID 28629370.

Mallal, S., Phillips, E., Carosi, G., Molina, J.M., Workman, C., Tomazic, J., et al., 2008. HLA-B*5701 screening for hypersensitivity to abacavir. N. Engl. J. Med. 358, 568–579. PMID 18256392.

Manolio, T.A., Collins, F.S., Cox, N.J., Goldstein, D.B., Hindorff, L.A., Hunter, D.J., et al., 2009. Finding the missing heritability of complex diseases. Nature 461, 747–753. PMID 19812666.

Martin, M.A., Klein, T.E., Dong, B.J., Pirmohamed, M., Haas, D.W., Kroetz, D.L., et al., 2012. guidelines for HLA-B genotype and abacavir dosing. Clin. Pharmacol. Ther. 91, 734–738. PMID 22378157.

Martin, M.A., Hoffman, J.M., Freimuth, R.R., Klein, T.E., Dong, B.J., Pirmohamed, M., et al., 2014. Clinical Pharmacogenetics Implementation Consortium guidelines for HLA-B genotype and abacavir dosing: 2014 update. Clin. Pharmacol. Ther. 95, 499–500. PMID 24561393.

McMahon, F.J., 2014. Prediction of treatment outcomes in psychiatry–where do we stand? Dialogues Clin. Neurosci. 16, 455–464. PMID 25733951.

McMahon, F.J., 2022. Supported in part by the NIMH intramural research program. Opinions expressed do not necessarily represent the NIMH, NIH, or Federal Government.

Mendlewicz, J., Rainer, J.D., 1977. Adoption study supporting genetic transmission in manic–depressive illness. Nature 268, 327–329. PMID 887159.

Michelson, D.J., Shevell, M.I., Sherr, E.H., Moeschler, J.B., Gropman, A.L., Ashwal, S., 2011. Evidence report: genetic and metabolic testing on children with global developmental delay: report of the quality standards subcommittee of the American Academy of Neurology and the Practice Committee of the Child Neurology Society. Neurology 77, 1629–1635. PMID 21956720.

Middeldorp, C.M., Wray, N.R., 2018. The value of polygenic analyses in psychiatry. World Psychiatr. 17, 26–28. PMID 29352547.

Miller, D.T., Adam, M.P., Aradhya, S., Biesecker, L.G., Brothman, A.R., Carter, N.P., et al., 2010. Consensus statement: chromosomal microarray is a first-tier clinical diagnostic test for individuals with developmental disabilities or congenital anomalies. Am. J. Hum. Genet. 86, 749–764. PMID 20466091.

Moeschler, J.B., Shevell, M., Committee On Genetics, 2014. Comprehensive evaluation of the child with intellectual disability or global developmental delays. Pediatrics 134, e903–e918. PMID 25157020.

Moriyama, B., Obeng, A.O., Barbarino, J., Penzak, S.R., Henning, S.A., Scott, S.A., et al., 2016. Clinical Pharmacogenetics Implementation Consortium (CPIC) guidelines for CYP2C19 and voriconazole therapy. Clin. Pharmacol. Ther. 102, 45–51. PMID: 27981572.

Muir, A.J., Gong, L., Johnson, S.G., Lee, M.T., Williams, M.S., Klein, T.E., et al., 2014. Clinical Pharmacogenetics Implementation Consortium (CPIC) guidelines for IFNL3 (IL28B) genotype and PEG interferon-alpha-based regimens. Clin. Pharmacol. Ther. 95, 141–146. PMID 24096968.

Nurnberger Jr., J.I., Austin, J., Berrettini, W.H., Besterman, A.D., Delisi, L.E., Grice, D.E., et al., 2018. What should a psychiatrist know about genetics? Review and recommendations from the Residency Education Committee of the International Society of Psychiatric Genetics. J. Clin. Psychiatr. 80, 17nr12046. PMID 30549495.

Pardinas, A.F., Holmans, P., Pocklington, A.J., Escott-Price, V., Ripke, S., Carrera, N., et al., 2018. Common schizophrenia alleles are enriched in mutation-intolerant genes and in regions under strong background selection. Nat. Genet. 50, 381–389. PMID 29483656.

Phillips, E.J., Sukasem, C., Whirl-Carrillo, M., Muller, D.J., Dunnenberger, H.M., Chantratita, W., et al., 2018. Clinical Pharmacogenetics Implementation Consortium guideline for HLA genotype and use of carbamazepine and oxcarbazepine: 2017 Update. Clin. Pharmacol. Ther. 103, 574–581. PMID 29392710.

Pindrik, J., Ye, X., Ji, B.G., Pendleton, C., Ahn, E.S., 2014. Anterior fontanelle closure and size in full-term children based on head computed tomography. Clin. Pediatr. (Phila) 53, 1149–1157. PMID 24920348.

Powis, L., Oliver, C., 2014. The prevalence of aggression in genetic syndromes: a review. Res. Dev. Disabil. 35, 1051-1071 PMID 24594523.

Press, K.R., Wieczorek, L., Hoover-Fong, J., Bodurtha, J., Taylor, L., 2016. Overview: referrals for genetic evaluation from child psychiatrists. Child Adolesc. Psychiatr. Ment. Health 10, 7. PMID 27022409.

Ramsey, L.B., Johnson, S.G., Caudle, K.E., Haidar, C.E., Voora, D., Wilke, R.A., et al., 2014. The Clinical Pharmacogenetics Implementation Consortium guideline for SLCO1B1 and simvastatin-induced myopathy: 2014 update. Clin. Pharmacol. Ther. 96, 423–428. PMID 24918167.

Reichenberg, A., Cederlof, M., McMillan, A., Trzaskowski, M., Kapra, O., Fruchter, E., et al., 2016. Discontinuity in the genetic and environmental causes of the intellectual disability spectrum. Proc. Natl. Acad. Sci. U. S. A. 113, 1098–1103. PMID 26711998.

Relling, M.V., Klein, T.E., 2011. CPIC: Clinical Pharmacogenetics Implementation Consortium of the Pharmacogenomics Research Network. Clin. Pharmacol. Ther. 89, 464–467. PMID 21270786.

Relling, M.V., Gardner, E.E., Sandborn, W.J., Schmiegelow, K., Pui, C.H., Yee, S.W., et al., 2013. Clinical Pharmacogenetics Implementation Consortium guidelines for thiopurine methyltransferase genotype and thiopurine dosing: 2013 update. Clinical Pharmacogenetics Implementation Consortium. Clin. Pharmacol. Ther. 93, 324–325. PMID: 23422873.

Relling, M.V., McDonagh, E.M., Chang, T., Caudle, K.E., McLeod, H.L., Haidar, C.E., et al., 2014. Clinical Pharmacogenetics Implementation Consortium (CPIC) guidelines for rasburicase therapy in the context of G6PD deficiency genotype. Clin. Pharmacol. Ther. 96, 169–174. PMID 24787449.

Relling, M.V., Schwab, M., Whirl-Carrillo, M., Suarez-Kurtz, G., Pui, C.H., Stein, C.M., et al., 2018. Clinical Pharmacogenetics Implementation Consortium guideline for thiopurine dosing based on TPMT and NUDT15 genotypes: 2018 update. Clin. Pharmacol. Ther. 105, 1095–1105. PMID: 30447069.

Relling, M.V., Klein, T.E., Gammal, R.S., Whirl-Carrillo, M., Hoffman, J.M., Caudle, K.E., 2020. The Clinical Pharmacogenetics Implementation Consortium: 10 years later. Clin. Pharmacol. Ther. 107, 171–175. PMID 31562822.

Saberzadeh-Ardestani, B., Karamzadeh, R., Basiri, M., Hajizadeh-Saffar, E., Farhadi, A., Shapiro, A.M.J., et al., 2018. Type 1 diabetes mellitus: cellular and molecular pathophysiology at a glance. Cell J. 20, 294–301. PMID 29845781.

Saito, Y., Stamp, L.K., Caudle, K.E., Hershfield, M.S., McDonagh, E.M., Callaghan, J.T., et al., 2016. Clinical Pharmacogenetics Implementation Consortium (CPIC) guidelines for human leukocyte antigen B (HLA-B) genotype and allopurinol dosing: 2015 update. Clin. Pharmacol. Ther. 99, 36–37. PMID 26094938.

Samanek, M., Voriskova, M., 1999. Congenital heart disease among 815,569 children born between 1980 and 1990 and their 15-year survival: a prospective Bohemia survival study. Pediatr. Cardiol. 20, 411–417. PMID 10556387.

Schiff, M., Benoist, J.F., Aissaoui, S., Boespflug-Tanguy, O., Mouren, M.C., De Baulny, H.O., et al., 2011. Should metabolic diseases be systematically screened in nonsyndromic autism spectrum disorders? PLoS One 6, e21932. PMID 21760924.

Scott, S.A., 2011. Personalizing medicine with clinical pharmacogenetics. Genet. Med. 13, 987–995. PMID 22095251.

Scott, S.A., Sangkuhl, K., Gardner, E.E., Stein, C.M., Hulot, J.S., Johnson, J.A., et al., 2011. Clinical Pharmacogenetics Implementation Consortium guidelines for cytochrome P450-2C19 (CYP2C19) genotype and clopidogrel therapy. Clin. Pharmacol. Ther. 90, 328–332. PMID 21716271.

Scott, S.A., Sangkuhl, K., Stein, C.M., Hulot, J.S., Mega, J.L., Roden, D.M., et al., 2013. Clinical Pharmacogenetics Implementation Consortium guidelines for CYP2C19 genotype and clopidogrel therapy: 2013 update. Clin. Pharmacol. Ther. 94, 317–323. PMID 23698643.

Shawky, R.M., Sadik, D.I., 2011. The role of genomics in prevention or reducing the impact of congenital malformations. Genet. Counsel. 22, 135–141. PMIC 21848005.

Shuldiner, A.R., Relling, M.V., Peterson, J.F., Hicks, J.K., Freimuth, R.R., Sadee, W., et al., 2013. The Pharmacogenomics Research network Translational Pharmacogenetics Program: overcoming challenges of real-world implementation. Clin. Pharmacol. Ther. 94, 207–210. PMID 23588301.

Stahl, E.A., Breen, G., Forstner, A.J., McQuillin, A., Ripke, S., Trubetskoy, V., Mattheisen, M., Wang, Y., Coleman, J.R.I., Gaspar, H.A., De Leeuw, C.A., Steinberg, S., Pavlides, J.M.W., Trzaskowski, M., Byrne, E.M., Pers, T.H., Holmans, P.A., Richards, A.L., Abbott, L., Agerbo, E., Akil, H., Albani, D., Alliey-Rodriguez, N., Als, T.D., Anjorin, A., Antilla, V., Awasthi, S., Badner, J.A., Baekvad-Hansen, M., Barchas, J.D., Bass, N., Bauer, M., Belliveau, R., Bergen, S.E., Pedersen, C.B., Boen, E., Boks, M.P., Boocock, J., Budde, M., Bunney, W., Burmeister, M., Bybjerg-Grauholm, J., Byerley, W., Casas, M., Cerrato, F., Cervantes, P., Chambert, K., Charney, A.W., Chen, D., Churchhouse, C., Clarke, T.K., Coryell, W., Craig, D.W., Cruceanu, C., Curtis, D., Czerski, P.M., Dale, A.M., De Jong, S., Degenhardt, F., Del-Favero, J., DePaulo, J.R., Djurovic, S., Dobbyn, A.L., Dumont, A., Elvsashagen, T., Escott-Price, V., Fan, C.C., Fischer, S.B., Flickinger, M., Foroud, T.M., Forty, L., Frank, J., Fraser, C., Freimer, N.B., Frisen, L., Gade, K., Gage, D., Garnham, J., Giambartolomei, C., Pedersen, M.G., Goldstein, J., Gordon, S.D., Gordon-Smith, K., Green, E.K., Green, M.J., Greenwood, T.A., Grove, J., Guan, W., Guzman-Parra, J., Hamshere, M.L., Hautzinger, M., Heilbronner, U., Herms, S., Hipolito, M., Hoffmann, P., Holland, D., Huckins, L., Jamain, S., Johnson, J.S., Jureus, A., Kandaswamy, R., Karlsson, R., Kennedy, J.L., Kittel-Schneider, S., Knowles, J.A., Kogevinas, M., Koller, A.C., Kupka, R., Lavebratt, C., Lawrence, J., Lawson, W.B., Leber, M., Lee, P.H., Levy, S.E., Li, J.Z., Liu, C., Lucae, S., Maaser, A., MacIntyre, D.J., Mahon, P.B., Maier, W., Martinsson, L., McCarroll, S., McGuffin, P., McInnis, M.G., McKay, J.D., Medeiros, H., Medland, S.E., Meng, F., Milani, L., Montgomery, G.W., Morris, D.W., Mühleisen, T.W., Mullins, N., Nguyen, H., Nievergelt, C.M., Adolfsson, A.N., Nwulia, E.A., O'Donovan, C., Loohuis, L.M.O., Ori, A.P.S., Öruc, L., Osby, U., Perlis, R.H., Perry, A., Pfennig, A., Potash, J.B., Purcell, S.M., Regeer, E.J., Reif, A., Reinbold, C.S., Rice, J.P., Rivas, F., Rivera, M., Roussos, P., Ruderfer, D.M., Ryu, E., Sánchez-Mora, C., Schatzberg, A.F., Scheftner, W.A., Schork, N.J., Weickert, C.S., Shehktman, T., Shilling, P.D., Sigurdsson, E., Slaney, C., Smeland, O.B., Sobell, J.L., Hansen, C.S., Spijker, A.T., St Clair, D., Steffens, M., Strauss, J.S., Streit, F., Strohmaier, J., Szelinger, S., Thompson, R.C., Thorgeirsson, T.E., Treutlein, J., Vedder, H., Wang, W., Watson, S.J., Weickert, T.W., Witt, S.H., Xi, S., Xu, W., Young, A.H., Zandi, P., Zhang, P., Zöllner, S., eQTLGen Consortium, BIOS Consortium, Adolfsson, R., Agartz, I., Alda, M., Backlund, L., Baune, B.T., Bellivier, F., Berrettini, W.H., Biernacka, J.M., Blackwood, D.H.R., Boehnke, M., Børglum, A.D., Corvin, A., Craddock, N., Daly, M.J., Dannlowski, U., Esko, T., Etain, B., Frye, M., Fullerton, J.M., Gershon, E.S., Gill, M., Goes, F., Grigoroiu-Serbanescu, M., Hauser, J., Hougaard, D.M., Hultman, C.M., Jones, I., Jones, L.A., Kahn, R.S., Kirov, G., Landén, M., Leboyer, M., Lewis, C.M., Li, Q.S., Lissowska, J., Martin, N.G., Mayoral, F., McElroy, S.L., McIntosh, A.M., McMahon, F.J., Melle, I., Metspalu, A., Mitchell, P.B., Morken, G., Mors, O., Mortensen, P.B., Müller-Myhsok, B., Myers, R.M., Neale, B.M., Nimgaonkar, V., Nordentoft, M., Nöthen, M.M., O'Donovan, M.C., Oedegaard, K.J., Owen, M.J., Paciga, S.A., Pato, C., Pato, M.T., Posthuma, D., Ramos-Quiroga, J.A., Ribasés, M., Rietschel, M., Rouleau, G.A., Schalling, M., Schofield, P.R., Schulze, T.G., Serretti, A., Smoller, J.W., Stefansson, H., Stefansson, K., Stordal, E., Sullivan, P.F., Turecki, G., Vaaler, A.E., Vieta, E., Vincent, J.B., Werge, T., Nurnberger, J.I., Wray, N.R., Di Florio, A., Edenberg, H.J., Cichon, S., Ophoff, R.A., Scott, L.J., Andreassen, A.O., Kelsoe, J., Sklar, P., Bipolar Disorder Working Group of the Psychiatric Genomics Consortium, 2019. Genome-wide association study identifies 30 loci associated with bipolar disorder. Nat. Genet. 51, 793–803. PMID 31043756.

Sullivan, A.K., Marcus, M., Epstein, M.P., Allen, E.G., Anido, A.E., Paquin, J.J., et al., 2005. Association of FMR1 repeat size with ovarian dysfunction. Hum. Reprod. 20, 402–412. PMID 15608041.

Van Der Horst, H.J., De Wall, L.L., 2017. Hypospadias, all there is to know. Eur. J. Pediatr. 176, 435–441. PMID 28190103.

Van Der Linde, D., Konings, E.E., Slager, M.A., Witsenburg, M., Helbing, W.A., Takkenberg, J.J., et al., 2011. Birth prevalence of congenital heart disease worldwide: a systematic review and meta-analysis. J. Am. Coll. Cardiol. 58, 2241–2247. PMID 22078432.

Verlingue, C., David, A., Audrezet, M.P., Le Roux, M.G., Mercier, B., Moisan, J.P., et al., 1993. Asymptomatic carrier of two CFTR mutations: consequences for prenatal diagnosis? Prenat. Diagn. 13, 1143–1148. PMID 7513889.

Wilcken, B., Wiley, V., Hammond, J., Carpenter, K., 2003. Screening newborns for inborn errors of metabolism by tandem mass spectrometry. N. Engl. J. Med. 348, 2304–2312. PMID 12788994.

Wilke, R.A., Ramsey, L.B., Johnson, S.G., Maxwell, W.D., McLeod, H.L., Voora, D., et al., 2012. The Clinical Pharmacogenomics Implementation Consortium: CPIC guideline for SLCO1B1 and simvastatin-induced myopathy. Clin. Pharmacol. Ther. 92, 112–117. PMID 22617227.

Zablotsky, B., Black, L.I., Maenner, M.J., Schieve, L.A., Blumberg, S.J., 2015. Estimated prevalence of autism and other developmental disabilities following questionnaire changes in the 2014 National Health Interview Survey. Natl. Health Stat. Report 87, 1–20. PMID 26632847.

Zanuso, S., Sacchetti, M., Sundberg, C.J., Orlando, G., Benvenuti, P., Balducci, S., 2017. Exercise in type 2 diabetes: genetic, metabolic and neuromuscular adaptations. A review of the evidence. Br. J. Sports Med. 51, 1533–1538. PMID 28501806.

Zuk, O., Schaffner, S.F., Samocha, K., Do, R., Hechter, E., Kathiresan, S., et al., 2014. Searching for missing heritability: designing rare variant association studies. Proc. Natl. Acad. Sci. U. S. A. 111, E455–E464. PMID 24443550.

Chapter 6

Genetic and Genomic Testing

Bo Yuan[1,2], Shen Gu[1,3], Amy Kimball[4], Ethylin Wang Jabs[5,*] and Christine M. Eng[1]

[1]*Department of Molecular and Human Genetics, Baylor College of Medicine, Houston, TX, United States;* [2]*Human Genome Sequencing Center, Baylor College of Medicine, Houston, TX, United States;* [3]*School of Biomedical Sciences, The Chinese University of Hong Kong, Hong Kong S.A.R., China;* [4]*Harvey Institute for Human Genetics, Greater Baltimore Medical Center, Baltimore, MD, United States;* [5]*Department of Genetics and Genomic Sciences, Department of Pediatrics, Department of Cell, Developmental and Regenerative Biology, Icahn School of Medicine at Mount Sinai, New York, NY, United States*

Abstract

Individuals with suspected genomic disorders can present clinically with a myriad of signs and symptoms. A differential diagnosis can be developed on a clinical basis, but definitive identification of the underlying genetic etiology often requires genetic testing of a specific nature. In this chapter, we will discuss the strategies currently employed to detect suspected genomic variant types from chromosomal abnormalities to copy number variations and from single gene to genomic disorders.

A variety of laboratory tests have been implemented to provide genetic information for individuals referred for various medical indications. These include but are not limited to pursuit of cytogenetic or molecular answer(s) for genetic disorders, screening for wellness or reproductive risks, assessment of drug metabolism and response to treatment, and personalized medicine. Genetic and genomic tests are designed to target-specific problems utilizing appropriate methodologies. In this chapter, the major genetic test methodologies and their utilities in patient care are discussed.

6.1 Cytogenetics and Cytogenetic Testing

Clinical cytogenetics is used to understand genetic changes at the chromosomal level. Chromosomal abnormalities may cause multisystematic anomalies, congenital defects, neurodevelopmental disorders and intellectual disabilities, and cancers. Technologies to study chromosomal abnormalities include chromosome analysis, fluorescence in situ hybridization (FISH), chromosome microarray analysis, and other conventional cytogenetic methods, each with their own advantages or limitations of clinical utilities.

6.1.1 Chromosome Analysis

Chromosomal abnormalities can be numerical involving the gain or loss of a whole chromosome or chromosome sets, or terminal or interstitial deletions or duplications involving part of a chromosome (Section 2.2.1) (Gardner et al., 2012). Balanced or unbalanced chromosomal structural changes such as translocations, inversions, and ring chromosomes can also be generated (Section 2.2.1). Notably, the majority of conceptions with chromosomal abnormalities result in spontaneous abortions (Table 2.1). Mosaic changes, those with a mixture of cells with normal or abnormal chromosome complements, can be more compatible with life and may be observed in less severely affected individuals (Sections 2.2.2, 11.13.2).

* Author EWJ was employed by Icahn School of Medicine at Mount Sinai at the time of writing the manuscript, but is currently employed by Mayo Clinic

Genomics in the Clinic. https://doi.org/10.1016/B978-0-12-816478-5.00006-6

Regardless of the types of chromosomal abnormalities, their pathogenicity may be linked to one or the combination of several of the following mechanisms (Section 2.2.1): (1) dosage effect due to a lack (deletion, or copy number loss) or an excess (duplication, or copy number gain) of chromosomal material; (2) disruption of gene(s) at the breakpoint junctions of a rearrangement; (3) position effect due to inappropriate regulation of gene expression in a new chromosomal environment; and (4) disrupted imprinting effect due to incongruent parental origin of a chromosome or chromosomal segment (Section 2.7.5).

Individuals with balanced structural rearrangements may be phenotypically normal other than his or her reproductive health. Alternatively, disease-causing gene(s) may be disrupted at the breakpoint junctions of the balanced events, or phenotypic consequence may arise due to position effect, leading to the balanced events being pathogenic. Notably, it is critical to distinguish true balanced rearrangements versus events that appear to be balanced, but are unbalanced. Apparently balanced events at the cytogenetic level may be unbalanced changes when studied at higher resolution with chromosomal microarrays (CMAs) (Section 6.1.3) or DNA sequence analysis (Sections 6.2.1, 6.2.2 and 6.3.1) (Kang et al., 2010; Neill et al., 2011; Gu et al., 2016).

The preparation steps of chromosome analysis, or the karyotype, generally include cell culture (stimulant may be added for selective cell growth), harvest (synchronization at metaphase and hypotonic solution treatment), cell fixation on slides, and Giemsa staining (requires prior trypsin treatment) (Section 2.1.3). The resolution attained with chromosome analysis increased with the development of different banding methodologies, with G-banding the most commonly used. In G-banding, chromosomes are treated with trypsin to provide access to staining material, and Giemsa staining is used to consistently produce a banding pattern of light (gene-rich regions that are guanine and cytosine (GC)-rich and less condensed chromatin) and dark (gene-poor regions that are adenine and thymine (AT)-rich and more condensed chromatin) (Section 2.1.3, Fig. 2.3). Chromosome analysis has also been facilitated by synchronizing the cell culture to metaphase of mitosis, allowing better separation of chromosomes and high-resolution chromosome G-banding to detect more subtle chromosome abnormalities, e.g., microscopic gain or loss and structural changes. The resolution of chromosome analysis also depends on sample types. Peripheral blood sample generally achieve the best resolution, which is followed by amniotic fluid, tissue, and bone marrow samples (Fig. 6.1).

Chromosome analysis is widely used to detect large chromosomal abnormalities (Section 2.2.1), including polyploidy, aneuploidy, structural rearrangements, and microscopic deletion/duplications for prenatal or postnatal genetic diagnosis. Chromosome analysis is useful in diagnosing common trisomy disorders (trisomy 21—Sections 11.1.3, 11.13.2; trisomy 13, trisomy 18—Section 11.1.5) and sex chromosome aneuploidies (Turner syndrome—Section 11.7.5, and Klinefelter syndrome—Section 11.12.5) as well as chromosomal translocations (Section 11.1.7). It has been commonly performed on: (1) peripheral blood samples from subjects with multiple congenital anomalies; (2) amniotic fluid samples (usually obtained by amniocentesis between weeks 16 and 20 of pregnancy) or chorionic villus samples (can be obtained by chorionic villus sampling [CVS] as early as 10 weeks of pregnancy, usually between 11 and 14 weeks) (Section 4.4.2) for prenatal diagnosis of pregnancies at-risk for chromosomal disorders; (3) products of conception (POC) samples that may help to elucidate genetic reasons of fetal demise; and (4) peripheral blood samples from couples who have had two or more spontaneous abortions (Section 11.1.7). Confined placental mosaicism in CVS can be confirmed by amniocentesis. Maternal cell contamination may be associated with the chromosome analysis of CVS and POC samples and may need to be ruled out to provide accurate results of the fetal sample. Chromosome studies of phenotypically normal parents, who are potential carriers of balanced translocations or inversions, may provide risk assessment and counseling for future pregnancies.

In addition to diagnostic workup, chromosome analysis has substantial clinical utility in acquired malignancies. It is often performed on samples with malignancies (bone marrow or solid tumor) to delineate chromosomal abnormalities and

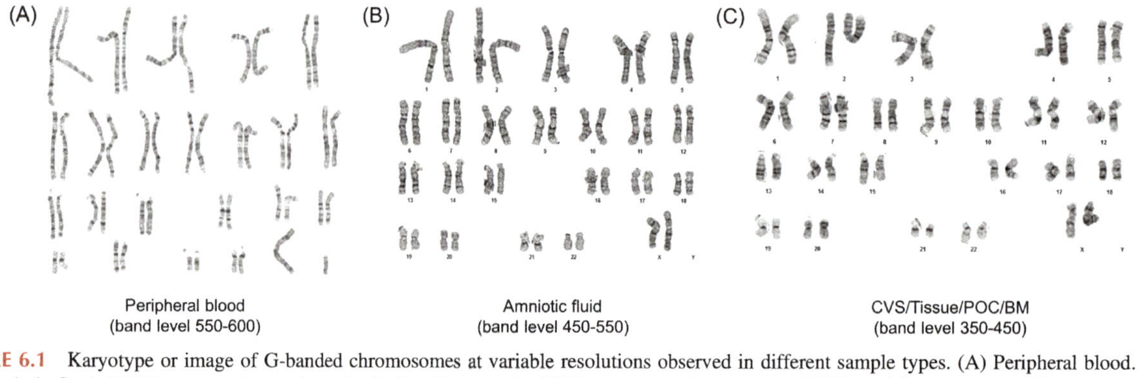

FIGURE 6.1 Karyotype or image of G-banded chromosomes at variable resolutions observed in different sample types. (A) Peripheral blood. (B) Amniotic fluid. (C) Tissue samples such as chorionic villus sample (CVS), products of conception (POC), and bone marrow (BM).

provide insight to help guide medical treatment. Variable stimulant and cell culture time may be used for advantageous growth of certain cell types, the enrichment of which helps analysis.

Although chromosome analysis has great value in detecting abnormalities such as aneuploidies, structural abnormalities, microscopic deletions/duplications, as well as oncology analysis, some of these may now fall in the detectable range of sequencing technologies with comparable sensitivity/specificity. Sequencing methods include next-generation sequencing (NGS), which is massive parallel sequencing of short-read fragments usually of few hundred base pairs (bp), or third-generation sequencing, long-read sequencing of 30,000–100,000 bp or epigenetic marker sequencing of methylated DNA and histones (Sections 2.6, 6.3.4). Complex karyotypes containing numerous clonal copy number and/or structural abnormalities may remain challenging for sequencing technologies to delineate a full picture of complex rearrangements and clonal evolution.

6.1.2 Fluorescence in situ Hybridization

FISH was developed in the early 1980s to detect the presence/absence or localization of a specific chromosome fragment. FISH is a targeted assay performed using a specifically designed, fluorescently labeled DNA probe with sequences complementary to the targeted chromosome region. The DNA probe is hybridized to the targeted chromosome region in the cell, and deletion/duplication could be informed by loss or gain of fluorescent signals, whereas abnormal organization of signals (e.g., split or fusion signals) may inform structural abnormalities (e.g., translocation or inversion). FISH can be performed on interphase cells (during S phase with less condensed chromosomes) or metaphase cells (during M phase with more condensed chromosomes) (Section 2.1.1), depending on different purposes of the analyses and whether fresh or preserved material is studied.

FISH is a frequently used test for prenatal specimens to provide a genetic diagnosis of specific common autosomal (trisomies 13, 18, or 21) or sex chromosomal (X, Y) aneuploidies (Fig. 6.2). Because no cell culture is need for interphase FISH studies, it can provide a quick prenatal diagnostic test for abnormal noninvasive prenatal screening (Section 4.4.1) results, pregnancies with abnormal ultrasound findings (Section 4.3.2), or a rapid screening test for advanced maternal age (Section 4.2.2). FISH also can be offered for detecting more subtle chromosomal abnormalities, such as microdeletions (e.g., Alagille syndrome—Section 5.1.4) or microduplications (e.g., Charcot–Marie–Tooth syndrome type 1A, a peripheral neuropathy often autosomal dominant inheritance), which sometimes may require metaphase FISH analysis.

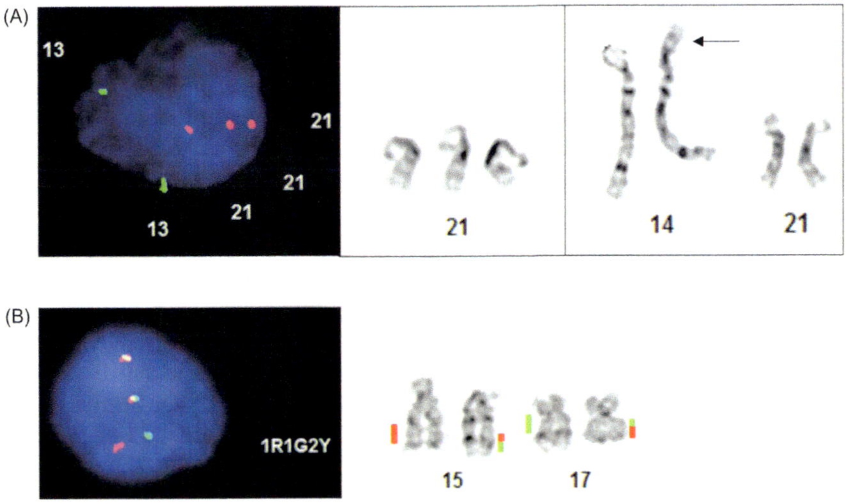

FIGURE 6.2 FISH analysis examples. A. FISH analysis for trisomy 13 and trisomy 21. The FISH probes of chromosomes 13 and 21 target the *RB1* locus (*green*) and *DSCR4* locus (*red*), respectively, on each chromosome. As illustrated on the *left panel*, the *green* signal represents two copies of the *RB1* locus, indicating two copies of chromosome 13 (no trisomy), and the *red* signal represents three copies of the *DSCR4* locus, indicating trisomy 21. However, such FISH analysis cannot illustrate any potential structural changes involving the trisomic event, for example, as shown in the *right panel*, trisomy 21 resulting from a Robertsonian translocation between chromosomes 14 and 21 (arrow shows translocated chromosome 21). B. FISH analysis for targeted detection of translocation between chromosomes 15 and 17 involving the *PML* and *RARA* genes, which is diagnostic of acute promyelocytic leukemia (APL). FISH probes are designed to target the *PML* locus (15q24, *red*) and *RARA* locus (17q21, *green*). As illustrated in the *left panel*, one *red* and one *green* signal, each indicates one intact copy of *PML* and *RARA*, respectively. A translocation between two loci will form fusion of two genes into two products, *PML/RARA* and *RARA/PML*; both are illustrated by two remaining fusion signals, each containing one *green* signal adjacent to one *red* signal.

FISH is an important assay in oncological studies to pinpoint disease-specific rearrangements that may provide diagnostic, prognostic, or therapeutic insights to guide patient care. Although FISH cannot be applied for a genome-wide detection, it targets the breakpoint more specifically than chromosome analysis. Such specificity allows detection of low-level clonal rearrangements. Again, oncological FISH analysis can be applied on uncultured cells, making fast turnaround time possible.

FISH only informs the abnormalities of the target region, meaning that the exact mechanism of chromosomal rearrangement and the possibility of abnormalities involving other chromosomal regions cannot be revealed by FISH analysis. For example, a breakpoint showing a pattern of a FISH signal targeting the immunoglobulin heavy locus (*IGH*) gene at chromosome 14q32 informs a rearrangement of the locus; however, the fusion product from either loss, gain, or change in orientation of DNA material resulting in the rearrangement may not be detected. Another important example may be illustrated by trisomy 21, which can be detected by FISH, but does not provide information of potential structural changes. Chromosome analysis can determine whether the trisomy 21 involves a Robertsonian translocation, which greatly increases the recurrence risk of trisomy 21 in the family (Section 2.2.1).

6.1.3 Chromosome Microarray Analysis

Genomic disorders, or contiguous gene syndromes (e.g., a deletion at Xp21 causing Duchenne muscular dystrophy—Section 11.13.5—with retinitis pigmentosa and chronic granulomatous disease with intellectual disability (Francke et al., 1985)), are conditions that result from copy number variants (CNVs) or rearrangements affecting a continuum of adjacent genes of the genome rather than from DNA sequence base changes (Harel and Lupski, 2018; Lupski, 2009).

Genomic disorders or CNVs can be recurrent or nonrecurrent (Lupski, 1998, 2009). Recurrent rearrangements involve the loss, gain, or change of orientation of the same chromosomal segment(s) that arises recurrently and independently with breakpoint junctions clustering in the same chromosomal regions in unrelated individuals (Edelmann et al., 1999; Bailey et al., 2001). During meiosis or mitosis, misalignment and unequal crossing-over between sister chromatids or homologous chromosomes, intermediated by low copy repeats (also known as segmental duplications) where the breakpoints cluster, lead to products with a deleted, or reciprocally a duplicated, chromosomal segment (Section 2.1.2). Nonrecurrent rearrangements involve different spans of chromosomal segments with variable breakpoints in each individual. They can be aligned to reveal the smallest region of overlap, which may contain the same set of dosage-sensitive gene(s) that potentially explain the overlapping clinical presentations among patients sharing nonrecurrent CNVs at a specific genomic locus (Zhang et al., 2010; Trivellin et al., 2014; Yuan et al., 2015).

Examples of common deletion and duplication syndromes are summarized in Table 6.1. Note that some reciprocal deletions and duplications could have traits from the opposite ends of the clinical spectrum in patients. For example, the syndromes of 5q35 deletion and duplication are associated with macrocephaly and microcephaly, respectively.

Chromosome microarray analysis (CMA, microarray) has been recommended as the first-tier test for patients with developmental disorders and congenital anomalies and collectively has a diagnostic yield of 15%–20% (Miller et al., 2010). There are two major platforms used for chromosome microarray analysis: comparative genomic hybridization array (aCGH) and single-nucleotide polymorphism (SNP) (Section 2.4.1) array.

aCGH detects relative copy number changes (gain or loss) by hybridizing differentially labeled patient and control DNA at the same time to the sequence-specific probes mounted on the same platform (Fig. 6.3). The DNA dosage difference between patient and control can be inferred by the comparative signal intensity from their probe-bound DNA fragments: an increased signal of the patient sample indicates a gain, and the opposite indicates a loss. SNP array uses a different technology by interrogating the ratio of two alleles (Section 2.4) (A allele and B allele) in the patient's DNA (Fig. 6.3). The probes target thousands of SNP loci across the genome, and the dosage change and B-allele frequency are calculated for each locus to infer copy number change or absence of heterozygosity (AOH), the latter of which cannot be revealed by aCGH. The B-allele frequency provides extra data that may be useful for indicating imprinted disorders such as Prader–Willi syndrome (Section 11.13.1) or Angelman syndrome.

The design of the SNP array depends on the distribution of useable SNPs and therefore the resolution may be limited. On the aCGH platform, the probe distribution can be flexibly adjusted to meet the desired resolution at certain regions, and therefore high probe density at specific regions can be achieved for high-resolution detectability. aCGH designs with a focus on exonic (coding) regions of disease-associated genes may achieve single-exon resolution (Yuan et al., 2020). However, neither SNP array nor aCGH can detect balanced structural variants such as balanced translocations or inversions (Section 2.2.1).

TABLE 6.1 Common deletion and duplication syndromes.

Deletion/duplication	Dosage-sensitive Gene[a]	Phenotype
1p36 deletion	*RERE*	ID, speech delay, behavioral difficulties, growth deficiency, hypotonia, characteristic faces, seizures, brain malformations, CHD
5q35 deletion	*NSD1*	Sotos syndrome phenotype, cleft palate, language delay, ID, macrocephaly
5q35 duplication	*NSD1*	Microcephaly, short stature, DD, delayed bone maturation
7q11.23 deletion (Williams syndrome)	*ELN*, others	Periorbital fullness, long philtrum, large mouth, stellate iris pattern, supravalvular aortic stenosis, ID, gregarious personality
7q11.23 duplication	–	Speech delay, mild craniofacial anomalies, DD, autism, aortopathy
8p23 deletion	*GATA4*	CHD
9q34 deletion (Kleefstra syndrome)	*EHMT1*	ID, speech, delay, hypotonia, seizures, characteristic facies, CNS abnormalities, CHD
11q deletion syndrome (Jacobsen syndrome)	*ETS1*, *FLI1*	ID, growth deficiency, characteristic facies, thrombocytopenia, CHD
15q11-q13 deletion (Prader–Willi syndrome; paternal)	*SNRPN*	ID, hypotonia, feeding difficulties (infancy) followed by hyperphagia, obesity, hypogonadotropic hypogonadism, small hands and feet
15q11-q13 deletion (Angelman syndrome; maternal)	*UBE3A*	Severe ID, significant language impairment, gait ataxia, microcephaly, seizures
15q11-q13 duplication (maternal)	–	ID, autism
16p11.2 deletion	–	ID, autism spectrum disorder, epilepsy, macrocephaly, chiari malformation, cerebellar tonsillar extopia, vertebral anomalies
16p11.2 duplication	–	ID, autism, schizophrenia, microcephaly, CNS abnormalities
17p11.2 deletion (Smith–Magenis syndrome)	*RAI1*	ID, behavioral problems, self-injury, disturbed sleep pattern, congenital anomalies
17p11.2 duplication (Potocki–Lupski syndrome)	*RAI1*	ID, autism, hypotonia, FTT, congenital anomalies
17p11.2 deletion (HNPP)	*PMP22*	Hereditary neuropathy with liability to pressure palsy
17p11.2 duplication (CMT1A)	*PMP22*	Charcot–Marie–Tooth disease, type 1A
17q11.2 deletion	*NF1*	Neurofibromatosis, DD, ID, facial dysmorphism
17q11.2 duplication	–	DD, ID, seizures
17q21.31 deletion (Koolen–De Vries syndrome)	*KANSL1*	DD, ID, hypotonia, friendly demeanor, characteristic facies, CHD, genitourinary anomalies, seizures
17q21.31 duplication	–	DD, ID, autism
22q11.2 deletion (Di George/velocardiofacial syndrome)	*TBX1*, others	Cardiac abnormalities, cleft palate, T cell dysfunction, learning difficulties, psychiatric issues
22q11.2 duplication	–	DD, ID, growth retardation, hypotonia
Xq28 deletion (Rett syndrome)	*MECP2*	Stereotypic behavior, ataxia, DD, autism, epilepsy, spastic paraparesis
Xq28 duplication (*MECP2* duplication syndrome)	*MECP2*	Males: DD, severe ID, hypotonia, seizures, progressive spasticity

CHD, congenital heart disease; *CMT1A*, Charcot–Marie–Tooth disease type 1A; *CNS*, central nervous system; *DD*, developmental delay; *FTT*, failure to thrive; *HNPP*, hereditary neuropathy with liability to pressure palsy; *ID*, intellectual disability.

[a]*Dosage-sensitive gene is provided if a single gene has been associated with most of the phenotypic features of the syndrome. Many microdeletion and microduplication syndromes are contiguous gene syndromes, with more than one dosage-sensitive gene involved.*

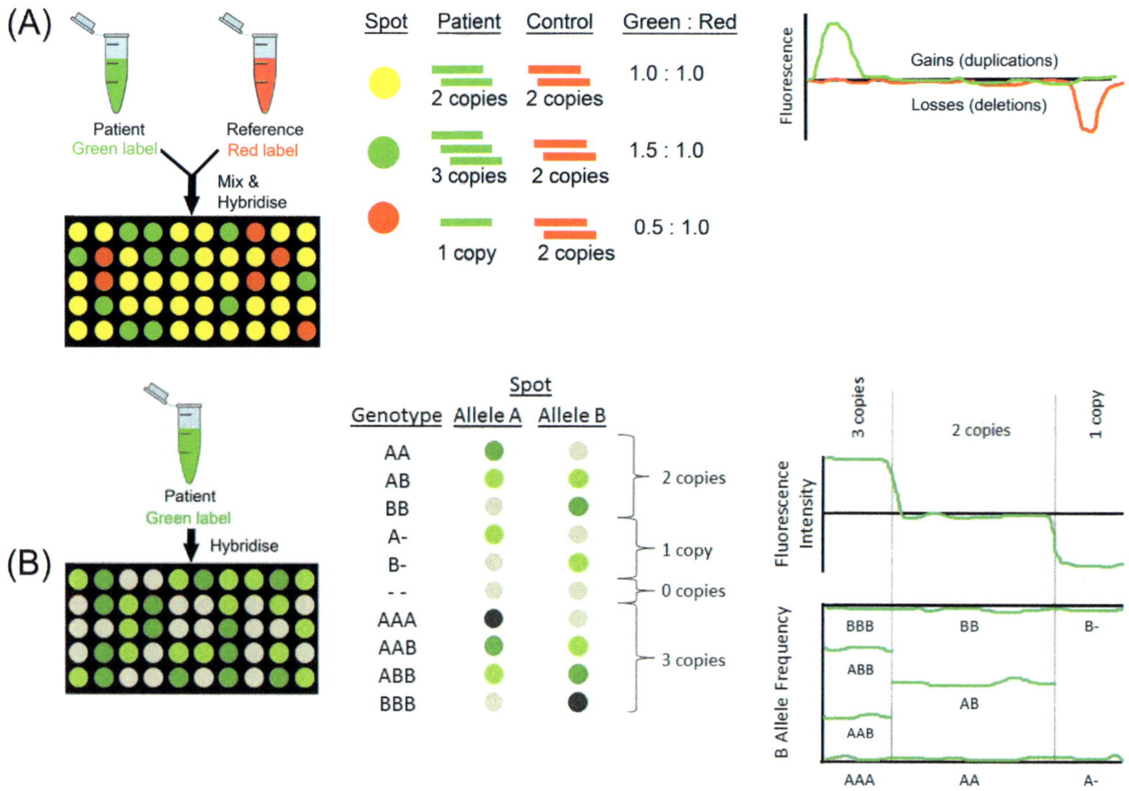

FIGURE 6.3 Chromosome microarray technology. A. Comparative genomic hybridization array (aCGH). DNA samples of patient and control reference are labeled by two different fluorescent colors, mixed and hybridized to the array platform; and ratio of *green* and *red* fluorescence is measured. *Yellow spot* indicates ratio of 1 with equal amounts patient and reference DNA. Fluorescent read out with increased *green or red* signals indicates gain or loss of DNA, respectively. B. Single-nucleotide polymorphism (SNP) array. DNA sample of patient is labeled by fluorescence, hybridized to array platform; and number of alleles A and B are measured. Fluorescent read out indicates ratio of copies or B allele frequency (e.g., 100% is top tracing, 0% is bottom tracing, and 50% for AB tracing). With absence of heterozygosity or loss of heterozygosity, the tracing will show only the top and bottom tracings (i.e., represented by the left most portion of plot with 100% B allele and 0% A allele). Source *Karampetsou, E., Morrogh, D., Chitty, L., 2014. Microarray technology for the diagnosis of fetal chromosomal aberrations: which platform should we use? J. Clin. Med. 3, 663–78, Figs. 1 and 2, via MDPI. Modified from original.*

6.1.4 Advantages and Limitations of Cytogenetic-Based Genetic Tests

Each cytogenetic approach has its advantages and limitations. All assays can be used in both constitutional studies for congenital anomalies and oncological studies for somatic variant detection; however, the sensitivity of each assay is variable for different types of variants (Table 6.2). Chromosome analysis is reliable for detecting large chromosome abnormalities including microscopically visible CNVs and structural variants (balanced or unbalanced), aneuploidies, and polyploidies. A karyotype is particularly useful in deciphering complex genomic rearrangements within a chromosome or involving multiple chromosomes and rare chromosome aberrations such as marker chromosome or ring chromosome.

FISH is designed in a way to target specifically one or multiple variants as in cancers, or to achieve rapid turnaround times in prenatal and neonatal settings. For example, a panel of probes can be designed for multiple loci critical to acute lymphoblastic leukemia, such as *BCR/ABL1*, *MYB*, *KMT2A*, *IGH*, and *CDKN2A*, to detect any potential rearrangement affecting those loci. Given hundreds of cells being examined in one test, FISH has ultra-high sensitivity of mosaicism (Section 2.2.2) detection for a given locus and therefore is exceptionally valuable in oncological studies targeting structural abnormalities. However, FISH cannot delineate the genome-wide picture of rearrangement, and therefore chromosome analysis and FISH are often performed concurrently, especially in oncological cases due to their complexity, to understand the overall picture of chromosome abnormalities.

Submicroscopic chromosome gains or losses as a result of chromosome rearrangement may not be visible by chromosome or FISH analysis if not targeted. Microarray, with probes covering the entire genome, can readily detect a net gain or loss with high breakpoint resolution. Microarray, depending on the design, can detect gains or losses as small as a few hundred base pairs in size (Fig. 6.4). Absence of heterozygosity (AOH or loss of heterozygosity [LOH]) or uniparental

TABLE 6.2 Comparison of chromosome analysis, FISH, and microarray.

	Chromosome analysis	FISH	Microarray
Resolution	>5–10 megabases (Mb)	>20–50 kilobases (kb)	Exonic (as small as hundreds of base pairs [bp])
Target	Genome-wide	Locus specific	Genome-wide or locus specific
Sample	Cultured cells	Cultured/uncultured cells	DNA
Detectable variants	Microscopic CNV, balanced/un-balanced SV, aneuploidy, triploidy	Microscopic CNV, balanced/un-balanced SV, aneuploidy, triploidy	Submicroscopic CNV, unbalanced SV, aneuploidy, triploidy (possible), AOH/UPD
Mosaicism sensitivity	Medium	High	Low
Breakpoint	Low	Medium	High

AOH, absence of heterozygosity; *CNV*, copy number variation; *SV*, structural variant; *UPD*, uniparental disomy.

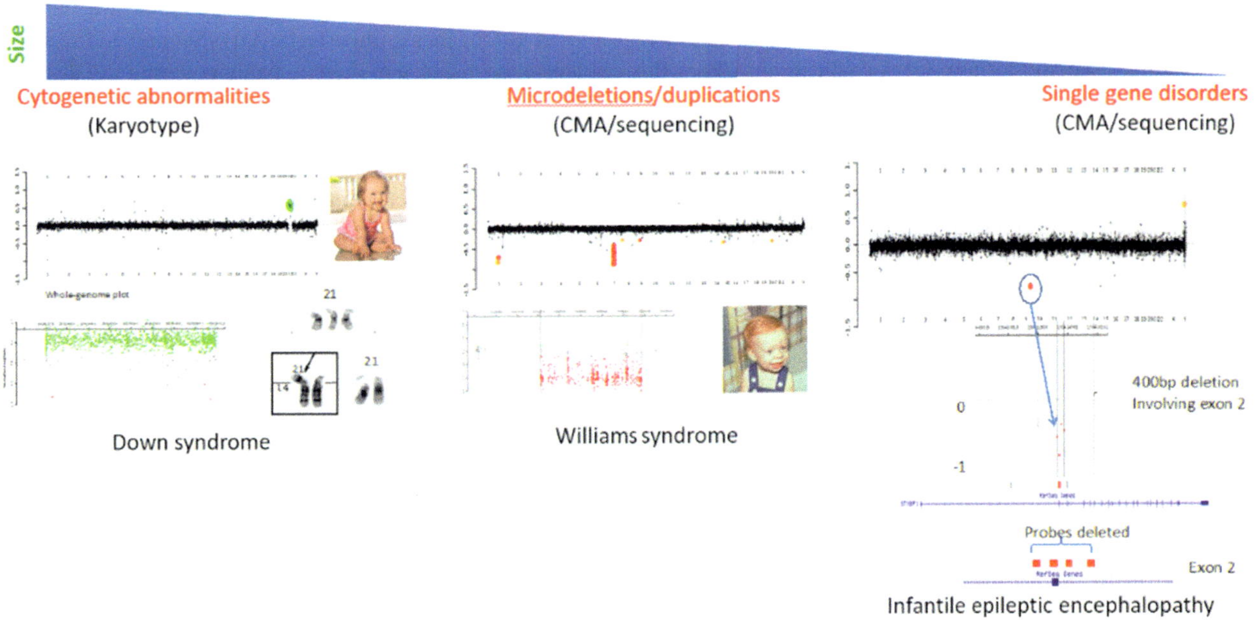

FIGURE 6.4 Increased clinical sensitivity of chromosome microarray analysis (CMA). In *left panel*, CMA and chromosome analysis have equivalent sensitivity for chromosome aneuploidies. Plots of log ratio on the Y axis and chromosome 1 to Y location on the X-axis (enlarged below) are shown. *Left panel* shows increased or gain of signals of chromosome 21 as green dots and karyotype with trisomy 21. However, CMA cannot differentiate if the aneuploidy is derived from a Robertsonian translocation rearrangement as a karyotype can (below). *Middle panel* shows how CMA can detect copy number variants below the detection limit of chromosome analysis, and the plot shows decreased or loss of signals with deletion of chromosome 7q11.2 as red dots. *Right panel* shows an exonic copy number variant (CNV) detected by CMA. When specifically designed for exonic coverage, CMA can readily detect heterozygous exonic losses (or gains) as small as a few hundred base pairs. Such events are possible but still remain challenging for sequencing-based methods.

isodisomy (UPD) (Section 2.7.5) can only be detected by SNP-based microarray, but not by chromosome, FISH, or aCGH analyses. In many clinical settings, microarray is the first choice option for the detection of genomic gains or losses when a specific syndrome is not clinically suspected.

6.2 Genotyping

There are numerous methods to detect disease-contributing variants underlying Mendelian or single gene disorders. The utility of different methods depends on the clinical differential diagnosis and the predicted variant spectrum of genes that cause disease. Variant-targeted panels are usually used for genes for which specific pathogenic variants have been identified in the majority of cases, or for population-specific or founder variants. Sanger sequencing, involving polymerase chain reaction (PCR) coupled with sequencing, is a commonly used targeted detection method. A PCR product generated by annealing complementary primers to amplify the specific DNA region containing the variant of interest is sequenced to detect the presence of the pathogenic variant. Sanger sequencing that utilizes extension of DNA fragments of various lengths terminated with dideoxynucleotides at the 3′ end and separation by capillary electrophoresis may be used for targeted variant(s), exon(s), or gene(s) that are highly suspected. With the development of NGS, multiple genes can be sequenced simultaneously by massive parallel sequencing with significantly high throughput and low cost. Although NGS has also been validated to detect copy number variants for a subset of genes, NGS and microarray combined analyses still serve as a comprehensive genetic evaluation for different variant types.

6.2.1 Targeted Gene Analysis

The single gene or Mendelian disorders are caused by defects of a single gene with different inheritance patterns, including autosomal dominant, autosomal recessive, and X-linked forms (Section 2.7). Pathogenic variants causing single gene disorders can be detected by sequencing technologies (Table 6.4). Gene-based or even variant-based molecular analysis may be applied when there is a specific differential diagnosis clinically. For example, the specific and common variant c.1138G>A in the *FGFR3* gene, leading to p.Gly380Arg, may be ordered if achondroplasia is highly suspected (Section 11.10.2). More often, the clinical presentation may be atypical or complex, resulting in a broad spectrum of differential diagnoses. For example, a patient has clinical features suggestive of osteogenesis imperfecta (OI) (Section 11.10.3), but pinpointing a specific form of OI is challenging. Under this circumstance, using a multigene panel covering the disorders within the phenotypic spectrum, such as an OI gene panel or a larger skeletal dysplasia gene panel, may achieve molecular diagnosis more efficiently. In some cases, where a specific single gene disorder is highly suspected yet no sequence variant is detected by sequencing methods, microarray or multiplex ligation–dependent probe amplification (MLPA) can be used concurrently as supplementary analyses focusing on potential copy number change(s) of the targeted gene as copy number changes may be difficult to detect with sequencing. MLPA is a variation of the multiplex PCR that permits amplification of multiple targets with only a single primer pair, detecting copy number changes.

Targeted variant analysis may be achieved by methods such as SNP genotyping arrays, pyrosequencing (sequencing by synthesis with detection of the order of nucleotides by a light reaction when a pyrophosphate is released), or Sanger sequencing with short turnaround time (Fig. 6.5). It is performed to detect major variant(s) of genes high on the differential diagnosis list, for example, the c.20T > A (p.Glu6Val) variant on the beta-hemoglobin gene for sickle cell disease. Targeted variant analysis also can be used for screening purposes, for example, carrier screening (Section 4.1) for a panel of population enriched common pathogenic variants of cystic fibrosis (CF) (Sections 11.4.1, 11.5.1), hemoglobinopathies, Tay–Sachs disease, and others. A multi-variant CF panel has been used for molecular confirmation of infants positive for

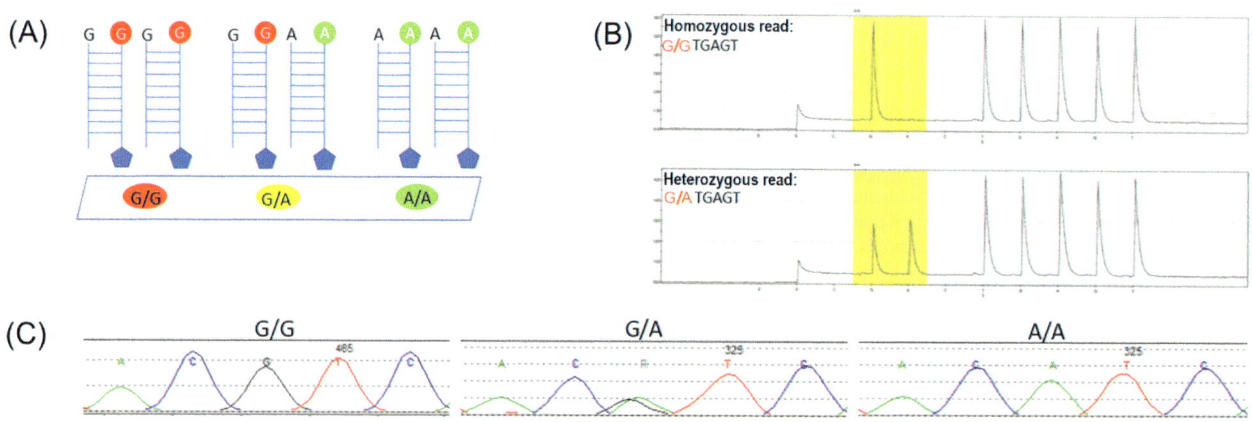

FIGURE 6.5 Examples showing genotyping by SNP array, pyrosequencing, and Sanger sequencing. (A) Single base extension is one of the common detection methods used in SNP array. An array has immobilized allele-specific oligonucleotide (ASO) probes (anchored in *blue, right DNA strand*). They hybridize to the DNA target(s) (*left strand*) and are extended by a single base that is complementary to the base being interrogated and is identified by fluorescent dyes. (B) Pyrosequencing is a sensitive and fast genotyping method. It differentiates zygosity based on the height of each peak, which represents the intensity of signal generated by pyrophosphate released from each nucleotide incorporation. Highlighted in *yellow* are the peaks of the homozygous SNP and the heterozygous SNPs that are half as high. (C) Sanger sequencing showing (from *left to right*) homozygous reference G/G, heterozygous G/A, or homozygous A/A variants at one nucleotide site.

CF on newborn screening and sweat chloride testing. Additionally, targeted variant analysis for one or two variants can be performed on individuals with a known family history of particular variant(s), such as with a family history of cancer (Section 11.16.2).

6.2.2 Single Gene Full-Sequence Analysis

All variants (benign or pathogenic) within the entire coding sequences of a gene can be revealed by Sanger sequencing or NGS, the latter of which may also provide copy number information. Compared to targeted variant genotyping (Section 6.2.1), full gene analysis interrogates the entire locus, which may achieve higher clinical sensitivity for some specific genes, and provides opportunity to discover novel disease-causing variants. On the other hand, full gene sequencing generates more data that lead to multiple variants and extensive curation, and variant interpretation is needed to determine the clinical significance of each identified variant. A recommended guideline has been developed by the American College of Medical Genetics and Genomics (ACMGG) and the Association for Molecular Pathology (AMP) to assist variant interpretation from massive sequencing data (Richards et al., 2015).

6.2.3 Multigene Panel Analysis

Since implemented in the mid-2000s, NGS has significantly advanced the sequencing throughput and cost, which offers superb clinical and research potentials (Goodwin et al., 2016). Multiple NGS platforms have been invented using different technologies including sequencing by ligation, sequencing by synthesis using cyclic reversible termination method, and sequencing by synthesis using single-nucleotide addition method. Sequencers generate different data of short-read pair-end sequences with read length from 50 bp to 150 bp, short-read single-end sequences with read length from 50 to 75 bp, and longer single-end sequences ranging from 200 to 600 bp in a much shorter runtime; however, with the latter, the error rate is higher for insertion/deletions (indels) (Section 2.4.1) and homopolymers (tract of purines or pyrimidines) compared to other platforms.

The NGS process consists of DNA library preparation, targeted region capture, sequencing, and postsequencing bioinformatic analyses. The library preparation step is to prepare a library of size-selected DNA fragments ligated with sequencing adaptors, which consist of sequencing indexes to differentiate samples, enabling multiplex sequencing. The DNA library will then be used for the capturing step where target enrichment occurs. Hybridization capture is one method that uses long, biotinylated oligonucleotide baits (probes) to hybridize to and isolate the regions of interest. Depending on the design, capturing uses target-specific probes to pull down targets-of-interest, for example, all exons (Section 2.3.4) of a group of genes associated with connective tissue disorders. This allows simultaneous analyses of multiple targets. After sequencing in multiplex, a bioinformatic pipeline is used to decode the data (Fig. 6.6). The components of a typical pipeline include demultiplexing, read alignment, variant calling, variant filtering, and annotation (Roy et al., 2018). Raw data generated by sequencing goes through bioinformatic analysis, which lead to generation of user-friendly variant files (annotated VCF). The critical components generated by the pipeline include FASTQ, BAM, and variant files (VCF). A series of quality control metrics (QC) are generated to monitor the process. A FASTQ file is a text-based format for storing nucleotide sequence and its corresponding quality scores. A BAM is a compressed binary file to represent aligned sequences. A VCF file contains variant calls annotated with Human Genome Variation Society (HGVS) calculations and customized annotations. Variant nomenclature, including nucleotide change and amino acid changes based on specific transcripts, is calculated by following HGVS recommendations. Databases of variant minor allele frequency, gene–disease association, prediction scores, and others may be used for annotation. Quality control is performed extensively during the pipeline to ensure data quality. The components should be validated individually for streamlined and clinical-grade production. Downstream of the bioinformatic processes, the variants will be interpreted with clinical and molecular genetic knowledge, and clinical reports will be issued.

NGS panel sequencing may focus on a group of genes associated with related phenotypes. For example, disease genes within the RAS-MAPK pathway causing Noonan-spectrum disorders/RASopathies (e.g., *BRAF, CBL, HRAS, KRAS, MAP2K1, MAP2K2, NRAS, PTPN11, RAF1, RIT1, SHOC2, SOS1,* and others) may be included in an NGS panel. This may provide more efficient molecular diagnosis for a patient suspected with Noonan or Noonan-like syndromes (Sections 11.4.3, 11.12.2).

Multiple variant types including single-nucleotide variants (SNVs) and small insertions or deletions (indels) are analyzed in NGS gene panels (Section 2.4). Copy number variants (CNVs) may be reported if the pipeline of a certain panel is clinically validated for reporting CNVs. Otherwise, MLPA or array CGH (Section 6.1.3) may be used as a complementary or confirmative method for the CNV analysis.

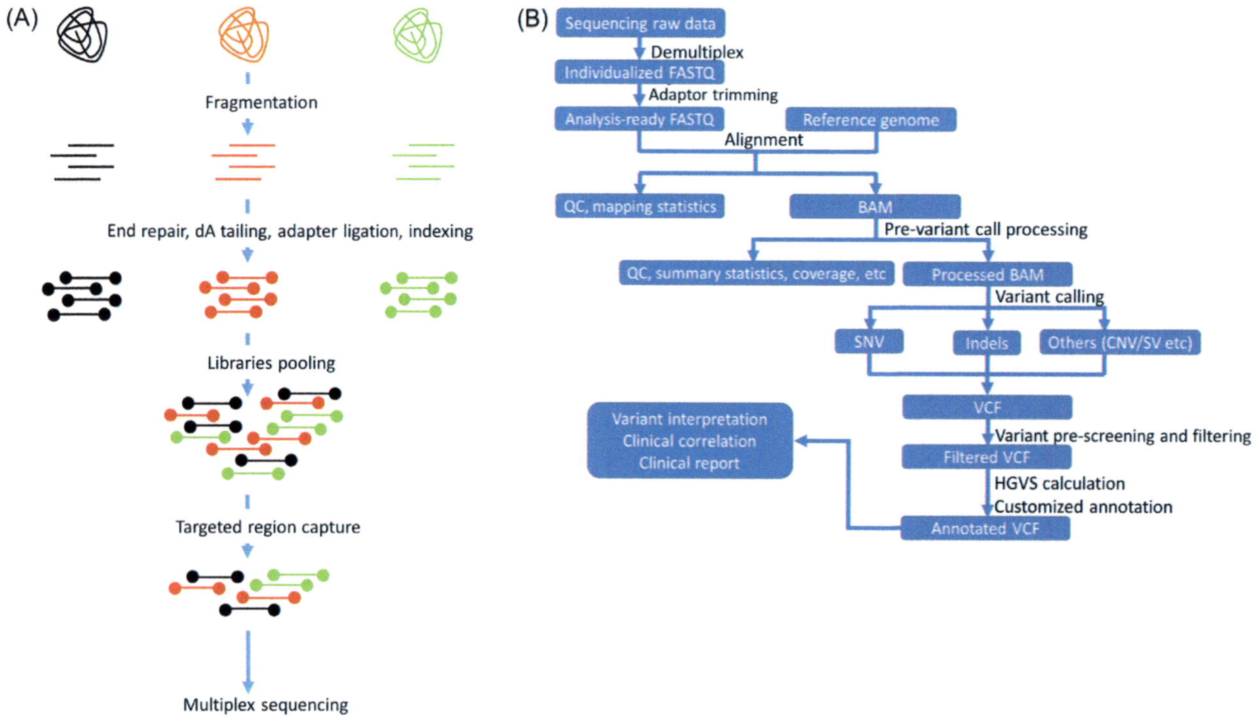

FIGURE 6.6 General next-generation sequencing (NGS) workflow. A. A simplified NGS wet-lab process. Multiple samples can be processed simultaneously, enabling multiplexing and increasing sequencing throughput. Two major steps are involved, including library preparation and targeted capture. Sequencing libraries are typically generated by fragmenting genomic DNA and adding specialized adapters with sample indexes to both ends of the DNA fragment. The indexes are used to distinguish among different libraries during data analysis. Targeted capture enriches the regions of interest, aiming for a more specific and focused study (e.g., targeted NGS panels or exome sequencing). B. Bioinformatic workflow for NGS data process, analysis, and interpretation. Raw data generated by sequencing goes through a bioinformatic pipeline. The major steps include raw sequencing data processing, sequence alignment, variant calling, variant annotation and filtering, and variant prioritization and interpretation. BAM, binary alignment map; QC, quality control; indel, insertion/deletion; SNV, single-nucleotide variant; CNV, copy number variant; SV, structural variant; HGVS, Human Genome Variation Society; VCF, variant call format. Source *Roy, S., Coldren, C., Karunamurthy, A., Kip, N.S., Klee, E.W., Lincoln, S.E., Leon, A., Pullambhatla, M., Temple-Smolkin, R.L., Voelkerding, K.V., Wang, C., Carter, A.B., 2018. Standards and guidelines for validating next-generation sequencing bioinformatics pipelines: a joint recommendation of the Association for Molecular Pathology and the College of American Pathologists. J. Mol. Diagn. 20, 4–27. PMID 29154853.*

NGS panels typically cover the coding exons of genes. Intronic and untranslated regions (5′-UTR or 3′-UTR) may not be covered unless disease-relevant variants have been reported in the literature for these noncoding regions and the assay is designed to capture these regions. NGS typically has limited detection ability for variants in pseudogenes, or nonfunctional regions of DNA, due to ambiguous mapping of short reads to highly homologous sequences (Section 2.3.4). Such variants may be initially identified by the NGS method, and a confirmation method is necessary to distinguish if the variant belongs to the functional gene. Long range PCR method or allele-specific PCR with primers uniquely annealed to the targeted region is usually used for this purpose. Although this makes NGS possible for differentiating a pseudogene and its corresponding functional gene, this is a gene-by-gene analysis that cannot be utilized universally for multigene interrogation.

6.2.4 Carrier Screening

While most gene panels are offered for diagnostic purposes, there also exist ones designed for apparently asymptomatic individuals, in particular carrier screening for reproductive risk (Section 4.1). Carrier screening is mostly used to provide reproductive assessments by detecting pathogenic variants in genes that potentially contribute to carrier status of recessive (autosomal or X-linked) disorders (Sections 2.7.1, 2.7.2). It is useful to assess reproductive risk for couples undergoing pregnancy or planning a pregnancy, having a family history of a recessive disorder, or belonging to a specific ethnicity with high risk of a recessive disorder (e.g., Tay–Sachs disease in Ashkenazi Jewish population). Among over 2000 recessive conditions, carrier screening mostly targets the most severe and/or prevalent diseases dependent or independent of a specific ethnicity.

Traditionally carrier screening uses targeted genotyping methods to detect a number of genetic variants for the limited group of disorders; for example, a *CFTR* mutation panel for the most common pathogenic variants causing CF (Sections 11.4.1, 11.5.1). This approach is effective in detecting population-specific variants. For example, the *CFTR* c.1521_1523delCTT, p.Phe508del (p.F508del) variant is the most common pathogenic allele for CF, contributing to 66% of pathogenic alleles overall; a panel consisting of 23 pathogenic variants offers detection rate of 97% in Ashkenazi Jewish, 88.3% in non-Hispanic whites, 69% in African Americans, and 57% in Hispanic Americans (Savant et al., 2001). However, the mutation targeted panels lack sensitivity for pan-ethnicity screening purposes. With the low-cost and high-throughput NGS technology, the targeted genotyping has now been transformed to a much broader screening of all exons for a significantly larger number of variants and genes. A list of conditions has been recommended for carrier screening by both The American College of Medical Genetics (ACMG) and The American College of Obstetricians and Gynecologists (ACOG) (Table 6.3). Beyond that, more conditions with high severity and prevalence regardless of ethnicity can be included in the carrier screening panel, i.e., expanded or universal carrier screening (Grody et al., 2013). Currently, comprehensive carrier screening panel can include hundreds of conditions. This also brings unprecedented challenges for reporting due to enormous data generated by such large panels.

TABLE 6.3 **ACOG/ACMG recommended carrier screening.**

Conditions	Genes
Alpha-Thalassemia	*HBA1/HBA2*
Beta hemoglobinopathies (Beta-thalassemia and sickle cell)	*HBB*
Bloom syndrome	*BLM*
Canavan disease	*ASPA*
Cystic fibrosis	*CFTR*
Familial dysautonomia	*IKBKAP*
Fanconi anemia	*FANCC*
Fragile X syndrome	*FMR1*
Gaucher disease	*GBA*
Mucolipidosis IV	*MCOLN1*
Niemann–Pick disease, type A	*SMPD1*
Spinal muscular atrophy	*SMN1*
Tay–Sachs disease	*HEXA*

Table created based on recommendations from the following sources: Gross, S.J., Pletcher, B.A., Monaghan, K.G., Professional Practice and Guidelines Committee, 2008. Carrier screening in individuals of Ashkenazi Jewish descent. Genet Med 10, 54–6. PMID 18197057 and Committee Opinion No. 691: Carrier Screening for Genetic Conditions. 2017. Obstet Gynecol, 129, e41–e55. PMID 28225426.

6.3 Genome Testing

6.3.1 Exome Sequencing

The exome is composed of the protein-coding sequences, or exons, of approximately 20,000 to 25,000 genes constituting 1%–2% of the human genome (Section 2.3.4). It may be extended to nonprotein-coding genes such as noncoding RNA (ncRNA) and microRNA. To date, more than 80% of known disease-contributing variants are identified within exomes, and therefore sequencing the exome may currently be the most efficient diagnostic method for Mendelian (single gene) disorders, especially for patients with broad differential diagnoses of suspected genetic conditions (Table 6.4). The exome represents a subset of the human genome, and the exome information can either be "pulled out" in solution from the DNA specimen prior to sequencing, or bioinformatically extracted from the entire genome sequencing data. Exome sequencing (ES) can be considered as a comprehensive NGS panel, where all genes are included for analysis regardless of their clinical significance or relationship to a disease state.

Since broadly implemented as a clinical test in 2011 (Lee et al., 2007), ES has provided a diagnostic yield at the range of 25%–40%, depending on the specific clinical phenotypes, for rare Mendelian disorders (Yang et al., 2013). Because ES

TABLE 6.4 Comparison of targeted, exome, and genome sequencing.

	Targeted sequencing	Exome sequencing	Genome sequencing
DNA sequenced	Specific region(s)	All coding regions and intronic splice sites	Entire genome
Number of genes	1–500, customized	20,000–25,000	All genes, translocations, and noncoding DNA
Depth of sequencing	High, >500x	Intermediate, >50–100x	Low, >30x
Mosaicism sensitivity	High	Intermediate	Low
Variant, number, and type detected	Small number dependent on gene and panel size; SNPs, indels	20,000; known and novel SNPs, indels, SVs	4,000,000; known and novel SNPs, indels, SVs
Cost	Lower	Intermediate	Higher
Turnaround time	Days to weeks	Weeks	Weeks to months
Limitations	Detection of variants only in preselected genes	Incomplete exome capture, detection of variants only in 2% of genome, coding regions	Complex analysis and interpretation, detection of many variants of unknown significance (VUS)

Indel, insertions/deletions; *SNP*, single-nucleotide polymorphism; *SV*, structural variant; *VUS*, variant of unknown significance.

also provides a look at the genes with unknown disease association, this enables the investigators to uncover many more new disease genes compared to the past several decades (Chen et al., 2022). The research on gene discovery informed by ES in turn contributes to the increasing diagnostic yield for new patients as well as previously studied patients who have their exomes reanalyzed (Liu et al., 2019). ES has also helped to identify a subset of patients who have a digenic or blended phenotype of two or more distinct Mendelian disorders (Posey et al., 2017).

Due to the larger-scale sequencing, as exemplified by ES or genome sequencing (Section 6.3.2), variants in disease genes unrelated to a patient's chief complaint for genetic testing may be identified. Some of these changes may provide opportunities to establish timely intervention and disease management to prevent or reduce morbidity and mortality. For example, identification of a pathogenic variant in *BRCA1 or BRCA2* or similar genes may help to care for the cancer risk of the individual in the future (Section 11.16.1). Therefore, such pathogenic findings are referred as incidental findings, mostly representing conditions that are medically actionable. As recommended by ACMG, pathogenic findings belong to the minimum list of at least 70 genes should be offered to be reported from clinical genomic sequencing (Section 7.2) (Kalia et al., 2017).

6.3.2 Genome Sequencing

While ES has been widely used in the clinical field, genome sequencing (WGS, referring to full, complete, or entire genome sequencing of an individual's DNA at one time as opposed to genomic sequencing of both exons and introns of a gene or segment of DNA) is at the early stage of clinical application while being used extensively in research studies. Compared to ES, much more genetic information, including variants from deep introns, 5′UTR and 3′UTR of a gene, noncoding sequences, and intergenic regions, can be revealed by WGS analysis (Table 6.4). The uniformity of genome-wide coverage provided by WGS has demonstrated a better copy number variant detection capability, which provides CNV analysis comparable to CMA. Moreover, structural variants and repeat expansions may also be detectable by WGS analysis, due to the fast-advancing analytical algorithms available to parse the WGS data.

Although the benefit of WGS is obvious, there are still numerous challenges for its application as a clinical test. First, due to the lower coverage (30–50X genomes vs. 100X all exomes sequenced), WGS may not be the optimal methodology to detect low-level mosaicism comparing to other more sensitive assays. Second, WGS generates a large amount of data, making data storage and computational analysis potentially challenging. More importantly, because our understanding of the genome outside the protein-coding regions is limited, the noncoding variants identified by WGS may impose a burden on clinical interpretation. However, our knowledge of noncoding variants has been gradually and steadily advanced by the publicly available genome databases (such as gnomAD, https://gnomad.broadinstitute.org/), the research discoveries promoted by WGS, and the multi-omic strategy integrating metabolomics, phenomics, proteomics, and transcriptomics.

6.3.3 Transcriptome

RNA studies may either target the expression level of the RNA production of a certain gene (e.g., increased or decreased RNA amount) (Section 2.3.2) or the abnormal RNA product altering the integrity of a normal RNA transcript (e.g., gene truncation or fusion); both events may be harmful to cells and lead to disorders. Transcriptomics is a tissue-specific profiling analysis of the entire RNA from a sample. Therefore, a transcriptomic study focuses on deciphering the totality of RNA abnormalities within the sample comparing to a reference control. A targeted RNA study reveals the change of the transcript of interest; in contrast, a transcriptomic profiling reveals not only the root cause but also the downstream alterations caused by the primary variation.

To prepare RNA analysis, complementary DNA (cDNA) is synthesized from the single-stranded mRNA template in a reaction catalyzed by the enzyme reverse transcriptase. Transcriptomics studies usually are carried out by microarray (Section 6.1.3) or RNA sequencing. Microarray uses an array of oligonucleotides ("probes") to hybridize to the RNA targets, and the abundance of each RNA target is quantitated by the fluorescence intensity at each probe location on the array. This is considered a targeted analysis of the expression level of the probe-targeted transcripts. RNA sequencing is the application of NGS (Section 6.2.3) on an RNA sample from a specimen. Like its use with DNA-sequencing, it serves as a more sensitive and high-throughput approach than microarray to determine the abundance and integrity of the gene transcripts. Studying RNA has been critical in the diagnostic field to detect aberrant transcription or fusion of certain gene transcripts. RNA studies, promoted by transcriptomics, have been widely used in the research field to provide supporting evidence for abnormal findings on the DNA level in poorly understood regions, such as introns, 5′- or 3′-UTRs, and intergenic regions. In the near future, RNA sequencing will be used as an adjunct to exome or genome sequencing to help interpret variants of unknown significance in clinical testing.

6.3.4 Third-Generation Sequencing

Long-read sequencing (third-generation sequencing) provides much longer reads (up to hundreds of kilobases). NGS may detect structural variants (SVs) (Section 2.4.1) by analyzing "clipped reads," a special group of reads with partial sequences divergently aligned to distinct locations of the genome. Identification of such "clipped reads" indicates a potential SV splitting the read. Intuitively, the alignment quality increases with more useable base pairs provided by long reads. In fact, third-generation sequencing has demonstrated superb capability in SV detection. Identification of epigenetic changes is also feasible by third-generation sequencing. This is based on the principle that the epigenetic markers (Section 2.6) change the property of the four nucleotides due to DNA modifications. However, the third-generation sequencing still faces critical challenges in clinical utility due to the much higher error as compared to NGS.

6.4 Other Specific Tests

6.4.1 Triplet Repeats (Trinucleotide Repeats)

Being dynamic mutations, the extent of triplet repeats or trinucleotide repeat expansion leads to variable disease severity (Paulson, 2018). In a dominantly inherited triplet repeat disorder, the transmission of the pathogenic triplet repeat allele may increase in copy number, which positively correlates with the increased disease manifestation or severity in the offspring. This is termed anticipation. Fragile X syndrome is one of the well-studied triplet repeat disorders, which is mostly associated with the CGG triplet repeat expansion at the promoter region of the *FMR1* gene on chromosome X. The normal allele has 5–44 repeats. A fully expanded allele with more than 200 CGG repeats (full mutation alleles), causing aberrant hypermethylation and loss of *FMR1* expression, leads to classic fragile X syndrome in males (Section 11.12.3). The smaller expansions between 55 and 200 (premutation alleles) do not cause fragile X syndrome, but increase risk for fragile X–associated tremor/ataxia syndrome (FXTAS) (Sections 11.12.3 and 11.13.11) and the fragile X-associated neurodevelopmental disorders (FXAND) in both male and female carriers, and *FMR1*-related primary ovarian insufficiency (FXPOI) in females, and may expand into full mutation alleles in offspring. Alleles with 45–54 repeats (intermediate alleles) do not cause disease but are at risk of expanding into premutation in the offspring. Anticipation has been observed in other loci, such as *DMPK* (myotonic dystrophy) (Section 11.13.4), *HTT* (Huntington disease) (Section 11.13.12), *FXN*

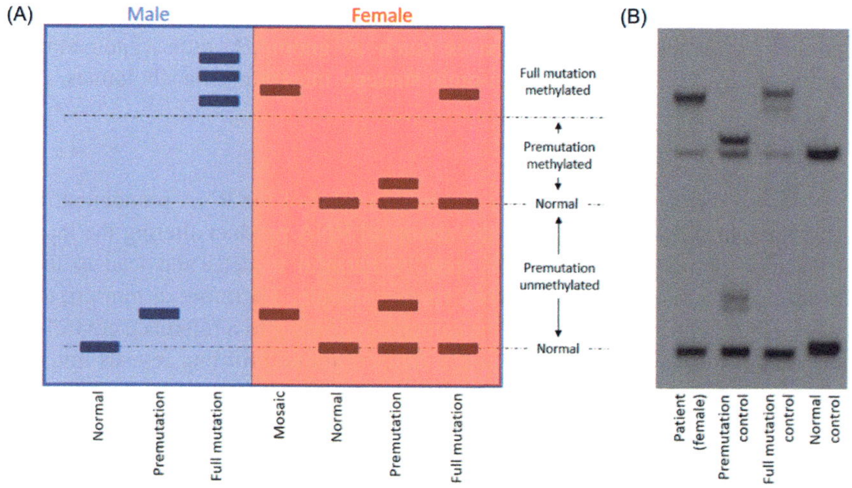

FIGURE 6.7 Southern blot provides diagnostic test for fragile X syndrome. (A) A schematic diagram showing the Southern blot patterns for normal, premutation, and full mutations in male or female. (B) A southern blot film showing the diagnostics for a female patient with fragile X syndrome. Control samples, including premutation, full mutation, and normal controls, are run together with the patient sample.

(Friedreich ataxia), and others. Most triplet repeat disorders have autosomal dominant or X-linked inheritance, except for Friedreich ataxia, which is inherited in an autosomal recessive manner.

Southern blot is traditionally used, and it still serves as the primary diagnostic test for triplet repeat disorders. Southern blot detects specific DNA sequences or fragments, which can be visualized by hybridizing to target-specific radiolabeled DNA probes. The experiment is designed to use site-specific restriction enzyme to generate DNA fragments with genomic content spanning the desired triplet repeat locus according to the human genome reference sequence. The size of the fragment increases as the triplet expands. The number of repeats can be determined by sizing the band of the patient sample and comparing it with the control. Specifically, a methylation-sensitive enzymatic digestion is introduced for fragile X syndrome to evaluate the methylation status (Fig. 6.7).

Triplet repeat-primed PCR has now been used widely as an alternative, and sometimes more accurate approach to determine a triplet repeat expansion. The assay is design by using a pair of unique sequence-targeted primers to amplify the entire region. In addition, the unique sequence-targeted primer is paired with a primer sitting on the triplet repeats, which will amplify from multiple repeat sites within in the triplet repeat region (Fig. 6.5). The PCR results in an admixture of products primed from each possible triplet site as well as the full length of the expanded region, and a capillary electrophoresis is used to differentiate the products based on their sizes. Southern blot sizes the expansion based on the size estimation of the DNA fragment blotted on the film, whereas PCR-based sizing method apparently provides a more accurate calculation of the repeat number. However, due to limitations in polymerization fidelity of the DNA polymerase particularly in a highly repetitive GC-rich region, PCR-based method may not be accurate for large expansions (e.g., >150 repeats), and Southern blot should be used to confirm the finding of such large expansions.

6.4.2 Mitochondrial Disease

Mitochondria are known to be the energy producers of our bodies and occur in thousands of copies in almost all cells (Section 2.5). Most of the proteins that form the respiratory chain are encoded by the nuclear genome, but 13 subunits are encoded by the mitochondrial genome. A hallmark of mitochondrial disease is the involvement of multiple organ systems, which can be observed at birth or later in life and may be caused by variants in the mitochondrial genome inherited from the mother or in the nuclear genome (Section 2.7.4), which are typically transmitted in an autosomal recessive manner (Section 2.7.1). Due to the presence of thousands of copies of the mitochondrial genome per cell, an important feature of mitochondrial genetics is the concept of homoplasmy and heteroplasmy. Homoplasmy reflects the identical nature of all mitochondria assayed while heteroplasmy refers to the existence of more than one mitochondrial genotype. Nuclear genome variants will be detected by standard constitutional molecular genetic approaches while mitochondrial genome variants need special approaches. With suspicion of mitochondrial disease, both the nuclear and mitochondrial genes may need to be sequenced (Sections 11.6.2, 11.6.5).

The mitochondrial genome is a double-stranded 16.6 kb circular molecule that encodes 13 proteins of the oxidative phosphorylation (OXPHOS) system as well as ribosomal RNAs and transfer RNAs. Like the nuclear genomes, pathogenic variants in the mitochondrial genome can be SNV, indels, or larger rearrangements such as deletions and duplications

(Section 2.4.1). NGS methods have been developed to sequence the entire mitochondrial genome with a high depth of coverage in order to detect low levels of heteroplasmy. Long range PCR is generally the initial step to target only the mitochondrial genome followed by NGS, referred to as amplicon NGS. Southern blotting can be employed to detect clinically significant deletions and duplications in the mitochondrial genome. To diagnose mitochondrial depletion syndromes, classified as myopathic, encephalomyopathic, hepatocerebral, or neurogastrointestinal phenotypes and caused by recessive mutations in nuclear genes, quantitative PCR is used to detect the mitochondrial genome content and compared to that in normal controls. The tissue chosen for mitochondrial mutation analysis is important to obtain the highest diagnostic yield. Muscle is often considered a tissue of choice; however, an invasive procedure is needed. Liver and cultured fibroblasts are also sent; however, usually blood and saliva are the primary sample types because of the relative ease to obtain these samples (Section 6.5).

6.4.3 Imprinting Disorders

Genomic imprinting is mediated by epigenetic modifications, e.g., DNA methylation or histone modification, of the human genome that causes a gene to be expressed on a parent-of-origin basis (Section 2.7.5). The imprinting is established during gametogenesis of egg or sperm formation during meiosis, and the locus-specific imprinting status remains unchanged during organogenesis. Several human chromosomal regions are subjected to imprinting, and genetic abnormality of these regions results in imprinting disorders through complex molecular mechanisms (Table 6.5) (Spence et al., 1988; Kotzot, 2008).

Disorders associated with genetic abnormalities of these regions may require a molecular diagnosis to support clinical diagnosis. Prader–Willi syndrome (PWS) (Section 11.13.1) and Angelman syndrome (AS) at the 15q11.2q13 chromosome region are two of the most studied imprinting disorders. PWS and AS are two distinct disease entities; however, the most common pathogenesis is due to a 5–6 Mb deletion at the same genomic region of 15q11.2q13, consisting of ~70% of the total molecular diagnoses for each disorder. The only difference is the molecular basis of the parental origin of the deletion, which can be differentiated by methylation studies, for example, by methylation-sensitive PCR and/or Southern blot using methylation-sensitive restriction enzymes.

6.4.4 Somatic Testing

Somatic mutations occur at the postzygotic stage affecting a subset of cells of an individual after conception (Sections 2.2.2, 2.4.2). As a result, the mutation may be harbored by part of the individual's body but absent in other body parts. These mutations, usually confined to specific tissues, may be detrimental to normal cell growth and cause disorders manifested at the somatic level. Malignancies, such as hematological cancers or solid tumors, may be good examples that show somatic mutation contributing to tumorigenesis. However, this should be differentiated from the cancer syndromes that are caused by inherited germline mutations in genes such as *BRCA1*, *BRCA2*, *PMS2*, and others (Chapter 8). Somatic mutations causing malignancies are usually confined to affected tissues and not able to be transmitted to offspring unless the mutation affects the gonad-producing germ cells; whereas germline mutations in genes associated with hereditary cancer can be inherited from parents and passed on to offspring.

Due to the tissue-specific nature of somatic mutations, the methodology of detection needs to be designed with high sensitivity, and the genetic tests are always recommended to be performed on affected tissues. For example, activating mutations in the *PIK3CA* gene cause somatic overgrowth syndromes with characteristic features including hemihypertrophy, megalencephaly, skeletal, and capillary malformation. Genetic tests performed on the affected tissue from the patient (e.g., affected skin biopsy) provide better diagnostic yield to detect the mutations, which may be absent or present at low level in the patient's germline that is beyond the detection limit of a specific test.

Chromosome analysis and FISH are two major cytogenetic technologies that have been widely used for decades to detect clonal chromosome rearrangements contributing to hematological malignancies or solid tumors (Sections 6.1.1, 6.1.2). Chromosome analysis focuses on the metaphase chromosome spread for large and microscopic visible aberrations; FISH is usually used to target the classic events for specific disorders, but it can detect submicroscopic rearrangements. Complementary to cytogenetic abnormalities, somatic tests built on higher resolution, molecular methods are used to detect base-pair level abnormalities underlying a somatic disorder. SNP-based microarray (Section 6.1.3) often is used to detect reliably submicroscopic copy number variants or loss of heterozygosity (LOH) in affected tissues. The detection of sequence variants, especially low abundance somatic SNPs and insertions/deletions, can be accomplished by sensitive NGS technologies with ultrahigh depth on the targeted region (Section 6.2.3). NGS also enables simultaneous interrogation, in a single assay, of millions of genetic changes of variable types potentially underlying malignancy, which include single-nucleotide variant/indels and copy number variants. NGS methods using mate-pair

TABLE 6.5 **Reported imprinted loci and their associated disorders in human.**

Imprinted regions	Disorder	Critical genes	Parental origin of imprint	Disease-causing variant	Molecular mechanism
6q24	Transient neonatal Diabetes Mellitus	*PLAGL1, HYMA1*	Maternal	Paternal UPD6 Paternal duplication Hypomethylation	*PLAGL1* and *HYMA1* overexpression
7q32	Russell-Silver syndrome	Unclear	Unclear	Maternal UPD7 (Section 11.7.4) Mosaic trisomy 7	Unclear
11p15.5-IC1	Russell–Silver syndrome	*H19, IGF2*	Paternal	Maternal UPD15 Maternal duplication Hypomethylation	*HG19* overexpression
	Beckwith–Wiedemann syndrome			Paternal UPD15 (Section 11.7.1) Paternal duplication Hypermethylation	*IGF2* overexpression
11p15.5-IC2	Russell–Silver syndrome	*CDKN1C, KCNQ10T1, KCNQ1*	Maternal	Maternal UPD15 Hypermethylation	*CDKN1C* overexpression
	Beckwith–Wiedemann syndrome			Paternal UPD15 (Section 11.7.1) Hypomethylation Maternal *CDKN1C* pathogenic variant	*CDKN1C* reduced expression
14q32.2	Temple syndrome	Unclear	Paternal	Maternal UPD14 Paternal deletion hypomethylation	Reduced expression of paternally expressed genes (*RTL1s, DLK1, DIO3*)
	Kagami–Ogata syndrome			Paternal UPD14 Maternal deletion hypermethylation	Reduced expression of maternally expressed genes (*MEG3s, MEG8, ncRNA*)
15q11.2q13	Prader–Willi syndrome	*SNURF-SNRPN UBE3-AS UBE3A*	Maternal	Maternal UPD15 Paternal deletion (Section 11.13.1) hypermethylation	*SNURF-SNRPN* reduced expression
	Angelman syndrome			Paternal UPD15 Maternal deletion Hypomethylation maternal *UBE3A* pathogenic variant	*UBE3A-AS* overexpression causing *UBE3A* reduced expression
20q13.22	Pseudohypoparathyroidism	*GNAS*	Paternal	Paternal UPD20 Methylation defect *GNAS* pathogenic variant Maternal *STX16* deletions Maternal *NESP* and/or *NESP-AS* deletions	*GNAS* reduced expression

IC, imprinting center; *UPD*, uniparental disomy.

sequencing (long-insert paired-end sequencing), designed for specific chromosomal rearrangements, have been developed for detection of somatic chromosomal abnormalities in hematological malignancies and solid tumors.

The comprehensive tumor profiling, including single-nucleotide variant/indel, copy number variant, LOH, chromosomal rearrangements, gene fusion, and tumor mutational burden (TMB, measurement of the number of mutations carried by tumor cells) detections that are enabled by high-throughput analytical methods, provides clinically actionable variants that may guide tailored treatment strategy and clinical trial options. Liquid biopsy, targeting on the circulating tumor DNA (ctDNA) released by tumors into the blood circulation, is a simple and noninvasive testing methodology that offers surveillance of tumor progression or treatment responses alternative to previous surgical biopsies for some specific tumors, such as *EGFR* in non−small cell lung cancer. Liquid biopsy is still at its early stage with many limitations to overcome before it gains clinical utility in a broader spectrum.

6.5 Source of Sample (Blood, Saliva, Tissue)

Many options are now available for the source of starting material acceptable for clinical genetic testing. While whole blood is still required for postnatal chromosome analysis and FISH, most other molecular-based assays can begin with less invasive sources of DNA including saliva, buccal swab, and dried blood spots. High-quality DNA in sufficient quantity can be extracted from these samples and yield high-quality results in the downstream processing. However, the method and technique of collection greatly impact the results; thus directions for optimal sample collection should be carefully followed or testing may fail and a recollection become necessary. DNA can be extracted from tissue samples such as tumor, muscle, or liver from fresh samples or formalin-fixed, paraffin-embedded (FFPE) samples. Prenatal samples for DNA extraction are generally obtained by invasive procedures such as chorionic villus sampling or amniocentesis (Section 4.4.2). Analysis can be done on the direct material or cultures can be established if larger quantities of DNA are required. Noninvasive genetic testing is increasingly utilized in the prenatal setting where fetal status can be ascertained against a maternal background by studying cell-free DNA (Section 4.4.1).

6.6 Ordering Genetic and Genomic Tests

Besides understanding the various methodologies for testing and the spectrum of tests available, there are many practical considerations when ordering genetic or genomic testing. Such considerations include not only which gene(s) to be tested, but also what tests are available, where to send testing, and how to minimize cost. Genetic testing is ordered most often clinically for a karyotype, microarray, and gene panel(s). Testing also is sent frequently for a specific known familial variant (single-site testing), and a single gene test. Comprehensive testing, like exome or genome sequencing or a specialized test, occurs less often but is becoming more frequent.

Familial variant testing is appropriate when a patient has a known family history of a genetic variant causing a specific condition (Section 11.16.2). For example, if a patient communicates that her sister or other family member has a breast cancer gene mutation, or *BRCA1/2* variant, and desires testing, testing for that specific variant is appropriate, depending on the age of the individual and the context (Section 11.18.1). It is important to provide the lab with any documentation of the family member's variant. The lab may perform single-site testing for this variant only. Single-site testing generally has a lower cost and conclusive results of whether the variant either is absent or present (Section 7.1). In some cases, a clinician may elect to include sequencing of the entire gene or include additional genes or a full panel as well, especially if there are other symptoms and disorders in the patient's personal medical history and/or extended family history. In this case, it is still important to consult with the lab to confirm that they include assessing for the known familial variant as well.

Single gene testing targets one specific gene. This is more commonly used when a specific diagnosis is likely, and the condition in question has one commonly associated gene. An example of this is a patient who has symptoms fulfilling the clinical criteria for Marfan syndrome (Section 11.10.4). Testing of the *FBN1* gene is often the first step. However, there are a number of conditions with overlapping symptoms, and sometimes using a panel of genes associated with Marfan syndrome and related disorders may be of benefit. Despite this, however, many insurance companies can have strict criteria for covering single gene testing versus gene panel tests as a first line of testing, and thus this may dictate how testing proceeds.

Generalized panel testing, exome sequencing and genome sequencing are used when findings are not suggestive of any one specific condition and the clinician desires broad coverage of many genes or conditions. Specialized tests are performed when specific conditions are considered. The clinician searches information about testing availability (https://www.ncbi.nlm.nih.gov/gtr/) and specimen requirements online and contacts CLIA-approved, clinical laboratories to discuss specifics of testing. Pretest and posttest counseling are important as additional information such as uncertain results, unrelated conditions, or consanguinity may be uncovered through this testing.

Cost of testing varies greatly depending on the type of testing performed. Most insurance carriers have specific criteria that dictate whether testing will be covered and a process for obtaining preauthorization. In some cases, this may require a statement of medical necessity and in others, genetic counseling as well. Communication with the insurance company prior to onset of genetic testing is recommended. In general, insurance companies usually approve karyotype analysis and microarray tests and are more likely to approve single-site testing or single gene, than panel or broader testing. Furthermore, many genetic labs will assist with the process and often offer patients alternative payment options if the out of pocket cost when using insurance is high. When availability or costs are prohibitive, clinicians can look for relevant research studies that may offer testing on a research basis. However, research test results should be confirmed in a CLIA-approved, clinical laboratory.

References

Bailey, J.A., Yavor, A.M., Massa, H.F., Trask, B.J., Eichler, E.E., 2001. Segmental duplications: organization and impact within the current human genome project assembly. Genome Res. 11, 1005–1017. PMID 11381028.

Chen, C.A., Lattier, J., Zhu, W., Rosenfeld, J., Wang, L., Scott, T.M., Du, H., Patel, V., Dang, A., Magoulas, P., Streff, H., 2022. Retrospective analysis of a clinical exome sequencing cohort reveals the mutational spectrum and identifies candidate disease–associated loci for BAFopathies. Genet. Med. 24 (2), 364–373. PMID 34906496.

Edelmann, L., Pandita, R.K., Spiteri, E., Funke, B., Goldberg, R., Palanisamy, N., Chaganti, R.S., Magenis, E., Shprintzen, R.J., Morrow, B.E., 1999. A common molecular basis for rearrangement disorders on chromosome 22q11. Hum. Mol. Genet. 8, 1157–1167. PMID 10369860.

Francke, U., Ochs, H.D., De Martinville, B., Giacalone, J., Lindgren, V., Disteche, C., Pagon, R.A., Hofker, M.H., Van Ommen, G.J., Pearson, P.L., Wedgwood, R.J., 1985. Minor Xp21 chromosome deletion in a male associated with expression of Duchenne muscular dystrophy, chronic granulomatous disease, retinitis pigmentosa, and McLeod syndrome. Am. J. Hum. Genet. 37, 250–267. PMID 4039107.

Gardner, R.J.M., Sutherland, G.R., Shaffer, L.G., 2012. Chromosome Abnormalities and Genetic Counseling. Oxford Monographs on Medical Genetics No 61, fourth ed. Oxford University Press, Oxford. https://doi.org/10.1093/med/9780195375336.001.0001.

Goodwin, S., Mcpherson, J.D., Mccombie, W.R., 2016. Coming of age: ten years of next-generation sequencing technologies. Nat. Rev. Genet. 17, 333–351. PMID 27184599.

Grody, W.W., Thompson, B.H., Gregg, A.R., Bean, L.H., Monaghan, K.G., Schneider, A., Lebo, R.V., 2013. ACMG position statement on prenatal/preconception expanded carrier screening. Genet. Med. 15, 482–483. PMID 23619275.

Gu, S., Szafranski, P., Akdemir, Z.C., Yuan, B., Cooper, M.L., Magrina, M.A., Bacino, C.A., Lalani, S.R., Breman, A.M., Smith, J.L., Patel, A., Song, R.H., Bi, W., Cheung, S.W., Carvalho, C.M., Stankiewicz, P., Lupski, J.R., 2016. Mechanisms for complex chromosomal insertions. PLoS Genet. 12, e1006446. PMID 27880765.

Harel, T., Lupski, J.R., 2018. Genomic disorders 20 years on-mechanisms for clinical manifestations. Clin. Genet. 93, 439–449. PMID 28950406.

Kalia, S.S., Adelman, K., Bale, S.J., Chung, W.K., Eng, C., Evans, J.P., Herman, G.E., Hufnagel, S.B., Klein, T.E., Korf, B.R., Mckelvey, K.D., Ormond, K.E., Richards, C.S., Vlangos, C.N., Watson, M., Martin, C.L., Miller, D.T., 2017. Recommendations for reporting of secondary findings in clinical exome and genome sequencing, 2016 update (ACMG SF v2.0): a policy statement of the American College of Medical Genetics and Genomics. Genet. Med. 19, 249–255. PMID 27854360.

Kang, S.H., Shaw, C., Ou, Z., Eng, P.A., Cooper, M.L., Pursley, A.N., Sahoo, T., Bacino, C.A., Chinault, A.C., Stankiewicz, P., Patel, A., Lupski, J.R., Cheung, S.W., 2010. Insertional translocation detected using FISH confirmation of array-comparative genomic hybridization (aCGH) results. Am. J. Med. Genet. 152A, 1111–1126. PMID 20340098.

Kotzot, D., 2008. Complex and segmental uniparental disomy updated. J. Med. Genet. 45, 545–556. PMID 18524837.

Lee, J.A., Carvalho, C.M., Lupski, J.R., 2007. A DNA replication mechanism for generating nonrecurrent rearrangements associated with genomic disorders. Cell 131, 1235–1247. PMID 18160035.

Liu, P., Meng, L., Normand, E.A., Xia, F., Song, X., Ghazi, A., Rosenfeld, J., Magoulas, P.L., Braxton, A., Ward, P., Dai, H., Yuan, B., Bi, W., Xiao, R., Wang, X., Chiang, T., Vetrini, F., He, W., Cheng, H., Dong, J., Gijavanekar, C., Benke, P.J., Bernstein, J.A., Eble, T., Eroglu, Y., Erwin, D., Escobar, L., Gibson, J.B., Gripp, K., Kleppe, S., Koenig, M.K., Lewis, A.M., Natowicz, M., Mancias, P., Minor, L., Scaglia, F., Schaaf, C.P., Streff, H., Vernon, H., Uhles, C.L., Zackai, E.H., Wu, N., Sutton, V.R., Beaudet, A.L., Muzny, D., Gibbs, R.A., Posey, J.E., Lalani, S., Shaw, C., Eng, C.M., Lupski, J.R., Yang, Y., 2019. Reanalysis of clinical exome sequencing data. N. Engl. J. Med. 380, 2478–2480. PMID 31216405.

Lupski, J.R., 1998. Genomic disorders: structural features of the genome can lead to DNA rearrangements and human disease traits. Trends Genet. 14, 417–422. PMID 9820031.

Lupski, J.R., 2009. Genomic disorders ten years on. Genome Med. 1, 42. PMID 19439022.

Miller, D.T., Adam, M.P., Aradhya, S., Biesecker, L.G., Brothman, A.R., Carter, N.P., Church, D.M., Crolla, J.A., Eichler, E.E., Epstein, C.J., Faucett, W.A., Feuk, L., Friedman, J.M., Hamosh, A., Jackson, L., Kaminsky, E.B., Kok, K., Krantz, I.D., Kuhn, R.M., Lee, C., Ostell, J.M., Rosenberg, C., Scherer, S.W., Spinner, N.B., Stavropoulos, D.J., Tepperberg, J.H., Thorland, E.C., Vermeesch, J.R., Waggoner, D.J., Watson, M.S., Martin, C.L., Ledbetter, D.H., 2010. Consensus statement: chromosomal microarray is a first-tier clinical diagnostic test for individuals with developmental disabilities or congenital anomalies. Am. J. Hum. Genet. 86, 749–764. PMID 20466091.

Neill, N.J., Ballif, B.C., Lamb, A.N., Parikh, S., Ravnan, J.B., Schultz, R.A., Torchia, B.S., Rosenfeld, J.A., Shaffer, L.G., 2011. Recurrence, submicroscopic complexity, and potential clinical relevance of copy gains detected by array CGH that are shown to be unbalanced insertions by FISH. Genome Res. 21, 535–544. PMID 21383316.

Paulson, H., 2018. Repeat expansion diseases. Handb. Clin. Neurol. 147, 105–123. PMID 29325606.

Posey, J.E., Harel, T., Liu, P., Rosenfeld, J.A., James, R.A., Coban Akdemir, Z.H., Walkiewicz, M., Bi, W., Xiao, R., Ding, Y., Xia, F., Beaudet, A.L., Muzny, D.M., Gibbs, R.A., Boerwinkle, E., Eng, C.M., Sutton, V.R., Shaw, C.A., Plon, S.E., Yang, Y., Lupski, J.R., 2017. Resolution of disease phenotypes resulting from multilocus genomic variation. N. Engl. J. Med. 376, 21–31. PMID 27959697.

Richards, S., Aziz, N., Bale, S., Bick, D., Das, S., Gastier-Foster, J., Grody, W.W., Hegde, M., Lyon, E., Spector, E., Voelkerding, K., Rehm, H.L., Committee, A.L.Q.A., 2015. Standards and guidelines for the interpretation of sequence variants: a joint consensus recommendation of the American College of medical genetics and genomics and the association for molecular Pathology. Genet. Med. 17, 405–424. PMID 25741868.

Roy, S., Coldren, C., Karunamurthy, A., Kip, N.S., Klee, E.W., Lincoln, S.E., Leon, A., Pullambhatla, M., Temple-Smolkin, R.L., Voelkerding, K.V., Wang, C., Carter, A.B., 2018. Standards and guidelines for validating next-generation sequencing bioinformatics pipelines: a joint recommendation of the association for molecular Pathology and the College of American Pathologists. J. Mol. Diagn. 20, 4–27. PMID 29154853.

Savant, A., Lyman, B., Bojanowski, C., Upadia, J., 2001. Cystic fibrosis. Seattle (WA). In: Adam, M.P., Ardinger, H.H., Pagon, R.A., Wallace, S.E., Bean, L.J.H., Mirzaa, G., Amemiya, A. (Eds.), GeneReviews((R)). University of Washington, Seattle (WA) 1993–2023 [updated March 9, 2023]. PMID 20301428.

Spence, J.E., Perciaccante, R.G., Greig, G.M., Willard, H.F., Ledbetter, D.H., Hejtmancik, J.F., Pollack, M.S., O'Brien, W.E., Beaudet, A.L., 1988. Uniparental disomy as a mechanism for human genetic disease. Am. J. Hum. Genet. 42, 217–226. PMID 2893543.

Trivellin, G., Daly, A.F., Faucz, F.R., Yuan, B., Rostomyan, L., Larco, D.O., Schernthaner-Reiter, M.H., Szarek, E., Leal, L.F., Caberg, J.H., Castermans, E., et al., 2014. Gigantism and acromegaly due to Xq26 microduplications and GPR101 mutation. N. Engl. J. Med. 371, 2363–2374. PMID: 25470569.

Yang, Y., Muzny, D.M., Reid, J.G., Bainbridge, M.N., Willis, A., Ward, P.A., Braxton, A., Beuten, J., Xia, F., Niu, Z., Hardison, M., Person, R., Bekheirnia, M.R., Leduc, M.S., Kirby, A., Pham, P., Scull, J., Wang, M., Ding, Y., Plon, S.E., Lupski, J.R., Beaudet, A.L., Gibbs, R.A., Eng, C.M., 2013. Clinical whole-exome sequencing for the diagnosis of mendelian disorders. N. Engl. J. Med. 369, 1502–1511. PMID 24088041.

Yuan, B., Harel, T., Gu, S., Liu, P., Burglen, L., Chantot-Bastaraud, S., Gelowani, V., Beck, C.R., Carvalho, C.M., Cheung, S.W., Coe, A., Malan, V., Munnich, A., Magoulas, P.L., Potocki, L., Lupski, J.R., 2015. Nonrecurrent 17p11.2p12 rearrangement events that result in two concomitant genomic disorders: the PMP22-RAI1 contiguous gene duplication syndrome. Am. J. Hum. Genet. 97 (5), 691–707. PMID: 26544804.

Yuan, B., Wang, L., Liu, P., Shaw, C., Dai, H., Cooper, L., Zhu, W., Anderson, S.A., Meng, L., Wang, X., Wang, Y., et al., 2020. CNVs cause autosomal recessive genetic diseases with or without involvement of SNV/indels. Genet. Med. 22, 1633–1641. PMID 32576985.

Zhang, F., Potocki, L., Sampson, J.B., Liu, P., Sanchez-Valle, A., Robbins-Furman, P., Navarro, A.D., Wheeler, P.G., Spence, J.E., Brasington, C.K., Withers, M.A., 2010. Identification of uncommon recurrent Potocki-Lupski syndrome-associated duplications and the distribution of rearrangement types and mechanisms in PTLS. Am. J. Hum. Genet. 86, 462–470. PMID 20188345.

Chapter 7

Genetic and Genomic Results and Management

Anne Slavotinek[1,2]

[1]*Medical Genetics and Genomics, Benioff Children's Hospital Oakland Division of Genetics, University of California, San Francisco, Benioff Children's Hospital, San Francisco, CA, United States;* [2]*Division of Medical Genetics, Department of Pediatrics, University of California, San Francisco, San Francisco, CA, United States*

Abstract

Genetic testing may be recommended after a clinical evaluation in various subspecialty areas within genetics. Ideally, testing is made after informed consent and through a genetic professional, but many specialists and some primary care providers increasingly are sending genetic testing. Results are interpreted for both the patient and referring physicians, and subsequent genetic counseling and recommendations can be made based on these findings. This chapter reviews genetic testing results, including variant classification and secondary findings.

Following a clinical evaluation, including physical examination or ultrasound of a fetus, genetic testing may be recommended. These results need to be interpreted and likely will dictate further management. Ideally testing is initiated with informed consent and through a genetic professional who will help with the interpretation and recommendations. In any case, the referring physicians will want to be able to understand genetic-related reports and have some indication of the next steps. In this chapter, we discuss the interpretation of laboratory results and provide examples of a typical report for a karyotype, single-nucleotide polymorphism (SNP) microarray, gene sequencing, and exome sequencing test. Factors that influence variant classification into pathogenic, benign, or a variant of uncertain significance are also explained, together with a description of secondary findings.

7.1 Interpretation of Results

The results of a genetic test are evaluated by the testing laboratory in the context of patient and family history data, prior medical evaluations, and examination findings. For the most reliable and relevant interpretation of genetic testing, complete and accurate clinical information must be supplied (Richards et al., 2015). Genetic testing can be performed for numerous clinical indications that involve different presentations and body structures. Clinical findings may be absent despite the presence of a pathogenic, or disease-causing variant, a situation in which the variant can be referred to as having reduced penetrance or age-dependent penetrance. Deleterious variants in the same family but in different individuals can also be associated with variation in the clinical presentation that is termed variable expressivity (Section 2.8). Thus, close correlation and expert interpretation of the genetic data (genotype) with clinical findings (phenotype) is critical for accurate interpretation of genetic testing results. Follow-up evaluations by the provider may also be useful to seek evidence that will support a genetic test result, or vice versa. Additional clinical testing to investigate for subtle findings that support disease-causing variants may be useful. Testing in other family members to establish whether a genetic variant is present only in the patient who is then the first in the family with the variant ('de novo' variant), or is inherited from either or both parents, is also

Genomics in the Clinic. https://doi.org/10.1016/B978-0-12-816478-5.00007-8

frequently needed for accurate report interpretation. It may also be necessary to determine if the genetic variant co-segregates with the disease in the family if more than one relative has similar clinical findings. In general, for complex genetic tests, such as next-generation sequencing, interpretation of results is more accurate when parental samples are included with the patient (trio testing), as de novo occurrence, and inheritance from both parents (biallelic inheritance) in autosomal recessive diseases, can be determined (Section 2.7.1). Also, parental testing can determine that a previously unreported, or novel, variant is more likely to be benign if an unaffected parent has the variant, or more likely to be pathogenic if the variant has arisen 'de novo.' Please also see Chapter 12, Communicating Genetics, for ways to share results with families.

7.1.1 How to Read a Genetic Report

In addition to demographic and test data, a genetic testing report will include the genetic testing result, an interpretation section, references, guidance for next steps, and a description of test methodology, together with appropriate disclaimers and information for specialists (Lubin et al., 2009; Rehm et al., 2013; Aziz et al., 2015; Richards et al., 2015). All genetic testing reports typically contain demographic information and patient identifiers, such as full name, date of birth, and sex. Sample details that are recorded include the type of specimen, the dates of sample collection and receipt by the testing laboratory, and a unique laboratory identifier. The name of the ordering provider and institution also is provided. Each laboratory will typically present this information, together with the report results, in a laboratory-specific format.

The following sections provide sample reports for three commonly ordered genetic tests including karyotype, SNP microarray, and exome sequencing.

Karyotype Report

A karyotype, or high-resolution chromosome analysis, will detect balanced chromosome aberrations and is thus complementary to a comparative genomic hybridization (CGH) microarray or SNP array (Section 6.1). The most frequent indication for a karyotype is suspicion of Down syndrome, also known as trisomy 21 (Sections 11.1.3, 11.13.2). An example of a karyotype report is provided in Fig. 7.1, Karyotype report example.

"Specimen type" (Fig. 7.1, superscript 1) indicates the type of tissue tested, for example, blood cells, skin cells (fibroblasts), or other tissue. The tissue type is relevant, as different cell types have different origins, and may show variation in karyotype even when obtained from the same person.

"Metaphases counted" (Fig. 7.1, superscript 2) indicates the number of cells in metaphase, or dividing cells, in which the total number of chromosomes were counted, but not analyzed in detail. The standard number counted is 20 cells. The number of metaphases is useful for determining whether mosaicism, or two different cell lines that can differ in the number or structure of chromosomes, is present (Section 2.2.2). A high number of cells with a normal chromosome complement make mosaicism less likely. If mosaicism has been suspected, the number of metaphases examined will be increased to 50 or higher. Number of "Metaphases analyzed" (Fig. 7.1, superscript 3) indicates the number of cells in metaphase, for which the technician carefully analyzed every banded chromosome to detect deletions, duplications, and other structural changes. The minimum number of analyzed cells is five. "Metaphases karyotyped" (Fig. 7.1, superscript 4) indicates the number of analyzed metaphase cells that were documented as digital, photographic, or drawing images, in which all the chromosomes were lined up in pairs following a specific order defined by chromosome size and number. The minimum standard number of karyotyped cells is two for non-mosaic and constitutional specimens.

"Number of cultures" (Fig. 7.1, superscript 5) indicates the number of separate dishes, or culture vessels, that were used to grow the cells. Cytogenetic results from two cultures are more reliable than results from one culture for detecting mosaicism and differentiating cultural artifact. In addition, one dish can provide a substitute if there is cell contamination or another type of error/incident. Prenatal specimens must be reported from at least two different cultures.

"Banding technique" (Fig. 7.1, superscript 6) indicates the technical process used to produce specific banding patterns on chromosomes, by which each chromosome can be clearly recognized. The commonest banding technique is G-banding using trypsin and Giemsa, typically referred to as G- or GTG-banding (Fig. 2.3). "Banding resolution" (Fig. 7.1, superscript 7) indicates the number of chromosome bands present in a single set of chromosomes (1–22, X and/or Y). Higher resolutions are more detailed and thus are more useful for detecting smaller chromosome aberrations. High-resolution chromosome analysis using peripheral blood is typically at the 550 band level or higher; prenatal specimens are more typically 400–550 band level (Fig. 6.1).

"Results" (Fig. 7.1, superscript 8) of a typical chromosome analysis will provide the numbers of autosomes and sex chromosomes, followed by a description of any aberration present. 46,XX indicates a normal female karyotype, whereas 46,XY is a normal male karyotype. Aneuploidy indicates gain or loss of an individual chromosome or chromosomes. For

Chromosome analysis

Patient name:
DOB:
Patient ID:
Sample ID:
Referring Physician:

Date collected:
Date received:
Specimen type[1]: Peripheral blood

Indication for study: Rocker bottom feet, growth retardation

Metaphases counted[2]:	20
Metaphases analyzed[3]:	5
Metaphases karyotyped[4]:	2
Number of Cultures[5]:	2
Banding technique[6]:	GTG
Banding resolution[7]:	550

RESULTS[8]: 47,XY,+18[9]
Abnormal male karyotype

INTERPRETATION[10]:
Cytogenetic analysis showed an abnormal male karyotype with three copies of chromosome 18 (trisomy 18) in all metaphase cells examined. This is consistent with a clinical diagnosis of Edwards syndrome (trisomy 18). Typical manifestations of Edwards syndrome include growth delays, cardiac defects, neural tube defects, facial anomalies and reduced survival . Clinical correlation is advised.

The risk for recurrence exists for this and other aneuploidies increases with maternal age.

No other chromosome abnormalities were observed. The standard cytogenetic methodology used in this analysis does not routinely detect small rearrangements and low-level mosaicism and cannot detect microdeletions.

RECOMMENDATIONS[11]:
Genetic counseling is recommended for this family.

REFERENCES:
Subramaniam et al., 2016 (PMID 26728763)

Signed: Laboratory Director/Clinical Cytogeneticist/Medical Consultant

FIGURE 7.1 Karyotype report example.

example, 47,XY,+21 is the karyotype of a male individual with Down syndrome due to trisomy 21 (Section 11.1.3 and 11.13.2). Mosaicism (Section 2.2.2) is indicated by a '/' that separates two distinct cell lines. If mosaicism is present, the karyotype will include the number of cells with each chromosome complement in brackets. For example, 47,XX,+21(25)/46,XX(25) describes two distinct karyotypes in a total of 50 cells, one with a normal male karyotype that was found in 25 cells and one with trisomy 21 that was present in 25 cells (Section 11.13.2). After the number of autosomes and sex chromosomes has been listed, any cytogenetic abnormalities that have been detected are described (Fig. 7.1, superscript 9). Reviews of chromosome nomenclature are available (Stevens-Kroef et al., 2017) and are based on a regularly updated International System for Human Cytogenomic Nomenclature (ISCN, 2016).

"Interpretation" (Fig. 7.1, superscript 10) is the most important section of the report, in which the result is described and interpreted, and management recommendations are provided. A diagnosis may be listed, together with a short description of the diagnosis and references. "Recommendations" (Fig. 7.1, superscript 11) typically include genetic counseling for atypical results. Parental testing or additional tests may also be recommended. References are provided, often including relevant cases with similar phenotypes and genotypes to support the interpretation and recommendations.

Single-Nucleotide Polymorphism Array and Comparative Genomic Hybridization Array Report

Copy number variants and chromosome aberrations (Sections 2.2.1 and 2.4.1) can be detected by a variety of technological methods. A SNP array (Fig. 7.2, superscript 1) or CGH array (Section 6.1.3) can detect chromosome aberrations that are below the microscopic level (submicroscopic) used for karyotype analysis and thus are preferred by many providers. A SNP array also will detect the presence of homozygosity, or two identical copies of an allele, and uniparental disomy, or inheritance of two copies of a chromosome from the same parent (Section 2.7.5). However, neither array can detect balanced chromosome rearrangements, even at the microscopic level, that can be detected with a karyotype. An example of an SNP array report is provided in Fig. 7.2, SNP array report example.

Patient Name:	Specimen type: Blood, peripheral, in EDTA
DOB:	Date specimen obtained:
Lab Accession:	Date specimen received:
Pedigree #:	Referring physician:
Gender:	Referring facility:
Race:	Referring facility MRN:

Clinical History: Cleft palate, speech delay, immunodeficiency

Test Performed[1]: SNP Array for blood analysis

RESULTS:
SNP Array
Copy Number Variations[2] Result: Loss[3], Pathogenic (22q11.21[5]), estimated 2.6 Mb[6]
Regions of Homozygosity[4] Result: Not detected

SNP Array Result:
ISCN: arr[hg19] 22q11.21(18,844,632-21,463,730)x1[7]

INTERPRETATION[8]:
Abnormal female result with an interstitial deletion within the 22q11.21 region consistent with DiGeorge syndrome (DGS; OMIM 188400)

COPY NUMBER VARIATIONS:
The results of this test indicate copy number LOSS at chromosome 22q11.21. This copy number loss is estimated 2.6 Mb in size. Within the ~2.6 Mb 22q11.21 region, which spans the typically deleted region (Breakpoints A-D) within the 22q11.2 locus, there are 44 OMIM genes, including the 22q11.2 deletion syndrome critical gene TBX1, and 39 other genes (see below). 22q11.2 deletions are known to be associated with congenital heart disease, cleft palate or velopharyngeal insufficiency, immune function dysregulation, hypocalcemia, facial anomalies, hyperparathyroidism, developmental delays and intellectual disability (McDonald McGinn et al., 2013). Overlapping copy losses have been reported in the literature in individuals presenting with 22q11.2 deletion syndrome. This deletion is consistent with the reported patient presentation and is considered to be pathogenic. Clinical correlation is recommended.

Abnormal Findings:
1) 22q11.21 Copy Loss
Genomic Coordinates:18,844,632-21,463,730

Estimated Size:	2.619 Mb
Number of Probes:	927 Probes
Significance:	Clinically significant
Inheritance:	Unknown

REGIONS OF HOMOZYGOSITY:
No regions of homozygosity were detected.

GUIDANCE FOR NEXT STEPS:
1) Parental analyses could be considered to clarify whether this copy number change was de novo or inherited from a parent.
2) Genetic counseling is recommended.
3) Clinical correlation is required.

Microarray parental variant study is available for the 22q11.21 copy loss at no additional charge to determine whether this alteration is inherited or de novo. Please use the microarray parental variant study requisition form to request this analysis.

RESOURCES FOR PHYSICIANS:
1) UCSC genome browser at http://genome.ucsc.edu/cgi-bin/hgGateway
2) Online Mendelian Inheritance in Man at http://www.ncbi.nlm.nih.gov/omim

RESOURCES FOR PATIENTS:
1) International 22q11.2 Foundation: www.22q.org

REFERENCES[9]: McDonald McGinn et al., 2013 (PMID 20301696)

OMIM genes and other genes[10]:
The 22q11.2 deletion contains: BCRP7, DGCR6, PRODH, DGCR5, DGCR9, DGCR10, CA15P1, DGCR2,DGCR11, TSSK1A, DGCR14, TSSK2, GSC2, SLC25A1, CLTCL1, KRT18P62, HIRA, C22orf39, RN7SL168P, MRPL40, UFD1L, CDC45, CLDN5, SEPT5, GP1BB, TBX1,GNB1L, C22orf29, TXNRD2, COMT, MIR4761, ARVCF, TANGO2, MIR185, DGCR8, MIR3618, MIR1306, TRMT2A, RANBP1, ZDHHC8, LINC00896, RTN4R, MIR1286, DGCR6L, GGTLC3, PI4KAP1, RN7SKP131, RIMBP3, CA15P2, PPP1R26P2, PPP1R26P3, FAM230A, USP41, ZNF74, RNU6-225P, SCARF2, KLHL22, KLHL22-IT1, RN7SL812P, KRT18P5, MED15, SMPD4P1, SLC9A3P2, ABHD17AP4, POM121L4P, BCRP5, TMEM191A, PI4KA, SERPIND1, SNAP29, CRKL, RN7SL389P, AIFM3, LZTR1, THAP7, THAP7-AS1, TUBA3FP, P2RX6, SLC7A4, MIR649, P2RX6P, TUBA3GP, BCRP2

METHODOLOGY:
Whole genome single nucleotide polymorphism-based cytogenomic array (SNP array) was performed using the Illumina CytoSNP-850K Platform with genome build hg19/GRCh37 to detect genome-wide genomic copy number changes (CNVs) and regions of homozygosity (ROH). This array platform contains over 850,000 SNPs in 15x redundancy throughout the genome with enriched coverage for 3,262 genes of known relevance in both constitutional and cancer applications. The average probe spacing across the whole array is approximately 1.8kb. CNVs in the backbone of the genome involving less than 16 contiguous probes are not reported. Therefore, the overall effective resolution across the whole array is approximately 29.0kb. Involvement of a minimum of 12 contiguous probes is required to report a CNV in a critical gene/locus. Benign and likely benign CNVs as well as CNVs less than 1Mb with unknown clinical significance but without involvement of annotated disease-causing genes are not reported for the prenatal testing. For postnatal testing, CNVs <200kb are not reported unless there is strong evidence of pathogenicity. Classification of clinical significance of CNVs is based on the ACMGG standards and guidelines. Data was analyzed using BlueFuse software (Illumina, San Diego, CA) and compared to available data in UC-Santa Cruz Genome Browser, Online Mendelian Inheritance in Man (OMIM),

FIGURE 7.2 Single-nucleotide polymorphism (SNP) array report example. *Adapted with permission using information from an SNP Array Report sample provided by the University of California San Francisco Clinical Cytogenetic Laboratory.*

Database of Genomic Variants (DGV), the Database of Chromosomal Imbalance and Phenotype in Humans Using Ensembl Resources (DECIPHER), Clinical Genome Resource (ClinGen), and ClinGen Dosage Sensitivity Map.

DISCLAIMER:
SNP array is designed to identify CNVs associated with genomic imbalances, as well as ROH. This test will detect aneuploidy, triploidy, deletions, duplications, and ROH of the loci represented on the microarray utilized. It will not detect balanced genomic rearrangements, imbalances and ROH of regions not represented on the microarray. It may not detect tetraploidy caused by endoduplication. The detection level for an alteration may vary depending on the locus. Failure to detect an alteration (gain or loss) at any disease-causing locus/gene does not exclude the diagnosis of any of the disorders associated with the locus/gene, since other mutation mechanisms may lead to the diagnosis. Some unrecognized potential pathogenic CNVs may be classified as unknown clinical significance or non-pathogenic at the time of analysis based on available data. This assay may detect mosaicism of >30% in non-mixed specimens with optimal DNA quality. Mosaicism may not be detected in specimens with mixed cells/tissues (such as prenatal specimens mixed with maternal cells) and specimens with suboptimal DNA quality. This assay may detect maternal cell contamination (MCC) or chimerism in prenatal specimens, which may result in non-informative results for copy number variants and ROH. Copy number variants and ROH below the cut-off thresholds established in this lab are not reported, unless evidence supports the variant or ROH is pathologic. The lab will report ROH detected in known imprinted regions/chromosomes, which may indicate pathogenic uniparental disomy (UPD). UPD suggested by ROH must be verified by a different method. In addition, ROH may indicate increased risk of expression of recessive mutations. However, it may also reveal consanguinity or inbreeding. The lab will not report ROH that is not related to the known imprinted regions/chromosomes unless a written request with a consent form for releasing such findings is signed by the patient, and the lab is not liable for any legal issues related to releasing such findings. The performance characteristics of this array assay was determined by the UCSF Clinical Laboratories as required by Clinical Laboratory Improvement Amendments of 1988 (CLIA-88) regulations. It has not been cleared or approved for specific uses by the US Food and Drug Administration (FDA). The FDA has determined that such clearance or approval is not necessary. This test is used for clinical diagnosis purposes. It should not be regarded as investigational or for research. Pursuant to the requirements of CLIA-88, this laboratory has established and verified the test accuracy and precision.

Signed: Laboratory Director/Clinical Molecular Geneticist/ Clinical Cytogeneticist/ Medical Consultant

FIGURE 7.2 cont'd

SNP arrays report copy number variants (Fig. 7.2, superscript 2), or imbalances in chromosome copy number, as a result of deletions (copy number loss, or monosomy; Fig. 7.2, superscript 3) or duplications (copy number gains, or trisomy – Section 2.2.1). In addition, SNP arrays will report regions of homozygosity when present (Fig. 7.2, superscript 4). Large regions of homozygosity that involve multiple chromosome regions are suggestive of relatedness between parents, or parental consanguinity (Section 3.2.4), because of a shared ancestor in common.

The chromosome region, or location of the copy number variant (Fig. 7.2, superscript 5), is provided in the report, together with the estimated size of the copy number variant (Fig. 7.2, superscript 6). Size is typically provided in million base pairs (megabases, or Mb). Smaller deletions or duplications may be described in 1000 base pairs (kilobases, or kb) (Sections 11.2.4, 11.4.5 and 11.13.5).

Copy number variants are described according to International System for Human Cytogenomic Nomenclature (ISCN, 2016) (Fig. 7.2, superscript 7). Array is abbreviated arr, and the specific reference build, or version, of the genome that was used is listed, for example, human genome assembly #19 (hg19) also named as Genome Reference Consortium Human Build 37 (GRCh37). A normal result would be listed as arr(1−22,X)x2 for a female or arr(1−22)x2,(X,Y)x1 for a male. If a copy number variant is present, after the variant is described, including the range of nucleotide positions on the chromosome for the abnormal probes, the copy number is provided, for example, as ×1 (copy number deletion with one copy rather than the typical two copies of an autosome) or ×3 (copy number duplication with three copies rather than the typical two copies of an autosome).

The "Interpretation" section (Fig. 7.2, superscript 8) discusses chromosome aberration(s) or homozygosity if detected. For copy number variants, size and gene content are noted, together with whether the copy number variant has been associated with disease or seen in healthy controls. The interpretation will also provide a conclusion as to whether the cytogenetic aberration is considered pathogenic (disease causing), likely pathogenic, a variant of uncertain significance, likely benign or benign (unrelated to disease or testing indication) (Riggs et al., 2020). Frequently, correlation with clinical findings and patient status is recommended. References may also be provided (Fig. 7.2, superscript 9), together with a list of all of the genes or transcripts involved in the variant (Fig. 7.2, superscript 10).

Next-Generation Sequencing Report

For genetic tests that use next-generation sequencing (Section 6.3), report information will include demographics, results, interpretation, and methodological sections. The following information is typically included (Fig. 7.3, Exome sequencing report, after Rehm et al., 2013, Supplementary Materials).

The name of the test performed, for example, exome sequencing (Section 6.3.1) or the name of the panel (Fig. 7.3, superscript 1), is listed first. If specific variants are analyzed, the laboratory will often note the variant(s) tested.

The indication for the test, together with a synopsis of clinical information from the healthcare provider, who ordered the test, is listed next (Fig. 7.3, superscript 2).

Patient Name:	Specimen type: Blood, peripheral
DOB:	Date specimen obtained:
Lab Accession:	Date specimen received:
Pedigree #:	Referring physician:
Gender:	Referring facility:
Race:	Referring facility MRN:

TEST PERFORMED:[1]
Exome sequencing

INDICATION FOR TEST:[2]
Speech delay, facial anomalies, short stature

RESULT[3]: Positive. A variant in a gene with an established role in the reported phenotype was identified

DNA VARIANTS:

Gene[4]	Disease[6]	Inheritance[7]	Variant[8]	Zygosity[9]	Inheritance[10]	Classification[11]
KMT2D	Kabuki Syndrome	Autosomal dominant	c.70G>T p.Glu24*	Heterozygous	*De novo*	Likely Pathogenic

INTERPRETATION SUMMARY[12]:
This individual is heterozygous for a missense substitution, c.70G>T, predicting p.Glu24* in the *KMT2D* gene. Parental studies did not demonstrate this variant in either parent, indicating that the variant is *de novo*. Pathogenic variants in the *KMT2D* gene are associated with Kabuki syndrome, an autosomal dominant disorder characterized by developmental delays, cardiac defects, cleft lip/palate, short stature, gastrointestinal anomalies and facial anomalies including long palpebral fissures, depressed nasal tip, prominent or cupped ears and down-turned corners of the mouth (Adam et al., 2019). This variant has not been seen before in individuals with Kabuki syndrome, but is predicted to interrupt gene function, and has been classified as likely pathogenic. Clinical correlation is recommended to determine whether this genetic disorder is consistent with the patient's condition.

INDIVIDUAL VARIANT INTERPRETATION[13]:
p.Glu24* in Exon 1 of *KMT2D* (NM_003482.3), Likely Pathogenic.
This missense change has not been previously reported as a disease-causing variant, but represents a stop codon change that is likely to lead to premature protein termination and nonsense medicated decay, indicating that the variant is likely to lead to loss of gene function. Software prediction programs were unanimous in designating this variant as likely to disrupt gene function. This variant is not listed in the NHLBI ESP Variant Server, 1000 Genomics or GnomAD databases, indicating it is not a common benign variant in these populations.

RECOMMENDATIONS[14]:
Genetic counseling is recommended for this individual and their family.

TEST METHOD[15]: Genomic DNA was extracted from the submitted specimen and additional familial specimens (mother and father) and the (Capture Reagent) kit was used to target the exon regions of their genomes. These targeted regions were sequenced using the (Instrument Model) sequencing system with 100bp paired-end reads. The DNA sequence was mapped to, and analyzed in comparison with, the published human genome build (UCSC hg19 reference sequence). The targeted coding exons and splice junctions of the known protein-coding RefSeq genes were assessed for the average depth of coverage and data quality threshold values*. Sequence changes in this individual were compared to the other provided family members. All reportable sequence variants were confirmed by Sanger sequence analysis using a separate DNA preparation.

*The values below represent metrics from this individual's exome sequencing:
Mean Depth of Coverage[16] 121X
Quality threshold[17] 97.2%

All reported variants are confirmed by independent Sanger sequencing. This test was developed and its performance characteristics determined by The Molecular Diagnostic Laboratory (CLIA#111111111). It has not been cleared or approved by the U.S Food and Drug Administration (FDA).

LIMITATIONS[18]: Absence of a plausible explanation for the reported phenotype by exome sequencing does not exclude a genetic basis of the patient's condition. Some types of genetic abnormalities, such as copy number changes, may not be detectable with the technologies performed by this exome analysis test. It is possible that the genomic region where a disease-causing mutation exists in the proband was not captured using the current technologies and therefore was not detected. Additionally, it is possible that a particular genetic abnormality may not be recognized as the underlying cause of the genetic disorder due to incomplete scientific knowledge about the function of all genes in the human genome and the impact of variants in those genes. Only variants in genes associated with the medical condition, or thought to be clinically relevant potentially for the proband's medical condition, are reported here.

REFERENCES[19]: Adam et al., 2019 (PMID 21882399)

Signed: Laboratory Director/Clinical Molecular Geneticist/ Clinical Cytogeneticist/ Medical Consultant

FIGURE 7.3 Exome sequencing report example. *Adapted with permission from Rehm, H. L., Bale, S. J., Bayrak-Toydemir, P., Berg, J. S., Brown, K. K., Deignan, J. L., Friez, M. J., Funke, B. H., Hegde, M. R., Lyon, E., Working Group of the American College of Medical, G. & Genomics Laboratory Quality Assurance, C., 2013. ACMG clinical laboratory standards for next-generation sequencing. Genet Med. 15, 733–47. PMID 23887774.*

The overall result, or outcome of the test (Fig. 7.3, superscript 3): Results (Section 7.1.3) can be listed as positive, indicating detection of a pathogenic or likely pathogenic sequence variant that is consistent with the diagnosis given on the test requisition, or negative, in that no sequence variants were identified that explain the primary diagnostic indication for ordering the test. The overall result can also be inconclusive, in that a clear explanation for the patient's condition was not found. Inconclusive results can arise if variants of unknown significance are identified, or if a single heterozygous pathogenic variant is identified for an autosomal recessive condition (Rehm et al., 2013; Section 4.4.1). Laboratories may also describe a sequencing result as a causative or "pathogenic" variant in a disease gene with an established association with the reported phenotype, a likely causative or "likely pathogenic" variant in a disease gene with a likely association with the reported phenotype, or as a secondary finding (Richards et al., 2015; Section 7.2). Other result interpretations may be appropriate for certain indications, such as carrier status for autosomal recessive carrier screening tests (Rehm et al., 2013). It is important to remember that when a causative sequence variant is not identified, it does not necessarily mean there is no genetic cause for the patient's condition, as most genetic testing does not have a 100% detection rate and the current understanding of the meaning of genetic variation is still incomplete.

The gene name (Fig. 7.3, superscript 4) in which the sequence variant has been identified is provided.

A reference transcript (Fig. 7.3, superscript 5) for each gene is also noted when the reported sequence variant is present in the coding sequence of a gene. The transcript used typically represents the longest known transcript or the most clinically relevant transcript (den Dunnen et al., 2016). This information can also be provided in the methodological section of the report.

The disease association (Fig. 7.3, superscript 6), or the disease or condition associated with the gene(s) mentioned in the report, is noted.

Mode of inheritance (Fig. 7.3, superscript 7), or the inheritance pattern of the condition (Section 2.7), is provided if known, for example, autosomal dominant, autosomal recessive, or X-linked inheritance.

Variant information (Fig. 7.3, superscript 8) is also noted. The sequence variant(s) is listed using nomenclature standardized by the Human Genome Variation Society (HGVS). Sequence variants are described at the nucleotide (cDNA) level and the protein level. The terminology may include the use of "g." to indicate the position of the sequence variant in the genomic sequence, "c." to indicate the position of the sequence variant in the coding DNA sequence, "p." for protein, and "m." for mitochondria. Stop codons are represented by "X", "*", or "Ter". Deletions are listed as "del" and insertions as "ins". In this example, at the 70th coding nucleotide of the DNA, a guanine has been changed to a thymidine, and at the protein level, the 24th amino acid, a glutamate, has been converted to a stop codon. The exon in which the variant is located may also be given. The genomic coordinates can be defined according to a standard genome build (e.g., hg19) or a genomic reference sequence that covers the entire gene (including the 5′ and 3′ untranslated regions and promoter).

The zygosity (Fig. 7.3, superscript 9) of the sequence variant is noted. Variants can be present in one copy of the double-stranded DNA (heterozygous, or monoallelic) or present in both copies of the double-stranded DNA (homozygous, or biallelic) (Section 2.3.3). In typical males, only one copy of the X chromosome is present and sequence variants on the X chromosome are described as hemizygous.

Parental origin of the sequence variant may also be included if parents have been tested (Fig. 7.3, superscript 10). If neither parent has the sequence variant found in their biological child, the variant is described as 'de novo'.

Variant classification (Fig. 7.3, superscript 11). Variants can be classified as pathogenic (Section 7.1.3), benign (Section 7.1.4), or as variants of unknown significance (Section 7.1.5). The terms 'mutation' and 'polymorphism' have been replaced by the term 'variant' with modifiers according to the five-tier system (see below and Sections 7.1.3–7.1.5; Richards et al., 2015). Next-generation sequencing tests (Section 6.3) can also reveal variants in genes with as yet unknown functions in humans, or variants related to another clinical condition that is different from the testing indication (Section 7.2, Recommendations for reporting secondary findings).

For targeted next-generation sequencing tests (Sections 6.2.1 and 6.2.2), or gene panels (Section 6.2.3), the report will contain a summary of the genes analyzed and can include the average sequence coverage achieved across the targeted region(s) (Rehm et al., 2013).

A summary of the interpretation and predicted effects of the reported sequence variant(s) is provided (Fig. 7.3, superscript 12). Many laboratories use a five-tiered terminology system for Mendelian diseases (Sections 2.7.1 and 2.7.2). Variants can be designated as (i) pathogenic, (ii) likely pathogenic, or greater than 90% certainty of a variant being disease causing, (iii) unknown significance, (iv) likely benign, or greater than 90% certainty of a variant being benign, and (v) benign (Richards et al., 2015). It is recommended that all assertions of pathogenicity (including "likely pathogenic") be reported with respect to a condition and inheritance pattern, for example, c.1521_1523delCTT, predicting p.Phe508del can be reported as pathogenic in relation to cystic fibrosis (Sections 11.4.1 and 11.5.1), a condition with autosomal recessive inheritance. It is also important to recognize that for many sequence variants, data to support a quantitative assignment to one of the above variant categories are not available (Richards et al., 2015).

The next section of the report contains evidence that supports variant classification and whether the variant(s) identified are considered likely to partially or fully explain the patient's clinical features (Fig. 7.3, superscript 13). Information as to whether the variant has been reported previously in disease or control databases or in the medical literature is also cited, together with a summary of the results from software prediction programs that analyze variant pathogenicity and evolutionary conservation (Richards et al., 2015). Comments on penetrance and variable expressivity of variants and genes may also be included (Section 2.8). Expert judgment can be required for variant interpretation. Some reports will also supply recommendations for the referring healthcare provider regarding suggested future testing, such as testing for the same sequence variant in other family members (Fig. 7.3, superscript 14). The results from these additional tests may be useful for variant interpretation.

Testing laboratories use computational, or in silico, predictive programs to aid with variant classification (Richards et al., 2015). These programs are typically directed at determining the effects of missense variants on the function of the protein and assessing whether sequence variants affect splicing (Sections 11.3.2, 11.3.5, and 11.5.5). Examples of these tools include PolyPhen2, SIFT, MutationTaster, and Combined Annotation Dependent Depletion (Kumar et al., 2009; Adzhubei et al., 2010; Schwarz et al., 2010; Kircher et al., 2014). However, algorithms for missense variant prediction have been estimated to be only 65%–80% accurate when examining known disease variants and thus the results of in silico prediction programs are typically combined with other data when used to classify variants (Thusberg et al., 2011). Splice site prediction tools are considered to have higher sensitivity (~90%–100%; Richards et al., 2015). Population databases are also available for determining the frequency of sequence variants in large control populations (Section 7.2). However, variants are entered into population databases without annotations as to pathogenicity, and it cannot be assumed that all of the individuals whose variant data were used were healthy (Richards et al., 2015). In addition, limited representation of specific ethnic groups in these large population databases may mean that unique polymorphisms are absent from "controls." Disease databases are also available and contain variants that have been reported in patients with disease, together with assessment of pathogenicity (Section 7.2). Both disease- and gene-specific databases often contain errors or incomplete variant classification and different databases vary in frequency and degree of curation, so that interpretation may not be entirely accurate. Many laboratories will also perform extensive searches using the medical literature prior to variant interpretation.

A summary of the methods used for variant detection is provided by the testing laboratory, together with the limitations of the assay (Fig. 7.3, superscript 15). The description of the methods generally describes the techniques used to obtain and analyze the nucleic acids, for example, polymerase chain reaction or exon capture and Sanger sequencing, or next-generation sequencing (Sections 6.2 and 6.3). Technical parameters regarding the level of coverage of the panel, exome or genome, should be provided. An important technical aspect is the mean depth of coverage (Fig. 7.3, superscript 16), which refers to the mean sequence read depth or mean number of sequencing reads across the regions targeted by the test. The quality threshold (Fig. 7.3, superscript 17) refers to the percentage of the defined target region where read depth has at least 10X coverage or 10 sequencing reads that permits high-quality base calling, annotation, and evaluation. Average quality thresholds may range from >90 to 95% of the targeted region and it is important to note that a small portion of the target region may not be covered with sufficient depth or quality to call variant positions confidently (Rehm et al., 2013).

Lastly, limitations of the report also are provided (Fig. 7.3, superscript 18), together with references used for variant interpretation or methods (Fig. 7.3, superscript 19).

Mitochondrial Testing

For mitochondrial testing (Section 6.4.2), the gene name and m. numbering (e.g., m.8993T > C) and p. (protein amino acid) numbering are provided from the current mitochondrial reference sequence (Falk et al., 2015; Shen et al., 2018). As the genes in the mitochondrial genome encode for transfer RNA as well as for protein, evaluating amino acid changes is relevant only for genes encoding proteins.

The occurrence of a variant in a subset of mitochondria, or heteroplasmy, and the occurrence of a variant in all mitochondria, or homoplasmy, will also be reported (Section 2.7.4). The percentage of heteroplasmy for the variant is noted if relevant, as varying levels of heteroplasmy will influence the penetrance and expressivity of clinical findings within families with a mitochondrial disease (Sections 11.6.2 and 11.6.5). Heteroplasmy can vary in different tissues. Low levels of heteroplasmy should be interpreted in the context of the tissue tested and may be more meaningful in an affected tissue, such as muscle (Richards et al., 2015). Mitochondrial variants are inherited from the maternal lineage and it can be difficult to determine if a variant is 'de novo' because of the possibility of a low degree of heteroplasmy or variation in heteroplasmy between different tissues.

Pharmacogenetic Testing

There is a wide range of sequence variation in the genes that determines responses to drugs (Sections 5.6 and 5.7.4). Interpreting sequence variants in these genes typically requires the analysis of a combination of variants. Nomenclature for pharmacogenetic variants typically uses star (*) alleles, which can represent a combination of variants on the same allele, or haplotype. For many genes involved in drug response, the overall phenotype is derived from a combination of variants on both alleles, or diplotype. Terms related to metabolism (rapid, intermediate, poor), efficacy (resistant, responsive, sensitive), or 'risk' are frequently used for describing pharmacogenetic testing results, rather than an assessment of sequence variant pathogenicity (Richards et al., 2015) (Section 11.15).

7.1.2 Review of the Medical Literature and Resources

Guidance documents for recommended reporting practices in genetic testing are available (Lubin et al., 2008, 2009; Scheuner et al., 2012; Rehm et al., 2013). These documents include full sample reports from the American College of Medical Genetics and Genomics (ACMG) clinical laboratory standards for next-generation sequencing (Rehm et al., 2013). Guidelines for interpretation of pathogenicity of sequence variants have also been published by the American College of Medical Genetics and Genomics and this has become the standard reference for many testing laboratories (Richards et al., 2015).

The following lists contain URLs for population and disease-associated variant databases. The list is not exhaustive and links may become outdated.

Population databases:

dbSNP Database: http://www.ncbi.nlm.nih.gov/snp
dbVar Database of structural variation: http://www.ncbi.nlm.nih.gov/dbvar
Exome Variant Server: http://evs.gs.washington.edu/EVS
GnomAD: https://gnomad.broadinstitute.org/

Disease-associated variant databases:

ClinVar: http://www.ncbi.nlm.nih.gov/clinvar
Human Gene Mutation Database: http://www.hgmd.org
Leiden Open Variation Database: http://www.lovd.nl
Online Mendelian Inheritance in Man: http://www.omim.org

Nomenclature for the description of sequence variants:

Human Genome Variation Society http://www.HGVS.org/varnomen

7.1.3 Pathogenic Variants

Typically, the classification of a variant as disease causing, or pathogenic, is determined from a collection of evidence that is synthesized (Richards et al., 2015). In general, variants are classified as pathogenic if they have been demonstrated or predicted to disrupt gene and protein function, have a strong association with the disease for which the test is being performed, and are rare in healthy individuals. After a detailed assessment of each pathogenic criterion, the resulting evidence for each variant is weighted as very strong, strong, moderate, or supporting (Richards et al., 2015). The criteria then are combined according to the scoring rules to choose a classification from the five-tier system of pathogenic, likely pathogenic, variant of uncertain significance, likely benign, or benign (see Richards et al., 2015, Table 5).

As pathogenic variants frequently result in changes to treatment, management, or surveillance of patients, accurate variant interpretation is critical. Segregation of a variant with other affected individuals in the same family does not prove that a sequence variant is the cause of the disease, as even sequence variants that are predicted to disrupt protein function sometimes do not cause disease (Richards et al., 2015). A careful clinical evaluation may be needed to rule out mild symptoms and determine if affected individuals have a non-genetic, or an alternative genetic, cause. For accurate variant interpretation, biological family relationships may need confirmation and decreased and/or age-dependent penetrance should be considered to ensure that asymptomatic family members are definitely without disease (Richards et al., 2015). In addition, the fact that a patient has a phenotype matching the known spectrum of clinical features for a gene may not constitute evidence for pathogenicity, as patients undergoing genetic testing may have similar clinical presentations (Richards et al., 2015).

Testing laboratories use information about a sequence variant, such as its prior association with disease, its predicted deleteriousness, and frequency in disease and control databases to classify variants as pathogenic. A summary of the evidence and criteria used to designate a sequence variant as pathogenic is available in Richards et al. (2015, Table 3). Evidence can be classified as very strong, strong, moderate, or supporting. Confirmatory studies using an alternative method, such as Sanger sequencing (Sections 6.2.1 and 6.2.2), have been recommended for all sequence variants that are considered to be pathogenic or likely pathogenic for a Mendelian disorder (Richards et al., 2015). These methods include Sanger sequencing for verification of variants identified with next-generation sequencing methodologies; this is sometimes performed by the testing laboratory prior to report generation. Selected criteria that primarily apply to very strong or strong evidence for pathogenicity are listed below.

Very Strong Evidence of Pathogenicity

Sequence variants that result in nonsense, frameshift, and canonical splice site variation (e.g., involving the nucleotides located within 1–2 base pairs of the splice site junctions) are frequently predicted to be disruptive to gene function and are thus considered as strong evidence of pathogenicity (Richards et al., 2015). Such variants are predicted to cause loss of function because of absence or partial absence of the gene product and nonsense-mediated decay of the messenger RNA (mRNA programmed destruction if mRNA has a stop codon at incorrect place). Variants that involve the start (initiation) codon and single- or multiexon deletions (Sections 2.41, 11.2.4 and 11.4.5) are also considered likely to be pathogenic (Richards et al., 2015). Caveats to the above generalization include genes for which loss of function is not known to cause disease in humans, splice site variants that may skip an exon, but leave the remainder of the protein intact, and variants at the 3′ end of the gene near the stop codon that may not undergo nonsense-mediated decay.

Strong Evidence of Pathogenicity

Variants that are present for the first time in an affected patient, or 'de novo', are strong evidence of pathogenicity. For most 'de novo' variants, the patient is the only member of the family to be clinically affected. As above, both maternity and paternity should be confirmed. It is also desirable that the phenotype, or clinical features of the patient, match those associated with the deleterious variant(s) in the gene (Richards et al., 2015). Variants for which there are well-established functional studies that support a damaging effect on the gene or gene product, such as enzyme assays in metabolic disease and assays that assess the impact of splice site variants at the mRNA level, also are considered as strong evidence of variant pathogenicity.

Moderate Evidence of Pathogenicity

An example of moderate evidence of pathogenicity includes the prevalence of a sequence variant in affected individuals that is increased significantly compared with the prevalence of the sequence variant in controls. If the variant is absent from (or below the expected carrier frequency, when recessive) a large, general population or a control cohort (>1000 individuals) and the population is ethnically matched to the patient harboring the identified variant, then this observation can be considered a moderate piece of evidence for pathogenicity. It is also important to consider the ethnicity of the patient and control population, as variants may be present at variable frequencies in individuals with different ancestries. Occurrence of a variant in a protein domain established to be critical for function in which all missense variants in that domain have been previously interpreted as pathogenic is considered moderate evidence of pathogenicity, as is the finding of a variant located in a mutational hotspot.

7.1.4 Benign Status

A sequence variant can be considered benign if there is sufficient evidence for the testing laboratory to conclude that it is not the cause of the patient's disorder or the indication for testing. Similar to the process for classifying pathogenic variants, criteria for designating a sequence variant as benign are chosen based on the evidence observed for the variant and then combined according to scoring rules to determine the classification from the five-tier system (Richards et al., 2015, Table 5). For benign variants, the evidence can be classified as stand-alone, strong, or supporting. Selected criteria that primarily apply to strong evidence are listed below; a summary of all criteria is available (Richards et al., 2015, Table 4). Benign variants can also be unique to individuals or families and absence in an ethnically-matched population is not considered evidence for pathogenicity.

Stand-alone Evidence of Benign Status

If the allele frequency for a variant is more than 5%, it is considered as stand-alone support for a benign interpretation (Richards et al., 2015).

Strong Evidence of Benign Status

An allele frequency in controls that is greater than expected for the disorder is considered strong support for a benign interpretation for a rare Mendelian disorder. If the disease under investigation is fully penetrant from an early age, observation of the variant in a well-documented healthy adult for a recessive (homozygous), dominant (heterozygous), or X-linked (hemizygous) condition is considered strong evidence for a benign interpretation (Richards et al., 2015).

Supporting Evidence of Benign Status

If a sequence variant is detected in a patient for whom there is a clear alternate genetic explanation for the disease, this is interpreted as supporting evidence of a benign status for that variant.

7.1.5 Variants of Uncertain Significance

Sequence variants that do not fulfill the criteria for either pathogenic or benign, or if the evidence is conflicting, are interpreted as variants of uncertain significance (VUSs; Richards et al., 2015). For reports that list VUSs, changes to treatment or medical management typically are not recommended for these variants due to the uncertainty in interpretation. However, healthcare providers should consider periodic reanalysis of the genetic test and regular contact with the testing laboratory to determine if classification of the variant changes toward a pathogenic or likely pathogenic result, in which case alterations to management may be appropriate, or toward a benign or likely benign status, in which case the patient and family can be reassured and an alternative etiology considered. Guidelines regarding physician responsibility regarding recontact have been established (David et al., 2019) and close follow-up of the patient and family is frequently recommended. Sometimes a VUS will appear to fit well with the patient's presentation and the clinical geneticist may choose to initiate further clinical studies to try to clarify the significance of the VUS, or contact a research laboratory for additional information.

7.2 Recommendations for Reporting Secondary Findings

The reporting of secondary findings, or incidental findings unrelated to the diagnostic indication for genetic testing, remains controversial, as for many genes and variants, an exact interpretation or a known role in disease remains elusive. Recommendations regarding the reporting of secondary findings (Section 10.3) were first issued for exome sequencing and genome sequencing (Section 6.3.1) by the American College of Medical Genetics (ACMG) in 2013 (Green et al., 2013). Other later papers of importance include the ACMG Board of Directors statement on secondary findings (ACMG Board of Directors, 2015; Miller et al., 2021a) and updated lists of the genes that are analyzed for secondary findings (Kalia et al., 2017; Miller et al., 2021b). Surveys of current practices in genetic test interpretation and reporting are also available (O'Daniel et al., 2017).

However, the broad scope of next-generation sequencing ensures that genetic variants unrelated to the purpose of the test may be identified that may have important medical implications for the individual sequenced and perhaps for other family members. In the first guidelines on reporting of secondary findings, laboratories performing exome or genome sequencing for any clinical indication were requested to specifically analyze 57 genes (later shortened to 56 genes) that were selected based on substantial clinical evidence that pathogenic variants are highly likely to result in severe disease that is preventable if identified before symptoms occur (Green et al., 2013). These recommendations were updated in the following year due to a general consensus among ACMG members and other relevant stakeholders. This stated that patients should be able to opt out of the analysis of genes unrelated to the indication for testing and that a decision as to whether the next-generation sequencing data would be analyzed for secondary findings should be made during the process of informed consent (Scheuner et al., 2015; ACMG Board of Directors, 2015). Parents or guardians also have the option to opt out of receiving secondary findings when exome or genome sequencing is performed on children (ACMG Board of Directors, 2015). In 2017, the list of genes which are analyzed for secondary findings was refined to 59 (Kalia et al., 2017) and updated in 2021 to 73 (Miller et al., 2021). The ACMG supports the reporting of known pathogenic or expected pathogenic variants but does not recommend reporting variants of uncertain significance as secondary findings (Kalia et al., 2017).

Overall, an estimated 1%–2% of unselected populations was found to have a secondary finding on the original ACMG list of 56 genes (Ormond et al., 2019). Many of the genes that are currently considered for reporting of secondary findings are relevant to cancer or heart disease, including cardiomyopathies (Section 11.4.5) or arrhythmogenic disorders (Section 11.4.4), and are thus likely to be actionable (Kalia et al., 2017). However, in practice, there is considerable variation in how laboratories practice and laboratory-specific judgments concerning clinical utility and patient benefit may influence reporting of secondary findings (Ackerman and Koenig, 2018). In addition, guidelines for secondary variant reporting vary, both nationally and internationally, and the decision as to whether to report secondary findings may depend on local policy and on how the patient has been counseled regarding this possibility (Claustres et al., 2014; Ormond et al., 2019).

7.3 Limitations of Genetic Testing

All healthcare providers should be aware of certain caveats and limitations when considering sending or receiving results of genetic testing.

7.3.1 Symptomatology Important in Genetic Test Choice and Interpretation

Close correlation between the clinical features of a patient and a genetic testing result is required for accurate test interpretation, as described above. In healthy or asymptomatic individuals, the chance of finding a pathogenic variant is decreased compared to testing individuals with a known condition, and more stringent criteria for implication of a variant in disease are required (Richards et al., 2015).

Many of the conditions for which panel testing is efficacious are genetically heterogeneous, meaning that deleterious, or damaging, variants in more than one gene can cause the disorder. In these conditions, clinical distinction between different causative genes is not always possible and thus a panel test that interrogates multiple genes simultaneously is preferred. Examples of conditions for which gene panels (Section 6.2.3) are commonly employed include hearing loss, cardiomyopathy, epilepsy, autism, and intellectual disability. Panel testing is most effective in clinical situations where a diagnosis either is known or highly likely; panels should not be used when the diagnosis is unclear, or when the presentation is atypical, as in these circumstances, the causative gene may not be included on the gene panel. In contrast, exome and genome sequencing (Sections 6.3.1 and 6.3.2) are used frequently when there is no suitable gene panel, if the underlying diagnosis is unknown, or if two conditions with distinct etiologies are suspected.

7.3.2 Limited Phenotype–Genotype Correlation

For many single gene disorders, a relationship between the clinical findings, or phenotype, in a patient and the underlying causative variant, or genotype, is absent or unclear (Orgogozo et al., 2015). The lack of a phenotype–genotype correlation implies that detection of a specific genetic variant may not allow prediction of the clinical outcome. Absence of a phenotype–genotype correlation can occur due to the rarity of the underlying condition, such that insufficient information is available in the medical literature, or the incomplete description of a patient's features. Underascertainment (Section 2.8) of clinical features despite a causative genotype may also hamper delineation of a phenotype–genotype correlation. Determination of a phenotype–genotype correlation is also more difficult for complex traits, in which more than one gene may be involved in pathogenesis and for conditions in which environmental factors interact with genotype.

7.3.3 DNA Variation and Results of Unknown Clinical Significance

Many genetic tests result in sequence variants or copy number variants (Section 2.4.1) of unknown clinical significance due to factors such as the large numbers of genes tested with next-generation sequencing technologies and difficulties with variant interpretation due to underrepresentation of some ethnicities in population databases. Current guidelines caution against using variants of unknown clinical significance for clinical decision-making and efforts to resolve the classification of the variant as pathogenic or benign should be undertaken. Additional monitoring of the patient for the disorder in question may also be prudent. It is therefore critical for referring healthcare providers to remain in communication with the testing laboratory regarding reanalysis and reinterpretation of results and to inform the laboratory about any relevant changes in health status for patients who have undergone genetic testing.

7.3.4 Cases With Better Clinical Testing

Although genetic testing is utilized increasingly, the availability of alternative forms of diagnostic testing that are highly sensitive and specific and sometimes less costly should not be forgotten. These tests may also be useful for confirmation of a genetic diagnosis, or for interpretation of unclear genetic testing results. Examples include nasal nitrous oxide measurements and nasal biopsies that can be used to diagnose primary ciliary dyskinesia (Lucas et al., 2017; Sections 5.1.3 and 5.1.4) and sweat chloride testing for cystic fibrosis (Sections 11.4.1 and 11.5.1) (de Boeck et al., 2017).

7.3.5 No Available Treatment or Clinical Trials, Especially for Orphan Diseases

For some genetic diseases, obtaining a precise molecular or cytogenetic diagnosis may not lead to meaningful options for improving patient care or outcomes, and thus a genetic diagnosis may have limited clinical utility. For many orphan diseases, the onset of the disease is during early childhood or even prior to birth, and thus detection and diagnosis may be delayed to an extent that curative interventions are no longer possible. However, as conditions continue to be delineated and more cases are reported in the medical literature, management guidelines can be formulated that are useful for optimizing patient care. In addition, obtaining an exact diagnosis enables a discussion of inheritance patterns and recurrence risks with the patient and other family members, as well as prognosis and expectations, support groups, and prenatal counseling. Finally, as causative genes and genetic pathways are elucidated together with improvements in our understanding of the mechanisms for diseases, treatment and clinical trials that are targeted toward the underlying molecular cause can be developed.

7.4 Management and Services Available with Genetic Diagnosis

It is critical for healthcare providers to use the results of genetic testing in an accurate and meaningful way. Sequence variants and copy number variants that are reported as pathogenic (Section 7.1.3) should always be interpreted in context of the patient's clinical findings. Follow-up testing to generate additional data and for medical surveillance is frequently required after a positive genetic test; in addition, consideration of the risks for future reproductive planning and for other family members also is important. In this section, we summarize some of the strategies and resources that are available after a positive genetic test result is obtained. We discuss genetic counseling, websites containing management guidelines, family follow-up, support groups, and involvement in clinical trials and research.

7.4.1 Genetic Counseling

Obtaining a positive genetic testing result provides a clinical diagnosis for an affected patient and thus is frequently considered the gateway to accurate and personalized medical care and relevant materials, including services, support, and research (Section 11.12.2). Depending on the prevalence of the condition and the availability of information, the benefits of a genetic diagnosis can be profound. A diagnosis confers access to descriptions of other affected patients that can be summarized in the genetic testing report, obtained from the medical literature, or available in disease-specific databases. These descriptions typically include the range of clinical features associated with the diagnosis and the inheritance pattern. Information regarding causation, or how the chromosomal or gene variant results in the presentation of the patient, may also be available. Finally, data that are specific to the exact, or to similar, sequence variants or copy number variants found in the patient may be ascertained.

The above information can be complex, and providers who are not specialized in genetics may feel uncomfortable being the first health professional to relay test information to patients and families (see Chapter 12 for examples of discussions). Genetic counseling (Patch and Middleton, 2018) has been shown to be highly useful for improving understanding of genetic testing results (Section 3.3). Among those who utilized genetic counseling services in direct-to-consumer personal genomic testing, over half reported that counseling improved their understanding of their results (54.5%) and many noted an improved understanding of genetics in general (43.9%; Darst et al., 2013).

Genetic counseling services can be obtained by several different methods. Patients can be referred to an academic medical center, as most tertiary care centers will have a genetic service that can provide access to genetic counseling by a clinical geneticist or genetic counselor. Counseling is frequently provided at an outpatient clinic appointment, but telegenetic services may also be available. Referral to a licensed professional through the company that performed the genetic testing may also be possible. The National Society of Genetic Counselors provides a website that enables searches for genetic services based on name, location, and specialization (https://www.nsgc.org/page/find-a-gc-search) and the American College of Medical Genetics also has a webpage enabling searches for genetic clinics (https://clinics.acmg.net/).

Finally, independent companies can provide genomic services, including genetic counseling, that also aid with the integration of genetic testing results into patient care.

7.4.2 Disease Management and Surveillance

Familiarity with the management strategies required for numerous genetic conditions is rarely attained by practicing physicians, as it is burdensome for providers to remain up to date with a myriad of disorders (Cody et al., 2018). In addition, providing healthcare information for genetic disorders may be challenging, as the clinical picture often includes a complex medley of medical and developmental issues that can present in a unique manner in an affected individual (Cody et al., 2018). Thus, curated online resources are typically consulted and utilized for management guidelines pertaining to genetic diagnoses.

Clinical guidelines covering surveillance and treatment have been produced for many commonly encountered genetic conditions that are caused by sequence variants in single genes or copy number variants (Section 2.4.1). A selection of currently available resources curated for common and rare genetic disorders is provided in Table 7.1.

Practice guidelines can also be designed by professional societies. The "ACT" sheets of the American College of Medical Genetics and Genomics (ACMG; https://www.acmg.net/ACMG/Medical-Genetics-Practice Resources/ACT_-Sheets_and_Algorithms.aspx) are resources for healthcare providers looking for information on genetic conditions, in particular those identified through newborn screening. The ACT sheets and algorithms provide clinical information and management steps for patients. Additional practice resources from the ACMG are also available for selected clinical conditions (https://www.acmg.net/ACMG/Medical-Genetics-Practice-Resources/ACT_Sheets_and_Algorithms.aspx).

It is important to be aware that the clinical information found online can vary in terms of accuracy and quality (Brütting et al., 2019) and that materials also differ in curation and readability. Information that is directed primarily at physicians may not be appropriate for patients and families and vice versa. There are specific guidelines to provide management information for chromosome aberrations (Cody et al., 2018), in addition to literature and database searches. Books and the medical literature available from Pubmed (https://pubmed.ncbi.nlm.nih.gov) can also serve as resources. For rare and novel conditions or genes, suitable information may not exist and may need to be collated from the medical literature. Follow-up over a period of months to years may also be required to ensure the family comes to terms with the diagnosis and the implications for healthcare and is informed and managed according to the most current guidelines.

TABLE 7.1 Information sources for healthcare after reaching a genetic diagnosis.

Guideline	Audience	Description from website	Scope and curation	URL
Pubmed	Healthcare providers	Citations for biomedical literature from MEDLINE, life science journals, and online books	More than 29 million; updated as new articles are released	https://www.ncbi.nlm.nih.gov/pubmed/
Genereviews	Healthcare providers	Expert-authored and peer-reviewed clinical information for inherited conditions covering diagnosis, management, and genetic counseling	More than 736 chapters; updated every 2–4 years	https://www.ncbi.nlm.nih.gov/books/NBK1116/
Orphanet	Healthcare providers and patients	Expert-authored and peer-reviewed articles that include clinical practice guidelines, fact-sheets, and summary information	Rare diseases; updated regularly	https://www.orpha.net/consor/cgi-bin/Disease_Search.php?lng=EN
Genetics home reference	Patient/family	Consumer-friendly information about the effects of genetic variation on human health, including educational resources	More than 1200 conditions and 1400 genes; updated	https://ghr.nlm.nih.gov/
Unique chromosome support group	Patient/family	Small charity supporting, informing, and networking with families living with a rare chromosome disorder	More than 19,500 members; information reviewed by healthcare professionals	https://www.rarechromo.org/

7.4.3 Family Follow-Up

A genetic diagnosis typically enables determination of the inheritance pattern (Section 2.7) for the patient and can be used to quantify the risk of a disorder in other family members (Section 11.12.2). Thus, along with disease-specific information, the pattern of inheritance should be explained to the patient, together with the recurrence risk for the patient and the implications for other family members, including children, siblings, and particularly the parents of children for prenatal decision-making. The need to test other "at-risk" family members should also be discussed if appropriate. Patients are typically encouraged to share their genetic testing results with family members that may benefit from the information. It is important to remember that healthcare providers typically should not contact at-risk relatives without the consent of their patient (Rothstein, 2018). However, physicians can support patients in making contact by providing written information about the condition and inheritance pattern, together with support group information. Genetic testing on additional family members, or segregation analysis, can then be employed to determine if at-risk relatives carry the same genetic variant(s). Often, targeted testing can be done.

7.4.4 Support Groups and Patient Advocacy Groups

In addition to consulting guidelines for clinical care, information that is relevant for patients and families should be offered after provision of a genetic diagnosis. In one study, 58% of more than 500 patients diagnosed with melanoma preferred to receive recommendations on reliable sources of information and support from their healthcare provider (Brütting et al., 2018, 2019). Patients preferred internet (63%) as media, followed by booklets (58%), whereas clinicians favored face-to-face communication and booklets (Lialiou and Mantas, 2016). However, online resources are still provided by a minority of clinicians (Brütting et al., 2019).

A large number of family support groups and patient advocacy groups have been created for patient and family support. Most patient support groups can be located using the internet and are often present on Facebook and other social media platforms. The Genetic Alliance is a nonprofit health advocacy organization for those with genetic conditions (http://www.geneticalliance.org/). Genetics home reference also provides lists of support and advocacy groups (https://ghr.nlm.nih.gov/about/support-advocacy-groups). However, it is important to recognize that support groups may be regionally based and thus may not always be suitable for in-person meetings, depending on patient location. In addition, support groups vary in the input they solicit from healthcare providers, although many function in combination with a medical advisory board. For further discussion of support organizations, see Section 3.5.

7.4.5 Clinical Trials and Research

After a genetic diagnosis, participation in an observational study, clinical trial, or research study relevant to the underlying condition may be appropriate, in particular if conventional treatments are not available or well defined. In an observational study, patients or healthy individuals are studied by researchers who gather information, group participants according to broad characteristics, and compare changes over time; this type of research or natural history study is conducted in a normal setting (https://www.nia.nih.gov/health/what-are-clinical-trials-and-studies). In contrast, a clinical trial is a research study that is aimed at evaluating a medical, surgical, or behavioral intervention. Clinical trials can be critical for comparing the efficacy of different treatments and interventions. The National Institutes of Health maintains a website that lists current and past clinical trials (https://clinicaltrials.gov/). This database lists available trials, together with contact information and adverse event data that is submitted by the study sponsor or principal investigator. Prior to drug approval, the United States Food and Drug Administration (FDA) typically requires Phases I, II, and III clinical trials to be conducted to determine if the drug can be approved for use (for summary, see https://www.nia.nih.gov/health/what-are-clinical-trials-and-studies). When discussing trial and research participation with patients, it is important to provide a balanced viewpoint that includes both potential advantages and disadvantages. Concerns related to patient participation in research include privacy and the risks of data sharing (Raza and Hall, 2017). Efforts to advance diversity in clinical trials and research by including studies with participants that are children, members of both sexes, and all ethnicities are also extremely important for reducing health disparities and enabling all groups to benefit from genomic advances (Avilés-Santa et al., 2017).

7.4.6 Orphan Drugs and Gene Therapies

The acquisition of a genetic diagnosis enables precision medicine, an approach for disease treatment and prevention that takes individual variability in genes, environment, and lifestyle for each person into account (Collins et al., 2016). In many cases, treatment for a genetic disorder comprises supportive measures that are directed at ameliorating the complications of the underlying condition. However, therapies that target the underlying genetic defect are increasingly being designed and utilized. In 2017, for example, the Food and Drug Administration (FDA) approved voretigene neparvovec (VN), marketed as Luxturna, as a gene therapy targeting inherited retinal diseases resulting from biallelic variants in *RPE65*. VN is an adeno-associated virus (AAV) vector containing modified human *RPE65* cDNA (AAV2-hRPE65v2) that is administered by injection into the subretinal space in the eye (Russell et al., 2017; Miraldi Utz et al., 2018). A Phase III clinical trial (ClinicalTrials.gov number NCT00999609) showed that bilateral injection of VN resulted in a significant improvement in multiluminance mobility testing that evaluated the ability of a patient to navigate a marked path while avoiding obstacles in varying levels of luminance (Russell et al., 2017). Several Phase I/II clinical trials for patients with *RPE65*-associated Leber's congenital amaurosis (Section 11.3.3) using subretinal AAV-mediated gene replacement have been completed (Miraldi Utz et al., 2018).

As another example, an antisense oligonucleotide therapy, Nusinersen (marketed as Spinraza) was approved for spinal muscular atrophy (SMA) in 2016 (Section 11.2.4). This drug is delivered by intrathecal injection and has a complex mechanism of action that involves binding to a splicing silencer region on SMN2 pre-mRNA that leads to augmentation of the translation of functional SMN protein (Rao et al., 2018). A double-blind, randomized, Phase III clinical trial involving 121 symptomatic infants with type 1 SMA (Section 11.2.4) was stopped early, as 41% of those receiving Nusinersen showed motor improvements compared to none of the infants who underwent sham treatment (Rao et al., 2018). AVXS-101 (onasemnogene abeparvovec; Zolgensma) is an AAV 9 vector carrying an SMN1 cDNA that has also been developed for the treatment of SMA (Dabbous et al., 2019). This therapy is designed to boost directly the levels of SMN protein (Lopes, 2018; Rao et al., 2018). Drawbacks for gene therapy include the need to obtain a precise molecular diagnosis prior to treatment, cost (Darrow et al., 2019), and patient access.

References

Ackerman, S.L., Koenig, B.A., 2018. Understanding variations in secondary findings reporting practices across U.S. genome sequencing laboratories. AJOB Empir. Bioeth. 9, 48–57. PMID 29131714.

Adam, M.P., Hudgins, L., Hannibal, M., 2011 Sep 1. Kabuki syndrome. In: Adam, M.P., Mirzaa, G.M., Pagon, R.A., Wallace, S.E., Bean, L.J.H., Gripp, K.W., Amemiya, A. (Eds.), GeneReviews®. University of Washington, Seattle, Seattle (WA), 1993–2023 (updated 2022 Sep 15). PMID 21882399.

Adzhubei, I.A., Schmidt, S., Peshkin, L., Ramensky, V.E., Gerasimova, A., Bork, P., Kondrashov, A.S., Sunyaev, S.R., 2010. A method and server for predicting damaging missense mutations. Nat. Methods 7, 248–249. PMID 20354512.

American College of Medical Genetics (ACMG) Board of Directors, 2015. ACMG policy statement: updated recommendations regarding analysis and reporting of secondary findings in clinical genome-scale sequencing. Genet. Med. 17, 68–69. PMID 25356965.

Aviles-Santa, M.L., Heintzman, J., Lindberg, N.M., Guerrero-Preston, R., Ramos, K., Abraido-Lanza, A.L., Bull, J., Falcon, A., McBurnie, M.A., Moy, E., Papanicolaou, G., Pina, I.L., Popovic, J., Suglia, S.F., Vazquez, M.A., 2017. Personalized medicine and Hispanic health: improving health outcomes and reducing health disparities - a National Heart, Lung, and Blood Institute workshop report. BMC Proc. 11, 11. PMID 29149222.

Aziz, N., Zhao, Q., Bry, L., Driscoll, D.K., Funke, B., Gibson, J.S., Grody, W.W., Hegde, M.R., Hoeltge, G.A., Leonard, D.G.B., Merker, J.D., Nagarajan, R., Palicki, L.A., Robetorye, R.S., Schrijver, I., Weck, K.E., Voelkerding, K.V., 2015. College of American Pathologists' laboratory standards for next-generation sequencing clinical tests. Arch. Pathol. Lab Med. 139, 481–493. PMID 25152313.

Brütting, J., Bergmann, M., Garzarolli, M., Rauschenberg, R., Weber, C., Berking, C., Tilgen, W., Schadendorf, D., Meier, F., NVKH Supporting Group, 2018. Information-seeking and use of information resources among melanoma patients of German skin cancer centers. J. Dtsch. Dermatol. Ges. 16, 1093–1101. PMID 30091517.

Brütting, J., Bergmann, M., Schadendorf, D., Weber, C., Tilgen, W., Berking, C., Meier, F., NVKH Supporting Group, 2019. Sources of information and support for melanoma patients: differences between patients' and clinicians' preferences. J. Dtsch. Dermatol. Ges. 17, 652–654. PMID 30933429.

Claustres, M., Kozich, V., Dequeker, E., Fowler, B., Hehir-Kwa, J.Y., Miller, K., Oosterwijk, C., Peterlin, B., Van Ravenswaaij-Arts, C., Zimmermann, U., Zuffardi, O., Hastings, R.J., Barton, D.E., European Society of Human Genetics, 2014. Recommendations for reporting results of diagnostic genetic testing (biochemical, cytogenetic and molecular genetic). Eur. J. Hum. Genet. 22, 160–170. PMID 23942201.

Cody, J.D., Hasi-Zogaj, M., Heard, P., Hill, A., Rupert, D., Sebold, C., Soileau, B., Hale, D.E., 2018. The chromosome 18 clinical resource center. Mol. Genet. Genomic Med. 6, 416–421. PMID 29603904.

Collins, H., Calvo, S., Greenberg, K., Forman Neall, L., Morrison, S., 2016. Information needs in the precision medicine era: how genetics home reference can help. Interact J. Med. Res. 5, e13. PMID 27122232.

Dabbous, O., Maru, B., Jansen, J.P., Lorenzi, M., Cloutier, M., Guerin, A., Pivneva, I., Wu, E.Q., Arjunji, R., Feltner, D., Sproule, D.M., 2019. Survival, motor function, and motor milestones: comparison of AVXS-101 relative to nusinersen for the treatment of infants with spinal muscular atrophy Type 1. Adv. Ther. 36, 1164–1176. PMID 30879249.

Darrow, J.J., 2019. Luxturna: FDA documents reveal the value of a costly gene therapy. Drug Discov. Today 24, 949–954. PMID 30711576.

Darst, B.F., Madlensky, L., Schork, N.J., Topol, E.J., Bloss, C.S., 2013. Perceptions of genetic counseling services in direct-to-consumer personal genomic testing. Clin. Genet. 84, 335–339. PMID 23590221.

David, K.L., Best, R.G., Brenman, L.M., Bush, L., Deignan, J.L., Flannery, D., Hoffman, J.D., Holm, I., Miller, D.T., O'Leary, J., Pyeritz, R.E., ACMG Social Ethical Legal Issues Committee, 2019. Patient re-contact after revision of genomic test results: points to consider-a statement of the American College of Medical Genetics and Genomics (ACMG). Genet. Med. 21, 769–771. PMID 30578420.

De Boeck, K., Vermeulen, F., Dupont, L., 2017. The diagnosis of cystic fibrosis. Presse. Med. 46, e97–e108. PMID 28576637.

Den Dunnen, J.T., Dalgleish, R., Maglott, D.R., Hart, R.K., Greenblatt, M.S., McGowan-Jordan, J., Roux, A.F., Smith, T., Antonarakis, S.E., Taschner, P.E.M., 2016. HGVS recommendations for the description of sequence variants: 2016 update. Hum. Mutat. 37, 564–569. PMID 26931183.

Falk, M.J., Shen, L., Gonzalez, M., Leipzig, J., Lott, M.T., Stassen, A.P.M., Diroma, M.A., Navarro-Gomez, D., Yeske, P., Bai, R., Boles, R.G., Brilhante, V., Ralph, D., DaRe, J.T., Shelton, R., Terry, S.F., Zhang, Z., Copeland, W.C., Van Oven, M., Prokisch, H., Wallace, D.C., Attimonelli, M., Krotoski, D., Zuchner, S., Gai, X., MSeqDR Consortium Participants. MSeqDR Consortium Participants, Bale, S., Bedoyan, J., Behar, D., Bonnen, P., Brooks, L., Calabrese, C., Calvo, S., Chinnery, P., Christodoulou, J., Church, D., Clima, R., Cohen, B.H., Cotton, R.G., de Coo, I.F.M., Derbenevoa, O., den Dunnen, J.T., Dimmock, D., Enns, G., Gasparre, G., Goldstein, A., Gonzalez, I., Gwinn, K., Hahn, S., Haas, R.H., Hakonarson, H., Hirano, M., Kerr, D., Li, D., Lvova, M., Macrae, F., Maglott, D., McCormick, E., Mitchell, G., Mootha, V.K., Okazaki, Y., Pujol, A., Parisi, M., Perin, J.C., Pierce, E.A., Procaccio, V., Rahman, S., Reddi, H., Rehm, H., Riggs, E., Rodenburg, R., Rubinstein, Y., Saneto, R., Santorsola, M., Scharfe, C., Sheldon, C., Shoubridge, E.A., Simone, D., Smeets, B., Smeitink, J.A., Stanley, C., Suomalainen, A., Tarnopolsky, M., Thiffault, I., Thorburn, D.R., Van Hove, J., Wolfe, L., Wong, L.-J., 2015. Mitochondrial Disease Sequence Data Resource (MSeqDR): a global grass-roots consortium to facilitate deposition, curation, annotation, and integrated analysis of genomic data for the mitochondrial disease clinical and research communities. Mol. Genet. Metab. 114, 388–396. PMID 25542617.

Green, R.C., Berg, J.S., Grody, W.W., Kalia, S.S., Korf, B.R., Martin, C.L., McGuire, A.L., Nussbaum, R.L., O'Daniel, J.M., Ormond, K.E., Rehm, H.L., Watson, M.S., Williams, M.S., Biesecker, L.G., American College of Medical Genetics & Genomics, 2013. ACMG recommendations for reporting of incidental findings in clinical exome and genome sequencing. Genet. Med. 15, 565–574. PMID 23788249.

International Standing Committee on Human Cytogenomic Nomenclature, McGowan-Jordan, J., Simons, A., Schmid, M., 2016. ISCN: An International System for Human Cytogenomic Nomenclature. Karger, Basel; New York. PMID 1678834.

Kalia, S.S., Adelman, K., Bale, S.J., Chung, W.K., Eng, C., Evans, J.P., Herman, G.E., Hufnagel, S.B., Klein, T.E., Korf, B.R., McKelvey, K.D., Ormond, K.E., Richards, C.S., Vlangos, C.N., Watson, M., Martin, C.L., Miller, D.T., 2017. Recommendations for reporting of secondary findings in clinical exome and genome sequencing, 2016 update (ACMG SF v2.0): a policy statement of the American College of Medical Genetics and Genomics. Genet. Med. 19, 249–255. PMID 27854360.

Kircher, M., Witten, D.M., Jain, P., O'Roak, B.J., Cooper, G.M., Shendure, J., 2014. A general framework for estimating the relative pathogenicity of human genetic variants. Nat. Genet. 46, 310–315. PMID 24487276.

Kumar, P., Henikoff, S., Ng, P.C., 2009. Predicting the effects of coding non-synonymous variants on protein function using the SIFT algorithm. Nat. Protoc. 4, 1073–1081. PMID 19561590.

Lialiou, P., Mantas, J., 2016. Online information seeking behaviour by nurses and physicians: a cross-sectional study. Stud Health Technol Inform 225, 33–37. PMID 27332157.

Lopes, J.M., 2018. #AAN2018 — Type 1 SMA Babies Treated with AVXS-101 Achieving New Milestones, Trials Show (Online). Philadelphia, PA: BioNews Services, LLC. Available at https://smanewstoday.com/2018/04/24/aan2018-type-1-sma-babies-avxs-101-continue-improve-clinical-trials/. Accessed November 2021.

Lubin, I.M., Caggana, M., Constantin, C., Gross, S.J., Lyon, E., Pagon, R.A., Trotter, T.L., Wilson, J.A., McGovern, M.M., 2008. Ordering molecular genetic tests and reporting results: practices in laboratory and clinical settings. J. Mol. Diagn. 10, 459–468. PMID 18669879.

Lubin, I.M., McGovern, M.M., Gibson, Z., Gross, S.J., Lyon, E., Pagon, R.A., Pratt, V.M., Rashid, J., Shaw, C., Stoddard, L., Trotter, T.L., Williams, M.S., Amos Wilson, J., Pass, K., 2009. Clinician perspectives about molecular genetic testing for heritable conditions and development of a clinician-friendly laboratory report. J. Mol. Diagn. 11, 162–171. PMID 19197001.

Lucas, J.S., Barbato, A., Collins, S.A., Goutaki, M., Behan, L., Caudri, D., Dell, S., Eber, E., Escudier, E., Hirst, R.A., Hogg, C., Jorissen, M., Latzin, P., Legendre, M., Leigh, M.W., Midulla, F., Nielsen, K.G., Omran, H., Papon, J.-F., Pohunek, P., Redfern, B., Rigau, D., Rindlisbacher, B., Santamaria, F., Shoemark, A., Snijders, D., Tonia, T., Titieni, A., Walker, W.T., Werner, C., Bush, A., Kuehni, C.E., 2017. European Respiratory Society guidelines for the diagnosis of primary ciliary dyskinesia. Eur. Respir. J. 49, 1601090. PMID 27836958.

Miller, D.T., Lee, K., Gordon, A.S., Amendola, L.M., Adelman, K., Bale, S.J., Chung, W.K., Gollob, M.H., Harrison, S.M., Herman, G.E., Hershberger, R.E., Klein, T.E., McKelvey, K., Richards, C.S., Vlangos, C.N., Stewart, D.R., Watson, M.S., Martin, C.L., ACMG Secondary Findings Working Group, 2021a. Recommendations for reporting of secondary findings in clinical exome and genome sequencing, 2021 update: a policy statement of the American College of Medical Genetics and Genomics (ACMG). Genet. Med. 23, 1391–1398. PMID: 34012069.

Miller, D.T., Lee, K., Chung, W.K., Gordon, A.S., Herman, G.E., Klein, T.E., Stewart, D.R., Amendola, L.M., Adelman, K., Bale, S.J., Gollob, M.H., Harrison, S.M., Hershberger, R.E., McKelvey, K., Richards, C.S., Vlangos, C.N., Watson, M.S., Martin, C.L., ACMG Secondary Findings Working Group, 2021b. ACMG SF v3.0 list for reporting of secondary findings in clinical exome and genome sequencing: a policy statement of the American College of Medical Genetics and Genomics (ACMG). Genet. Med. 23, 1381–1390. PMID: 34012068.

Miraldi Utz, V., Coussa, R.G., Antaki, F., Traboulsi, E.I., 2018. Gene therapy for RPE65-related retinal disease. Ophthalmic Genet. 39, 671–677. PMID 30335549.

O'Daniel, J.M., McLaughlin, H.M., Amendola, L.M., Bale, S.J., Berg, J.S., Bick, D., Bowling, K.M., Chao, E.C., Chung, W.K., Conlin, L.K., Cooper, G.M., Das, S., Deignan, J.L., Dorschner, M.O., Evans, J.P., Ghazani, A.A., Goddard, K.A., Gornick, M., Farwell Hagman, K.D., Hambuch, T., Hegde, M., Hindorff, L.A., Holm, I.A., Jarvik, G.P., Knight Johnson, A., Mighion, L., Morra, M., Plon, S.E., Punj, S., Richards, C.S.,

Santani, A., Shirts, B.H., Spinner, N.B., Tang, S., Weck, K.E., Wolf, S.M., Yang, Y., Rehm, H.L., 2017. A survey of current practices for genomic sequencing test interpretation and reporting processes in US laboratories. Genet. Med. 19, 575–582. PMID 27811861.

Orgogozo, V., Morizot, B., Martin, A., 2015. The differential view of genotype-phenotype relationships. Front. Genet. 6, 179. PMID 26042146.

Ormond, K.E., O'Daniel, J.M., Kalia, S.S., 2019. Secondary findings: how did we get here, and where are we going? J. Genet. Couns. 28, 326–333. PMID 30821867.

Patch, C., Middleton, A., 2018. Genetic counselling in the era of genomic medicine. Br. Med. Bull. 126, 27–36. PMID 29617718.

Rao, V.K., Kapp, D., Schroth, M., 2018. Gene therapy for spinal muscular atrophy: an emerging treatment option for a devastating disease. J. Manag. Care Spec. Pharm. 24, S3–S16. PMID 30582825.

Raza, S., Hall, A., 2017. Genomic medicine and data sharing. Br. Med. Bull. 123, 35–45. PMID 28910995.

Rehm, H.L., Bale, S.J., Bayrak-Toydemir, P., Berg, J.S., Brown, K.K., Deignan, J.L., Friez, M.J., Funke, B.H., Hegde, M.R., Lyon, E., Working Group of the American College of Medical, Genetics and Genomics Laboratory Quality Assurance Committee, 2013. ACMG clinical laboratory standards for next-generation sequencing. Genet. Med. 15, 733–747. PMID 23887774.

Richards, S., Aziz, N., Bale, S., Bick, D., Das, S., Gastier-Foster, J., Grody, W.W., Hegde, M., Lyon, E., Spector, E., Voelkerding, K., Rehm, H.L., ACMG Laboratory Quality Assurance Committee, A.L.Q.A., 2015. Standards and guidelines for the interpretation of sequence variants: a joint consensus recommendation of the American College of Medical Genetics and Genomics and the Association for Molecular Pathology. Genet. Med. 17, 405–424. PMID 25741868.

Riggs, E.R., Andersen, E.F., Cherry, A.M., Kantarci, S., Kearney, H., Patel, A., Raca, G., Ritter, D.I., South, S.T., Thorland, E.C., Pineda-Alvarez, D., Aradhya, S., Martin, C.L., ACMG, 2020. Technical standards for the interpretation and reporting of constitutional copy-number variants: a joint consensus recommendation of the American College of Medical Genetics and Genomics (ACMG) and the Clinical Genome Resource (ClinGen). Genet. Med. 22, 245–257. PMID 31690835.

Rothstein, M.A., 2018. Reconsidering the duty to warn genetically at-risk relatives. Genet. Med. 20, 285–290. PMID 29388945.

Russell, S., Bennett, J., Wellman, J.A., Chung, D.C., Yu, Z.F., Tillman, A., Wittes, J., Pappas, J., Elci, O., McCague, S., Cross, D., Marshall, K.A., Walshire, J., Kehoe, T.L., Reichert, H., Davis, M., Raffini, L., George, L.A., Hudson, F.P., Dingfield, L., Zhu, X., Haller, J.A., Sohn, E.H., Mahajan, V.B., Pfeifer, W., Weckmann, M., Johnson, C., Gewaily, D., Drack, A., Stone, E., Wachtel, K., Simonelli, F., Leroy, B.P., Wright, J.F., High, K.A., Maguire, A.M., 2017. Efficacy and safety of voretigene neparvovec (AAV2-hRPE65v2) in patients with RPE65-mediated inherited retinal dystrophy: a randomised, controlled, open-label, phase 3 trial. Lancet 390, 849–860. PMID 28712537.

Scheuner, M.T., Hilborne, L., Brown, J., Lubin, I.M., Members of the RAND Molecular Genetic Test Report Advisory Board, 2012. A report template for molecular genetic tests designed to improve communication between the clinician and laboratory. Genet. Test. Mol. Biomarkers 16, 761–769. PMID 22731646.

Scheuner, M.T., Peredo, J., Benkendorf, J., Bowdish, B., Feldman, G., Fleisher, L., Mulvihill, J.J., Watson, M., Herman, G.E., Evans, J., 2015. Reporting genomic secondary findings: ACMG members weigh in. Genet. Med. 17, 27–35. PMID 25394173.

Schwarz, J.M., Rodelsperger, C., Schuelke, M., Seelow, D., 2010. MutationTaster evaluates disease-causing potential of sequence alterations. Nat. Methods 7, 575–576. PMID 20676075.

Shen, L., Attimonelli, M., Bai, R., Lott, M.T., Wallace, D.C., Falk, M.J., Gai, X., 2018. MSeqDR mvTool: a mitochondrial DNA Web and API resource for comprehensive variant annotation, universal nomenclature collation, and reference genome conversion. Hum. Mutat. 39, 806–810. PMID 29539190.

Stevens-Kroef, M., Simons, A., Rack, K., Hastings, R.J., 2017. Cytogenetic nomenclature and reporting. Methods Mol. Biol. 1541, 303–309. PMID 27910032.

Subramaniam, A., Jacobs, A.P., Tang, Y., Neely, C., Philips 3rd, J.B., Biggio, J.R., Robin, N.H., Edwards, R.K., 2016. Trisomy 18: a single-center evaluation of management trends and experience with aggressive obstetric or neonatal intervention. Am. J. Med. Genet. 170A, 838–846. PMID 26728763.

Thusberg, J., Olatubosun, A., Vihinen, M., 2011. Performance of mutation pathogenicity prediction methods on missense variants. Hum. Mutat. 32, 358–368. PMID 21412949.

Chapter 8

Cancer Genetics Referrals and Management

Takae M. Brewer[1], Rebekah Moore[1,*], Shao-Tzu Li[2], Joanne Ngeow[2,3] and Charis Eng[4,5]

[1]*Genomic Medicine Institute, Cleveland Clinic, Cleveland, OH, United States;* [2]*Cancer Genetics Service, Division of Medical Oncology, National Cancer Centre Singapore, Singapore;* [3]*Lee Kong Chian School of Medicine, Nanyang Technological University, Singapore;* [4]*Genomic Medicine Institute, Center for Personalized Genetic Healthcare, Cleveland Clinic Lerner Research Institute, and Taussig Cancer Institute, Cleveland, OH, United States;* [5]*Department of Genetics and Genomic Sciences, Germline High Risk Focus Group, Case Comprehensive Cancer Center, Case Western Reserve University School of Medicine, Cleveland, OH, United States*

Abstract

Identifying individuals with hereditary cancer syndromes is an important aspect of comprehensive cancer care and prevention. Primary care providers should be aware of the possibility of hereditary cancer syndromes present in patients and families and should consider referral to genetics. This chapter discusses the general indications for referrals to cancer genetic specialists, implications of the genetic test results, utility of multigene panels, and where to refer to ensure management informed by genetic results, as well as issues with direct-to-consumer genetic testing.

The pathogenesis of cancer is complex. Identifying individuals with hereditary cancer syndromes is an important aspect of comprehensive cancer care and prevention. The primary care physician has a pivotal role in making referrals to genetics when hereditary cancer syndromes are suspected. This chapter is focused on hereditary cancer syndromes associated with adult-onset cancers. In addition to adult-onset hereditary cancer syndromes, there are hereditary syndromes that result in pediatric cancers, including retinoblastoma and neurofibromatosis. Some of these are discussed in more detail in the cancer-related cases in Section 11.16.

8.1 Hereditary Cancer Risk Assessment

Hereditary cancer syndromes account for an average of 5%–10% of all cancer diagnoses, with a broad range of 4% for acute myeloid leukemia to more than 50% for pheochromocytoma/paraganglioma (Lu et al., 2015; Neumann et al., 2019). Genetic evaluation is a crucial part of comprehensive cancer care for management and prevention today. However, genetic services are currently underutilized in the primary care setting for cancer risk assessment (Childers et al., 2017).

Cancer is a heterogeneous disease and often multifactorial (Section 2.7.3). Cancer, however, is fundamentally a genetic disease with molecular mechanisms underlying its development and progression. Cancer occurs when there are changes in genes, which either promote uncontrolled cell proliferation (oncogene) or suppress tumor development (tumor suppressor gene). According to the two-hit hypothesis, cancer arises when there are two pathogenic alterations in a cancer-associated gene. Sporadic cancers tend to occur later in life since it takes time for two insults in the same gene to occur. In case of a

* Author RM was employed by Cleveland Clinic at the time of writing the manuscript, but is currently employed by AstraZeneca Pharmaceuticals

Genomics in the Clinic. https://doi.org/10.1016/B978-0-12-816478-5.00008-X

hereditary cancer syndrome, however, the first hit is germline (inherited) and the second hit is acquired in a specific tissue (due to spontaneous or environmental effects), leading to cancers with earlier onset.

Cancer risk assessment is essential for two main reasons:

(1) To identify patients with a personal history of cancer who may be at an increased risk for additional primary cancers so that cancer surveillance and risk-reduction and therapeutic strategies can be tailored based on the risk.
(2) To identify asymptomatic individuals who may be at an increased risk of developing cancer based on their family history so that individualized surveillance and risk-reduction strategies can be formulated.

8.1.1 Indications for Referrals in the Primary Care Setting

The primary care physician is one of the first contacts that individuals with hereditary conditions may encounter in the healthcare system. As part of population health and preventative medicine, it is important to make referrals to the cancer genetic service when hereditary cancer syndromes are suspected. A careful cancer risk assessment in combination with personalized interpretation of genetic test results is crucial in order to deliver the most appropriate medical management options for high-risk groups.

The hereditary cancer risk assessment begins with a thorough review of personal and family history of cancer. As a rule of thumb, hereditary cancer syndromes should be suspected if the personal or family history includes any of the following:

(1) Cancer diagnosed at a younger age (≤45−50 years of age)
(2) Two or more primary cancers (including bilateral cancers in paired organs) in the same individual
(3) Three or more close blood relatives on one side of the family diagnosed with the same type of cancer, regardless of age of diagnosis
(4) Rare types of cancer (such as male breast cancer, ovarian cancer, pancreatic cancer, and pheochromocytoma)
(5) The presence of a known pathogenic germline mutation in the family

Family history may be absent in the case of a de novo mutation in a patient with cancer. A de novo mutation is a mutation in a gene arising in a germ cell of an individual for the first time in the family and has not been passed down. In such instances, the clinical red flags of heritable cancer described above and/or characteristic presentations (e.g., macrocephaly, breast cancer, thyroid cancer, and skin manifestations in *PTEN* hamartoma tumor syndrome) may serve as clues for possible hereditary cancer syndromes that warrant further evaluation by a cancer genetic specialist.

Some hereditary cancer syndromes occur in the absence of typical characteristics such as cancer of early age of onset, family history, and/or known pathogenic mutations in the family. Sections 8.4.1 and 8.4.3 list types of cancer, which warrant special attention, as well as consideration of referral to genetics regardless of the age of onset or family history.

Ethnicity should be inquired as some populations have a higher likelihood of carrying a "founder mutation" (for example, Ashkenazi Jewish or Icelandic ancestry) (Section 3.2.3). The presence of consanguinity, especially between biological parents, should be assessed as well since the incidence of recessive conditions is increased with consanguinity; for example, *MUTYH*-associated polyposis is autosomal recessive (Section 2.7.1).

Those who meet criteria as outlined by National Comprehensive Cancer Network (NCCN) should be referred to a cancer genetic specialist, such as a cancer geneticist or a cancer genetic counselor, for further analysis of personal and family history of cancer, possible physical examination, discussion of hereditary cancer syndromes and testing options, and coordination of genetic testing, if indicated. Posttest counseling by the genetic specialist typically includes results disclosure, a discussion of appropriate medical management recommendations based on the genetic test results, and the identification of family members who may also benefit from genetic testing.

As illustrations, the following two sections discuss two main components in the NCCN guidelines in genetic/familial high-risk assessment, which are most frequently encountered in the primary care setting: (1) hereditary breast and ovarian cancer syndrome, and related genetic differential diagnoses, and (2) hereditary colorectal cancer syndromes, and related genetic differential diagnoses. In addition, risk-assessment recommendations for multiple endocrine neoplasia (MEN) and leukemia are discussed.

8.1.2 Breast and Ovarian Cancer Risk Assessment

According to the American Cancer Society Cancer Facts & Figures, breast cancer is the most frequent type of female cancer and second-leading cause of cancer death among women in the United States (Siegel et al., 2019). Ovarian cancer is rare, accounting for approximately 1.3% of all cancer cases in the United States. However, ovarian cancer remains a challenging diagnosis, accounting for 5% of all cancer deaths, which is the highest death rate among gynecologic cancers (Siegel et al., 2019). Hereditary cancer syndromes due to predisposition genes account for approximately 10% of breast (Couch et al., 2017; Tung et al., 2016) and 15%–20% of ovarian cancer cases (Carter et al., 2018; Walsh et al., 2011).

NCCN outlines criteria for genetic consultation, which provides practical guidance to screen for individuals at risk for hereditary breast and ovarian cancer syndrome (Sections 11.16.1, 11.16.2) (NCCN, 2022b). Men also should be referred to genetics if any of the NCCN criteria is met because of two important reasons: (1) mutation-positive men can pass down the gene mutation to their offspring regardless of gender; (2) *BRCA1/2* is associated with cancers that occur in prostate and pancreas (Sections 11.16.4, 11.16.5). Key elements in the current NCCN recommendations are:

(1) Presence of a known germline mutation in a breast and/or ovarian cancer susceptibility gene within the family or in the patient's tumor
(2) Patient with early-onset breast cancer (≤45 years of age)
(3) Two or more breast cancer primaries in the same individual [46 to 50 years of age]
(4) Diagnosis of the following at any age:
Patient with triple-negative breast cancer (Tested negative for estrogen receptors, progesterone receptors, and hormone epidermal growth factor receptor 2. Common treatments such as hormone therapy and drugs that target estrogen, progesterone, and HER-2 are ineffective. Using chemotherapy is still an effective option.)
- Lobular breast cancer with personal or family history of diffuse gastric cancer
- Ovarian cancer
- Pancreatic cancer
- Metastatic prostate cancer
- High-grade prostate cancer (Gleason score ≥7; Gleason score reflects a grading system to determine the aggressiveness of the cancer based on the histological appearance on biopsy. There are two grades, from the largest and next largest area, and the total score is equal to the sum of both grades) and Ashkenazi Jewish ancestry
- Male breast cancer

(5) Breast cancer at any age and:
- Ashkenazi Jewish ancestry
- One or more close blood relative* with any of the below:
  - o Breast cancer ≤45 years of age
  - o Triple negative breast cancer
  - o Lobular breast cancer with personal or family history of diffuse gastric cancer
  - o Ovarian cancer
  - o Male breast cancer
  - o Pancreatic cancer
  - o High-grade (Gleason score ≥7) or metastatic prostate cancer
- Two or more close relatives with breast cancer at any age [*Close relatives include first-, second-, and third-degree relatives: first-degree (parents, full siblings, children); second-degree (grandparents, grandchildren, aunts, uncles, nephews, nieces, or half-siblings); third-degree (first cousins, great-grandparents, or great-grandchildren)]

For an individual who does not meet the above criteria, a referral to genetics should be considered if there is a first- or second-degree relative who meets any of the following (Sections 11.16.2, 11.16.3, 11.16.4, 11.16.5) (NCCN, 2022b):

(1) Breast cancer diagnosed ≤45 years of age
(2) Ovarian cancer
(3) Male breast cancer
(4) Pancreatic cancer
(5) Metastatic prostate cancer
(6) Two or more breast cancer primary tumors in a single individual
(7) Two or more individuals with breast cancer primaries on the same side of family at any age

The criteria above are outlined primarily to screen for hereditary breast and ovarian cancer syndrome caused by *BRCA1, BRCA2, CDH1 and PALB2* mutations. To facilitate identification of other high-penetrance breast cancer–predisposition genes including *TP53 and PTEN*, NCCN further recommends genetic referrals if the criteria below are met (NCCN, 2022b):

(1) An individual with a personal and/or family history on the same side of the family with the following:
- Presence of a known germline mutation in *TP53* or in *PTEN*
- Individuals meeting the classic Li-Fraumeni syndrome criteria or Chompret criteria*
- Hisory of Bannayan-Riley-Ruvalcaba syndrome (BRRS), adult Lhermitte-Duclos disease (cerebellar tumors), at least two biopsy-proven trichilemmomas, and some features listed in the clinical diagnostic criteria for Cowden syndrome (CS)/*PTEN* hamartoma tumor syndrome (PHTS)*

8.1.3 Colorectal Cancer Risk Assessment

According to the American Cancer Society Cancer Facts & Figures, colorectal cancer is the third most common type of cancer and third-leading cause of cancer-related deaths among both men and women in the United States (Siegel et al., 2019). Approximately 5%–10% of colorectal cancer cases are hereditary in nature (Lynch and De La Chapelle, 2003). Hereditary colorectal cancer syndromes include Lynch syndrome (Section 11.16.3), familial adenomatous polyposis (FAP), *MUTYH*-associated polyposis, Peutz–Jeghers syndrome, Juvenile polyposis syndrome, Cowden/*PTEN* hamartoma tumor syndrome, and serrated polyposis syndrome. Current NCCN guidelines recommend formal hereditary colorectal cancer evaluation by genetics if any of the following is met (NCCN, 2022c):

(1) Presence of a known genetic mutation in the individual or family
(2) Personal history of any of the following:
- Ten or more adenomatous polyps
- Five or more serrated polyps in proximal or sigmoid colon
- Two or more hamartomatous polyps

(3) Family history of one or more relative with polyposis

Lynch syndrome is the most common genetic cause of adult-onset colorectal cancer, accounting for approximately 3% of cases, which may be an underestimate (Moreira et al., 2012). Colorectal cancer in Lynch syndrome occurs earlier than sporadic colorectal cancer (average age of diagnosis is 45 years for Lynch syndrome, compared to 67 years for sporadic cases), thus necessitating initiation of clinical surveillance at a younger age (Lynch and De La Chapelle, 2003; National Cancer Institute, DCCPS, Surveillance Research Program). The type of surgery differs for early stage (stages I–III) colorectal cancers in the sporadic versus Lynch syndrome setting (subtotal/hemicolectomy vs. total colectomy, respectively). Furthermore, Lynch syndrome predisposes mutation-positive individuals to multiple types of extracolonic cancers, including endometrium, ovarian, prostate, stomach, and genitourinary (Section 11.16.3).

Lynch syndrome is caused by germline mutations in any of the Lynch syndrome–associated DNA mismatch repair (MMR) genes (*MLH1*, *MSH2*, *MSH6*, *PMS2*) or the *EPCAM* gene. Loss of the DNA MMR gene function, termed mismatch repair deficiency (dMMR), is detected by the absence of staining for DNA MMR proteins expressed by the DNA MMR genes with the immunohistochemistry technique. Furthermore, cancers arising in Lynch syndrome demonstrate the microsatellite high cellular phenotype. A microsatellite is a series of repetitive DNA sequences that exists at multiple locations in the human genome (Section 2.4.1). During DNA replication, the microsatellite area is susceptible to replication errors, which are normally detected and repaired by the DNA MMR system. When the DNA MMR system is defective, as in the case of Lynch syndrome, the DNA mismatch errors cannot be corrected, leading to the retention of the unstable and error-prone microsatellite regions. This is known as microsatellite instability-high (Tautz and Renz, 1984). The degree of microsatellite instability (high or low) is detected by amplifying the target region with the polymerase chain reaction (PCR).

The NCCN recommends a genetic referral for further evaluation of Lynch syndrome when any of the following criteria is met (NCCN, 2022c):

(1) A known Lynch syndrome–associated mutation in the family: Germline mutation in any of the Lynch syndrome–associated DNA MMR genes (*MLH1*, *MSH2*, *MSH6*, *PMS2*) or *EPCAM*
(2) An individual with colorectal or endometrial cancer plus any of the following:
- Diagnosed <50 years of age
- Another synchronous or metachronous Lynch syndrome–related cancer†

* Refer to the latest NCCN guidelines for further details

† Lynch syndrome–related cancers include: colorectal, endometrial, gastric, ovarian, pancreas, ureter and renal pelvis, brain (usually glioblastoma), biliary tract, small intestinal cancers, sebaceous adenomas, sebaceous carcinomas, keratoacanthomas

- One or more first-degree or second-degree relative with Lynch syndrome—related cancer diagnosed at < 50 years of age
- Two or more first-degree or second-degree relatives with Lynch syndrome—related cancers regardless of age

(3) An individual with colorectal or endometrial cancer at any age with tumor showing dMMR or microsatellite instability-high histology[§]

(4) Family history of any of the following:
- One or more first-degree relative with colorectal or endometrial cancer diagnosed <50 years of age
- One or more first-degree relative with colorectal or endometrial cancer and another synchronous or metachronous Lynch syndrome—related cancer regardless of age
- Two or more first-degree or second-degree relatives with Lynch syndrome—related cancer, including one or more diagnosed <50 years of age
- Three or more first-degree or second-degree relative with Lynch syndrome—related cancers, regardless of age

(5) An individual with a Lynch syndrome—related cancer or unaffected individuals with a 5% risk or greater of having a DNA MMR gene mutation based on predictive models (PREMM5, MMRpro, MMRpredict)

8.1.4 Multiple Endocrine Neoplasia Syndromes

MEN is a group of syndromes associated with high risks of developing multiple types of cancer. Although rare, cancers in MEN tend to be multifocal, early-onset, and aggressive. Thus, it is important to recognize the clinical features of these syndromes so that appropriate and prompt referrals are made to genetics and other subspecialists for the patient as well as at-risk family members. As a rule, we recommend any patient with personal diagnosis of pheochromocytoma/paraganglioma or medullary thyroid carcinoma should be referred to genetics.

In this section, two main types of MEN syndromes are discussed. They are inherited both in an autosomal dominant manner. Multiple endocrine neoplasia type 1 (MEN 1) is caused by pathogenic variants in the *MEN1* gene while multiple endocrine neoplasia type 2 (MEN 2) is caused by pathogenic variants in the *RET* gene.

MEN 1 is characterized by three primary tumors: parathyroid, pancreatic islet cell, and pituitary tumors. A clinical diagnosis of MEN 1 is made when two or more MEN 1—associated tumors are present in an individual. MEN 1 is not limited to tumors mentioned above and individuals may also develop adrenal cortical tumor, pheochromocytoma (catecholamine-secreting tumors in adrenal glands; termed paraganglioma when extra-adrenal), bronchopulmonary neuroendocrine tumor, gastric neuroendocrine tumor, thymic neuroendocrine tumor, lipomas, angiofibromas, collagenomas, and meningiomas (Thakker et al., 2012). Current guidelines recommend individuals with a clinical diagnosis or suspicion of MEN 1, or for at-risk individuals with an *MEN1* mutation identified in a family, should be referred to genetics for further evaluation of MEN 1 (NCCN, 2022d).

MEN 2 is classified into three clinical subtypes; MEN 2A, MEN 2B, and familial medullary thyroid carcinoma, all of which are caused by a pathogenic mutation in the *RET* gene. They are associated with a high risk of developing medullary thyroid cancer (70%—100% risk by age 70 years). Pheochromocytomas are also commonly seen in MEN 2A and MEN 2B (50%). Familial medullary thyroid carcinoma, by operational definition, presents with medullary thyroid cancer alone. This entity is currently considered as a variant of MEN 2A (Eng, 2019).

MEN 2A—associated tumors include medullary thyroid cancer, adrenal pheochromocytoma, and parathyroid neoplasia. A clinical diagnosis of MEN 2A is made when two or more of these tumors are present in an individual or in close relatives. Clinical exam findings which are suggestive of MEN 2A include lichen planus amyloidosis and Hirschsprung disease. Familial medullary thyroid carcinoma, a variant of MEN 2A, is clinically diagnosed in families with four or more cases of medullary thyroid carcinoma without family history of pheochromocytoma or parathyroid neoplasms (Eng, 2019).

In contrast to MEN 2A, parathyroid neoplasia is absent in MEN 2B. MEN 2B is clinically diagnosed with the presence of MEN 2B—associated tumors or clinical manifestations, which include: (1) medullary thyroid cancer, (2) pheochromocytoma, (3) mucosal neuromas of the lips and tongue, (4) medullated corneal nerve fibers, (5) ectopic lenses, (6) Marfanoid body habitus (Section 11.10.4), (7) distinctive facial features with enlarged lips, and (8) the absence of tear production upon crying. Intestinal ganglioneuroma are also highly associated with MEN 2B (NCCN, 2022d).

§. i.e., presence of tumor-infiltrating lymphocytes, Crohn-like lymphocytic reaction, mucinous/signet ring differentiation, or medullary growth pattern.

For MEN 2, the NCCN recommends a genetic referral for further evaluation when any of the following criteria is met (NCCN, 2022d):

(1) An individual with a diagnosis of medullary thyroid carcinoma, clinical diagnosis of MEN 2, or primary C-cell hyperplasia
(2) At-risk blood relatives of an individual with a known germline *RET* mutation, including those at a very young age

For MEN syndromes, cancer genetic testing is indicated for at-risk young children due to extremely early onset of cancer. For MEN 1 and MEN 2A, genetic testing should be offered to at-risk individuals by age 5. Children with MEN 2A as young as 5 years have been reported to develop medullary thyroid carcinoma. In MEN 2B, metastatic medullary thyroid cancer may occur at an early age unless the mutation-positive individual undergoes total thyroidectomy at less than a year old. Thus, all children who are at risk for MEN 2B should be tested as soon as after birth (Eng, 2019).

8.1.5 Leukemia

After the World Health Organization revised its leukemia diagnostic classifications in 2016 to include germline genetic status, greater emphasis has been placed on the identification of patients with leukemia-predisposing conditions (Arber et al., 2016). Leukemia-predisposing conditions are either syndromic or nonsyndromic (Sections 11.16.6, 11.16.7). Well-established syndromic causes include (but are not limited to):

(1) Inherited bone marrow failure syndromes
   - Fanconi anemia
   - Diamond–Blackfan anemia
   - Shwachman–Diamond syndrome
   - Short telomere syndromes, including dyskeratosis congenita
(2) Down syndrome/trisomy 21 (Section 11.13.2)
(3) Noonan syndrome (Section 11.4.3)
(4) Li–Fraumeni syndrome
(5) *GATA2* deficiency
(6) Constitutional dMMR syndrome
(7) Bloom syndrome

Genes associated with nonsyndromic leukemia-predisposing conditions include, but are not limited to: *ANKRD26, CEBPA, DDX41X, ELANE, ETV6, HAX1, RUNX1, SAMD9, SAMD9L,* and *SRP72* (Porter et al., 2016). Given the significant heterogeneity among these genetic conditions, careful attention to the clinical history, physical exam, and laboratory workup is required in order to guide the identification of these individuals.

A patient with a hematologic malignancy with a suggestive personal and/or family history as below should be referred for a genetic assessment (Bochtler et al., 2018; University of Chicago Hematopoietic Malignancies Cancer Risk Team, 2016):

(1) Multiple cancer diagnoses of any type, presenting in early adulthood
(2) Physical anomalies, unexpected short stature, failure to thrive, endocrinopathies, immunodeficiencies, premature gray hair, skin pigmentation, infertility, or other characteristic features suggestive of inherited bone marrow failure syndromes or other syndromic conditions
(3) Leukemia, long-standing cytopenia, bleeding tendency, hematopoietic abnormality such as macrocytosis/dysplasia diagnosed in two or more individuals within a three-generation pedigree
(4) Excessive toxicities with chemotherapy or radiation, including protracted or nonrecovering cytopenia, or a history of poor stem cell mobilization
(5) Cytogenetic and molecular genetic aberrations in leukemia cells typical of a specific genetic condition

Leukemia-predisposing conditions may present with bone marrow failure, hematologic malignancies, myelodysplastic syndrome, or nonhematologic manifestations. However, the absence of characteristic clinical features or a negative family history does not exclude genetic diagnoses because they can be caused by de novo germline pathogenic variants. Additionally, these conditions are characterized often by marked inter- and intrafamilial variations regarding penetrance and expressivity. These factors make the identification of leukemia-predisposing conditions challenging, and although most are recognized during childhood, some patients reach adulthood before receiving a genetic diagnosis (Bochtler et al., 2018).

8.2 Referral for Cancer Genetic Testing

Once a hereditary cancer syndrome or risk is suspected, a referral to the cancer genetic service should be made for comprehensive risk assessment, further recommendations, and coordination of genetic testing. The primary care physician or oncologist should ensure that the patient has a clear understanding of why an evaluation by a genetic specialist is important, which will aid in compliance with the referral (Section 3.1). The patient should understand that the goal of pursuing genetic counseling and testing is to determine if there is a hereditary cancer predisposition and to understand the importance for his/her primary care physician/oncologist to guide management in a gene-informed manner.

To prepare the patient for a genetic counseling session, the patient should be made aware that the session will involve detailed discussion of family history (at least three generations) (Section 3.2) and focus on cancer diagnoses, including cancer type, severity (metastatic or high-grade prostate cancer), and ages of onset. Some patients find it helpful to invite a close relative to accompany the session or to assist in providing pertinent family history. Furthermore, if family members have previously pursued genetic testing, a copy of their test results, if available, should be brought to the session, even if they are negative because there may be limitations to the previous test used (Section 7.3).

While the patient may meet criteria for a hereditary cancer syndrome referral to genetics, it may be recommended that a family member with a personal history of cancer undergo genetic counseling and testing first. Individuals with a personal history of cancer have the highest likelihood to test positive for mutations in cancer susceptibility genes. Thus, the referring physician should discuss with an unaffected patient that the affected family members are the most informative candidates for testing, and ideally should pursue genetic testing first. Based on the genetic test results on affected family members, the risk stratification strategy may largely change for the unaffected individual. In the event that the affected family members have passed away or are not interested in pursuing genetic testing, then genetic testing can certainly be coordinated for the unaffected patient.

The primary care physician or oncologist should educate the patient that the results may have impact on medical management recommendations for other family members as well. This may include screening or risk-reduction options for unaffected family members and reproductive or family planning considerations for family members of childbearing age (as some hereditary cancer syndromes are associated with being carriers for other rare disorders). These recommendations are discussed in more detail later in this chapter.

For reference, a specimen for genetic testing is typically obtained by blood draw or saliva/buccal swab. In case of underlying hematologic malignancies, cultured skin fibroblasts are considered a preferable specimen.

The cancer genetic specialist will discuss topics including basic cancer genetics (multifactorial etiology of cancer, two-hit hypothesis, etc. [section 8.1]), inheritance patterns, penetrance, variable expressivity, benefits and limitations of genetic testing, choices of testing, interpretation of possible results, surveillance strategies, and/or treatment implications. The specialist will also touch upon possible impact on close blood relatives. Information regarding genetic discrimination laws, as well as financial considerations such as insurance coverage and self-payment options, will be discussed. Referrals to multidisciplinary cancer specialists will be made as appropriate.

8.2.1 Utility of Multigene Panels

There are multiple genes that have been linked with breast, ovarian, colon, and other cancers. Some genes are associated with an increased risk for multiple cancers, and there are multiple genes that are associated with similar phenotypes. Therefore, multigene panels (Section 6.2.3) are widely utilized today when there are no known deleterious mutations previously identified in the family. Furthermore, multigene testing by next-generation sequencing has become important in retesting those who previously tested negative for a limited number of genes. Studies show pathogenic variants in non-*BRCA1/2* predisposition genes occur in 4%–7% of breast cancer cases, which can be missed if only *BRCA1/2* testing is performed (Couch et al., 2017; Tung et al., 2016).

In addition to including multiple genes associated with a high risk of cancer, panel tests also typically include moderate-risk genes for breast cancer (*ATM, CHEK2*, and *PALB2*) and lower-risk genes (*NF1, BARD1*, and others). While there are not specific testing criteria for all these genes, NCCN guidelines do include recommendations for medical management based on a pathogenic mutation identified in a gene other than *BRCA1/2* or the Lynch syndrome genes.

With the multigene panels becoming a predominant testing of choice in many circumstances, pretesting consultation is crucial and understanding the caveats of multigene testing is important in the informed consent process. The patient needs to be aware that multigene panel testing may reveal unexpected findings in a gene, which may or may not be related to the presenting phenotype. For example, panel testing may show a gene associated with colorectal cancer in a patient with personal and strong family history of breast cancer, but not colorectal cancer.

8.2.2 Impact of Genetic Testing Results

Genetic testing results often provide useful information, which guides the clinician to formulate a personalized plan for each patient. When a pathogenic or deleterious mutation is detected, enhanced surveillance, medical and/or surgical prevention, and targeted therapeutic options may be offered. National guidelines as well as expert consensus exist regarding medical management for many genes with pathogenic mutations associated with hereditary cancer syndromes. Specialists with experiences treating individuals with hereditary cancer syndromes (e.g., high-risk breast specialists, gynecologic oncologists), if available, should be involved in further discussion for optimum care.

Pathogenic mutations in cancer susceptibility genes do not cause cancer in all mutation-positive individuals (incomplete penetrance) (Section 2.8). While some mutations can increase the risk of multiple types of cancer, the types of cancer seen in hereditary cancer syndromes may vary among individuals and even in the same family carrying the same mutation (variable expressivity) (Section 2.8). A positive genetic test result provides information to identify close relatives who are at risk and may benefit from cascade testing. Thus, the mutation-positive individuals should be encouraged to share the information with their close relatives. Genetic counselors can facilitate genetic testing for at-risk family members or help find cancer genetic specialists in other cities for nonlocal family members. See also Section 7.1 for detailed discussion of genetic testing results.

A negative testing result is considered an "informative" or "true" negative when a mutation associated with a hereditary cancer syndrome has previously been identified in the family. For instance, when an asymptomatic woman is negative for the familial *BRCA1* mutation previously detected in her mother with breast cancer, then this woman's result is a true negative, which may reduce her breast and ovarian cancer risk to the same level as the general population. This scenario, however, applies only when there is no family history of breast cancer on the other side of the family. If she has strong family history of breast cancer on her father's side, for example, further testing in affected relatives on her father's side may be useful.

A negative testing result is considered "uninformative" when there are no previously detected mutations in the family. In this situation, a hereditary cancer syndrome cannot be completely ruled out because there could be cancer susceptibility mutations in the family that cannot be detected at this time. This scenario happens frequently when an asymptomatic individual who is the first person in the family to pursue testing has a negative result. For instance, hereditary cancer syndromes are suspected in an asymptomatic woman whose sister was diagnosed with breast cancer at a young age and subsequently died. With her sister no longer available for testing, this woman has pursued genetic testing and is found to be negative for pathogenic mutations. In this case, the women's result is "uninformative" and increased cancer screening may still be indicated based on her family history and other factors. Genetic testing is recommended typically for other family members after an "uninformative negative" test result, as there may still be a pathogenic mutation carried by other family members that would explain the family history of cancer. Furthermore, for individuals with negative testing, a genetic reevaluation in the future may be indicated as new cancer susceptibility genes are expected to be identified. Patients are encouraged to contact their primary care physician or genetic specialist every few years to determine if additional testing is available and indicated.

A variant of unknown significance (VUS) is encountered often in genetic testing results (Section 7.1.5). A VUS is a change in DNA (variant) for which there is not enough evidence to conclude it as pathogenic or benign. The frequency of a VUS has been reported to be 5%–20% and may depend on other factors, including the prevalence of the VUS in the tested population (Eccles et al., 2015). Medical or surgical recommendations should not be made based solely on a VUS result. Instead, recommendations should be carefully individualized based on the personal risk (ancestry, environmental, etc.) as well as the pattern of cancer occurrence in the family. It is not recommended typically to test family members for a VUS. In some instances, a VUS may seem to be more like a likely pathogenic variant based on amino acid change or phenotypic information. We recommend that medical management recommendations in the setting of a VUS be genetic-driven as increased screening still may be indicated for the patient and/or their family members in these situations. As genetic testing laboratories accumulate more data about a variant, the VUS may be upgraded to pathogenic/likely pathogenic or downgraded to benign/likely benign. Patients should be notified of the recategorization of their test results as clinicians receive updates from the testing laboratory.

Breast cancer risk evaluation tools such as the Tyrer–Cuzick IBIS model (Tyrer et al., 2004) should be utilized to risk stratify and identify women who should undergo increased breast cancer surveillance in the setting of negative or inconclusive genetic test results. The American Cancer Society recommends high-risk breast cancer screening (annual mammograms and breast magnetic resonance imaging (MRI)) for women whose estimated lifetime risk of breast cancer is 20% or greater (Saslow et al., 2007). A referral to a high-risk breast specialist will be beneficial if available.

The polygenic risk score is an emerging risk stratification tool that utilizes quantitative genetics to predict disease risks. Recent efforts with genome-wide association studies have identified common genetic variants (single-nucleotide polymorphisms) in the general population, which are associated with higher cancer risks (Al Olama et al., 2014; Michailidou et al., 2015). Currently, the polygenic risk score does not add clinically useful information beyond a good personal and family health history; as such, there are no consensus guidelines to incorporate the polygenic risk score to change management or surveillance strategies at present. Further studies are warranted to evaluate its clinical utility (see also Section 5.7.5).

8.3 Management Recommendations for Mutation-Positive Individuals

There are multiple hereditary breast cancer syndromes and hereditary colon cancer syndromes, each of which is associated with different cancer risks and screening recommendations. As the field of cancer genetics continues to evolve, management guidelines for mutation-positive individuals will continue to be updated. Below we discuss the most common recommendations for *BRCA1/2*, polyposis syndromes, Lynch syndrome, and multiple neuroendocrine neoplasia, as well as leukemias, as outlined by the NCCN. However, providers should refer to the current NCCN guidelines, as these recommendations are updated annually.

8.3.1 Management Recommendations for BRCA1/2 Mutation-Positive Individuals

Currently, recommendations for hereditary breast and ovarian cancer syndrome as well as most other hereditary breast cancer syndromes include mammograms and breast MRIs on an annual basis, sometimes compared to each other. However, the age at which to begin breast cancer screening varies depending on the lifetime risk of breast cancer and the age of breast cancer onset for each gene. The discussion of risk-reducing mastectomy should be individualized and a breast surgeon with expertise in the high-risk population management should be involved when possible (Section 11.16.1).

Because of difficulty in screening for ovarian cancer with currently available modalities including imaging and blood test (serum CA-125), risk-reducing bilateral salpingo-oophorectomy (RRSO) is recommended typically for *BRCA1/2* mutation-positive women. The age at which to pursue this procedure varies from ages 35–45 years and is dependent upon the risk of developing ovarian cancer and the age of onset for the disease based on the gene mutation detected. RRSO should be discussed in the setting of the patient's age, current health, and family planning. When women do not elect to undergo this procedure, annual transvaginal ultrasound combined with serum CA-125 screening may be offered as an alternative, although the benefit is uncertain and is not routinely recommended. This screening strategy should be considered at the discretion of the clinician in consultation with the genetics specialist because not all genes confer high enough risks to warrant risk-reducing surgeries.

8.3.2 Management Recommendations for Polyposis Syndromes

For adenomatous polyposis syndromes, the management is individualized based on the type of mutation and the number and histology of polyps observed. FAP and its variants are caused by germline mutations in the Adenomatous Polyposis Coli (*APC*) gene (Burt et al., 1995). Clinical manifestations, the degree of polyp burden, and risk for colorectal cancer vary depending on the mutation location in the *APC* gene (Nieuwenhuis and Vasen, 2007). *MUTYH*-associated polyposis is an autosomal recessive disorder and is caused by biallelic *MUTYH* gene mutations (mutations in both copies of the *MUTYH* gene). In many cases, more frequent colonoscopy screening is recommended. However, the ages of the initial screening and follow-up intervals vary by gene.

While proctocolectomy or colectomy is recommended for classic FAP, the management for attenuated FAP and *MUTYH*-associated polyposis depends on the patient's age and the burden of adenoma. Colonoscopy remains the gold standard for colorectal cancer screening and is initiated at a younger age and performed at more frequent intervals for this population. In general, colonoscopy should be performed every 1–2 years in cases of small adenoma burden. However, surgery is usually recommended for cases of large adenoma burden and when screening with colonoscopy is considered inadequate. In such instances, surgical interventions with prophylactic colectomy or proctocolectomy become necessary. Gastroenterology specialists with expertise in high-risk colorectal cancer syndromes should be involved.

8.3.3 Management Recommendations for Lynch Syndrome

Since Lynch syndrome (Section 11.16.3) is associated with many different types of cancer, screening strategies are often complex. To provide comprehensive care, involving high-risk specialists who are familiar with screening guidelines and management specifically for Lynch syndrome would be beneficial (the multidisciplinary approach). The multidisciplinary team often involves specialties such as gastroenterology, gynecology, urology, and neurology. In general, colorectal cancer screening for Lynch syndrome starts at age 20–25 years or 2–5 years prior to the youngest age of diagnosis in the family (whichever comes first). The NCCN recommends that colonoscopy should be repeated every 1–2 years (NCCN, 2022c). Some data suggest an association between aspirin administration and a decreased risk of colorectal cancer in this population (Burn et al., 2011). Thus, aspirin may be used for preventative purposes for Lynch syndrome–associated cancers. However, there are no established guidelines regarding the most appropriate aspirin dosing and/or duration at present.

Surveillance for extracolonic cancers in Lynch syndrome needs to be individualized, and most recommendations stem from expert opinions. In general, surveillance is considered for gastric, small bowel, and urethral cancers for those with family history of these cancers. For endometrial and ovarian cancer prevention, decisions for hysterectomy and RRSO should be individualized for those who have completed childbearing. There is currently insufficient evidence regarding the association between Lynch syndrome and breast and prostate cancers, and screening for these cancers should be determined based on the general population guidelines and family history.

Mutations in the Lynch syndrome–associated genes as well as several other genes, including *BRCA1*, *BRCA2*, *PALB2*, *ATM*, and possibly *STK11*, are known to increase the risk of pancreatic cancer in both men and women. There is no consensus guideline for pancreatic cancer screening based on family history or positive genetic test results. When screening is considered, it needs to be individualized and performed at experienced medical centers that have multidisciplinary teams familiar with surveillance strategies for this unique population. The American College of Gastroenterology guidelines recommend that endoscopic ultrasound and/or MRI should be performed annually, starting at age 50 years, or 10 years younger than the earliest age of pancreatic cancer diagnosis in the affected family (Syngal et al., 2015).

8.3.4 Management Considerations for Multiple Endocrine Neoplasia Syndromes

Once MEN is identified, enhanced long-term surveillance with clinical, biochemical, and radiological screening with and without prophylactic surgery by the multidisciplinary team (including endocrinology, surgery, gastroenterology, and oncology) is indicated. Management for individuals with MEN syndromes is complex, and we recommend these patients to be treated and followed at centers with expertise in MEN.

Current NCCN guidelines recommend individuals with MEN 1 should be screened for parathyroid disease, pancreas neuroendocrine tumors, pituitary tumors, and bronchial and thoracic neuroendocrine tumors (NCCN, 2022d). Due to early onset of tumors in MEN 1, some experts recommend the age to begin screening should be as early as 5 years old for certain cancers (e.g., pituitary tumor) (Thakker et al., 2012). Individuals with MEN 2 should be screened for medullary thyroid cancer with biochemical testing, starting at 3–5 years old for MEN 2A and at 6 months old for MEN 2B. In general, prophylactic thyroidectomy is recommended by age 5 or at diagnosis for individuals with MEN 2A and during the first year of life for MEN 2B. Timing of prophylactic thyroidectomy, however, should be based on the aggressiveness of *RET* mutations detected. For individuals with high-risk mutations, prophylactic thyroidectomy should be performed at an early age. In contrast, delaying thyroidectomy may be appropriate for those with low-risk mutations. Furthermore, individuals with MEN 2 syndromes should undergo annual screening for pheochromocytoma (MEN 2A and MEN 2B) and hyperparathyroidism (MEN 2A) (Eng, 2019; NCCN, 2022d).

8.3.5 Management Considerations for Hereditary Leukemic Conditions

Many of the hereditary leukemic conditions (Sections 11.16.6, 11.16.7) are associated with an increased risk for severe chemotherapy/radiation-related toxicities and therefore, although rare, an early identification is crucial. It is vital to uncover germline genetic etiologies in both syndromic and nonsyndromic hematologic malignancies before initiating chemotherapy or whole-body irradiation prior to bone marrow transplantation. The high-intensity induction therapy may lead to permanent bone marrow failure or lack of bone marrow recovery if not accounted for prior to treatment. Prompt recognition may be helpful to optimize the induction therapy with reduced-intensity regimens as well as to formulate tailored surveillance and management. Furthermore, a genetic diagnosis allows for subsequent cascade testing of at-risk relatives, which can direct tailored risk management to those with the condition and alleviate those unaffected. Additionally, since close relatives may be considered as potential bone marrow donors, an established genetic diagnosis can help guide the donor selection.

8.3.6 Management Recommendations for Other Hereditary Cancer Syndromes

As the field of hereditary cancer genetics has transitioned away from single-gene tests to multigene panels, the NCCN guidelines have also evolved to include medical management recommendations for genes other than the ones listed in detail above.

For some genes, the inclusion of breast MRI to breast cancer screening is recommended, but risk-reducing bilateral mastectomy is listed as a "consideration" based on a patient's personal and family history rather than a recommendation (e.g., *ATM*, *CHEK2*, *PALB2*, *NF1*). The age of onset for mammogram and breast MRI screening varies among these genes as well, depending on penetrance of breast cancer at younger ages. Likewise, recommendation for an RRSO may be given for genes with a higher risk for ovarian cancer (e.g., *BRIP1)* (Section 11.16.2).

Additional recommendations may be made to address increased risks of cancer beyond breast and ovarian cancer. In the setting of a *CDH1* mutation, risk-reducing gastrectomy is recommended for patients over screening via as random gastric biopsies. In individuals with *PTEN* mutations, increased screening is recommended for thyroid, colon, kidney, and skin in addition to breast and uterus. For individuals with *NF1* mutations, they also receive a diagnosis of neurofibromatosis type 1 and will need also to follow standard guidelines for this neurocutaneous condition (Section 11.11.2).

Please note that this list is not inclusive and a review of current NCCN guidelines for any pathogenic variant in a cancer predisposition gene is strongly recommended before making any changes to a patient's medical and surgical management.

8.4 Follow-up after Cancer Genetic Testing

There are other considerations following cancer genetic counseling and testing, in addition to screening for associated cancers.

8.4.1 Autosomal Recessive Considerations for Cancer Predisposition Syndromes

Of note, there are some genes associated with hereditary cancer syndromes where biallelic pathogenic/likely pathogenic variants (pathogenic variants on both copies of the same gene) result in rare autosomal recessive disorders. In these cases, each of the parents carries a pathogenic or likely pathogenic variant in the same gene (does not have to be the same variant). In instances where both parents have pathogenic variants in the same gene, each child has a 25% chance of having the autosomal recessive disorder, a 50% chance to have one pathogenic variant and have the hereditary cancer syndrome, and a 25% chance to not have inherited a pathogenic variant from either parent. Table 8.1 lists the hereditary cancer genes commonly included on multigene cancer panels that are associated with autosomal recessive disorders; it is not an all-inclusive list.

For individuals of childbearing age who test positive for one of the genes listed in Table 8.1, consideration of carrier testing is recommended for their partner so that informed decisions regarding family planning and preimplantation genetic testing can be made. If positive, a hereditary cancer risk assessment would also be recommended. Family planning discussions are another avenue for primary care providers and obstetrician/gynecologists to review patients' family histories and refer to a cancer genetic specialist if indicated.

8.4.2 Direct-to-Consumer Genetic Testing for Hereditary Cancer Syndromes

Direct-to-consumer (DTC) genetic testing (Chapter 9) has recently become more common and may be pursued in the hope of learning more about genetic risks for medical conditions including cancer. It is important to be aware that DTC testing has limitations in this setting. Tandy-Connor et al. (2018) examined 49 DTC-positive cases received for confirmatory testing by a clinical laboratory. Approximately 40% of variants called in DTC testing were false positives in cancer-related genes including *BRCA1*, *BRCA2*, *MLH1*, and *TP53*. Consumers and healthcare professionals should not use the DTC test results to make any cancer-related treatment or surveillance decisions, including enhanced screening strategies and prophylactic surgeries. Such decisions should be made only with confirmatory clinical testing results and genetic counseling by a genetic expert.

Negative DTC results also do not rule out hereditary cancer syndromes. For example, DTC tests may only test for selected variants, such as specific *BRCA1/BRCA2* mutations found most commonly in the Ashkenazi Jewish population, although there are more than 1000 *BRCA* mutations known to increase cancer risks. Thus, negative results in limited *BRCA* testing do not rule out the presence of hereditary breast and ovarian cancer syndromes, especially in the general population. Furthermore, there are genes other than *BRCA* that are clinically important, but the DTC testing may omit.

Thus, the primary care physician or oncologist should recommend a referral to a genetic specialist for patients who report prior DTC testing or are considering this testing related to cancer. Both pretest genetic counseling, important for facilitating a well-informed decision-making including the risks and benefits of genetic testing, and posttest counseling, important for accurate interpretation of both positive and negative results, are absent in the majority of the DTC testing processes.

TABLE 8.1 Cancer Predisposition Genes Associated with Autosomal Recessive Disorders.

Gene	Autosomal recessive disorder
ATM	Ataxia telangiectasia
BRCA2	Fanconi anemia
BRIP1	Fanconi anemia
EPCAM	Constitutional mismatch repair deficiency
MLH1	Constitutional mismatch repair deficiency
MSH2	Constitutional mismatch repair deficiency
MSH6	Constitutional mismatch repair deficiency
NBN	Nijmegen breakage syndrome
PALB2	Fanconi anemia
PMS2	Constitutional mismatch repair deficiency
RAD51C	Fanconi anemia

Reference: NCCN, 2022b; NCCN, 2022c.

8.4.3 Follow-up in the Primary Care Setting

A patient's personal and family history should be updated regularly. Any significant change in history may provide new information suggestive of hereditary cancer syndromes. Table 8.2 lists cancers for which a genetic referral should be considered for all patients, regardless of family history and/or age of onset. Novel and clinically important cancer predisposition genes are expected to be discovered and more comprehensive genetic testing options will be available as a

TABLE 8.2 Cancers for which a genetic referral should be considered regardless of family history or age of onset.

Adrenocortical cancer
Breast cancer and of Ashkenazi Jewish ancestry
Triple negative breast cancer
Colorectal cancer or endometrial cancer with evidence of DNA mismatch repair deficiency by either microsatellite instability or loss of DNA mismatch repair protein expression by immunohistochemistry staining
Giant cell astrocytoma (subependymal giant cell astrocytoma)
Male breast cancer
Medullary thyroid cancer
Ovarian cancer
Pancreatic cancer
Pancreatic neuroendocrine tumor (both nonfunctioning and functioning)
Prostate cancer, metastatic
Prostate cancer with high-grade histology (gleason score 7 or higher) and of Ashkenazi Jewish ancestry
Pheochromocytoma/paraganglioma
Rhabdomyosarcoma of embryonal anaplastic subtype
Two primary cancers

Credit: NCCN, 2022a, b, c, d, and e.

result. In addition, referral and testing criteria continue to expand. When appropriate, patients should be re-referred to genetics for update testing as well as reevaluation for hereditary cancer syndromes.

In addition, the primary care physician has repeated opportunities to advocate for identifying at-risk family members once a mutation has been identified in a family. First-degree relatives have a 50% risk to carry the identified mutation and second-degree relatives have a 25% risk. Once an individual has tested positive for a cancer susceptibility gene, it is recommended that all at-risk family members pursue genetic testing, a process that is called cascade testing. This type of cascade testing is useful to identify asymptomatic individuals at risk for a hereditary condition.

References

Al Olama, A.A., Kote-Jarai, Z., Berndt, S.I., Conti, D.V., Schumacher, F., Han, Y., Benlloch, S., Hazelett, D.J., Wang, Z., Saunders, E., Leongamornlert, D., Lindstrom, S., Jugurnauth-Little, S., Dadaev, T., Tymrakiewicz, M., Stram, D.O., Rand, K., Wan, P., Stram, A., Sheng, X., Pooler, L.C., Park, K., Xia, L., Tyrer, J., Kolonel, L.N., Le Marchand, L., Hoover, R.N., Machiela, M.J., Yeager, M., Burdette, L., Chung, C.C., Hutchinson, A., Yu, K., Goh, C., Ahmed, M., Govindasami, K., Guy, M., Tammela, T.L.J., Auvinen, A., Wahlfors, T., Schleutker, J., Visakorpi, T., Leinonen, K.A., Xu, J., Aly, M., Donovan, J., Travis, R.C., Key, T.J., Siddiq, A., Canzian, F., Khaw, K.T., Takahashi, A., Kubo, M., Pharoah, P., Pashayan, N., Weischer, M., Nordestgaard, B.G., Nielsen, S.F., Klarskov, P., Roder, M.A., Iversen, P., Thibodeau, S.N., McDonnell, S.K., Schaid, D.J., Stanford, J.L., Kolb, S., Holt, S., Knudsen, B., Coll, A.H., Gapstur, S.M., Diver, W.R., Stevens, V.L., Maier, C., Luedeke, M., Herkommer, K., Rinckleb, A.E., Strom, S.S., Pettaway, C., Yeboah, E.D., Tettey, Y., Biritwum, R.B., Adjei, A.A., Tay, E., Truelove, A., Niwa, S., Chokkalingam, A.P., Cannon-Albright, L., Cybulski, C., Wokolorczyk, D., Kluzniak, W., Park, J., Sellers, T., Lin, H.Y., Isaacs, W.B., Partin, A.W., Brenner, H., Dieffenbach, A.K., Stegmaier, C., Chen, C., Giovannucci, E.L., Ma, J., Stampfer, M., Penney, K.L., Mucci, L., John, E.M., Ingles, S.A., Kittles, R.A., Murphy, A.B., Pandha, H., Michael, A., Kierzek, A.M., Blot, W., Signorello, L.B., Zheng, W., Albanes, D., Virtamo, J., Weinstein, S., Nemesure, B., Carpten, J., Leske, C., Wu, S.-Y., Hennis, A., Kibel, A.S., Rybicki, B.A., Neslund-Dudas, C., Hsing, A.W., Chu, L., Goodman, P.J., Klein, E.A., Zheng, S.L., Batra, J., Clements, J., Spurdle, A., Teixeira, M.R., Paulo, P., Maia, S., Slavov, C., Kaneva, R., Mitev, V., Witte, J.S., Casey, G., Gillanders, E.M., Seminara, D., Riboli, E., Hamdy, F.C., Coetzee, G.A., Li, Q., Freedman, M.L., Hunter, D.J., Muir, K., Gronberg, H., Neal, D.E., Southey, M., Giles, G.G., Severi, G., Breast and Prostate Cancer Cohort Consortium (BPC3), PRACTICAL (Prostate Cancer Association Group to Investigate Cancer-Associated Alterations in the Genome) Consortium, COGS (Collaborative Oncological Gene-environment Study) Consortium, GAME-ON/ELLIPSE Consortium, Cook, M.B., Nakagawa, H., Wiklund, F., Kraft, P., Chanock, S.J., Henderson, B.E., Easton, D.F., Eeles, R.A., Haiman, C.A., 2014. A meta-analysis of 87,040 individuals identifies 23 new susceptibility loci for prostate cancer. Nat. Genet. 46, 1103–1109. PMID 25217961.

Arber, D.A., Orazi, A., Hasserjian, R., Thiele, J., Borowitz, M.J., Le Beau, M.M., Bloomfield, C.D., Cazzola, M., Vardiman, J.W., 2016. The 2016 revision to the World Health Organization classification of myeloid neoplasms and acute leukemia. Blood 127, 2391–2405. PMID 27069254.

Bochtler, T., Haag, G.M., Schott, S., Kloor, M., Kramer, A., Muller-Tidow, C., 2018. Hematological malignancies in adults with a family predisposition. Dtsch. Arztebl. Int. 115, 848–854. PMID 30722840.

Burn, J., Gerdes, A.M., Macrae, F., Mecklin, J.P., Moeslein, G., Olschwang, S., Eccles, D., Evans, D.G., Maher, E.R., Bertario, L., Bisgaard, M.L., Dunlop, M.G., Ho, J.W.C., Hodgson, S.V., Lindblom, A., Lubinski, J., Morrison, P.J., Murday, V., Ramesar, R., Side, L., Scott, R.J., Thomas, H.J.W., Vasen, H.F., Barker, G., Crawford, G., Elliott, F., Movahedi, M., Pylvanainen, K., Wijnen, J.T., Fodde, R., Lynch, H.T., Mathers, J.C., Bishop, D.T., CAPP2 Investigators, 2011. Long-term effect of aspirin on cancer risk in carriers of hereditary colorectal cancer: an analysis from the CAPP2 randomised controlled trial. Lancet 378, 2081–2087. PMID 22036019.

Burt, R.W., DiSario, J.A., Cannon-Albright, L., 1995. Genetics of colon cancer: impact of inheritance on colon cancer risk. Annu. Rev. Med. 46, 371–379. PMID 7598472.

Carter, N.J., Marshall, M.L., Susswein, L.R., Zorn, K.K., Hiraki, S., Arvai, K.J., Torene, R.I., McGill, A.K., Yackowski, L., Murphy, P.D., Xu, Z., Solomon, B.D., Klein, R.T., Hruska, K.S., 2018. Germline pathogenic variants identified in women with ovarian tumors. Gynecol. Oncol. 151, 481–488. PMID 30322717.

Childers, C.P., Childers, K.K., Maggard-Gibbons, M., Macinko, J., 2017. National estimates of genetic testing in women with a history of breast or ovarian cancer. J. Clin. Oncol. 35, 3800–3806. PMID 28820644.

Couch, F.J., Shimelis, H., Hu, C., Hart, S.N., Polley, E.C., Na, J., Hallberg, E., Moore, R., Thomas, A., Lilyquist, J., Feng, B., McFarland, R., Pesaran, T., Huether, R., LaDuca, H., Chao, E.C., Goldgar, D.E., Dolinsky, J.S., 2017. Associations between cancer predisposition testing panel genes and breast cancer. JAMA Oncol. 3, 1190–1196. PMID 28418444.

Eccles, B.K., Copson, E., Maishman, T., Abraham, J.E., Eccles, D.M., 2015. Understanding of BRCA VUS genetic results by breast cancer specialists. BMC Cancer 15, 936. PMID 26608569.

Eng, C., September 27, 1999. Multiple endocrine neoplasia type 2. Adam, M.P., Mirzaa, G.M., Pagon, R.A., Wallace, S.E., Bean, L.J.H., Gripp, K.W., Amemiya, A. (editors). In: GeneReviews® [Internet]. University of Washington, Seattle (WA), 1993–2023 [updated August 15, 2019]. PMID 20301434.

Lu, C., Xie, M., Wendl, M.C., Wang, J., McLellan, M.D., Leiserson, M.D.M., Huang, K.L., Wyczalkowski, M.A., Jayasinghe, R., Banerjee, T., Ning, J., Tripathi, P., Zhang, Q., Niu, B., Ye, K., Schmidt, H.K., Fulton, R.S., McMichael, J.F., Batra, P., Kandoth, C., Bharadwaj, M., Koboldt, D.C., Miller, C.A., Kanchi, K.L., Eldred, J.M., Larson, D.E., Welch, J.S., You, M., Ozenberger, B.A., Govindan, R., Walter, M.J., Ellis, M.J., Mardis, E.R., Graubert, T.A., Dipersio, J.F., Ley, T.J., Wilson, R.K., Goodfellow, P.J., Raphael, B.J., Chen, F., Johnson, K.J., Parvin, J.D., Ding, L., 2015. Patterns and functional implications of rare germline variants across 12 cancer types. Nat. Commun. 6, 10086. PMID 26689913.

Lynch, H.T., De La Chapelle, A., 2003. Hereditary colorectal cancer. N. Engl. J. Med. 348, 919–932. PMID 12621137.

Michailidou, K., Beesley, J., Lindstrom, S., Canisius, S., Dennis, J., Lush, M.J., Maranian, M.J., Bolla, M.K., Wang, Q., Shah, M., Perkins, B.J., Czene, K., Eriksson, M., Darabi, H., Brand, J.S., Bojesen, S.E., Nordestgaard, B.G., Flyger, H., Nielsen, S.F., Rahman, N., Turnbull, C., Breast and Ovarian Cancer Susceptibility (BOS) Study, Fletcher, O., Peto, J., Gibson, L., Dos-Santos-Silva, I., Chang-Claude, J., Flesch-Janys, D., Rudolph, A., Eilber, U., Behrens, S., Nevanlinna, H., Muranen, T.A., Aittomaki, K., Blomqvist, C., Khan, S., Aaltonen, K., Ahsan, H., Kibriya, M.G., Whittemore, A.S., John, E.M., Malone, K.E., Gammon, M.D., Santella, R.M., Ursin, G., Makalic, E., Schmidt, D.F., Casey, G., Hunter, D.J., Gapstur, S.M., Gaudet, M.M., Diver, W.R., Haiman, C.A., Schumacher, F., Henderson, B.E., Le Marchand, L., Berg, C.D., Chanock, S.J., Figueroa, J., Hoover, R.N., Lambrechts, D., Neven, P., Wildiers, H., Van Limbergen, E., Schmidt, M.K., Broeks, A., Verhoef, S., Cornelissen, S., Couch, F.J., Olson, J.E., Hallberg, E., Vachon, C., Waisfisz, Q., Meijers-Heijboer, H., Adank, M.A., Van Der Luijt, R.B., Li, J., Liu, J., Humphreys, K., Kang, D., Choi, J.Y., Park, S.K., Yoo, K.Y., Matsuo, K., Ito, H., Iwata, H., Tajima, K., Guenel, P., Truong, T., Mulot, C., Sanchez, M., Burwinkel, B., Marme, F., Surowy, H., Sohn, C., Wu, A.H., Tseng, C.C., Van Den Berg, D., Stram, D.O., Gonzalez-Neira, A., Benitez, J., Zamora, M.P., Arias Perez, J.I., Shu, X.-O., Lu, W., Gao, Y.-T., Cai, H., Cox, A., Cross, S.S., Reed, M.W.R., Andrulis, I.L., Knight, J.A., Glendon, G., Mulligan, A.M., Sawyer, E.J., Tomlinson, I., Kerin, M.J., Miller, N., kConFab Investigators, AOCS Group, Lindblom, A., Margolin, S., Teo, S.H., Yip, C.H., Taib, N.A.M., Tan, G.-H., Hooning, M.J., Hollestelle, A., Martens, J.W.M., Collée, J.M., Blot, W., Signorello, L.B., Cai, Q., Hopper, J.L., Southey, M.C., Tsimiklis, H., Apicella, C., Shen, C.-Y., Hsiung, C.-N., Wu, P.-E., Hou, M.-F., Kristensen, V.N., Nord, S., Alnaes, G.I.G., NBCS, Giles, G.G., Milne, R.L., McLean, C., Canzian, F., Trichopoulos, F., Peeters, P., Lund, E., Sund, M., Khaw, K.-T., Gunter, M.J., Palli, D., Mortensen, L.M., Dossus, L., Huerta, J.-M., Meindl, A., Schmutzler, R.K., Sutter, C., Yang, R., Muir, K., Lophatananon, A., Stewart-Brown, S., Siriwanarangsan, P., Hartman, M., Miao, H., Chia, K.S., Chan, C.W., Fasching, P.A., Hein, A., Beckmann, M.W., Haeberle, L., Brenner, H., Dieffenbach, A.K., Arndt, V., Stegmaier, C., Ashworth, A., Orr, N., Schoemaker, M.J., Swerdlow, A.J., Brinton, L., Garcia-Closas, M., Zheng, W., Halverson, S.L., Shrubsole, M., Long, J., Goldberg, M.S., Labrèche, F., Dumont, M., Winqvist, R., Pylkäs, K., Jukkola-Vuorinen, A., Grip, M., Brauch, H., Hamann, U., Brüning, T., , GENICA Network, Radice, P., Peterlongo, P., Manoukian, S., Bernard, L., Bogdanova, N.V., Dörk, T., Mannermaa, A., Kataja, V., Kosma, V.-M., Hartikainen, J.M., Devilee, P., Tollenaar, R.A.E.M., Seynaeve, C., Van Asperen, C.J., Jakubowska, A., Lubinski, J., Jaworska, K., Huzarski, T., Sangrajrang, S., Gaborieau, V., Brennan, P., McKay, J., Slager, S., Toland, A.E., Ambrosone, C.B., Yannoukakos, C., Kabisch, M., Torres, D., Neuhausen, S.L., Anton-Culver, H., Luccarini, C., Baynes, C., Ahmed, S., Healey, C.S., Tessier, D.C., Vincent, D., Bacot, F., Pita, G., Alonso, M.R., Álvarez, N., Herrero, D., Simard, J., Pharoah, P.P.D.P., Kraft, P., Dunning, A.M., Chenevix-Trench, G., Hall, P., Easton, D.F., 2015. Genome-wide association analysis of more than 120,000 individuals identifies 15 new susceptibility loci for breast cancer. Nat. Genet. 47, 373–380. PMID 25751625.

Moreira, L., Balaguer, F., Lindor, N., De La Chapelle, A., Hampel, H., Aaltonen, L.A., Hopper, J.L., Le Marchand, L., Gallinger, S., Newcomb, P.A., Haile, R., Thibodeau, S.N., Gunawardena, S., Jenkins, M.A., Buchanan, D.D., Potter, J.D., Baron, J.A., Ahnen, D.J., Moreno, V., Andreu, M., Ponz De Leon, M., Rustgi, A.K., Castells, A., EPICOLON Consortium, 2012. Identification of Lynch syndrome among patients with colorectal cancer. JAMA 308, 1555–1565. PMID 23073952.

National Cancer Institute, DCCPS, Surveillance Research Program. Surveillance, Epidemiology, and End Results (SEER) Program 18 Research Data (2011-2015). Online: http://www.seer.cancer.gov (Accessed 07.09.2023).

National Comprehensive Cancer Network, 2022a. NCCN Clinical Practice Guidelines in Oncology (NCCN Guidelines): Thyroid Carcinoma. Version 2.2022-May 5, 2022. [Online]. https://www.nccn.org/professionals/physician_gls/pdf/thyroid.pdf.

National Comprehensive Cancer Network, 2022b. NCCN Clinical Practice Guidelines in Oncology (NCCN Guidelines): Genetic/Familial High-Risk Assessment: Breast, Ovarian, and Pancreatic. Version 2.2022-March 9, 2022. [Online]. https://www.nccn.org/professionals/physician_gls/pdf/genetics_bop.pdf.

National Comprehensive Cancer Network, 2022c. NCCN Clinical Practice Guidelines in Oncology (NCCN Guidelines): Genetic/Familial High-Risk Assessment: Colorectal. Version 1.2022-June 8, 2022. [Online]. https://www.nccn.org/professionals/physician_gls/pdf/genetics_colon.pdf.

National Comprehensive Cancer Network, 2022d. NCCN Clinical Practice Guidelines in Oncology (NCCN Guidelines): Neuroendocrine and Adrenal Tumors. Version 1.2022-May 23, 2022. [Online]. https://www.nccn.org/professionals/physician_gls/pdf/neuroendocrine.pdf.

National Comprehensive Cancer Network, 2022e. NCCN Clinical Practice Guidelines in Oncology (NCCN Guidelines): Central Nervous System Cancers. Version 1.2022-June 2, 2022. [Online]. https://www.nccn.org/professionals/physician_gls/pdf/cns.pdf.

Neumann, H.P.H., Young Jr., W.F., Eng, C., 2019. Pheochromocytoma and paraganglioma. N. Engl. J. Med. 381, 552–565. PMID 31390501.

Nieuwenhuis, M.H., Vasen, H.F., 2007. Correlations between mutation site in APC and phenotype of familial adenomatous polyposis (FAP): a review of the literature. Crit. Rev. Oncol. Hematol. 61, 153–161. PMID 17064931.

Porter, C.C., 2016. Germ line mutations associated with leukemias. Hematology Am. Soc. Hematol. Educ. Program 2016, 302–308. PMID 27913495.

Saslow, D., Boetes, C., Burke, W., Harms, S., Leach, M.O., Lehman, C.D., Morris, E., Pisano, E., Schnall, M., Sener, S., Smith, R.A., Warner, E., Yaffe, M., Andrews, K.S., Russell, C.A., American Cancer Society Breast Cancer Advisory Group, 2007. American Cancer Society guidelines for breast screening with MRI as an adjunct to mammography. CA Cancer J. Clin. 57, 75–89. PMID 17392385.

Siegel, R.L., Miller, K.D., Jemal, A., 2019. Cancer statistics, 2019. CA Cancer J. Clin. 69 (1), 7–34.

Syngal, S., Brand, R.E., Church, J.M., Giardiello, F.M., Hampel, H.L., Burt, R.W., American College of Gastroenterology, 2015. ACG clinical guideline: genetic testing and management of hereditary gastrointestinal cancer syndromes. Am. J. Gastroenterol. 110, 223–262. PMID 25645574.

Tandy-Connor, S., Guiltinan, J., Krempely, K., LaDuca, H., Reineke, P., Gutierrez, S., Gray, P., Tippin Davis, B., 2018. False-positive results released by direct-to-consumer genetic tests highlight the importance of clinical confirmation testing for appropriate patient care. Genet. Med. 20, 1515–1521. PMID 29565420.

Tautz, D., Renz, M., 1984. Simple sequences are ubiquitous repetitive components of eukaryotic genomes. Nucleic Acids Res. 12, 4127–4138. PMID 6328411.

Thakker, R.V., Newey, P.J., Walls, G.V., Bilezikian, J., Dralle, H., Ebeling, P.R., Melmed, S., Sakurai, A., Tonelli, F., Brandi, M.L., Endocrine Society, 2012. Clinical practice guidelines for multiple endocrine neoplasia type 1 (MEN1). J. Clin. Endocrinol. Metab. 97, 2990–3011. PMID 22723327.

Tung, N., Lin, N.U., Kidd, J., Allen, B.A., Singh, N., Wenstrup, R.J., Hartman, A.R., Winer, E.P., Garber, J.E., 2016. Frequency of germline mutations in 25 cancer susceptibility genes in a sequential series of patients with breast cancer. J. Clin. Oncol. 34, 1460–1468. PMID 26976419.

Tyrer, J., Duffy, S.W., Cuzick, J., 2004. A breast cancer prediction model incorporating familial and personal risk factors. Stat. Med. 23, 1111–1130. PMID 15057881.

University of Chicago Hematopoietic Malignancies Cancer Risk Team, 2016. How I diagnose and manage individuals at risk for inherited myeloid malignancies. Blood 128 (14), 1800–1813. PMID 27471235.

Walsh, T., Casadei, S., Lee, M.K., Pennil, C.C., Nord, A.S., Thornton, A.M., Roeb, W., Agnew, K.J., Stray, S.M., Wickramanayake, A., Norquist, B., Pennington, K.P., Garcia, R.L., King, M.C., Swisher, E.M., 2011. Mutations in 12 genes for inherited ovarian, fallopian tube, and peritoneal carcinoma identified by massively parallel sequencing. Proc. Natl. Acad. Sci. U. S. A. 108, 18032–18037. PMID 22006311.

Chapter 9

Direct-to-Consumer Testing

Angela E. Scheuerle[1,2,3]

[1]Department of Pediatrics, Division of Genetics and Metabolism, University of Texas Southwestern Medical Center, Dallas, TX, United States; [2]Department of Pathology Division of Genetic Diagnostics, University of Texas Southwestern Medical Center, Dallas, TX, United States; [3]McDermott Center for Human Growth and Development, University of Texas Southwestern Medical Center, Dallas, TX, United States

Abstract

Direct-to-consumer genetic testing gives the general public individual information about relationships, ethnic backgrounds, single gene diseases, and complex traits without counseling by a medical professional. There are various testing technologies, most of which are used in clinical diagnostics; however, interpretation of test results can be complex and direct-to-consumer genetic testing may not provide medically actionable information. Direct-to-consumer genetic testing should not be considered a substitute for diagnostic testing performed by a Clinical Laboratory Improvement Amendment (CLIA)-certified laboratory with professional genetic oversight.

Another form of genetic laboratory evaluation is direct-to-consumer (DTC) genetic testing. Identity and lineage testing are sought commonly, and more detailed testing is also marketed. Results are sent directly to the individual. Some companies will provide genetic counseling information upon inquiry, but other tests are conducted outside the realm of trained genetic professionals. The results can be helpful, but likewise can be misunderstood and potentially falsely reassuring.

9.1 Direct-to-Consumer Testing and Limitations

The phrase "Direct-to-Consumer" (DTC) came into general medical parlance in 1981 with the first print advertisement for Merck's pneumococcal pneumonia vaccine. Pharmaceutical DTC advertising in the United States is now ubiquitous (Ventola, 2011). DTC genetic testing began in 1997 when Identigene began selling paternity testing services online. This was and is DNA "fingerprinting" or identity testing. It does not generate data that can be analyzed for disease risk. The early 2000s saw a leap in genetic testing technology, making it possible to identify disease markers and pathogenic genetic variants on a large scale. At that time, about 2007, DTC genetic testing exploded. Today, DTC genetic testing is marketed, accurately or not, as answering questions about ancestry, disease risk (Sections 11.17.1, 11.17.3), recessive carrier status, pharmacogenomic risks (Sections 5.6, 5.7.4), and a variety of things not recognized as genetically diagnosable, such as prediction of athletic ability. Some companies claim to analyze DNA and predict risk of multifactorial disorders that have a significant environmental component such as obesity, which is not a validated clinical approach. Other "tests" are actually scams that claim to do DNA analysis to optimize one's diet and nutrition or make other equally questionable predictions (Section 11.17.6) (Gardner et al., 2018).

9.1.1 Identity Testing

This method is used for paternity, zygosity, and tissue or person identification. It is so-called "DNA fingerprinting" or "genetic fingerprinting" and takes advantage of microsatellites, highly variable but easily analyzed areas of the genome.

Genomics in the Clinic. https://doi.org/10.1016/B978-0-12-816478-5.00009-1

Microsatellites (Section 2.4.1) are stretches of DNA in short repeated patterns of 2–4 base pairs. The number of repeats at any microsatellite is highly variable. By measuring the number of repeats at a few dozen microsatellites, it is possible to say statistically that sample A matches sample B because they share enough of the same "fingerprint." A and B might be twins, or a parent and child, or a tissue sample and a crime suspect. This is also one way to do ancestry analysis as the relation between two (or more) individuals can be calculated based on the proportion of microsatellites shared. The result from this testing is a statistical estimate because it queries only a proportion of the DNA.

9.1.2 Lineage Testing

Mothers pass on mitochondria to all their children (Section 2.5) and fathers pass on Y chromosomes to their sons (Section 2.7.2). Using this information it is possible to trace a maternal lineage back by analyzing mitochondrial DNA and, likewise, a paternal lineage looking at the Y chromosome. The mutation rate for mitochondrial DNA has been estimated and comparison of different individuals can suggest the phylogeny of mutations, the chronologic sequence in which the variations occur generation to generation (van Oven and Kayser, 2009). Similar data exist for the Y chromosome (van Oven et al., 2014). By comparing the derived chronology of variation to the mitochondria or Y-chromosome sequence in existing (or archeologic) populations, it is possible to follow the variations backward in both time and geography. Thus, it is theoretically possible to describe, essentially, how a particular mitochondrial genome or Y chromosome was inherited in populations that moved around the world from their origin in Africa. This technique was used by the National Genographic Project, an anthropological study by the National Geographic Society that looked at human migration patterns over history (Behar et al., 2007). This method also can be used in a general way as ancestry or genealogy testing (Section 11.17.3).

9.1.3 Disease Risk Testing

A haplotype is a contiguous set of genes or definable DNA sequence inherited as a block because it has not been altered by recombination. Haplotype analysis can be used for ancestry and ethnicity estimates. Because haplotypes reflect specific lineages, they may be helpful in differentiating or linking population groups. At a population level, haplotype "mapping" is a highly developed branch of anthropology and paleontology. It is less useful at the individual level.

Within a particular population group, a haplotype can be associated with disease risk. For example, some haplotypes in some population groups are statistically associated with adult-onset conditions such as cancer or heart arrhythmias. This is not the same as finding a pathogenic sequence variant. Rather, it is a statistical recognition that a defined haplotype is seen with unusual frequency in people of a population group who have a particular condition, i.e., individuals in population group A who have a "haplotype #5" are at higher risk of having disease X. One example of this is the entities associated with HLA haplotypes (Section 5.7.4).

The important point is that this disease risk estimate only works within a population group. The DNA sequence designated "haplotype #5" in the example may occur in multiple groups. However, it is only statistically associated with disease risk in one of them and not in others (Section 11.17.2).

9.1.4 Mutation Scanning

The most sensitive broad testing is done via single-nucleotide polymorphism (SNP) or single-nucleotide variant (SNV) (Section 2.4.1) analysis. This evaluates DNA at the single base pair level and can identify true pathogenic variants (Section 7.1.3). Testing SNPs of many genes looks at multiple specific sites in the genome quickly. Medically this is used in preconception and prenatal testing (Sections 4.4 and 10.1.1) to see if parents are carriers of certain variants for recessive conditions (Section 2.7.1). It is not the same as gene sequencing. It does not analyze for all possible pathogenic variants in a gene, just the most common. For example, in the gene associated with cystic fibrosis, *CFTR*, a set of 40 pathogenic variants account for disease in about 95% of the population of European descent. Analyzing for these 40 variants is considered sufficient for screening in patients with a negative family history, but it reduces the carrier risk rather than eliminating it (Section 6.2.4). There may be a pathogenic variant elsewhere in the gene that is not detected by this method.

In some conditions, there may be more than one potential gene of interest. Mutation scanning may evaluate for pathogenic variants across the more common genes, but not all of them, for a given condition. This is somewhat broader testing, but, again, "no pathogenic variant found" is not the same thing as "normal" because it depends on the number of variants investigated and the detection rate of the multigene screening method.

9.1.5 Genotyping

DTC testing that purports to evaluate disease risk is done by SNP genotyping (Section 2.4.1) or next-generation sequencing (Section 6.2). As such, there is the real potential to detect variants that are truly pathogenic; however, there are caveats. First, this is a highly developed technology that, even though much simplified from the 1990s, requires skilled technicians and trained interpreters. Second, as with haplotypes, knowing the individual's population background is important. A DNA sequence common in one population group may be rare in another. It may be misinterpreted as pathogenic when present because it is rare, even though it is actually benign. Third, DTC companies interpret SNP variant pathogenicity based in part on their specific internal databases; likely skewed in their population groups. There are public, more widely based databases easily available, though these also have deficits and cannot yet be considered population-based. Fourth, DTC companies may not include all the genes of interest to a particular condition. For example, testing the breast cancer genes *BRCA1* and *BRCA2* is insufficient to identify a risk of breast cancer, as there are now as many as 45 genes tested in diagnostic labs for the same condition (Genetic Testing Registry Online, 2022) (Section 11.17.1).

9.1.6 Traits and "Wellness"

DTC testing for minor physical traits and lifestyle advice has become more in demand. This marketing and testing usually falls in the "entertainment" realm because results do not affect clinical management or treatment. Many human characteristics are not determined by a single gene, or even by genes alone. As such, using genetics to predict or explain the presence of these traits is not medically interpretable. Any human feature that is a spectrum, such as height, hair color, or musical talent, results from an interplay of many genes and/or of genes and environment. Height is mathematically estimable in a child based on parental heights, though "normal" has a wide range and is dependent on genes involved in bone length, growth hormone production, etc. Further, final adult height is also dependent upon general health and nutritional status. There are DTC testing companies that claim to do genetic testing for some of these traits. However, a reading of one company's website finds many disclaimers about trait testing: that it is a statistical prediction, that it depends on other factors (race, age, sex, etc.) and that some people will have a contradictory genetic result.

In the same vein are "Wellness" reports (Section 11.17.6). These are focused on traits that relate to lifestyle in topics such as diet, exercise, and sleep. There are companies that do these tests exclusively, claiming, for example, to predict someone's ideal diet based on their genetics. There is no basis in science for such claims. While appropriate nutritional advice is not, per se, bad, there is the real risk of a consumer being caught up in the testing and taking unnecessary, expensive, or even medically contraindicated supplements.

9.1.7 Ancillary Interpretation

DTC genetic testing companies can release results in various formats, from very general verbal reports to the specific base sequence data. There are companies that allow uploading via the internet of the DTC DNA testing data files for secondary interpretation. As with the DTC companies themselves, these secondary evaluations may be of variable quality and should be used with caution.

9.2 Ethical Considerations

Availability of DNA-based testing marketed directly to the general public raises both ethical (Wasson et al., 2006) and scientific (Helgason and Stefánsson, 2010; Tandy-Connor et al., 2018) concerns. Physicians order diagnostic tests in the context of objective patient findings and professional sample handling. DTC testing is collected by the consumer who also provides the history and medical information. While actual problems are likely the minority, there is a risk of sample mislabeling and misinterpretation. Some of the DTC labs have Clinical Laboratory Improvement Amendment (CLIA) certification, but not all, and those without are operating with no oversight. Additionally, the protections against testing minor children without consent or assent are widely variable (Moray et al., 2017) (Sections 10.1.2, 11.17.1, 11.17.3, and 11.18.1).

In both the United States (ACMG Board of Directors, 2016) and the United Kingdom (Human Genetics Commission, 2007), professional genetic organizations recommend that genetic testing be done through appropriate health professionals who can provide adequate pre- and posttest counseling (Section 3.3). The goal is to assist the patient in considering the implications of testing as well as its benefits and limits.

Some consumers may choose DTC genetic testing as a way to keep genetic information out of the medical record and away from their insurance company. While it will provide this avenue, the DTC companies themselves may become the owners of both the DNA sample and any information derived from it. Ownership depends upon consent obtained via the companies' specific consent forms. This information can then be shared with pharmaceutical companies (Glaxo Smith Kline, 2018) and with law enforcement (Haag, 2019), abrogating the consumers' desired privacy (Seife, 2013). It has been suggested that there has now been enough DTC genetic testing of persons of Northern European descent that in the United States even someone who has not been tested may be identifiable to at least a third cousin level (Erlich et al., 2018). Conversely, if DTC testing identifies a health risk such as a high risk of colon cancer, the consumer may then have a legal obligation to disclose that on health and insurance forms. Existing laws that protect consumers from misuse of genetic information cover most health insurances but do not provide protection in life or disability insurance (Genetics Home Reference, 2020). To our knowledge, this has not been tested in court as of this time. See also Chapter 10 for further ethical considerations.

Offers of DTC genetic testing through Medicare is a problem that has arisen in the past few years. While there are a number of examples, in 2019 the US Federal Bureau of Investigation charged a DTC testing agency with Medicare fraud (Lynch and Nouvelage, 2019). The typical scam is carried out at health fairs or other community events (Section 11.17.6) and usually targets the elderly. The consumer is told that there will be genetic testing done by cheek swab that can test them for a number of conditions like cancer and that cost will be covered by Medicare. The testing, if done, can be substandard, and the lab can pocket the profit from the Medicare payment or can use access to the Medicare account for other fraudulent purposes. If Medicare is charged and the charge is denied, a bill, potentially for thousands of dollars, could be sent to the consumer (Kinney, 2019).

9.3 Medical and Scientific Considerations

The test result itself is not always straightforward. CLIA-certified molecular genetic labs report a variant finding on a five-point scale: pathogenic, likely pathogenic, variant of uncertain significance, likely benign, and benign (Richards et al., 2015). The most common finding in a DNA-based clinical diagnostic test is "variant of uncertain significance (VUS)." This is as it sounds: a statistically unusual finding for the patient's background about which there is insufficient information to label it either pathogenic or benign (Section 7.1).

There are two major concerns. The first is that the accumulated data change over time. Variants that were initially labeled at one point on the scale may be reclassified as more information becomes available. Pathogenic variants may be relabeled as benign and initially benign variants may come to be recognized as pathogenic. This is not malpractice or malfeasance but simply the progress of genetic science. For this reason, the current recommendation is that nondiagnostic genome data should be reinterrogated every few years (Das et al., 2014; Mersch et al., 2018; Sorelle et al., 2019). This is not a service commonly or currently offered by DTC testing programs except as submission of a new sample.

The second concern is about false-positive or false-negative results. First, medical decisions about diagnostic testing, treatments, or other aspects should never be based solely on positive DTC testing results. Instead, all DTC testing needs to be confirmed in a CLIA-approved lab if the company itself does not use one. It is best if the result is reviewed by the person's physician, possibly with retesting in a primary diagnostic lab, prior to carrying out any medical or surgical actions. The accuracy of results from the DTC test may itself be uncertain. One study found that DTC testing had a 40% false-positive rate (Tandy-Connor et al., 2018). This is a significant problem if patients choose to act on the results immediately.

False-negative results are even more difficult. Genetic testing, even formal diagnostic testing, is unambiguous only when clearly abnormal. Negative and "nondiagnostic" results should not be interpreted as ruling out a diagnosis or risk of disease. There is high person-to-person variability, as well as significant limits of testing technology and a large number of genes of uncertain clinical significance. Consumers may inappropriately interpret a low risk or "negative" result as a guarantee of health (Wasson et al., 2006) (Section 11.17.1).

9.4 Risk and Benefits

There are some benefits of DTC testing to consumers, science, and other interested parties. To the consumer undergoing the testing, theoretical benefits include: the entertainment value of "identifying" a purported ethnic background; the possible identification of distant family members, who have also submitted samples to the same DTC company; the possible identification and location of birth family by an adoptee (Section 11.17.3); and the potential determination of disease risk via identification of pathogenic genetic variants (Section 11.17.5). Benefits for science may include pooling large numbers of samples to build DNA databases, thus allowing better definition of benign versus pathogenic variants, and also collaboration with pharmaceutical companies for susceptibility analysis and drug discovery.

Other potential benefits from DTC genetic testing might include facilitation of criminal investigations. Law enforcement access to the DTC databases may lead to increased identification of criminals.

There are also risks of DTC testing to both the consumers and to others, as well as to science. Risks to the consumer include direct privacy violations such as exposure of nonpaternity (Section 10.3.2) or other nonbiologic relationship; violation of birth mother rights and protections in closed adoptions; and loss of control over use of the DNA sample and any data derived from it. There is the proxy risk to the privacy of individuals who do not participate in the testing process and who may not want their DNA information shared. In addition to the risk of poor consumer sample handling and lack of laboratory certification and oversight, there may be laboratory misinterpretation or overinterpretation of data leading to a false-positive or false-negative report of disease risk. This can be complicated by the consumer's misinterpretation of nondiagnostic DTC testing results as "normal." And there is the potential legal concern if DTC testing identifies an increased disease risk that creates an obligation of disclosure to insurance and/or employer.

Risks to science with DTC testing include accumulation of data with unclear background. Information about ethnicity, health, and other characteristics is self-reported by the consumer rather than derived from the medical record and is otherwise unvalidated.

9.5 Conclusions

DTC genetic testing companies take advantage of the "genetic mystique"—the idea that DNA-based information is an ultimate answer. Direct access also circumvents the physician as gatekeeper and supports, or appears to support, patient autonomy. This is obviously appealing to a large segment of the population. Unfortunately, the lay understanding of genetic science and the risks associated with testing may be insufficient. For physicians, the better approach is that DTC genetic testing should be considered at best supplemental or preliminary, rather than standard, clinical testing. Any DTC testing results suggesting diagnosis of disease risk should be followed up with a standard medical investigation including appropriate DNA testing in a certified diagnostic lab (Tandy-Connor et al., 2018) (Section 11.17.5). Consumers also should be aware of the potential limitations and risks for this type of testing (Section 7.3). Undoubtedly, in the future there will be more DTC genetic testing, and hopefully this will be accompanied with regulatory oversight to increase the benefits and reduce risks to the public.

References

ACMG Board of Directors, 2016. Direct-to-consumer genetic testing: a revised position statement of the American College of Medical Genetics and Genomics. Genet. Med. 18, 207–208. PMID 26681314.

Behar, D.M., Rosset, S., Blue-Smith, J., Balanovsky, O., Tzur, S., Comas, S., et al., 2007. The Genographic project public participation mitochondrial DNA database. PLoS Genet. 3, e104. PMID: 17604454.

Das, K.J., Ingles, J., Bagnall, R.D., Semsarian, C., 2014. Determining pathogenicity of genetic variants in hypertrophic cardiomyopathy: importance of periodic reassessment. Genet. Med. 16, 286–293. PMID 24113344.

Erlich, Y., Shor, T., Pe'er, I., Carmi, S., 2018. Identity inference of genomic data using long-range familial searches. Science 362, 690–694. PMID 30309907.

Gardner, C.D., Trepanowski, J.F., Del Gobbo, L.C., Hauser, M.E., Rigdon, J., Ioannidis, J.P.A., et al., 2018. Effect of low-fat vs low-carbohydrate diet on 12-month weight loss in overweight adults and the association with genotype pattern or insulin secretion: the DIETFITS Randomized Clinical Trial. JAMA 319, 667–679. PMID 29466592.

Genetics Home Reference, 2020. Can the results of direct-to-consumer genetic testing affect my ability to get insurance? (Online). Bethesda, MD. Available: https://medlineplus.gov/genetics/understanding/dtcgenetictesting/dtcinsurancerisk/ (Accessed May 19, 2023).

Genetic Testing Registry (Online), 2022. National Center for Biotechnology Information. National Library of Medicine. Available: https://www.ncbi.nlm.nih.gov/gtr/all/tests/?term=Breast%20cancer (Accessed May 19, 2023).

Glaxo Smith Kline (GSK), 2018. GSK Press Release: GSK and 23andMe Sign Agreement to Leverage Genetic Insights for the Development of Novel Medicines (Online). Available: https://www.gsk.com/en-gb/media/press-releases/gsk-and-23andme-sign-agreement-to-leverage-genetic-insights-for-the-development-of-novel-medicines/ (Accessed August 21, 2020).

Haag, M., 2019. FamilyTreeDNA Admits To Sharing Genetic Data with F.B.I. (Online). New York Times, New York. Available: https://www.nytimes.com/2019/02/04/business/family-tree-dna-fbi.html.

Helgason, A., Stefansson, K., 2010. The past, present, and future of direct-to-consumer genetic tests. Dialogues Clin. Neurosci. 12, 61–68. PMID 20373667.

Human Genetics Commission, 2007. More Genes Direct (Online). London, UK nationalacademies.org. Available: http://sites.nationalacademies.org/cs/groups/pgasite/documents/webpage/pga_053238.pdf (Accessed September 2019).

Kinney, R., September 27, 2019. Beware Of a Growing Medicare Scam: "Free Genetic Testing" Administration For Community Living. https://acl.gov/news-and-events/acl-blog/beware-growing-medicare-scam-free-genetic-testing (Accessed October 12, 2021).

Lynch, S.N., Nouvelage, E., September 27, 2019. U.S. agents raid genetic testing labs, charge 35 in Medicare fraud probe. https://www.reuters.com/article/us-usa-fraud-genetics/us-agents-raid-genetic-testing-labs-charge-35-in-medicare-fraud-probe-idUSKBN1WC1PH (Accessed October 12, 2021).

Mersch, J., Brown, N., Pirzadeh-Miller, S., Mundt, E., Cox, H.C., Brown, K., et al., 2018. Prevalence of variant reclassification following hereditary cancer genetic testing. JAMA 320, 1266–1274. PMID 30264118.

Moray, N., Pink, K.E., Borry, P., Larmuseau, M.H., 2017. Paternity testing under the cloak of recreational genetics. Eur. J. Hum. Genet. 25, 768–770. PMID 28272533.

Richards, S., Aziz, N., Bale, S., Bick, D., Das, S., Gastier-Foster, J., et al., 2015. Standards and guidelines for the interpretation of sequence variants: a joint consensus recommendation of the American College of Medical Genetics and Genomics and the Association for Molecular Pathology. Genet. Med. 17, 405–424. PMID 25741868.

Seife, C., 2013. 23andMe Is Terrifying, but Not for the Reasons the FDA Thinks (Online). Scientific American. https://www.scientificamerican.com/article/23andme-is-terrifying-but-not-for-the-reasons-the-fda-thinks/.

Sorelle, J.A., Thodeson, D.M., Arnold, S., Gotway, G., Park, J.Y., 2019. Clinical utility of reinterpreting previously reported genomic epilepsy test results for pediatric patients. JAMA Pediatr. 173, e182302. PMID 30398534.

Tandy-Connor, S., Guiltinan, J., Krempely, K., Laduca, H., Reineke, P., Gutierrez, S., et al., 2018. False-positive results released by direct-to-consumer genetic tests highlight the importance of clinical confirmation testing for appropriate patient care. Genet. Med. 20, 1515–1521. PMID 29565420.

Van Oven, M., Kayser, M., 2009. Updated comprehensive phylogenetic tree of global human mitochondrial DNA variation. Hum. Mutat. 30, E386–E394. PMID 18853457.

Van Oven, M., Van Geystelen, A., Kayser, M., Decorte, R., Larmuseau, M.H., 2014. Seeing the wood for the trees: a minimal reference phylogeny for the human Y chromosome. Hum. Mutat. 35, 187–191. PMID 24166809.

Ventola, C.L., 2011. Direct-to–consumer pharmaceutical advertising: therapeutic or toxic? Pharm. Therapeutics 36, 669–674, 681-684. PMID: 22346300.

Wasson, K., Cook, E.D., Helzlsouer, K., 2006. Direct-to-consumer online genetic testing and the four principles: an analysis of the ethical issues. Ethics Med. 22, 83–91. PMID 17115513.

Further Reading

Donohue, J.M., Cevasco, M., Rosenthal, M.B., 2007. A decade of direct-to-consumer advertising of prescription drugs. N. Engl. J. Med. 357, 673–681. PMID 17699817.

Chapter 10

Ethical and Psychosocial Issues

Lynn Wein Bush[1], Amy Kritzer[2] and Cheryl D. Lew[3]

[1]*Division of Genetics and Genomics, Department of Pediatrics, Boston Children's Hospital, Pediatrics, Harvard Medical School and Harvard Medical School Center for Bioethics, Boston, MA, United States;* [2]*Division of Genetics and Genomics, Department of Pediatrics, Boston Children's Hospital, Pediatrics, Harvard Medical School, Boston, MA, United States;* [3]*Clinical Professor of Pediatrics (Emerita), Pediatrics and Emeritus Faculty, Center for Bioethics. Children's Hospital Los Angeles, Pediatrics, Keck School of Medicine at University Southern California, Los Angeles, CA, United States*

Abstract

It is critical to consider a host of challenging ethical and psychosocial issues that can arise with genomic testing. The informed consent process requires more extended communication and counseling than is typical in clinical medicine. Preconception, prenatal, pediatric, and adult issues need to be considered. A continuing conversation is essential between healthcare professionals with expertise in the patient's clinical phenotypes as well as the appropriate application, and interpretation, of advanced genetic testing.

It is critical to consider a host of challenging ethical and psychosocial issues that can arise with genomic testing and are crucial to address during genetic consultations. The informed consent process for genome-based sequencing, as well as targeted genetic and exome testing, requires more extended communication and counseling than typical in clinical medicine. Preconception, prenatal, pediatric, and adult issues need to be considered, highlighting the importance of understanding the benefits, risks, and limitations inherent with current genomic testing relevant to an individual and family members. A continuing conversation is essential between healthcare professionals with expertise in the patient's clinical phenotypes as well as the appropriate application, and interpretation, of advance genetic testing.

10.1 The Informed Consent Process

Essential to medical decision-making in most countries, the informed consent process is a means of demonstrating respect for persons, with voluntariness and freedom from coercion being key characteristics (ACOG Committee on Ethics, 2009; Katz et al., 2016). Several significant contextual nuances exist regarding genome-based sequencing, with ethical and psychosocial implications that warrant inclusion in the consent process. These distinctions tend to necessitate more extended communication and counseling than typical in clinical medicine so the patient may gain a sufficient appreciation of the benefits, risks, and limitations of the genomic testing and recognize that the information generated may influence their blood relatives as well as themselves. Furthermore, due to the complexity of interpretation and variable predictability of results from genomic technologies, the need to respect the autonomy of the patient in either choosing for or declining genetic testing is crucial to the consent process.

In the domain of clinical genomics, the informed consent process is optimally a continuing conversation between healthcare professionals with expertise in the clinical phenotypes in question as well as the appropriate application of predictive and targeted genetic testing. The informed consent process for clinical exome or genome sequencing and large panel genetic testing begins prior to ordering the test and includes conversations about patients' preferences for learning

Genomics in the Clinic. https://doi.org/10.1016/B978-0-12-816478-5.00010-8

results of secondary or incidental findings (Sections 7.2, 10.3). Then, during the disclosure and discussion of results, which is to include negative findings, the potential for future reanalysis of data and their preference for recontact is again raised.

For children and individuals with limited cognitive capacity, other responsible individuals have to be involved in the consent process, such as parents, other recognized or designated family members or legal guardians, or other healthcare surrogate decision-makers. While assents are not mandatory in many countries, they are generally legally required for minor children participating in clinical research in the United States, but not for clinically indicated procedures, testing, or treatments. However, an ethical process in the clinical setting includes some degree of participation by most patients even if the surrogate party is the legal decision-maker (Section 10.1.2).

Genomic testing becomes increasing complicated when parental trio testing or sibling testing is done (Section 10.1.3). Further, disclosure of actionable findings to extended members of a family for purposes of anticipatory reproductive or other healthcare decision-making raises the complexity to an even higher level. Offering testing to extended family members may raise difficult issues of genetic privacy, identity, and potential assignment of blame for the patient's condition or other forms of shaming and stigma. Because psychosocial factors can emerge with some individuals in the context of targeted and genomic testing, participation of a clinician closely familiar with the patient's situation and skilled in providing comprehensive support to patients and families may be beneficial if included in the conversation.

While the informed consent process is typically conducted in person, the means of communication can vary and include telephone or telemedicine video conferencing. The remote modes of communication require, if at all possible, the presence of a knowledgeable and supportive clinician with the patient and family when indicated. The clinician may take the role of facilitating understanding, reframing the communication, ensuring that the patient and/or other decision-makers are heard by the geneticist. Supplementary genomic education via written material, the internet, or video in the clinic has the potential to aid in decision-making in advance of finalizing whether to provide consent.

Genomic testing is still at an early stage with uncertainty about the correlation between a large assortment of genetic variants and often highly pleiomorphic phenotypes as well as limited population diversity in databases. The informed consent process must acknowledge this uncertainty as well as the potential of not receiving a diagnosis for the condition that testing was sought, or having results reinterpreted and revised as scientific knowledge expands. In addition, prior to consenting, competent patients and/or surrogates need to be informed of the possibility for disclosure of "secondary" or "incidental" findings (findings that are unrelated to the purpose for testing, Sections 7.2, 10.3) that may assume a larger significance than warranted in the minds of some patients and/or their decision-makers or extended family members, while underappreciated by others. Individuals have a right to refuse to know aspects of their incidental or secondary findings, including if these individuals believe that knowledge will be more distressful than beneficial to them. A particularly ethically, psychologically, and potentially legally, challenging situation occurs when a patient reconsiders the desire to "know everything" and decides to opt-out of learning secondary findings just before the clinician was going to disclose actionable or important medical results. Consistent with other areas of medicine, patients may decline at the last moment; however, they must be advised if the lab report is already in their medical record. It is also important that the clinician document in the medical record that the competent patient actively declined learning about information which the clinician and lab may have deemed actionable or important.

There is always an ethical obligation on the part of healthcare providers to present, in a transparent way, the benefits versus risks of any intervention or provision of information. Patients must weigh whether or not the additional information of uncertain significance (Section 7.3.3) represents a net benefit or a net risk to them and decide accordingly what information they wish to receive. Nevertheless, there is an obligation to inform patients of the potential of new knowledge in the future which may bring significance to some of these finding, thus leaving open the possibility of future communication.

10.1.1 Preconception and Prenatal Considerations

Although there are professional guidelines in most countries for the appropriate use of genetic tests in the preconception and prenatal stages of gestation (International Society of Prenatal Diagnosis et al., 2018), for the nongeneticist healthcare professional, early consultation with genetic specialists and/or other subspecialists who have disease-specific genetic knowledge will minimize likelihood of ordering incorrect testing. Early consultation will be helpful in minimizing violations of ethical obligations to avoid doing harm to patients while maximizing the likelihood of benefiting them. As such, even in countries with resource limitations, all healthcare providers who care for women considering pregnancy or already pregnant, even in the absence of known risk factors, family history, or observed fetal anomalies, can benefit in their practices from the education and collaboration of early genetic consultation, potentially accessed via electronic communication (Best et al., 2018).

One cannot genuinely engage in an appropriate informed consent conversation with the patient without sufficient skills in pedigree assessment (Section 3.2.1) and risk assessment. Without having genetic expertise, failure to obtain accurate and complete data can result in poor or incorrect testing strategies which can result in significant misinformation for the patient with resultant medical harm as well as psychological harm (Section 11.18.2).

Common errors in the counseling of patients around prenatal diagnoses (Section 4.4.2) which can generate ethical errors of practice include (Bianchi and Chiu, 2018): failure to advise about limitations of many noninvasive screening approaches (Section 4.4.1), i.e., that the screening is not diagnostic; and failure to inform about the specific information which can be acquired from testing, and that some results may be indeterminate or reflective of an etiology unrelated to the health of the fetus (such as maternal malignancy).

If counseling is inadequate, women may not be informed of the need to confirm positive screening tests with more invasive yet diagnostic procedures such as chorionic villus sampling or amniocentesis (Section 4.4.2). An especially serious example of a counseling and diagnostic error would be pregnancy termination on the basis of a false-positive test for trisomy 18 on noninvasive prenatal screening (Section 4.4.1) without confirmatory tissue diagnosis. There is a common misunderstanding or erroneous belief that abnormal findings on prenatal evaluation should always be followed by strong advice for pregnancy or termination and can be seen by the lay public as well as unsophisticated healthcare providers as an unethical form of paternalism. Rather, the healthcare provider's central ethical obligation is to provide accurate information in an unbiased manner which respects the moral valuation and psychological framework of the individual gestating mother and her family about the goals of their family planning. Failure to do so is a breach of the healthcare provider's obligation to act in the best interests of the mother and her fetus.

Should there be a pre- or postnatally confirmed, genetically determined, pathological process, there may be a desire on the part of the mother or other family members for genomic evaluation of already born siblings. The goal of such evaluations is usually to identify carrier or other disease states to enable early interventions for the older children. However, phenotypic expression in siblings may be subtle, and pleiomorphic in nature. Therefore, this sort of clinical evaluation should be referred to medical geneticists or to subspecialists with specific expertise in the diagnosis and management of certain diseases, particularly for rare diseases. An example might be the fetus or neonate suspected to have congenital central hypoventilation syndrome, an extremely rare dysregulation of breathing control. Because of the rarity of this condition, counseling and evaluation of family members should be carried out by one of the few pediatric pulmonologists who are acknowledged experts in this disease. Inappropriate counseling and misdiagnosis has delayed treatment with serious consequences for the overall neurocognitive development of children with this disorder and other very rare diseases.

Lack of equitable and timely access to appropriately indicated genetic testing is a significant social justice issue. Even if competent genetic consultation is available and can be done, testing remains expensive, and particularly for patients living in countries without universal health coverage, many families lack the financial means to pursue recommended testing in the absence of insurance approval or other financial healthcare benefits which include genetic testing. Certainly, the informed consent process must include disclosures about the potential out of pocket costs so that families may consider what options they have for either seeking other sources of funding or for foregoing testing completely (Section 6.6).

Trained and certified genetic experts may be rare in some geographic locations. This scarcity of expertise also impacts negatively access to clinically indicated genetic assessment and testing. In such circumstances, primary care providers may have a moral obligation to obtain for themselves basic education and training in first steps in anticipatory genetic diagnosis as well as to develop strong collaborative liaisons with regional genetic experts for further guidance, as telemedicine becomes more feasible across the globe.

10.1.2 Pediatric Considerations

In recognition of the predictive power of sequencing information and significant amount of uncertainty in findings generated as the technology accelerates into pediatrics, the ethical, legal, and psychosocial implications of genetic testing in children and adolescents are especially notable and were reconsidered jointly by the American Academy of Pediatrics (AAP) and the American College of Medical Genetics and Genomics (ACMG) (Ross et al., 2013) and then by the American Society of Human Genetics (ASHG) (Botkin et al., 2015). Followed in part by Canada, the United Kingdom, and several other countries, importantly they both:

1) highlighted continuing challenges throughout the informed assent process that includes parental decision-making based on "the best interests of the child," (Bush, 2014) and
2) set policy permitting some flexibility with predictive testing in children for adult-onset conditions, such as if applied to diminish significant psychological distress or for facilitating important family planning decisions, provided the child's interest is not negatively impacted.

In 2016, the American Academy of Pediatrics issued a stronger statement to their 1995 policy (reaffirmed consistently) with the aim of boosting implementation of their original concept: the informed consent process for children and adolescents in the clinical setting includes both parental permission as well as the assent of children 8 years of age and over who are developmentally capable cognitively and psychologically to understand basic concepts of any procedure to be undertaken. This allows children and adolescents cognitively capable to express their viewpoints, their understanding of their own health status, as well as any concerns they may have (Katz et al., 2016).

With the increased application of comprehensive genome-based technologies in the pediatric patient population across the globe, an interprofessional workgroup with bioethics expertise developed a statement with the American College of Medical Genetics and Genomics to parallel the best interest standards of the American Academy of Pediatrics and to emphasize engagement of the older child—adolescent in conversation about relatively distinct features of exome or genome sequencing so they may share their feelings and opinions (Bush et al., 2018a). Important points regarding some unique aspects of genome-based testing specific for the older pediatric patient to consider, discussed in age and developmentally appropriate language, include: (1) Some results have no medical relevance in childhood and most likely will not be reported to the pediatric patient but might be disclosed to the parent or legal guardian; (2) information about potential adult-onset conditions not shared by the parent might be accessible once the child reaches 18 years, if the report placed in a medical record will be still available. Several other points are common to adults as well, such as: (1) An answer to the medical question being tested for may not be found; (2) unanticipated information may be given which may not be related to the reason being tested; (3) some testing results may have uncertain or unknown significance because there is not sufficient information so they are not yet well understood; (4) new information is being learned each day and therefore interpretation of test results may change over time.

While active engagement and assent are optimal in pediatric clinical care, that is not standard practice for many countries. The American College of Medical Genetics and Genomics notes there are circumstances when these conversations are not attainable in the clinical setting, such as when parents prohibit these conversations citing discordant cultural, religious, or personal values. This tends to cause moral distress for the clinical team who should still strive to foster some level of dialogue, although it may be a lengthy process. Parents should be cautioned, however, that trust can be eroded if children or adolescents ultimately discover that they were excluded from all facets of an informed consent process regarding a comprehensive test on their DNA.

See Sections 11.17.1 and 11.18.1 for examples of ethical challenges surrounding predictive *BRCA* and other cancer gene testing, particularly when involving minors (Bush et al., 2018a; Fenwick et al., 2017). As with adults, the ethical implications are minimized if identification of a variant in an affected relative is possible to enable targeted testing to be done rather than more comprehensive panels which brings with it the potential added complexity of secondary/incidental findings (Section 7.2).

10.1.3 Trios, Siblings

While many genetic tests are done solely on one individual, for broad genome-based testing, such as exome sequencing, it is optimal, if possible, to send family member samples to serve as reference samples. This is most often done as a trio (mother, father, and child). This allows identification not only of genetic changes that are de novo in a child but of variants that may be inherited. Although parents may be asymptomatic, given the variability in the presentation of many genetic disorders, sometimes potential variants with medical consequences may be discovered in them as well. Ethical complexities can ensue, such as consanguinity (Rehder et al., 2013) (Sections 3.2.4, 10.3.3), misattributed parentage (Eno et al., 2019) (Section 10.3.2), and other issues of inheritance; these need to be addressed with each blood relative being tested, in advance of the testing during the informed consent process, as well as separate counseling during the return of findings.

An example of the value and ethical challenges of parental trio exome testing: Ahmad is a 6-year-old boy with seizures and a vascular malformation in his brain. He had trio exome sequencing (Section 6.3.1) which included his mother and father. Ahmad was identified as having a variant of uncertain significance in a gene associated with vascular malformations that typically do not present until the third or fourth decade of life, but can sometimes present in children. Ahmad's father, who has been healthy and never had brain imaging, was also identified as carrying this variant. This result, although classified as a variant of uncertain significance (Section 7.1.5), raises numerous questions for the father. Should the father be offered further diagnostic evaluation as well as clinical follow-up for any asymptomatic vascular malformation which could lead to CNS hemorrhage? A positive result on workup of the father could result in changing the child's result status. What consequences are there for the father regarding obtaining life insurance or long-term care insurance, particularly relevant in countries lacking universal coverage?

In cases when sibling testing is clinically appropriate, the ethical justification rests on the benefits vs. risks of the testing for the sibling. There are two circumstances of clinical appropriateness of sibling testing: (1) to determine if the sibling shares the same genetic abnormality as the proband and may benefit from further clinical evaluation and treatment; (2) to provide insight as to whether the proband has a shared genetic defect or has a de novo variant if parent(s) is not available or child is adopted. The benefits versus risks for the sibling are clear for situation one, where a previously unsuspected condition emerges which may be amenable to anticipatory treatment. For situation 2, the benefit would be the reassurance that the sibling does not share the proband's genetic condition. A deeper bioethical justification rests on the benefits associated with sibling relationships and the value of the unaffected sibling (even if a minor) being able to benefit his/her sibling.

The following is an example of sibling testing: Ryan, a 3-year-old boy with autism spectrum disorder, came to clinic with his mother and twin brother, Jason, who also has autism. Their mother explained that Ryan is much more severely affected and notes that although both had significant early intervention and school based services, Jason experienced remarkable progress and is now talking in full sentences, yet Ryan remained "in his own world" with no words. Their mother wanted to understand if there was an underlying genetic explanation. The geneticist recommended an autism-specific gene panel (Section 6.2.3) containing approximately 2000 genes and explained that it is best done with family samples since the panel is so large. Ryan's mother clarified that her wife was Ryan's biological mother and they used a sperm donor to conceive the pregnancy. Because that testing is usually offered as a trio, Jason could be used as the other reference sample. The mother agreed, and testing revealed a variant in Ryan not present in either his mother or brother. The gene is specifically known to cause significant speech delay. His mother acknowledges this explanation makes them understand the differences between their sons better and this knowledge is of benefit to the entire family.

10.1.4 Variants of Uncertain Significance

When doing genetic testing, results often are not clear cut, with many variants reported by the clinical sequencing lab as indeterminant regarding whether they are disease causing or pathogenic, or benign. These are referred to as variants of uncertain significance (VUSs) (Section 7.1.5). The clinical significance of a VUS requires careful interpretation and often will remain uncertain until further evidence emerges, especially by sharing clinical-phenotypic information (Bush et al., 2018b). The psychological implications of VUS are notable for some patients and families due to varying tolerances for uncertainty of genetic outcomes. Therefore, astute clinicians will recognize and anticipate the need for necessary psychosocial and clinical support with continuity of care for patients and families.

The case of Anya is an example. She was a 22-year-old presenting to the genetic clinic for evaluation of hypermobility after her third shoulder dislocation and second ankle dislocation. A connective tissue gene panel was sent and revealed a variant of uncertain significance in a gene associated with aortic aneurysm. Anya was sent for cardiac evaluation and echocardiogram, which were found to be normal. Anya asked what this means for her future. Would Anya need echocardiograms and ongoing follow-up with cardiology? Could there be other family members at risk and should her parents and siblings be tested? Genetic counseling can be very helpful to the patient and her family in considering these many aspects.

10.1.5 Data Sharing, Biobanking

There are significant potential benefits and potential risks associated with exome and genome data sharing and DNA biobanking. Cumulative genomic data with clinical phenotypes are invaluable for ascertaining the genetic variations which may account for the pleomorphisms of some complex and puzzling syndromes. Furthermore, accumulation of case reports along with their variants of uncertain significance ultimately lead to determination if they are pathogenic or benign and identification of new diseases (Bush et al., 2018b). However, the risks of data sharing and biobanking include the potential loss of genetic identity, privacy, and loss of autonomous control of individuals' genetic information especially with respect to potential commodification of that material for biological products when the benefits of these products are not distributed equitably. There are also potential concerns for some individuals regarding commodification of regenerative medicine biological products which do not benefit the donors of the genetic materials. Donors of genetic materials may confer benefits to others in addition to the greater society without the prospect of reaping any equivalent benefits from themselves. This would violate the accepted standards of ethical clinical and health research activities where risks of interventions as well as the benefits thereof should be shared equitably.

Existing professional guidelines, including those from the American College of Medical Genetics and Genomics which pertain to responsible collection, sharing, and dissemination of genetic data (Bush et al., 2018B), are based on beneficence for the patient as well as future patients, with the premise of fair distribution of both risks and benefits in clinical and clinical research activities.

An example of the benefits of data sharing: Serena, a 31-year-old with neurologic findings, had exome sequencing (Section 6.3.1) and two variants were found in an autosomal recessive (Section 2.7.1) gene for which there was no reported disorder caused by mutations in this gene. The physician who ordered the testing considered this a potential diagnosis due to the suspected role of the gene; however, there was not enough scientific evidence at that time to confirm the diagnosis. After confirming that Serena had agreed in the informed consent process to the sharing, Serena's genetic information was entered into a computer database meant to match patients around the globe with candidate genes (for example, https://www.matchmakerexchange.org/*). One year later, the physician was contacted to collaborate and share data about the Serena's clinical presentation. After comparing information among physicians, it was determined that multiple individuals with genetic variants in this gene had similar anatomical abnormalities plus medical and developmental complications. They were then all requested, through a research protocol's informed consent process, to allow biobanking of additional DNA specimens so an international consortium of scientists could further study this disorder in the future, in order to establish which variants are pathogenic through functional studies, and determine phenotype–genotype correlations for a research publication.*

10.2 Cost and Insurance Concerns

While the cost of some aspects of clinical exome sequencing has been greatly reduced, a considerable number of skilled, expert hours are still required, such as the informed consent process, ongoing counseling, and detailed analysis of results with correlation to gene variant patterns and phenotypes. Therefore, the costs of such testing remain beyond the financial means of most patients and their families, and especially since many insurers in the United States to date do not cover, or place significant limitations on, exome sequencing as a benefit. Patients and families bear financial and time factor burdens of a diagnostic odyssey using sequential rounds of less-comprehensive testing, and greater ethical consideration is needed given an oft-times greater diagnostic yield for phenotypically complex children.

And even if prior authorization for testing is given, authorization doesn't always guarantee coverage, particularly in countries without universal health coverage. Given the high cost of testing and lack of insurance coverage too, many times patients find themselves in a position of not being able to get genomic testing recommended by a clinician. This is exemplified in Section 11.18.2 in which insurance would not cover preconceptual genetic testing and in vitro fertilization for the known family history of ornithine transcarbamylase deficiency, an enzyme disorder of the urea cycle, and the patient had to forgo those preconception procedures because the cost surpassed her finances. Cognizant that the cost of prenatal genetic testing is significantly less, she had to select less optimal timing based on her values and await results from her chorionic villous sampling (Section 4.4.2).

An example in which insurance rules and burdensome copays compromise best care involves José, a 23-year-old with increasingly refractory seizures who tried multiple medications yet experienced significant medication side effects. Reluctant to keep changing medications, he asked if further genetic testing might help determine if there is an underlying genetic etiology for his seizures. The physician recommended a seizure gene panel (Section 6.2.3). Though an adult, José was still listed as a dependent on his parent's medical insurance plan due to his disability. The request was initially denied but, after appeal, prior authorization was obtained. Concerned about cost, José called the insurance company and was informed that even though testing is covered they still must pay a $2500 deductible. José's parents could not afford this, especially considering high copays they already pay for his hospitalizations and medications. He asked the doctor if there were any other options; unfortunately, José learned no options exist given his circumstances in the United States. However, in countries with universal health coverage, these fees may not be relevant.

10.3 Secondary and Incidental Findings

Another significant ethical and psychosocial challenge regarding exome and genome sequencing centers are the discovery and reporting of secondary findings, those unrelated to the clinical indication for testing and typically unexpected. The

American College of Genetics and Genomics curates some findings that they deem "medically actionable" (Kalia et al., 2017), and when disclosed to the patient and/or pertinent family member should include recommendations for appropriate counseling, medical follow-up, and referrals (Sections 7.2, 11.4.5). The American College of Genetics and Genomics also provides guidelines about the identification and management of more potentially "delicate" secondary findings, such as consanguinity (Rehder et al., 2013) and misattributed paternity (Eno et al., 2019), both explored below.

The President's Commission for Bioethics (Presidential Commission, 2013) and American Society for Human Genetics (Botkin et al., 2015) are other sources of guidance that also stress the critical necessity to discuss these difficult, often unanticipated findings that may be found, particularly during family-based (duo or trio, patient with one parent or both parents, respectively) sequencing. Besides the need to explicitly address secondary findings during the early phase of the informed consent process to facilitate understanding and decision-making as to whether or not to pursue exome and genome sequencing, should secondary genomic findings be revealed. It is imperative to provide good counseling.

Although psychological distress has long been argued as a reason not to permit predictive testing of adult-onset sequencing for children or even for younger adults, recent reviews of the literature found serious psychological distress to be uncommon. To date, empirical research has largely not supported long-term negative psychological impact; however, more research is needed to better identify who may be at most risk, including for secondary findings (Wakefield et al., 2016).

10.3.1 Identity

Individuals develop a narrative about themselves, with perspectives constructed based on varying levels of reality that we use to define who we are. There is the potential that after genetic testing one's identity can be in turmoil, raising existential angst (e.g., So who am I?).

Consider the following example: Jennifer is a 14-year-old girl presenting for initial evaluation due to primary amenorrhea and scant pubic and axillary hair. Her mother reported she was born full term and had no significant health problems or learning difficulties. Jennifer's two older sisters had their first menstrual periods at age 12, and her mother assumed Jennifer was a "late bloomer" and was surprised to be referred for further evaluation. At that visit an abdominal ultrasound was ordered and revealed an absent uterus. Genetic testing was sent and revealed a karyotype of 46,XY. A geneticist was consulted and, considering androgen insensitivity syndrome (Section 11.8.1), sent further testing, which confirmed the diagnosis. Sensitive counseling followed, explaining to Jennifer and her parents that Jennifer was genetically male, with some physical traits of a female. Recommendations were made for medical follow-up with other specialists, but also for the teen and family to see a psychologist as well.

10.3.2 Misattributed Parentage

Because the consent for language in commercial labs and academic institutions varies significantly regarding (1) whether families will be informed if duo or trio testing detects misattributed parentage and (2) whether these secondary findings will be reported in the medical record or final lab report, standardization would benefit all involved (Eno et al., 2019). It is also critical to clearly articulate expectations prior to test requisition, since some women need to consider potential harms and may choose to decline testing, particularly if they feel at risk for partner abuse following disclosure. Current recommendations from professional organizations differ whether to report misattributed paternity at all, or if so, to whom; some disclose it confidentially to the mother only, while some routinely report misattributed parentage to both parents, to the lab and medical professionals involved as well as the families.

Ethical complexities involving misattributed paternity are challenging for clinicians as well as the families, as exemplified in this case scenario: Jasmine was a 1-month-old girl identified on newborn screening as having sickle cell disease. She presented for evaluation and her parents asked about recurrence risks. Because sickle cell disease is an autosomal recessive disorder, each parent is most likely a carrier of a variant in the gene, and the risk is 25% for future pregnancies. The father requested carrier testing. As they were leaving, the mother asked to speak in private and then explained that she did not want to go through with carrier testing because her husband was not Jasmine's biological father and she did not want him to find out. Was it acceptable for the healthcare professional to deliberately lie in this circumstance? On the other hand, if the social father is not the biological father, is there any particular obligation to disclose to him that he is not the father? Optimally there are guidelines set forth by the medical institution or a given country's professional organization. In any event, these concerns need to be discussed with the mother and preferably also clinical peers and/or ethical professionals.

10.3.3 Consanguinity

Exome sequencing (Section 6.3.1) and other genome-wide assays (Section 6.3.2) can detect long stretches of homozygosity (Section 2.7.1) on chromosomes, thus lacking the variation typically expected in individuals. When there are lengthy segments without heterozygosity (human variability) in many or all of the chromosomes, the findings are typically interpreted as reflecting that the parents are close biological relatives, termed "consanguinity" (Fenwick et al., 2017). Importantly, extensive regions of homozygosity in a minor should raise a red flag in the clinician for a suspicion of incest. Suspected child sexual abuse is a mandatory reportable responsibility for physicians and other clinicians or teachers in the United States and many other countries. However, with child adoptions or gamete donation for assisted reproduction, inadvertent consanguinity should also be considered. Consanguinity (Section 3.2.4) in consenting adults is not, however, a reportable duty for clinicians in the United States, even if deemed illegal in certain regions. In an increasingly global society, we see many families from societies and cultures where cousin consanguinity is not only frequent but expected. Therefore, great sensitivity and a nonjudgmental demeanor must be employed in the consent process so as to elicit trust and respect within the relationship between provider and family.

A case example of these principles: Kwame is a 7-year-old boy with autism. On a detailed history, his mother and father both noted they did not know their family history as they were both adopted as infants. They related they come from the same small town and met in high school. It is explained that a chromosomal microarray (Section 6.1.3) is a genetic test looking very broadly across the chromosomes for missing or extra pieces or areas of the chromosomes that are exactly the same, known as areas of homozygosity. The parents were eager to know if there was an underlying genetic etiology and wanted to proceed with the recommended genetic testing. The microarray returned with no deletions or duplications but showed areas of homozygosity at 25%. This level of homozygosity indicates a close familial relationship between the parents. Since they were both adopted, there is concern that they could be half-siblings. Further sensitive discussion is indicated in follow-up.

10.4 Genetic Information Nondiscrimination Act

To address significant concerns that individuals undergoing genetic-genomic testing in the clinical or research setting may face discrimination by employers and/or health insurers based on their genetic information, a federal law was enacted in the United States in 2008, the Genetic Information Nondiscrimination Act (GINA) (see further readings). However, life insurance and long-term disability insurance are not protected by GINA, and notably, businesses with less than 15 employees do not have to abide by GINA terms. Moreover, GINA does not offer protections for preexisting conditions. The possibility for misuse of genetic information by insurers and employers has not been adequately resolved by GINA in the United States and is a particular ethical concern for most countries lacking some form of a national universal insurance policy. This also poses significant ethical and psychosocial implications in some countries with national coverage who mandate testing and patients have no choice, such as China, hence precluding voluntariness from the consent process. Although in 2013 the United States finalized their HIPPA ruling to further clarify that health information protections should include genetic information to be in compliance with GINA, there remains insufficient protections in our society and unethical discriminatory practices are rampant throughout the globe.

10.5 Reinterpretation of Findings and Patient Recontact

Given that genomic technologies and their application in the clinical and research setting are evolving far faster than our ability to understand the significance of the information, many ethical challenges currently arise regarding patient recontact after revision of genomic test results. Interpretation and classification of variants, ranging from highly pathogenic to benign, and inclusive of a significant number of unknowns, often shift over time as bioinformatics, curation, and population databases iteratively get refined. Results from sequencing provided on one day may be reanalyzed with its significance reinterpreted a month or year or decade later. A very real ethical and legal conundrum exists as to how often reinterpretation ought to be done, who has responsibility for reviewing seemingly daily updates to the scientific literature that may modify findings previously reported, and when and how is the revised information relayed back to clinician and patient if at all? For example, if a patient is informed genetic testing results show a variant of uncertain significance (VUS), because no mention had appeared in the literature search, but 3 years later a journal reports on increased lung cancer in four individuals with that variant, many questions instantly arise: Who will update the ordering clinician? How will the patient(s) learn they need surveillance, especially if they moved without forwarding you their address?

To emphasize concerns with these implications in the context of diagnostic sequencing in the clinical setting, recently, the American College of Medical Genetics and Genomics has provided points to consider policy statement and laboratory guidelines regarding reinterpretation of clinical genomic information (David et al., 2019). Recommendations include that (1) the clinical laboratory inform the ordering physician of new or revised gene-variant-disease discovery even if a referral was made to a medical geneticist for counseling of test results and (2) "reasonable effort" ought to made by the ordering physician to recontact the patient (David et al., 2019). The concept of shared responsibilities in decision making, was also raised as an essential component for effective patient recontact, notably the importance of patient's updating their contact information on a regular basis.

Shortly thereafter, the American Society of Human Genetics issued guidelines in a policy statement primarily focused in the research domain (Bombard et al., 2019) and a policy commentary from an NIH grant consortium exploring legal challenges appeared in Science (Cousin-Frankel, 2019). What is the standard of care in part shapes legal risk. That too is evolving, as depicted recently when the South Carolina Supreme Court ruled for the first time that a testing laboratory may be deemed a "healthcare provider" in response to a case. Simplified here, the case involved the death of a boy with a severe seizure disorder whose mother accused the lab of failing to appropriately reclassify the boy's particular pathogenic variant as Dravet syndrome, reportedly known to clinical laboratory's key scientist but not disclosed at the time. Lacking that information resulted in antiepileptic medication prescribed to the boy that was contraindicated for his specific mutation, for which the family claims directly led to the son's fatality. Ethical and legal scholars predict many other cases will follow in an attempt to establish precedent for standard of care as it exists today with constantly revised databases involving genome-based testing.

References

ACOG Committee on Ethics, 2009. ACOG Committee Opinion No. 439: informed consent. Obstet. Gynecol. 114, 401–408. PMID 19623005.

Best, S., Wou, K., Vora, N., Van Der Veyver, I.B., Wapner, R., Chitty, L.S., 2018. Promises, pitfalls and practicalities of prenatal whole exome sequencing. Prenat. Diagn. 38, 10–19. PMID 28654730.

Bianchi, D.W., Chiu, R.W.K., 2018. Sequencing of circulating cell-free DNA during pregnancy. N. Engl. J. Med. 379, 464–473. PMID 30067923.

Bombard, Y., Brothers, K.B., Fitzgerald-Butt, S., Garrison, N.A., Jamal, L., James, C.A., et al., 2019. The responsibility to recontact research participants after reinterpretation of genetic and genomic research results. Am. J. Hum. Genet. 104, 578–595. PMID 30951675.

Botkin, J.R., Belmont, J.W., Berg, J.S., Berkman, B.E., Bombard, Y., Holm, I.A., et al., 2015. Points to consider: ethical, legal, and psychosocial implications of genetic testing in children and adolescents. Am. J. Hum. Genet. 97, 6–21. PMID 26140447.

Bush, L., 2014. In the best interest of the child: psychological and ethical reflections on traditions, contexts, and perspectives in pediatric clinical genomics. Am. J. Bioeth. 14, 16–18. PMID 24592832.

Bush, L.W., Bartoshesky, L.E., David, K.L., Wilfond, B., Williams, J.L., Holm, I.A., et al., 2018a. Pediatric clinical exome/genome sequencing and the engagement process: encouraging active conversation with the older child and adolescent: points to consider-a statement of the American College of Medical Genetics and Genomics (ACMG). Genet. Med. 20, 692–694. PMID 29565417.

Bush, L.W., Beck, A.E., Biesecker, L.G., Evans, J.P., Hamosh, A., Holm, I.A., et al., 2018b. Professional responsibilities regarding the provision, publication, and dissemination of patient phenotypes in the context of clinical genetic and genomic testing: points to consider-a statement of the American College of Medical Genetics and Genomics (ACMG). Genet. Med. 20, 169–171. PMID 29323668.

Couzin-Frankel, J., 2019. Medical DNA Sequencing Leads to Lawsuits and Legal Questions [Online]. Science Magazine. Available: https://www.sciencemag.org/news/2019/04/medical-dna-sequencing-leads-lawsuits-and-legal-questions [Accessed May 2019].

David, K.L., Best, R.G., Brenman, L.M., Bush, L., Deignan, J.L., Flannery, D., et al., 2019. Patient re-contact after revision of genomic test results: points to consider-a statement of the American College of Medical Genetics and Genomics (ACMG). Genet. Med. 21, 769–771. PMID 30578420.

Eno, C., Bayrak-Toydemir, P., Bean, L., Braxton, A., Chao, E.C., El-Khechen, D., et al., 2019. Misattributed parentage as an unanticipated finding during exome/genome sequencing: current clinical laboratory practices and an opportunity for standardization. Genet. Med. 21, 861–866. PMID 30214068.

Fenwick, A., Plantinga, M., Dheensa, S., Lucassen, A., 2017. Predictive genetic testing of children for adult-onset conditions: negotiating requests with parents. J. Genet. Counsel. 26, 244–250. PMID 27680566.

International Society for Prenatal Diagnosis, Society for Maternal Fetal Medicine & Perinatal Quality Foundation, 2018. Joint position statement from the International Society for Prenatal Diagnosis (ISPD), the Society for Maternal Fetal Medicine (SMFM), and the Perinatal Quality Foundation (PQF) on the use of genome-wide sequencing for fetal diagnosis. Prenat. Diagn. 38, 6–9. PMID 29315690.

Kalia, S.S., Adelman, K., Bale, S.J., Chung, W.K., Eng, C., Evans, J.P., et al., 2017. Recommendations for reporting of secondary findings in clinical exome and genome sequencing, 2016 update (ACMG SF v2.0): a policy statement of the American College of Medical Genetics and Genomics. Genet. Med. 19, 249–255. PMID 27854360.

Katz, A.L., Webb, S.A., Committee on Bioethics, 2016. Informed consent in decision-making in pediatric practice. Pediatrics 138, e20161485. PMID 27456510.

Presidential Commission, 2013. Presidential Commission For the Study of Bioethical Issues: Anticipate and Communicate: Ethical Management of Incidental and Secondary Findings in the Clinical, Research, and Direct-To-Consumer Contexts [Online]. Washington,D.C. Available: https://bioethicsarchive.georgetown.edu/pcsbi/node/3183.html [Accessed May 2019].

Rehder, C.W., David, K.L., Hirsch, B., Toriello, H.V., Wilson, C.M., Kearney, H.M., 2013. American College of Medical Genetics and Genomics: standards and guidelines for documenting suspected consanguinity as an incidental finding of genomic testing. Genet. Med. 15, 150–152. PMID 23328890.

Ross, L.F., Saal, H.M., David, K.L., Anderson, R.R., 2013. American Academy of Pediatrics, American College of Medical Genetics and Genomics Technical report: ethical and policy issues in genetic testing and screening of children. Genet Med 15, 234–245. PMID 23429433.

Wakefield, C.E., Hanlon, L.V., Tucker, K.M., Patenaude, A.F., Signorelli, C., Mcloone, J.K., et al., 2016. The psychological impact of genetic information on children: a systematic review. Genet. Med. 18, 755–762. PMID 26741411.

Further Reading

Genetic Discrimination, 2019. Genetic Discrimination [Online]. National Human Genome Research Institute. Available: https://www.genome.gov/about-genomics/policy-issues/Genetic-Discrimination. [Accessed April 2019].

Chapter 11

Case Scenarios

Abstract

Referrals to genetic clinics and inpatient management and consultation encompass the practice of clinical genetics. The interactions among the geneticist, genetic counselor, primary care physician, and specialists involve a complex process of comprehensive integration of family history, review of medical records and past medical and surgical history, physical examination, and genetic testing. Based on genetic test results, patients and potentially other family members are counseled and recommendations are made for their care. This chapter presents more than 100 clinical cases of patient and family referrals to genetics, including a spectrum of the most common presentations and diagnoses. Cases give examples of patients with symptomatology of various organ systems or topics such as prenatal diagnosis, newborn screening, clinical genetics of syndromic and nonsyndromic conditions, multifactorial conditions, cancer, and pharmacogenetics. Also issues related to the practice of direct-to-consumer testing and ethics are discussed.

This chapter includes 102 cases, organized by topics in genetics, paralleling the first 10 chapters, or by primary body system. These cases describe common scenarios for referral to Genetics, written by geneticists, genetic counselors, and/or specialists, representing their personal clinical experience. Each case is presented initially from the primary care provider or specialist point of view, describing The Case, and then The Issues with differential diagnosis and a preliminary evaluation and workup. There is The Referral to Genetics (geneticist and/or genetic counselor), describing the genetic evaluation, including family history, genetic counseling, testing, results, management and recommendations, any implications for family members, caveats (if any), and outcome(s). Both the clinical and genetic diagnoses are listed at the end of each case. Because of the limited number of genetic centers, in general, the genetic workup for each case can be used as a recommended evaluation and testing guide for the primary care provider or specialist. Many laboratories that perform genetic testing will be able to provide remote genetic counseling to interpret results. In conjunction with these cases, there are lay descriptions and images of genetic concepts, tests, and results in Chapter 12 that can be used for counseling families. More recently, genetic centers are able to provide remote care which also may be useful. We hope that these cases are helpful in considering genomic evaluation and testing in the clinic.

Section 11.1

Prenatal Cases

Case 11.1.1

Prenatal Cases: Positive Carrier Screening in Pregnancy

Antonie D. Kline

Harvey Institute for Human Genetics, Greater Baltimore Medical Center, Baltimore, MD, United States

Abstract

Prenatal carrier screening early in pregnancy reveals the mother to have sickle cell trait carrier and to be a carrier for cystic fibrosis, and the father to be a carrier for spinal muscular atrophy.

The Case

Angie, a 25-year-old female, and her partner, Roger, are seen by you, her obstetrician, for their first pregnancy at 6 weeks gestation. They have been wondering about any disorders for which their fetus could be at risk. You note Angie's African American ethnicity and ask about her hemoglobinopathy status. Angie says that she thinks she has sickle cell trait, but is not sure. You ask if Angie and her partner would be interested in carrier screening. They say they are interested in all screening available. You recommend initial lab work on Angie, including a hemoglobin electrophoresis, as well as expanded carrier screening (Section 4.1). You also recommend noninvasive prenatal screening (Section 11.1.4). The hemoglobin electrophoresis shows that Angie has sickle cell trait. Her carrier screening returns with a variant in the gene for cystic fibrosis as well as the pathogenic variant or mutation known to be associated with sickle cell disease (Sections 2.7.1, 7.1.3).

The Issues

- **Is the developing fetus at risk for sickle cell disease?** Sickle cell disease is an autosomal recessive condition (Section 2.7.1) and the children are at risk only if both parents are carriers of the mutation, which causes the formation of hemoglobin S (SA). If Roger has sickle cell trait carrier (SA), their children would be at risk for sickle cell disease (SS) or if he is a beta-thalassemia carrier, their children would be at risk for sickle-thalassemia disease (SA2) (Section 4.1).
- **What is the risk to the developing fetus for cystic fibrosis?** If Roger is found to be a carrier for a gene variant (Section 2.4.1) in the gene for cystic fibrosis, their children would also be at risk for this. Angie and Roger are starting to get very nervous about their baby. You refer them for prenatal genetic counseling.

Referral for genetic counseling is helpful to review genetic testing results. The genetic counselor has more time available in which to discuss ongoing studies and also can help the couple evaluate what they want to find out about the current pregnancy, and what they might do with future results.

Genomics in the Clinic. https://doi.org/10.1016/B978-0-12-816478-5.00011-X

- **How accurate are these results?** Hemoglobin electrophoresis is a diagnostic test that will determine the risk for any hemoglobinopathy disorder, and include hemoglobin A2. Carrier screening (Section 4.1) is performed by gene sequencing (Section 6.2.2) and should pick up almost all variants in the gene DNA sequence; however, there could be variants in areas of the gene not sequenced, so it is not 100% sensitive if it is called "negative." If a variant is found, a "positive" test, then this is definitive and the individual is a carrier. The number of genes, which are tested on carrier screening, varies greatly, from several genes to hundreds depending on the panel used by a laboratory.

It is helpful to send all prenatal testing results to the genetic clinic prior to the couple's appointment in order for the genetic counselor to prepare for the visit.

The Referral

- **Genetic Evaluation:** Angie and Roger meet with the genetic counselor at 9 weeks gestation. A detailed three-generation pedigree (Sections 2.7, 3.2.1) is obtained. Angie has several cousins with sickle cell disease, but there are no other genetic conditions on either side. Both Angie and Roger are of African American ancestry. The genetic counselor notes Angie's sickle cell trait status as well as her positive carrier screening for cystic fibrosis.
- **Genetic Counseling:** Although sickle cell disease is common in the black population, cystic fibrosis is quite rare, but can occur. Both sickle cell disease and cystic fibrosis are autosomal recessive conditions (Section 2.7.1), so that Roger also has to be a gene mutation carrier in order for their child to have either condition. The genetic counselor offers expanded carrier screening (Section 4.1) to Roger for autosomal or X-linked conditions as recommended by American College of Medical Genetics and Genomics/American College of Obstetricians and Gynecologists (ACMGG/ACOG) guidelines. This expanded carrier screening will screen for hundreds of conditions, depending on the laboratory. The couple decides that Roger will have both the hemoglobin electrophoresis and the expanded carrier screening.
- **Management and Recommendations:** If Roger is found to be positive for either of these conditions, and if the couple is interested, fetal material can be tested subsequently by prenatal diagnostic testing. The genetic counselor discusses that this could be done by chorionic villous sampling (CVS), at 10–14 weeks gestation, or amniocentesis, on or after 16 weeks gestation (Section 4.4.2). The couple would want to think about the potential for positive results and whether they would want to know these and/or act upon them.

Although not in great detail, the genetic counselor would bring up for discussion with the couple the possibility of finding out that the fetus is affected. The counselor also would discuss whether the couple would want to think about continuing or terminating the pregnancy if affected, or considering adoption.

- **Results:** Two weeks later, Roger's results show that he is negative on the hemoglobin electrophoresis, indicating no increased risk for sickle cell disease in the fetus. On the expanded carrier screening, he is found to be a carrier for spinal muscular atrophy (Section 11.2.4) by having only one copy of the gene SMN1, and is not a carrier for cystic fibrosis. The genetic counselor is able to reassure the couple that their fetus should not be at any increased risk for sickle cell disease, sickle-thalassemia disease, or cystic fibrosis. The counselor also can reassure Angie and Roger about spinal muscular atrophy because all prenatal screening carrier panels assess for spinal muscular atrophy, per the ACMGG/ACOG guidelines, and Angie was not missing a copy.
- **Caveats:** The genetic counselor points out that these are all screening panels, and they will never identify 100% of the gene variants associated with all conditions, but these tests are very reassuring overall. She reminds them about the detailed prenatal ultrasound at midgestation, which will be done to assess their fetus for any physical findings.
- **Outcome:** The couple leaves, greatly reassured. Seven months later they give birth to a healthy baby girl.

The Diagnosis

Clinical Diagnosis: Patient and Partner—future pregnancies with no increased risk for sickle cell disease, cystic fibrosis, or spinal muscular atrophy.

Genetic Diagnosis: Patient—sickle cell disease, heterozygous carrier; cystic fibrosis, heterozygous carrier. Partner—spinal muscular atrophy, heterozygous carrier.

Further Reading

https://www.acmg.net/PDFLibrary/Sickle-Cell-Carrier-ACT-Sheet.pdf.

Committee Opinion No. 691: Carrier screening for genetic conditions, 2017. Obstet. Gynecol. 129, e41-e55. PMID 28225426.
Grody, W.W., Thompson, B.H., Gregg, A.R., Bean, L.H., Monaghan, K.G., Schneider, A., Lebo, R.V., 2013. ACMG position statement on prenatal/preconception expanded carrier screening. Genet. Med. 15, 482e483. PMID 23619275.

Case 11.1.2

Prenatal Cases: Maternal Diabetes Exposure

Shama Jari
Harvey Institute for Human Genetics, Greater Baltimore Medical Center, Baltimore, MD, United States

Abstract
A woman with childhood onset type 1 diabetes has elevated hemoglobin A1c and is found to have a fetus with a ventricular septal defect.

The Case

Stephanie, a 32-year-old G0 female, is seen by you, her primary care physician. She has a history of type 1 diabetes diagnosed at age 15 and is considering attempting to conceive. She asks about the potential issues associated with diabetes in pregnancy. She is currently taking insulin for glucose control.

The Issues

- **What are the risks associated with maternal diabetes?** You discuss that uncontrolled diabetes in the preconception period and first trimester is strongly associated with pregnancy loss. Miscarriage rates range from normal with very good glucose control to 44% with poor control. There is also a 4- to 8-fold increased risk of birth defects, which is largely dependent on preconception glycemic control. Associated malformations include congenital heart defects, neural tube defects, renal agenesis, and caudal dysplasia. Finally, there is also an increased risk of pregnancy complications. There is an increased risk for preterm labor, preeclampsia, IUGR, or macrosomia, and a two-fold increased risk of perinatal mortality.
- **What is the recommended course prior to conception?** Optimal glucose control prior to conception and during pregnancy is essential to minimizing the risks. You instruct the patient to optimize her glucose control prior to conceiving to reduce her risk. The blood hemoglobin A1c can be followed to assess glycemic control, with values below 6.5% indicating good control. Six months later Stephanie presents to her obstetrician at 8 weeks gestation for a new obstetrics visit. Her HgbA1c is 7.2% at that time. You inform her that her elevated HgbA1c is associated with an increased risk for miscarriage as well as congenital malformations.
- **What is the next step?** Stephanie is sent for noninvasive prenatal screening (NIPS) (Section 4.4.1) at 12 weeks, with reassuring results. In addition to a detailed fetal anatomic survey, a fetal echocardiogram is ordered in the second trimester (Section 4.3.2). The anatomy screen at 18 weeks gestation reveals a ventricular septal defect. The patient is referred for genetic counseling.

Genomics in the Clinic. https://doi.org/10.1016/B978-0-12-816478-5.00012-1

The Referral

- **Genetic Evaluation**: Stephanie and her partner Ed meet with the genetic counselor at 18 weeks gestation. The genetic counselor obtains a detailed three-generation pedigree, which is noncontributory (Sections 2.7, 3.2.1). Specifically, there is no family history of congenital heart defects. The counselor notes Stephanie's history of type 1 diabetes, management with insulin, and elevated hemoglobin A1c.
- **Genetic Counseling**: The counselor discusses that the ventricular septal defect can be associated with chromosomal anomalies, genetic syndromes, as well as maternal diabetes. A fetal echocardiogram can evaluate the heart in more detail than on a level III ultrasound. The counselor reiterates that there is an increased risk for birth defects especially with poor glycemic control prior to conception.
- **Management and Recommendations**: The option of prenatal diagnostic testing, specifically amniocentesis, is discussed for the detection of chromosomal abnormalities (Section 4.4.2). Risks and benefits are discussed, including the increased risk of miscarriage. In addition to performing a karyotype to test for chromosomal anomalies, the possibility of performing a prenatal microarray is discussed (Section 6.1.3). This testing can detect smaller changes in the chromosomes than can be detected by karyotype and may detect genetic syndromes that cannot be detected by karyotype. The couple is counseled that positive results may aid in their decision about pregnancy termination, or alternatively, may help to guide treatment postnatally. The couple elects to proceed with amniocentesis for karyotype and will reflex to microarray if karyotype is normal.
- **Results**: A fetal echocardiogram confirms the ventricular septal defect. No other abnormalities are seen. There is no contraindication to delivery at the community hospital. Follow-up echocardiogram is recommended after delivery. Amniocentesis reveals a normal female karyotype (Section 6.1.1). Microarray testing yields normal results as well.
- **Caveats**: Karyotype and microarray cannot rule out all genetic syndromes. The fetus could have an underlying condition not detectable with these tests. In addition, prenatal ultrasound cannot rule out all birth defects or intellectual disability.
- **Outcome**: The patient is reassured by her testing results. The heart defect is most likely isolated and may be related to her diabetes. During the third trimester, Stephanie's diabetic control is improved and her hemoglobin A1c is 6.9. The remainder of the pregnancy is uneventful. She goes on to deliver a baby girl at term with an apparently isolated ventricular septal defect.

If newborn exam reveals any features suggestive of a genetic syndrome, then consultation with a pediatric geneticist would be appropriate.

The Diagnosis

Clinical Diagnosis: Fetus—exposure to maternal diabetes, with ventricular septal defect.
Genetic Diagnosis: Fetus—none.

Further Reading

Goto, M.P., Goldman, A.S., 1994. Diabetic embryopathy. Curr. Opin. Pediatr. 6, 486–491. PMID 7951674.
Wilmot, E.G., Mansell, P., 2014. Diabetes and pregnancy. Clin. Med. 14, 677–680. PMID 25468858.

Case 11.1.3

Prenatal Cases: Advanced Maternal Age

Antonie D. Kline
Harvey Institute for Human Genetics, Clinical Genetics, Greater Baltimore Medical Center, Baltimore, MD, United States

Abstract
A fetus is found to have an increased nuchal translucency; noninvasive prenatal screening is positive for Down syndrome; and a chorionic villous sampling confirms trisomy 21.

The Case

Brianna, a 36-year-old G2 P1001 female, and her partner, Tyrell, are seen by you, her obstetrician, for their second pregnancy at 7 weeks gestation. They ask about all first trimester screening for any significant disorders for the fetus, including Down syndrome, but say that they would not be interested in invasive testing (Section 4.3).

The Issues

- **What are the increased risks to this pregnancy?** Women 35 years and older during pregnancy are considered to be at advanced maternal age. In addition to potential maternal and obstetrical complications, including increased incidence of pregnancy loss and difficulties with delivery, there is a higher incidence of aneuploidy (Section 2.2.1). This is due to meiotic nondisjunction (Section 4.2.2) with higher chances of abnormal chromosome number correlating with increasing maternal age (Table 4.1). Fetuses with abnormal chromosomes will often have increased mortality, growth delays and/or malformations. You review these risks with the couple.
- **What are the screening tests available during pregnancy?** Brianna is scheduled for an ultrasound (Section 4.3.1) at 12 weeks gestation at your office. You also discuss combined first trimester screening and noninvasive prenatal screening (NIPS) (Section 4.4.1). The ultrasound shows an increased nuchal translucency (Section 4.3.2) of 3.5 mm on ultrasound, which is over the 95th percentile for gestational age. The couple is very concerned.
- **How accurate are these tests?** The ultrasound is highly accurate, depending on the level of the provider. It is likely that a high-resolution ultrasound at the Prenatal Diagnostic Center will be more detailed than your in-office ultrasound. The first trimester screening tests (Section 4.4.1) can detect underlying conditions to varying degrees of predictability, but remain screening, not diagnostic, tests. Invasive prenatal diagnostic testing, which may not be covered by insurance, can assess the fetus for specific conditions, is highly accurate, and can be precisely diagnostic (Section 4.3.2).
- **What is the next step?** Because there is an increased risk for Down syndrome with the increased nuchal translucency, the couple choose to have noninvasive prenatal screening (NIPS) test, which returns positive for Down syndrome (Sections 2.2.1, 6.1.2). The couple is now extremely worried about their baby. At this point, you refer them to genetic counseling as soon as possible.

It would have been appropriate to refer for genetic counseling after the initial ultrasound, if you do not have enough time for in-depth discussion in the office and for the couple to have counseling prior to sending the NIPS. An initial genetic

Genomics in the Clinic. https://doi.org/10.1016/B978-0-12-816478-5.00013-3

counseling visit would include discussion of risks for positive testing and options for positive results, and there would not be a need for an emergency appointment. It is helpful to send all prenatal testing results to the genetic clinic prior to the couple's appointment in order for the genetic counselor to prepare for the visit.

The Referral

- **Genetic Evaluation**: Brianna and Tyrell meet with the prenatal genetic counselor at 13 weeks gestation. The ultrasound is repeated at the Prenatal Diagnostic Center, confirming the enlarged nuchal translucency (Section 4.4.2). The genetic counselor obtains a detailed three-generation pedigree (Sections 2.7, 3.2.1) which is noncontributory.
- **Genetic Counseling**: The counselor discusses that an enlarged nuchal translucency can be seen with congenital heart defects, genetic syndromes including Down syndrome, and maternal prenatal exposures to certain organisms. The counselor also discusses the results of the NIPS screening and the increased risk for Down syndrome. Further investigative studies are mentioned, including a detailed anatomy ultrasound at midgestation, a fetal echocardiogram, and additional testing for other conditions (Section 4.3).
- **Management and Recommendations**: The option of prenatal diagnostic testing to obtain fetal material to assess chromosomes (Section 6.1.1) is raised, if the couple would want to confirm the screening results, and if the benefits of having these results at this time outweigh the risk of the procedure. The genetic counselor discusses that this could be done by chorionic villous sampling (CVS) at 10–14 weeks gestation, or amniocentesis on or after 15 weeks gestation (Section 4.4.2). The genetic counselor discusses the risks of each procedure, including increased risk of miscarriage, higher with the CVS. The couple would want to think about potential positive results and whether they would want to know these results and/or act upon them. If so, the results could help with a decision about continuing or terminating the pregnancy, managing the pregnancy appropriately, or preparing emotionally ahead of time for the potential outcome (Section 4.4.3). The couple also can choose no testing and find out after delivery if their baby has Down syndrome. They decide on having a CVS, and arrangements are made for the procedure that afternoon. Chorionic villi are sent to the Cytogenetic Laboratory to be cultured and grown. The couple waits for the results.

Which laboratory to use will be likely dependent on the woman's insurance, and they may need preauthorization for the procedure and testing, regarding insurance coverage (Section 6.6). Approval may have been obtained prior to the visit.

- **Results**: The karyotype performed on the chorionic villus sample shows that the fetus has Down syndrome due to an extra chromosome 21 in all cells. Brianna and Tyrell return for further discussion with the genetic counselor. They discuss management options for the pregnancy. Would they consider termination of the pregnancy (Section 4.4.3)? Would they want to continue the pregnancy and find out more about what to expect after birth? Would they consider putting the baby up for adoption? The couple is offered a visit with a pediatric geneticist to discuss Down syndrome and to learn about neonatal recommendations.

Many clinical centers may not have pediatric subspecialists available for live consultation, and the couple can be referred to an outside center. Otherwise, there may be availability by telemedicine.

- **Caveats**: Although the prenatal diagnostic testing is very accurate, no test is 100% sensitive, especially if mosaicism is present (Section 2.2.2). In addition, a fetus could have an additional underlying condition or a congenital malformation undetected by the ultrasound.
- **Outcome**: The couple elects to continue the pregnancy under your service. Arrangements are made for a detailed midgestation ultrasound to better assess the anatomy of the fetus, and for a fetal echocardiogram, which is normal (Section 4.3). Six months later a baby boy is born with features of Down syndrome.

The Diagnosis

Clinical Diagnosis: Fetus—increased nuchal fold, confirmed to have Down syndrome.
Genetic Diagnosis: Fetus—47,XY,+21.

Further Reading

Carlson, L.M., Vora, N.L., 2017. Prenatal diagnosis: screening and diagnostic tools. Obstet. Gynecol. Clin. North Am. 44, 244–256. PMID 28499534.

Grieco, J., Pulsifer, M., Seligsohn, K., Skotko, B., Schwartz, A., 2015. Down syndrome: cognitive and behavioral functioning across the lifespan. Am. J. Med. Genet. C Semin. Med. Genet. 169, 135–149. PMID 25989505.

Seror, V., L'Haridon, O., Bussieres, L., Malan, V., Fries, N., Vekemans, M., Salomon, L.J., Ville, Y., SAFE 21 Study Group, 2019. Women's attitudes toward invasive and noninvasive testing when facing a high risk of fetal Down syndrome. JAMA Netw. Open 2, e191062. PMID 30924894.

Case 11.1.4

Prenatal Cases: Noninvasive Prenatal Screening Positive for Sex Chromosome Abnormality

Shama Jari

Harvey Institute for Human Genetics, Greater Baltimore Medical Center, Baltimore, MD, United States

Abstract

A fetus is found to have a positive prenatal screening for triple X syndrome, but is negative on diagnostic testing, and on ultrasound is found to have a unilateral clubfoot.

The Case

Elizabeth, a 35-year-old G3 P2002, and her partner, Ben, are seen by you, her obstetrician, at 12 weeks gestation. After a discussion of screening and testing options for aneuploidy, they elect to proceed with noninvasive prenatal aneuploidy screening (NIPS) using circulating cell-free DNA. This shows an increased risk for triple X syndrome. The couple is very concerned about this result.

The Issues

- **How accurate are the results?** While NIPS has a high sensitivity and specificity for autosomal trisomies, the test performance for sex chromosome abnormalities is much lower (Section 4.4.1). The positive predictive value for this result depends on the prevalence of the disorder in the population, and since triple X syndrome, also referred to as trisomy X, is relatively uncommon, the positive predictive value for this result is low. The couple is counseled that this is a screening test, and false positives as well as false negatives can occur. Confirmation of results with diagnostic testing is recommended, if the couple is interested.
- **What are the diagnostic tests available during pregnancy?** Prenatal diagnosis can be performed with chorionic villus sampling (CVS) in the first trimester (10−14 weeks) or amniocentesis in the second trimester (15−22 weeks) (Section 4.4.2). Both procedures carry a small risk for miscarriage, with a slightly higher risk for CVS. Chorionic villi or amniotic fluid are sent to the cytogenetic laboratory to be cultured and grown. Karyotype analysis will determine if there are any missing or extra chromosomes, and if there are any structural rearrangements or smaller deletions or duplications large enough to be detected microscopically (Section 2.2).
- **What is the next step?** Given the increased risk for a false-positive result of triple X syndrome on noninvasive prenatal screening, referral for genetic counseling is recommended. An appointment for an ultrasound is scheduled for immediately after the counseling to look for any associated anomalies.

Genomics in the Clinic. https://doi.org/10.1016/B978-0-12-816478-5.00014-5

It would have been appropriate to refer the patient for genetic counseling (Section 3.3) prior to sending the NIPS, for an in-depth discussion of the benefits and limitations of this screening test, as well as to go over other screening and testing options in detail. In addition, a first trimester ultrasound (Section 4.3.1) at 12 weeks is generally recommended even if NIPS is being sent for aneuploidy screening for the detection of fetal anomalies (Section 4.3.2) which may or may not be associated with chromosomal abnormalities (Section 2.2.1).

The Referral

- **Genetic Evaluation:** Elizabeth and Ben meet with the prenatal genetic counselor at 13 weeks gestation. A detailed three-generation pedigree is obtained (Sections 2.7, 3.2.1), which is only remarkable for some mild learning disabilities in the patient. Her NIPS results are reviewed in detail, including the sensitivity, specificity, and positive predictive value of the test.
- **Genetic Counseling:** Since NIPS is performed by evaluating cell-free DNA circulating in the maternal circulation, it is possible that the extra X chromosomes may be maternal in origin. The patient is therefore informed that these results could also reflect a change in her chromosomes. The natural history of triple X syndrome is discussed.

  Diagnostic testing options are reviewed. The availability of chorionic villous sampling or amniocentesis (Section 4.4.2) for the precise detection of chromosomal abnormalities is reviewed. The risk of miscarriage is discussed, which is slightly higher for CVS, as well as the benefit of earlier detection with CVS. The genetic counselor also has a discussion with the couple about the potential findings on CVS, and what the couple would want to do with the information obtained. In some cases, a precise diagnosis can help with decisions about continuing or terminating a pregnancy. Having a diagnosis can also help the parents to be prepared and can help with postnatal management. The couple can also choose no testing, or plan to evaluate the baby after birth. The couple decides to see how the ultrasound looks before deciding whether to undergo diagnostic testing.
- **Management and Recommendations:** The patient undergoes an ultrasound following her genetic counseling appointment at 13 weeks gestation, which reveals a unilateral clubfoot. The couple is very worried at this point. In a follow-up meeting with the genetic counselor after the ultrasound, the couple is informed that clubfoot may be isolated, or may be due to chromosomal or genetic syndromes. However, it is not a typical finding seen with triple X syndrome. The possibility of an underlying genetic syndrome is discussed with the couple. Given the abnormal ultrasound finding and the increased risk for triple X syndrome on NIPS, the couple elects to proceed with CVS. The procedure is performed that day. The couple anxiously awaits the results.

CVS results generally take 10–12 days. A direct analysis of the villi can be performed to provide preliminary information on the number of chromosomes (Section 6.1.2). Some deletions, duplications, and translocations (Section 2.2.1) can be detected as well, depending on the size of the change. Direct analysis results should be considered preliminary and should always be confirmed with final results on cultured villi.

- **Results:** Final CVS results reveal a normal female karyotype. The patient is again counseled about the possibility that her NIPS results may have been due to the maternal presence of an extra X chromosome. The patient is offered a karyotype (Section 2.1.3) on her blood. Elizabeth accepts this testing, and a sample of her blood is sent for chromosome analysis. The results reveal that she does indeed have triple X syndrome. She is counseled that this finding may explain her learning disabilities. The couple is relieved that the chromosomes are normal, but is still anxious about the clubfoot. The couple is referred to a pediatric orthopedic specialist to discuss management of the clubfoot, who reviews that cases of isolated clubfoot generally have a good prognosis, and most are correctable with stretching and casting after birth, followed by physical therapy.
- **Caveats:** Although the fetal chromosomes are normal, this testing cannot rule out other possible causes for the clubfoot. This would be more likely if additional anomalies are seen on the detailed anatomic survey in the second trimester. However, a normal anatomy screen also cannot rule out the possibility of a genetic syndrome.

Prenatal microarray can be performed on the villi to detect smaller changes in the chromosomes that are not visible with standard chromosome analysis (Section 6.1.3, Table 6.2). Some of these changes can be associated with certain genetic syndromes. However, no test is 100% sensitive. A normal microarray result cannot rule out all genetic syndromes.

- **Outcome:** Elizabeth undergoes a detailed fetal anatomic survey (Section 4.3) at 20 weeks that again reveals a unilateral clubfoot. No other anomalies are detected by ultrasound. A fetal echocardiogram is performed which is within normal limits. The couple is reassured by the lack of additional findings and declines additional testing with microarray. The patient is counseled that isolated clubfoot on prenatal ultrasound generally has a good prognosis. However, in about 7%

of cases, additional abnormalities can be seen at birth. Continued ultrasound surveillance is recommended during the pregnancy, as well as a thorough postnatal evaluation to rule out associated anomalies (Section 4.3.1). Serial prenatal ultrasounds reveal appropriate interval growth and no additional findings are seen. Elizabeth goes on to deliver a healthy baby girl at term. Evaluation of the newborn reveals what appears to be an isolated clubfoot, with no additional findings on physical exam.

The Diagnosis

Clinical Diagnosis: Patient—triple X syndrome. Fetus—clubfoot, nonsyndromic.
Genetic Diagnosis: Patient—47,XXX. Fetus—none.

Further Reading

Lauson, S., Alvarez, C., Patel, M.S., Langlois, S., 2010. Outcome of prenatally diagnosed isolated clubfoot. Ultrasound Obstet. Gynecol. 35, 708–714. PMID 20069547.

Ramdaney, A., Hoskovec, J., Harkenrider, J., Soto, E., Murphy, L., 2018. Clinical experience with sex chromosome aneuploidies detected by noninvasive prenatal testing (NIPT): accuracy and patient decision-making. Prenat. Diagn. 38, 841–848. PMID 30068017.

Wigby, K., D'Epagnier, C., Howell, S., Reicks, A., Wilson, R., Cordeiro, L., Tartaglia, N., 2016. Expanding the phenotype of triple X syndrome: a comparison of prenatal versus postnatal diagnosis. Am. J. Med. Genet. A. 170, 2870–2881. PMID 27644018.

Case 11.1.5

Prenatal Cases: Fetal Anomalies

Antonie D. Kline

Harvey Institute for Human Genetics, Greater Baltimore Medical Center, Baltimore, MD, United States

Abstract

A couple discovers that their fetus has several anomalies, including single umbilical artery, bilateral choroid plexus cysts, and a ventricular septal defect, and an amniocentesis reveals trisomy 18.

The Case

Carrie, a 30-year-old G1, has had an uneventful pregnancy, with normal early ultrasound and first trimester screening. She has arrived in your office with her partner, Georgina, for her midgestation ultrasound at 18 weeks gestation. You, her obstetrician, are called by the ultrasound technician that there are anomalies. You confirm the findings, which include a single umbilical artery, large bilateral choroid plexus cysts, and an abnormal four-chamber view of the heart, and speak with Carrie and Georgina.

The Issues

- **How accurate are these findings?** Prenatal ultrasound is a highly accurate method of assessing the fetus for structural anomalies, particularly at midgestation at 18–20 weeks gestation (Section 4.3.2). There is greater accuracy at a specialized center than in the obstetrician office, but it is very likely that these findings are present.
- **When would these findings have occurred?** Many congenital anomalies occur prior to the woman discovering that she is pregnant, usually in the first 6 weeks of gestation. Organogenesis occurs within the first 8 weeks of gestation, although the central nervous system continues developing later than this, even after birth. You reassure them that choroid plexus cysts usually are normal and do not harm the fetus, and recommend more detailed ultrasound studies.
- **Was anything done by the mother prenatally to cause the findings?** There are some medications and other exposures that can lead to congenital anomalies (Section 4.2.1), but most are due to multifactorial causes (partly environmental, partly genetic) (Section 2.7.3) or underlying genetic syndromes that could be inherited or de novo in that pregnancy. In Carrie's case, she has only been taking prenatal vitamins, with no other prenatal exposures, and you are able to reassure her that nothing she did or did not do led to these findings.

Maternal exposures that can lead to fetal anomalies include medications the pregnant woman is taking (e.g., insulin for diabetes), organisms the pregnant woman has been exposed to (e.g., Zika virus), recreational substances (e.g., alcohol), and other entities (Section 11.3.1).

- **What is the next step?** You discuss the findings, which are not specific for a known diagnosis, but concerning. You say that you would like to refer the couple to the Prenatal Diagnostic Center in the Genetics Division at the University

Genomics in the Clinic. https://doi.org/10.1016/B978-0-12-816478-5.00015-7

Hospital for more detailed ultrasound and genetic counseling, to discuss possible options including prenatal diagnosis by amniocentesis if they want to pursue possible cause(s) further. Carrie and Georgina are very interested in this. You refer them to Genetics; they are able to be seen in 2 days.

It is helpful for you to send a copy of the results of the abnormal ultrasound and any first trimester testing that has been done in your office prior to the couple being seen by Genetics.

The Referral

- **Genetic Evaluation**: Carrie and Georgina meet with the prenatal genetic counselor 2 days later, following a high-resolution ultrasound done in the office. The reproductive geneticist confirms the single umbilical artery and the choroid plexus cysts and also has noted a likely ventricular septal defect (VSD). A three-generation pedigree is obtained (Sections 2.7, 3.2.1); the couple used a sperm donor and has only limited paternal family information, but there are no contributory findings.
- **Genetic Counseling**: The genetic counselor discusses the abnormal ultrasound (Section 4.3.2), the implications, and the potential causes. Any of these as a single finding would raise much less concern. A group of malformations can be associated with aneuploidy (Section 2.2.1), particularly trisomy 18, but also trisomy 13, as well as underlying specific single gene disorders (Section 5.1). Trisomy 18, with an extra chromosome 18 in all cells, typically presents with small size, mildly dysmorphic facial features (Fig. 11.1.5.1), characteristic overlapped fingers with index finger over third and fifth over fourth, prominent heels with "rocker bottom" feet, and abnormal neurologic exam. There are often malformations including cleft lip and palate, congenital heart disease, and ambiguous genitalia. Survival is poor, generally less than 10% survive past 1 year, and there is global developmental delay, usually with subsequent severe intellectual disability if surviving.

  The ultrasound findings are discussed in detail. Choroid plexus cysts are small fluid-filled bubbles within the choroid plexus lining the ventricles of the brain. They are usually an incidental finding, occurring in about 1% of normal pregnancies, and most resolve spontaneously with no sequelae. There is a slightly increased chance of trisomy 18 in the fetus when these are seen, and 50%–70% of fetuses with trisomy 18 will have choroid plexus cysts. If at least one other anomaly is present, as in this case, there is a high association with trisomy 18.

  VSD involves an opening between the lower two chambers of the heart. It can be isolated or accompanied by other cardiac defects (Section 5.1.3), and a fetal echocardiogram is recommended for further assessment. It can be sporadic, associated with other malformations, or due to an underlying syndrome. Congenital heart disease occurs in about 1 in 100 pregnancies. In trisomy 18, 85% of fetuses will have a congenital heart defect, of which a VSD is one of the most common. It may close spontaneously or require surgical intervention.

  The umbilical cord typically has three blood vessels, with two being arteries and one a vein. A single umbilical artery is found in about 1% of all fetuses, and about 10% of fetuses with aneuploidy, and up to 30% have been associated with structural defects. When in combination with other anomalies, as in this case, it raises the suspicion for an underlying condition, and prenatal diagnostic studies should be offered, and a fetal echocardiogram obtained. When isolated and in a case with a normal heart on the initial ultrasound, no further recommendations are needed.

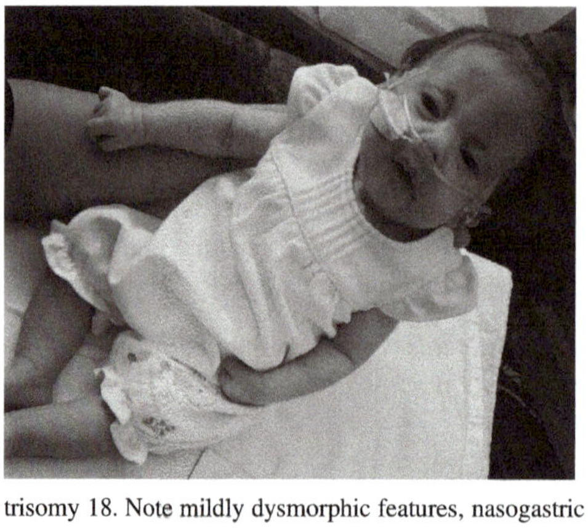

FIGURE 11.1.5.1 A 2-month-old infant with trisomy 18. Note mildly dysmorphic features, nasogastric feeding tube, and clenched fingers *[publishing permission obtained]*.

- **Management and Recommendations**: Fetal material could be tested for chromosomes by prenatal diagnostic testing (Sections 4.4.2, 6.1.1). The genetic counselor discusses that this could be done by amniocentesis. The couple would want to think about potential positive results and if they would want to know these and/or act upon them. Carrie and Georgina say that they do want to know about the possibility of trisomy 18, or other chromosomal changes. The genetic counselor discusses the options of keeping the pregnancy and also pregnancy termination (Section 4.4.3). Carrie undergoes an amniocentesis that afternoon. The couple says they will think about what they might want to do if the fetus is found to have trisomy 18.
- **Results**: The amniocentesis fluid is sent for cytogenetic analysis. Two days later, the genetic counselor calls the couple with the results of the fluorescence in situ hybridization (FISH) (Section 6.1.2) for chromosomes 13, 18, and 21 on uncultured cells. There are three signals for chromosome 18. They tell the counselor that they want to wait for the final results. Two weeks later, the amniocentesis confirms trisomy 18, with the chromosomes showing an extra chromosome 18 in every cell analyzed from a male fetus. The couple say that they have been struggling with the decision and want to return to discuss the options. On their return, the genetic counselor reviews clinical findings seen in trisomy 18 and again discusses the options of keeping the pregnancy, and also pregnancy termination.
- **Caveats**: Ultrasounds are very accurate but can miss many smaller or internal anomalies. The amniocentesis should be very accurate, but could potentially miss mosaicism (Section 2.2.2) or tiny rearrangements below the resolution of the microscope.
- **Outcome**: Carrie and Georgina decide to terminate the pregnancy, now at 20½ weeks gestation. The genetic counselor provides them with recommendations for local centers. She gives them a copy of "A Time to Decide, A Time to Heal," and a referral for a counselor if they are interested in short-term counseling related to this decision.

The Diagnosis

Clinical Diagnosis: Fetus—trisomy 18 with multiple anomalies, including single umbilical artery, bilateral choroid plexus cysts, and ventricular septal defect.
Genetic Diagnosis: Fetus—47,XY,+18.

Further Reading

Ebrashy, A., Kurjak, A., Adra, A., Aliyu, L.D., Wataganara, T., De Sa, R.A., Pooh, R., Sen, C., Stanojevic, M., 2016. Controversial ultrasound findings in mid trimester pregnancy. Evidence based approach. J. Perinat. Med. 44, 131–137. PMID 26506099.

Gurram, P., Figueroa, R., Sipusic, E., Kuhnly, N., Clark, S., Janicki, M.B., 2018. Isolated single umbilical artery and fetal echocardiography: a 25-year experience at a tertiary care city hospital. J. Ultrasound Med. 37, 463–468. PMID 28850695.

Minnick, M.A., Kelp, K.J., Ciotti, M.C., 1992. A Time to Decide, a Time to Heal. Pineapple Press.

Watson, W.J., Miller, R.C., Wax, J.R., Hansen, W.F., Yamamura, Y., Polzin, W.J., 2008. Sonographic findings of trisomy 18 in the second trimester of pregnancy. J. Ultrasound Med. 27, 1033–1038. PMID 18577667.

Case 11.1.6

Prenatal Cases: Clubfeet and Decreased Fetal Movement

Judith G. Hall

Departments of Pediatrics and Medical Genetics, British Columbia Children's Hospital, University of British, Vancouver, BC, Canada

Abstract

A newborn with a prenatal diagnosis of decreased fetal movement and multiple contractures is found to have arthrogryposis at birth and diagnosed with amyoplasia.

The Case

At the time of a 17-week ultrasound, a question of clubfeet is raised. Tanis, the mother, is 28 years old, and Dakota, the father, is 26 years old, and they had had one previous miscarriage at 8 weeks, about 8 months ago. Tanis had noted some spotting at 7 weeks in the pregnancy and had a mild URI at 10 weeks. Fetal measurements are at the 10th percentile. The family lives way out in the country, about an 8 hour drive from your small town family practice. You refer them to a tertiary care center for further evaluation.

Because of the distance and being harvest time, the family does not get to the tertiary care center until 21 weeks gestation. The ultrasound evaluation at the tertiary care center indicates generalized decreased fetal movement and more extensive contractures (Sections 4.3.2, 5.1.6), including contractures of the knees, equinovarus deformities of the feet, extension of the elbows, and flexed wrists. However, the lungs are developing normally and the fetus continues to be at the 10th percentile. By the time you receive the report, the family has returned home and mother realizes that she has not really noticed fetal movement. The pregnancy is now at 22 weeks.

The Issues

- **How big of a problem are the contractures in this fetus?** The contractures could be very extensive. With decreased fetal movement, there could be an underlying neurologic, neuromuscular, and/or genetic condition. Although the degree of involvement of this unknown potential condition is unclear, you realize that it could be severe, and survival may even be an issue.
- **What management options are there for the contractures?** Usually a multidisciplinary team, particularly orthopedist and physical therapist, is required to address all of the issues after a baby with contractures is born. There also could be additional congenital contractures that are unable to be detected prenatally, such as trismus, dislocated hips, and scoliosis. The primary care provider will need to learn about management as well.

Genomics in the Clinic. https://doi.org/10.1016/B978-0-12-816478-5.00016-9

- **Can a specific diagnosis be made?** Discussion by the parents with pediatric experts seems indicated at this point, so they can understand the potential range of the involvement. Additional ultrasounds and prenatal diagnostic testing may be suggested. You decide to refer the couple to the pediatric geneticist.

The Referral

- **Genetic Evaluation**: The geneticist is able to do a long-distance video online consultation or telehealth visit with the family. A full pregnancy history, medical history, and family history are obtained. The geneticist is able to see the parents and observe that both have full range of motion of joints, particularly hands and feet with normal creases, and no dysmorphic facial features. A three-generation pedigree is obtained (Sections 2.7, 3.2.1), with no significant family history, and the couple are not related. They are both Indigenous.

  The geneticist discusses with the family that contractures can be single or multiple; when multiple congenital contractures in different body areas are present, the descriptive term "arthrogryposis" is used. If picked up prenatally, a careful search for additional contractures should be made. There are over 400 specific genes that are known to have variants that cause arthrogryposis. Arthrogryposis is heterogeneous and a multidisciplinary team is required for care. The geneticist also indicates that a third of cases will die in utero or at the time of birth because of severe fetal akinesia sequence, or nonmovement in the fetus leading to secondary deformations, usually related to pulmonary hypoplasia (Section 4.3.2). Another 25% have intellectual disability, but the rest tend to do well, but require a great deal of physical therapy and often orthopedic surgery. Because of the extended elbows and good pulmonary size, a diagnosis of amyoplasia is suspected.
- **Genetic Counseling**: Because of the time delay, prenatal diagnosis and gene testing will not be possible prior to 24 weeks. The geneticist discusses possible continuation versus termination of the pregnancy (Section 3.1.2). The couple indicates this is a highly desired pregnancy and understands there is an overall 25% risk for intellectual disability. They want to continue the pregnancy, and they want to do all they can for the child's welfare.

  The geneticist discusses gene panel testing at the time of birth, and/or later, because of the relevance for medical management, future children, and their extended family. The couple needs to check whether their health insurance would cover such testing. The geneticist offers to meet again in a week to answer additional questions. The geneticist indicates the baby should be delivered in a tertiary care center and by Cesarean section, since children with arthrogryposis are often in breech or transverse position in utero, and 10% will fracture long bones at birth or in the perinatal period, presumably because of decreased mineralization secondary to decreased fetal movement.

  A week later during the follow-up online consultation, the couple said they had been searching the web for information on arthrogryposis and found the parent support group (Section 3.5) website as well as the National Organization of Rare Disorders (NORD) Website, and had found the suggestion that increasing fetal movement might lead to the contractures being less severe. They also read many family reports about children with amyoplasia, who seemed to be bright and doing well in spite of lots of therapy and surgeries. After discussing with the geneticist to be sure there was no potential harm (and with the knowledge that no relevant controlled clinical studies or trials have been performed), Tanis begins moderate exercise and drinking three cups containing caffeine a day. She notes fetal movement for the first time at 26 weeks.

  The fetus continues to grow along the 10th percentile. At 34 weeks, the mother moves into the larger town where the tertiary care center was located and stays with her sister. At 37 weeks after confirming that the lungs were mature, Cesarean section is performed with the baby in breech position. At the time of delivery, the placenta is examined and is noted to have a small area of thickening, but no suggestion of amniotic rupture or infarcts that can be associated with contractures.

  The baby boy is transferred to the NICU and has some mild breathing problems for the first 3 days. In addition, he has some trouble with breastfeeding for the first week. On exam, the baby boy, Jacy, is noted to be 3rd percentile for weight and 10th percentile for length, with a normal head circumference. There are symmetrical contractures with internal rotation of shoulders, extended elbows, flexed wrists, no creases on the fingers with severe camptodactyly (Fig. 11.1.6.1). The limbs are very thin with little palpable muscle. There are severe equinovarus deformities of the feet, fixed flexed knee contractures, and a question of dislocated hips. He is also noted to have dimples over the joints and a midline stork mark on the forehead.
- **Management and Recommendations**: The geneticist examines Jacy and confirms the prenatal suspicion of amyoplasia. This is a clinical diagnosis and the most common type of arthrogryposis, often requiring 2 or 3 years to observe the

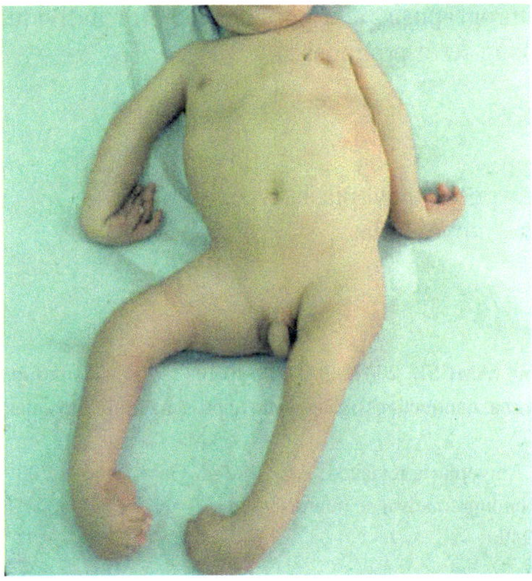

FIGURE 11.1.6.1 An infant with amyoplasia.

natural history. Birth history, positioning of extended elbows, and four limb symmetric involvement and severe equinovarus deformities are important clues to the diagnosis. It is typically nongenetic with no recurrences observed and no affected individuals having had affected children. Curiously, amyoplasia occurs frequently in only one of monozygotic twins. The placenta pathology and spotting during early pregnancy are suggestive of loss of a monozygotic twin in this case. The geneticist orders a gene panel (Section 6.2.3) to rule out a genetic cause for arthrogryposis.

A neurologist, orthopedist, and physical therapist are consulted. The physical therapist begins stretching exercises immediately under the direction of the orthopedist. The multidisciplinary team working with the geneticist begins physical therapy and stretching exercises and serial casting.

- **Results**: The arthrogryposis gene panel results return showing no relevant pathogenic variants or mutations. Jacy is discharged to your care. The mother becomes involved in the therapy and learns to do the stretching, being careful not to fracture the osteoporotic bones. He continues weekly physical therapy. The family is in contact with the parent support group, Arthrogryposis Multiplex Congenita Support, Inc. (AMCSI), and attends a national meeting, where they find an additional number of resources. They begin to recognize the many different types of arthrogryposis and network with families, who also have had children with amyoplasia. They join the childhood registry at Shriner's Hospital, shrinershospitalforchildren.org. They find at the meetings that there are talks related to the children, the parents, the unaffected children, and the families. They are encouraged to see that an additional child might actually be stimulatory for their little boy. There is also information concerning school readiness, and what one needs to do to obtain an aide for school, and about the use of computers rather than writing by hand.
- **Differential Diagnosis**: The only well-defined genetic disorders with extended elbows are autosomal recessive types of lethal arthrogryposis, with severe fetal akinesia and pulmonary hypoplasia not sustaining survival.
- **Caveats**: During prenatal diagnosis, when there are multiple congenital contractures (arthrogryposis) it is important to evaluate pulmonary status since it determines whether survival is possible. Fractures at the time of delivery or in the newborn period occur in about 10% of individuals with arthrogryposis. This is because the bones are gracile and osteopenic; therefore it is very important to be careful doing stretching in the newborn period. Fractures heal perfectly in almost all forms of arthrogryposis except Bruck syndrome, which is a variant of osteogenesis imperfecta (Section 11.10.3) with contractures.
- **Outcome**: You keep in contact with the members of the multidisciplinary team. By 5 years of age, Jacy is walking with braces and is noted in preschool to be very bright, to be a problem-solver, and to be very gregarious and interacting with the other children. By 6 years of age, there is a second pregnancy, which is monitored by ultrasound, and normal movement of the limbs is seen and felt by Tanis. When this child is born, she seems to stimulate the affected child, and the family is delighted. Through their contacts with the parent support group, they began to think about adopting a child with arthrogryposis from a developing country. By that time, there is a great deal of information about the adult outcomes (Adults with Arthrogryposis Multiplex Congenita (AMC) registry, contact bonita.sawatzky@ubc.ca), long-term

prognosis for different types of arthrogryposis, and possible jobs for individuals with arthrogryposis, and the family feels quite comfortable about the long-term prognosis.

The Diagnosis

Clinical Diagnosis: Fetus—arthrogryposis, amyoplasia type.
Genetic Diagnosis: Fetus/Newborn—amyoplasia.

Further Reading

Arthrogryposis Multiplex Congenita Support, Inc (AMCSI). 2005–2020. Available at: https://amcsupport.org/.

Hall, J.G., 2014. Arthrogryposis (multiple congenital contractures): diagnostic approach to etiology, classification, genetics, and general principles. Eur. J. Med. Genet. 57, 464–472. PMID 24704792.

Hall, J.G., Aldinger, K.A., Tanaka, K.I., 2014. Amyoplasia revisited. Am. J. Med. Genet. 164A, 700–730. PMID 24459070.

Hall, J.G., Dahan-Oleil, N. (Eds.), 2019. Collaborating to advance interdisciplinary care for individuals with arthrogryposis [Special Issue]. Am. J. Med. Genet. C 181 (3), 273–276. PMID 31471949.

Case 11.1.7

Prenatal Cases: Recurrent Pregnancy Loss

Shama Jari
Harvey Institute for Human Genetics, Greater Baltimore Medical Center, Baltimore, MD, United States

Abstract
A couple with three prior spontaneous abortions is found to have positive findings on testing. The woman has a polymorphism in the *MTHFR* gene, and the man is found to have a balanced translocation.

The Case

Anne, a 30-year-old G3 P0030 female, and her husband, Joseph, are seen by you, her obstetrician and gynecologist, after a recent miscarriage 2 weeks ago. They have now had three previous miscarriages at 8, 17, and 6 weeks gestation. Their last pregnancy was karyotyped and found to have an extra chromosome 16. The couple is anxious about their chances to have a viable pregnancy given the previous losses and ask if there is any identifiable cause for their pregnancy losses. You briefly discuss the possible causes for recurrent miscarriage and start the workup by ordering a hysterosalpingogram to rule out a structural defect. You also send bloodwork for a thrombophilia workup (including anticardiolipin antibodies), as well as a thyroid-stimulating hormone (TSH), prolactin, and hemoglobin A1c. All of this testing comes back negative, with the exception of a polymorphism in the methylenetetrahydrofolate reductase (*MTHFR*) gene c.677C > T (commonly referred to as C677T).

The Issues

- **How common is recurrent pregnancy loss?** Sporadic pregnancy loss is relatively common, affecting approximately 15%–25% of clinically recognized pregnancies. However, fewer than 5% of women have two consecutive miscarriages, and only about 1% of women will have three or more.
- **What are the causes of recurrent pregnancy loss?** There are many possible factors that can contribute to recurrent pregnancy loss. These include: genetics, antiphospholipid antibodies, uterine anomalies, thrombophilias, hormonal or metabolic disorders, infection, autoimmune disorders, sperm quality, and lifestyle issues.
- **What testing is recommended for recurrent pregnancy loss?** Recommended screening includes a karyotype to check for chromosomal rearrangements (Section 2.2.1), anticardiolipin antibodies, hysterosalpingogram or sonohysterography, TSH, prolactin, and hemoglobin A1c. In general, screening for thrombophilia (other than anticardiolipin antibodies) is not recommended for recurrent pregnancy loss.
- **What is the next step?** You offer the couple genetic counseling to discuss her *MTHFR* gene polymorphism and to send off additional testing if indicated.

Genomics in the Clinic. https://doi.org/10.1016/B978-0-12-816478-5.00017-0

Recurrent pregnancy loss is defined as a history of three or more consecutive losses with or without previous live births, but it would be appropriate to initiate a clinical evaluation after a patient experiences two consecutive miscarriages.

The Referral

- **Genetic Evaluation**: Anne and Joseph meet with the genetic counselor to discuss their test results prior to attempting to conceive again. They are anxious to find out what this means for a future pregnancy. The genetic counselor obtains a detailed three-generation pedigree (Sections 2.7, 3.2.1). This is significant for the three miscarriages to the couple, a paternal grandmother with a history of a miscarriage, and a paternal aunt with a congenital heart defect and mild intellectual disability.
- **Genetic Counseling**: The genetic counselor informs the couple that the polymorphism found in the *MTHFR* gene is relatively common in people of European descent. Patients who have homozygous variants (two abnormal copies) in this gene can have elevated levels of homocysteine in the blood, which can be a risk factor for venous thromboembolism (blood clot formation). However, a heterozygous variant (one abnormal and one normal copy) is not associated with an increased risk for thromboembolism (Section 2.7). Therefore, Anne's test results are not of any clinical significance and should not increase her risk for blood clots. Studies looking at the association between inherited thrombophilias and recurrent pregnancy loss have revealed inconsistent results. Routine testing of women with recurrent pregnancy loss for inherited thrombophilias (in the absence of a personal history of or first-degree relative with venous thromboembolism) is not currently recommended.

  The genetic counselor informs the couple that 2%–5% of couples with recurrent miscarriages carry a balanced reciprocal translocation or Robertsonian translocation (Section 2.2.1). Therefore, a peripheral blood karyotype (Section 6.1.1) is recommended for both Anne and Joseph in order to detect any balanced structural chromosomal rearrangements.
- **Results**: Karyotype results for Anne reveal normal chromosomes. However, Joseph's results reveal that he is a carrier for a balanced translocation (Section 2.2.1) between chromosomes 2 and 13. Translocation carriers can produce offspring with normal chromosomes, balanced translocations, or various combinations of unbalanced translocations. Some unbalanced translocations are more likely to cause miscarriage, whereas others can result in live-born offspring with intellectual disabilities, developmental delay, and/or congenital anomalies. The outcome depends on the specific segments involved and whether material is deleted or duplicated. This may explain Joseph's family history of miscarriage in his grandmother (balanced carrier) as well as his aunt with intellectual disability and congenital heart defect (unbalanced result).
- **Management and Recommendations**: The counselor discusses the option of prenatal diagnosis using chorionic villous sampling (CVS) or amniocentesis in a future pregnancy (Section 4.4.2). The alternate options of in vitro fertilization with preimplantation genetic screening and sperm donor are also discussed (Section 4.5.1).
- **Implications for Other Family Members**: The genetic counselor also discusses that this balanced translocation may be present in other members of Joseph's family. He may want to share this information with his siblings and cousins of child-bearing age, as it may affect their pregnancy risks as well (Section 3.4). If the family is not local, the genetic counselor can help refer to the nearest genetic centers.
- **Caveats**: This couple had testing of one of the products of the spontaneous abortions, which was found to have trisomy 16, the most common chromosomal abnormality found in this situation. However, likely the other two losses were due to unbalanced products of the balanced translocation.
- **Outcome**: Six weeks later, the couple find out that they have conceived spontaneously. They elect to proceed with CVS for the detection of chromosome abnormalities. This reveals that the pregnancy has normal chromosomes. The couple are relieved and go on to deliver a healthy baby girl at term.

The Diagnosis

Clinical Diagnosis: Patient—recurrent spontaneous abortions; polymorphism, clinically insignificant. Partner—balanced translocation carrier leading to unbalanced products.

Genetic Diagnosis: Patient—*MTHFR* gene, heterozygous polymorphism. Partner—46,XY,t(2;13).

Further Reading

ACOG Practice Bulletin No. 197, 2018. Inherited thrombophilias in pregnancy. Obstet. Gynecol. 132, e18−e34. PMID 29939939.

Alijotas-Reig, J., Garrido-Gimenez, C., 2013. Current concepts and new trends in the diagnosis and management of recurrent miscarriage. Obstet. Gynecol. Surv. 68, 445−466. PMID 23942472.

Practice Committee of the American Society for Reproductive Medicine, 2012. Evaluation and treatment of recurrent pregnancy loss: a committee opinion. Fertil. Steril. 98, 1103−1111. PMID 22835448.

Case 11.1.8

Prenatal Cases: Preimplantation Genetic Testing for a Single Gene Disorder

Virginia L. Corson

Gynecology and Obstetrics and Pediatrics, Johns Hopkins University School of Medicine, The Academy at Johns Hopkins, East Baltimore Campus, Welch Medical Library, Baltimore, MD, United States

Abstract

A couple whose previous fetus was affected with multiple anomalies secondary to a ciliopathy, Meckel syndrome, elect preimplantation genetic testing with the eventual transfer of an embryo free of the pathogenic variants identified in the affected fetus.

The Case

Angela and Joe present to your office to discuss their options and concerns about attempting another pregnancy. Six months ago the couple elected a termination of pregnancy at 20 weeks gestation following the diagnosis by ultrasound of multiple fetal anomalies, including polycystic kidneys, an occipital encephalocele, and polydactyly of the hands. An amniocentesis was performed prior to the termination to obtain a fetal sample for genetic testing. The possible diagnosis of Meckel syndrome, also known as Meckel–Gruber syndrome, was suspected by the genetics group based on the constellation of findings. A microarray analysis (Section 6.1.3) was performed, and no cytogenetic abnormalities were detected. Cultured amniotic fluid cells were sent out for molecular analysis of a panel of the genes (Section 6.2.3) known to cause Meckel syndrome, and two pathogenic variants or mutations in the *MKS1* gene on chromosome 17 were identified, confirming the clinical diagnosis.

The Issues

- **What is the risk of recurrence for Meckel syndrome?** Meckel syndrome is an autosomal recessive condition with a 25% recurrence risk in each pregnancy (Section 2.7.1). Each fetus/offspring would have a 50% chance of being a carrier and a 25% chance of inheriting neither parental mutation.
- **When is the best time to refer this couple for genetic counseling?** Referral prior to pregnancy would enable the couple to consider all of their options, including preimplantation genetic testing (Section 4.5.1) and the use of donor sperm or egg.

The counselor can explore the couple's understanding about the diagnosis and the possibility of a recurrence and describe the prenatal and preimplantation procedures in detail to facilitate the couple's decision-making.

Genomics in the Clinic. https://doi.org/10.1016/B978-0-12-816478-5.00018-2

The Referral

- **Genetic Counseling**: The genetic counselor reviews the autosomal recessive inheritance (Section 2.7.1) of Meckel syndrome and the clinical features of this condition, which is a ciliopathy disorder. For this and similar disorders, defects in the structure or function of cilia are thought to disrupt important signaling pathways during early development, often leading to fetal anomalies. Because previous testing identified the gene mutations in the affected fetus, prenatal diagnosis through molecular testing of a fetal sample obtained by chorionic villus sampling or amniocentesis (Section 4.4.2) is possible and is described to the couple. The option of ultrasound surveillance to monitor for fetal anomalies also is discussed as a noninvasive alternative. Preimplantation genetic testing−monogenic disease (PGT-M) (Section 4.5.1) and conception by donor sperm or egg (Section 4.5.2) are also discussed.

More than 35 conditions, including polycystic kidney disease, are now known to be caused by mutations in genes that cause cilia dysfunction. Issues for the couple's consideration include the benefits and risks associated with the procedures, costs and insurance coverage for the various options, and their feelings about a possible recurrence in light of their previous experience.

- **Management and Recommendations**: Angela and Joe elect to discuss these options for several weeks and then contact the counselor with their decision to proceed with PGT-M (Section 4.5.1). The counselor refers the couple to a reproductive endocrinologist who works with a laboratory that can perform the genetic testing on embryo biopsies. The patient is seen in consultation, at which time screening laboratory tests are performed. In addition, blood samples from the patient and her partner, as well as a stored sample from the previous affected pregnancy, are sent to the genetics laboratory so that a test can be designed for this couple. Results of the genetic test developed take 2−3 months. When results are available, the patient undergoes ovarian stimulation prior to egg retrieval. In vitro fertilization (IVF) utilizing intracytoplasmic sperm injection with a sample from the partner yields 11 fertilized eggs which are suitable for biopsy after 5 days. The embryos are then cryopreserved while results of the genetic testing on the biopsies are pending.
- **Results**: PGT-M reveals that three of the embryos have inherited both parental mutations and are predicted to be affected with Meckel syndrome. Preimplantation genetic testing−aneuploidy (PGT-A) indicates that one of these embryos and four additional embryos have a cytogenetic abnormality (Section 2.2.1). In addition, one embryo has an apparent mosaic chromosome abnormality (Section 2.2.2). Three embryos reportedly are unaffected with Meckel syndrome and unlikely to have a chromosome abnormality. These results are reported to the reproductive endocrinologist, who discusses them with the patient and sends a copy of the report to the genetic counselor. After some consideration, Angela and Joe elect to transfer one of the embryos with normal test results.

Mosaic results are not uncommon during PGT-A and are difficult to interpret, as the mosaicism can be confined to the biopsy sample obtained or could reflect true chromosomal mosaicism (Section 2.2.2) of the embryo. Many factors, such as maternal age (Section 4.2.2) and pregnancy history, may influence how many embryos to transfer.

- **Caveats**: The results of PGT-M and PGT-A are highly accurate in experienced laboratories; however, errors in diagnosis may occur. Patients should be offered screening and/or diagnostic testing for confirmation of the results. In addition, there is thought to be some increase in the risk for birth defects and rare genetic conditions influenced by imprinting (Sections 2.7.5, 6.4.3) in pregnancies conceived with IVF.
- **Outcome**: Angela elects to have noninvasive prenatal screening (Section 4.4.1) for aneuploidy, which yields normal results. In addition, her 20 week anatomy survey reveals no anomalies. She gives birth to a healthy daughter at 39 weeks gestation.

The Diagnosis

Clinical Diagnosis: Patient and Partner—future pregnancies with 25% risk for Meckel syndrome. Prior fetus—Meckel syndrome.

Genetic Diagnosis: Patient and Partner—*MKS1* gene, heterozygous pathogenic variant. Prior fetus—*MKS1* gene, compound heterozygous pathogenic variants.

Further Reading

MedlinePlus [Internet]. Bethesda (MD): National Library of Medicine (US); [updated 2020; cited 2020]. Meckel syndrome. Available from: https://medlineplus.gov/genetics/condition/meckel-syndrome.

Reiter, J.F., Leroux, M.R., 2017. Genes and molecular pathways underpinning ciliopathies. Nat. Rev. Mol. Cell Biol. 18, 533−547. PMID 28698599.

Treff, N.R., Zimmerman, R.S., 2017. Advances in preimplantation genetic testing for monogenic disease and aneuploidy. Annu. Rev. Genom. Hum. Genet. 18, 189−200. PMID 28498723.

Section 11.2

Newborn Screening Cases

Case 11.2.1

Newborn Screening Cases: False Positive Newborn Screening

Melissa P. Wasserstein

Division of Pediatric Genetic Medicine, The Children's Hospital at Montefiore, The University Hospital for Albert Einstein College of Medicine, Bronx, NY, United States

Abstract

A newborn screen for mucopolysaccharidosis type 1 is positive falsely because of pseudodeficiency alleles.

The Case

You receive a call from your state's newborn screening lab about Baby Girl Houlihan. Baby Girl Houlihan was born at term to a 32-year-old mother 4 days ago. The lab reports that her newborn screening test, collected at 24 h of age, was abnormal. Specifically, the α-iduronidase (IDUA) activity was low. IDUA is the screening analyte for mucopolysaccharidosis type 1 (MPS 1), which you know was recently added to your state's newborn screening panel.

The Issues

- **How accurate is the newborn screening program?** Newborn screening is a public health program designed to identify children with rare, disabling, or life-threatening preventable conditions at birth, often before symptoms are apparent. In general, all newborn screening tests can have false positive and false negative results, although there are ongoing efforts to enhance accuracy.
- **How are new disorders added on to a state's newborn screening panel?** The Recommended Uniform Screening Panel (RUSP) is a list of disorders recommended by the Secretary of the Department of Health and Human Services for inclusion on states' newborn screening panels. Disorders on the RUSP are chosen based on an extensive review of evidence about the potential benefit of screening, the availability and outcome of treatment, and the ability of states to implement screening. As of August 2022, there are 36 core conditions on the RUSP. MPS 1 was added to the RUSP in 2015 (Section 5.4), and many states have included it on their newborn screening panels.
- **Does the baby have MPS 1?** You know that this is a screening test and could be a false positive, but you are concerned that Baby Girl Houlihan may have MPS 1 and needs treatment. You call the parents to let them know about the abnormal newborn screen, and tell them that they must bring the baby to see the state's designated metabolic disease specialist right away.

Genomics in the Clinic. https://doi.org/10.1016/B978-0-12-816478-5.00019-4

The Referral

- **Genetic Evaluation:** Baby Girl Houlihan's parents bring her to see the metabolic geneticist the next morning. The geneticist notes that the baby is well appearing and does not have organomegaly or an umbilical or inguinal hernia but knows that most babies with MPS 1 appear typically healthy at birth. The geneticist discusses that MPS 1 is a rare disorder of glycosaminoglycan (GAG) metabolism. The intralysosomal accumulation of partially degraded GAGs results in multisystem involvement. MPS 1 has a broad phenotypic spectrum ranging from severe, early-onset, rapidly progressive disease (formerly called "Hurler syndrome") to an attenuated form with later onset presentation (formerly called "Scheie syndrome"). Children with severe MPS 1 have disease manifestations in the musculoskeletal system, brain, eyes, liver, spleen, heart, and lungs.
- **Genetic Counseling:** The metabolic geneticist says that in general, all newborn screening tests can have false positive and false negative results, although there are ongoing efforts to minimize these using second-tier biomarkers, DNA sequencing, and computational algorithms to enhance accuracy. For certain disorders like MPS 1, the false positive rate is high due to the high prevalence of pseudodeficiency alleles, which are specific variants in the α-iduronidase gene that result in low enzyme activity in vitro, but do not cause clinical symptoms. For MPS 1, pseudodeficiency alleles are more common than pathogenic variants, leading to a high false positive rate in newborn screening programs that rely on enzyme activity as a sole screening test. Thus, although low α-iduronidase on a baby's newborn screen can be indicative of MPS 1, it is very possible that Baby Girl Houlihan's results were falsely positive.
- **Management and Recommendations:** The parents are anxious for Baby Girl Houlihan's confirmatory testing and inquire about starting treatment even before the result is reported. Treatment options for MPS 1 include enzyme replacement therapy, which is used to manage the somatic manifestations, or bone marrow transplantation, which is used to manage the somatic and neurologic manifestations in severe MPS 1. Bone marrow transplantation is most effective when performed in children less than 2 years of age. You advise that it is safe to wait a few weeks to see if the baby is truly affected before initiating treatment.
- **Genetic Testing:** The metabolic geneticist sends a blood sample to the biochemical genetic laboratory to measure leukocyte IDUA activity and a urine specimen to measure GAGs. The geneticist also sends a blood test to sequence (Section 6.2.2) *IDUA*, the gene that encodes α-iduronidase. The baby's leukocyte α-iduronidase activity is in the low normal range at 1.2 nmol/h/mL (normal ≥ 1). The urine GAGs are negative. The genetic testing reveals two known pseudodeficiency alleles, and no pathogenic variants (Section 7.1).
- **Outcome:** Baby Girl Houlihan's parents are delighted with the news that she does not have MPS 1 and that the newborn screening result was, in fact, a false positive. Baby Girl Houlihan is discharged from the metabolic clinic.

The Diagnosis

Clinical Diagnosis: Healthy newborn infant, no evidence of metabolic disease.
Genetic Diagnosis: Alpha-iduronidase enzyme, false positive deficiency.

Further Reading

Clarke, L.A., October 31, 2002. Mucopolysaccharidosis type I. Adam, M.P., Everman, D.B. Mirzaa, G.M., Pagon, R.A., Wallace, S.E., Bean, L.J.H., Gripp, K.W., Amemiya, A., (editors). In GeneReviews. University of Washington, Seattle (WA), 1993–2023 [updated February 11, 2016]. PMID 20301341.

Clarke, L.A., Atherton, A.M., Burton, B.K., Day-Salvatore, D.L., Kaplan, P., Leslie, N.D., Scott, C.R., Stockton, D.W., Thomas, J.A., Muenzer, J., 2017. Mucopolysaccharidosis type I newborn screening: best practices for diagnosis and management. J. Pediatr. 182, 363–370. PMID 27939258.

Recommended Uniform Screening Panel [Online], August 2020. Health Resources and Services Administration, Rockville, MD. Available from: https://www.hrsa.gov/advisory-committees/heritable-disorders/rusp/index.html.

Taylor, J.L., Clinard, K., Powell, C.M., Rehder, C., Young, S.P., Bali, D., Beckloff, S.E., Gehtland, L.M., Kemper, A.R., Lee, S., Millington, D., Patel, H.S., Shone, S.M., Woodell, C., Zimmerman, S.J., Bailey, D.B., Muenzer, J., 2019. The North Carolina experience with mucopolysaccharidosis type I newborn screening. J. Pediatr. 211, 193–200 e2. PMID 31133280.

Case 11.2.2

Newborn Screening Cases: Abnormal Newborn Metabolic Screening

Melissa P. Wasserstein

Division of Pediatric Genetic Medicine, The Children's Hospital at Montefiore, The University Hospital for Albert Einstein College of Medicine, Bronx, NY, United States

Abstract

A newborn has an elevated phenylalanine on newborn screening and is found to have classic phenylketonuria.

The Case

You receive a call from your state's newborn screening lab about Baby Boy Pierce. Baby Boy Pierce was born at term to a 28-year-old, gravida 1 mother 4 days ago. The lab reports that his newborn screening test, collected at 27 hour of age, was abnormal; the phenylalanine was elevated at 378 μmol/L [normal below 150] and the phenylalanine/tyrosine ratio was elevated at 6 [normal below 3]. You recall that several years ago, you received a similar call from the state lab, but that time they simply asked you to send a repeat sample to the lab. This time, however, you are more concerned as the lab specifically asked you to ensure that the baby has urgent follow-up with a metabolic center to rule out phenylketonuria (PKU).

The Issues

- **How are newborn screening results reported?** The vast majority of newborn specimens will have normal results. A much smaller number will have abnormal results. Abnormal results can be "presumptive positive," which means that there is a high probability that the infant may have a disorder and should be urgently referred to a metabolic center. Abnormal results can also be "borderline," which means that the result was slightly abnormal. In many states, a borderline result means that the clinician is instructed to send a repeat newborn screening sample to the state lab for another analysis. The last possible result is "unsatisfactory," which indicates that the sample was not collected, shipped, or handled properly; in that case, the clinician will be asked to send another sample to the newborn screening lab.
- **Is newborn screening diagnostic?** No, newborn screening is not diagnostic. If the newborn screen test is abnormal, it does not necessarily mean that the infant has the disorder. It does mean, however, that an infant with a presumptive positive needs to be referred to a specialist who can confirm if the baby is truly affected with the disorder.
- **What is the next step?** You are concerned that Baby Boy Pierce, now given the first name of Benjamin, may really have PKU. You call the parents to let them know about the abnormal newborn screening and tell them that they must bring the baby to see the state's designated metabolic disease specialist, a geneticist, right away.

Genomics in the Clinic. https://doi.org/10.1016/B978-0-12-816478-5.00020-0

The Referral

- **Genetic Evaluation:** Benjamin's parents bring him to the metabolic geneticist the next morning. She obtains a three-generation pedigree (Sections 2.7 and 3.2) and on medical history discovers that the baby is feeding well. She discusses that elevated phenylalanine on a baby's newborn screen can be indicative of classic PKU, variant PKU, or hyperphenylalaninemia ("hyperphe"). These three disorders are all caused by deficient activity of phenylalanine hydroxylase (PAH) activity, which results in a reduced ability to break down the essential amino acid phenylalanine. Classic PKU is the most severe phenotype with the least residual PAH activity, and results in severe, irreversible intellectual disability if untreated. Milder forms also exist, including variant PKU and hyperphenylalaninemia, both of which may require treatment, although usually less stringent than that required for classic PKU. On physical exam, the metabolic geneticist notes that Benjamin is well appearing and has regained his birth weight while exclusively breast fed, but cautions that babies with any form of PKU usually look and act normally at birth. The geneticist tells them that it is important to first see if the baby's phenylalanine is still elevated, and if so, it is essential to determine the exact cause, as the treatment options differ greatly.
- **Genetic Testing:** The metabolic geneticist sends a plasma amino acid panel to the biochemical laboratory to measure phenylalanine, as well as specimens to check for a disorder of tetrahydrobiopterin (BH4) metabolism. The baby's plasma phenylalanine level is 1350 μmol/L [normal below 91], and the BH4 results are normal. The metabolic geneticist tells Benjamin's parents that based on the biochemical results, the baby has classic PKU. She tells them that although the biochemical results are enough to confirm the diagnosis, sequencing of the *PAH* gene may guide treatment options and permit reproductive planning in the future. Benjamin is found to be compound heterozygote (Section 2.7.1) for c.1222C>T (p.Arg408Trp) and c.1066-11G>A (p.Gln355_Tyr356insGlyLeuGln), two common *PAH* pathogenic variants associated with classic PKU.
- **Genetic Counseling:** The genetic counselor discusses that PKU is an autosomal recessive condition (Section 2.7.1), and each parent is a carrier for a pathogenic variant in the *PAH* gene. Any subsequent pregnancy will be at 25% risk for having the same condition. Prenatal diagnostic testing (Section 4.4.2) or preimplantation genetic testing (Section 4.5.1) is available to test ahead of time, if parents are interested.
- **Management and Recommendations:** Fortunately, treatment with a low phenylalanine diet is an effective way to maintain normal cognitive function, and newer treatment options including oral cofactor and injectable enzyme substitution therapy are available (Section 5.4.3). Benjamin's family meets with the metabolic dietician. The dietician reassures the mother that she can continue to breastfeed her son, although the amount of breast milk must be restricted and supplemented with a phenylalanine-free formula. The metabolic geneticist and the dietician tell the parents that he will need frequent follow-up with them so they can regularly adjust his diet to keep his phenylalanine within a target range.
- **Differential Diagnosis:** More rarely, an elevated phenylalanine level can result from a disorder of BH4 metabolism. In addition to affecting phenylalanine metabolism, the BH4 deficiency disorders also impact neurotransmitter metabolism, resulting in a significant neurologic phenotype. Treatment of the BH4 disorders usually involves a combination of medications.
- **Outcome:** Benjamin's phenylalanine level drops nicely on his new diet. As he gets older, he works closely with the metabolic team and sticks to his special diet, with the result that his phenylalanine level generally remains within target range. He grows and develops appropriately for age and eventually gets accepted to medical school where he hopes to become a surgeon.

The Diagnosis

Clinical Diagnosis: Phenylketonuria, classic form.
Genetic Diagnosis: *PAH* gene, compound heterozygous pathogenic variants.

Further Reading

Blau, N., Hennermann, J.B., Langenbeck, U., Lichter-Konecki, U., 2011. Diagnosis, classification, and genetics of phenylketonuria and tetrahydrobiopterin (BH4) deficiencies. Mol. Genet. Metab. 104 (Suppl), S2–S9. PMID 21937252.

Regier, D.S., Greene, C.L., January 10, 2000. Phenylalanine hydroxylase deficiency. Adam, M.P., Everman, D.B., Mirzaa, G.M., Pagon, R.A., Wallace, S.E., Bean, L.J.H., Gripp, K.W., Amemiya, A., (editors). In: GeneReviews® [Internet]. University of Washington, Seattle (WA), 16, 1993–2023 [updated January 5, 2017]. PMID 20301677.

Vockley, J., Andersson, H.C., Antshel, K.M., Braverman, N.E., Burton, B.K., Frazier, D.M., Mitchell, J., Smith, W.E., Thompson, B.H., Berry, S.A.; American College of Medical Genetics & Genomics Therapeutics Committee, 2014. Phenylalanine hydroxylase deficiency: diagnosis and management guideline. Genet. Med. 16, 188–200. PMID 24385074.

Case 11.2.3

Newborn Screening Cases: Abnormal Newborn Hearing Screening

Jessica Ogawa[1], Daniah Albokhari[2] and Joann Bodurtha[1,3]

[1]*McKusick-Nathans Department of Genetic Medicine, Johns Hopkins University School of Medicine, Baltimore, MD, United States;* [2]*Department of Pediatrics, Taibah University College of Medicne, Medina Saudi Arabia;* [3]*Departments of Pediatrics and Oncology, Johns Hopkins University School of Medicine, Baltimore, MD, United States*

Abstract

A newborn with congenital deafness/hard of hearing found to have homozygous pathogenic variants in the *GJB2* gene.

The Case

A newborn female, Gina, is in your clinic for her first well child checkup at 1 week of age. She is a full-term baby, born to a gravida 1 mother without complication and was discharged from the newborn nursery with her mother. Gina's mother tells you that she is exclusively breastfeeding, and she has no current concerns. You see in the discharge summary that Gina failed her initial hearing screening, and she had a rescreen attempt before her discharge from the hospital which she also failed (Section 5.1.2). You review with Gina's mother the prenatal and delivery histories, which seem uneventful, including no maternal infection, drugs, or medication exposures. On exam, you notice that Gina gained weight, and her length and head circumference are age-appropriate. She is moving her arms and legs, keeping her hands in tight fists, and able to focus on your face. When you make a loud noise, she does not startle or turn toward the sound.

The Issues

- **Is it possible that the screening test result is false?** There is about a 2.5%–8% chance that Gina has no deafness/hard of hearing even though she did not pass the screening test. This might be due to fluid in her ears, or perhaps she was crying during the testing. As Gina did not pass her screening, you refer her to an audiologist for more confirmatory testing to determine if she has hearing loss.
- **Can you repeat the screening test in your office today?** You explain to Gina's mother that it is not recommended for Gina to be screened more than two times because if Gina has true deafness/hard of hearing, she may pass falsely by chance.

Most pediatricians will not have the capability to do an otoacoustic emissions test, the standard newborn screening hearing test typically performed in the hospital, not in their office.

- **Is it better to test Gina when she is older for a more accurate result?** If Gina has a hearing disability, it is important to confirm it sooner rather than later, so you can start the appropriate intervention as early as possible to avoid

Genomics in the Clinic. https://doi.org/10.1016/B978-0-12-816478-5.00021-2

preventable sequelae. The audiologist will use the auditory brainstem response to test Gina's hearing. This is a gold standard test for assessment of hearing sensitivity in newborns and infants and does not require a response from Gina. Two weeks later you receive the report from the audiologist, which reveals that Gina has a profound bilateral sensorineural hearing loss.

- **Were there any prenatal clues to the situation?** Gina's prenatal history was benign, but because cytomegalovirus (CMV) is a common cause of congenital hearing loss, you obtain a CMV urine test, and it comes back negative (Section 4.2.1).
- **What are your concerns regarding her hearing loss?** Gina is at risk of deficits in speech development because of her deafness. This can negatively affect her social interactions and academic performance, but with early intervention mainly before 6 months of age, these risks can be ameliorated. You refer Gina for the early intervention program to initiate the evaluation and start intervention, and also you refer her to a geneticist to determine the cause of her hearing loss.

The Referral

- **Genetic Evaluation**: The geneticist and the genetic counselor meet with Gina and her parents in the genetic clinic 2 months later. They obtain a complete medical history including prenatal, birth, and developmental histories. They also review with Gina's parents three generations of the family medical history (Sections 2.7, 3.2). The parents report no family history of deafness/hard of hearing, or medical problems involving the major organs. Growth is at the 50th percentile. Gina's development is appropriate for her age: she holds her head unsupported, and she is smiling spontaneously, reaching for a toy with one hand, and following the toy from side to side. Her parents note she has begun to coo. She has a normal physical exam, not associated with dysmorphic features, any visible external ear malformations, ear tags or pits, or abnormal pigmentation.
- **Genetic Counseling**: In the setting of Gina's profound deafness, the absence of any specific features in Gina's physical exam, and the negative family history, the geneticist and the genetic counselor say that they suspect that Gina has nonsyndromic hearing loss and deafness (DFNB1). They explain to the parents that this is the most common cause of isolated hearing loss and deafness inherited in an autosomal recessive pattern (Section 2.7.1) and is characterized by congenital bilateral and nonprogressive sensorineural hearing loss. The condition can range from mild to profound in severity, with no other associated medical findings. They tell the parents that the diagnosis of the most common form of nonsyndromic hearing loss and deafness is established by identifying a mutation on sequencing (Section 6.2.2) the *GJB2* gene encoding the connexin 26 protein. Mutations in this gene account for deafness in an estimated 50% of all autosomal recessive, nonsyndromic hearing loss cases and 15%–40% of all deaf individuals.

  They discuss with the parents that if Gina is found to have *GJB2* pathogenic variants, then each of her future full siblings will have a one in four chance of having deafness. They also say that because Gina has a profound hearing loss, the chance for the siblings with the same *GJB2* pathogenic variants to present with severe-to-profound hearing loss is ~91% while the chance to have mild-to-moderate hearing loss is ~9%. Furthermore, they review with the parents the availability of prenatal diagnostic testing (Section 4.4.2) once the pathogenic variants have been identified in Gina.
- **Management and Recommendations**: Because children who have hearing loss rely heavily on their other senses, the geneticist recommends ophthalmologic evaluation for Gina and repeat audiometry regularly to confirm the stability of hearing loss. The geneticist also emphasizes the importance of enrollment in the appropriate infant services and educational programs for children with hearing disability. Speech therapy will be important starting when she is a little older. Cochlear implantation is an important option for Gina due to her profound hearing loss. Furthermore, the parents learn from the geneticist about the Individuals with Disabilities Education Act (IDEA). This Act protects Gina's right to receive an individualized education program (IEP) to ensure appropriate progress in her learning, whether she is going to be in a public school or if she needs to enroll in the school for the deaf.
- **Genetic Testing**: The next step for testing Gina is looking specifically for variants in *GJB2*. There is also a comprehensive deafness-specific genetic panel (Section 6.2.3) that includes many genes implicated in nonsyndromic as well as syndromic deafness and hard of hearing.
- **Differential diagnosis**: Deafness and hard of hearing (mild-to-moderate hearing loss) can be divided into syndromic or nonsyndromic. Many genetic conditions represent complex syndromes, of which hearing loss is only one of many findings. This is why a detailed physical examination is essential to recognize the other findings that may lead to the diagnosis of a specific syndrome. Syndromes that cause deafness/hard of hearing include Waardenburg syndrome (most commonly presenting with pigmentary changes and telecanthus) (Section 11.3.7), branchiootorenal syndrome (presents

with structural ear changes such as ear tags or pits) (Section 11.3.8), Usher syndrome (retinitis pigmentosa) (Section 11.3.9), Pendred syndrome (goiter), and Jervell and Lange-Nielsen syndrome (cardiac conduction defects).

- **Caveats**: The genetic counselor emphasizes that if current single gene testing does not identify an etiology for Gina's hearing loss, the possibility of genetic etiology remains, and further genetic testing may be pursued.
- **Outcome**: Genetic testing reveals homozygous pathogenic variants in the *GJB2* gene. Gina is further evaluated and receives a cochlear implant. Gina most likely will have more typical speech development due to the early identification and the appropriate intervention for her hearing loss.

The Diagnosis

Clinical Diagnosis: Congenital nonsyndromic sensorineural hearing loss.
Genetic Diagnosis: *GJB2* gene, homozygous pathogenic variants.

Further Reading

Gifford, K.A., Holmes, M.G., Bernstein, H.H., 2009. Hearing loss in children. Pediatr. Rev. 30, 207–215. PMID 19487429.

Li, M.M., Tayoun, A.A., DiStefano, M., Pandya, A., Rehm, H.L., Robin, N.H., et al., 2022. Clinical evaluation and etiologic diagnosis of hearing loss: A clinical practice resource of the American College of Medical Genetics and Genomics (ACMG). Genet. Med. 24 (7), 1392–1406. PMID 35802133.

Smith, R.J.H., Jones, M.K.N., September 28, 1998. Nonsyndromic hearing loss and deafness, DFNB1. Adam, M.P., Everman, D.B. Mirzaa, G.M., Pagon, R.A., Wallace, S.E., Bean, L.J.H., Gripp, K.W., Amemiya, A., (editors). In GeneReviews. University of Washington, Seattle (WA), 1993–2023 [updated August 18, 2016]. PMID 20301449.

Case 11.2.4

Newborn Screening Cases: Abnormal Newborn Neuromuscular Screening

Tara Newcomb[1], Russell J. Butterfield[2] and Lynne M. Kerr[2]

[1]*Division of Pediatric Neurology, University of Utah Health, School of Medicine, Papillion, NE, United States;* [2]*Division of Pediatric Neurology, Department of Pediatrics, University of Utah Health, Shriners Hospital for Children, Salt Lake City, UT, United States*

Abstract

A newborn screens positive for spinal muscular atrophy, and this newborn screening and the availability of treatment have changed the evaluation and outcome.

The Case

You first meet baby Jimmy a few days after birth. This is the first child to this couple, although you are also the pediatrician for the mother's older children from another union. The mother is annoyed as her insurance had made her travel to a hospital in northern Utah, instead of allowing her to deliver closer to home in Idaho, where you are Jimmy's pediatrician. Overall, Jimmy looks good. Four days after Jimmy was born, you receive a call from the Utah Department of Health, as Jimmy's newborn screen tested positive for spinal muscular atrophy (SMA). Your nurse calls Jimmy's mother and asks the family to come in for a visit. Jimmy seems to be eating and growing well and is moving appropriately for his age of one week.

The Issues

- **What do the newborn screening test results mean?** Jimmy tested positive for SMA on newborn screening, which has only recently been added to the newborn screen in Utah and is still not available in many other states. If Jimmy had been born in Idaho, this diagnosis would not be apparent at this point because at one week of age he is thriving.
- **Does the fact that Jimmy looks so good mean that he will have a milder course?** You are not certain, but you tell the family that it is hard to predict based on the initial exam, since it is a progressive condition.
- **What is the next step?** You briefly discuss the diagnosis and arrange for Jimmy and his family to visit the Neuromuscular Clinic at a multispecialty Children's Hospital in the next few days.

The Referral

- **Genetic Evaluation**: Jimmy is seen by the geneticist and genetic counselor at the Children's Hospital 3 days later. They review the medical history that Jimmy had an uneventful prenatal course. His mother had declined prenatal carrier

Genomics in the Clinic. https://doi.org/10.1016/B978-0-12-816478-5.00022-4

screening (Section 4.1). Jimmy is still doing well with feeding and growing. A three-generation pedigree (Sections 2.7, 3.2.1) does not elicit any positive findings. A physical exam is performed, and Jimmy's neurologic exam is noted to be normal.

The American College of Obstetrics and Gynecology has recommended that all pregnant women be offered screening for SMA.

- **Genetic Counseling**: They discuss the following with the family: SMA is an autosomal recessive (Section 2.7.1) neuromuscular disorder that causes decreased survival of the anterior horn cells in the spinal cord (lower motor neurons), which innervate voluntary muscles. This results in progressive muscle atrophy and weakness. In the most severe form, type I, infants begin to show signs of weakness prior to 6 months old. They never gain the ability to sit independently. Eventually the majority of voluntary muscle control is lost, including muscles for breathing and swallowing. Without treatment, which has only been available recently, most children do not survive beyond the 2nd year without extraordinary measures.

  As SMA is autosomal recessive, both mother and father have contributed an abnormal survival motor neuron gene (*SMN1*) to Jimmy. Missing or defective survival motor neuron protein (SMN protein) results in SMA. Most cases of SMA are caused by deletions (Section 2.4.1) in various combinations of exons 7 and/or 8 in both copies of *SMN1* on chromosome 5. In addition, there is a modifying gene, *SMN2*, also on chromosome 5, which is similar to *SMN1*, but not completely functional in most people (Section 2.8). In children with SMA, the number of copies (copy number) of the *SMN2* gene appears to control severity by a complex mechanism allowing some production of the survival motor neuron protein.

  The copy number of *SMN2* can range from one to four or more. Babies with one or two copies of *SMN2* genes usually have type 1 SMA. If more *SMN2* copies are present, the baby most likely will present later with a milder form of SMA. Type 2 patients usually have 3 copies of *SMN2* and they sit independently but will not be able to walk. Type 3 patients usually have four or more copies of *SMN2*; in type 3 SMA the child will sit and walk, although walking does not always continue through adolescence.
- **Genetic Testing**: Confirmatory genetic testing is sent, including deletion/duplication testing of *SMN1* and *SMN2* (Section 6.2.1). Jimmy is found to have deletions in both copies of his *SMN1* gene and to have two copies of *SMN2*, which is most likely to manifest in SMA type 1.
- **Management and Recommendations**: Multiple FDA-approved treatments have become available for SMA in the last few years and data show that treatment given before the child has symptoms is more helpful than after symptoms are obvious. One such medication, nusinersen, needs to be given by lumbar puncture with loading doses over the first few months and then every 4 months indefinitely. Nusinersen is an antisense oligonucleotide (ASO) targeted to *SMN2* pre-messenger ribonucleic acid (pre-mRNA, Section 2.3.3) that increases production of full-length SMN protein. Another treatment option, onasemnogene abeparvovec-xioi, is a one-time IV administration of a gene replacement therapy that uses an adeno-associated virus (AAV9) to deliver normal copies of the *SMN1* gene. Both treatment options are carefully discussed with the parents and they are interested in pursuing the gene replacement therapy. Prior to treatment, Jimmy will need one more test to determine if he has antibodies to the virus used to deliver the gene replacement therapy. If so, he will not be able to get the gene replacement therapy, but nusinersen may be a possibility. If he receives the gene replacement therapy, he will need to be closely monitored for complications, especially for effects on the liver.

  SMA is now included on the national recommended newborn screening priority list and at least 47 states now include SMA in newborn screening tests with more states in progress of adding it (Section 5.4).

Credit for the availability of newborn screening for SMA and support of potential therapies rests with many tireless parent advocacy groups.

- **Implications for Other Family Members:** Jimmy's parents would like to know the chance that subsequent children may also have SMA. This is an autosomal recessive condition (Section 2.7.1). As both mother and father are carriers for this condition, there is a 1 in four chance that another child would have SMA, a two in four chance that the child will be a carrier, and a 1 in four chance that the child does not have a *SMN1* mutation in either allele. Prenatal testing (Section 4.4.2) is available for future pregnancies. The family also has the option to pursue preimplantation genetic testing (Section 4.5.1) in conjunction with in vitro fertilization to screen embryos for SMA. The mother's other children have a 50% chance of also being a carrier.
- **Differential Diagnosis**: Although Jimmy was diagnosed because of positive newborn screening, he would have presented with early muscle weakness to allow for a clinical diagnosis to be made. The differential diagnosis for findings

seen in SMA type 1 includes infant botulism, metabolic myopathies such as mitochondrial myopathy or Pompe disease (glycogen storage disease type 2), and neuropathies.

- **Caveats**: Although it is exciting to have a choice of potential treatments for SMA, these medicines are not curative and have significant side effects, and also long-term effects of treatment are unknown at this time. Newborn screening cannot detect all cases of SMA because it only detects deletions of exon 7 in *SMN1*. Newborns who have one point mutation and a deletion or two point mutations are missed. This accounts for up to 5% of cases. In these cases, sequencing of the *SMN1* gene is available to confirm the diagnosis.
- **Outcome**: Jimmy receives the gene replacement therapy intravenously and is closely monitored by his neuromuscular team, his pediatrician, and blood testing. At his 8-month-old well child visit, when severe symptoms of SMA would have been present without treatment, Jimmy is healthy and achieving his milestones consistent with his age.

The Diagnosis

Clinical Diagnosis: Spinal muscular atrophy type 1.
Genetic Diagnosis: *SMN1* gene, homozygous deletion exon 7. *SMN2* gene, two normal copies.

Further Reading

Aguti, S., Malerba, A., Zhou, H., 2018. The progress of AAV-mediated gene therapy in neuromuscular disorders. Expert Opin. Biol. Ther. 18, 681–693. PMID 29781327.

De Vivo, D.C., Bertini, E., Swoboda, K.J., Hwu, W.L., Crawford, T.O., Finkel, R.S., et al., 2019. Nusinersen initiated in infants during the presymptomatic stage of spinal muscular atrophy: interim efficiency and safety results from the phase 2 NURTURE study. Neuromuscul. Disord. 29, 842–856. PMID 31704158.

Keinath, M.C., Prior, D.E., Prior, T.W., 2021. Spinal muscular atrophy: mutations, testing, and clinical relevance. Appl. Clin. Genet. 14, 11–25. PMID 33531827.

Prior, T. W., Leach, M. E., Finanger, E., February 24, 2000. Spinal muscular atrophy. Adam, M.P., Everman, D.B. Mirzaa, G.M., Pagon, R.A., Wallace, S.E., Bean, L.J.H., Gripp, K.W., Amemiya, A., (editors). In GeneReviews. University of Washington, Seattle (WA), 1993–2023 [updated December 3, 2020]. PMID 20301526.

Scoto, M., Finkel, R.S., Mercuri, E., Muntoni, F., 2017. Therapeutic approaches for spinal muscular atrophy (SMA). Gene Ther. 24, 514–519. PMID 28561813.

Section 11.3

Craniofacial Cases

Case 11.3.1

Craniofacial Cases: Congenital Microcephaly

Dhruv K. Patel[1] and Sonja A. Rasmussen[2]

[1]*University of Florida College of Medicine, Gainesville, FL, United States;* [2]*Departments of Pediatrics and Epidemiology, University of Florida College of Medicine and College of Public Health and Health Professions, Gainesville, FL, United States*

Abstract

A newborn with microcephaly is found to have congenital toxoplasmosis. Genetic workup and evaluation for other possible prenatal exposures (e.g., alcohol) are discussed.

The Case

Natalie is a 2-day-old female born at 37 weeks gestation to a G2 P1001 mother who is admitted to your service in the newborn nursery. She reportedly had an uneventful prenatal and delivery history. Her birth parameters are as follows: weight 25th percentile, length 10th percentile, head circumference below 3rd percentile. Your physical examination is unremarkable, except for microcephaly.

The Issues

- **How concerned are you by microcephaly in this patient?** Microcephaly is defined as an occipitofrontal head circumference measurement that is below a certain value, typically more than 2 standard deviations below the median, although different cut-offs are sometimes used, for babies of the same gestational age and sex. Severe microcephaly is usually defined as more than 3 standard deviations below the median. Microcephaly is associated with an increased risk for intellectual disability, seizures, and hearing and vision problems. Head circumference measurements can be incorrect, so repeating the measurement is important, but if the measurement is confirmed, this finding is of significant concern.
- **Could Natalie have a genetic condition?** Yes, microcephaly (both isolated and when associated with other findings) can be due to a genetic cause. This baby has congenital (present at birth) microcephaly. Microcephaly can also be of later onset (acquired). Congenital microcephaly can be due to genetic or nongenetic causes. Genetic causes include single-gene disorders and chromosome abnormalities. Congenital microcephaly can also be caused by particular exposures during pregnancy (Section 4.2.1), including infectious causes (e.g., cytomegalovirus, herpes simplex virus, *Toxoplasma gondii*, Zika virus) and noninfectious causes (e.g., fetal alcohol syndrome, prenatal exposure to methylmercury). Microcephaly can be isolated (no other physical findings) or syndromic (associated with other physical findings in systems outside the central nervous system). Associated findings often can suggest a particular

Genomics in the Clinic. https://doi.org/10.1016/B978-0-12-816478-5.00023-6

diagnosis. A physical examination looking for associated findings is used to guide additional laboratory testing and evaluation.

- **How helpful is prenatal history in determining the cause of her microcephaly?** Prenatal history is sometimes helpful in determining the cause of microcephaly; thus, you consider possible teratogenic exposures (Section 4.2.1), including exposure to specific infections, alcohol, medications, or chemical exposures. Several prenatal infections can cause microcephaly. You ask about exposures to cat feces or undercooked meat (risk factors for *T. gondii* infection) and exposures to the urine of young children (risk factor for cytomegalovirus). Rubella virus infection during pregnancy can also cause microcephaly, but congenital rubella syndrome is rarely seen in the United States because nearly all women of childbearing age have been vaccinated, so you confirm that her mother received all recommended vaccines in childhood. You also ask about travel history, important since Zika virus infection is primarily transmitted through the bite of a Zika-infected mosquito, and mosquito-borne transmission is limited to certain geographic regions. Sexual transmission is also possible, so you even ask about the partner's travel. However, you know that even when the prenatal history is unremarkable, an exposure during pregnancy might be the cause of this newborn's findings. Because your additional history and physical examination do not provide specific clues as to the cause of Natalie's microcephaly, you request a genetic consultation for Natalie.

The Referral

- **Genetic Evaluation**: Natalie is seen by a clinical geneticist the next day (at 3 days of age). A prenatal history is obtained, which is mostly unremarkable. The mother reports no illnesses during pregnancy. She notes that the family has two cats, but that she did not change the litter during her pregnancy. Neither the mother nor her partner traveled during the pregnancy. The mother is rubella immune, according to prenatal testing. A three-generation family history (Sections 2.7, 3.2) is also obtained by the genetic counselor: no family members with microcephaly, developmental delay, or other congenital anomalies are noted. The geneticist asks specifically whether the parents might be related, since consanguinity increases the risk of autosomal recessive conditions, and congenital microcephaly can be inherited in an autosomal recessive manner. The parents deny consanguinity. The geneticist conducts a careful physical examination of the infant, including repeating the head circumference measurement, and looking for minor findings (e.g., epicanthal folds or abnormally shaped ears) that could provide clues into the cause of the baby's microcephaly. The only abnormalities noted on the exam are microcephaly and intermittent abnormal eye movements consistent with nystagmus. She is otherwise nondysmorphic. Based on the eye findings, the geneticist recommends a pediatric ophthalmology consult. In addition to the ophthalmology consult, the geneticist recommends a chromosome microarray (Section 6.1.3) to look for extra or missing genetic material.
- **Results**: The eye exam shows bilateral chorioretinitis, suggesting an infectious cause, although chorioretinal abnormalities can be seen in some rare genetic disorders. The microarray comes back as normal. Because of the ophthalmology findings, an infectious disease consult is requested. The infectious disease specialists recommend testing for infectious causes of microcephaly, including antibody (IgM and IgG) testing for cytomegalovirus and *T. gondii.* The infectious disease consultant notes that IgG at this age reflects maternal antibodies, but that IgM supports prenatal infection, and that it is important to test for infectious diseases early in life because IgM antibody levels can wane, making diagnosis more challenging. Antibody testing shows IgG positive for cytomegalovirus, with a negative IgM, and both IgG and IgM positive for *Toxoplasma.* Given the microcephaly and findings on ophthalmologic exam, brain imaging (e.g., head CT scan) is also recommended to look for intracranial calcifications, hydrocephalus, or other structural abnormalities often seen with prenatal infection. Head CT scan shows intracranial calcifications, consistent with congenital toxoplasmosis.
- **Management and Recommendations**: Treatment is available for some (e.g., congenital toxoplasmosis and cytomegalovirus), but not all (e.g., Zika) congenital infections. These treatments have been shown to improve outcomes, which emphasize the importance of making a diagnosis. If no infectious cause can be identified, the next step would be to test for variants in a large number of genes that cause microcephaly using a gene panel (Section 6.2.3) from one of several genetic testing companies.

  Based on recommendations from the infectious disease specialist, treatment for congenital toxoplasmosis (pyrimethamine, sulfadiazine, and folinic acid [leucovorin] for 1 year) is initiated. Because of the risk for hearing loss, hearing testing (brainstem auditory evoked response) is performed, which is normal. Abdominal ultrasonography is also performed to look for intrahepatic calcifications or hepatosplenomegaly and is normal. Children with congenital toxoplasmosis are at increased risk of developmental delay, so referral to early intervention services is also recommended.

- **Differential Diagnosis**: Another possible cause of microcephaly is alcohol exposure during pregnancy. Children with effects of their mother's alcohol intake during pregnancy are said to have the fetal alcohol spectrum disorder (FASD), which can range from mild to severe. This condition likely occurs in about 1%–2% of all children. Sometimes children have a severe form of FASD called fetal alcohol syndrome (FAS), characterized by dysmorphic facial features (Section 5.1.2), problems with growth, and central nervous system abnormalities, including microcephaly, developmental delay, and problems with attention and hyperactivity. There is no lab test to diagnose fetal alcohol syndrome, so this diagnosis should be considered after other causes have been ruled out. The diagnosis depends on physical findings that are consistent with the diagnosis. Confirmation of maternal alcohol consumption can strengthen the case for diagnosis, although is not needed if the child meets other criteria. Early diagnosis and implementation of interventions have been shown to improve outcomes for children with FASD.
- **Caveats**: Prenatal history is often negative for exposures, even when a teratogenic exposure is responsible for the findings in a child. Thus, it should not be used to exclude a teratogenic exposure as cause of a child's findings. Toxoplasma IgM antibodies obtained before 5 days of age could represent contamination of the infant's blood with maternal blood; thus testing should be repeated after 5 days. Receipt of a transfusion with blood products or of immune globulin could also result in false positive results. False negative results are also possible. Consultation with an infectious disease specialist regarding diagnostic testing and treatment is recommended.
- **Outcome**: Natalie continues to receive treatment for congenital toxoplasmosis for her first year of life. She has limited vision and mild developmental delay, but her hearing remains normal.

The Diagnosis

Clinical diagnosis: Congenital microcephaly, due to prenatal exposure to toxoplasmosis.
Genetic diagnosis: Not applicable.

Further Reading

Ashwal, S., Michelson, D., Plawner, L., Dobyns, W.B., Quality Standards Subcommittee of the American Academy of Neurology & The Practice Committee of the Child Neurology, Society. 2009. Practice parameter: evaluation of the child with microcephaly (an evidence-based review): report of the quality standards subcommittee of the American Academy of Neurology and the Practice Committee of the Child Neurology Society. Neurology 73, 887–897. PMID 19752457.

Del Campo, M., Jones, K.L., 2017. A review of the physical features of the fetal alcohol spectrum disorders. Eur. J. Med. Genet. 60, 55–64. PMID 27729236.

Devakumar, D., Bamford, A., Ferreira, M.U., Broad, J., Rosch, R.E., Groce, N., Breuer, J., Cardoso, M.A., Copp, A.J., Alexandre, P., Rodrigues, L.C., Abubakar, I., 2018. Infectious causes of microcephaly: epidemiology, pathogenesis, diagnosis, and management. Lancet Infect. Dis. 18, e1–e13. PMID 28844634.

Maldonado, Y.A., Read, J.S., Committee on Infectious Diseases, 2017. Diagnosis, treatment, and prevention of congenital toxoplasmosis in the United States. Pediatrics 139, e20163860. PMID 28138010.

Woods, C.G., Parker, A., 2013. Investigating microcephaly. Arch. Dis. Child. 98, 707–713. PMID 23814088.

Case 11.3.2

Craniofacial Cases: Misshapen Head

Ethylin Wang Jabs[*]

Department of Genetics and Genomic Sciences, Department of Pediatrics, Department of Cell, Developmental and Regenerative Biology, Icahn School of Medicine at Mount Sinai, New York, NY, United States

Abstract

A 2-month-old twin presents with an abnormal head shape, prominent eyes, and midface retrusion on clinical exam and is found to have a de novo heterozygous, pathogenic variant in the *FGFR2* gene.

The Case

A 2-month-old infant, Jake, is accompanied by his parents to see you, his family primary care physician, for his well-child care visit. He was the product of a twin pregnancy. You notice his prominent forehead and measure his head circumference, which is normal for his age, but you palpate that his anterior fontanelle is closed. His parents think he looks fine, but the father relays that Jake's paternal grandmother is concerned that he looks different from his twin brother, Vince, because she thinks his eyes appear prominent. His mother becomes defensive and quickly says that Jake looks fine and his vision is fine too because he pays attention to her face, recognizes her at a distance, and is starting to follow objects with his eyes. You also note that he is able to hold his head up and is beginning to push up when lying prone. He smiles and coos at his mother. Although your assessment is that his development is normal, you are concerned that his anterior fontanelle is closed.

The Issues

- **What is concerning about the anterior fontanelle closing early?** The anterior fontanelle normally closes between 9 and 18 months, a few percent are closed by 3 months, and almost all are closed at 24 months (Section 5.1.1). The differential diagnosis for a prematurely closed anterior fontanelle typically includes hyperthyroidism, hypophosphatemia, and hyperparathyroidism, as well as vitamin D deficiency. Your evaluation, the results of his newborn screen and additional laboratory tests rule out these conditions. You wonder if he may have craniosynostosis. The definition of craniosynostosis is the premature fusion of skull bones that leads to an abnormal head shape. You decide to refer him to the craniofacial multidisciplinary clinic at the hospital, where he will be seen not only by the neurosurgeon, but by other specialists.

Multidisciplinary teams are the mechanism for organizing and coordinating health and care services to meet the individual's complex care needs. The teams bring together the expertise and skills of different professionals to screen,

[*] Author EWJ was employed by Icahn School of Medicine at Mount Sinai at the time of writing the manuscript, but is currently employed by Mayo Clinic

Genomics in the Clinic. https://doi.org/10.1016/B978-0-12-816478-5.00024-8

assess, plan, and manage care jointly. The American Cleft Palate-Craniofacial Association standards for team care include a patient care coordinator and speech-language pathology, surgery, and orthodontic specialties. There is access to professionals in psychology for neurodevelopmental and cognitive assessment, social work, audiology, general and pediatric dentistry, otolaryngology, and pediatrics/primary care. The team can refer to a neurosurgeon, an ophthalmologist, a radiologist, and, most importantly, a geneticist because about a third of all cases seen in craniofacial clinics will have a syndrome.

- **Does he have craniosynostosis?** The next week Jake is seen by the craniofacial multidisciplinary team. The pediatrician on the team concurs that his head circumference and growth curves are normal, at the 20th percentile, and his anterior fontanelle is closed, while his posterior fontanelle is still open and normal in size. The developmental psychologist also agrees that he has reached appropriate milestones for a 2–3 month-old. The neurologist, neurosurgeon, and ophthalmologist decide that if he has a craniosynostosis condition, he is not at immediate risk for increased intracranial pressure as his eye exam is normal, and he has no symptoms secondary to neurological involvement. The medical geneticist on the team notes that he has dysmorphic features with brachycephaly and refers the family to her genetic clinic for a more in-depth evaluation. The team recommends that you continue to follow him closely for changes in head growth, development, and neurologic signs while awaiting his genetic evaluation.
- **Can this be due to an underlying genetic condition?** The parents do not understand why they should see the clinical geneticist. They wonder if his misshapen head is due to crowding in the womb as his mother is very petite and carried the twin pregnancy to term. Could his head shape become better with time? Does having an abnormal head shape affect his brain? They also are worried whether the other twin may have the same condition, but Vince, as well as other members of the family, appears to have a normal skull shape. You explain that Jake seems to be doing well at this time and that you will follow his head growth and development and see him in a month. You explain that a referral for Jake to genetics will assist in determining the cause of his abnormal head shape, and deciding if any other body systems should be evaluated, and that Vince can be assessed as well.

The Referral

- **Genetic Evaluation**: Jake and Vince are 3 months of age when they are seen by the clinical geneticist, who obtains a full medical history, review of systems, developmental history, social history, and physical examination. His mother reports that she was 24 years old, and the father was 40 years old at the twins' birth. She had routine prenatal care, including ultrasounds, during the pregnancy that were normal. The twins were a product of a normal full term, Cesarean section delivery. The genetic counselor obtains a detailed three-generation pedigree (Sections 2.7, 3.2) and no one else in the family has symptoms of craniosynostosis, any other birth defects, miscarriages, stillbirths, early infant deaths, or consanguinity. On examination, Jake is found to have normal growth parameters including his head circumference. On physical exam, he has a prominent forehead, closed anterior fontanelle, midface retrusion or flattening, and flat occiput. The geneticist also examines Vince, whose exam is remarkable for relatively large, big toes and mild syndactyly of his second and third toes, bilaterally (Fig. 11.3.2.1). She informs the family of the clinical findings in both boys.
- **Genetic Counseling**: The clinical geneticist begins by explaining the features found in Jake and Vince, and that Jake has craniofacial features that may be consistent with a craniosynostosis syndrome. Vince does not have the same features but has large big toes and mild syndactyly that may be found in some forms of craniosynostosis syndromes. Craniosynostosis can be associated with a small fontanel or early fontanel closure. It is a common birth defect, occurring in about 1 in 2500 live births. Eighty-five percent of cases have craniosynostosis as the only clinical finding and thus are nonsyndromic. The other 15% of cases are syndromic with other clinical features, especially involving the bones of the hands and feet. There are more than 200 different, rare syndromes with craniosynostosis as a feature. About 30% of all cases have identifiable genetic pathogenic variants, and there are more than 100 different genes with variants that can cause craniosynostosis. Craniosynostosis can occur for the first time in a member of a family or be inherited. Although environmental causes have been reported to be associated with craniosynostosis, such as intrauterine constraint as in the case of twinning, parity, prematurity, and maternal tobacco or nitrosatable drug use, evidence for any of these being causative is not conclusive.
- **Genetic Testing**: The clinical geneticist explains to the parents that molecular genetic testing can be ordered for individuals in whom a craniosynostosis syndrome is suspected. The genetic counselor discusses using a multigene panel test (Section 6.2.3) for craniosynostosis that sequences and performs deletion/duplication testing of genes and chromosomal regions known to be involved in craniosynostosis. The different test results that can come back include a positive result with a known disease-causing variant, a negative result, and a variant of uncertain significance (VUS) (Section

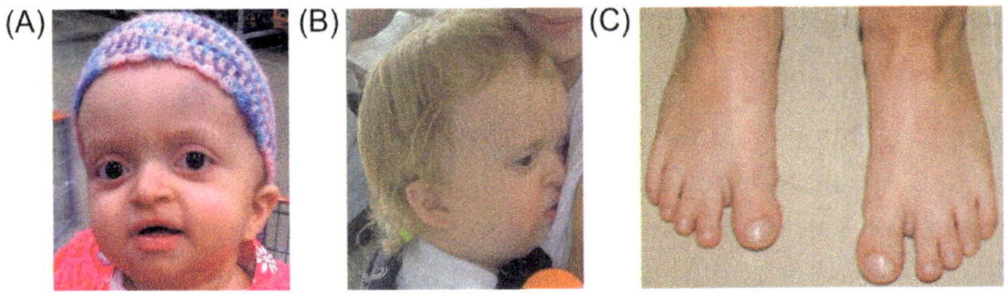

FIGURE 11.3.2.1 Crouzon or Pfeiffer syndrome. (A), (B) Prominent forehead, midface flattening, and flat occiput. *Photos used with permission from patient.* (C) Pfeiffer syndrome with large, big toes and syndactyly of second and third toes. *Reprinted from FGFR2 mutation in a patient without typical features of Pfeiffer syndrome—The emerging role of combined NGS and phenotype based strategies, Flottmann, R., Knaus, A., Zemojtel, T., Robinson, P.N., Mundlos, S., Horn, D., Spielmann, M. Eur J Med Genet. Aug; 58(8):376–80, Figure 1C and 2015, with permission from Elsevier.*

7.1.5). A positive test result will give a molecular diagnosis, which may be beneficial for knowing if the twins have a diagnosis of a craniosynostosis syndrome and assisting with management, treatment, prognosis, and family planning. A negative result means that the test was not successful in identifying a pathogenic variant in one of the genes tested. It does not mean that the twins do not have a genetic condition. If a variant of uncertain significance is found, parental samples can aid in the interpretation because if one of the parents has the variant and by examination does not have symptoms of craniosynostosis, then the variant is likely not to be pathogenic. The parents express understanding and decide that they would like to pursue testing, but worry about the cost because the twins have Medicaid health coverage. The counselor tells them that the panel which will test the 10 genes most commonly associated with craniosynostosis will be covered by their insurance, and other genomic testing could be considered later depending on these results. The parents sign a consent for the testing.

Determining which multigene panel to order can be challenging because various clinical molecular laboratories have different sets of genes to be tested (Section 6.2.3). Some have only genes associated with craniosynostosis, others have genes associated with several different craniofacial disorders including craniosynostosis, and some have genes associated with skeletal disorders including craniosynostosis. Some geneticists believe it is better to order a panel with the most common craniosynostosis genes, while others select the panel with the greatest total number of genes especially if they are not sure of the clinical diagnosis. Sometimes it is necessary to order a panel and customize the order by adding additional single gene(s) (Section 6.2.3) not included on the standard panel if the geneticist believes them most likely to be associated with the clinical symptoms or syndrome found upon examining the patient. Other geneticists order exome sequencing (Section 6.3.1) with a microarray (Section 6.1.3) of the genome to maximize the analysis. Often, the higher the number of genes assessed, the more likely there will be a variant of uncertain significance. What ultimately is ordered also may depend on which laboratory accepts the family's insurance and what the family is willing to pay out of pocket (Section 10.2).

- **Results**: Jake's test comes back with a heterozygous, missense variant in the fibroblast growth factor receptor 2 gene (*FGFR2*), a synonymous change, c.1032G > A, p.A344A. Although there is a nucleotide change from a "G" to an "A" that does not change the encoded amino acid alanine, the variant creates a cryptic splice site (Section 2.4.1). This new donor splice site sequence causes a 51 bp or 17 amino acid deletion, which results in a stable transcript and altered receptor function. Vince undergoes site-specific, targeted testing (Section 6.2.1) and also is found to have the same pathogenic variant.

  This variant and other pathogenic variants (missense, small deletions and insertions, and splice site—Section 2.4.1) in *FGFR2* have been found in patients with Crouzon, Pfeiffer, and Jackson–Weiss syndromes. They are gain of function variants causing a new or enhanced activity for the *FGFR2* protein. Also, all three conditions have clinical findings of craniosynostosis, especially synostosis of the coronal suture, ocular proptosis, midface retrusion, and mandibular and dental dysgenesis. Pfeiffer and Jackson–Weiss syndromes also have skeletal limb abnormalities of broad and large first toes and thumbs, syndactyly, and radioulnar synostosis. These conditions with overlapping craniofacial features were initially considered distinct, clinical conditions, but now because a common molecular cause has been identified, they represent conditions along the spectrum of *FGFR2*-related craniosynostosis syndromes.
- **Management and Recommendations**: Jake and Vince need to be monitored for risk of developing moderate or severe craniosynostosis, which would require surgical intervention to remove the fused bones to allow continued brain growth and development. If surgery is needed, it usually is done within the first year of life. Some children undergo a series of surgical procedures to assure normal brain and midface growth. They should be monitored by their family doctor and regularly followed by the multidisciplinary craniofacial clinic and relevant specialists. Generally, intelligence and global development is normal for most craniosynostosis syndromes with appropriate management. However, because

long-term studies suggest that there may be associated learning disabilities, development needs to be followed carefully.

- **Implications for Other Family Members**: To determine if this is a de novo or inherited variant, both parents undergo site-specific, targeted testing for the missense variant and are not found to have the change. This raises the possibility that Jake and Vince are identical twins with the same de novo variant. However, the parents always thought they looked very different from each other because Jake looks like his mother and Vince looks like his father. Genetic testing for identity is ordered, and the results show that they differ in multiple DNA polymorphisms, so they are fraternal twins. Collectively, the test results from this nuclear family suggest that one of the parents has germline mosaicism (Section 2.4.2) for the *FGFR2* variant. For the parent who is mosaic for the variant, the risk that a future offspring may be affected with this autosomal dominant condition may range from an empiric risk of as low as 1%–5% to as high as 50%. The clinical geneticist also mentions to the parents that advanced paternal age can be associated with an increased risk, particularly for missense variants to occur in a father's germline (Sections 2.4.2, 4.2.2). The parents, if they plan to have additional children, may want to have prenatal counseling to discuss their options, which can include targeted testing for the missense variant in future pregnancies and offspring (Sections 4.4, 4.5). The genetic counselor also explains that each of the boys will have a 50% risk of passing on the variant to each of their offspring, and the clinical manifestations can be variable, even among parents and offspring within a family, as in the case of Jake and Vince. The optimal time to discuss risks to offspring and reproductive options is when they are of childbearing age (Section 11.18.1)
- **Caveats**: Genetic tests can have different sensitivity levels for the detection of mosaicism (Sections 6.1.4, 6.3.2) and therefore, a negative result cannot rule out mosaicism, especially in the case of germline mosaicism in a parent because a DNA sample is taken from blood or saliva and not germ cells. Also, when using site-specific, single gene targeted testing as compared to other genetic tests that analyze multiple genes or sequences, nonpaternity may not be detected as a reason for a negative result in parents because only a few short DNA segments are analyzed for the possibility of polymorphisms present in the father that are not present in either twin.
- **Outcome**: The family understands the genetic risk for *FGFR2*-related craniosynostosis in their family, and the twins follow up with the multidisciplinary team to monitor for symptoms related to condition. At 6 months of age, Jake's head circumference becomes relatively smaller, crossing percentiles to the fifth percentile, and on CT scan imaging, the coronal suture is fused. He undergoes surgery and at 3 years of age is doing fine with his development on target when he enters pre-kindergarten. Vince's head shape changes, with his forehead becoming more prominent with flattening of the back of his skull, but by 3 years of age he has not required craniofacial surgery because his sutures have remained open by CT scan evaluation. The parents are given information how to contact a support group (Sections 3.5, 7.4), and they become actively involved in helping families with babies with craniosynostosis, sharing their experiences with them.

The Diagnosis

Clinical Diagnosis: Patients—*FGFR2*-related craniosynostosis. Parent—germline mosaicism.
Genetic Diagnosis: Patients—*FGFR2* gene, heterozygous pathogenic variant.

Further Reading

Dias, M.S., Samson, T., Rizk, E.B., Governale, L.S., Richtsmeier, J.T., Section on Neurologic Surgery, Section on Plastic and Reconstructive Surgery, 2020. Identifying the misshapen head: craniosynostosis and related disorders. Pediatrics 146 (3) e2020015511 PMID 32868470.

Li, X., Park, W.J., Pyeritz, R.E., Jabs, E.W., 1995. Effect on splicing of a silent FGFR2 mutation in Crouzon syndrome. Nat. Genet. 9 (3), 232–233. PMID: 7773284.

McCarthy, J.G., Warren, S.M., Bernstein, J., Burnett, W., Cunningham, M.L., Edmond, J.C., Figueroa, A.A., Kapp-Simon, K.A., Labow, B.I., Peterson-Falzone, S.J., Proctor, M.R., Rubin, M.S., Sze, R.W., Yemen, T.A., Craniosynostosis Working Group, 2012. Parameters of care for craniosynostosis. Cleft Palate Craniofac. J. 49 (Suppl), 1S–24S. PMID 21848431.

Wilkie, A.O.M., Johnson, D., Wall, S.A., 2017. Clinical genetics of craniosynostosis. Curr. Opin. Pediatr. 29 (6), 622–628. PMID 28914635.

Case 11.3.3

Craniofacial Cases: Congenital Nystagmus

Vaidehi Dedania[1], Alexander Bottini[2], Lediana Goduni[3] and Scott Brodie[1]

[1]*New York University School of Medicine, New York University Langone Medical Center, New York University Eye Center, New York, NY, United States;* [2]*The Retina Institute, St. Louis, MO, United States;* [3]*Bascom Palmer Eye Institute, Miami, FL, United States*

Abstract

A 6-month-old boy with congenital nystagmus and poor vision is found to have Leber congenital amaurosis.

The Case

A 6-month-old boy, James, is noted to be visually inattentive to his surroundings at his regular well-baby examination. His mother notes that his eye movements have been "disorganized" since early infancy, and, indeed, a coarse, horizontal nystagmus has been evident since 2 or 3 months of age. Most neurological milestones are normal, but James does not seem to recognize people or toys. His mother worries that his development is impaired compared to his healthy older sister when she was the same age.

The Issues

- **Is there a problem with the vision?** An infant should be able to fixate and track a penlight or other object of interest by a few months of age. This visual inattention should raise the question of visual or cognitive impairment.
- **Could the nystagmus be a normal variant?** Nystagmus (rhythmic oscillation of eye position) is never normal. It may indicate disruption of the afferent visual apparatus ("sensory nystagmus"), motor system impairment (such as cerebellar abnormalities), or be idiopathic ("primary motor nystagmus"). It can be the first sign of albinism (Section 11.11.1). But taken together with visual inattention, there is suspicion for a primary eye problem.
- **What is the next step?** These findings should prompt referral to an ophthalmologist, preferably a pediatric ophthalmologist if one is available, or to a pediatric neurologist if other neurologic issues, such as generalized developmental delays, are present. For James, there are no other issues. You refer James to a pediatric ophthalmologist.

Ophthalmologic evaluation will include assessment of visual attention (ability to fixate a penlight) and ability to "follow" or "track" a visual target; assessment of the quality and magnitude of the nystagmus, and integrity of the extraocular muscles; evaluation of the pupillary reactions to direct and consensual stimulation by light; examination of the anterior segment structures (cornea, iris, crystalline lens); and dilated fundus examination, usually with the indirect ophthalmoscope. This will help distinguish between disorders which are "ophthalmoscopically visible" (e.g., corneal, lens or vitreous opacity, optic nerve hypoplasia or aplasia, coloboma, retinal dysplasia, traction or detachment, or albinism) and those which are "ophthalmoscopically variable" (e.g., Leber congenital amaurosis, achromatopsia, congenital stationary night blindness) in which differentiation is frequently facilitated by electroretinography testing (ERG).

Genomics in the Clinic. https://doi.org/10.1016/B978-0-12-816478-5.00025-X

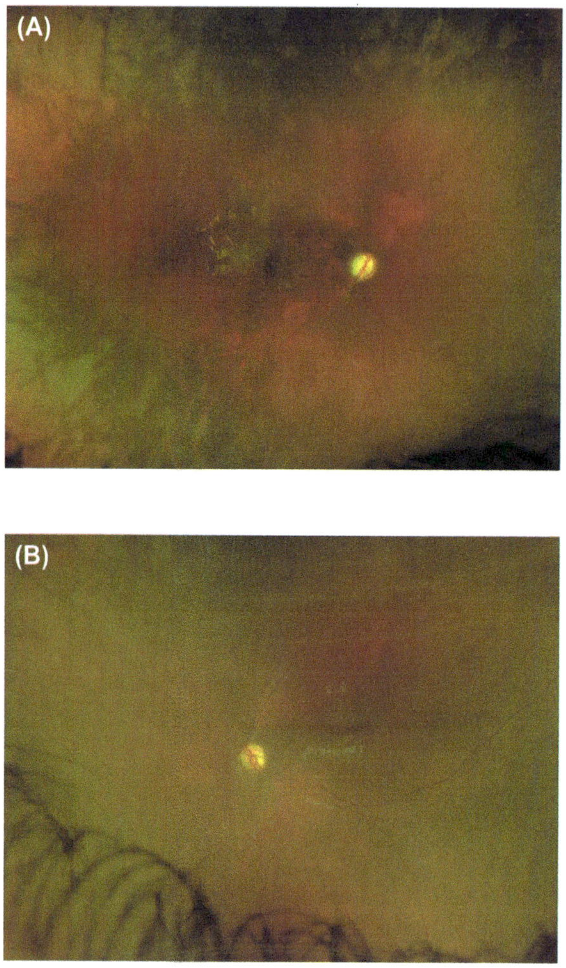

FIGURE 11.3.3.1 Fundus photographs of the (A) *right* and (B) *left* eye illustrating the optic nerve and retina in both eyes. Notable retinal findings include irregular reflections from the internal limiting membrane centrally and irregular coloration of the retinal pigment epithelium; optic disc pallor; and near-invisibility of the retinal blood vessels.

- **What does the ophthalmologic evaluation show?** Five days later, James is seen by the pediatric ophthalmologist. Dilated fundus examination (Fig. 11.3.3.1) confirms narrowing of the vessels. Abnormal wrinkling of the internal limiting membrane is noted in the posterior poles. ERG testing shows essentially "extinguished" responses—no recordable signals in response to flash stimulation of the eyes under light-adapted and dark-adapted conditions. A diagnosis of Leber congenital amaurosis is made. You refer James for a genetic evaluation to help counsel the family.

The Referral

- **Genetic Evaluation**: The geneticist and genetic counselor meet with James and his parents when he is 7 months of age. A medical history is taken and developmentally, James is found to be on target other than visually. A three-generation family history (Sections 2.7, 3.2) is obtained, which finds no evidence for any other family member with eye abnormalities. A physical exam is normal.
- **Genetic Counseling**: The genetic team discusses the genetic implications of his clinical diagnosis with the family. Leber congenital amaurosis is a group of autosomal recessive genetic eye conditions. It is part of the spectrum of early-onset retinal dystrophies. It is characterized by nystagmus and poor vision from birth, variable retinal findings, and extinguished or near-extinguished electroretinographic responses to light stimulation. Affected infants may be drawn to light and are sometimes seen to press on their eyes with their fingers (the "oculo-digital sign") as if attempting to generate retinal stimulation. James's parents have not seen him do this. At least 16 different causative genes have been identified. James's family requests genetic testing.

- **Results**: In recent years, genetic testing has played an increasing role in clarifying the diagnosis and visual prognosis. A blood sample is sent for single-gene testing (Section 6.2.2) and returns negative for pathogenic variants in the *RPE65* gene, one of the most common genes associated with Leber congenital amaurosis, and the only such gene for which there is an approved treatment.

If clinically the diagnosis is not clear, a gene panel (Section 6.2.3) could be sent for genes associated with similar eye findings. Leber congenital amaurosis is genetically heterogeneous (Section 2.8), and pathogenic variants are most commonly in CEP290, CRB1, GUCY2D, RDH12, and RPE65 genes. In about 30% of all people with Leber congenital amaurosis, the cause of the disorder is unknown. In this case, single-gene sequencing is being done first because mutations in RPE65 are treatable by gene therapy.

- **Management and Recommendations**: Testing should be done to confirm the diagnosis of Leber congenital amaurosis and to assess the possibility of gene replacement therapy (Section 7.4.6). A viral-vector-based gene replacement therapy for the *RPE65*-mediated variant of Leber congenital amaurosis, voretigene neparvovec, was approved by the FDA as of December 2017. This was the first FDA-approved gene replacement therapy.
- **Differential Diagnosis**: Other inherited retinal dystrophies include achromatopsia, Alstrom syndrome, congenital stationary night blindness, and other disorders of retinal dysfunction. These can be differentiated by an examination, family history, and testing, most importantly by the ERG findings.
- **Outcome**: James is not a candidate for *RPE65* gene replacement therapy and gene panel testing also is negative. He is enrolled in occupational therapy and vision therapy classes, where he and his family are able to better learn how to function with his limited vision, which may remain as is or may slowly degenerate.

The Diagnosis

Clinical Diagnosis: Leber congenital amaurosis.
Genetic Diagnosis: Unknown genetic cause.

Further Reading

Carr, R.E., Siegel, I.M., 1990. Electrodiagnostic Testing of the Visual System: A Clinical Guide. FA Davis, Philadelphia.

Miraldi Utz, V., Coussa, R.G., Antaki, F., Traboulsi, E.I., 2018. Gene therapy for RPE65-related retinal disease. Ophthalmic Genet. 39, 671−677. PMID 30335549.

Robson, A.G., Nilsson, J., Li, S., Jalali, S., Fulton, A.B., Tormene, A.P., Holder, G.E., Brodie, S.E., 2018. ISCEV guide to visual electrodiagnostic procedures. Doc. Ophthalmol. 136, 1−26. PMID 29397523.

Stephen, H.T., Tarun, S., 2018. In: Atlas of Inherited Retinal Diseases. Springer Nature, Switzerland.

Case 11.3.4

Craniofacial Cases: Loss of Central Vision

Vaidehi Dedania[1], Alexander Bottini[2], Lediana Goduni[3] and Scott Brodie[1]

[1]*Department of Ophthalmology, New York University School of Medicine, New York University Langone Medical Center, New York University Eye Center, New York, NY, United States;* [2]*The Retina Institute, St. Louis, MO, United States;* [3]*Bascom Palmer Eye Institute, Miami, FL, United States*

Abstract

A 7-year-old girl with worsening visual acuity and history of consanguinity and family history of Stargardt disease is found to have the same *ABCA4* pathogenic variants as the other family members.

The Case

A 7-year-old girl, Sara, accompanied by her mother, presents to you, her new pediatrician, with complaints of gradually worsening visual acuity. There is no history of night blindness. She otherwise is doing well. Her mother reports that Sara's father and she are second cousins and that Sara's older brother has Stargardt disease, with genetic testing revealing a pathogenic variant or mutation in the *ABCA4* gene. Visual acuity was only mildly impaired for Sara (20/50 in the right eye, 20/60 in the left eye) until several months ago, when more severe impairment has been noted. You refer Sara back to the pediatric ophthalmologist.

The Issues

- **What are her eye findings on follow-up evaluation?** Best-corrected visual acuity measures 20/400 in the right eye, 20/200 in the left eye. Color vision is severely impaired. Dilated fundus examination (Fig. 11.3.4.1) shows exaggerated granularity of the retinal pigment epithelium in the central macula, with apparent widening of the foveal depressions. Fundus autofluorescence imaging demonstrates small bull's-eye lesions (Fig. 11.3.4.2). Spectral mode optical coherence tomography (OCT) imaging, a noninvasive imaging modality that allows each layer of cells in the retina to be seen in cross section, confirms severe central retinal thinning with loss of the central outer segment ellipsoid lines (Fig. 11.3.4.3).
- **Does her brother's diagnosis have any implications for Sara?** Yes. Stargardt disease is a type of macular dystrophy that is genetically inherited in an autosomal recessive fashion (Section 2.7.1). Therefore, Sara has the same risk of 25% having inherited the disease as her brother since their parents are heterozygous carriers.
- **If Sara has Stargardt disease, will her vision continue to decline?** Usually, patients with Stargardt disease present with loss of central vision and disturbances of color vision. Generally, visual acuity does continue to decline slowly in these patients. The severity and rate of vision loss varies both within families and also between patients with the disease. Some patients can continue to retain relatively good functional vision while others may become legally blind, with an acuity of 20/200 or less in the better seeing eye. Additionally, there is also visual field loss to variable degrees. Sara's electroretinograms (ERGs), representing electric responses of the retina to different stimuli, show significant reduction

Genomics in the Clinic. https://doi.org/10.1016/B978-0-12-816478-5.00026-1

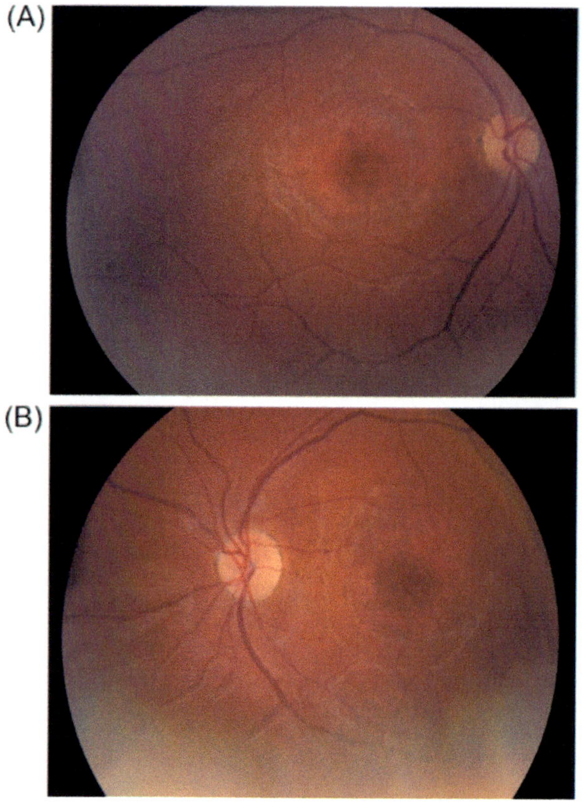

FIGURE 11.3.4.1 Fundus photographs of the optic nerves and macula of the (A) *right* and (B) *left* eye showing enlarged foveal depressions, with enhanced granularity of the underlying retinal pigment epithelium (*centrally*).

in both low light (scotopic) function and good light (photopic) and cone (color) function, compatible with either a diagnosis of Stargardt disease or cone-rod dystrophy (Fig. 11.3.4.4). The ophthalmologist recommends a referral to a geneticist for molecular diagnostic testing.

Prior to referring the patient to a geneticist, the ophthalmologist can use many different types of tools to make the diagnosis of Stargardt disease. Although this disease can have a varied presentation on clinical examination, other tests and imaging modalities such as visual field testing, fluorescein angiography, OCT, fundus autofluorescence, and electroretinography (ERG) can help the ophthalmologist make the diagnosis.

The Referral

- **Genetic Evaluation**: Stargardt disease is a disorder of retinal degeneration. It is important to note that it can present at essentially any age—it can be diagnosed in patients as young as 5 years old and in those greater than 50 years old. The peak incidence of disease is bimodal with greatest incidence in childhood and in early adulthood. Patients who present at younger ages tend to have more severe disease than those presenting at later ages.

  Pathogenic variants in *ABCA4* cause 95% of the cases of Stargardt disease, and they are also responsible for other retinal degenerations such as cone-rod dystrophies and retinitis pigmentosa. The other 5% of cases of Stargardt are caused by pathogenic variants in other genes, including *ELOVL4*, *PROM1*, *PRPH2*, and *BEST1*. Given such, retinal degenerations are becoming more frequently defined and named by the gene variant instead of a particular disease name (i.e., *ABCA4* macular dystrophy instead of Stargardt disease). The *ABCA4* gene is a very large gene that has 50 exons with more than 1000 reported disease-associated variants. The polymorphic nature of this gene and the wide allelic heterogeneity make it very difficult to assign genotype–phenotype associations. Overall, Stargardt disease is very heterogeneous, both clinically and genetically (Section 2.8). The highly variable nature of this disease suggests that there may be other genetic factors and/or environmental influences on the course of the disease.
- **Genetic Counseling**: Obtaining a thorough family history is also very important in making the diagnosis because the symptoms and inheritance pattern observed can be very valuable in making the diagnosis. Since Stargardt disease is an autosomal recessive disease (Section 2.7.1), the parents are consanguineous (Section 3.2.4), and one of the children already has a diagnosis of Stargardt disease; this strongly suggests that the parents are at least carriers of the disease.

(A)
(B)
(C)

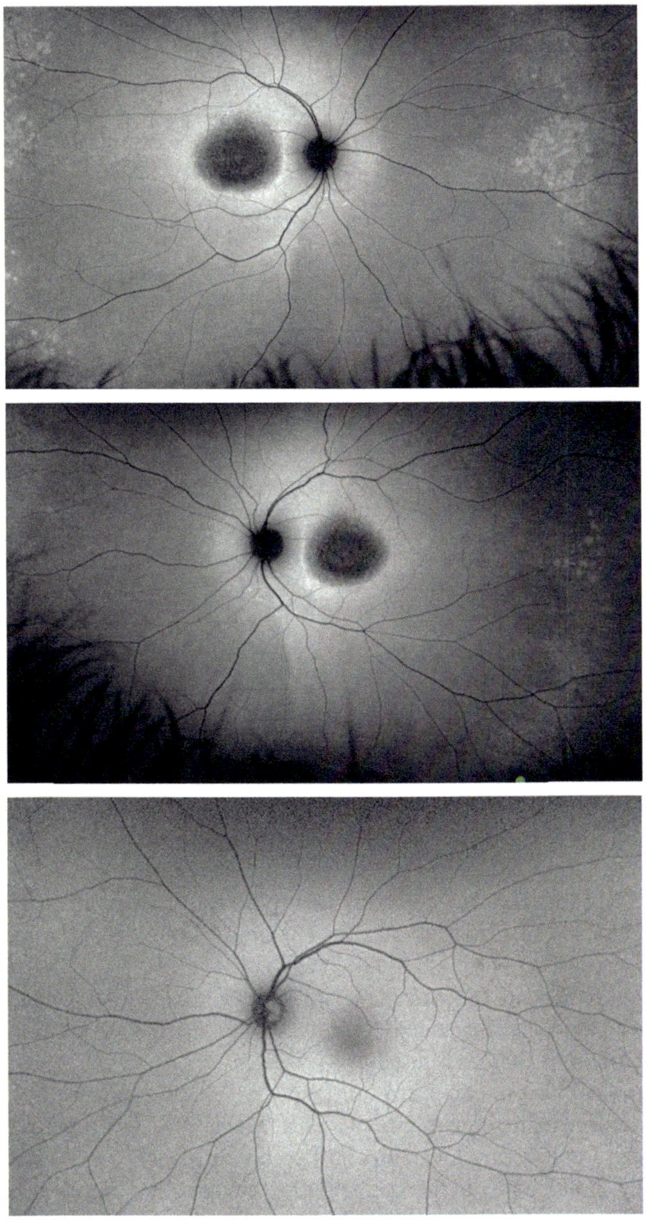

FIGURE 11.3.4.2 Fundus autofluorescence images of the (A) *right* and (B) *left* eye showing hypoautofluorescence (*darkness*) in the posterior poles (*centrally*), with central zones of hypofluorescence surrounded by hyperfluorescent halos. (C) Normal autofluorescence image for comparison.

If they are both carriers, each child they have has a 25% chance of having the disease, 50% of being a carrier and 25% of not having the disease and not being a carrier. If one parent has the disease and the other is a carrier, there is a 50% chance that each child will have the disease and a 50% chance that each child will be a carrier only. Of course, if each parent has two mutant copies of the gene (i.e., if each parent has the disease), then each child will also have the disease.

- **Management and Recommendations**: Genetic testing is recommended. Some of the current genetic testing methods do not detect all of the disease-causing alleles of *ABCA4*. Next-generation sequencing (Section 6.2.2) has been shown to be more effective and may also detect novel *ABCA4* variants. Furthermore, some pathogenic variants of the *ABCA4* gene are located outside of the promoter and can be missed by the current genetic testing methods so it is important to keep this in mind when interpreting the results of the genetic testing. There are some laboratories that give a pathogenicity score for the variant or mutation, which may be helpful when counseling patients.

  There is no treatment or cure for this disease. There is extensive research in the field at this time, including gene replacement therapies, stem cell therapies, and pharmacological interventions. At this time, it has been recommended

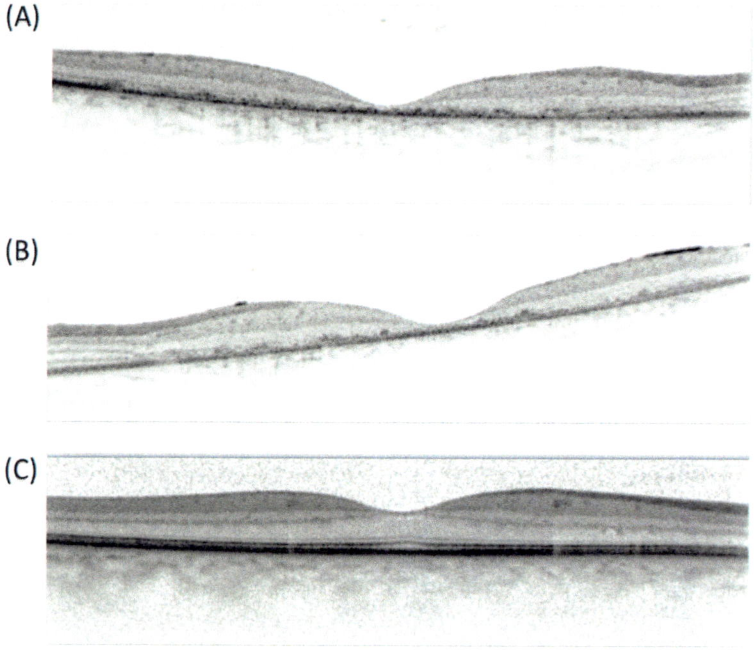

FIGURE 11.3.4.3 Spectral mode optical coherence imaging (OCT) of the (A) *right* and (B) *left* eye showing severe thinning of the central foveal floors, with disruption of outer segment ellipsoid lines. (C) Normal OCT image for comparison.

on theoretical grounds that patients with Stargardt disease protect themselves from extensive exposure to bright light by wearing sunglasses and hats when outside for prolonged periods of time. Smoking also should be avoided as are high dose vitamin A supplements. Clinical confirmation of preservation of vision by these means is lacking.

Treatment for cone and cone-rod dystrophies, which also cause progressive worsening of vision, is lacking as well and mostly focused on low vision aids and psychosocial support. It is important, therefore, that patients affected by all of these conditions be connected with low vision services and specialists to maximize their function. It is also important for them to contact a parent group or support organization (Sections 3.5, 7.4.4).

- **Results**: Gene testing identifies compound heterozygous variants (Sections 2.4.1, 2.7) in Sara's *ABCA4* alleles, confirming the diagnosis of autosomal recessive, *ABCA4*-associated macular dystrophy. In the absence of retinal flecks on retinal exam, the case is considered an atypical variant of Stargardt disease.
- **Implications for Other Family Members**: It is important also to emphasize to the parents that they are both at least carriers of the disease. The frequency of being a carrier is actually quite high—perhaps as many as 1 in 20 people. It is also possible that, since the age of presentation for Stargardt can vary (even up to 50 years) the parents may be not only carriers, but may also have the disease. Therefore, they should undergo thorough ophthalmological examination as well as genetic testing. Other family members should also be evaluated for Stargardt disease or other disorders based on initial evaluation.
- **Differential Diagnosis**: The history and testing modalities listed above are also important to rule out other causes of macular dysfunction as well, such as abnormalities of photoreceptor function (cone and cone-rod dystrophies). The cone and cone-rod dystrophies can be inherited in an autosomal recessive, autosomal dominant, or X-linked recessive fashion (Section 2.7), making family history critical to identification. Furthermore, it is also important to know that the cone and cone-rod dystrophies, which are a heterogeneous group of diseases of photoreceptor dysfunction, can present similarly to Stargardt disease with decreased vision, central scotomas, and color vision loss. These patients, unlike the Stargardt patients, may also have photophobia, which can help distinguish the entities. Similarly to Stargardt disease, these disorders also present in the first few decades of life. Some photoreceptor disorders are stationary (i.e., achromatopsia, blue cone monochromacy, and Bornholm eye disease) while others (i.e., cone and cone-rod dystrophies such as Alström syndrome) are progressive in nature, like Stargardt disease. In young children, the disease known as neuronal ceroid lipofuscinosis must also be considered as it can present similarly to Stargardt disease, but is fatal. One way to make the distinction between these is ERG. The ERG is very abnormal in neuronal ceroid lipofuscinosis, but less likely to be severely abnormal in young children with Stargardt disease. The differential diagnosis for Stargardt disease also includes drug-induced macular toxicity and Stargardt-like dominant macular dystrophy.

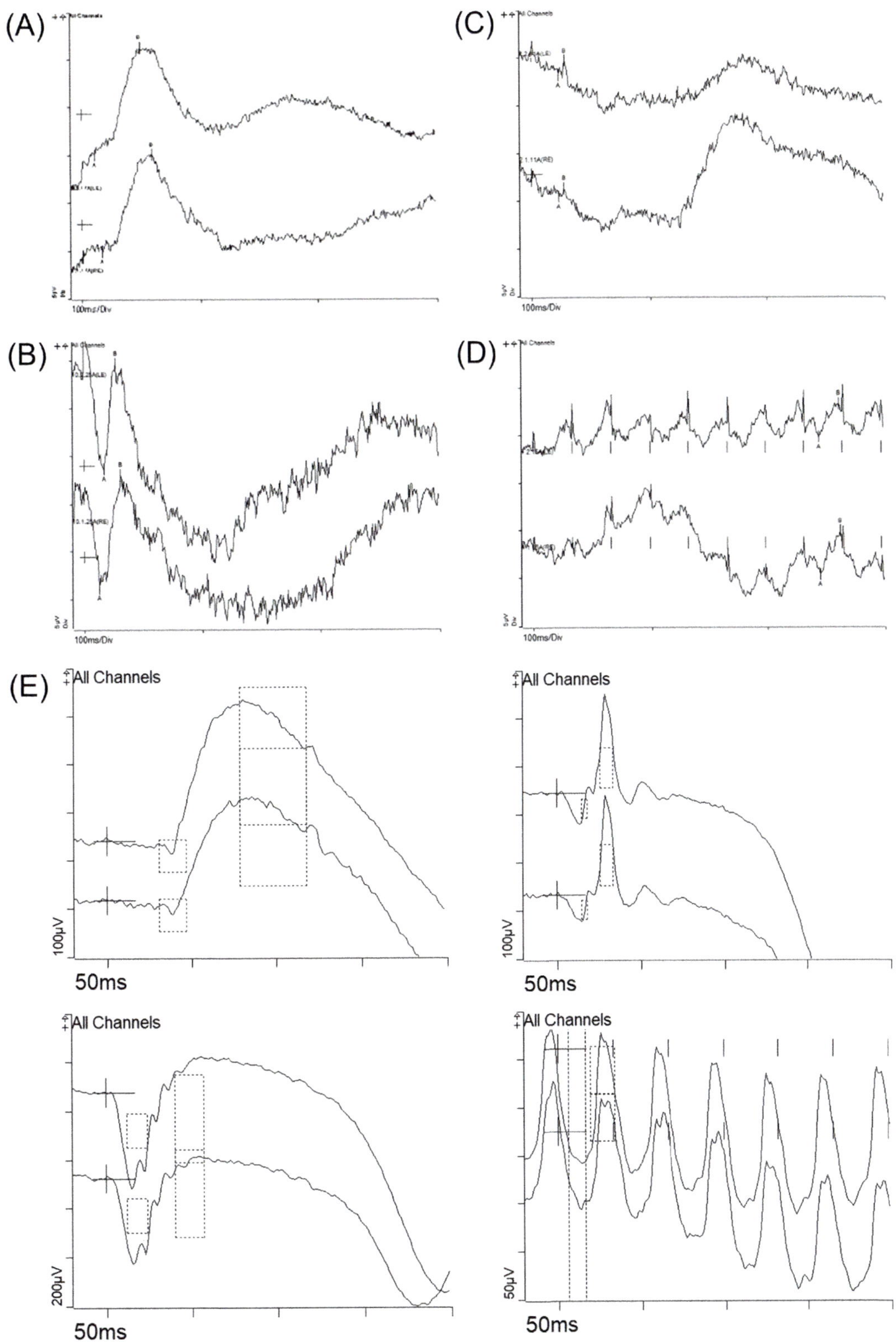

FIGURE 11.3.4.4 Full-field ERGs: Responses are significantly reduced to scotopic (*dark-adapted*) dim flash stimuli (A), scotopic bright flash (B), photopic (*light-adapted*) bright flash (C), and photopic 30-Hz flicker (D). The bottom four panels (E) show normal ERG response waveforms for comparison.

- **Caveats**: *ABCA4* pathogenic variants are implicated in other retinal diseases, including those affecting areas outside of the macula. These include cone and cone-rod dystrophies, as well as retinitis pigmentosa. The type of disease the patient will have depends on how much function the mutated *ABCA4* gene has remaining. Of these diseases, Stargardt disease is the mildest manifestation of the *ABCA4* associated conditions.
- **Outcome**: Sara is put in contact with a low vision specialist who can help her navigate low vision aids for daily life functions. Also, photoprotection is recommended and ultraviolet blocking glasses are encouraged when in direct sunlight.

The Diagnosis

Clinical Diagnosis: Patient and Brother—Stargardt disease.
Genetic Diagnosis: Patient—*ABCA4* gene, compound heterozygous pathogenic variantss.

Further Reading

Baldwin, A.N., Robson, A.G., Moore, A.T., Duncan, J.L., 2018. Abnormalities of rod and cone function. In: Schachat, A.P., Wilkinson, C.P., Hinton, D.R., Sadda, S.R., Wiedemann, P. (Eds.), Ryan's Retina, sixth ed. Elsevier.

Descartes, M., Royal, S.A., Franklin, J., Goodin, K., Mancuso, M., Mikhail, F.M., Holt, L., 2009. A new syndrome with Stargardt macular degeneration, abnormalities of the corpus callosum, mental retardation, and dysmorphic features: a case report of two siblings. Clin. Dysmorphol. 18, 178–180. PMID 19451813.

Gill, J.S., Georgiou, M., Kalitzeos, A., Moore, A.T., Michaelides, M., 2019. Progressive cone and cone-rod dystrophies: clinical features, molecular genetics and prospects for therapy. Br. J. Ophthalmol. 103, 711–720. PMID 30679166.

Sohn, E.H., Mullins, R.F., Stone, E.M., 2018. Macular dystrophies. In: Schachat, A.P., Wilkinson, C.P., Hinton, D.R., Sadda, S.R., Wiedemann, P. (Eds.), Ryan's Retina, sixth ed. Elsevier.

Tanna, P., Strauss, R.W., Fujinami, K., Michaelides, M., 2017. Stargardt disease: clinical features, molecular genetics, animal models and therapeutic options. Br. J. Ophthalmol. 101, 25–30. PMID 27491360.

Case 11.3.5

Craniofacial Cases: Bilateral Retinal Detachment

Vaidehi Dedania[1], Alexander Bottini[2], Lediana Goduni[3] and Scott Brodie[1]

[1]New York University School of Medicine, New York University Langone Medical Center, New York University Eye Center, New York, NY, United States; [2]The Retina Institute, St. Louis, MO, United States; [3]Bascom Palmer Eye Institute, Miami, FL, United States

Abstract

An 11-year-old girl with bilateral retinal detachments and high myopia and family history of similar features is found to have Stickler syndrome.

The Case

An 11-year-old girl, Trisha, was referred to you, a pediatric ophthalmologist, by her optometrist for a retinal detachment in both eyes. She reports that the vision in her right eye has decreased for at least 4 months, and she did not notice a change in her vision in the left eye. On examination, she has a total retinal detachment in the right eye, with the ability to see only her hand in front of her face and an inferior retinal detachment in her left eye with excellent vision. You note that her glasses have very thick lenses (Fig. 11.3.5.1), consistent with high myopia. She also has mildly hypermobile joints. On further history, her mother reports that the father has had retinal detachments in both eyes at the age of 19 years and mild spondyloepiphyseal dysplasia. You consider Stickler syndrome in your differential diagnosis.

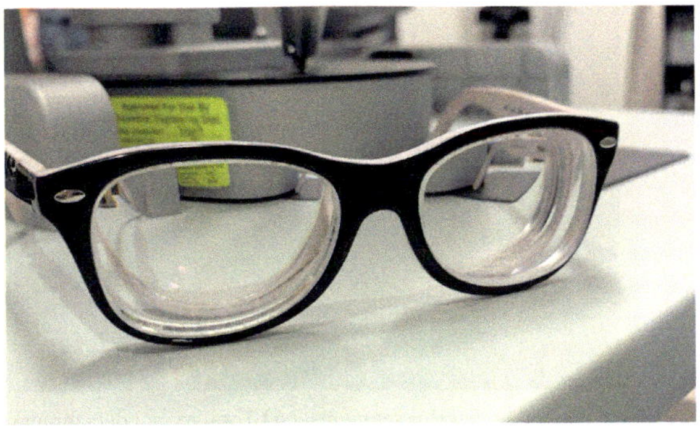

FIGURE 11.3.5.1 The patient's strong myopic glasses, evident by the thickness of the lens and minified image through the lenses.

Genomics in the Clinic. https://doi.org/10.1016/B978-0-12-816478-5.00027-3

The Issues

- **What is the likely cause of the retinal detachments?** Retinal detachments are uncommon in childhood and account for between 1.7% and 5.7% of all retinal detachments. While there are several salient inherited disorders that are associated with a predisposition for the development of retinal detachments, it is important to first consider other nonhereditary etiologies of your patient's retinal detachments, such as trauma or infection. You refer this patient to a retina specialist who performs a careful ocular history, ophthalmic examination, and multimodal ocular imaging. The retina specialist strongly suspects, given the family history of retinal detachments, that there is a hereditary vitreoretinal degenerative process. The retinal detachments could also be the result of trauma, retinopathy of prematurity, intraocular infection, inflammation, or a neoplastic process, but the retina specialist finds no evidence of these.
- **What does the retina specialist find on exam?** The retina specialist reports that your patient's retinal detachments are rhegmatogenous detachments associated with giant retinal tears (Fig. 11.3.5.2). Retinal detachments traditionally have been classified into rhegmatogenous, tractional, and exudative. Rhegmatogenous retinal detachment occurs when a tear in the retina leads to fluid accumulation with a separation of the neurosensory retina from the underlying retinal pigment epithelium. This is the most common type of retinal detachment. The specialist also notes your patient has an "optically empty vitreous" and a "membranous" appearing vitreous anomaly. Trisha will need surgery to repair these.
- **Could there be a hereditary cause of the retinal detachments?** The anatomic presentation of the retinal detachments and, in particular, the appearance of the vitreous gel are suggestive of a diagnosis of Stickler syndrome. Stickler syndrome features the clinical manifestations of inherited disorders of collagen synthesis. The ophthalmic findings, combined with her history of high myopia (near-sightedness) and musculoskeletal anomalies, are highly suggestive of Stickler syndrome type 1. The retina specialist, while planning for surgical repair of the patient's retinal detachments, also recommends referral to a clinical geneticist for further assessment. Trisha will be seen in clinical genetics.

(A)

(B)

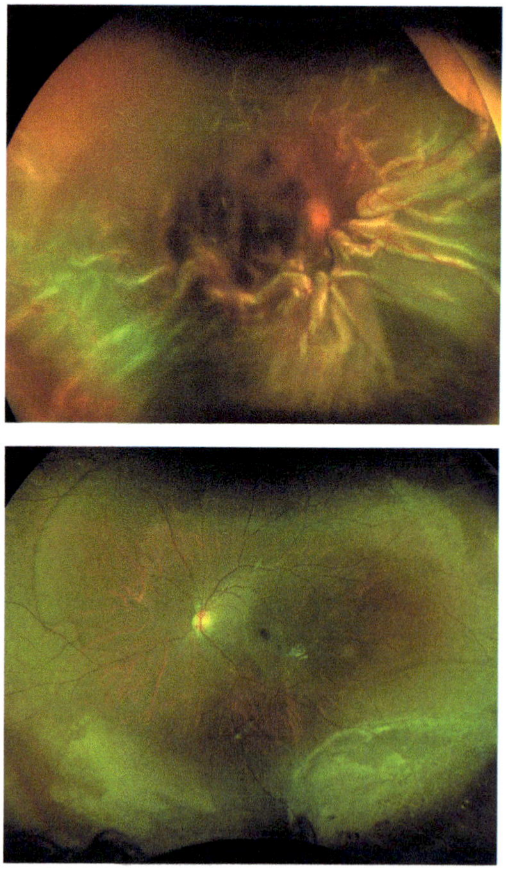

FIGURE 11.3.5.2 Wide-angle color photos of the patient's posterior segments. (A) Total retinal detachment with retinal folds approaching the optic disc is evident in the *right eye*, and (B) an inferotemporal (*lower right corner*) retinal detachment is present in the *left eye*.

It would be helpful for the family to bring any documentation of the retinal detachments in Trisha's father to the genetic visit, as well as any other relevant familial information.

The Referral

- **Genetic Evaluation**: The clinical geneticist performs a thorough past medical history, review of systems, developmental history, social history, and physical examination. Review of medical records reveals an unremarkable prenatal course and a normal, full-term, spontaneous vaginal delivery. Prior to the discovery of her retinal detachments, Trisha was followed closely by an optometrist for significant myopia and began wearing −7.50 diopter glasses at age 18 months. From parental interview, the genetic counselor creates a thorough three-generation pedigree (Sections 2.7, 3.2). This confirms the presence of bilateral retinal detachments in her father and also reveals monocular vision loss in a paternal uncle in childhood. Trisha has two healthy siblings. On examination, Trisha is found to have a small midline, U-shaped cleft of her soft palate and a small chin. She has a depressed nasal bridge, anteverted nares, and mild joint hypermobility with a Beighton score of 6 (Section 11.10.5).

  The geneticist and genetic counselor review the findings with Trisha's parents and discuss the strong suspicion for Stickler syndrome. They explain that there are many different types of Stickler syndrome and they can be inherited as an autosomal dominant or recessive condition (Section 2.7.1). Given Trisha's clinical presentation including the careful exams by the retina specialist and clinical geneticist, Stickler syndrome type 1 with autosomal dominant inheritance is the most likely diagnosis. The phenotype of the Stickler syndromes is characterized by ocular, facial, musculoskeletal, and auditory anomalies. The ocular findings include high myopia in a majority of cases, occasional cataracts, and an abnormal vitreous gel. The assessment of the vitreous gel by the ophthalmologist can not only suggest a diagnosis of Stickler syndrome but may also help identify the specific type. The vitreous gel in Stickler syndrome type 1 often has a characteristic membranous and beaded appearance. Retinal detachments, the most dire ocular complication of the Stickler syndromes, have an incidence of over 50%, with a significant proportion of patients having bilateral retinal detachments. The facial findings can include clefts of the hard or soft palate and midface hypoplasia. Pierre Robin sequence or complex (Section 5.1.2), with posterior displacement of the tongue, small chin, and airway difficulty with or without a U-shaped cleft palate, is associated with Stickler syndrome, and Trisha does exhibit these findings, although she did not have airway obstruction as an infant. Over 80% of patients with a diagnosis of Stickler syndrome report musculoskeletal symptoms, with common findings including spinal abnormalities, and early arthritis in adulthood. Auditory findings include subtle to severe sensorineural or conductive hearing loss. There is also a possible association between Stickler syndrome and mitral valve prolapse.
- **Genetic Counseling**: Stickler syndrome is an inherited disorder of collagen found in the vitreous gel and connective tissues. The syndrome is genetically heterogeneous with pathogenic variants or mutations in specific types of collagen, either types 2, 9, or 11 collagens. Stickler syndrome type 1 with *COL2A1* pathogenic variants is the most prevalent type of Stickler syndrome and is associated with a higher rate of retinal detachment than the other subtypes. An ocular-only subtype of Stickler syndrome type 1, with no systemic manifestations, also exists.

  The geneticist and genetic counselor recommend pursuing a molecular genetic diagnosis of Stickler syndrome type 1 with single gene testing by sequencing the *COL2A1* gene that codes for the collagen type 2 alpha 1 chain. The parents are informed that molecular genetic analysis now can confirm a diagnosis 95% of the time for this type.

  Should the single gene testing be negative for pathogenic variants, the genetic counselor explains that further testing can be done for the other collagen genes associated with the Stickler syndromes. The possibility of other rare collagenopathies affecting the vitreous and connective tissue is reviewed with the parents. Trisha's parents agree to begin genetic testing with single gene sequencing (Section 6.2.2) of the *COL2A1* gene.
- **Results**: Variant analysis of the *COL2A1* gene reveals a splice site mutation (Section 2.4.1) causing an alternatively spliced transcript resulting in a loss-of-function mutation (Section 7.1.3). Trisha's clinical diagnosis of Stickler syndrome type 1 is confirmed.
- **Management and Recommendations**: Trisha's genetic referral and evaluation have not delayed the treatment of her bilateral detachments. Her retinal specialist first undertakes repair of the inferior retinal detachment in her left eye with placement of a scleral buckle. The undetached areas of her peripheral left retina are also treated with prophylactic laser retinopexy to prevent additional detachments. The retina specialist next performs surgery to repair the total retinal detachment of her right eye with a scleral buckle and pars plana vitrectomy with placement of intraocular silicone oil. The retina specialist will see Trisha for regular follow-up for the rest of her life.

The clinical geneticist refers Trisha for audiology testing. Her audiogram reveals mild high-frequency hearing loss. She is asymptomatic but will be monitored annually for progressive hearing loss. Her primary care physician will monitor her musculoskeletal anomalies closely and consider consultation with an orthopedist if significant exacerbation occurs. She is evaluated by a cardiologist, and no valvular anomalies were noted. She is referred to the local Cleft Lip and Palate Team for evaluation of the newly discovered, cleft palate (Section 11.3.10). Trisha's parents are counseled that although her intelligence is expected to be normal, she may face difficulties in school due to her limited vision and hearing.

- **Implications for Other Family Members**: Her father and the two children of Trisha's paternal uncle in England undergo targeted mutation testing. Her father and one cousin are found to have the same *COL2A1* mutation. This 4-year-old cousin is referred to an ophthalmologist and found to have ocular and systemic findings of Stickler syndrome. The cousin undergoes bilateral 360-degree prophylactic laser of his peripheral retinas.
- **Caveats**: Given the variable expression of Stickler syndrome (Section 2.8), genetic evaluation with targeted mutation testing is warranted in the proband's siblings (Section 6.2.1).
- **Differential Diagnoses**: While Stickler syndromes are the most common hereditary causes of retinal detachments, they can occur with other hereditary disorders of connective tissue. Marfan syndrome (Section 11.10.4) and the Ehlers–Danlos syndromes (Section 11.10.5) are both associated with an increased risk of retinal detachment, as well as characteristic body habitus and musculoskeletal anomalies. The ocular exam may reveal thin scleral tissue or a dislocation of the intraocular lens in these syndromes. X-linked retinoschisis can also be a heritable cause of retinal detachments and is suggested by a characteristic splitting of the retinal layers in the macula noted on ocular examination and imaging. Other less common chondrodysplasias associated with vitreoretinal degeneration and retinal detachment include Kniest dysplasia, spondyloepiphyseal 1 dysplasia, spondyloepiphyseal dysplasia congenita, Knobloch syndrome, and Wagner syndrome. They can also occur with other causes of severe myopia, such as retinopathy of prematurity.
- **Outcome**: Four years after genetic diagnosis, Trisha has 20/400 vision in her right eye and 20/20 vision in her left eye. Her audiometric testing results are stable. She has mild arthritic symptoms that are well controlled with medication. She underwent repair of the cleft palate. Trisha plans to visit with her clinical geneticist again when she reaches child-bearing age to discuss prenatal testing. Her siblings are negative for the splice-site mutation in *COL2A1*.

The Diagnosis

Clinical Diagnosis—Patient, Father, Cousin, and Paternal Uncle—Stickler syndrome type 1.
Genetic Diagnosis—Patient, Father, and Cousin—*COL2A1* gene, heterozygous pathogenic variant.

Further Reading

Fincham, G.S., Pasea, L., Carroll, C., McNinch, A.M., Poulson, A.V., Richards, A.J., Scott, J.D., Snead, M.P., 2014. Prevention of retinal detachment in Stickler syndrome: the Cambridge prophylactic cryotherapy protocol. Ophthalmology 121, 1588–1597. PMID 24793526.

Poulson, A.V., Snead, M.P., 2006. Childhood retinal detachment. In: Lorenz, B., Moore, A. (Eds.), Essentials in Ophthalmology: Pediatric Ophthalmology, Neuro-Ophthalmology, Genetics. Springer, Berlin; New York.

Snead, M.P., McNinch, A.M., Poulson, A.V., Bearcroft, P., Silverman, B., Gomersall, P., Parfect, V., Richards, A.J., 2011. Stickler syndrome, ocular-only variants and a key diagnostic role for the ophthalmologist. Eye 25, 1389–1400. PMID 21921955.

Snead, M.P., Yates, J.R., 1999. Clinical and molecular genetics of Stickler syndrome. J. Med. Genet. 36, 353–359. PMID 10353778.

Tang, S., Ding, X., Luo, Y., 2018. Hereditary vitreoretinal degenerations. In: Schachat, A.P., Wilkinson, C.P., Hinton, D.R., Sadda, S.R., Wiedemann, P. (Eds.), Ryan's Retina, sixth ed. Elsevier, Amsterdam, Netherlands.

Case 11.3.6

Craniofacial Cases: Acquired Night Blindness

Vaidehi Dedania[1], Alexander Bottini[2], Lediana Goduni[3] and Scott Brodie[1]

[1]*New York University School of Medicine, New York University Langone Medical Center, New York University Eye Center, New York, NY, United States;* [2]*The Retina Institute, St. Louis, MO, United States;* [3]*Bascom Palmer Eye Institute, Miami, FL, United States*

Abstract

A 25-year-old woman with a history of consanguinity develops night blindness and is found to have retinitis pigmentosa.

The Case

Hannah, a 25-year-old healthy woman, presents to you, her primary care provider, with complaints of recent onset night blindness. There is no known family history of retinal degeneration, but her parents are first cousins. You refer her to a local ophthalmologist.

The Issues

- **What are the possible causes of the night blindness?** The differential diagnosis of acquired night blindness includes: hereditary retinal degenerative disorders, such as retinitis pigmentosa, choroideremia, and gyrate atrophy, vitamin A deficiency due to malabsorption syndromes, disorders of vitamin A transport, iatrogenic malabsorption (intestinal resections, bariatric surgery), and dietary vitamin A deficiency.
- **Could she also have decreased day vision?** It is important to differentiate acquired night blindness from congenital impairment of vision in dim light, which is typically nonprogressive. The latter generally represents one of the variants of congenital stationary night blindness (CSNB). As the terminology implies, patients with CSNB retain their level of vision in both daylight and dim lighting environments throughout life.
- **What does the ophthalmologist find?** Dilated fundus examination is notable for bone-spicule retinal pigmentation and retinal vascular narrowing. A tentative diagnosis of retinitis pigmentosa is made (Fig. 11.3.6.1). The bone-spicule pigmentation is a hypertrophic response of the retinal pigment epithelium (RPE) to the breakdown of the overlying photoreceptors, not the primary disease process. Variations in the retinal pigment response include a normal appearance (retinitis pigmentosa *sine pigmento*) or small spots of depigmentation (retinitis punctata albescens). Visual fields show mild—moderate defects with considerable constriction (Fig. 11.3.6.2). The ophthalmologist recommends an evaluation with a geneticist.

Genomics in the Clinic. https://doi.org/10.1016/B978-0-12-816478-5.00028-5

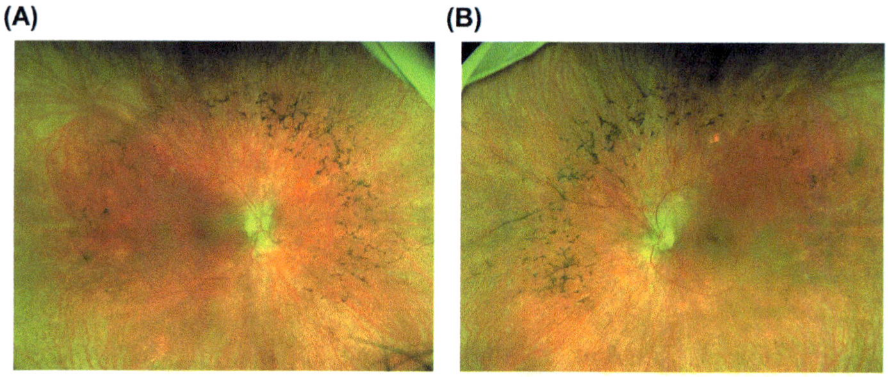

FIGURE 11.3.6.1 Fundus photographs of the (A) *right* and (B) *left* eye show bone-spicule pigmentation and vascular narrowing.

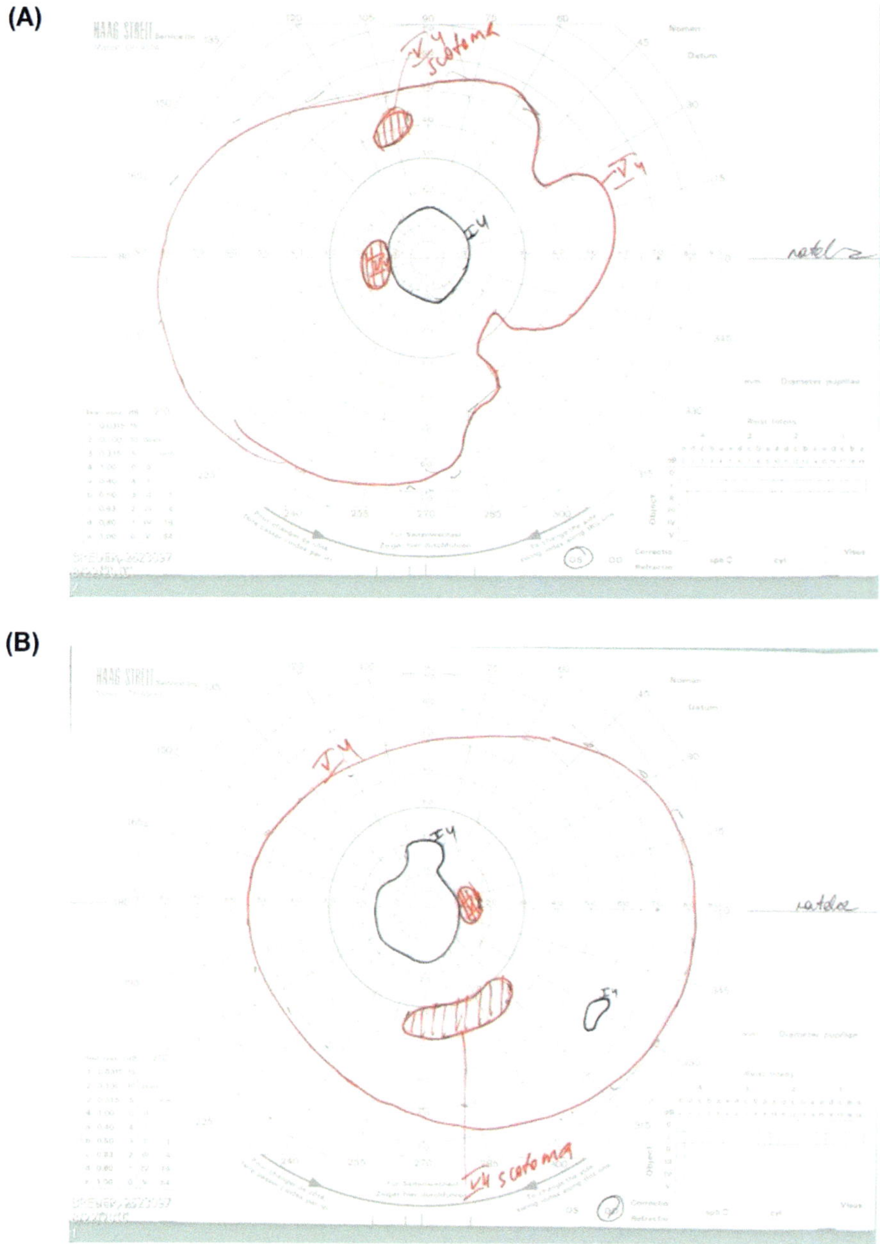

FIGURE 11.3.6.2 Goldmann visual fields. There is some constriction of the (A) *left* and (B) *right* visual fields.

The Referral

- **Genetic Evaluation**: Hannah is seen by a geneticist and genetic counselor. Medical history reveals no other abnormal findings. Hearing is normal. A three-generation pedigree shows that her parents are first cousins (Section 3.2.4), with her father's mother and her mother's mother being siblings, but no other significant findings. A physical exam is normal. They discuss that retinitis pigmentosa denotes a large class of inherited degenerative retinal disorders, characterized initially by night blindness, with later development of visual field constriction ("tunnel vision"), and, ultimately, loss of central vision. The most common fundus findings include apparent retinal vascular attenuation, "bone-spicule" retinal pigmentation, and mottling and atrophy of the retinal pigment epithelium, all of which she has.
- **Genetic Counseling**: There are over 60 genes in which pathogenic variants or mutations (Section 7.1.3) are known to cause nonsyndromic retinitis pigmentosa. More than 20 of these genes are associated with the autosomal dominant (Section 2.7.1) form of the disorder. Mutations in the gene for the visual pigment, rhodopsin, are the most common cause of autosomal dominant retinitis pigmentosa, accounting for 20%–30% of all cases. Most cases of X-linked (Section 2.7.2) retinitis pigmentosa are due to pathogenic variants in the *RGPR* and *RP2* genes. At least 35 genes have been associated with the autosomal recessive (Section 2.7.1) form of the disorder.

  A presumed autosomal recessive inheritance pattern is assigned to this case based on the family history of first-cousin parents, even in the absence of any other known affected family members. Hannah is advised that the risk of affected children is minimal, unless she marries a close relative. Her future children can be evaluated clinically by fundus examination and ERG testing if desired. She is advised that her children are presumed to be obligatory carriers for autosomal recessive retinitis pigmentosa.

  The role of genetic testing in this context is unclear at present, as there are no clinical decisions which hinge on the results. Hannah was informed that she can always be tested at a later date should she become a possible candidate for gene-based therapy in the future, or if genetic counseling questions arise. If she is thinking about having children, multigene panel testing (Section 6.2.3) may determine if her condition is indeed autosomal recessive and the associated risk of retinitis pigmentosa in future offspring and other family members. However, she elects not to undergo genetic testing or other laboratory studies, including vitamin A level, syphilis testing, ornithine level to rule out gyrate atrophy, or phytanic acid level to rule out Refsum disease.
- **Management and Recommendations**: Hannah is referred to a retina specialist, and the diagnosis of retinitis pigmentosa is confirmed on the basis of ERG testing (note near-extinguished light-adapted (left) and dark-adapted (right) ERGs (Fig. 11.3.6.3). Fundus autofluorescence imaging demonstrates bull's eye pattern suggesting metabolic stress on the retinal pigment epithelium underneath the damaged retinas (Fig. 11.3.6.4). Optical coherence tomography (OCT) imaging, cross sections of the macula, shows extensive intraretinal clefts (casually referred to as "cystoid edema," though unrelated to the cystoid edema seen in diabetes and after cataract surgery), indicating loss of interstitial

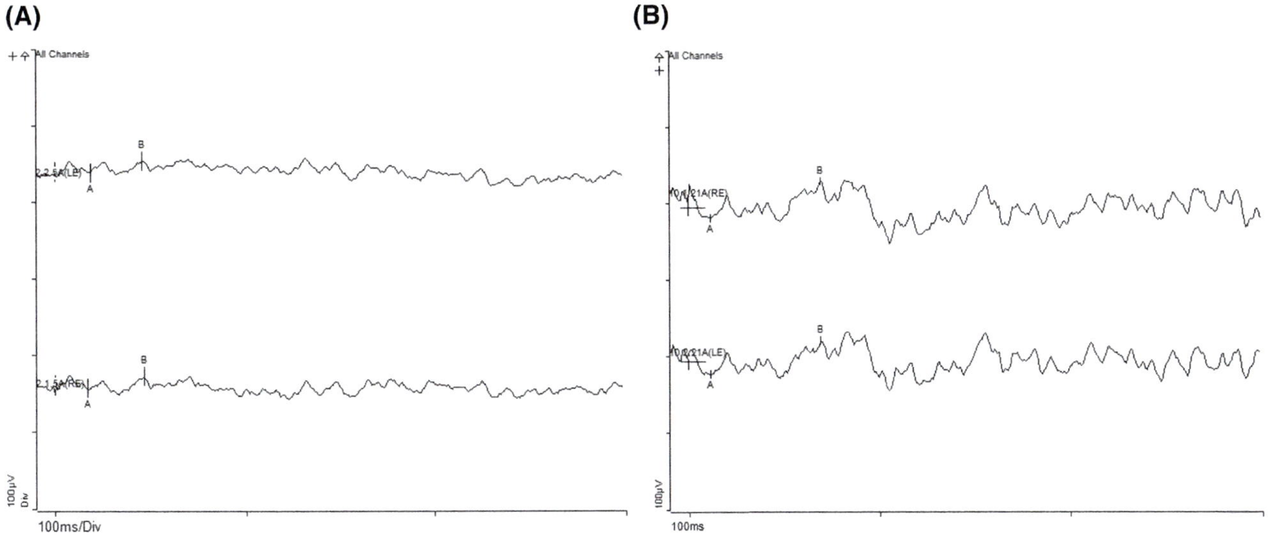

FIGURE 11.3.6.3 (A) Photopic (*light*-adapted) and (B) scotopic (*dark*-adapted) ERGs are nearly extinguished (*Right* eye, *top* tracings; *Left* eye, *bottom* tracings).

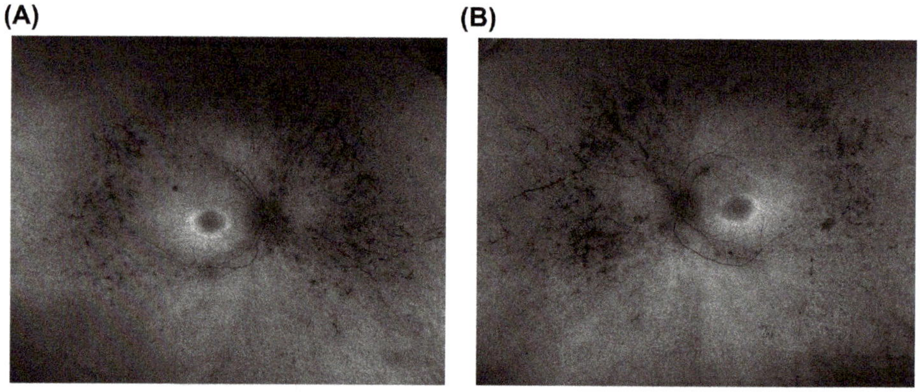

FIGURE 11.3.6.4 Fundus autofluorescence imaging of (A) *right* eye and (B) *left* eye shows bone-spicule pigmentation (*dark*) in the periphery, and central hyperfluorescent (*bright*) macular halos. Please see Figure 11.3.4.2 (C) for normal image.

nerve tissue, as well as loss of the "outer segment ellipsoid line" outside the central macular region in both eyes, indicating damage to the outer retinal layers (Fig. 11.3.6.5).

A trial of oral Diamox therapy for the "cystoid edema" is recommended and poorly tolerated. Hannah is treated instead with dorzolamide (Trusopt) eyedrops.

Nutritional supplementation treatment for retinitis pigmentosa, including high-dose vitamin A, lutein, and omega-3 fatty acids, remains somewhat controversial. Reports of slowing of loss of electroretinographic amplitude observed in the initial trials have not been confirmed with long-term follow-up studies showing palpable visual benefit.

For patients with end-stage disease, an electronic subretinal implant device which stimulates the inner retina according to input from a head-mounted video camera has been FDA approved, but Hannah sees far too well to be a candidate for this treatment for the foreseeable future.

- **Caveats**: Identification of vitamin A deficiency in acquired night blindness is important, as it is the one readily treatable diagnosis. In affluent societies, dietary insufficiency is nearly impossible except for bizarre diets, such as children who eat only potato chips, or ascetics who eat only brown rice. Bariatric surgery is essentially an exercise in deliberate iatrogenic malabsorption, and patients are frequently in a very mild negative vitamin A balance and deplete their tissue stores over a decade or so, presenting with night blindness many years after the surgical procedure. Massive vitamin A supplementation may reverse the symptoms in a matter of days—if there is any question as to the adequacy of intestinal absorption, parenteral injections may be more reliable.

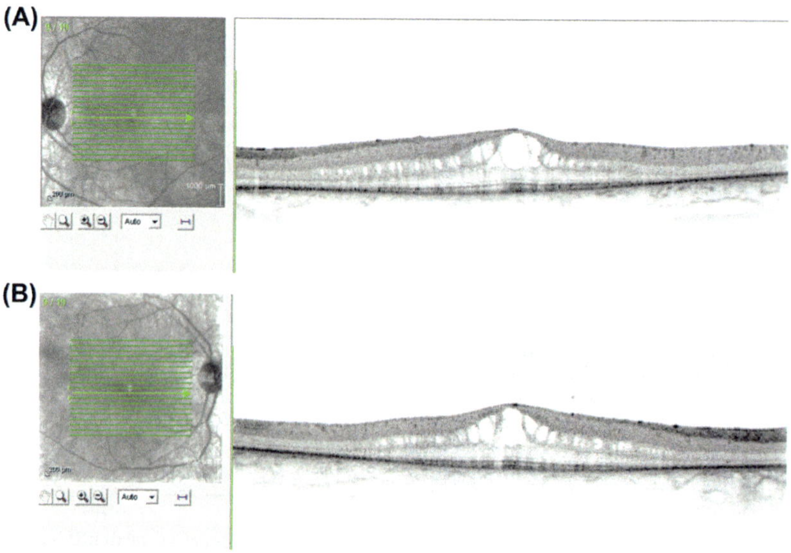

FIGURE 11.3.6.5 Optical coherence tomography (OCT) imaging through the foveal centers of the (A) *left* eye and (B) *right* eye, showing large cavities of fluid in the retina and midlevel retinoschisis (splitting of the retina). *Arrow* points to level of the macula. Please see Figure 11.3.4.3 (C) for normal image.

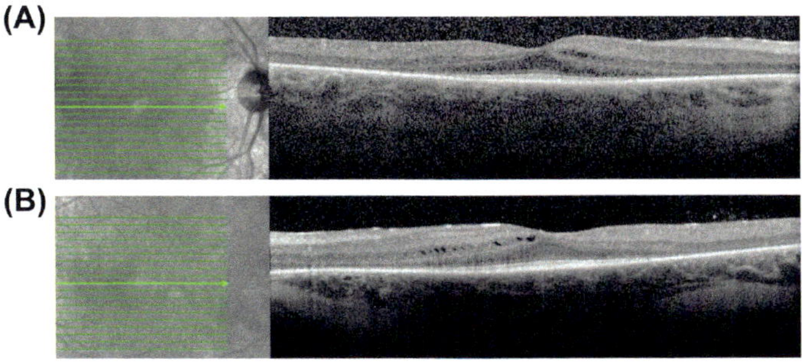

FIGURE 11.3.6.6 OCT images through the foveal centers of the (A) *right* and (B) *left* eye showing marked reductions in interstitial fluid and retinoschisis after treatment with topical dorzolamide eyedrops.

- **Outcome**: Two years later, Hannah returns with visual acuity significantly improved (20/30 OU improved to 20/25 OD, 20/20 OS), and marked reductions in retinal swelling (Fig. 11.3.6.6). For the time being, Hannah is able to function well visually, except for difficulties in dim light at night. The long-term prognosis, however, is for gradual deterioration in the visual fields. Ongoing care will include visual field testing every year or two. Determination of the rate of visual field loss is the most useful clinical tool for estimating the long-term visual prognosis in retinitis pigmentosa.

  Now she has time to revisit whether she should pursue genetic testing. She kept the contact information of the genetic clinic to make an appointment to review the latest testing options and how they may impact her medical care and other family members.

The Diagnosis

Clinical Diagnosis: Retinitis pigmentosa.
Genetic Diagnosis: Not tested.

Further Reading

Berson, E.L., Rosner, B., Sandberg, M.A., Weigel-Difranco, C., Brockhurst, R.J., Hayes, K.C., et al., 2010. Clinical trial of lutein in patients with retinitis pigmentosa receiving vitamin A. Arch. Ophthalmol. 128, 403–411. PMID 20385935.

Berson, E.L., Rosner, B., Sandberg, M.A., Weigel-Difranco, C., Willett, W.C., 2012. Omega-3 intake and visual acuity in patients with retinitis pigmentosa receiving vitamin A. Arch. Ophthalmol. 130, 707–711. PMID 22332205.

McBain, V.A., Egan, C.A., Pieris, S.J., Supramaniam, G., Webster, A.R., Bird, A.C., et al., 2007. Functional observations in vitamin A deficiency: diagnosis and time course of recovery. Eye 21, 367–376. PMID 16341129.

MedlinePlus [Internet]. Bethesda (MD): National Library of Medicine (US); [updated 2010; cited 2023]. Retinitis pigmentosa. Available from: https://medlineplus.gov/genetics/condition/retinitis-pigmentosa/#genes.

Rachitskaya, A.V., Yuan, A., 2016. Argus II retinal prosthesis system: an update. Ophthalmic Genet. 37, 260–266. PMID 26855177.

Case 11.3.7

Craniofacial Cases: Congenital Hearing Loss

Daniah Albokhari[1], Jessica Ogawa[2] and Joann Bodurtha[2,3]

[1]*Department of Pediatrics, Taibah University College of Medicine, Medina, Saudi Arabia;* [2]*McKusick-Nathans Department of Genetic Medicine, Johns Hopkins University School of Medicine, Baltimore, MD, United States;* [3]*Departments of Pediatrics and Oncology, Johns Hopkins University School of Medicine, Baltimore, MD, United States*

Abstract

A 2-month-old girl with a paternal family history of hearing loss is diagnosed with Waardenburg syndrome.

The Case

Emily is a 2-month-old former full-term baby girl who is being seen by you for her well child check. Mom mentions that Emily does not startle to loud noises. You note that she does not orient to your voice, but she is cooing. Otherwise, she is growing well, is tracking objects, and can hold her head up while prone. Particularly, you note that her head circumference is appropriate for her length and weight. Her mother had a home birth and did not have any issues with her pregnancy or delivery. When inquired about Emily's family history, the mother states that Emily's father has unilateral hearing loss and his side of the family has a history of hearing issues. On exam you note that Emily has bright blue eyes and a patch of white hair.

The Issues

- **What are you most concerned about?** Emily is showing signs of a hearing decrease as she doesn't respond to loud sound or orient to voice. It is important to note that babies can still make cooing noises even if they have profound hearing loss. Some babies who are born at home are evaluated at a hospital or are referred by their primary care doctor to an audiologist in order to assess their hearing; however, there may be some babies who do not receive a neonatal hearing screen. It is important to identify hearing issues early on as early intervention can lead to better communication outcomes (Section 5.1.2).
- **How is the father having unilateral hearing loss and Emily showing signs of hearing loss significant?** There are multiple family members on the father's side with hearing loss and Emily also has some physical findings such as a white patch of hair that together with their hearing loss suggest a genetic condition (Section 5.1.7).
- **What is the next step?** You discuss that Emily should be evaluated by an audiologist and, based on the exam and family history, a geneticist. You refer her to an audiologist for an audiology exam as soon as possible and have those results sent to the geneticist. As her father has hearing issues, it would be helpful if her father were present for the genetic visit.

Genomics in the Clinic. https://doi.org/10.1016/B978-0-12-816478-5.00029-7

It is important to share with the geneticist information ahead of the visit. This includes history and exam findings as well as supplementary data such as audiology testing or any pertinent blood work or workup.

The Referral

- **Genetic Evaluation**: Emily and both of her parents are seen in the genetic clinic. They bring her audiology report, which states she has bilateral profound sensorineural hearing loss. Her prenatal and postnatal history are reviewed and a detailed family history is taken. Emily's paternal grandfather, father, and paternal aunt have hearing loss and premature graying of hair (Section 5.1.7). On physical exam, she has a white forelock, dystopia canthorum with the inner canthi of eyes laterally displaced in the setting of normal interpupillary distance, bright blue eyes, high nasal root, and hypoplastic nasal alae. Her father is also examined. He has early graying, a hypopigmented skin spot on the forehead at his hairline, a unibrow (synophrys), and one brown iris, one bright blue iris (heterochromic irides) (Fig. 11.3.7.1, Section 5.1.2).

At times, pictures of affected relatives can be revealing.

- **Genetic Counseling**: The geneticist and genetic counselor discuss with the family that Emily clinically meets the criteria for Waardenburg syndrome type 1 based on her congenital sensorineural hearing loss, bright blue irides, dystopia canthorum, and white forelock. Waardenburg syndrome type 1 is an autosomal dominant condition (Section 2.7.1) that is characterized by sensorineural hearing loss, pigmentation defects, and dystopia canthorum. Waardenburg syndrome type 1 is differentiated from the other types of Waardenburg syndrome by the dystopia canthorum. Waardenburg type 1 is most commonly caused by pathogenic variants (Section 7.1.3) in the *PAX3* gene with approximately 6% of cases due to a deletion in *PAX3*. The majority of patients with Waardenburg syndrome type 1 inherit the variant from an affected parent. Each of Emily's future children has a 50% chance of inheriting the pathogenic variant.

There can be phenotypic variability (Section 2.8) within families with some members having profound congenital hearing loss and others with progressive postlingual hearing loss. For example, within a family as in this one, one family member can have bilateral hearing loss and other family members can have unilateral hearing loss.

- **Management and Recommendations**: Treatment is largely based on the severity of hearing loss. Patients with severe or profound sensorineural hearing loss have done well with cochlear implants. They should be referred to otolaryngology to determine which hearing aid is appropriate. Although the level of hearing loss in Waardenburg syndrome

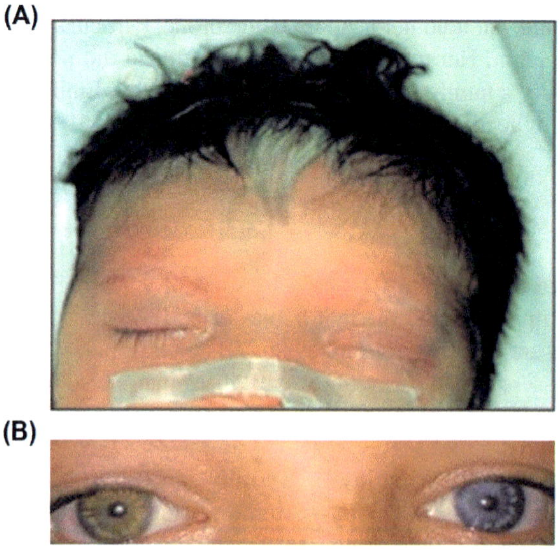

FIGURE 11.3.7.1 (A) Face showing white forelock and white left eyebrow. *(Images courtesy of Seidahmed, M.Z., Abdelbasit, O.B., Shaheed, M.M., Alhussein, K.A., Miqdad, A.M., Samadi, A.S., Khalil, M.I., Al-Mardawi, E., Salih, M.A., 2014. Genetic, chromosomal, and syndromic causes of neural tube defects. Saudi Med. J., 35 (Suppl. 1), S49–S56. PMID 25551112).* (B) Full heterochromia. *(Images courtesy of Haj Kassem, L., Ahmado, M.F., Sheikh Alganameh, M., 2018. A rare case of seven siblings with Waardenburg syndrome: a case report. J. Med. Case Rep., 12, 192. PMID 29973257. http://creativecommons.org/licenses/by/4.0/).*

type 1 typically does not progress, regular hearing tests are recommended. Any at-risk family member, such as siblings, should have a low threshold to obtain an audiology test.

Emily should also be enrolled in early intervention under the Individuals with Disabilities Education Act (IDEA) Part C. IDEA was passed in 1975 in order to enable children with disabilities to have adequate public education and assists states with funding in order to implement services such as early intervention. Under this law, states must provide appropriate public education up to 21 years of age to children with disabilities and eligible children with disabilities will receive an individualized education plan (IEP). Parents are also encouraged to collaborate with the school system to help design related service needs. Some examples of children who qualify for these services include those with hearing loss, visual impairment, intellectual disability, speech impairment, and orthopedic impairment.

Although visual acuity in Waardenburg syndrome type 1 is not thought to differ from the general population, an ophthalmology exam should be considered as unrecognized limited visual acuity could have further impact on development and would require different educational strategies.

- **Genetic Testing**: As this is an autosomal dominant condition, any sibling future or present is at 50% risk of having Waardenburg syndrome type 1. Sequencing and testing for deletions in *PAX3* in Emily or her father would be the first step (Section 6.2.2). If a pathogenic mutation is found, then it would be possible to have targeted prenatal testing or test other at-risk relatives.
- **Differential Diagnosis**: There are several other syndromes that share clinical features with Waardenburg syndrome type 1. Overlap of the pigmentary features and hearing loss is seen with Waardenburg syndrome type II and Waardenburg syndrome type IV, which also can include Hirschsprung disease (Section 5.1.4). Piebaldism can have a white forelock, but sensorineural hearing loss is not a common feature.
- **Caveats**: The white forelock can disappear and then reappear in the teenage years or later in some patients with Waardenburg syndrome type 1. It can also first appear at any age.
- **Outcome**: The family meets with the otolaryngologist and together, they decide on an appropriate hearing aid. When she is closer to 12 months old, Emily receives a cochlear implant. A pathogenic variant in *PAX3* is identified on genetic testing.

The Diagnosis

Clinical diagnosis: Patient and father—congenital sensorineural hearing loss due to Waardenburg syndrome type 1.
Genetic diagnosis: Patient—*PAX3* gene, heterozygous pathogenic variant.

Further Reading

Cullen, R.D., Zdanski, C., Roush, P., Brown, C., Teagle, H., Pillsbury 3rd, et al., 2006. Cochlear implants in Waardenburg syndrome. Laryngoscope 116, 1273–1275. PMID 16826074.

Milunsky, J.M., July 30, 2001. Waardenburg syndrome type I. Adam, M.P., Everman, D.B., Mirzaa, G.M., Pagon, R.A., Wallace, S.E., Bean, L.J.H., Gripp, K.W., Amemiya, A., (editors), In GeneReviews® [Internet]. University of Washington, Seattle, Seattle (WA), 1993–2023 [updated October 20, 2022]. PMID 20301703.

Program to Enhance the Health & Development of Infants and Children, 2020 (PEHDIC) [Online]. American Academy of Pediatrics. [Accessed March 2020]. Available: https://www.aap.org/en-us/advocacy-and-policy/aap-health-initiatives/PEHDIC/Pages/Early-Hearing-Detection-and-Intervention.aspx.

Song, J., Feng, Y., Acke, F.R., Coucke, P., Vleminckx, K., Dhooge, I.J., 2016. Hearing loss in Waardenburg syndrome: a systematic review. Clin. Genet. 89, 416–425. PMID 26100139.

Case 11.3.8

Craniofacial Cases: Hearing Loss and Renal Agenesis

Daniah Albokhari[1], Jessica Ogawa[2] and Joann Bodurtha[2,3]

[1]*Department of Pediatrics, Taibah University College of Medicine, Medina, Saudi Arabia;* [2]*McKusick-Nathans Department of Genetic Medicine, Johns Hopkins University School of Medicine, Baltimore, MD, United States;* [3]*Departments of Pediatrics and Oncology, Johns Hopkins University School of Medicine, Baltimore, MD, United States*

Abstract

An 11-year-old boy with unilateral hearing loss as well as renal agenesis is diagnosed with branchiootorenal syndrome.

The Case

Boris is an 11-year-old boy who comes to you for a sick visit because he has decreased hearing in one ear. He does not have any rhinorrhea, congestion, or recent illness. He thinks it might have been present for several months and has progressively worsened. He has needed to increase the volume of his headphones to the point where it is then too loud for his other ear. He has not had any history of trauma, meningitis, or recurrent ear infections. He previously received the varicella vaccine and has not had any rashes or burning sensation around his ear. He had passed his newborn hearing screening. The family history is notable for his mother having hearing loss and end-stage renal disease that were diagnosed in her 20s. She recently had a kidney transplant and has been doing well. Examination of Boris's ears reveals that his external canal and tympanic membranes are within normal limits, but his ears are asymmetric. He has small, superficial preauricular pits bilaterally. Based on the history and exam, you refer Boris to audiology.

This case is a good example of the importance of the contribution of a good family history, and keeping an up-to-date family history as signs and symptoms of many conditions may present later (Section 3.2).

The Issues

- **What should the workup include?** A renal ultrasound should be obtained. In the newborn period if there are isolated ear pits (Section 5.1.2), a renal ultrasound is not mandatory; however, in the setting of hearing loss and a family history of hearing loss, a renal ultrasound should be obtained. This is to detect any malformation and is based on recommendations published in Pediatrics in 2001 that any patient with preauricular pits, cup ears, or an ear anomaly with another feature including a family history of hearing loss or ear anomalies, teratogenic exposure, polydactyly, or imperforate anus should have a renal ultrasound.
- **Has the patient had hearing issues since birth that were missed**? Boris did not have any developmental delay or signs that he was hard of hearing prior to this visit; however, if at any point in time there had been any concern

Genomics in the Clinic. https://doi.org/10.1016/B978-0-12-816478-5.00030-3

from the parents, teachers, or his pediatrician, he would have had a hearing test earlier based on the American Academy of Pediatrics hearing assessment recommendations.

With the introduction of universal hearing screening in 2001, approximately 97% of newborns in the United States are screened with the goal of identifying hard of hearing infants by 3 months of age. At birth, 6 in 1000 neonates are found to have some degree of hearing loss. As there are various causes of later onset hearing loss, the prevalence increases to 17 in 1000 individuals by the end of adolescence. This has prompted the recommendation by the Academy of Pediatrics Bright Futures for another hearing screening for children between 15 and 17 years old.

- **What does the workup show**? The renal ultrasound shows unilateral renal agenesis (Section 5.1.5). You refer him to pediatric nephrology based on these findings. On audiology exam, he is found to have moderate unilateral sensorineural hearing loss. Because there are multiple findings, you also refer him for a genetic evaluation.

The Referral

- **Genetic Evaluation**: Boris and his parents are seen in the genetic clinic. His history and imaging are reviewed and a family history taken. Boris has normal vision and has had normal development throughout. Boris's mother says that her brother died in his 40s and had end-stage renal disease, deafness, and some type of surgery in his neck. She cannot recall more specifics on her brother's findings other than that he had issues with "lumps" in his neck that were removed. She herself had bilateral renal hypoplasia and mild sensorineural hearing loss, but no issues with neck masses. On exam, Boris is found to have asymmetrically positioned ears that are cup shaped, preauricular pits bilaterally, a long narrow face, and retrognathia. No palpable cysts or evidence of a branchial cleft remnant are found on his neck. The remainder of his exam is unremarkable.
- **Genetic Counseling**: The geneticist and genetic counselor discuss their findings with the family. They believe that Boris and his mother have branchiootorenal (BOR) syndrome, and most likely his uncle does as well. This is an autosomal dominant condition (Section 2.7.1) characterized by hearing loss, branchial fistulae and cysts, preauricular pits, malformations of the inner, outer, and middle ear, and renal anomalies. Patients with this syndrome can have conductive, sensorineural, or mixed hearing loss. The onset of hearing loss varies from early childhood to adulthood and the severity of hearing loss also varies from mild to profound. The side of unilateral hearing loss can vary within families. While most patients will have preauricular pits, they can also have preauricular tags, microtia, or a lop ear where there is folding of the upper portion of the ear inferiorly and anteriorly (Fig. 11.3.8.1). Branchial cleft cysts or fistulae are common and could appear as a lump, dimple, or small opening that can drain or get infected (Section 5.1.6). The renal anomalies (Section 5.1.5) seen in BOR vary from renal agenesis to dysplasia and can be unilateral or bilateral. While renal agenesis is the most common renal anomaly seen in BOR, renal hypoplasia, ureteropelvic junction obstruction, calyceal cysts, pelviectasis, hydronephrosis, and vesicoureteral reflux can be seen.

  The known genes associated with BOR are *EYA1*, *SIX1*, and *SIX5*. The majority of cases have a parent who is affected. Approximately 10% of individuals have a de novo pathogenic variant (Sections 2.4.1, 7.1.3).
- **Management and Recommendations**: Renal issues are the most common cause of morbidity and mortality in patients with BOR, so it is important to have a nephrologist follow Boris and to obtain regular lab work including BUN, creatinine, and urinalysis. A computed tomography (CT) to evaluate for temporal bone anomalies is needed. A careful examination of his neck does not reveal any branchial cleft cysts or fistulae. If a neck mass is palpated, a CT of the neck is warranted. Any branchial cleft cysts or fistulae would need to be assessed and likely removed by otolaryngology. A hearing aid is fitted with semiannual monitoring to ensure that it is not progressing or fluctuating. The family is advised to avoid using ototoxic and renal toxic medications like aminoglycoside antibiotics. A letter is sent to the school asking for accommodations for Boris, such as preferential seating closer to the teacher.
- **Genetic Testing**: There is gene panel testing (Section 6.2.3) for the genes that are associated with BOR. If a pathogenic variant (Section 7.1) is found, then other family members who are at risk can be assessed, and prenatal testing can be performed, looking for the known pathogenic variant.
- **Differential Diagnosis**: Townes–Brocks syndrome can also have preauricular pits or tags, as well as hearing and renal impairment with dysplastic ears, but often has imperforate anus and thumb anomalies. Other conditions that have ear pits and/or tags with hearing issues are the cat eye syndrome (duplication 22q) and the oculo-auriculo-vertebral spectrum (OAVS) (Section 5.1.2), hemifacial microsomia, and other craniofacial microsomia syndromes. Patients with craniofacial microsomia have a wide range of phenotypic variability with some having mild facial asymmetry with a preauricular skin tag to others having severe mandibular hypoplasia with microtia, but the majority of patients

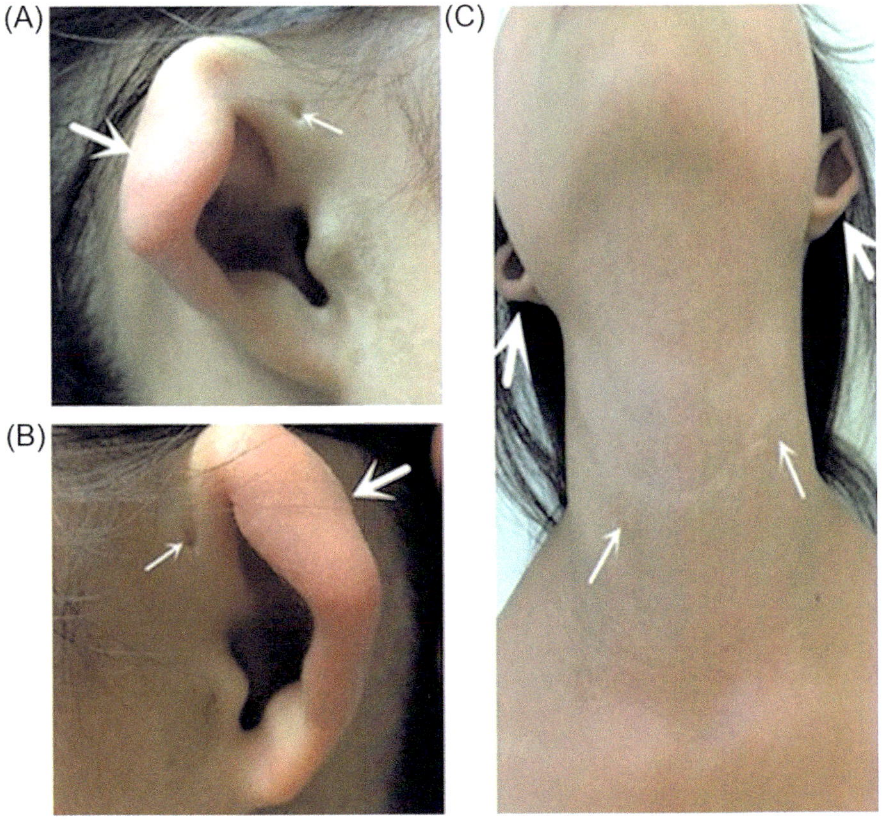

FIGURE 11.3.8.1 Ears and neck of a patient with branchiootorenal (BOR) syndrome. (A) and (B) Preauricular pits indicated by the smaller arrows; microtia, the larger arrows. (C) Branchial cleft fistulae indicated by the smaller arrows; microtia, the larger arrows. *Adapted Figure from Wang, Y.G., Sun, S.P., Qiu, Y.L., Xing, Q.H., Lu, W., 2018. A novel mutation in EYA1 in a Chinese family with Branchio-oto-renal syndrome. BMC Med. Genet., 19, 139. PMID 30086703 used under CC BY (https://creativecommons.org/licenses/by/4.0/).*

have hearing loss. Hearing and renal issues are also present in Kabuki syndrome, but these patients have intellectual disability and long palpebral fissures with eversion of the lower eyelid. Other syndromes on the differential include Alport syndrome, Muckle–Wells syndrome, and Barakat syndrome.

- **Caveats**: Even with genetic testing, a pathogenic variant is not identified in a little over half of the cases.
- **Outcome**: Boris does well with a behind the ear hearing aid. Due to his unilateral renal agenesis, he will continue to be monitored for hypertension, proteinuria, and renal failure. Testing shows a known pathogenic variant in *EYA1* that was also found in his mother. His mother reaches out to her brother's wife so that Boris's cousins who are at risk can be offered testing after confirmation in their aunt.

The Diagnosis

Clinical diagnosis: Patient, mother, and maternal uncle—branchiootorenal syndrome.

Genetic diagnosis: Patient and mother—*EYA1* gene, heterozygous pathogenic variant.

Further reading

Grindle, C.R., 2014. Pediatric hearing loss. Pediatr. Rev. 35, 456–463. PMID 25361905.

Smith, R.J.H., March 19, 1999. Branchiootorenal spectrum disorder. Adam, M.P., Everman, D.B., Mirzaa, G.M., Pagon, R.A., Wallace, S.E., Bean, L.J.H., Gripp, K.W., Amemiya, A., (editors), In GeneReviews® [Internet]. University of Washington, Seattle, Seattle (WA), 1993–2023 [updated September 6, 2018]. PMID 20301554.

Wang, R.Y., Earl, D.L., Ruder, R.O., Graham Jr., J.M., 2001. Syndromic ear anomalies and renal ultrasounds. Pediatrics 108, E32. PMID 11483842.

Case 11.3.9

Craniofacial Cases: Hearing Loss and Night Blindness in a Teenager

Jessica Ogawa[1], Daniah Albokhari[2] and Joann Bodurtha[1,3]

[1]*McKusick-Nathans Department of Genetics Medicine, Johns Hopkins University School of Medicien, Baltimore, MD, United States;* [2]*Department of Pediatrics, Taibah University College of Medicine, Medina, Saudi Arabia;* [3]*Departments of Pediatrics and Oncology, Johns Hopkins University School of Medicine, Baltimore, MD, United States*

Abstract

A 16-year-old boy with a history of bilateral moderate sensorineural hearing loss develops night blindness and is found to have Usher syndrome with retinitis pigmentosa.

The Case

Manuel, a 16-year-old male, is in your office for his annual checkup. Manuel tells you that he is having difficulty seeing at night. He denies any pain or redness of the eyes. When you ask him about light sensitivity, he tells you he has been noticing that bright light is bothering him lately. He is not taking any medications, and he denies any history of eye trauma. School eye checks and office screening have been normal. You know Manuel's medical history is significant for bilateral moderate, sensorineural hearing loss diagnosed at 13 months of age, for which he is using hearing aids, with good adaptation. Other than his new vision problem, he has no complaints. He is in 11th grade, doing well at school. You perform a physical examination, including neurological, and cranial nerve exam. You try to perform fundus examination, but his photophobia limits your exam.

Now with universal hearing screening, most cases of congenital hearing loss should be detected after birth in the United States, and often the etiology determined early.

The Issues

- **How concerned are you by his eye findings?** His eye findings are concerning, especially because you cannot get a good exam of his eyes. You refer him to an ophthalmologist who finds retinal pigmentary changes and arteriolar narrowing in the fundus consistent with the diagnosis of retinitis pigmentosa (Sections 5.1.2, 11.3.6). You are aware that this type of abnormality in the retina can worsen over time and lead to visual loss. You see him again after the ophthalmological evaluation.

Night blindness can also be due to primary eye findings, such as myopia, cataracts, or glaucoma, as well as systemic causes such as poorly controlled diabetes, vitamin A deficiency, and exposure to some medications.

Genomics in the Clinic. https://doi.org/10.1016/B978-0-12-816478-5.00031-5

- **Do you think his visual problem is related to his hearing loss?** You tell the mother that whenever there is more than one organ system involvement, a syndromic condition should be considered. You think that Manuel's visual findings and hearing loss are connected and could be a part of a syndrome. You discuss that you think Manuel should be assessed by a geneticist.
- **Why is it important to be evaluated by a geneticist?** Seeing a geneticist may help to clarify if there is an exact underlying diagnosis for Manuel, in order to provide him with targeted specific management and determine the likely course or severity of the disease.
- **Who else could be at risk?** It would be helpful to review the family history with the geneticist who may identify at-risk individuals and provide the family with a specific recurrence risk.

The Referral

- **Genetic Evaluation**: The geneticist and the genetic counselor see Manuel and his mother in the genetic clinic 2 months later. They review Manuel's medical history including gestational, perinatal, postnatal, and developmental histories. They complete a three-generation family medical history (Sections 2.7, 3.2). Manuel has two siblings, a brother, and a sister, who do not have any known vision or hearing disabilities. There is no known extended family history of eye or hearing issues, kidney disease, dystonia, ataxia, arrhythmias, or sudden death. He walked at approximately 1 year of age and has had no balance issues. A physical exam is performed, and growth parameters are noted to be at the 50th percentile. He has no dysmorphic features, including normal dentition, and no goiter, or abnormal pigmentation.
- **Genetic Counseling**: The geneticist and the genetic counselor state that, based on Manuel's medical history and physical exam, he could have a condition called Usher syndrome type 2 (USH2). They explain to Manuel and his mother that Usher syndrome (Section 5.1.2) is classified into three forms. This classification depends on the severity of the hearing loss, the presence and absence of balance problems, and the age at which signs and symptoms appear. Individuals with USH2 may develop moderate to severe sensorineural hearing loss along with progressive blindness due to retinitis pigmentosa, but they will have a normal balance. The other two forms of Usher syndrome are characterized by vestibular dysfunction and balance issues along with the hearing loss and retinitis pigmentosa.

  Usher syndrome accounts for approximately 3%–5% of hearing loss among children and about 50% of the individuals who are deaf-blind in the United States. USH2 is due to pathogenic variants in one of three genes and inherited in an autosomal recessive manner (Section 2.7.1). Thus, recurrence risks for Manuel's parents to have a future child with USH2 are 1 in 4 (25%).

There are multiple causes for becoming deaf-blind, only some of which are genetic. Teratogenic causes include prenatal viral exposures (e.g., TORCH infections, HIV) and prenatal alcohol exposure (Section 11.3.1). Genetic causes include chromosomal disorders and other complex syndromes (see below). Nongenetic causes are often acquired accidentally (e.g., head trauma, infection).

- **Management and Recommendations**: The genetic team discusses with Manuel and his mother the importance of regular auditory evaluation to detect any changes that may require modifications to hearing aids and recommends ongoing ophthalmologic evaluations to detect potentially treatable complications such as cataracts. They also tell Manuel to avoid competitive sports because they require a full range of vision, and people with retinitis pigmentosa will likely develop tunnel vision, which would put him at high risk of accidental injury.
- **Genetic Testing**: There are multigene panels (Section 6.2.3) that test for the three genes involved in USH2, also multigene panels that include genes known to cause deafness and hard of hearing. Prenatal diagnosis (Section 4.4.2) for future children is feasible for the family if a specific mutation has been identified in Manuel.
- **Results**: Manuel is found to have biallelic pathogenic variants (Section 7.1) in the *USH2A* gene. Currently, there is no treatment available for the Usher syndromes, but various treatment strategies for retinitis pigmentosa are under investigation for future therapy including the use of antioxidants, stem cells, retinal prostheses, and gene therapy. Treatment with vitamin A palmitate has been suggested to slow the retinitis pigmentosa progression, but no formal studies have approved its benefit in USH2.
- **Differential Diagnosis**: There are other syndromes characterized by comorbidities and produce concurrent symptoms of hearing loss and retinitis pigmentosa such as Alstrom syndrome (cone-rod dystrophy, obesity, and cardiomyopathy), PHARC (polyneuropathy, hearing loss, ataxia, retinitis pigmentosa, and cataract), RHYNS syndrome (retinitis pigmentosa, hypopituitarism, nephronophthisis, and mild skeletal dysplasia), and deafness-dystonia-optic neuronopathy. Retinal dystrophy with vestibular dysfunction and, more rarely, hearing loss can occur in the Bardet–Biedl syndromes (Section 11.5.4). Retinitis pigmentosa can also present alone without hearing loss (Section 11.3.6). If there had been other findings in Manuel, a more detailed search for etiology would have occurred.

- **Caveats**: Although the majority of the pathogenic variants in *USH2A* are detected by sequence analysis method, a small percentage of the pathogenic variants can be missed and thus require target gene deletion/duplication analysis to be identified.
- **Outcome**: Manuel continues to have regular auditory and ophthalmologic evaluations. Manuel's hearing remains stable, with no progression in the hearing loss over time. However, ophthalmological surveillance reveals that his retinopathy continues to progress each year.

The Diagnosis

Clinical diagnosis: Usher syndrome type 2.
Genetic diagnosis: *USH2A* gene, compound heterozygous pathogenic variants.

Further Reading

Castiglione, A., Busi, M., Martini, A., 2013. Syndromic hearing loss: an update. Hear. Bal. Commun. 11, 146–159. https://doi.org/10.3109/21695717.2013.820514.

Koenekoop, R., Arriaga, M., Trzupek, K.M., Lentz, J., December 10, 1999. Usher syndrome type II. Adam, M.P., Ardinger, H.H., Pagon, R.A., Wallace, S.E., Bean, L.J.H., Gripp, K.W., Mirzaa, G.M., Amemiya, A., (editors). In GeneReviews. University of Washington, Seattle (WA), 1993–2001 [updated Octobeter 22, 2020]. PMID 20301515.

Musarella, M.A., MacDonald, I.M., 2011. Current concepts in the treatment of retinitis pigmentosa. J. Ophthalmol. 2011, 753547. PMID: 21048997.

Case 11.3.10

Craniofacial Cases: Cleft Palate

Ethylin Wang Jabs[*]

Department of Genetics and Genomic Sciences, Department of Pediatrics, Department of Cell, Developmental and Regenerative Biology, Icahn School of Medicine at Mount Sinai, New York, NY, United States

Abstract

A 17-year-old mother has a child with nonsyndromic cleft palate, likely multifactorial in etiology.

The Case

Cassandra is a 17-year-old patient of your family practice. She is a single mother who just gave birth to her baby girl with a cleft palate (Fig. 11.3.10.1A). Baby Kyra was the product of a normal 39 week gestation and was vaginally delivered with no complications. Cassandra was G2P0010 with a previous miscarriage and had hypothyroidism treated with synthroid and gestational diabetes treated with diet during the pregnancy. Ultrasounds performed during the pregnancy were normal. Kyra's birth weight, length, and head circumference were between the 50th–75th percentiles. Her Apgar scores were 9/9. She was noted to have jaundice and hypoglycemia that resolved. Kyra had difficulties sucking because of her cleft palate, and the nurses gave Cassandra a special bottle for Kyra that allows her to suck and feed while avoiding nasal regurgitation and choking. Kyra was seen by a pediatric plastic surgeon who discussed future surgical repair of the palate at about one year of age. She passed her hearing screen and was discharged home with Cassandra after 3 days. Cassandra is overwhelmed and worried about caring for Kyra.

The Issues

- **Should the infant be referred to other clinical specialists for her care?** You note Kyra's normal growth parameters and ask about feeding, which is going well with the special bottle according to Cassandra. You know that one of the first manifestations associated with cleft palate is difficulty with feeding. Aerophagia is a problem in these infants with cleft palate and requires more frequent burping and slower feeding. The inability to generate intraoral breath pressure due to nasal air emission in children with cleft palate frequently manifests later as articulation difficulties, particularly consonant weakness, and unintelligible speech. Hearing disorders are prevalent among individuals with oral clefting, as a result of chronic otitis media with effusion due to eustachian tube dysfunction.

It is recommended that parents of children with cleft lip and/or palate bring them to a multidisciplinary cleft care team. There are teams of health professionals that work together to help parents coordinate care for children with oral clefting, including cleft palate as well as those with cleft lip alone or with cleft palate. Teams include experts in pediatrics, surgery,

[*] Author EWJ was employed by Icahn School of Medicine at Mount Sinai at the time of writing the manuscript, but is currently employed by Mayo Clinic

Genomics in the Clinic. https://doi.org/10.1016/B978-0-12-816478-5.00032-7

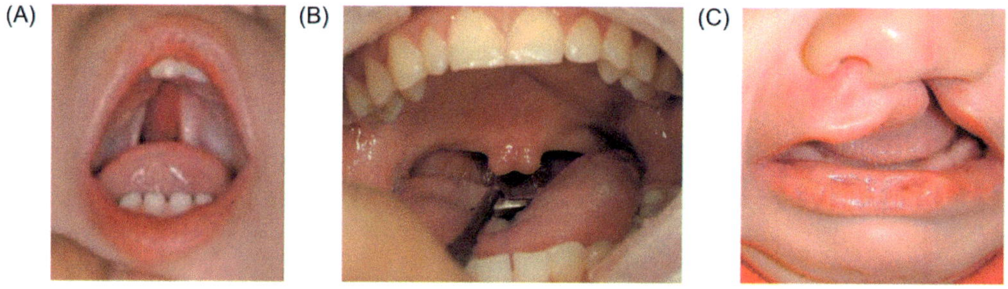

FIGURE 11.3.10.1 Various forms of oral clefting. (A) Cleft palate. *Courtesy of Mary L. Marazita, Center for Craniofacial and Dental Genetics, University of Pittsburgh.* (B) Bifid uvula. *Source Yun-Ha Park, Hyun-Jun Jo, In-Seok Hong, Dae-Ho Leem, Jin-A Baek and Seung-O Ko, Treatment of velopharyngeal insufficiency in a patient with a submucous cleft palate using a speech aid: the more treatment options, the better the treatment results, Maxillofac Plast Reconstr Surg. 2019 May; 41(1): 19. http://creativecommons.org/licenses/by/4.0.* (C) Unilateral cleft lip and palate with lower lip pits. *Source Su Yin Htun, Kurt-W Büto, Hanlie Engelbrecht and Engela Honey, Interlink of two syndromes—Demarquay-van der Woude and popliteal pterygium in one family group, J Interdiscipl Med Dent Sci 2016, 4:188, Figure 1. http://creativecommons.org/licenses/by/4.0/.*

dentistry with orthodontics, otolaryngology with audiology, speech pathology and/or occupational therapy, feeding/nutrition, social work, and psychology. Teams also have a clinical geneticist and genetic counselor. Together they develop a personalized protocol to address individual patient's developmental, medical, and surgical needs. Children with oral clefting can have problems in feeding, ear infections, dental misalignment and malocclusion, and speech. If a child has a cleft lip, surgery to close the lip usually is performed within the first 3 to 6 months of life and if a child has a cleft palate, surgery to close the palate is performed at about 6 months to 18 months of age.

- **Does the infant have isolated cleft palate or a syndrome with cleft palate as one of a constellation of features?** You perform a complete physical exam on Kyra, and you find no additional abnormal findings. You still are not sure if she has isolated cleft palate because, although she does not have additional congenital anomalies, there are syndromes that present with features that are discovered with additional diagnostic testing.
- **Could the mother have a mild case of cleft palate?** Although you also have followed Cassandra since she was a child and you have never noticed an abnormal palate or a bifid uvula (Fig. 11.3.10.1B), you double check, but you see no evidence of any form of oral clefting. You decide to refer Kyra to a multidisciplinary team with a clinical geneticist, to determine if she has isolated cleft palate or a syndromic diagnosis, and to be followed longitudinally by the team.

A bifid uvula may be an isolated, benign finding, or it may be related to a submucous cleft palate. It can be referred to as a forme fruste of a cleft palate. Individuals may be unaware of their bifid uvula and the possibility of an increased risk for familial cleft palate.

The Referral

- **Genetic Evaluation**: Kyra and Cassandra are seen by the geneticist and genetic counselor as part of the team visit 2 weeks later. A past medical history, review of systems, developmental history, social history, and physical examination are obtained by the clinical geneticist. The family history is taken by the genetic counselor and is notable for a distant maternal cousin with a cleft lip and palate. No additional family members have birth defects, or consanguinity; however, Cassandra knows little about the father's side of the family. Kyra undergoes a detailed physical examination to rule out additional dysmorphic findings. A survey of measurements are taken of clinical features such as inner and outer canthal distances, ear length, chest circumference, internipple distance, and palm, middle finger, and feet lengths to document that they are within normal range for age and there are no objective dysmorphisms. Other than the cleft palate, the geneticist finds no other abnormalities. Cassandra also is examined and found not to have a mild form of cleft palate or even a bifid uvula.
- **Genetic Counseling**: Cleft palate occurs in about 2000 newborns each year in the United States. About 50% of all children with cleft palate are nonsyndromic. The etiopathogenesis of isolated cleft palate is not completely understood and generally thought to be multifactorial. Oligogenic or multigenic models with reduced penetrance and the risk related to environmental factors have been reported (Section 2.7.3). Although environmental factors, such as maternal smoking, alcohol consumption, diabetes, and intake of certain drugs during pregnancy, have been associated with cleft palate, conflicting results also have been published.

There are more than 300 syndromes with oral clefting as a feature. The more common genetic syndromes that can present with a cleft palate include 22q11.2 deletion syndrome (also called velocardiofacial syndrome, or chromosome 22q11.2 deletion, and presenting often with congenital heart disease; Sections 11.9.2, 11.14.2), Treacher Collins syndrome (mandibulofacial dysostosis with bilateral downslanting eyes, dysplastic ears, and hypoplastic zygomatic and mandibular bones), and Stickler syndrome (eye findings including myopia, hearing loss, midface flattening, and spondyloepiphyseal dysplasia; Section 11.3.5). For all patients with cleft palate only, 2%–11% will have 22q11.2 deletion syndrome. Each of these conditions can be de novo in the individual, and there can be phenotypic variability (Section 2.8). Each exhibits autosomal dominant inheritance (Section 2.7.1), so there is a 50% risk for an affected individual that his/her offspring may inherit the condition.

The geneticist does not think Kyra has Pierre Robin complex. Pierre Robin complex that can present with cleft palate is thought to be due to a sequence of events, beginning with the mandibular dysmorphology of micrognathia leading to displacement of the tongue upward, resulting in failure of the palate to close and then respiratory obstruction. This complex can be a feature of Stickler syndrome as well. Rare cases of Pierre Robin complex have been associated with a heterozygous variant in a noncoding regulatory region of the *SOX9* gene that can be inherited as an autosomal dominant condition, but most cases appear to be multifactorial (Section 2.7.3) with its genetic etiology unknown.

Cleft palate can occur with cleft lip (CLP) and is different developmentally from cleft palate alone. CLP is the most common form of oral clefting occurring in about 1 in 700 live births with a variable incidence among races and ethnicity; CLP is least common in those of African ancestry, intermediate in those of European ancestry and most common in those of Asian, Native American and Hispanic ancestry. If CLP is a feature of a syndrome, it is mostly associated with lower lip pits or lip mounds, which can be diagnosed as van der Woude syndrome (Fig. 11.3.10.1C). Lip pits and mounds can be easily missed on examination but are important to detect since they can be present as the sole finding in a parent. Van der Woude syndrome is autosomal dominant (Section 2.7.1) caused by *IRF6* heterozygous variants, so that the recurrence risk is 50%, rather than 1%–5% for multifactorial CLP, for an affected parent. *IRF6* pathogenic variants are identified in 0.47% of individuals with nonsyndromic orofacial cleft.

- **Genetic Testing**: The geneticist and genetic counselor discuss with Cassandra that Kyra does not appear to have an underlying syndrome. Testing could be done for the most common syndromes, as in some cases the phenotypic presentation can be very mild. The multisystem disorder, 22q11.2 deletion syndrome can be detected using chromosomes and FISH testing (Section 6.1.2) or microarray testing (Section 6.1.3) with about a 95% detection rate. The other conditions would require a targeted gene (Sections 6.2.1, 6.2.2) or gene panel testing (Section 6.2.3). All of these tests would require a blood or buccal sample.

  Cassandra understands the risks and benefits of the testing but does not want to have genetic testing at this time because she is worried that genetic testing may yield information that Kyra may have other medical issues unrelated to cleft palate, and she already is stressed and preoccupied with arranging Kyra surgical care as soon as possible to improve her oral function. She wants to focus on her feeding and growth.

- **Management and Recommendations**: The geneticist and genetic counselor express understanding about Cassandra's concerns and encourage her to continue Kyra's care with the multidisciplinary team of specialists. They can discuss genetic testing in the future during other visits to the team. If she plans to have additional children, she may seek prenatal counseling. Cassandra states at this time she hopes not to have more children as she is not prepared to handle more!

  Because of the presence of high myopia in Stickler syndrome, the geneticist recommends that Kyra undergo a pediatric ophthalmology evaluation in the next 6 months to make sure she does not have signs of this condition. Other recommendations for Kyra from the other team members include that she should continue using the special bottle as her interval growth has been good and that her feeding mechanics look appropriate. Because she passed the newborn hearing screen, repeat audiology testing will not be done for a year. The social worker also meets with Cassandra to make sure she has a strong support system and to see if obtaining the state insurance for children with complex healthcare needs is indicated. Cassandra says that her mother's family is local and provides some support, and the state insurance would be very helpful. She also is given information about the American Cleft Palate Association, which provides families with information and resources (Section 7.4).

- **Outcome**: Cleft palate repair surgery is planned for Kyra at 12 months. Prior to this, speech pathology works with Kyra to introduce cup feeding to be used after surgery to decrease tension on the wound. When Kyra undergoes cleft repair, tympanostomy ear tubes are placed to relieve her inner ear effusions. Kyra does well, and by 2 years of age she is speaking sentences with good clarity. Cassandra returns with Kyra for genetic testing once she feels the initial stressors and complex care for a child with a cleft are manageable. She consents for Kyra to have a microarray and an oral clefting gene panel performed, which are both negative. Cassandra has a second child, who does not have oral clefting.

The Diagnosis

Clinical diagnosis: Cleft palate, likely multifactorial.
Genetic diagnosis: Unknown genetic cause.

Further Reading

Dixon, M.J., Marazita, M.L., Beaty, T.H., Murray, J.C., 2011. Cleft lip and palate: understanding genetic and environmental influences. Nat. Rev. Genet. 12 (3), 167–178. PMID 21331089.

Katsanis, S.H., Jabs, E.W., July 20, 2004. Treacher Collins syndrome. Adam, M.P., Everson, D.B., Mirzaa, G.M., Pagon, R.A., Wallace, S.E., Bean, L.J.H., Gripp, K.W., Amemiya, A., (editors). In GeneReviews® [Internet]. University of Washington, Seattle, Seattle (WA), 1993–2023 [updated August 20, 2020]. PMID 20301704.

Robin, N.H., Moran, R.T., Ala-Kokko, L., June 9, 2000. Stickler syndrome. Adam, M.P., Everman, D.B., Mirzaa, G.M., Pagon, R.A., Wallace, S.E., Bean, L.J.H., Gripp, K.W., Amemiya, A., (editors). In GeneReviews® [Internet]. University of Washington, Seattle, Seattle (WA), 1993–2023 [updated May 6, 2021]. PMID 20301479.

Schutte, B.C., Saal, H.M., Goudy, S., Leslie, E.J., October 30, 2003. *IRF6*-related disorders. Adam, M.P., Everman, D.B., Mirzaa, G.M., Pagon, R.A., Wallace, S.E., Bean, L.J.H., Gripp, K.W., Amemiya, A., (editors). In GeneReviews® [Internet]. University of Washington, Seattle, Seattle (WA), 1993–2023 [updated March 4, 2021]. PMID 20301581.

Varadarajan, S., Balaji, T.M., Raj, A.T., Gupta, A.A., Patil, S., Alhazmi, T.H., et al., 2021. Genetic mutations associated with Pierre Robin syndrome/ sequence: a systematic review. Mol. Syndromol. 12 (2), 69–86. PMID 34012376.

Section 11.4

Cardiovascular and Pulmonary Cases

Case 11.4.1

Cardiovascular and Pulmonary Cases: Recurrent Pneumonias

Maksym Puliaiev[1] and Joel N. Kline[1,2]
[1]*Pulmonary and Critical Care Medicine, University of Iowa School of Medicine, University of Iowa Health Care, Iowa City, IA, United States;*
[2]*Professor of Internal Medicine, Pulmonary, Critical Care and Occupational Medicine Environmental Health, University of Iowa School of Medicine, University of Iowa Health Care, Iowa City, IA, United States*

Abstract
A 27-year-old male presenting with recurrent respiratory tract infections and chronic sinus congestion is found to have bronchiectasis and infertility due to cystic fibrosis.

The Case

A 27-year-old male, Hendrix, is seen by you, his primary care provider for recurrent respiratory tract infections. The patient mentions history of "chest colds" and pneumonias since childhood. He experiences these three to four times per year and has been hospitalized at least three times in his lifetime for intravenous antibiotic treatment. He recently established care with you. Prior to now, he went to urgent care clinics and emergency rooms every time he had severe symptoms and generally received short courses of oral antibiotics which were helpful. The patient also reports chronic sinus congestion, occasional shortness of breath and productive cough, fatigue, intermittent constipation, and diarrhea.

On physical examination, patient is in no respiratory distress. He is thin and relatively short. Examination of the nose shows polyps. Lung auscultation reveals scattered expiratory rhonchi in both lower lung fields.

Hendrix has been married for 6 years. He shares with you that he and his wife have been trying to have a child but thus far have been unsuccessful. He is employed as a bank teller, and previously worked in the food industry as a server. He admits to smoking up to one pack of cigarettes per day for 4 years in his early 20s, but quit 2 years ago since he felt it made his breathing worse. The patient denies any history of alcohol abuse or illicit drug use. He has never traveled outside of the United States.

The patient's parents and two siblings have no history of lung disease. His maternal grandfather, who was a heavy smoker, suffered from emphysema and died of lung cancer.

The Issues

- **What is the most likely cause of his symptoms?** Hendrix most likely has obstructive lung disease. In a young adult, asthma is the most common cause of airflow obstruction. Although he has a history of smoking, this is fairly minimal,

[a] Current affiliation: Pulmonary and Critical Care Medicine, MercyOne Healthcare System, Des Moines, IA, United States

Genomics in the Clinic. https://doi.org/10.1016/B978-0-12-816478-5.00033-9

and in any case, chronic obstructive pulmonary disease (COPD) is highly unlikely at a young age. Perhaps he could have some primary pulmonary problem, given the recurrent pulmonary infections and chronic sinusitis.

- **What is the next step in the evaluation?** You obtain a chest X-ray, which demonstrates a tram-tracking appearance of the airways in the bilateral lower lobes. This pattern suggests bronchiectasis, a chronic pulmonary condition in which the airways are dilated and their walls thickened due to persistent inflammation and infection. You refer him to a pulmonologist.
- **Can all of his findings be tied together?** The pulmonologist evaluates the patient and performs pulmonary function tests in the office. These show obstructive lung disease pattern with decrease in forced expiratory volume 1 (FEV1) to 2.8 L (71% of predicted), a reduced FEV1/FVC (forced vital capacity) ratio of 0.66 (normal greater than 0.70) and no significant response to bronchodilators. The lack of response to bronchodilators makes asthma less likely. Hendrix's symptoms include recurrent pulmonary infections and chronic sinusitis, as well as gastrointestinal symptoms and likely infertility. This constellation of findings is suggestive of cystic fibrosis (CF), a genetic disorder caused by pathogenic variants or mutations (Section 2.4.1) in the cystic fibrosis transmembrane conductance regulator (*CFTR*) gene. The disease is characterized by abnormal function of the epithelial cells responsible for ion conductance, mucus production, and mucociliary transport. This leads to difficulty in mucus clearance in the upper and lower respiratory tract and other organs (i.e., pancreas, biliary ducts, vas deferens), persistent inflammation, and bacterial colonization, which results in recurrent pneumonias and structural changes of the airways long-term.
- **Is not cystic fibrosis diagnosed only in children?** Although cystic fibrosis is classically a pediatric disorder, presenting with pulmonary complications, chronic diarrhea, and failure to thrive early, an increasing number of adults are now being diagnosed with it. This has been attributed to more widespread physician awareness of the variability in disease presentation, availability of genotyping techniques, and clearer diagnostic criteria. Patients with adult-diagnosed disease represent a unique cohort of patient with cystic fibrosis. They usually have a variant which preserves some residual CFTR function, leading to slower disease progression, less severe symptoms, and improved survival compared to the patients with classic cystic fibrosis who are diagnosed shortly after birth.
- **What testing should be done to confirm the diagnosis?** For this patient, it would be reasonable to start with advanced chest imaging, such as computed tomography (CT) of the chest, to further characterize the abnormalities noted on the chest X-ray. CT scan will add information regarding any concomitant parenchymal lung disease. Further diagnostic steps may include bronchoscopy with bronchoalveolar lavage and microbiologic studies (i.e., bacterial and fungal cultures) to evaluate for an active lung infection or bacterial colonization. Regarding the diagnosis of cystic fibrosis, current guidelines recommend the sweat chloride test as a first step for all patients in whom the disease is suspected. The concentration of above 60 mmol/L confirms the diagnosis, while below 29 mmol/L makes cystic fibrosis unlikely. If sweat chloride testing is indeterminate, genetic testing for *CFTR* gene pathogenic variants is the next step in evaluation. Of note, patients with adult-diagnosed cystic fibrosis occasionally exhibit negative sweat chloride testing; therefore if a high suspicion for the diagnosis exists, genetic testing should be performed regardless of the sweat chloride test results. Hendrix undergoes sweat chloride testing which yields a concentration of 57 mmol/L, highly suggestive but not diagnostic of cystic fibrosis. His CT scan confirms bronchiectasis. Hendrix is referred for genetic testing.

If sweat chloride testing is indeterminate and genetic testing does not reveal cystic fibrosis causing variants, cystic fibrosis functional testing may be needed (including nasal potential difference and intestinal current measurement). This testing should always be carried out at a center with considerable experience and expertise in the diagnosis and management of cystic fibrosis.

The Referral

- **Genetic Evaluation**: A full medical history and a family history are obtained, and a detailed physical examination is performed by the geneticist. Cystic fibrosis (CF) is the most common life-threatening, autosomal recessive (Section 2.7.1) genetic disorder among those of European descent. There are over 30,000 individuals with cystic fibrosis in the United States with a proportion of patients diagnosed as adults. Diagnosis of cystic fibrosis is critical both for appropriate management of the affected individual as assessing and counseling family members about the risk of cystic fibrosis in themselves or their offspring.

  The geneticist explains to the patient that there are six classes of *CFTR* gene mutations, which affect the production of CFTR protein, its processing function, and stability at the cell membrane. There are over 1700 known mutations associated with cystic fibrosis, and the most common mutation, Phe508del (also known as delta F508), leads to a deletion (Section 2.4.1) with loss of the amino acid phenylalanine at the 508th position of the protein. This causes the CFTR protein to fold abnormally and results in destruction of the CFTR by cellular proteases and complete loss of protein function. Individuals homozygous (Section 2.7.1) for this variant exhibit the classic disease phenotype which is apparent shortly after birth. Of note, a small percentage (below 5% according to some studies) of patients with adult-diagnosed cystic fibrosis have been found to be homozygous for Phe508del. Patients with adult-diagnosed cystic fibrosis usually have variants which preserve some residual CFTR function (such as class III-V variants). They are either homozygous for these types of variants or are compound heterozygotes (Section 2.7.1), having one of these variants and a variant with complete loss of function. It is recommended that Hendrix undergo genetic testing.

- **Genetic Counseling**: The genetic counselor discusses the autosomal recessive nature of cystic fibrosis, in that all patients inherit a mutation in the *CFTR* gene from each parent. Carrier parents have a 25% chance of having an affected child. The counselor recommends that Hendrix discuss genetic testing for cystic fibrosis with his wife, and notify his parent and siblings. Each of his siblings has a two-thirds chance of being a carrier for one of the known mutations and could have an affected child. The counselor also educates the patient about congenital bilateral absence of vas deferens (CBAVD), a condition that causes obstructive azoospermia and infertility in male patients with cystic fibrosis. It is extremely common, affecting up to 98%–99% of patients with cystic fibrosis (Section 5.1.5). Physical examination, ultrasonographic evaluation of seminal vesicles and testicles, and measurement of semen quantity and pH can be used to confirm the diagnosis. Patients with congenital bilateral absence of vas deferens can father children by means of assisted reproductive therapy (Section 4.5.1), but need to be educated about the fact that if their partner is also a carrier, 25% of offspring will be affected. If their partner is not a carrier, all of their children will be carriers of one of the known mutations and this will need to be discussed when they reach childbearing age. The carrier rate in the general population varies from 1 in 22 to much higher, depending on the ethnic background.
- **Genetic Testing**: Subsequent genetic testing reveals that Hendrix is positive for two heterozygous (Section 2.4.1) mutations leading to Arg117His and Gly551Asp, class IV and III mutations, respectively, confirming cystic fibrosis.
- **Management and Recommendations**: Management of patients with cystic fibrosis aims to address the pathophysiologic mechanisms of the disease. Airway clearance, either by breathing techniques or use of accessory devices, is helpful in improving the mucociliary transport and mobilizing secretions. Hendrix is referred to one of the regional Cystic Fibrosis centers, where a multidisciplinary team assists him with management of his disease, nutritional and psychosocial support, and genetic counseling. He is taught airway clearance breathing techniques. Mucolytic agents such as inhaled dornase alfa (a proteolytic enzyme) and hypertonic saline are almost universally recommended to augment the airway clearance and decrease the frequency of disease exacerbations. Prophylactic use of azithromycin is advocated in most patients due to its antiinflammatory effects and its ability to slow down airway remodeling. In patients colonized with *P. aeruginosa* and other microorganisms, inhaled tobramycin and aztreonam are used to reduce bacterial load. Disease exacerbations frequently require hospitalization and intravenous antibiotics as well as occasional oxygen supplementation and ventilatory support. His workup reveals colonization with *P. aeruginosa.* Hendrix is prescribed inhaled dornase alfa and hypertonic saline, as well as inhaled tobramycin and oral azithromycin.

In all patients with cystic fibrosis, nutritional deficits should be assessed and corrected. Pancreatic enzyme replacement is important if there is evidence of pancreatic insufficiency. Insulin replacement may be needed in patients with cystic fibrosis–associated diabetes.

In addition to the aforementioned therapies, recent advances in research of the cystic fibrosis genetic mechanisms have led to development of novel targeted therapies (Section 7.4.6) which modulate the CFTR protein and restore functional ion transport. Among these, two agents, ivacaftor and lumacaftor, are currently recommended for use in patients with cystic fibrosis. They have been shown to decrease sweat chloride concentration, increase lung function, improve quality of life, and reduce the frequency of pulmonary exacerbations. Hendrix starts treatment with ivacaftor.

Patients with progressive disease unresponsive to conventional therapies should be considered for a referral for lung transplantation. Timely referral should be arranged for patients with progressively declining lung function (FEV1 $<$ 50% with rapid decline by 20% or more in preceding 12 months; FEV1 $<$ 40% with 2 or more exacerbations per year, episodes of hemoptysis or pneumothorax), or individuals with significant hypoxia, hypercapnia, or pulmonary hypertension irrespective of FEV1.

As mentioned previously, adult-diagnosed cystic fibrosis patients generally have a milder form of the disease than those diagnosed in childhood. At times, only one organ system is affected, most commonly the respiratory tract. Pancreatic insufficiency or cystic fibrosis–associated diabetes is less prevalent in this cohort of patients and occur in up to 15% of cases per some studies. Ten and fifteen year transplant-free survival for patients with adult-diagnosed cystic fibrosis can be as high as 88% and 86%, respectively. Advanced age at diagnosis, presence of diabetes, and decreased lung function are important predictors of worse survival.

- **Differential**: The differential diagnosis for cystic fibrosis includes primary ciliary dyskinesia, lung disease due to alpha-1 antitrypsin deficiency, chronic cough, bronchial asthma, and chronic obstructive pulmonary disease (including chronic bronchitis and emphysema).
- **Outcome**: Six months later, Hendrix's FEV1 improves to 78%, he appears symptom-free, and he has had no exacerbations to date. Hendrix's wife undergoes carrier testing for cystic fibrosis and finds that she is a carrier for the common Phe508del mutation. He and his wife are thinking about adoption, since they worry about the risk of their children having the disease. His brother and his new wife will be meeting with a genetic counselor to discuss carrier testing (Section 4.1) for their future pregnancies. His sister has three children, all healthy, and is not planning on any further pregnancies, but she should be aware that she could be a carrier and thus, her children could be as well.

The Diagnosis

Clinical Diagnosis: Cystic fibrosis, adult onset; congenital bilateral absence of the vas deferens.
Genetic Diagnosis: *CFTR* gene, compound heterozygous pathogenic variants.

Further Reading

Desai, S., Wong, H., Sykes, J., Stephenson, A.L., Singer, J., Quon, B.S., 2018. Clinical characteristics and predictors of reduced survival for adult-diagnosed cystic fibrosis. Analysis of the Canadian CF Registry. Ann. Am. Thorac. Soc. 15, 1177–1185. PMID 29944384.

Elborn, J.S., 2016. Cystic fibrosis. Lancet 388, 2519–2531. PMID 27140670.

Farrell, P.M., White, T.B., Ren, C.L., Hempstead, S.E., Accurso, F., Derichs, N., et al., 2017. Diagnosis of cystic fibrosis: consensus guidelines from the Cystic Fibrosis Foundation. J. Pediatr. 181S, S4–S15 e1. PMID 28129811.

Gilljam, M., Ellis, L., Corey, M., Zielenski, J., Durie, P., Tullis, D.E., 2004. Clinical manifestations of cystic fibrosis among patients with diagnosis in adulthood. Chest 126, 1215–1224. PMID 15486385.

Stoltz, D.A., Meyerholz, D.K., Welsh, M.J., 2015. Origins of cystic fibrosis lung disease. N. Engl. J. Med. 372, 351–362. PMID 25607428.

Case 11.4.2

Cardiovascular and Pulmonary Cases: Dyspnea with History of Neonatal Jaundice

Scott M. Schecter[1] and Joel N. Kline[2]

[1]*Division of Pulmonary and Critical Care Medicine, University of Virginia Health, Charlottesville, VA, United States;* [2]*Internal Medicine, Pulmonary, Critical Care and Occupational Medicine, Environmental Health, University of Iowa, C33GH, University of Iowa Health Care, Iowa City, IA, United States*

Abstract

A 42-year-old man with a history of neonatal jaundice presents with increasing dyspnea and is found to have severe emphysema and liver dysfunction, consistent with alpha-1 antitrypsin deficiency.

The Case

A 42-year-old male, Erik, notes to you, his primary care physician, that he has been experiencing increasing dyspnea on exertion. Previously he was able to jog approximately 1 mile several times a week, but recently he has not been able to run as far without having to stop due to shortness of breath. On history he reports that he has smoked approximately half a pack a day for 20 years. He is employed as an accountant and has no unusual inhalation exposures in his home or work life. On physical exam, the patient is noted to have reduced breath sounds in all lung fields and scattered end expiratory wheezes. Additionally, tender subcutaneous nodules are noted on his lower extremities. You order pulmonary function testing (PFT) which shows an obstructive pattern with forced expiratory volume 1 (FEV1) of 2.2 L (52% of predicted), forced vital capacity (FVC) of 3.5 L (74% of predicted), and a FEV1/FVC ratio of 0.62 (normal greater than 0.70). There is no significant bronchodilator response (improvement in airflow after inhalation of albuterol). The diffusion capacity (DLCO) is significantly reduced to 45% of predicted. Basic lab work reveals mildly increased for liver enzymes for AST and ALT and is otherwise unremarkable.

The Issues

- **What is the next step?** Shortness of breath is one of the more common presenting symptoms to primary care providers who care for adults. The differential diagnosis is broad, with asthma, chronic obstructive pulmonary disease (COPD), and heart failure being some of the most common causes. The practitioner must keep an open mind about the etiology of her patient's diagnosis. Pulmonary function testing is a critical aspect of the diagnostic approach, and referral may be

Genomics in the Clinic. https://doi.org/10.1016/B978-0-12-816478-5.00034-0

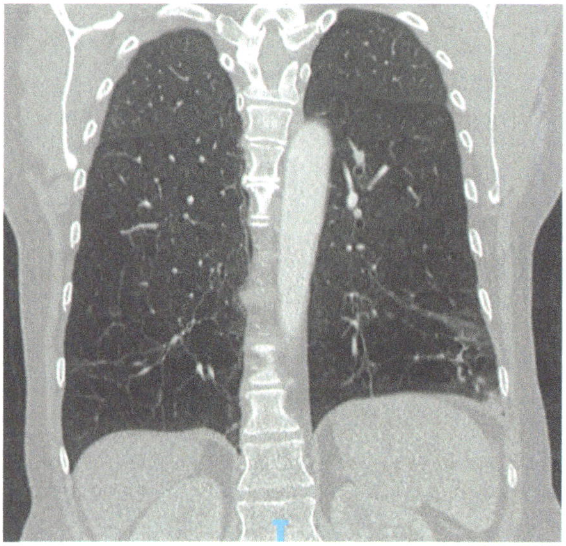

FIGURE 11.4.2.1 Coronal view of chest CT scan demonstrating emphysema throughout all lobes of both lungs and some markings consistent with scarring in the left lower lobe.

warranted when an atypical presentation is encountered. You then order a CT scan to evaluate his lungs (Fig. 11.4.2.1). The scan is significant for emphysema, which is seen in all lobes of the lungs.

- **How should the pulmonary function tests be interpreted?** You discuss the PFTs with the pulmonologist. While both the FEV1 and FVC are reduced below 80% of the value predicted for a 42-year-old male of Erik's height and weight, the FEV1 is reduced out of proportion to the FVC, leading to a low FEV1/FVC ratio, the hallmark of airflow obstruction. The diffusion lung capacity is measured by having the subject inhale a gas mixture containing a very low concentration of carbon monoxide (CO), and then measuring the CO concentration in the exhaled breath. Since CO easily passes across the walls of the alveoli and the blood vessels, then is tightly bound to red blood cells, the reduction in CO concentration between the inhaled and exhaled gases correlates with a subject's alveolar-capillary surface area.
- **What is the likely etiology of this patient's shortness of breath?** The pulmonologist says that given the obstructive nature of the patient's pulmonary function tests, in addition to the radiographic findings of emphysema, a diagnosis of COPD would be the most likely explanation. COPD includes chronic bronchitis, emphysema, and bronchiectasis. Asthma is another common cause of obstructive lung disease, but several features argue toward COPD in this case. Erik did not have childhood symptoms, does not have a bronchodilator response, and demonstrated a reduced diffusion lung capacity, suggesting a reduced alveolar-capillary interface as is seen in emphysema, and he also has parenchymal changes on CT scan (Fig. 11.4.2.1). Other diseases which commonly present with obstructive patterns such as cystic fibrosis (Section 11.4.1), bronchiectasis, or bronchiolitis are less compatible with this presentation, especially the radiographic findings.
- **Is this presentation of COPD typical?** Several features of this case are atypical for the presentation of COPD. The presentation of disease before age 45 is early especially in light of the patient's 10 pack-year smoking history. In the developed world, most COPD is caused by tobacco smoking, whereas in developing countries inhalation of biomass fuels used for cooking and heating are important etiologic factors. While CT imaging was not necessary in the initial workup of this patient, lower lobe emphysema is less commonly seen in comparison to upper lobe disease in COPD caused by cigarette smoking. Additionally, the findings of liver dysfunction and skin findings suggestive of panniculitis are not commonly seen in COPD patients. On further history, neonatal jaundice is reported. This, together with the atypical features, is associated with the later presentation of emphysema caused by alpha-1 antitrypsin (AAT) deficiency. These atypical disease features point to the possibility of alpha-1 antitrypsin deficiency in this patient.
- **When is alpha-1 antitrypsin testing warranted?** Although most COPD is due primarily to cigarette smoking, alpha-1 antitrypsin deficiency is a genetic cause of emphysema, albeit one whose presentation is heavily influenced by a history of cigarette smoking. According to the 2016 guidelines, all adults with COPD should be tested for alpha-1 antitrypsin deficiency. Additionally, patients with unexplained abnormalities in liver function, necrotizing panniculitis, or unexplained bronchiectasis should be tested. Alpha-1 antitrypsin levels should be measured but these are not sufficient to make a diagnosis on their own as they can vary with inflammation. Confirmatory testing should be sought by either protease typing, or *SERPINA1* genotyping. You refer Erik for a genetic evaluation.

The Referral

- **Genetic Evaluation**: A full medical history and a family history are obtained on Erik, and a detailed physical examination is performed by the geneticist. The patient was told by his mother that he was jaundiced at birth and needed a blood transfusion. He has been healthy until he recently noted shortness of breath. His family history is significant for COPD in his father, who smoked for much of his adult life. His exam is remarkable for reduced breath sounds and wheezing.
- **Genetic Counseling**: Alpha-1 antitrypsin deficiency is an autosomal recessive condition (Section 2.7.1). It may be diagnosed by genetic testing. The gene that encodes the AAT protein is the *SERPINA1* gene located on the long arm of chromosome 14. Several different alleles (Section 2.4.1) have been characterized and the phenotypes are designated by different letters. M refers to the normal protein product while S and Z are the most common variants. Of the typical variants, patients with the homozygous ZZ phenotype have the lowest levels and are at high risk for loss of lung function. Heterozygous individuals with the MZ phenotype are at increased risk of emphysema as well.
- **Testing Strategy**: Sequencing of the *SERPINA1* gene can be done (Section 6.2.2), but usually alpha-1 antitrypsin levels and protease (protein) typing are the first-line tests. A serum alpha-1 antitrypsin level is sent on Erik and returns with a low level of 19 mg/dL [normal 100–300 mg/dL]. Additional confirmatory testing with protein typing is pursued. The patient is found to have the ZZ phenotype.
- **Management and Recommendations**: The patient is informed of the diagnosis and counseled about his condition and educated about the importance of smoking cessation. In addition to this, it is important that he try to avoid all forms of secondhand smoke, as well as dust and fume exposure as well. In addition to standard of care COPD treatments, the patient is referred to an alpha-1 antitrypsin specialist. Alpha-1 antitrypsin generally should be managed by a pulmonary specialist, especially if replacement therapy is utilized. Current recommendations support the use of alpha-1 antitrypsin intravenous augmentation therapy in patients with FEV1 between 30% and 65% of predicted, as Erik's was. This has been shown to decrease the progression of emphysema as measured by loss of lung density on CT scan. These changes were only found to be significant in former smokers as opposed to never and current smokers.

  Regarding the mild hepatic involvement, recommendations include avoiding alcohol and monitoring his liver function at least twice a year. Erik is also told about the Alpha-1 Foundation and given their education information (Section 3.5). This can help him get linked with a local support group.
- **Implications for Other Family Members**: It is important for family members to understand the disease, especially the need for tobacco avoidance. The genetic counselor recommends that all immediate and extended family members of the individual who are affected be tested. Testing of his parents reveals that they each have the MZ phenotype. Each parent contributed a Z allele to the patient, making him a homozygote; his siblings are at an equal risk of 25%. His father is at increased risk for emphysema and is recommended to try and stop smoking.
- **Differential Diagnosis**: The differential diagnosis for alpha-1 antitrypsin deficiency includes asthma, COPD, and occupational lung disease.
- **Caveats**: Patients with loss of function or null mutations (Section 2.4.1) in the *SERPINA1* gene will have severe emphysema but will not have hepatic manifestations. This is because the protein does not accumulate in the liver.
- **Outcome**: Erik stops smoking and enters a pulmonary rehabilitation program. He elects to undergo augmentation therapy to preserve lung function.

The Diagnosis

Clinical Diagnosis: Patient—alpha-1 antitrypsin deficiency. Parents—increased risk for emphysema.
Genetic Diagnosis: Patient—protease inhibitor, homozygous ZZ alleles. Parents—protease inhibitor, heterozygous MZ alleles.

Further Reading

Chapman, K.R., Burdon, J.G., Piitulainen, E., Sandhaus, R.A., Seersholm, N., Stocks, J.M., et al., 2015. Intravenous augmentation treatment and lung density in severe alpha1 antitrypsin deficiency (RAPID): a randomised, double-blind, placebo-controlled trial. Lancet 386, 360–368. PMID 26026936.

Sandhaus, R.A., Turino, G., Brantly, M.L., Campos, M., Cross, C.E., Goodman, K., et al., 2016. The diagnosis and management of alpha-1 antitrypsin deficiency in the adult. Chronic. Obstr. Pulm. Dis. 3, 668–682. PMID 28848891.

Townsend, S., Newsome, P., Turner, A.M., 2018. Presentation and prognosis of liver disease in alpha-1 antitrypsin deficiency. Expet Rev. Gastroenterol. Hepatol. 12, 745–747. PMID 29768056.

Case 11.4.3

Cardiovascular and Pulmonary Cases: Pulmonic Stenosis and Short Stature

Bruce D. Gelb

Mindich Child Health and Development Institute, Departments of Pediatrics and Genetics and Genomic Sciences, Icahn School of Medicine at Mount Sinai, New York, NY, United States

Abstract

A toddler with pulmonary valve stenosis, modest short stature, and borderline neurodevelopment is found to have Noonan syndrome.

The Case

A 3-year-old girl, Charlene, whose family recently moved to town, comes to see you, her pediatrician, for her first well-child care visit. Her parents mention that Charlene's previous pediatrician heard a cardiac murmur when Charlene was an infant. A cardiac evaluation revealed mild pulmonary valve stenosis, which has been stable over time. In reviewing Charlene's history, you learn that she had some mild feeding difficulties during the first year of life, which resolved. Her growth has been a bit slow with height assessments around the third percentile. A review of Charlene's development reveals that she began walking at 14 months, and she was a bit late to talk. She is now running, climbing, and speaking in short sentences. She is not yet toilet trained. On examination, you feel that she might have mild facial dysmorphic features and the beginnings of a pectus deformity. Her neurologic exam appears normal.

The Issues

- **What do you make of all of Charlene's findings?** You note multiple findings both on history and physical exam and feel that it is possible that she could have an underlying genetic condition. She has a history of pulmonic stenosis, feeding problems initially, borderline late developmental milestones, and, on exam, she is mildly dysmorphic and short and may have a pectus deformity. You consider Noonan syndrome in your differential diagnosis, but wonder about a mild skeletal dysplasia or a rarer genetic condition.
- **Does Charlene have a genetic diagnosis?** You review Charlene's family history to determine if anyone else has unusual manifestations or any genetic diagnosis. You discover that no other relative has pulmonary valve stenosis or other forms of congenital heart disease, short stature, or a pectus deformity or has been diagnosed with Noonan syndrome. You are not sure if Charlene has Noonan syndrome. Although she has mild pulmonary valve stenosis and mild short stature with her height measurements taken at different ages following just below the third percentile growth curve, her facial features are not obvious and classic for the syndrome, and she does not clearly manifest a pectus deformity yet. Similarly, Charlene's neurodevelopment is fairly normal but consistently at the outer limits as she attained milestones.

Genomics in the Clinic. https://doi.org/10.1016/B978-0-12-816478-5.00035-2

- **Are there any recommendations for Charlene's development being on the later end of normal?** Now that Charlene is meeting her milestones; her development only bears monitoring. She likely would not be approved for early intervention therapy through the state program, which only provides therapy if the child is more than 25% behind the expected, although if you felt very concerned, private therapy could be considered and might be covered by insurance.
- **Should Charlene be referred to a specialist for her short stature?** While you could refer Charlene to a pediatric endocrinologist due to her short stature, that input would not likely result in a specific genetic diagnosis. At the age of 3 years with growth sustaining a stable height percentile, therapeutic intervention for her short stature would not yet be indicated. Based on her findings, you refer Charlene to a clinical geneticist to determine if she has Noonan syndrome.

The Referral

- **Genetic Evaluation**: The clinical geneticist and genetic counselor obtain a full present illness, past medical and surgical history, review of systems, developmental history, family history with three-generation pedigree (Sections 2.7.1, 3.2), and physical examination. The mother reports that she was 27 years old at the time of Charlene's birth and that her father was 35 years old then. Her first trimester ultrasound (Section 4.3.1) during her pregnancy with Charlene revealed a mild increase in nuchal translucency (Section 4.3.2). Noninvasive prenatal testing screening (Section 4.4.1) for aneuploidies was negative. The increased nuchal translucency resolved on subsequent imaging studies, and no further evaluation was undertaken. The remainder of the pregnancy was unremarkable as was the perinatal course. The family history shows that neither parent has features of Noonan syndrome. Charlene has two older siblings, both of whom are also without features of Noonan syndrome. The mother is of African American ancestry and the father is of Afro-Caribbean ancestry. Consanguinity is denied.

  On physical examination, the clinical geneticist notes dysmorphic facial features with a triangular-shaped face, high anterior hairline, diamond-shaped eyebrows, and ears that are normally set but have mildly thickened outer helices. In examining the eyes, the palpebral fissures are neutral, and there is mild ptosis bilaterally. The irides are brown. The midpupillary distance is normal. The hair is normal. The neck is short, but there is no evidence of pterygia. The nipples are widely spaced but, as with the pediatrician's examination, the presence of a pectus deformity is not definitive. The genitalia are normal. There is mild cubitus valgus. There is no evidence of scoliosis. Muscle tone is normal. Examination of the skin reveals a couple of simple nevi, one small café au lait spot but no lentigines.

  Based on the history and physical examination, the clinical geneticist and genetic counselor are suspicious of the diagnosis of Noonan syndrome but feel unable to make a definitive diagnosis. In addition, they feel that Noonan syndrome with multiple lentigines (NSML; formerly, LEOPARD syndrome) must also be part of the differential diagnosis as lentigines may appear after Charlene's current age in patients with Noonan syndrome with multiple lentigines. For the diagnosis of Noonan syndrome, there is a scoring system available but not widely used. If applied, Charlene has a major feature (pulmonary valve stenosis) and two minor features (suggestive facial dysmorphology and a broad thorax) (Fig. 11.4.3.1), which would be deemed sufficient to assign the diagnosis of Noonan syndrome.

- **Genetic Counseling**: The clinical geneticist and genetic counselor discuss with Charlene's parents that they suspect, but are not certain, that Charlene has Noonan syndrome. They explain that Noonan syndrome is an autosomal dominant condition (Section 2.7.1), but that in approximately 50% of cases, it arises in a child for the first time in the family. Molecular genetic testing (Section 6.2) is indicated for individuals like Charlene for whom the diagnosis is suspected and would be informative with a mutation positive result, but a negative result would not rule out the condition. Such testing is often performed even when the clinical diagnosis is certain, particularly for young infants, because the prognosis for certain features associated with Noonan syndrome may depend on the specific causal genetic variant (i.e., genotype–phenotype correlations) (Section 2.8). In addition, the geneticist explains that she is also considering a second diagnosis, Noonan syndrome with multiple lentigines, which has overlapping features so can be difficult to differentiate from Noonan syndrome in infants and toddlers.
- **Genetic Testing**: They discuss the gene testing panel (Section 6.2.3) for Noonan syndrome and the related disorders (called the RASopathies). For individuals who have a definitive diagnosis of Noonan syndrome, the test identifies the causal variant in about 85–90% of cases; for Noonan syndrome with multiple lentigines, the sensitivity is even better, establishing a molecular diagnosis in about 95% of cases. The genetic counselor explains the different possible outcomes for the testing: the finding of a pathogenic or likely pathogenic variant (Section 7.1.3) would formally establish the diagnosis of Noonan syndrome or Noonan syndrome with multiple lentigines, and because the test does not detect

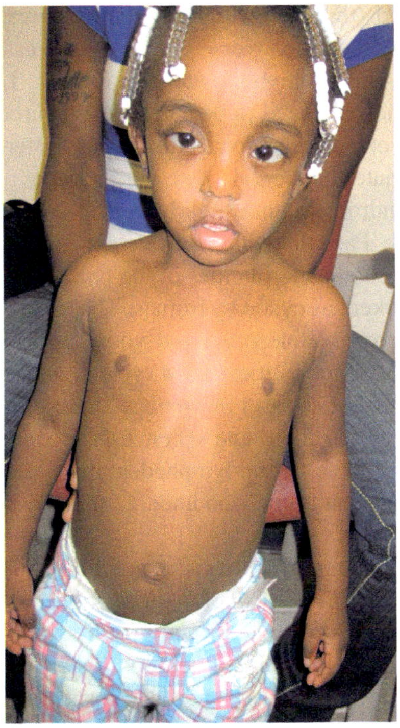

FIGURE 11.4.3.1 A 3-year-old girl with Noonan syndrome due to a pathogenic variant in the *PTPN11* gene.

all cases of those disorders, a negative test could not formally exclude either. She also points out that there is a gray area, variants of unknown significance (VUSs), which are results that the laboratory is unable to assign as causal or benign. Of note, such uncertain findings are less likely for Noonan syndrome and Noonan syndrome with multiple lentigines, particularly for the major gene *PTPN11,* because the range of different causal variants has been established more than most Mendelian genetic disorders. Because Noonan syndrome and Noonan syndrome with multiple lentigines are completely penetrant disorders (Section 2.8), the status of a VUS may be able to be further delineated through genotyping of the parents, and variants inherited from an unaffected parent can be safely excluded from being causal. Charlene's parents, having absorbed the information that the geneticist and counselor have presented, opt to proceed with RASopathy panel genetic testing, for which they give their consent.

- **Results**: A blood sample is taken from Charlene and sent to a commercial laboratory offering RASopathy genetic testing. Three weeks later, the results become available, showing that Charlene harbors a known pathogenic variant in the gene, *PTPN11*, most commonly associated with RASopathies. Moreover, the mutation that Charlene has is causal for Noonan syndrome, not Noonan syndrome with multiple lentigines. There are several genotype–phenotype correlations for genetic variants underlying Noonan syndrome so counseling after return of all positive results should reflect them.
- **Management and Recommendations**: Charlene needs to be monitored for changes in the severity of her current manifestations of Noonan syndrome as well as for the emergence of ones not yet present. For the pulmonary valve stenosis, the fact that the obstruction has remained mild is encouraging, but it can become worse later in childhood and adolescence. In fact, while natural history studies of pulmonary valve stenosis in general suggest that, if the obstruction is mild at age 5 years, then progression is unlikely, the experience with patients with Noonan syndrome shows that progression can occur even in adolescence. She should be followed by pediatric cardiology.

  Short stature is a frequent feature in Noonan syndrome. Untreated, only about one-third of patients will achieve an adult height in the normal range, albeit one typically shorter than what would be expected based on parental heights. The FDA has approved the use of growth hormone therapy for Noonan syndrome. Referral to a pediatric endocrinologist is often undertaken around age 6–8 years.

  Children with Noonan syndrome can have a broad range of neurocognitive outcomes, including typically developing; disabilities in the mild range are most commonly diagnosed. When there are disabilities, there is not a single pattern that is observed. Among those with Noonan syndrome caused by *PTPN11* pathogenic variants, the range of neurodevelopmental outcomes is broad (this is in contradistinction to variants in other RASopathy genes—Section 11.12.2—for which outcomes are either better or worse than for those associated with *PTPN11* mutations). Children with Noonan syndrome are also more likely to exhibit attention deficits. Taken together, pediatricians caring for children with the diagnosis of Noonan syndrome need to be attentive to neurodevelopmental issues and, if present or suspected, either refer for evaluation or work with the local educational system to assure that one is undertaken and then, depending on the results, that appropriate interventions are put in place.

While most features of Noonan syndrome manifest early in life, there are two aspects that tend to present later, in adolescence or even early adulthood. Patients with Noonan syndrome are believed to generally have abnormal lymphatic vascular systems. Developmental abnormalities of the lymphatic system manifest in utero with the increased nuchal translucency or, when more marked, as cystic hygromas. These may lead to the webbed neck (pterygia) in some affected individuals. For reasons unknown, adolescents and young adults may develop peripheral lymphedema as a further manifestation of their lymphatic abnormalities. Most typically, these are merely lymphedema of the lower extremities, a problem that is generally modest and can be managed with pressure stockings. In rarer cases, there is more marked lymphedema with risk of skin breakdown and secondary infection. Lymphatic problems can also present in more serious and/or disabling fashion such as scrotal lymphedema, chylous abdominal ascites, and chylous pleural effusions. In general, patients with Noonan syndrome who undergo thoracic surgery such as cardiac surgery are at higher risk for postoperative chylous effusions.

The second later onset problem in Noonan syndrome is obstruction of the cerebral spinal fluid (CSF) circulation due to a Chiari malformation. The exact prevalence of Chiari malformations in Noonan syndrome has not been established. If a patient with the diagnosis of Noonan syndrome presents with symptoms consistent with CSF obstruction, most commonly recurrent headaches, imaging of the central nervous system is needed to exclude this possibility prior to initiating therapy for more common types of recurrent headaches such as migraines.

- **Implications for Other Family Members:** Because the parents are still considering the possibility of having additional children, they opt to be tested for the specific *PTPN11* variant that Charlene has. This site-specific, targeted testing (Section 6.2.1) is negative. The geneticist explains that this makes the chances of a recurrence of Noonan syndrome due to this variant in a future pregnancy extremely low but still slightly higher than the general population due to the small possibility of gonadal mosaicism (Section 2.4.2). Testing of peripheral blood does not detect gonadal mosaicism. If fetal imaging for a subsequent pregnancy revealed findings that might be consistent with Noonan syndrome, most commonly increased nuchal translucency, then site-specific, targeted testing could be performed after an amniocentesis or chorionic villus sampling (Section 4.4.2) if the parents elect to have prenatal testing.

  While the recurrence risk (Section 3.4) for Charlene's parents is low, Charlene herself is expected to have a 50% chance of transmitting her Noonan syndrome–causing *PTPN11* variant to any offspring. As Noonan syndrome is completely penetrant, a child of Charlene's inheriting that variant would expect to manifest the condition. However, Noonan syndrome has variable expressivity (Section 2.8) so the precise features in an affected child might not match those in Charlene, both with respect to presence/absence and severity.

- **Differential Diagnosis**: In addition to Noonan syndrome with multiple lentigines, there are other possible genetic conditions to consider with overlapping findings. These would include other RASopathies such as Costello syndrome, cardiofaciocutaneous syndrome, and Noonan syndrome with loose anagen hair. Other genetic conditions can have overlapping features such as short stature and neck webbing in females with Turner syndrome (Section 11.7.5) in females, and short stature, some facial features and cardiac defects in males with Aarskog syndrome, an X-linked condition, also called faciodigitalgenital dysplasia.
- **Caveats**: Variants in the *PTPN11* gene for Noonan syndrome with multiple lentigines are nearly mutually exclusive from those causing Noonan syndrome.
- **Outcome**: Charlene's parents understand Noonan syndrome is the cause of their daughter's ongoing medical issues and the recurrence risks for her offspring, themselves, and their other children. They also understand the issues related to the diagnosis for which she should be monitored. They consider whether or not to pursue growth hormone therapy for Charlene when she is older and decide they will consult an endocrinologist. Finally, the parents, and later Charlene, have access to resources such as the Noonan Syndrome Foundation, which is a parent-led organization with the mission of supporting, educating, and advocating for families coping with Noonan syndrome (Section 3.5).

The Diagnosis

Clinical Diagnosis: Noonan syndrome.
Genetic Diagnosis: *PTPN11* gene, heterozygous pathogenic variant.

Further Reading

DYSCERNE—Noonan Syndrome Guideline Development Group, 2010. Management of Noonan Syndrome: A Clinical Guideline [Online]. University of Manchester, Manchester, UK. Available: https://rasopathiesnet.org/wp-content/uploads/2014/01/265_Noonan_Guidelines.pdf.

Gelb, B.D., Tartaglia, M., November 30, 2007. Noonan syndrome with multiple lentigines. Adam, M.P., Everman, D.B., Mirzaa, G.M., Pagon, R.A., Wallace, S.E., Bean, L.J.H., Gripp, K.W., Amemiya, A., (editors). In GeneReviews. University of Washington, Seattle (WA), 1993–2023 [updated June 30, 2022]. PMID 20301557.

Roberts, A.E., Allanson, J.E., Tartaglia, M., Gelb, B.D., 2013. Noonan syndrome. Lancet 381, 333–342. PMID 23312968.

Van Der Burgt, I., Berends, E., Lommen, E., Van Beersum, S., Hamel, B., Mariman, E., 1994. Clinical and molecular studies in a large Dutch family with Noonan syndrome. Am. J. Med. Genet. 53, 187–191. PMID 7856646.

Case 11.4.4

Cardiovascular and Pulmonary Cases: Severe Palpitations

Brittney Murray and Cynthia A. James
Division of Cardiology, Johns Hopkins University, School of Medicine, Baltimore, MD, United States

Abstract
An 11-year-old girl with severe palpitations is found to have long QT syndrome.

The Case

Amanda, age 11 years, has palpitations while at her desk at school. The school nurse finds her heart rate to be over 150 beats per minute. Her parents pick her up at school and bring her to you, her pediatrician. Amanda reports having palpitations so disturbing at night that they wake her up. Her heart rate and rhythm are unremarkable during your physical exam. You send her for an electrocardiogram (ECG), which shows sinus rhythm at 81 beats per minute with a corrected QT interval (QTc) of 473 milliseconds (ms) (borderline: 441−460 ms, and abnormal: >460 ms in females 1−15 years; >470 ms in adult females). You review Amanda's medical history and confirm she is not taking any potentially QT-prolonging medications.

The Issues

- **What is the etiology of Amanda's palpitations?** While palpitations in preteens are not uncommon, you are concerned about the ECG results. You suspect long QT syndrome (LQTS), an inherited arrhythmia syndrome associated with an increased risk of sudden cardiac death. You order a stress test which shows normal exercise capacity but prolonged QTc in recovery, which you know is also consistent with a long QT syndrome diagnosis.
- **Is there any clue to an inherited arrhythmia?** The family reports that there is no family history of sudden death or syncope below age 50.
- **What is the next step?** You are aware that while long QT syndrome is inherited, not all individuals with long QT-associated variants have symptoms or evident QT prolongation. As you suspect Amanda has long QT syndrome, you refer the family to a multidisciplinary inherited heart disease clinic for further evaluation, genetic testing, and management.

Genomics in the Clinic. https://doi.org/10.1016/B978-0-12-816478-5.00036-4

The Referral

- **Genetic Evaluation**: A full medical and family history is obtained. Amanda has two healthy siblings and parents with no symptoms in their 40s. Her paternal grandmother has a history of unspecified dizziness. Her paternal great-grandmother died suddenly in her sleep at age 74 years.

  Amanda's physical exam by a pediatric electrophysiologist is normal. The electrophysiologist reviews the ECG and exercise stress test tracings and confirms the QTc prolongation reported in both. An echocardiogram is ordered which shows normal cardiac chamber size and function.
- **Genetic Counseling**: The genetic counselor meets with Amanda and her parents and explains that long QT syndrome is typically an autosomal dominant (Section 2.7.1) condition with reduced penetrance (Section 2.8). The risk of life-threatening arrhythmic events in long QT syndrome is associated with both the extent of QTc prolongation and the genotype. Long QT syndrome is caused by pathogenic variants in genes encoding ion channels in the heart. While at least 17 genes have been associated with long QT syndrome most patients have pathogenic variants (Section 7.1.3) in one of three genes: *KCNQ1* (LQT1), *KCNH2* (LQT2), or *SCN5A* (LQT3). The counselor explains that should Amanda have a long QT syndrome pathogenic variant it will confirm her diagnosis.
- **Genetic Testing**: Based on Amanda's history and clinical findings, a targeted long QT panel of 12 genes (Section 6.2.3) is ordered. Ordering a targeted long QT syndrome genetic test in this situation where the diagnosis is clear rather than a large arrhythmia panel or combined arrhythmia/cardiomyopathy panel decreases the likelihood of the family receiving a variant of uncertain significance (Section 7.1.5) in a gene not associated with long QT syndrome. Amanda provides a saliva sample and the genetic test is sent.
- **Results**: Genetic testing reveals a heterozygous *KCNH2* deletion (c.220_233del, p.Thr74CysfsTer66), a known pathogenic variant, confirming a diagnosis of long QT syndrome in Amanda. *KCNH2* variants are associated with LQTS2. This subtype is amenable to management with beta-blockers.
- **Management and Recommendations**: Amanda is started on nadolol, the recommended beta-blocker for LQT2 patients. Amanda and her parents are also counseled about the importance of avoiding QT-prolonging drugs. The family is advised that they expect Amanda to do well with this regimen but that they should seek care immediately should she experience syncope or recurrent severe palpitations. The combination of her QTc interval, female sex, and genotype indicates she has a low to moderate risk of a life-threatening arrhythmia. The team also provides the family with information to share with the school about Amanda's diagnosis. The team has a call with the school nurse to help her understand Amanda's diagnosis and to develop an emergency response plan should Amanda experience recurrent symptoms at school.
- **Implications for Other Family Members**: The genetic counselor and pediatric cardiologist also discuss the importance of cascade screening, or single variant targeted testing in at-risk family members (Section 6.2.1), to identify other individuals at risk. Amanda's father has a screening ECG in his primary care doctor's office which shows a QTc of 490 ms (abnormal >450 ms in adult males). Based on this finding, he is diagnosed with long QT syndrome as well, and targeted single variant testing (Section 6.2.1) confirms that he also carries the *KCNH2* variant. Amanda's father is also started on nadolol, counseled about avoiding QT-prolonging drugs, and, as he is an avid marathon runner, recommended to curtail his exercise habits. Subsequent family screening identifies long QT syndrome in a paternal uncle, and his two sons (Amanda's cousins) as well. Finally, the family is connected with a patient support organization (SADS) (Section 3.5).
- **Differential Diagnosis**: The differential diagnosis could include anxiety, as well as idiopathic premature ventricular contractions.
- **Caveats**: A significant proportion of carriers of pathogenic long QT syndrome variants have a QTc interval in the normal range. A normal screening ECG does not exclude the possibility a patient has an LQTS-associated variant. Indeed, the three most common genes mentioned above are among the American College of Medical Genetics and Genomics actionable genes that individuals can choose to find out if they have a variant or secondary finding when doing exome or genome sequencing for other clinical reasons (Sections 6.3.1, 6.3.2, 7.2).
- **Outcome**: Amanda is maintained on the beta-blocker. Her palpitations lessen and she no longer feels them. She now sleeps well through the night and has no further events at school.

The Diagnosis

Clinical Diagnosis: Patient and Father—long QT syndrome.

Genetic Diagnosis: Patient, Father, Paternal Uncle, and Cousins—*KCNH2* gene, heterozygous pathogenic variant.

Further Reading

Ackerman, M.J., Priori, S.G., Willems, S., Berul, C., Brugada, R., Calkins, H., et al., 2011. HRS/EHRA expert consensus statement on the state of genetic testing for the channelopathies and cardiomyopathies: this document was developed as a partnership between the Heart Rhythm Society (HRS) and the European Heart Rhythm Association (EHRA). Europace 13, 1077–1109. PMID 21810866.

Adler, A., Novelli, V., Amin, A.S., Abiusi, E., Care, M., Nannenberg, E.A., et al., 2020. An international, multicentered, evidence-based reappraisal of genes reported to cause congenital long QT syndrome. Circulation 141 (6), 418–428. PMID 31983240.

Mazzanti, A., Maragna, R., Vacanti, G., Monteforte, N., Bloise, R., Marino, M., et al., 2018. Interplay between genetic substrate, QTc duration, and arrhythmia risk in patients with long QT syndrome. J. Am. Coll. Cardiol. 71, 1663–1671. PMID 29650123.

Case 11.4.5

Cardiovascular and Pulmonary Cases: Exercise-Induced Syncope

Brittney Murray and Cynthia A. James
Division of Cardiology, Johns Hopkins University, School of Medicine, Baltimore, MD, United States

Abstract
A 15-year-old boy faints while swimming and is found to have arrhythmogenic right ventricular cardiomyopathy.

The Case

Billy is a 15-year-old competitive swimmer who was practicing when he lost consciousness in the water. He was immediately pulled from the pool and transported to the emergency room by ambulance. He was found to be in sustained ventricular tachycardia with a left bundle superior axis morphology at 250 beats per minute. The arrhythmia was terminated by cardioversion and Billy now is admitted to your service for evaluation. Billy's physical exam is unremarkable. You order an electrocardiogram (ECG) and cardiac troponins.

The Issues

- **What is the cause of sustained ventricular tachycardia in this young athlete?** ECG monitoring reveals frequent premature ventricular contractions. His ECG is also notable for T wave inversions in leads V1−V3. Cardiac troponins are mildly elevated. The differential diagnosis includes idiopathic right ventricular tachycardia, myocarditis, arrhythmogenic cardiomyopathy, premature coronary artery disease, and several inherited arrhythmia syndromes. As Billy is an athlete, you are unsure whether the anterior precordial T wave inversions on ECG are abnormal or are physiologic adaptions to exercise (e.g., athlete's heart). However, you recognize that arrhythmogenic right ventricular cardiomyopathy (ARVC) frequently occurs in athletes and is consistent with the morphology of his sustained ventricular tachycardia.
- **Can imaging help determine the diagnosis?** You order a cardiac MRI to assess his heart structure and function. The results reveal a dilated and hypokinetic right ventricle with regional wall motion abnormalities. Billy now meets diagnostic 2010 Task Force Criteria for arrhythmogenic right ventricular cardiomyopathy based on the T wave inversions on his ECG, his ventricular tachycardia, and his imaging findings. However, you remain concerned that these findings could still be consistent with myocarditis in an athlete.
- **How should this patient be treated?** Based on his exercise-induced cardiac syncope, abnormal imaging, and presumptive arrhythmogenic right ventricular cardiomyopathy diagnosis you recommend placement of an implantable cardioverter defibrillator (ICD) to prevent sudden cardiac death. Arrhythmogenic right ventricular cardiomyopathy is

Genomics in the Clinic. https://doi.org/10.1016/B978-0-12-816478-5.00037-6

associated with frequent ventricular arrhythmias and a high risk of sudden cardiac death. A prior history of sustained ventricular tachycardia and cardiac syncope is class I indications for implantable cardioverter defibrillator placement. After discussing the risks, benefits, and alternatives with Billy and his family, a transvenous implantable cardioverter defibrillator is implanted. Billy is also started on beta blockers.

- **Can the diagnosis be confirmed using genetic testing?** While Billy is an inpatient you order arrhythmogenic right ventricular cardiomyopathy genetic testing from your in-house academic laboratory. The lab tests six genes associated with arrhythmogenic right ventricular cardiomyopathy. You are somewhat surprised when no arrhythmogenic right ventricular cardiomyopathy-associated genetic variant is identified. You refer the family to a cardiogenetic program for further genetic evaluation, a diagnostic second opinion, and to discuss management.

The Referral

- **Genetic Evaluation**: A three-generation pedigree (Section 3.2) focused on cardiac history is taken by the genetic counselor. Billy and his family report no family history of sudden death, arrhythmias, or cardiomyopathy. However, Billy's 45-year-old mother, Mary, describes a 10-year history of increasingly frequent palpitations. She has not been evaluated for these symptoms.
- **Genetic Counseling**: The genetic counselor explains that arrhythmogenic right ventricular cardiomyopathy is usually an autosomal dominant condition (Section 2.7.1). Inheritance is characterized by incomplete, age-related penetrance and variable expressivity (Section 2.8). Therefore, even though neither of Billy's parents have had significant symptoms they may still be at risk for arrhythmogenic right ventricular cardiomyopathy.
- **Genetic Testing**: The genetic test result from the in-house academic laboratory is reviewed. The genetic counselor explains that it is not terribly surprising that the local genetic test identified no pathogenic or likely pathogenic ARVC-associated variant since the genetic basis of arrhythmogenic right ventricular cardiomyopathy cannot be identified in nearly half of probands. However, the initial genetic test did not include all arrhythmogenic right ventricular cardiomyopathy-associated genes and also did not include deletion/duplication analysis (Sections 6.1.4, 6.2) of the six genes that were sequenced. Therefore, a full panel (Section 6.2.3) of 60 arrhythmia genes with deletion/duplication analysis is ordered. Ordering a broader panel is important in this situation due to the lingering diagnostic uncertainty and the fact that a small proportion of arrhythmogenic right ventricular cardiomyopathy patients have multiple pathogenic variants (Sections 2.8, 7.1) contributing to the phenotype. Patients presenting at younger ages are more likely to have multiple variants. As arrhythmogenic right ventricular cardiomyopathy rarely occurs prior to puberty, Billy is a relatively young ARVC patient.
- **Results**: The genetic test results identify a deletion of exons 2 and 3 of the *PKP2* gene. *PKP2* encodes PLAKOPHILIN 2, a protein of the cardiac desmosome. Pathogenic and likely pathogenic variants in *PKP2* are the most common genetic causes of ARVC. While *PKP2* was sequenced as part of the initial genetic test, the in-house lab did not identify this deletion because deletion/duplication analysis was not performed. Deletions and duplications are common in arrhythmogenic right ventricular cardiomyopathy.

  This genetic test result strongly supports Billy's clinical arrhythmogenic right ventricular cardiomyopathy diagnosis. Identification of a pathogenic desmosomal variant is part of the diagnostic criteria for this condition (major criterion).
- **Management and Recommendations**: The genetic counselor explains to Billy and his family that it is likely he inherited the *PKP2* deletion from one of his parents since de novo variants in arrhythmogenic right ventricular cardiomyopathy are rare. It is recommended that Billy not participate in competitive or frequent high intensity aerobic exercise. Such exercise is associated with both arrhythmias and heart failure risk in patients. The family also is connected with other patients with arrhythmogenic right ventricular cardiomyopathy and support groups (Section 3.5). A referral for counseling is also offered.
- **Implications for Other Family Members**: Cascade genetic testing (Section 6.2.1) identifies that Billy's *PKP2* variant was inherited from his mother. Mary's cardiac testing reveals an abnormal ECG. Holter monitoring shows frequent premature ventricular contractions. Together these findings result in Mary also meeting arrhythmogenic right ventricular cardiomyopathy diagnostic criteria.

  The cardiac electrophysiologist offers Mary an implantable cardioverter defibrillator to prevent sudden cardiac death. Decision-making for an implantable cardioverter defibrillator in the absence of a history of cardiac syncope or sustained ventricular arrhythmias is challenging. A "risk calculator" has recently been developed to help physicians and patients predict individualized arrhythmic risk in these patients. Mary undergoes an electrophysiology study with high-dose adrenaline (isoproterenol) which is able to induce arrhythmias. After discussing the risks, benefits, and limitations of implantable cardioverter defibrillator placement with the electrophysiologist, she elects implantation of a primary prevention ICD. She is also cautioned against vigorous exercise.

- **Caveats**: When considering genetic testing for arrhythmogenic right ventricular cardiomyopathy it is critical that the test chosen be capable of identifying duplications and deletions, as these types of variants are common in ARVC-associated genes. Of note, the *PKP2* gene is included in the defined set of genes considered medically actionable by the American College of Genetics and Genomics (Section 7.2). Patients can decide to learn their results for variants in this gene, even when unrelated to the primary medical reason for exome or genome sequencing, as a secondary finding.
- **Outcome**: Quitting competitive swimming has been challenging for Billy, as his identity and social life was strongly associated with being an athlete. His mother Mary reports feeling guilty about passing the *PKP2* variant to Billy. For both, adjusting to life with an ICD has had its challenges. Billy and his mother both do well, however. Billy has one episode of sustained ventricular tachycardia appropriately treated by his implantable cardioverter defibrillator. With adjustment of his medication, he has no further episodes. His mother has no arrhythmias. Mary reaches out to her relatives to inform them of the diagnosis and their risk of also carrying a *PKP2* variant. Cascade genetic testing is ongoing. The family becomes a valued part of the arrhythmogenic right ventricular cardiomyopathy (ARVC) support community.

The Diagnosis

Clinical Diagnosis: Patient and Mother—arrhythmogenic right ventricular cardiomyopathy.
Genetic Diagnosis: Patient and Mother—*PKP2* gene, heterozygous pathogenic variant.

Further Reading

Cadrin-Tourigny, J., Bosman, L.P., Nozza, A., Wang, W., Tadros, R., Bhonsale, A., et al., 2019. A new prediction model for ventricular arrhythmias in arrhythmogenic right ventricular cardiomyopathy. Eur. Heart J. 40, 1850–1858. PMID 30915475.

Corrado, D., Wichter, T., Link, M.S., Hauer, R.N., Marchlinski, F.E., Anastasakis, A., et al., 2015. Treatment of rrhythmogenic right ventricular cardiomyopathy/dysplasia: an international Task Force consensus statement. Circulation 132, 441–453. PMID 26216213.

Groeneweg, J.A., Bhonsale, A., James, C.A., Te Riele, A.S., Dooijes, D., Tichnell, C., et al., 2015. Clinical presentation, long-term follow-up, and outcomes of 1001 arrhythmogenic right ventricular dysplasia/cardiomyopathy patients and family members. Circ. Cardiovasc. Genet. 8, 437–446. PMID 25820315.

Hoorntje, E.T., Te Rijdt, W.P., James, C.A., Pilichou, K., Basso, C., Judge, D.P., et al., 2017. Arrhythmogenic cardiomyopathy: pathology, genetics, and concepts in pathogenesis. Cardiovasc. Res. 113, 1521–1531. PMID 28957532.

Case 11.4.6

Cardiovascular and Pulmonary Cases: Hyperlipidemia

Emily E. Brown and Cynthia A. James
Division of Cardiology, Johns Hopkins University, Baltimore, MD, United States

Abstract
A 27-year-old man with a history of early hyperlipidemia is diagnosed with familial hypercholesterolemia.

The Case

Gregory, a 27-year-old male, was first diagnosed with hyperlipidemia in college. He was previously taking rosuvastatin but had to stop due to costs. He returned to your practice during the past year. Within the last 2 months he began taking atorvastatin 20 mg and niacin. A recent lipid panel revealed total cholesterol 227 mg/dL (normal adult male <200 mg/dL), low density lipoprotein cholesterol (LDL-C) 174 mg/dL (normal <100 mg/dL), high density lipoprotein (HDL) 31 mg/dL (normal adult male >40 mg/dL, adult female >60 mg/dL), and triglycerides 109 mg/dL (normal <150 mg/dL). Prior to restarting medications, his untreated lipid panel had revealed a total cholesterol 396 mg/dL, LDL-C 300 mg/dL, HDL 34 mg/dL, and triglycerides 416 mg/dL. He does not have a history of coronary artery disease and denies chest pain, palpitations, and syncope.

The Issues

- **What are Gregory's risk factors?** You review Gregory's family history and learn that Gregory's father and sister have high cholesterol. Asking further about his diet and lifestyle, you learn Gregory does not regularly exercise and drinks a lot of soda pop. You are not sure if his high triglycerides are due to his lifestyle or are hereditary.
- **Could there be a genetic cause for his hyperlipidemia?** Gregory's positive family history and personal history of LDL-C levels >190 mg/dL are suggestive of familial hypercholesterolemia. However, Gregory does not meet diagnostic criteria using the Dutch Lipid Clinic Score. You know that familial combined hyperlipidemia, which is characterized by elevated triglycerides and LDL-C levels, is also a hereditary condition which should be on the differential diagnosis.
- **Is additional therapy for his dyslipidemia needed?** You review the latest American Heart Association guidelines which indicate the goal LDL-C level for patients with severe hypercholesterolemia and no history of coronary artery disease would be <100 mg/dL with therapy. Based on this, you believe Gregory needs additional therapy but are not sure what to prescribe. You refer him to a lipid specialty clinic for discussion of additional medical management for hyperlipidemia and evaluation for genetic causes.

Genomics in the Clinic. https://doi.org/10.1016/B978-0-12-816478-5.00038-8

The Referral

- **Genetic Evaluation:** A full medical history and family history are obtained and a physical exam is performed by the lipidologist. A family history is obtained by a genetic counselor. Gregory has a significant family history of premature coronary artery disease and hyperlipidemia. His sister recently had a lipid panel which revealed a total cholesterol around 350 mg/dL. Gregory's father has been treated for high cholesterol for many years and is currently taking ezetimibe and a statin. On treatment, his total cholesterol measured 220 mg/dL. Gregory's paternal grandfather died from congestive heart failure secondary to a heart attack at 50 years old.

  Gregory's physical exam is normal. He does not have any physical sequelae of hyperlipidemia including tendon xanthomas, xanthelasmas, or corneal arcus.

  Applying Gregory's personal and family history to the Dutch Lipid Clinic Network Criteria, Gregory has "Probable Familial Hypercholesterolemia."
- **Genetic Counseling:** The genetic counselor discusses with Gregory that his personal and family histories are suggestive of a hereditary dyslipidemia. The two most common hereditary forms are familial hypercholesterolemia which is characterized by an elevated LDL-C level and familial combined hyperlipidemia which is characterized by elevated LDL-C *and* triglycerides. While both conditions are known to be genetic, the genetics of familial combined hyperlipidemia is not well understood. The genetic counselor reviews that familial hypercholesterolemia is an autosomal codominant disorder (Section 2.7.1). The heterozygous form is due to a single pathogenic variant while the homozygous form is due to two pathogenic variants (Section 7.1.3). First-degree relatives of an individual with heterozygous familial hypercholesterolemia have a 50% chance to also carry the variant and subsequently be at an increased risk to develop the condition (Section 3.4). It is reviewed that genetic testing for familial hypercholesterolemia can aid in making the diagnosis.
- **Genetic Testing:** Genetic testing for familial hypercholesterolemia includes sequencing (Section 6.2.2) and deletion/duplication analysis (Section 6.2) of four genes: *PCSK9, LDLR, APOB,* and *LDLRAP1.* Around 70% of individuals with a definitive clinical diagnosis of familial hypercholesterolemia will have a positive genetic test result. However, around 30% of patients who do not meet definitive clinical criteria but have suggestive histories will receive a positive result. A saliva sample is sent off for analysis.
- **Results:** Genetic testing reveals a pathogenic variant in *LDLR* (p.Tyr188*). This confirms Gregory has familial hypercholesterolemia. His elevated triglycerides are likely related to lifestyle factors and not an underlying biological mechanism.
- **Management and Recommendations:** The genetic counselor and lipidologist discuss further information regarding familial hypercholesterolemia with Gregory. Individuals with familial hypercholesterolemia are at up to 20 times greater risk to develop coronary artery disease compared to the general population. Early, effective treatment is imperative to preventing coronary heart disease. The lipidologist reviews the lab results, agrees that additional therapy is recommended, and prescribes ezetimibe. If Gregory's LDL-C levels remain >100 mg/dL on ezetimibe, a PCSK9 inhibitor will likely be the next step. This medication inhibits the PCSK9 enzyme which results in more LDL receptors on the liver. Consequently, LDL-C uptake into the liver increases reducing the concentration in the blood stream which in turn reduces the risk to develop coronary artery disease.
- **Implications for Other Family Members:** It is also reviewed that all of Gregory's first-degree relatives have a 50% chance to also carry this pathogenic variant (Sections 2.7.1 and 3.4) and be at risk to develop familial hypercholesterolemia. It is recommended that relatives have their cholesterol levels checked periodically and consider targeted genetic testing (Section 6.2.1) for the familial variant given the reduced penetrance and variable expressivity associated with familial hypercholesterolemia.
- **Caveats:** Negative physical exam findings do not rule out familial hypercholesterolemia. Similarly, although hypertriglyceridemia is typically associated with familial combined hyperlipidemia, elevated triglycerides do not rule out this diagnosis. Since there is reduced penetrance (Section 2.8) with this condition it is important to correlate genetic test results with clinical findings before assigning a diagnosis.
- **Outcome:** Despite taking a high intensity statin and ezetimibe, Gregory's cholesterol levels remain elevated (total cholesterol 200 mg/dL and LDL-C 146 mg/dL). Given that Gregory's positive genetic test results confirm his diagnosis of familial hypercholesterolemia, his insurance company approves a PCSK9 inhibitor for further treatment. The positive genetic test results also allow cascade screening within the family and subsequent appropriate treatment for individuals identified with familial hypercholesterolemia. Gregory does well with the medicine and trying to alter his diet, and a year later his panel reveals LDL-C 67 mg/dL and total cholesterol 105 mg/dL.

The Diagnosis

Clinical Diagnosis: Patient, Father, and Sister—hyperlipidemia.
Genetic Diagnosis: Patient—*LDLR* gene, heterozygous pathogenic variant.

Further Reading

Civeira, F., Ros, E., Jarauta, E., Plana, N., Zambon, D., Puzo, J., de Esteban, J.P.M., Ferrando, J., Zabala, S., Almagro, F., Gimeno, J.A., Masana, L., Pocovi, M., 2008. Comparison of genetic versus clinical diagnosis in familial hypercholesterolemia. Am. J. Cardiol. 102, 1187−1193. 1193.e.1. PMID 18940289.

Goldberg, A.C., Hopkins, P.N., Toth, P.P., Ballantyne, C.M., Rader, D.J., Robinson, J.G., Daniels, S.R., Gidding, S.S., de Ferranti, S.D., Ito, M.K., McGowan, M.P., Moriarty, P.M., Cromwell, W.C., Ross, J.L., Ziajka, P.E., 2011. Familial hypercholesterolemia: screening, diagnosis and management of pediatric and adult patients: clinical guidance from the National Lipid Association Expert Panel on Familial Hypercholesterolemia. J. Clin. Lipidol. 5, 133−140. PMID 21600517.

Grundy, S.M., Stone, N.J., Bailey, A.L., Beam, C., Birtcher, K.K., Blumenthal, R.S., Braun, L.T., de Ferranti, S., Faiella-Tommasino, J., Forman, D.E., Goldberg, R., Heidenreich, P.A., Hlatky, M.A., Jones, D.W., Lloyd-Jones, D., Lopez-Pajares, N., Ndumele, C.E., Orringer, C.E., Peralta, C.A, Saseen, J.J., Smith, Jr., S.C., Sperling, L., Virani, S.S., Yeboah, J., 2019. 2018 AHA/ACC/AACVPR/AAPA/ABC/ACPM/ADA/AGS/APhA/ASPC/NLA/PCNA guideline on the management of blood cholesterol: a report of the American College of Cardiology/American Heart Association task force on clinical practice guidelines. Circulation 139 (24), e1082−1143. PMID 30586774.

Nordestgaard, B.G., Chapman, M.J., Humphries, S.E., Ginsberg, H.N., Masana, L., Descamps, O.S., et al., 2013. Familial hypercholesterolaemia is underdiagnosed and undertreated in the general population: guidance for clinicians to prevent coronary heart disease: consensus statement of the European Atherosclerosis Society. Eur. Heart J. 34, 3478−3490a. PMID 23956253.

Sturm, A.C., Knowles, J.W., Gidding, S.S., Ahmad, Z.S., Ahmed, C.D., Ballantyne, C.M., Baum, S.J., Bourbon, M., Carrie, A., Cuchel, M., de Ferranti, S.D., Defesche, J.C., Treiberger, T., Hershberger, R.E., Hovingh, G.K., Karayan, L., Kastelein, J.J.P., Kindt, I., Lane, S.R., Leigh, S.E., Linton, M.F., Mata, P., Neal, W.A., Nodestgaard, B.G., Santos, R.D., Harada-Shiba, M., Sijbrands, E.J., Stitziel, N.O., Yamashita, S., Wilemon, K.A., Ledbetter, D.H., Rader, D.J.; Convened by the Familial Hypercholesterolemia Foundation, 2018. Clinical genetic testing for familial hypercholesterolemia: JACC scientific expert panel. J. Am. Coll. Cardiol. 72 (6), 662−680. PMID 30071997.

Case 11.4.7

Cardiovascular and Pulmonary Cases: Myocardial Infarction in a Healthy Adult

Reed Pyeritz

Department of Medicine, Department of Genetics, Perelman School of Medicine, University of Pennsylvania, Smilow Center for Translational Research, Philadelphia, PA, United States

Abstract

A 45-year-old healthy man has a myocardial infarction with no risk factors other than a grandfather with a similarly aged cardiac death and a daughter with Moyamoya disease and is found to have an aortopathy due to a pathogenic variant in the *ACTA2* gene.

The Case

Otis, a 45-year-old previously healthy man, develops sudden-onset crushing anterior chest pain radiating to his left arm. Taken to the emergency room, he is found to have a transmural anteroseptal myocardial infarction. He undergoes coronary angiography that shows a near-totally occluded anterior descending artery. He agrees to percutaneous coronary artery angioplasty that is successful. He is discharged without symptoms in the care of a cardiologist. On the follow-up visit, you, the nurse practitioner, review his medical status.

The Issues

- **What were his risk factors?** Otis had been asymptomatic until that event. Routine primary care had not revealed any risk factors for occlusive arterial disease: normal LDL, above average HDL, normal blood pressure, and a nonsmoker. A family history focused on myocardial infarction shows no one in the immediate family affected.
- **Is there anything significant in the family history?** You take a detailed, inclusive family history, reviewing risk factors for occlusive arterial disease in detail. He relates various disorders in a number of relatives, but you emphasize cardiovascular disorders. He recalls a history of his maternal grandfather dying of some sort of an aortic disorder. You initially assume this was likely the abdominal aneurysm relatively common in older men, but nonetheless inquire about his age. Your patient states that his grandfather died so young that he never knew him, probably in his 40s. This would be distinctly uncommon for an atherosclerotic abdominal aortic aneurysm. Otis says his parents are still alive in their late 60s and to his knowledge have never been evaluated for any cardiovascular problems. He has two children. He also mentions that his young adult daughter had a sudden stroke-like episode a few years previously from which she recovered. It was caused by a vascular malformation with a strange name. You ask, 'Was it Moyamoya?'. He says, 'That's it'.
- **What is the next step?** You immediately counsel Otis about the possibility of an inherited condition and recommend a referral to a cardiovascular or general clinical geneticist.

Genomics in the Clinic. https://doi.org/10.1016/B978-0-12-816478-5.00039-X

The Referral

- **Genetic Evaluation:** Otis is seen by a cardiovascular geneticist and a genetic counselor. The history is reviewed and a detailed three-generational family history is obtained, as above. A physical exam is negative, showing no signs of a possible underlying connective tissue disorder such as vascular Ehlers–Danlos syndrome, lipid disorder, or any arterial disease associated with an overt syndrome. The skin exam reveals livedo reticularis, but is otherwise negative. They concur that he likely has a hereditary condition affecting blood vessels.
- **Genetic Testing:** Although the geneticist and genetic counselor suspect pathogenic variants in the gene smooth muscle alpha-2 actin, *ACTA2*, they recommend a full arteriopathy gene panel (Section 6.2.3), with over 24 genes associated with aortopathy or arteriopathy. Alternatively, exome sequencing (Section 6.3.1) or genome sequencing (Section 6.3.2) could be offered, but these are much more difficult to have approved by the insurance company (Section 10.2).

The cost and time for analysis of one gene or many are similar. So, on the possibility that a pathogenic variant is not present in ACTA2, a full panel of genes should be ordered and analyzed.

- **Results:** A specific pathogenic variant in the *ACTA2* gene is found. Although any pathogenic variant (Section 7.1.3) can predispose to aortopathy, those that lead to a substitution of one of several amino acids for arginine at position 179 are particularly detrimental. They predispose to aortopathy as well as arteriopathy. Otis is found to have this type, namely, the arginine has been changed to a histidine.
- **Genetic Counseling:** Pathogenic variants in the gene *ACTA2* predispose to a variety of arterial disorders, including aortic aneurysm, particularly of the thoracic portion, dissection of the aorta and its major branches, and a cerebral vascular congenital malformation termed Moyamoya. The latter appears as a tangled collection of blood vessels in the brain and can predispose to transient ischemic attacks or outright strokes, as was seen in Otis' daughter. Having only one copy of the variant (heterozygosity—Section 2.3.4) is sufficient to produce the syndrome, so the syndrome is inherited as an autosomal dominant (Section 2.7.1) trait. Otis' myocardial infarction was likely due to a dissection of a coronary artery. The arteriopathy due to a variation in *ACTA2* affects vascular smooth muscle cells, in distinction to defects of the extracellular matrix such as those seen in Marfan syndrome (Section 11.10.4).
- **Management and Recommendations:** The patient is counseled about his own risks. His aorta should be imaged to establish baseline diameters, and imaging should be repeated periodically. If no dilatation is present, this can be done annually because of his specific variant, but should be done more frequently depending on the diameter of the dilated portion. Since the risk of aortic dissection is greater at a smaller diameter than that in Marfan syndrome and other aortopathies, he should be followed by both a cardiologist and cardiothoracic surgeon familiar with familial aortopathies. He should also have imaging of the cerebral vasculature to determine if an asymptomatic Moyamoya defect is present. A person with any aortopathy should be counseled against vigorous activity, isometric exertion, and contact sports.

  No specific medical therapy is available currently. For someone with an *ACTA2* pathogenic variant, treatment with a beta-adrenergic blocker can delay dilatation and reduce the risk of dissection.
- **Implications for Other Family Members:** Otis should also be counseled about the risks to his parents and two children. Since it is more likely that a predisposition was inherited from his mother, because of her father's history, she should be referred to a cardiologist for screening of her aorta and determination if she has any subtle manifestations of a generalized arteriopathy. The same recommendation should be offered to his two children, especially his daughter with Moyamoya. His unaffected daughter should also undergo imaging of the cerebral vasculature to determine if an asymptomatic Moyamoya defect is present. Genetic testing would be offered to all of these family members, but particularly his unaffected daughter.
- **Caveats:** Determining how often to image most of the arterial circulation is unclear. Similarly, a definite diameter at which to recommend surgical intervention for any artery is uncertain.
- **Outcome:** Otis recovers from his myocardial infarction. Echocardiography shows that his aortic root dimension is 42 mm, which is mildly enlarged. He is referred to a cardiologist for routine follow-up, to include echocardiography every 6 months to follow the diameter of his aortic root. He is also referred to a cardiothoracic surgeon expert in prophylactic repair of the ascending aorta using a valve-sparing technique. There will be a continuing dialogue between Otis, his cardiologist, and his surgeon in the years ahead.

  His daughter with Moyamoya tests positive using a targeted variant analysis for the same *ACTA2* variant as Otis, and she also will be referred to a cardiologist for echocardiography and regular follow-up. His other daughter tests negative for the variant and needs no further evaluation.

The Diagnosis

Clinical Diagnosis: Patient—familial thoracic aortic aneurysm. Daughter—Moyamoya disease.
Genetic Diagnosis: Patient and Daughter—*ACTA2* gene, heterozygous pathogenic variant.

Further Reading

Guo, D.C., Papke, C.L., Tran-Fadulu, V., Regalado, E.S., Avidan, N., Johnson, R.J., Kim, D.H., Pannu, H., Willing, M.C., Sparks, E., Pyeritz, R.E., Singh, M.N., Dalman, R.L., Grotta, J.C., Marian, A.J., Boerwinkle, E.A., Frazier, L.Q., LeMaire, S.A., Coselli, J.S., Estrera, A.L., Safi, H.J., Veeraraghavan, S., Muzny, D.M., Wheeler, D.A., Willerson, J.T., Yu, R.K., Shete, S.S., Scherer, S.E., Raman, C.S., Buja, L.M., Milewicz, D.M., 2009. Mutations in smooth muscle alpha-actin (ACTA2) cause coronary artery disease, stroke, and Moyamoya disease, along with thoracic aortic disease. Am. J. Hum. Genet. 84, 617–627. PMID 19409525.

Regalado, E.S., Guo, D.C., Prakash, S., Bensend, T.A., Flynn, K., Estrera, A., Safi, H., Liang, D., Hyland, J., Child, A., Arno, G., Boileau, C., Jondeau, G., Braverman, A., Moran, R., Morisaki, T., Morisaki, H.; Montalcino Aortic Consortium; Pyeritz, R., Coselli, J., LeMaire, S., Milewicz, D.M., 2015. Aortic disease presentation and outcome associated with ACTA2 mutations. Circ. Cardiovasc. Genet. 8, 457–464. PMID 25759435.

Regalado, E.S., Mellor-Crummey, L., De Backer, J., Braverman, A.C., Ades, L., Benedict, S., Bradley, T.J., Brickner, M.E., Chatfield, K.C., Child, A., Feist, C., Holmes, K.W., Iannucci, G., Lorenz, B., Mark, P., Morisaki, T., Morisaki, H., Morris, S.A., Mitchell, A.L., Ostergaard, J.R., Richer, J., Sallee, D., Shalhub, S., Tekin, M.; Montalcino Aortic Consortium; Estrera, A., Musolino, P., Yetman, A., Pyeritz, R., Milewicz, D.M., 2018. Clinical history and management recommendations of the smooth muscle dysfunction syndrome due to ACTA2 arginine 179 alterations. Genet. Med. 20, 1206–1215. PMID 29300374.

Case 11.4.8

Cardiovascular and Pulmonary Cases: Cardiomyopathy and Peripheral Neuropathy

Jaya Ganesh

Department of Genetics and Genomic Sciences, The Icahn School of Medicine at Mount Sinai, New York, NY, United States

Abstract

A 55-year-old man with third-degree atrioventricular block, bradycardia, and neuropathy is found to have hereditary transthyretin amyloidosis.

The Case

Jared, a 55-year-old man, presents to you in the emergency department with dizziness, chest pain, and weakness. His wife reports that he has been having intermittent symptoms over the last 10 months and usually would feel better after resting briefly, but this episode is severe and more prolonged. He had seen a cardiologist 8 months before and was diagnosed with thickening of the heart muscle; he has not yet been seen for a follow-up. Subsequent assessment reveals a complex medical history that started almost a decade ago with pain and paresthesia of the feet. This was variously attributed to heavy smoking, Raynaud's phenomenon, and possible diabetic neuropathy. He failed a glucose tolerance test previously and had been overweight, but he has had a lot of weight loss, abdominal discomfort, and poor appetite recently. Of late he has difficulty walking, poor balance, and feels he is clumsy when using his hands. He had an abnormal nerve conduction study and EMG, but no definite diagnosis could be established. A brain MRI done a year ago was normal. His brother died recently from a "heart attack" and had similar nerve problems, and his father had died from "heart and nerve issues." On examination, you note that he appears tired and ill. There is evidence of weight loss and atrophy of muscles and weakness of the extremities in a glove and stocking distribution. He has bradycardia.

The Issues

- **What does your initial workup show?** Laboratory testing reveals mild hypochromic anemia and mild proteinuria on urinalysis, but unremarkable liver and kidney function tests. Total creatine kinase was sent and is normal. Echocardiogram confirms the cardiomyopathy. EKG shows third-degree atrioventricular block. You admit him for monitoring, cardiac consultation, and further management.

Genomics in the Clinic. https://doi.org/10.1016/B978-0-12-816478-5.00040-6

- **Can the findings be tied together?** The bradycardia and AV block explain his more recent symptoms of dizziness and chest pain and are linked to the cardiomyopathy. However, the progressive neurological symptoms are not explained by the cardiac history. The laboratory findings do not support diabetic neuropathy.
- **What is the next step?** It is possible that the neurological issues are distinct and unrelated to the cardiac illness, but the family history is compelling and points to a unifying diagnosis for both cardiac and neurologic symptoms with genetic underpinnings. You wonder if this could be Fabry disease because of pain, paresthesia, cardiac involvement, and mild proteinuria. You recommend referral to neurology, nephrology, and clinical genetics.

The Referral

- **Genetic Evaluation:** A medical history including the family history is obtained by the genetic counselor and clinical geneticist, and a detailed physical examination is performed. Jared reports normal early development and was in good health until onset of an intermittent burning sensation of both feet in his mid-40s. Symptoms progressed gradually to include weakness of both hands and clumsiness while walking. Abdominal discomfort with intermittent diarrhea and bloating were noted, along with erectile dysfunction. He has been evaluated by two neurologists and had a workup that was inconclusive. His physical examination confirms the bradycardia, muscle atrophy, and weakness noted in the emergency department.

  Jared is one of four children born to a nonconsanguineous couple of Swedish descent. His father died of a "heart attack" in his early 50s. Jared recalls that his father had been laid off work due to nerve damage from "carpal tunnel syndrome" for several years previously. His brother started having nerve pain and weakness in his 40s and had a nerve biopsy that was abnormal, but Jared does not have more details. The brother sustained a fall while climbing stairs, was known to have an enlarged heart, and died recently at 60. His two older sisters are alive and well and his mother died at 98 years of age following a fall but had been overall healthy. He has a paternal first cousin who is demonstrating some weakness. He himself has no children.
- **Genetic Counseling:** The geneticist and genetic counselor tell Jared that they suspect an autosomal dominant (Section 2.7.1) disorder based on the family history, as there is evidence of male-to-male transmission in the pedigree, and thus X-linked disorders (Section 2.7.2) are excluded. Hereditary amyloidosis due to pathogenic variants in the transthyretin (*TTR*) gene is a strong possibility. Hereditary transthyretin amyloidosis (ATTR or hATTR) is inherited in an autosomal dominant manner, and the disorder is characterized by peripheral sensorimotor neuropathy that is gradually progressive. Symptoms of autonomic neuropathy such as orthostatic hypotension, impotence, urinary incontinence or retention, anhidrosis, and GI dysmotility may be seen or even dominate the clinical picture. Nonneurological findings include cardiomyopathy and conduction defects, nephropathy, and eye involvement such as vitreous opacities and glaucoma. Leptomeningeal amyloidosis can manifest as headache, visual impairment, seizures, pyramidal and extrapyramidal symptoms, dementia, and intracranial hemorrhage.

  The onset of the disease may be early in the third to fifth decade of life, especially in patients from Japan and Portugal, where the prevalence of the disease is higher, while other populations report a later onset of symptoms. Some affected individuals are homozygous (Section 2.3.4) with two copies of the same *TTR* pathogenic variant. Cardiac amyloidosis is present in two-thirds of the patients and is particularly common in the African American population, due to the prevalence of a specific pathogenic variant in this cohort. Pathogenic variants in the *APOA1* gene and the gelsolin gene are less common causes of hereditary neuropathic amyloidosis.
- **Genetic Testing:** Molecular testing of the *TTR* gene (the gene underlying hereditary transthyretin amyloidosis) is included in various commercially available genetic testing panels for hypertrophic cardiomyopathy. Isolated *TTR* gene sequencing (Section 6.2.2) or targeted variant analysis if the familial *TTR* pathogenic variant is known is also widely available. Jared undergoes sequencing of the *TTR* gene.
- **Results:** Molecular testing reveals that Jared is heterozygous (Sections 2.3.4 and 2.7) for the common pathogenic variant c.148G > A (p.Val30Met) in the *TTR* gene. The geneticist and genetic counselor discuss with Jared and his wife that his clinical symptoms and test results are consistent with hereditary transthyretin amyloidosis. Pathogenic variants in *TTR* that encode transthyretin (formerly called prealbumin) cause misfolding of the protein and degradation of the tetrameric molecule into monomers. This leads to tissue deposition and damage, which induces systemic deposition of amyloid. Progressive amyloid accumulation can cause multiorgan dysfunction and death. The average life expectancy, untreated, from symptom onset is 3–20 years. The presence of cardiomyopathy is associated with a worse prognosis.

- **Management and Recommendations:** Symptomatic evaluation and treatment include surveillance for cardiac disease and management of arrhythmia including pacemaker/defibrillator placement, CNS imaging and symptomatic management, ophthalmologic assessment for vitreous opacities and glaucoma, renal evaluation for proteinuria and chronic renal failure, and nutritional management. Jared undergoes all of these multidisciplinary evaluations. He has nuclear scintigraphy scanning with SPECT using technetium pyrophosphate (Tc-99m PYP) that shows a Grade 3 tracer uptake, a heart-to-contralateral ratio of 1.5, and serum and urine immunofixation and serum-free light chains all negative for any sign of plasma cell dyscrasia. All of this, in the context of a *TTR* pathogenic variant, left ventricular hypertrophy, and absence of other risk factors, is considered to be diagnostic in Jared for cardiac amyloidosis due to hereditary amyloidosis TTR (ATTR-CM).

  Orthotopic liver transplantation can provide a source of normal *TTR* gene product and was a consideration in patients who were younger (age below 60 years), diagnosed early with peripheral neuropathy limited to lower extremities, and had no significant cardiac, renal, or CNS disease. However, progressive organ involvement is known to occur even after transplantation, and orthotopic liver transplant is no longer the preferred option for the management of these patients. Liver transplantation to treat ATTR has dramatically decreased with the advent of novel TTR-stabilizing therapy. TTR tetramer stabilizers include tafamedis, which stabilizes the tetramer by binding to it and reducing the degradation and misfolding of the protein (approved by the FDA in 2019 for management of ATTR-CM), and diflunisal, a nonsteroidal antiinflammatory drug (NSAID) used off-label for ATTR as it may improve TTR tetramer stability.

  There are also gene silencing therapies. This strategy has led to the development of FDA-approved drugs to treat hATTR polyneuropathy but not ATTR-CM. These include patisiran, a double-stranded small interfering ribonucleic acid molecule (siRNA, Section 2.3.3) that binds to a genetically conserved sequence in the 3′ untranslated region of mutant and normal *TTR* messenger RNA causing RNA interference, resulting in a reduction of serum TTR protein and TTR protein deposits in tissues. Inotersen is a modified antisense oligonucleotide that inhibits hepatic production of transthyretin. It has been approved by the FDA in 2018 for treatment of ATTR polyneuropathy. Clinical trials looking at dual therapy with stabilizer and silencer together are in progress (Section 7.4).
- **Implications for Family Members:** Each child of an affected individual (who is heterozygous (Section 2.3.4) for one *TTR* pathogenic variant) has a 50% chance of inheriting the *TTR* variant. The penetrance (Section 2.8) of hereditary transthyretin amyloidosis is not 100% and can be variable depending on the pathogenic variant, ethnicity, and even geographic regions. Some individuals may be asymptomatic until late adulthood. Thus, both of Jared's sisters have a 50% chance of being affected, even though they are not yet symptomatic. Targeted variant analysis (Section 6.2.1) can be offered to them. Jared's brother would likely to have tested positive for the variant, so his children may have the same risk.
- **Differential Diagnosis:** Fabry disease, a lysosomal storage disorder which can cause cardiomyopathy, bradyarrhythmia, and neuropathic pain, is excluded since it cannot be an X-linked condition with male-to-male transmission. Also pain and paresthesia in Fabry disease are early findings and do not account for muscle wasting, loss of balance, and other neurological symptoms. Charcot–Marie–Tooth disease, a group of neurologic disorders affecting peripheral nerves, may explain the peripheral neuropathy, and autosomal dominant forms of the disorder are known, but cardiac involvement is not seen in this condition. Similarly, some forms of spinocerebellar ataxia (Section 11.13.10) have an autosomal dominant inheritance and peripheral neuropathy, but associated pyramidal and extrapyramidal signs are typically more evident and cardiac involvement is not a feature. Systemic amyloidoses are acquired disorders either due to a plåsma cell disorder (Primary AL amyloidosis) or due to chronic inflammation or infectious processes (Secondary AA amyloidosis).
- **Caveats:** Tissue biopsies that were previously done to demonstrate the presence of amyloid deposition may be useful if variants of uncertain significance (Section 7.1.5) are detected in *TTR* molecular testing, especially if the symptoms are still evolving, and the clinical picture is unclear. While a positive family history is supportive, absence of affected relatives does not exclude the diagnosis.
- **Outcome:** The decision for what medication to use for treatment depends on a number of factors. These include: the predominance of symptoms (neurologic vs. cardiac), patient preference (infusion vs. subcutaneous vs. oral), and side effect profile of the medications (patisiran needs pretreatment with Benadryl, steroid, and tylenol; inotersen has the potential for renal and platelet issues so needs frequent lab monitoring at least in the beginning of therapy; tafamadis is generally well tolerated but very expensive). Given predominant cardiac disease, Jared is started on tafamidis and will be closely monitored.

The Diagnosis

Clinical Diagnosis: Patient, Father, and Brother - hereditary transthyretin amyloidosis.
Genetic Diagnosis: Patient - *TTR* gene, heterozygous pathogenic variant.

Further Reading

Gertz, M., Adams, D., Ando, Y., Beirao, J.M., Bokhari, S., Coelho, T., et al., 2020. Avoiding misdiagnosis: expert consensus recommendations for the suspicion and diagnosis of transthyretin amyloidosis for the general practitioner. BMC Fam. Pract. 21, 198. PMID 32967612.

Luigetti, M., Romano, A., Di Paolantonio, A., Bisogni, G., Sabatelli, M., 2020. Diagnosis and treatment of hereditary transthyretin amyloidosis (hATTR) polyneuropathy: current perspectives on improving patient care. Therapeut. Clin. Risk Manag. 16, 109–123. PMID 32110029.

Ruberg, F.L., Grogan, M., Hanna, M., Kelly, J.W., Maurer, M.S., 2019. Transthyretin amyloid cardiomyopathy: JACC state-of-the-art review. J. Am. Coll. Cardiol. 73, 2872–2891. PMID 31171094.

Sekijima, Y., November 5, 2001. Hereditary transthyretin amyloidosis. Adam, M.P., Everman, D.B. Mirzaa, G.M., Pagon, R.A., Wallace, S.E., Bean, L.J.H., Gripp, K.W., Amemiya, A., (editors). In GeneReviews. University of Washington, Seattle (WA), 1993–2023 [updated June 17, 2021]. PMID 20301373.

Section 11.5

Gastrointestinal Cases

Case 11.5.1

Gastrointestinal Cases: Bowel Obstruction in an Infant

Antonie D. Kline

Harvey Institute for Human Genetic, Greater Baltimore Medical Center, Baltimore, MD, United States

Abstract

A newborn with history of an echogenic focus in the bowel develops a bowel obstruction after delivery, which is found to be due to meconium plug syndrome from cystic fibrosis. Her parents are consanguineous, and prenatally both were found to be carriers of cystic fibrosis.

The Case

A newborn female infant, Amira, was born prematurely at 37 weeks gestation to a gravida 2 para 0010 mother, and she was admitted to your service in the nursery. She had an uneventful prenatal and delivery history. On her initial physical exam, she is average size for gestational age at birth, with no unusual features, and has generalized hirsutism, a normal abdominal exam and a normal neurologic exam. By day 2 of life, you return to the nursery because she still has not passed meconium. On exam, her abdomen is distended and there are very minimal upper bowel sounds. She is urinating normally.

The Issues

- **Why has the baby not passed meconium?** You request an abdominal X-ray, which shows several dilated loops of bowel. It is possible that the baby has a bowel obstruction. This could be due to an intestinal stenosis or atresia, Hirschsprung disease, meconium plug syndrome, or other primary bowel issue (Section 5.1.4).
- **Were there any prenatal clues to the situation?** You look in more detail at the prenatal records. A second trimester ultrasound showed an echogenic focus in the bowel (Section 4.3.2). You speak with the parents. They mention that the prenatal carrier screening (Section 4.1) was abnormal. Both parents were found to be carriers for cystic fibrosis, but they declined prenatal diagnostic testing. No records are available. You remember that cystic fibrosis has an increased risk for meconium plug syndrome and meconium ileus. You are not sure if Amira has cystic fibrosis, but it is in the differential. You are uncertain about taking the parents' word that their screening test was positive, and you do not know what type of testing to send on the newborn. It is very appropriate to call genetics. You consult genetics to help evaluate the newborn.
- **Should she be seen by pediatric surgery for the possible bowel obstruction?** You discuss the case with the pediatric surgeon, who recommends treating the baby with enemas and watching her carefully with repeat X-rays. You transfer the baby to the NICU.

Genomics in the Clinic. https://doi.org/10.1016/B978-0-12-816478-5.00041-8

The Referral

- **Genetic Evaluation**: A full pregnancy history, a medical history, and a family history are obtained, and a detailed physical examination is performed by the pediatric geneticist and genetic counselor. The parents report that the mother underwent carrier screening (Section 4.1) and was found to have a variant in the gene for cystic fibrosis. They do not know the specific variant. The rest of the carrier testing was normal. The obstetrician discussed the case with the parents, and carrier testing only for cystic fibrosis was sent on the father. He was also found to have a variant, but neither parent knows their specific results. Records will be obtained from the obstetrician.

  The Family history is obtained. There is consanguinity (Section 3.2.4). The parents are first cousins: both of their mothers are sisters. The families are originally from Saudi Arabia. They state that no member of their families has ever had cystic fibrosis. Amira has a normal physical exam by the geneticist, other than the slightly distended abdomen and generalized hirsutism.
- **Genetic Counseling**: The geneticist and genetic counselor discuss with the parents that, based on their prenatal screening, there is a 25% chance that the baby could have cystic fibrosis, since it is autosomal recessive (Section 2.4.1), assuming that their recollection is accurate. The fact that no one has had cystic fibrosis on either side does not rule out this possibility. It is likely that both parents are carrying the same pathogenic variant or mutation because of the consanguinity (Section 3.2.4).

  Cystic fibrosis causes pulmonary, pancreatic, digestive, and reproductive abnormalities due to abnormal and thick mucus secretion. It can be a severe lifelong condition with most individuals only reaching middle age, although there is a spectrum of clinical involvement. Lung complications include progressive bronchiectasis, airway obstruction, and chronic damage and lead to the greatest morbidity and mortality. Sinusitis and nasal polyps may occur. Pancreatic insufficiency is found in most affected individuals with malabsorption, steatorrhea, and subsequent poor growth. Failure to thrive in infants can be a presenting sign. With Amira's presentation of a possible bowel obstruction, and the family history, there is a strong chance that this is meconium plug syndrome due to cystic fibrosis, and the baby has developed meconium ileus, which occurs in up to 20% of newborns with cystic fibrosis. Other organ systems may be involved as well. The majority of males have congenital absence of the vas deferens and are infertile. In cystic fibrosis, the chloride content of the sweat will be elevated, to more than 60 mEq/L. This measurement should be performed only after the infant is 6 months of age, and when elevated should prompt molecular testing. Cystic fibrosis is due to pathogenic variants in the *CFTR* gene, which codes for a chloride channel protein that helps transport chloride into cells. Pathogenic variants in the gene cause defects in the channel and production of the abnormal mucus.
- **Genetic Testing**: The obstetrician's records are received. Both parents had been found to have the Phe508del or F508del (formerly called the delta F508) mutation, the most common mutation in the *CFTR* gene leading to cystic fibrosis (Section 11.4.1). A blood sample is sent on the baby now that the result of the parental screening is available. Newborn screening (Section 5.4) was previously sent on the baby. An immunoreactive trypsinogen (IRT) is included on the newborn screening in all of the United States. If elevated, molecular testing is recommended. This will likely return before any molecular testing.
- **Results**: Amira continues to have worsening dilated bowel loops. The enemas produce some mucous plugs but no meconium. The next morning, there is free air in the bowel, indicated a bowel perforation. She is taken to the operating room for bowel resection and repair. The IRT returns elevated, further indicating that cystic fibrosis is the likely diagnosis.
- **Management and Recommendations**: Additional information is shared with the parents. Homozygous F508del mutations cause more classic, severe cases of the spectrum of cystic fibrosis. All of their affected children will likely have both pulmonary and gastrointestinal involvement because of the severity of the F508del mutation. If they have an affected boy, he will be infertile (Section 5.5.1). Referral will be made to the Cystic Fibrosis clinic at the tertiary care center for expert care and preventative management. Pulmonary care will include medications to help dilate airways as well as antibiotics for infections when needed, chest physical therapy, and other respiratory care. Treatment for pancreatic insufficiency will be started including fat-soluble vitamin supplementation, oral pancreatic enzymes, and monitoring and support for nutrition.

  Newer therapies have been approved for cystic fibrosis following clinical trials, depending on the mutations present and the age of the child. Ivacaftor is an FDA-approved medication that improves function of the chloride channel in the lungs and is recommended for children with F508del variants either alone or in combination with other medications. *CFTR* modulators are undergoing trials currently. Most recently, the median age of survival has increased into the low 50s. There is a parent support organization, to which the family is referred. Each of Amira's siblings will have a 25% chance of having cystic fibrosis. Future prenatal testing is briefly discussed, with molecular testing available on invasive prenatal diagnostic testing, and the option of preimplantation genetic testing (Section 4.5.1) also possible. They can be seen in genetics in the future to discuss these options in greater detail.

- **Caveats**: Even with negative prenatal carrier screening, an individual could be carrying a mutation. Different laboratories screen for a different number of variants, and many of them may be unique to the family. For infants who are found to have an elevated immunoreactive trypsinogen, a single elevated sweat chloride is sufficient for a diagnosis of cystic fibrosis.
- **Outcome**: Amira is found to be affected, with homozygous F508del status. She slowly improves. Upon discharge she is referred to the Cystic Fibrosis clinic.

The Diagnosis

Clinical Diagnosis: Meconium plug syndrome due to cystic fibrosis.

Genetic Diagnosis: Patient—*CFTR* gene, homozygous pathogenic variants. Parents—*CFTR* gene, heterozygous pathogenic variant.

Further Reading

Naehrlich, L., 2020. The changing face of cystic fibrosis and its implications for screening. Int. J. Neonatal. Screen. 6, 54. PMID 33123635.

Savant, A., Lyman, B., Bojanowski, C., Upadia, J., March 26, 2001. Cystic fibrosis. Adam, M.P., Everman, D.B. Mirzaa, G.M., Pagon, R.A., Wallace, S.E., Bean, L.J.H., Gripp, K.W., Amemiya, A. (Eds.), In GeneReviews, Seattle (WA), 1993–2023 [updated March 9, 2023]. PMID 20301449.

Sparbel, K.J.H., Tluczek, A., 2011. Patient and family issues regarding genetic testing for cystic fibrosis: a review of prenatal carrier testing and newborn screening. Annu. Rev. Nurs. Res. 29, 303–329. PMID 22891510.

Harrison, P.T., 2022. CFTR RNA- and DNA-based therapies. Curr. Opin. Pharmacol. 65, 102247. PMID 35709547.

Case 11.5.2

Gastrointestinal Cases: Elevated Liver Transaminases in an Infant

Hilary J. Vernon
Department of Genetic Medicine, Johns Hopkins University School of Medicine, Baltimore, MD, United States

Abstract
A 7-month-old girl with failure to thrive, hepatomegaly and elevated liver transaminases, is found to have glycogen storage disease type III.

The Case

Mollie, a 7-month-old girl with medical history significant for failure to thrive, is noted to have an enlarged and firm liver on physical examination. Laboratory studies are sent and show an elevation in alanine aminotransferase (ALT) of 650 IU/L and in aspartate aminotransferase (AST) of 560 IU/L. Gamma-glutamyl transferase (GGT) and total bilirubin are normal. Creatine kinase is mildly elevated at 305 U/L. Mollie is sent to pediatric gastroenterology for an urgent evaluation.

The Issues

- **Why does this child have hepatomegaly?** An abdominal ultrasound is performed and shows an enlarged liver with homogeneous echogenicity (Section 5.4.1). No masses or abnormal vascularity are observed. There is mild splenomegaly. The kidneys appear normal.
- **What other lab work can be done to investigate liver dysfunction?** Follow-up 4 hour (h) fasting labs are sent, and Mollie is found to have elevated triglycerides of 300 mg/dL and mild hypoglycemia with a blood glucose of 60 mg/dL. Prothrombin time (PT) and partial thromboplastin time (PTT) are normal. Uric acid and lactate are normal. Urinalysis shows moderate ketones.
- **Are there any clues in Mollie's history of present illness to aid in her diagnosis?** On obtaining further history, Mollie's parents tell you that she often becomes irritable if her bottle-feeding is delayed. Additionally, she is not sleeping through the night, and still awakens every 3–4 h throughout the night for a bottle. In retrospect, this is a sign of her intolerance to fasting. Because of these findings, Mollie is referred for a genetic evaluation to rule out an inborn error of metabolism.

The Referral

- **Genetic Evaluation**: A pregnancy history and medical history are obtained by the geneticist, and a detailed physical examination is performed. The parents report that the pregnancy was uncomplicated. Family history is obtained. Mollie

Genomics in the Clinic. https://doi.org/10.1016/B978-0-12-816478-5.00042-X

is the only child to her parents, who are both healthy. There is no known extended family history of multiple miscarriages, stillbirths, or infant or childhood deaths. Parental consanguinity is denied. Mollie is found to have length and weight both below the 3rd percentile and her head circumference at the 10th percentile. Her facial features appear normal and hepatomegaly is noted. The rest of her exam is normal.

- **Genetic Counseling**: The geneticist and genetic counselor discuss with the family that Mollie's clinical symptoms and laboratory test results are consistent with a glycogen storage disorder, specifically glycogen storage disease type III (GSD III, also known as Cori Disease). GSD III is caused by a deficiency of the glycogen debrancher enzyme (amylo-1,6-glucosidase, 4-alpha-glucanotransferase deficiency), which is a key enzyme in glycogen degradation. Sequencing of the *AGL* gene that encodes for glycogen debrancher enzyme is recommended.

  Mollie's parents ask about any recurrence risk for future children if this is the correct diagnosis. GSD III is an autosomal recessive disorder in which both carrier parents are asymptomatic but have a 25% chance of recurrence for each future child they have together (Section 2.7.1).
- **Genetic Testing**: Mollie is found to have two pathogenic variants (Section 7.1.3) in *AGL*. These pathogenic variants are known to be associated with GSD IIIa. Her parents are found to each carry one variant, indicating that the variants occur on separate chromosomes or are in "trans" rather than on the same chromosome and gene or are in "cis" (Section 2.4.1).
- **Management and Recommendations**: A multispecialist team approach is employed for Mollie's long-term care, including genetics, gastroenterology, neurology, and cardiology. Involvement of a skilled metabolic dietician is key for long-term management (Section 5.4.3) of GSD III. Avoidance of fasting, with regular feeds every 3–4 h, is recommended throughout infancy. After Mollie is 1 year of age, she can start taking weight-based dosing of uncooked cornstarch, which is a slowly digestible carbohydrate source and can be used to avoid hypoglycemia. Glycosade® (an extended-release cornstarch) can be employed after 5 years of age. A high protein diet is also recommended.

  An emergency plan is also provided to the family describing action to be taken in the setting of hypoglycemia, possible emergency room management, and precautions to be taken in the setting of planned fasting (i.e., provision of dextrose containing IV fluids before surgery).

  Regular screening for other complications that can be seen with GSD III is recommended including: echocardiograms for cardiomyopathy and electrocardiograms to monitor for signs of cardiac hypertrophy, regular liver ultrasounds to screen for adenomas or signs of cirrhosis, and regular measurement of liver transaminases, bilirubin, PT, creatine kinase, and lipid profiles.
- **Differential Diagnosis**: Some overlapping clinical features can be seen in glycogen storage disorder type I (GSD I), type VI (GSD VI), and type IX (GSD IX).
- **Caveats**: Cardiac involvement is seen in the majority of patients, but usually is not seen in infancy. Mollie will need regular cardiac monitoring and management of cardiomyopathy if it develops.
- **Outcomes**: With regular treatment and monitoring, Mollie's growth rate begins to normalize. As she gets older, her hepatomegaly improves and she is more tolerant to short periods of fasting. She undergoes regular screening, and other complications of GSD III have not yet manifested as of age 7 years.

The Diagnosis

Clinical Diagnosis: Glycogen storage disease type III.
Genetic Diagnosis: *AGL* gene, compound heterozygous pathogenic variants.

Further Reading

Kishnani, P.S., Austin, S.L., Arn, P., Bali, D.S., Boney, A., Case, L.E., et al., 2010. Glycogen storage disease type III diagnosis and management guidelines. Genet. Med. 12, 446–463. PMID 20631546.

Sentner, C.P., Hoogeveen, I.J., Weinstein, D.A., Santer, R., Murphy, E., McKiernan, P.J., et al., 2016. Glycogen storage disease type III: diagnosis, genotype, management, clinical course and outcome. J. Inherit. Metab. Dis. 39, 697–704. PMID 27106217.

Case 11.5.3

Gastrointestinal Cases: Failure to Thrive

Antonie D. Kline
Harvey Institute for Human Genetics, Greater Baltimore Medical Center, Baltimore, MD, United States

Abstract
A 17-month-old female with intrauterine growth restriction (IUGR), subsequent failure to thrive, and delayed development, and with minor dysmorphic features, small hands and feet, and hirsutism, is found to have Cornelia de Lange syndrome.

The Case

Lydia is seen by you, her nurse practitioner, at 17 months of age. She had been born small for gestational age, with intrauterine growth restriction (IUGR) noted in the third trimester. TORCH titers and HIV testing were negative. You had treated her for reflux when she had persistent vomiting as an infant, and this has continued intermittently. She has had some constipation as well. You had noted a drop-off in her growth parameters before 12 months, with her head circumference below the second percentile, and have not seen her since. Her development has been borderline normal to delayed. She achieved a pincer grasp at 12 months, walked at 15 months, and although said first word at 10 months, is just now starting to say her second word. Her mother says that when she herself was younger, she was always the smallest in the class. On exam, you note a gain of only one pound since her 12 month visit, and all of her measurements have plateaued. You notice she appears different from family members, including a small nose and chin as well as small feet, although no specific diagnosis comes to mind. You discuss that there are enough findings in Lydia that you feel she should be evaluated by a pediatric geneticist.

The Issues

- **How concerned are you by her growth?** Her drop-off in growth is quite concerning, since toddlers are expected to have continued growth. You discuss that although she started off small and was making forward progress, she has plateaued in her size and now has failure to thrive. You could send some bloodwork, but because you have noted some dysmorphic features, the slow growth, and small feet, with history of IUGR, microcephaly, and persistent vomiting, you prefer to defer the workup and see what the geneticist recommends. In the meantime, you discuss increasing calories with a specific toddler formula and food additions. You also discuss increasing fiber to help with the constipation.

There are a number of causes of failure to thrive which are treatable and important not to miss. These include hypothyroidism, growth hormone deficiency, some inborn errors of metabolism, cystic fibrosis, and chronic infections, although these less commonly include small head size. An appropriate workup should be recommended if etiology is unclear including: complete blood count (CBC), comprehensive metabolic panel (CMP), thyroid function tests at a minimum, possibly with plasma amino acids, urine organic acids, a sweat chloride test, growth hormone, and/or HIV testing if

Genomics in the Clinic. https://doi.org/10.1016/B978-0-12-816478-5.00043-1

mother has not had this. If length or height also has dropped off, a bone age should be done if child is older than one year. These tests could be done prior to the genetic evaluation.

- **How concerned are you by her development?** You discuss that her development shows variable delay, more in language than the other areas. She should be evaluated developmentally and referred for early intervention therapy through the state program. You refer the family to the state program for infants and toddlers.
- **Does she have something genetic**? It is unclear at this point. The more findings, the more likely there is a single underlying cause. She could have something inherited or new in herself, and it is best to wait for the genetic evaluation and then consider additional testing.
- **Could the family history have anything to do this**? On brief review, no other family member has had slow growth. It would be good to discuss a detailed family history with the geneticist. You refer the family to the local hospital pediatric geneticist.

It is helpful for you to send a copy of the growth charts and records prior to the child being seen by genetics, as well as any studies that you have done. Also, if this were an infant, you could call and request an urgent visit, or admit to the hospital if very severe. For a toddler, it may be less urgent, but it may be helpful to call to get a timely visit.

The Referral

- **Genetic Evaluation:** Lydia and her parents are seen in the pediatric genetic clinic by the pediatric geneticist and genetic counselor 3 weeks later. History is reviewed and a family history (Section 3.2) taken. Lydia tends to vomit frequently after meals. Her bowel movements are sporadic, usually only every 3–4 days. No history of delayed growth or development in other relatives is noted on family history; although the mother grew slowly, she is 5′ 2″ in height, and the father is 5′ 10″ in height. A physical exam is performed, and the patient is photographed. On exam, growth parameters are noted to be below the third percentile, and the head circumference is below the second with flattening of the right occipital region. Mildly dysmorphic features are noted, including arched synophrys, long lashes, epicanthal folds, small mouth and chin, and high palate. She also has small hands and feet, clinodactyly of fifth fingers, hirsutism on the back, cutis marmorata, and a normal neurologic exam.

It may be helpful for the family to know ahead of time that this will be a long visit, with very thorough discussion of history and family history of three generations, as well as a very detailed examination with lots of measurements, and findings documented by photography.

- **Genetic Counseling:** The geneticist and genetic counselor confer and return to discuss the findings with the family. They say that they feel that Lydia has a rare condition called Cornelia de Lange syndrome (CdLS). Individuals with CdLS usually have proportionate small stature, some aspects of developmental delays particularly language, and often have some similar minor findings and small hands and feet (Fig. 5.3G–I). Facial features can be distinctive (Fig. 5.2D). CdLS has a broad range of involvement, and some children can be very severely involved. Lydia appears to be milder and has been making good progress.

  CdLS is due to pathogenic variants or mutations in at least seven genes related to the protein complex, cohesin. The most commonly involved is the *NIPBL* gene, which has pathogenic variants found in about 65% of cases. Most of the time CdLS is a sporadic (new to family) autosomal dominant (Section 2.7.1) condition. Other causal dominant genes include *SMC3, RAD21, ANKRD11*, and *BRD4*, but some of the genes are X-linked dominant (*SMC1A, HDAC8*), and also there can be gonadal mosaicism (Section 2.4.2) in one of the parents. Recurrence risk for future children is about 1.5% empirically.
- **Management and Recommendations:** Growth is typically slow in CdLS, and there are specific growth charts for CdLS that should be used. Lydia may not benefit from increased calories, but this can be monitored. Gastroesophageal reflux disease is very common in most children with CdLS, and because of Lydia's history of emesis, an endoscopy is recommended to rule this out. Constipation is common, and often children may need more than just increased fiber, but also a laxative regularly. Malrotation can occur in 10% of children with CdLS, and an upper GI series through to the duodenum is important to identify this. Overall a referral to pediatric gastroenterology is useful. An echocardiogram and a renal ultrasound will detect congenital heart defects and renal malformations, each found in about 20% of affected children. Regular eye evaluations and audiology testing are recommended. Development needs to be monitored. Early intervention therapy is very important, especially speech therapy. It is helpful to teach the children methods of communication before they get frustrated and develop behavioral issues.

- **Genetic Testing:** If the parents are interested in prenatal diagnosis (Section 4.4.2) for future children, it is important to test Lydia, ideally before they become pregnant. There is gene panel (Section 6.2.3) testing for CdLS, and there is also exome sequencing (Section 6.3.1) to detect variants in genes for all cohesin-related conditions. It is likely that chromosomes (Section 6.1.1) and microarray (Section 6.1.3) will be normal. Lydia's parents decide to pursue panel testing, and the genetic counselor investigates insurance coverage (Section 10.2). The insurance denies coverage for this test.
- **Caveats:** For individuals with CdLS, genetic testing can still be negative about 20% of the time. There can be mosaicism for the pathogenic variant (Section 2.4.2) in the affected individual, in which case a buccal swab or skin biopsy (Section 6.5) may detect a variant when the blood test has been negative.
- **Differential Diagnosis:** Although CdLS is often a diagnosis made clinically, there are specific features that are helpful with this. Other syndromes can overlap, with growth and developmental delay and some of the same features. These include Rubenstein–Taybi syndrome (which has different facial features, very short broad thumbs, and great toes—Figs. 5.2C and 5.3E, F), KBG syndrome (which has less dysmorphic facial features and can have macrodontia, large fontanelles, but does have significant overlap, and is due to variants also in the *ANKRD11* gene), Nicolaides–Baraitser syndrome (which has sparse scalp hair, hirsute body, coarse facial features, and severe seizures), Coffin–Siris syndrome (which has sparse scalp hair, hirsute body, hypotonia, and hypoplastic fifth fingernails), and fetal alcohol spectrum disorder (which has different facial features and less speech delays—Section 11.3.1). Other genetic syndromes with proportionate small stature but less severe developmental delays and different facial features include Russell–Silver syndrome (which tends to have normal head circumference and can have hemihyperplasia—Section 11.7.4), 22q11 deletion syndrome (which can have more normal growth—Section 11.9.2, Fig. 5.2B), and other microdeletion syndromes.
- **Outcome:** Lydia does well with therapy and starts to use sign language. She is started on stronger medications for the gastroesophageal reflux and constipation, and both improve. The rest of her workup is negative for organ system malformations. The family is referred to the parent support group for CdLS (Section 3.5), whose resources are extremely helpful for the family. Two years later, genetic testing is approved by the insurance company. She is found to have a heterozygous missense variant in the *SMC1A* gene, confirming the clinical diagnosis of milder CdLS. Her parents undergo single site variant testing (Section 6.2.1) and they are both negative. They understand that there is still the residual risk due to the possibility of gonadal mosaicism.

The Diagnosis

Clinical Diagnosis: Cornelia de Lange syndrome.
Genetic Diagnosis: *SMC1A* gene, heterozygous pathogenic variant.

Further Reading

Ansari, M., Poke, G., Ferry, Q., Williamson, K., Aldridge, R., Meynert, A.M., et al., 2014. Genetic heterogeneity in Cornelia de Lange syndrome (CdLS) and CdLS-like phenotypes with observed and predicted levels of mosaicism. J. Med. Genet. 51, 659–668. PMID 25125236.

Deardorff, M.A., Noon, S.E., Krantz, I.D., September 16, 2005. Cornelia de Lange Syndrome. Adam, M.P., Mirzaa, G.M., Pagon, R.A., Wallace, S.E., Bean, L.J.H., Gripp, K.W., Amemiya, A., (editors). In GeneReviews® [Internet]. University of Washington, Seattle, Seattle (WA), 1993–2023 [updated October 15, 2022]. PMID 20301283.

Kline, A.D., Moss, J.F., Selicorni, A., Bisgaard, A.M., Deardorff, M.A., Gillett, P.M., et al., 2018. Diagnosis and management of Cornelia de Lange syndrome: first international consensus statement. Nat. Rev. Genet. 19, 649–666. PMID 29995837.

Case 11.5.4

Gastrointestinal Cases: Obesity, Hypotonia, and Peripheral Vision Loss

Kim M. Keppler-Noreuil

Division of Genetics and Metabolism, Department of Pediatrics, University of Wisconsin School of Medicine and Public Health, Madison, WI, United States

Abstract

An 8-year-old boy with history of obesity, enlarged fatty liver, night vision loss, bicuspid aortic valve, enlarged cystic kidney, dysmorphic features, hypotonia, decreased smell, and learning problems is found to have Bardet–Biedl syndrome.

The Case

Thomas, your 8-year-old patient, comes for his annual well-child check. You have not seen him since he was 1 year of age because his family had relocated out-of-country in the interim, but now they have returned to live in the area. You review his past medical history. He was a normally grown infant born at term gestation, who had an echogenic, possible cystic kidney on prenatal ultrasound at 7 months gestation. Postnatal abdominal ultrasound showed left fetal lobulations and nephromegaly, but normal function. After birth he was noted to have primary hypospadias, which was repaired at age 1 year. Other past medical issues have included bicuspid aortic valve with mild aortic stenosis detected on echocardiogram 2 years ago because of a heart murmur detected on auscultation, and enlarged liver with moderate fatty infiltration seen on ultrasound done yearly for follow-up of his earlier noted kidney abnormality by his pediatrician. His parents are concerned about progressive night vision loss, first noticed in-between 2–3 years of age, along with more recent decrease in his peripheral vision. He has a prescription for glasses, but they have not helped much. He has learning problems, especially with speech delay and low muscle tone. Thomas also has had increased appetite with significant weight gain since 1 year of age. He drinks a lot of fluids and has increased urination. On your exam, Thomas's weight is 69.3 kg (greater than 97th percentile), height 143 cm (95th percentile), and you notice left exotropia, small testes, short fingers and toes, and metatarsus varus.

The Issues

- **How concerned are you about his reported progressive night vision and peripheral vision loss?** You know that these symptoms can indicate retinal disease, and Thomas has only had an evaluation with an optometrist. Sometimes these symptoms are difficult for a young child to describe and could be inaccurate. However, the differential diagnosis for these symptoms includes retinal cone-rod dystrophies that can lead to severe vision loss. You refer him to pediatric ophthalmology.

  Thomas is seen by a pediatric ophthalmologist at 8 years 9 months of age, who elicits history that Thomas has reduced vision at dusk and with bright light, and that his side vision is decreased. On exam, he has intermittent exotropia, left eye preference, and reduced visual acuity (OD Snellen 20/50; OS Snellen 20/63) consistent with myopic

Genomics in the Clinic. https://doi.org/10.1016/B978-0-12-816478-5.00044-3

astigmatism. His retinal exam shows macular granular pigmentary changes with loss of foveal reflex, and mildly attenuated vessels. An electroretinogram shows reduced and delayed responses.

- **What about the obesity?** Thomas's parents have normal BMIs. His weight is significantly over the 97th percentile, and he has fatty liver (associated with obesity), along with symptoms of polyuria and polydipsia, which raise concern about diabetes mellitus or diabetes insipidus. They report that he is always hungry, even after eating a big meal. He also has small testes for his age with the obesity that may indicate a functional abnormality in his hypothalamic–pituitary axis. You refer him to pediatric endocrinology.

  The pediatric endocrinologist obtains testing for thyroid function, cortisol, prolactin, and glucose metabolism, which are normal. Because of his history of polyuria and polydipsia, a urinalysis and urine osmolality are done with no evidence of diabetes insipidus. The endocrinologist recommends routine annual follow-up.
- **What is the next step?** You are concerned that, in addition to his obesity, he has multiple congenital anomalies including: bicuspid aortic valve with mild aortic stenosis, kidney enlargement with prior cysts, hypospadias and small testes, short fingers and toes, and developmental problems. With multiple anomalies, neurodevelopmental issues, and obesity, it is likely that he has an underlying genetic disorder. You refer to pediatric genetics.

The Referral

- **Genetic Evaluation**: Thomas and his parents are seen in the pediatric genetic clinic by the pediatric geneticist and genetic counselor. History is reviewed, and a family history taken. No history of vision or developmental problems or obesity is noted on family history. A physical exam is performed, and the patient is photographed. On exam, Thomas's BMI is 34 kg/m^2. His head circumference is at the 25th percentile. He is noted to have mildly dysmorphic features: bitemporal narrowing, low anterior hairline, bilateral epicanthal folds, strabismus with left exotropia, prominent ear lobes, inverted nipples, phallus embedded in suprapubic fat, and small testes. He has brachydactyly, mild fifth finger clinodactyly, metatarsus varus, and mild second and third toes syndactyly (Fig. 11.5.4.1A and B). His neurologic exam shows muscle tone mildly decreased, slow and inaccurate rapid alternating movements, imbalance on tandem gait, wide-based gait, and mild dysmetria, limb tremor on finger-to-nose testing. He has reduced sense of smell on the UPENN smell identification test (UPSIT).
- **Genetic Counseling**: The geneticist and genetic counselor review Thomas's findings and return to speak with the parents. They discuss that they think Thomas has a diagnosis of Bardet–Biedl syndrome (BBS). BBS is characterized by obesity with onset soon after birth, pigmentary retinal dystrophy, cognitive disabilities, renal defects, and skeletal abnormalities, including brachydactyly, and/or postaxial polydactyly. It is caused by abnormal primary cilia, with more than 19 BBS genes identified. Diagnosis is suspected when a child has symptoms of night vision and peripheral vision loss and obesity. Ophthalmology and genetic evaluations in combination with molecular testing can confirm the diagnosis. Thomas has rod-cone dystrophy, obesity, learning disabilities, hypogonadism, hyposmia, renal abnormalities, brachydactyly, polyuria/polydipsia, strabismus, hepatic involvement, and ataxia/poor coordination/imbalance. The diagnosis of BBS can be made clinically by the presence of four primary features [of rod-cone dystrophy, polydactyly (postaxial), truncal obesity, learning disabilities, hypogonadism in males or genital abnormalities in females, partial or complete anosmia, renal abnormalities, or specific craniofacial features] or three primary and two secondary features [of speech disorder or delay, strabismus/cataracts/astigmatism, brachydactyly/syndactyly, developmental delay, hepatic involvement, polyuria/polydipsia, ataxia/poor coordination/imbalance, mild hypertonia (especially of the lower limbs), diabetes mellitus, dental crowding/hypodontia/high-arched palate, cardiovascular abnormalities, or other (psychiatric problems, autism spectrum disorder, hearing loss)].

BBS is inherited in an autosomal recessive manner (Section 2.7.1). Interfamilial and intrafamilial variability in phenotypic features can occur. Genetic counseling is provided: each sibling of an affected individual has a 25% chance of being affected, a 50% chance of being an asymptomatic carrier, and a 25% chance of being unaffected and not a

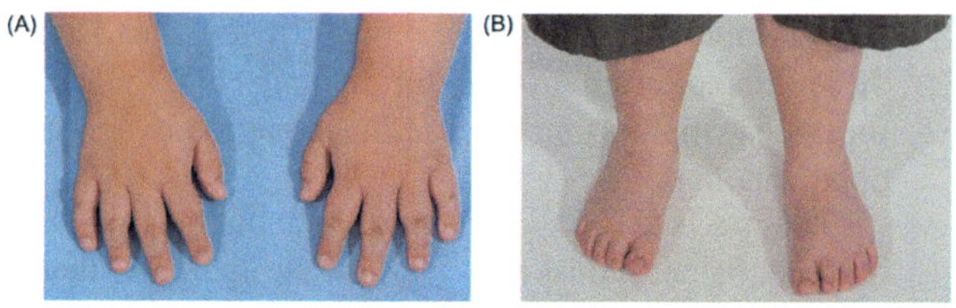

FIGURE 11.5.4.1 (A) Typical hand findings in Bardet–Biedl syndrome, including broad hands with brachydactyly and fifth finger clinodactyly. (B) Note broad feet with metatarsus varus, and mild syndactyly of second and third toes.

carrier. Carrier testing is possible if the pathogenic variants (Section 7.1.3) in a family are known. The geneticist and counselor offer testing and describe to the parents that approximately 80% of BBS is attributed to pathogenic variants in 14 genes: *BBS1*, *BBS2*, *ARL6* (*BBS3*), *BBS4*, *BBS5*, *MKKS* (*BBS6*), *BBS7*, *TTC8* (*BBS8*), *BBS9*, *BBS10*, *TRIM32* (*BBS11*), *BBS12*, *MKS1* (*BBS13*), and *CEP290* (*BBS14*), with *BBS1* and *BBS10* being the most common (each accounting for ~20% of BBS). Approximately 5% of BBS is attributed to pathogenic variants in at least five additional genes to date. Identification of pathogenic variants will confirm the diagnosis, provide information about natural history and associated complications reported with the specific genetic variant, and help with management and surveillance recommendations.

- **Management and Recommendations**: The geneticist recommends regular ophthalmologic evaluation, monitoring of renal function, endocrine status and lipid profile, and annual blood pressure measurement. In particular, the following surveillance and management guidelines for Thomas's care are made:
  (1) Visual aids and educational programs for the visually impaired, including preferential seating and lighting as well as mobility training and learning Braille at school;
  (2) Obesity is managed with diet, exercise, and behavioral therapies; hypercholesterolemia and diabetes mellitus are treated as in the general population;
  (3) Early intervention and special education address cognitive disability; speech therapy for speech delay/impairment; physical and occupational therapy to discuss strategies for compensating with decreased peripheral vision and visual field;
  (4) Follow-up of renal anomalies and risk for hypertension;
  (5) Possible need for hormone replacement therapy for hypogonadism;
  (6) Cardiac abnormalities are treated as in the general population. Antibiotic prophylaxis for surgical and dental procedures is indicated for individuals with structural cardiac anomalies;
  (7) Prenatal diagnosis using second trimester ultrasound examination (Section 4.3) to detect anomalies associated with BBS such as renal cysts can be done. Prenatal diagnostic testing (Section 4.4.2) for the pathogenic variants found in Thomas for future pregnancies and preimplantation genetic testing (Section 4.5.1) can be offered.

  It is also important for Thomas to be referred to developmental pediatrics for a thorough neuropsychological assessment, which will help with providing potential services in school. He is 8 years old going into the third grade. Weschler Intelligence Scale for Children-IV (WISC-IV), Rey Auditory Verbal Learning Test (RAVLT), and other tests are administered. The pediatric neuropsychologist reviews the results with the parents. Thomas has a full scale IQ of 86, which is in the low average range of intellectual ability. However, there is a significant difference between Thomas's verbal abilities and his working memory and processing speeds. Speech therapy is recommended, and he will need an IEP in school.

  The geneticist also counsels that testing by gene sequencing (Section 6.2.2) for *BBS1* and *BBS10* could be done because these are the two most commonly with pathogenic variants in persons of northern European descent. However, because Thomas does not have the typical associated finding of polydactyly, the geneticist recommends a multigene panel (Section 6.2.3), which is also an effective way to molecularly diagnose a person with BBS.
- **Results**: Compound heterozygous (Section 2.7.1) pathogenic variants in *BBS12* are identified. Mutation 1: c.del1092, p.E365RfsX18, and Mutation 2: c.1627G > A, p.E543K. These results confirm that Thomas has inherited two autosomal recessive (Section 2.7.1) mutations causing BBS. The geneticist reviews the results of the genetic testing, which confirms the diagnosis of BBS.
- **Implications for Other Family Members**: Based on Thomas's family history, his 5-year-old brother is healthy with normal growth and development, and unlikely to be affected with BBS. You discuss availability of carrier testing for the brother for future reproductive risk counseling. His parents are most likely carriers, and prenatal diagnostic testing (Section 4.4.2) could be offered for future pregnancies.
- **Differential Diagnosis**: Other syndromes can overlap with obesity and developmental delay, and some of the same features. There are additional distinctive features for these syndromes, which include: Prader–Willi syndrome (feeding difficulties and hypotonia in the first year of life followed by hyperphagia and obesity at 18–24 months—Section 11.13.1), Alström syndrome (early onset retinopathy, sensorineural hearing loss, kidney dysfunction, insulin resistance, dilated cardiomyopathy), Cohen syndrome (prominent central incisors, truncal obesity, severe myopia, neutropenia, short stature), Smith–Magenis syndrome (short stature, small hands and feet, square facies, sleep disturbances), and Albright hereditary osteodystrophy (short stature, microcephaly, short fourth and fifth metacarpals, metatarsals, subcutaneous calcifications). Each of these syndromes has different identifiable genetic etiologies that can be detected with genetic testing. Conditions with obesity and without developmental delay include: congenital leptin deficiency, leptin

receptor deficiency, and pro-opiomelanocortin. There are other primary retinal dystrophies, such as Stargardt disease (Section 11.3.4) and neuronal ceroid lipofuscinosis.

- **Caveats**: It is important to recognize that individuals with BBS may have variable or incomplete clinical findings, like Thomas, who did not have one of the primary features of polydactyly (but rather had brachydactyly). In addition, most of the symptoms may not be present at birth, but appear and progressively worsen during the first and second decades of life. Also, there are individuals who have clinical diagnosis of BBS, but have a negative genetic test result, i.e., they do not have one of the tested or known causative gene variants for BBS. Finally, there is noticeable phenotypic overlap with some other ciliopathy disorders, e.g., Alström syndrome, Joubert syndrome, Meckel syndrome, and McKusick–Kaufman syndrome, which complicates clinical and genetic diagnosis of BBS.
- **Outcome**: Thomas will likely have progressive retinal degeneration and associated worsening vision loss. He will need regular follow-ups with ophthalmology and visual aids and educational programs at home and school. Thomas will receive speech and learning assistance at school. He is in a regular program of diet and exercise to help with his weight control. He is monitored regularly by endocrinology for annual thyroid function, liver function, lipid profile, fasting glucose level or glucose tolerance testing, and pituitary function. His renal function will be monitored annually with BUN and creatinine levels. We anticipate that his cardiac status will be stable.

The Diagnosis

Clinical Diagnosis: Bardet–Biedl syndrome.
Genetic Diagnosis: *BBS12* gene, compound heterozygous pathogenic variants.

Further Reading

Forsythe, E., Gunay-Aygun, M., July 14, 2003. Bardet-biedl syndrome overview. Adam, M.P., Mirzaa, G.M., Pagon, R.A., Wallace, S.E., Bean, L.J.H., Gripp, K.W., Amemiya, A., (editors). In GeneReviews® [Internet]. University of Washington, Seattle, Seattle (WA), 1993–2023 [updated March 23, 2022]. PMID 20301537.

Genetic and Rare Diseases Information Center (GARD), 2023. Bardet-Biedl syndrome [Online]. Maryland: NIH National Center for Advancing Translational Sciences. Available: https://rarediseases.info.nih.gov/diseases/6866/bardet-biedl-syndrome. (Accessed May 2023).

Rump, P., Jazayeri, O., Van Dijk-Bos, K.K., Johansson, L.F., Van Essen, A.J., Verheij, J.B., et al., 2016. Whole-exome sequencing is a powerful approach for establishing the etiological diagnosis in patients with intellectual disability and microcephaly. BMC Med. Genom. 9, 7. PMID 26846091.

Case 11.5.5

Gastrointestinal Cases: Acute Abdominal Pain

Ayuko Iverson and Manisha Balwani

Departments of Genetics and Genomic Sciences and Medicine, Icahn School of Medicine at Mount Sinai, New York, NY, United States

Abstract

A 25-year-old presenting with recurrent abdominal pain is found to have acute intermittent porphyria.

The Case

Rebecca is a 25-year-old female with a history of recurrent severe abdominal pain which is unexplained. She has presented three times over the last year to the emergency room with severe generalized abdominal pain, nausea, vomiting, and weakness. She has required admissions for these symptoms, where she was treated with IV fluids and pain medications and discharged home. During her first admission, appendicitis was suspected and she had an appendectomy. On a subsequent admission, a urinary tract infection was suspected and she was given Bactrim which made her symptoms worse. After her most recent admission, you are now seeing her in follow-up. Her prior inpatient workup included an upper endoscopy which showed mild gastritis and a normal colonoscopy. Her abdominal pain is typically accompanied by nausea, muscle weakness, and fatigue and usually occurs after she has had two to three alcoholic drinks. She is on no medications at present. She has a normal physical examination.

The Issues

- **Are her recurrent episodes of concern?** You note that she appears genuinely in pain, upset that she is unable to return to work until you have evaluated her. A question of malingering has been raised in the emergency room, but you suspect that she has true pathology. Given her clinical picture and prior negative workup, you decide to refer her to gastroenterology.
- **What does the gastroenterologist diagnose?** Based on the above history, normal exam, and negative workup, the gastroenterologist decides to test her for acute intermittent porphyria. An elevated urine porphobilinogen (PBG) from a random urine sample is needed to establish this diagnosis. The urine PBG results a few days after Rebecca's visit and is elevated. You refer Rebecca to a clinical geneticist to confirm the diagnosis of acute intermittent porphyria and to discuss management.

The Referral

- **Genetic Evaluation**: Rebecca meets with the clinical geneticist, who obtains a medical history and completes a physical exam. She is currently asymptomatic and her physical exam is normal. A three-generation pedigree (Section 3.2)

Genomics in the Clinic. https://doi.org/10.1016/B978-0-12-816478-5.00045-5

is obtained and there are no other family members with acute intermittent porphyria; however, she had a maternal grandmother who also had recurrent unexplained abdominal pain and was diagnosed with hepatocellular carcinoma in her 60s.

- **Genetic Counseling**: The clinical geneticist reviews the diagnosis of acute intermittent porphyria with Rebecca. This autosomal dominant (Section 2.7.1) condition is caused by a pathogenic variant within one copy of the *HMBS* gene which encodes hydroxymethylbilane synthase. Approximately 90% of individuals who have pathogenic variants (Section 7.1.3) in *HMBS* do not develop acute attacks and are characterized as having latent, or asymptomatic, acute intermittent porphyria. This incomplete penetrance (Section 2.8) explains why Rebecca's pedigree does not fit the typical autosomal dominant picture.

  The disease, which is more common in women than men, is characterized by acute neurovisceral attacks of severe abdominal pain, which can be accompanied by tachycardia, hypertension, nausea, vomiting, constipation, and hyponatremia. Neurological findings, such as seizures, altered mental status, and peripheral neuropathy, may also occur. Clinical manifestations are thought to be due to toxicity from elevated 5-aminolevulinic acid (ALA) levels. The clinical geneticist reviews that hepatocellular carcinoma, hypertension, and chronic kidney disease are long-term complications. Family history details are also reviewed. It is possible to be carrying a variant and have no symptoms.

There can be symptomatic and latent cases of acute intermittent porphyria within families. Often cases are undiagnosed for many years. Patients with acute intermittent porphyria have an increased risk of hepatocellular carcinoma.

- **Management and Recommendations**: The clinical geneticist gives recommendations to Rebecca on how to prevent an acute attack, including avoiding certain medications, alcohol, illicit drugs, exogenous hormones, fasting, and stress. Given that Rebecca has attacks when she drinks alcohol, the clinical geneticist recommends that she avoid drinking. The geneticist also recommends she wear a medical alert bracelet with her diagnosis and use barrier methods or a copper intrauterine device for birth control, as hormonal contraception can precipitate attacks.

  Severe attacks require admission for intravenous hemin administration, pain control with opiates, caloric support with dextrose, treatment of hypertension with beta blockers, and correction of hyponatremia if needed. If the patient has four or more attacks a year, prophylactic hemin infusions can be considered. A new RNAi therapy, givosiran, which targets ALA synthase 1 and decreases ALA and PBG levels is FDA approved for patients with recurrent attacks. Rebecca is a candidate for this treatment if she continues to have attacks after eliminating known triggers.

  The geneticist also discusses doing basic lab work, including a complete blood count, comprehensive metabolic panel, testing of urine PBG and ALA for a baseline level, and molecular testing of *HMBS*.
- **Results**: Rebecca's urine PBG and ALA are both elevated to 20 times and 10 times the upper level of normal, respectively, which is common in acute intermittent porphyria patients even between acute attacks. Molecular testing finds a splice variant (Sections 2.3.4, 2.4.1, 7.1.3) within *HMBS*, confirming a diagnosis of acute intermittent porphyria. A comprehensive metabolic panel shows normal renal and liver function.
- **Implications for Other Family Members**: Once the pathogenic variant is determined, parents and siblings can be screened to identify latent patients so they can institute lifestyle changes to prevent acute attacks. It is likely that her mother will carry the same variant and that her grandmother was also affected.
- **Caveats**: As symptoms for acute intermittent porphyria are often nonspecific, the diagnosis can be missed or not considered. Recurrent acute attacks of abdominal pain are sometimes dismissed as malingering which can also lead to a delay in diagnosis and appropriate treatment. In some affected patients, measurement of urinary PBG and ALA may be normal in between acute attacks.
- **Differential Diagnosis**: Recurrent abdominal pain can be seen in other acute hepatic porphyrias, such as hereditary coproporphyria and variegate porphyria. While both of these conditions can also increase urine PBG levels, they can also present with cutaneous manifestations such as blistering. In addition to molecular testing, plasma, urine, and fecal porphyrin analysis can help to differentiate between these three acute porphyrias.
- **Outcome**: An emergency hemin infusion plan is created for Rebecca in the event of an acute attack and a follow-up appointment is scheduled in 3 months. She avoids alcohol and other known triggering factors and does well. She reports no acute attacks at her follow-up appointment and has had no further episodes. She should be monitored long term for recurrence of symptoms and chronic kidney disease. To monitor for the increased risk for hepatocellular carcinoma, after the age of 50 years she should undergo hepatic ultrasound and alpha-fetoprotein levels every 6 months.

The Diagnosis

Clinical Diagnosis: Acute intermittent porphyria.
Genetic Diagnosis: *HMBS* gene, heterozygous pathogenic variant.

Further Reading

Anderson, K.E., Bloomer, J.R., Bonkovsky, H.L., Kushner, J.P., Pierach, C.A., Pimstone, N.R., et al., 2005. Recommendations for the diagnosis and treatment of the acute porphyrias. Ann. Intern. Med. 142, 439–450. PMID 15767622.

Balwani, M., Wang, B., Anderson, K.E., Bloomer, J.R., Bissell, D.M., Bonkovsky, H.L., Network., et al., 2017. Acute hepatic porphyrias: recommendations for evaluation and long-term management. Hepatology 66, 1314–1322. PMID 28605040.

Bissell, D.M., Anderson, K.E., Bonkovsky, H.L., 2017. Porphyria. N. Engl. J. Med. 377, 862–872. PMID 28854095.

Sardh, E., Harper, P., Balwani, M., Stein, P., Rees, D., Bissell, D.M., et al., 2019. Phase 1 trial of an RNA interference therapy for acute intermittent porphyria. N. Engl. J. Med. 380, 549–558. PMID 30726693.

Case 11.5.6

Gastrointestinal Cases: Abdominal Pain and Arthralgias

Ayuko Iverson

Departments of Genetics and Genomic Sciences and Medicine, Icahn School of Medicine at Mount Sinai, New York, NY, United States

Abstract

A 50-year-old male presenting with arthralgias, abdominal pain, and fatigue is found to have elevated liver enzymes, serum ferritin and transferrin saturation levels, and the molecular diagnosis of hemochromatosis due to homozygous pathogenic variants in the *HFE* gene.

The Case

Josiah is a 50-year-old male with a history of arthralgias and fatigue who is presenting to you, his primary care physician, for evaluation. He began feeling fatigue, weakness, and joint stiffness and pain in his hands a year ago, after a camping trip. Evaluations for Lyme disease and rheumatoid arthritis at that time were negative. He reports that in addition to these symptoms, he now has developed intermittent abdominal pain. He drinks one to two cans of beer every week. His physical exam reveals hepatomegaly and joint swelling around his second and third metacarpophalangeal joints.

The Issues

- **What is the initial workup?** Due to his abdominal complaints and hepatomegaly, liver function testing is obtained and reveals transaminitis. A right upper quadrant ultrasound confirms hepatomegaly, but no other abnormalities. You wonder about an infectious cause, but viral hepatitis serologies are negative. You refer him to hepatology for further evaluation.
- **What does the hepatologist diagnose?** Based on the above history and workup, the hepatologist is concerned about the possibility of hemochromatosis and decides to test for iron overload. Serum iron, serum ferritin, transferrin saturation, and total iron-binding capacity are obtained and reveal elevated serum ferritin and transferrin saturation levels. Josiah is referred to a clinical geneticist to confirm the diagnosis of hemochromatosis and to discuss management.

The Referral

- **Genetic Evaluation**: Josiah meets with the genetic counselor and clinical geneticist, who obtains a medical history and completes a physical exam. He denies dyspnea on exertion, loss of libido, or erectile dysfunction. There is no evidence of heart failure on exam and skin pigmentation is normal. A three-generation pedigree (Section 3.2) is obtained and there are no family members with a known diagnosis of hemochromatosis; however, he has an older brother who

Genomics in the Clinic. https://doi.org/10.1016/B978-0-12-816478-5.00046-7

was recently diagnosed with cirrhosis. He has three children, reportedly healthy, and his family is of Northern European background. *HFE* sequencing is recommended.

- **Genetic Counseling**: The genetic counselor and clinical geneticist review the diagnosis of *HFE* hemochromatosis with Josiah. This autosomal recessive (Section 2.7.1) condition is caused by biallelic pathogenic variants within *HFE*, which encodes a protein that regulates hepcidin, an iron regulatory hormone. In *HFE* hemochromatosis, dysregulation of hepcidin leads to inappropriately increased iron absorption in the small intestine.

  The disease is characterized by excessive iron deposition within multiple organ systems, including the liver, pancreas, heart, skin, and joints. Early symptoms include fatigue, weakness, weight loss, abdominal pain, and arthralgias. In untreated individuals, symptoms can progress to skin hyperpigmentation ("bronze diabetes"), hypogonadism, diabetes mellitus, cardiomyopathy, cirrhosis, and liver cancer.

  There is a range of phenotypic manifestations of *HFE* hemochromatosis due to incomplete penetrance (Section 2.8). Most individuals only have biochemical evidence of iron overload with increased ferritin concentrations and transferrin saturation levels, but no end-organ damage. Other individuals have neither clinical nor biochemical evidence of iron overload. Clinical hemochromatosis is more common in men than women, as menstruation delays iron accumulation. Approximately 75%–90% of individuals with hemochromatosis are asymptomatic.
- **Genetic Testing**: *HFE* sequencing shows homozygous (Section 2.7.1) p.Cys282Tyr pathogenic variants (Section 7.1.3). Approximately one-third of northern Europeans are heterozygous (Section 2.7.1) for either p.Cys282Tyr or p.His63Asp. The prevalence of individuals homozygous for p.Cys282Tyr is approximately 1:150 to 1:300 in European populations and accounts for 90% of cases. Clinical hemochromatosis occurs in 10% of those who are homozygous for the p.Cys282Tyr variants, and Josiah is among them. Homozygotes for p.Cys282Tyr are at greater risk of developing iron overload (relatively high, but not 100%) than p.Cys282Tyr/p.His63Asp compound heterozygotes (0.5%–2.0%), who, in turn, are at a greater risk than homozygotes for p.His63Asp.
- **Management and Recommendations**: The clinical geneticist recommends that Josiah begin weekly phlebotomy treatment to achieve a serum ferritin concentration of ≤50 ng/mL. Once this concentration is achieved, serum ferritin should be monitored every 3 months and kept at less than 300 ng/mL through maintenance phlebotomy.

  The clinical geneticist also reviews lifestyle modifications including avoiding exogenous iron and vitamin C supplements, which can increase intestinal absorption of iron. Complete alcohol cessation is recommended, as well as avoiding uncooked seafood, which can lead to severe infections with *Vibrio vulnificus*.

  The geneticist orders thyroid function testing, and a hemoglobin A1C level, which are normal. An abdominal MRI is recommended to evaluate for iron deposition and shows iron overload. Transient elastography (TE), a specialized noninvasive ultrasound, is also recommended to screen for liver cirrhosis, given Josiah's elevated transaminases. Screening results that are concerning for cirrhosis should be confirmed by a liver biopsy. The geneticist reviews that if cirrhosis is found, then Josiah must be monitored at least every 6 months for liver cancer by abdominal ultrasound and alpha fetoprotein levels. The TE results are not concerning for fibrosis.
- **Implications for Other Family Members**: Once an individual is diagnosed with *HFE* hemochromatosis, family members can be screened with iron studies and *HFE* sequencing to identify individuals with biallelic pathogenic variants and to determine carrier status. Although hemochromatosis is an autosomal recessive condition, the carrier rate is high in certain populations, and children of affected individuals also can be affected by inheriting a pathogenic variant (Section 7.1.3) from the other parent, thus making it appear as a dominant condition in certain families. This is known as pseudodominance (Section 2.7.1). Josiah's older brother should be offered testing, as he may also be affected.
- **Differential Diagnosis**: Iron overload can be due to pathogenic variants in other proteins that regulate iron levels, such as ferroportin and ceruloplasmin. In general, these disorders have an earlier age of onset for clinical symptoms and higher serum ferritin and transferrin saturation levels compared to individuals with *HFE* hemochromatosis.
- **Caveats**: As early symptoms for hemochromatosis are nonspecific, the diagnosis may be missed or not considered. In some patients, there may be no clinical symptoms of iron overload, or only biochemical abnormalities.
- **Outcome**: Josiah begins weekly therapeutic phlebotomy and after a few sessions, his serum ferritin levels fall below 50 ng/mL. He stops drinking alcohol and stops his daily multivitamin which contained both iron and vitamin C. His repeat liver function testing normalizes and his abdominal pain improves, although he continues to experience arthropathy. He continues to receive maintenance phlebotomy every 2–4 months and his primary care physician monitors his iron studies and complete blood counts at each phlebotomy session. Josiah's brother undergoes *HFE* sequencing and is also found to be affected. None of his children have abnormal iron studies or biallelic pathogenic variants.

The Diagnosis

Clinical Diagnosis: Patient and Brother—hemochromatosis.
Genetic Diagnosis: Patient and Brother—*HFE* gene, homozygous pathogenic variants.

Further Reading

Adams, P.C., Reboussin, D.M., Barton, J.C., McLaren, C.E., Eckfeldt, J.H., McLaren, G.D., et al., 2005. Hemochromatosis and iron-overload screening in a racially diverse population. N. Engl. J. Med. 352, 1769–1778. PMID 15858186.

Allen, K.J., Gurrin, L.C., Constantine, C.C., Osborne, N.J., Delatycki, M.B., Nicoll, A.J., et al., 2008. Iron-overload-related disease in HFE hereditary hemochromatosis. N. Engl. J. Med. 358, 221–230. PMID 18199861.

Bacon, B.R., Adams, P.C., Kowdley, K.V., Powell, L.W., Tavill, A.S., American Association for the Study of Liver Diseases, 2011. Diagnosis and management of hemochromatosis: 2011 practice guideline by the American Association for the Study of Liver Diseases. Hepatology 54, 328–343. PMID 21452290.

European Association for the Study of the Liver, 2010. EASL clinical practice guidelines for HFE hemochromatosis. J. Hepatol. 53, 3–22. PMID 20471131.

Section 11.6

Metabolic Cases

Case 11.6.1

Metabolic Cases: Sweet Smell in an Infant

George A. Diaz

Department of Genetics and Genomic Sciences, Icahn School of Medicine at Mount Sinai, New York, NY, United States

Abstract

An 8-month-old infant with sweet smell to urine is found to have negative testing for maple syrup urine disease.

The Case

Isabelle is a thriving 8-month-old infant in your pediatric practice whose parents are worried because they have noted an occasional odor coming from her diaper that they think smells sweet like maple syrup. Her parents have done some investigation by computer search and are worried that Isabelle might have maple syrup urine disease. You ask if Isabelle has had any changes in her behavior or level of alertness in association with these episodes and her parents report that she has not had any such changes. You note in your records that Isabelle has been breast fed since birth and has been weaned to baby food jars and purees without difficulties. She is the third child for this family and her parents have never noted a similar odor with their other children. She was delivered uneventfully and had a negative newborn screen. You do not note any smell of maple syrup from her diaper.

The Issues

- **How should you evaluate the smell?** You order a urine dipstick test for ketones, which is negative. You confirm with the state newborn screening program that Isabelle had a negative newborn screening result for maple syrup urine disease.
- **Does a negative newborn screen preclude the possibility of maple syrup urine disease (MSUD) in Isabelle?** You recall that newborn screening is a screening test, not diagnostic testing, and there can be false negatives (Section 5.4). It would be possible to miss a milder case of maple syrup urine disease by newborn screening.
- **What is the next step?** Based on your assessment of Isabelle, you do not believe that she has a milder form of maple syrup urine disease but to be certain, you refer her to a metabolic specialist for definitive evaluation.

The Referral

- **Genetic Evaluation**: Isabelle is assessed by a metabolic geneticist who notes her negative medical history, normal development, and normal physical examination with normal growth, who concurs with your assessment based on

Genomics in the Clinic. https://doi.org/10.1016/B978-0-12-816478-5.00047-9

history. The geneticist also asks the family if there is any history of ingestion of herbal tea or any other compound containing fenugreek. The family denies that either Isabelle or her mother knowingly ingested any such items.

- **Genetic Counseling**: The metabolic geneticist reviews that maple syrup urine disease is a disorder of branched chain amino acid metabolism that is a primary target of the Recommended Uniform Screening Panel and screened for in all states of the United States (Section 5.4). Patients with untreated maple syrup urine disease will present in the neonatal period with neurological findings progressing from irritability and poor feeding to lethargy and finally to coma. Neurological symptoms may include opisthotonic posturing, focal dystonia, and bicycling movements. Cerebral edema may lead to death from central respiratory failure. During episodes of hyperleucinemia, the clinical symptoms of the milder disease variants can be identical to those seen in classical maple syrup urine disease (ataxia, altered mental status, progression to coma).

  The disorder is caused by pathogenic variants in the *BCKDHA*, *BCKDHB*, or *DBT* genes, which encode the components of the branched chain ketoacid dehydrogenase enzyme complex. Variants that result in less than 3% residual enzyme activity will cause classic maple syrup urine disease and will be flagged for a newborn screening referral due to elevation of leucine and isoleucine. Variants that result in higher residual activity (up to 40%) cause a less severe form of disease called intermediate maple syrup urine disease. Rarely, patients may present with the intermittent form that does not have the characteristic biochemical markers of maple syrup urine disease except during a period of illness that triggers a metabolic crisis. This intermittent form of maple syrup urine disease, which can result from variants in the aforementioned genes or in a gene encoding a regulatory phosphatase (*PPM1K*), may be missed by newborn screening. It is possible that Isabelle has this form.

  Genetic Testing: Evaluation of fasting plasma amino acids on Isabella is recommended. They reveal normal branched chain amino acids and no alloisoleucine, a pathognomonic marker for maple syrup urine disease. These findings exclude the classic and intermediate forms of maple syrup urine disease but are not able to exclude intermittent MSUD. DNA is collected from her to assess whether she has any pathogenic variants in the *BCKDHA*, *BCKDHB*, *DBT*, or *PPM1K* genes by sequencing (Section 6.2.2). This testing does not identify any variants in these genes. The family is counseled that negative results are not definitive but nonetheless, in the absence of clinical symptoms, the negative results are highly reassuring.
- **Management and Recommendations**: Classical and intermediate maple syrup urine disease will require dietary restriction of branched chain amino acids from infancy but intermittent maple syrup urine disease patients do not require dietary restriction when they are well (Section 5.4.3).
- **Caveats**: Negative results are not definitive because changes in introns or regulatory regions are not well covered by the available technology and as yet uncharacterized genes that lead to an intermittent maple syrup urine disease phenotype may be discovered in the future.
- **Outcome**: Based on the evaluation, there is no evidence for an intermittent form of maple syrup urine disease in Isabelle. Her parents are reassured that she does not have maple syrup urine disease. Despite the assessments that she was unlikely to have maple syrup urine disease, they had been watching her protein intake due to concerns for triggering a crisis while they awaited the results of genetic testing. Isabelle grows and thrives and the parents note the sweet smell only rarely as she gets older.

The Diagnosis

Clinical Diagnosis: Healthy newborn infant, no evidence metabolic disease.
Genetic Diagnosis: None.

Further Reading

Korman, S.H., Cohen, E., Preminger, A., 2001. Pseudo-maple syrup urine disease due to maternal prenatal ingestion of fenugreek. J. Paediatr. Child Health 37, 403–404. PMID 11532065.

Oyarzabal, A., Martinez-Pardo, M., Merinero, B., Navarrete, R., Desviat, L.R., Ugarte, M., et al., 2013. A novel regulatory defect in the branched-chain alpha-keto acid dehydrogenase complex due to a mutation in the PPM1K gene causes a mild variant phenotype of maple syrup urine disease. Hum. Mutat. 34, 355–362. PMID 23086801.

Sewell, A.C., Mosandl, A., Bohles, H., 1999. False diagnosis of maple syrup urine disease owing to ingestion of herbal tea. N. Engl. J. Med. 341, 769. PMID 10475807.

Strauss, K.A., Puffenberger, E.G., Carson, V.J., January 30, 2006. Maple syrup urine disease. Adam, M.P., Mirzaa, G.M., Pagon, R.A., Wallace, S.E., Bean, L.J.H., Gripp, K.W., Amemiya, A., (editors). In GeneReviews [Internet]. University of Washington, Seattle, Seattle (WA), 1993–2023 [updated April 23, 2022]. PMID 20301495.

Case 11.6.2

Metabolic Cases: Apparent Life-Threatening Event

Pankaj Prasun

Department of Genetics and Genomic Sciences, Icahn School of Medicine at Mount Sinai, New York, NY, United States

Abstract

An 8-month-old girl brought in for an apparent life-threatening event with a history of motor delay and eye twitches is found to have Leigh syndrome.

The Case

An 8-month-old female infant, Iona, presented to the emergency department after she was found to have cyanosis of the lips. Her parents had noted she was gasping for air while asleep at home and suddenly saw her lips turn blue. They woke her up by patting her on her back and she returned to a normal color. Her past medical history has been unremarkable. She was awake and alert after being brought to the emergency department and was breathing comfortably. The emergency physician sent laboratory studies and admitted her for evaluation of this apparent life-threatening event (ALTE). After admission to your service, you evaluate the infant and find her to be hypotonic. After further inquiry, Iona's parents mention that her pediatrician has been concerned that she is not sitting without support at 8 months of age. By then, initial blood work is available, which shows a high lactate of 6.3 mmol/L (normal 0.5−2.2 mmol/L).

The Issues

- **Is this an isolated episode or is there an underlying cause?** An infant with ALTE is usually evaluated for gastro-intestinal reflux disease, upper respiratory tract infection, pneumonia, seizure disorder, cardiac arrhythmia, heart failure, hypoglycemia, hypocalcemia, or child abuse as the underlying cause. However, often no cause is found. In this case, you are worried that there may be an underlying cause because of motor delay, hypotonia, and elevated lactate on initial evaluation. You request a pediatric neurology consultation.
- **What does the neurology evaluation show**? The pediatric neurologist is also concerned because of high lactate and generalized hypotonia. Apart from Iona's history of motor delay, her past medical history is also significant for intermittent eye twitching episodes for which her pediatrician had recommended a pediatric neurology evaluation. This appointment was scheduled but she was admitted for the ALTE before the appointment occurred. An EEG and MRI of brain are recommended.
- **What is the next step?** The EEG is unremarkable, but the MRI of brain shows bilateral ischemic lesions in basal ganglia and midbrain. The radiologist raises the possibility of a metabolic disorder. Based on the neuroimaging findings and a high lactate, pediatric genetics is consulted.

Genomics in the Clinic. https://doi.org/10.1016/B978-0-12-816478-5.00048-0

The Referral

- **Genetic Evaluation**: A detailed antenatal, birth, postnatal, and family history are obtained by the geneticist and genetic counselor. Iona is the first child of this couple, who are nonconsanguineous. Pregnancy was complicated by oligohydramnios. Birth history was significant for meconium aspiration. In addition, Iona was treated by phototherapy after birth for neonatal jaundice. After discharge from the hospital, she was well, but developed the eye twitching at 2 months of age. On examination, her weight is in the 5th percentile and length in the 10th percentile; however, she has a small head below the 2nd percentile (Z score—2.5). There is generalized hypotonia and mild head lag. There is no facial dysmorphism, and eye movements appear normal. The laboratory data are reviewed. In addition to the elevated lactate, there is also metabolic acidosis, with increased anion gap.
- **Genetic Counseling**: The geneticist and genetic counselor meet with Iona's family and discuss the possible causes. The presence of a small head, low muscle tone, motor delay, high lactate in blood, and abnormal MRI of brain are all concerning for a mitochondrial disease (Section 5.4.1), particularly Leigh syndrome (LS). They review this condition with the family.

  LS is a rare mitochondrial disease. The usual presentation is developmental regression and hypotonia precipitated by a viral illness in a child below 2 years of age. The condition is progressive and often results in death by 3 years of age. There are characteristic bilaterally symmetrical lesions in the brainstem and basal ganglia. These lesions are seen on the MRI of the brain as hyperintense regions in T2-weighted images. There are associated signs and symptoms of basal ganglia and brainstem involvement such as dystonia and abnormal breathing. Other findings sometimes associated with LS are retinitis pigmentosa, optic atrophy, hypertrophic cardiomyopathy, hepatomegaly, and renal tubulopathy. Often the child's development is normal before the onset of illness or there may be only mild delays.

  LS is genetically heterogeneous. Because the main function of mitochondria is to produce ATP, which is the main source of energy for cells, in mitochondrial diseases there is energy failure. Tissues with high energy demand such as brain, muscle, and heart are more commonly affected in mitochondrial diseases. Although each mitochondrion has its own DNA inherited from the mother, its genetic material is not sufficient for its function. Mitochondria depend on the nucleus for maintenance and function (Section 2.5). Hence, variants either in mitochondrial DNA or in nuclear genes can cause a mitochondrial disease (Section 6.4.2). More than 75 genes have been implicated in the etiology of LS. Pathogenic variants or mutations in nuclear genes comprise about 80% of LS cases, while mitochondrial DNA variants contribute about 20%. The inheritance pattern of LS can be maternal or autosomal depending on the genetic etiology (Section 2.7.4). Hence, a confirmatory genetic diagnosis is essential for accurate genetic counseling.
- **Genetic Testing**: A mitochondrial gene panel (Section 6.2.3), which consists of sequencing the entire mitochondrial DNA as well as about 200 nuclear genes implicated in mitochondrial disease, is recommended for diagnosis.
- **Management and Recommendations**: Unfortunately, no curative treatment is available for LS at present. Management is mainly symptomatic, avoidance of mitotoxic medications, and physical and occupational therapy. Mitochondrial supplements such as coenzyme Q10, L-carnitine, thiamine, and riboflavin can be tried but their efficacy is not proven. Periodic evaluations for cardiac, endocrine, renal, and audiological complications are necessary. A multidisciplinary approach (audiologist, ophthalmologist, neurologist, pulmonologist, cardiologist, physical therapy, and rehabilitation specialist) is needed for optimal management. Iona is referred for cardiac evaluation, physical therapy, occupational therapy, and hearing evaluation. Her parents are trained in basic life support.

  Physical therapy is started. Due to suspicion of LS, mitochondrial supplements (thiamine, riboflavin, and coenzyme Q10) are started and Iona is discharged.
- **Results**: About a month later, Iona and her parents return for follow-up in the genetic clinic. By this time, the mitochondrial gene panel test is done. It is positive for the m.8993 T > G mutation in the mitochondrial gene *MT-APT6*, associated with LS, thus confirming the diagnosis. The child is homoplasmic (Section 2.5) for this mutation. When all of the mitochondrial DNA in a cell have the same mutation, there is homoplasmy.

  Because mitochondrial DNA is inherited in a maternal fashion, Iona's mother is recommended to have targeted genetic testing (Section 6.2.1). The results show that she has a heteroplasmy (Section 2.5) level of 47% for the same mutation. There are thousands of mitochondria in a single cell. A single cell may harbor both mitochondrial DNA (mtDNA) with a pathogenic mutation and wild-type mitochondrial DNA together. This is called heteroplasmy and is measured as a percentage of the mitochondrial DNA with the mutation. The heteroplasmy level may vary between different cells of a tissue, between different tissues of an individual, and between different individuals carrying same mitochondrial DNA mutation. A higher heteroplasmy level may lead to clinical symptoms while a lower level of heteroplasmy may be clinically silent. Typically, clinical and/or biochemical manifestations from a pathogenic mitochondrial DNA mutation are seen when the heteroplasmy level exceeds 60%–80%. This is called the "threshold effect." However, the threshold varies for different mutations and between different tissues. Hence, the clinical manifestation and severity of a mitochondrial disease is highly variable among individuals carrying the same pathogenic mutation.

- **Implications for Other Family Members**: Iona's mother is currently asymptomatic. However, she plans to have another child in the future. Mitochondrial disease caused by a pathogenic mitochondrial DNA point mutation is maternally inherited, as the mitochondrial DNA in a zygote is almost exclusively derived from the egg. The sperm contains relatively fewer mitochondria and those are removed before fertilization. During oogenesis, as the primordial germ cell divides to form primary oocytes, there is rapid reduction in the number of mitochondria per cell. The primary oocyte then undergoes maturation to form a mature oocyte, which is associated with replication of the mitochondria. This reduction followed by replication in mitochondrial number leads to a different proportion of mitochondrial DNA with the mutation in the mature oocyte compared to the primordial germ cell. A mutation below the threshold level in the mother may reach above the threshold in the fetus. Hence, all the mother's future children will inherit this mutation but they may or may not be affected based on the level of the mitochondrial mutation in their mitochondrial DNA.

  Prenatal diagnosis is challenging at present. Chorionic villous sampling (CVS) (Section 4.4.2) or preimplantation genetic testing (PGT) (Section 4.5.1) can provide information on the heteroplasmy level of the fetus/embryo. Prevention of transmission of mitochondrial disease can be achieved by in vitro fertilization of a donor oocyte with sperm from the male partner and implantation of the embryo in the recipient (oocyte donation). Adoption is another option. Iona's mother is referred to a reproductive medicine clinic for detailed discussion about these options.
- **Differential Diagnosis**: The differential diagnosis includes the following conditions: mitochondrial DNA depletion syndrome, mitochondrial translation defects, pyruvate dehydrogenase deficiency, biotinidase deficiency, biotin–thiamine-responsive basal ganglia disease, and dihydrolipoamide dehydrogenase deficiency. All of these metabolic disorders can clinically present as LS and may present indistinguishably. Molecular studies are needed to confirm.
- **Caveats**: The possibility of giving birth to an affected child when a low heteroplasmy level is detected by CVS, amniocentesis, and preimplantation genetic testing cannot be completely excluded, as the heteroplasmy level may vary between fetus and placenta (CVS) and between cells of embryo (amniocentesis and PGT). The interpretation becomes even more complicated when there is an intermediate level of heteroplasmy found by these methods.
- **Outcome**: Unfortunately, Iona dies at the age of 13 months due to respiratory failure. Her mother follows up at the reproductive clinic. She has also been enrolled in a research protocol for mitochondrial donation for mothers with mitochondrial DNA mutation. This technique involves removal of the nuclear genome from an oocyte carrying the mitochondrial DNA mutation followed by transfer to an enucleated donor oocyte with healthy mitochondria. Thus, the resulting embryo will have healthy mitochondria from the donor, but nuclear genetic materials from both parents. This procedure is not approved in the United States at present. However, some centers are enrolling patients and freezing their eggs for the future when this procedure will be legal in the United States.

The Diagnosis

Clinical Diagnosis: Leigh syndrome.

Genetic Diagnosis: Patient—*MT-ATP6* gene, homoplasmic pathogenic variant. Mother—*MT-ATP6* gene, heteroplasmic pathogenic variant.

Further Reading

Gerards, M., Sallevelt, S.C., Smeets, H.J., 2016. Leigh syndrome: resolving the clinical and genetic heterogeneity paves the way for treatment options. Mol. Genet. Metabol. 117, 300–312. PMID 26725255.

Lake, N.J., Compton, A.G., Rahman, S., Thorburn, D.R., 2016. Leigh syndrome: one disorder, more than 75 monogenic causes. Ann. Neurol. 79, 190–203. PMID 26506407.

Thorburn, D.R., Rahman, J., Rahman, S., October 30, 2003. Mitochondrial DNA-associated Leigh syndrome and NARP. Adam, M.P., Mirzaa, G.M., Pagon, R.A., Wallace, S.E., Bean, L.J.H., Gripp, K.W., Amemiya, A., (editors). In GeneReviews ® [Internet]. University of Washington, Seattle, Seattle (WA), 1993–2023 [updated May 4, 2022]. PMID 20301352.

Case 11.6.3

Metabolic Cases: Hypoglycemia in a Child

Hilary J. Vernon
Department of Genetic Medicine, Johns Hopkins University School of Medicine, Baltimore, MD, United States

Abstract
A 4-year-old girl with a first time seizure is found to have hypoglycemia due to MCAD deficiency.

The Case

Alice, a 4-year-old girl with no significant past medical history, presents to you in the emergency room via ambulance with a first-time seizure. In the ambulance, her blood glucose level is found to be 20 mg/dL. She is given two boluses of IV dextrose and started on a 10% dextrose IV drip. On examination in the emergency room, she is irritable, but awake, with a normal cardiac, pulmonary, and neurological exam. There is mild hepatomegaly. Laboratory testing (drawn while the patient was on dextrose containing IV fluids) reveals a complete blood count (CBC) with a moderate neutrophilia, and a comprehensive metabolic panel with a normal sodium level, bicarbonate of 17 meq/L, elevations in liver enzyme levels AST and ALT, and a blood glucose of 105 mg/dL. Total creatine kinase is sent and is normal.

The Issues

- **Why does this child have hypoglycemia?** The differential diagnosis for pediatric hypoglycemia is wide and can include multiple metabolic and endocrine etiologies. You send a plasma lactate, which is normal at 1.8 mmol/dL, and thyroid stimulating hormone (TSH) and thyroxine, which are normal. You do not send insulin levels at this time because the child is not currently hypoglycemic. You send a urinalysis to determine if this child is dehydrated. The urinalysis shows trace ketones and a specific gravity of 1.026, indicating highly concentrated urine.
- **Are there any clues in Alice's history of present illness to aid in her diagnosis?** On further history, it is found that Alice had developed a gastrointestinal illness the day before and had vomited several times. She also was refusing to eat or drink for the past 24 h, with the exception of small sips of water.

  You find it striking that in this clinical setting the urine ketones are only "trace". You recall that inappropriately low ketone excretion can occur with fatty acid oxidation disorders, which is a group of inborn errors of metabolism caused by deficient mitochondrial fatty acid metabolism (Section 5.4.1). You also recall that fasting can instigate metabolic crises in affected patients. You consult genetics to assist with diagnosis and management.

Genomics in the Clinic. https://doi.org/10.1016/B978-0-12-816478-5.00049-2

The Referral

- **Genetic Evaluation**: A pregnancy history and medical history are obtained by the geneticist, and a detailed physical examination is performed. The parents report that the pregnancy was uncomplicated. Alice was born at home, and newborn screening was declined. She had typical early development and is a very active child, with no history of fatigue, muscle pain, or weakness. She is nondysmorphic and well developed with an average height and weight for age.

  Family history is obtained. Alice is the only child to her parents, who are both healthy. There is no known extended family history of multiple miscarriages, stillbirths, or infant or childhood deaths. Parental consanguinity is denied.

  Based on the patient's medical history, the following tests are ordered (Section 5.4.2): free and total carnitine, acylcarnitine profile, and urine organic acids. The free and total carnitine is normal. The acylcarnitine profile shows abnormal elevations in the C6–C10 species, with C8 being the highest. Urine organic acids show elevations of hexanoylglycine and suberylglycine. These findings are characteristic of medium-chain acyl-CoA dehydrogenase (MCAD) deficiency.
- **Genetic Counseling**: The geneticist and genetic counselor discuss with the family that Alice's clinical symptoms and test results are consistent with MCAD deficiency. This is a disorder caused by deficient oxidation of medium-chain fats and results in deficient liver production of ketones from these medium-chain fats. Symptoms can arise when there is depletion of the body's glycogen (carbohydrate stores), often in the setting of an acute illness. Affected individuals are at risk for hypoglycemia, coma, and death if not treated appropriately.

  MCAD deficiency is an autosomal recessive disorder (Section 2.7.1) where both carrier parents are asymptomatic but have a 25% chance of each child they have together being affected.
- **Genetic Testing**: Molecular testing of the *ACADM* gene (the gene underlying MCAD deficiency) reveals that Alice is homozygous (Section 2.3.4) for one of the most common pathogenic variants (Section 7.1.3): c.985A > G (p.Lys329-Glu). Her parents are found to each carry one of the mutations.
- **Management and Recommendations**: The key to management of MCAD deficiency is avoidance of fasting, and provision of intravenous glucose when adequate oral intake is not possible. Patients and their families should be provided with instructions for when to seek medical attention, and given an emergency letter that describes necessary medical care.
- **Differential Diagnosis**: Overlapping clinical features, including hypoglycemia with fasting, can be seen in disorders of gluconeogenesis (e.g., phosphoenolpyruvate carboxykinase deficiency), ketotic hypoglycemia, and disorders of ketone synthesis or utilization.
- **Caveats**: MCAD deficiency is included on the newborn screen in all 50 states (Section 5.4) and has a high sensitivity for detection of this disease. However, this child's parents declined newborn screening, and thus the MCAD deficiency was not diagnosed earlier. Additionally, while false-negative newborn screening for MCAD deficiency is very rare, it can occur in milder cases.
- **Outcome**: Alice is kept on dextrose containing IV fluids until she is able to eat again and her emesis has resolved. Her parents are now aware of the medical precautions that must be taken in the setting of this disease, including avoidance of fasting and emergency administration of dextrose containing IV fluids in the setting of intercurrent illness (Section 5.4.3). An emergency plan and letter for management of future episodes are provided to the family, and continued care is provided in collaboration with Alice's pediatrician.

The Diagnosis

Clinical Diagnosis: Medium-chain acyl-coA dehydrogenase deficiency.

Genetic Diagnosis: Patient—*ACADM* gene, homozygous pathogenic variants. Parents—*ACADM* gene, heterozygous pathogenic variant.

Further Reading

Jager, E.A., Kuijpers, M.M., Bosch, A.M., Mulder, M.F., Gozalbo, E.R., Visser, G., et al., 2019. A nationwide retrospective observational study of population newborn screening for medium-chain acyl-CoA dehydrogenase (MCAD) deficiency in the Netherlands. J. Inherit. Metab. Dis. 42, 890–897. PMID 31012112.

Merritt, J.L., Chang, I.J., April 20, 2000. Medium-chain acyl-coenzyme A dehydrogenase deficiency. Adam, M.P., Mirzaa, G.M., Pagon, R.A., Wallace, S.E., Bean, L.J.H., Gripp, K.W., Amemiya, A., (editors). In GeneReviews ® [Internet]. University of Washington, Seattle, Seattle (WA), 1993–2023 [updated June 27, 2019]. PMID 20301597.

Rhead, W.J., 2006. Newborn screening for medium-chain acyl-CoA dehydrogenase deficiency: a global perspective. J. Inherit. Metab. Dis. 29, 370–377. PMID 16763904.

Case 11.6.4

Metabolic Cases: Recurrent Vomiting and Language Delay

George A. Diaz

Department of Genetics and Genomic Sciences, Icahn School of Medicine at Mount Sinai, New York, NY, United States

Abstract

A 4-year-old girl with recurrent vomiting and lethargic episodes, and language delay is found to have ornithine transcarbamylase deficiency.

The Case

Esme is a 4-year-old girl who is brought to you by her parents for recurrent episodes of vomiting. They have come to see you because she had an episode of vomiting last week and has continued to have occasional nausea in the past few days. You ask about her medical history and learn that Esme has been seen several times in the emergency department over the past 2 years for vomiting and lethargy in association with viral infections. She has always responded well to intravenous fluids given for dehydration and antiemetic medication, recovering within several hours. Her parents are concerned that her symptoms seem more persistent this time. In reviewing Esme's medical history with her parents, you learn that she has a history of language delay and that her growth has been steady but tracking at the third percentile for weight. Her parents tell you that she is a picky eater who eats vegetables but not meat or fish.

The Issues

- **Could Esme have an underlying condition?** Because she has several seemingly unrelated, positive findings on history, it is possible that she has an underlying genetic condition. It could be a metabolic disorder because of the vomiting and lethargy and recurrent nature, as well as her avoidance of protein. Even though many inborn errors of metabolism are screened for on newborn screening, this does not rule out milder forms or partial deficiencies.
- **How should you screen for an underlying metabolic disorder?** You obtain an ammonia level and liver function tests. The ammonia is just above the normal range and the liver function tests reveal a mildly elevated AST and ALT. Her BUN is below the normal range. You refer Esme to a metabolic specialist for evaluation for concern for an inborn error of metabolism. The history you have elicited and the low Her blood urea nitrogen (BUN) makes you concerned that Esme might have an underlying urea cycle disorder.

Genomics in the Clinic. https://doi.org/10.1016/B978-0-12-816478-5.00050-9

The Referral

- **Genetic Evaluation**: Esme is assessed by a metabolic geneticist who also is concerned about the history of recurrent encephalopathy, developmental delay, and protein aversion. The ammonia level was only slightly elevated but as Esme was well at the time of the visit, so intermittent hyperammonemia is not excluded. At the time of this visit with the metabolic specialist, she is well. The geneticist thinks that Esme could have a urea cycle defect (Section 5.4.1).
- **Genetic Counseling**: Urea cycle disorders result from deficiency in the enzymes or transporters that carry out the detoxification of ammonia generated from dietary protein. The first steps of this process occur in the mitochondrial matrix and result in low levels of the urea cycle intermediate citrulline. Deficiencies in the cytoplasmic enzymes result in elevations of citrulline, argininosuccinic acid, or arginine. Accordingly, detection of citrullinemia, argininosuccinic lyase deficiency, or arginase deficiency is possible on newborn screening. Unfortunately, detection of N-carbamyl glutamate synthetase (NAGS) deficiency, carbamoylphosphate synthetase (CPS1) deficiency, or ornithine transcarbamylase (OTC) deficiency is not yet feasible by newborn screening. Urea cycle disorders (UCDs) can present catastrophically in the newborn period, with onset of vomiting, encephalopathy, and progression to hyperammonemic coma occurring in the first 2–3 days of life in severe cases. The more severe conditions (NAGS, CPS1, and OTC deficiency) are not amenable to detection by newborn screening. The conditions that are amenable to newborn screening (citrullinemia, ASA lyase deficiency, and arginase deficiency) may present in the neonatal period but also can present in later life.

  In particular, ornithine transcarbamylase deficiency, the X-linked (Section 2.7.2) and most common form of urea cycle disorders, can manifest in females who have randomly inactivated the X-chromosome (Section 2.6) carrying the functional copy of the *OTC* gene in a large fraction of their liver cells. Later-onset forms of urea cycle disorders have a broad spectrum of presentations but common symptoms include a history of protein aversion, history of developmental issues, recurrent vomiting, and recurrent encephalopathy. Symptoms may be triggered by infectious illnesses, stressors such as surgery or prolonged fasting, or use of corticosteroids or other medications that induce catabolism.
- **Genetic Testing**: Both biochemical and molecular testing are recommended. Evaluation of fasting plasma amino acids (Section 5.4.2) reveals an elevation of glutamine consistent with hyperammonemia, a low citrulline level, and a low arginine level. These findings are suggestive of a deficiency in one of the mitochondrial urea cycle enzymes (NAGS, CPS1, and OTC). Analysis of a urine sample detects elevated orotic acid, consistent with a diagnosis of ornithine transcarbamylase deficiency. Based on the results of the evaluation, Esme is diagnosed with ornithine transcarbamylase deficiency. Molecular testing is recommended. A sample is also obtained from her mother.

Urine organic acids will not routinely detect orotic acid so this test has to be ordered specifically instead of organic acid analysis.

- **Management and Recommendations**: Esme is started on a protein restricted diet, citrulline supplementation, and ammonia scavenger medication in order to reduce her risk of future hyperammonemic crises.
- **Results**: A blood sample is collected from Esme and testing is ordered for a panel (Section 6.2.3) of urea cycle enzyme and transporter genes. The testing identifies a heterozygous variant in the *OTC* gene that has been described in cases of neonatal onset OTC deficiency in several males.
- **Implications for Other Family Members**: Analysis of the maternal sample reveals that Esme has inherited the mutation from her mother. Biochemical testing in Esme's mother is negative, suggesting that her functional *OTC* gene is expressed on her active X-chromosome in more of her liver cells to give her a lower risk for hyperammonemic events than Esme. Although she is not treated, she is counseled that she should be monitored carefully for signs of hyperammonemia if she is administered corticosteroids or after surgical procedures or other extreme stressors. Esme's parents are also counseled that the risk of having a boy with a potentially catastrophic form of the disease is 25%.
- **Differential Diagnosis**: Intermittent vomiting and encephalopathy can be a feature of a number of disorders. Nonmetabolic disorders should be excluded including a seizure disorder or cyclic vomiting syndrome due to abdominal migraine. Non-urea cycle metabolic disorders that can present with recurrent vomiting and encephalopathy include organic acidurias, fatty acid oxidation disorders, glycogen storage disorders, mitochondrial disorders, and intermittent maple syrup urine disease, among others.
- **Caveats**: Mild cases of OTC deficiency can be very difficult to diagnose until the condition is unmasked by a severe stress such as surgery or use of corticosteroids. The nonspecific nature of the symptoms makes this a challenging condition to identify in the absence of a severe decompensation event.
- **Outcome**: Additionally, Esme's parents had been considering getting pregnant. The couple elect to pursue reproductive genetic counseling and are considering preimplantation genetic diagnosis (Section 4.5.1) for their next pregnancy.

Esme is started on a low-protein diet, medical formula supplementation, and ammonia scavenger medication. Her family is provided with an emergency illness protocol focusing on carbohydrate caloric support and ammonia monitoring to reduce her risk of a severe hyperammonemic event. Her family reports that her developmental language delay improves and her growth curve also improves with the changes in her nutrition.

The Diagnosis

Clinical Diagnosis: Patient—ornithine transcarbamylase deficiency. Mother—asymptomatic.
Genetic Diagnosis: Patient and Mother—*OTC* gene, heterozygous pathogenic variant.

Further Reading

Bennett, E.E., Hummel, K., Smith, A.G., Longo, N., 2019. Acute presentation and management of the encephalopathic child with an undiagnosed inborn error of metabolism. J. Emerg. Med. 56, e5–e8. PMID 30420308.

Ganetzky, R.D., Bedoukian, E., Deardorff, M.A., Ficicioglu, C., 2017. Argininosuccinic acid lyase deficiency missed by newborn screen. JIMD Rep. 34, 43–47. PMID 27515243.

Lichter-Konecki, U., Caldovic, L., Morizono, H., Simpson, K., Mew, N.A., MacLeod, E., August 29, 2013. Ornithine transcarbamylase deficiency. Adam, M.P., Mirzaa, G.M., Pagon, R.A., Wallace, S.E., Bean, L.J.H., Gripp, K.W., Amemiya, A., (editors). In GeneReviews ® [Internet]. University of Washington, Seattle, Seattle (WA), 1993–2023 [updated May 26, 2022]. PMID 24006547.

Summar, M.L., Mew, N.A., 2018. Inborn errors of metabolism with hyperammonemia: urea cycle defects and related disorders. Pediatr. Clin. North Am. 65, 231–246. PMID 29502911.

Case 11.6.5

Metabolic Cases: Lactic Acidosis

Bryn D. Webb[1,2] and Hilary J. Vernon[3]

[1]*Department of Genetics and Genomic Sciences, Icahn School of Medicine at Mount Sinai, New York, NY, United States;* [2]*Department of Pediatrics, University of Wisconsin School of Medicine and Public Health, Madison, WI, United States;* [3]*Department of Genetic Medicine, Johns Hopkins University School of Medicine, Baltimore, MD, United States*

Abstract

A 5-year-old male with acute hemiparesis following a viral illness, history of short stature, developmental delay, and hypotonia is found to have lactic acidosis due to MELAS.

The Case

Juan, a 5-year-old boy with a past medical history significant for short stature, developmental delay, and hypotonia, presents to you in the emergency room with acute onset of weakness in his right arm and leg in the setting of a fever and viral upper respiratory tract infection. On examination, the child is nondysmorphic, with a normal cardiac, pulmonary, and abdominal exam. He has mild right-sided hemiparesis and generalized hypotonia. Laboratory testing reveals a normal CBC and a normal comprehensive metabolic panel with the exception of a low bicarbonate of 13 meq/L.

The Issues

- **Why does this child have a metabolic acidosis?** You request a venous blood gas and determine that the child has a blood pH of 7.29, in the setting of a primary metabolic acidosis with a moderate respiratory compensation. A lactate level is sent and found to be 8.2 mmol/L (normal level is 0.5−2.0 mmol/L).
- **Why did this child have an acute onset of focal neurologic symptoms?** You are concerned about a focal brain abnormality and request a brain MRI. T2-weighted imaging reveals increased signal in the left parietal, temporal, and occipital lobe regions. Although these abnormalities are not in a vascular distribution, you request a follow-up MRA/MRV. This shows no vascular abnormalities.
- **Are there any clues in Juan's past medical history to aid in his diagnosis?** You learn that Juan had typical early developmental milestones, including sitting up and rolling over, but that he did not walk until 24 months of age. Since "late bloomers" are common in the family, his parents were not particularly concerned. Speech delay was also detected at this time, and a hearing test revealed moderate bilateral sensorineural hearing loss. He measures at less than the fifth percentile for height with other growth parameters normal, but other family members are not particularly tall. He also has history of episodes of headaches with vomiting of unknown cause.
- **What is the next step?** You recall that focal neurologic deficits in the setting of physiologic stress (such as fever) can be seen in mitochondrial disorders. In addition, elevated serum lactate is also a marker of mitochondrial dysfunction (Section 5.4.1). You consult genetics to assist with diagnosis and management.

Genomics in the Clinic. https://doi.org/10.1016/B978-0-12-816478-5.00051-0

The Referral

- **Genetic Evaluation**: The family meets with the metabolic geneticist and the genetic counselor. A pregnancy history and medical history are obtained by the geneticist, and a detailed physical examination is performed. The parents report that the pregnancy was uncomplicated other than significant maternal fatigue, and that Juan had an uncomplicated neonatal course. Juan's physical exam reveals the generalized hypotonia and mild right-sided hemiparesis as seen earlier. Follow-up labs are requested, and a mild elevation in creatine kinase is identified. Blood pyruvate is high at 0.24 mmol/L, and lactate/pyruvate ratio is also high at 34. Plasma amino acids are also sent and show moderately elevated alanine and proline, both of which are consistent with lactic acidosis.

  Family history is obtained. Juan's mother has a history of diabetes and hearing loss, and two maternal siblings died in childhood. Juan's older sister has a history of migraine headaches. His maternal grandmother also had diabetes and migraine headaches. His maternal uncle died at age 37 years of cardiomyopathy. There is no known consanguinity. Most family members live outside of the United States and none have received a genetic evaluation.
- **Genetic Counseling**: The geneticist and genetic counselor discuss with the family that Juan's clinical symptoms could be consistent with mitochondrial encephalomyopathy, lactic acidosis, and stroke-like episodes (MELAS), a maternally inherited mitochondrial disorder.

  Mitochondrial disorders are a heterogeneous group of conditions with often multisystemic presentation involving high energy requiring organ systems. Mitochondria are double-membrane cellular organelles that house the electron transport chain, which functions to generate cellular ATP (energy). Mitochondrial diseases result from impaired oxidative phosphorylation (decreased synthesis of ATP by the electron transport chain) (Section 2.5). Mitochondria contain more than 1500 proteins that are encoded by either the mitochondrial genome or the nuclear genome. Thus, primary mitochondrial disease may be due to pathogenic variants in mitochondrial DNA or nuclear DNA and can have maternal, autosomal recessive, autosomal dominant, or X-linked inheritance (Section 2.7). Secondary mitochondrial dysfunction may also result from a variety of diseases including neurodegenerative disorders, cancer, obesity, and other conditions.

  Regarding MELAS, the maternal family history of diabetes, hearing loss, cardiomyopathy, and migraine headaches is consistent with this diagnosis. MELAS is a multisystem disorder and most patients develop symptoms between ages 2 and 40 years, with childhood-onset being most common. The most common initial presenting symptoms in more than 25% of cases are seizures, recurrent headaches, stroke-like episodes, cortical vision loss, muscle weakness, recurrent vomiting, and short stature. Other initial clinical manifestations include altered consciousness, impaired mentation, hearing impairment, diabetes mellitus (type 1 or 2), developmental delay, and fever. In Juan's case, he presented with symptoms of stroke-like episodes, muscle weakness, short stature, headaches, and developmental delay.

  MELAS is caused by a pathogenic variant in mitochondrial DNA. It displays maternal inheritance and exhibits heteroplasmy. Heteroplasmy refers to the fact that a cell may have some mitochondria with a pathogenic variant and some without. The proportion of mitochondrial DNA with the pathogenic variant is the percent heteroplasmy. The percentage of heteroplasmy of the pathogenic variant plays an important role in the penetrance (Sections 2.5, 2.8) and severity of the disease. If Juan is found to have a pathogenic variant (Section 7.1.3) causative of MELAS, other members of his maternal family are at risk for carrying this variant, and women carrying this variant are at risk of passing it on to all of their offspring. Written informed consent is obtained for mitochondrial DNA sequencing (Section 6.4.2), given the strong suspicion for MELAS based on Juan's presentation and family history.
- **Results**: Juan is found to have 90% heteroplasmy in blood for the pathogenic variant (Section 7.1.3) *MT-TL1*, m.3243A > G. *MT-TL1* encodes mitochondrial leucine tRNA, and the m.3243A > G variant is the most common cause of MELAS, accounting for approximately 80% of MELAS cases. Juan's sister is tested and is found to have 65% heteroplasmy for the m.3243A > G pathogenic variant in her blood, and his mother is found to have 75% heteroplasmy in her blood.

  The combination of deafness, diabetes, and the mtDNA mutation in Juan's mother is consistent with a diagnosis of Maternally Inherited Diabetes and Deafness (MIDD), which has the same genetic etiology as MELAS but possibly associated with a lower mutation heteroplasmy level. It is not clear at this time if Juan's sister has a clinical diagnosis of mitochondrial disease, although the migraines are suspicious as a clinical presentation.
- **Management and Recommendations**: There is no cure for MELAS. Juan will be treated supportively. The disease progresses with episodic deterioration. Febrile illness and stress may trigger exacerbations.

  L-arginine has been shown to decrease severity of stroke-like symptoms, enhance microcirculation, and decrease ischemic damage in patients with MELAS. In the acute setting, L-arginine should be given intravenously. It can also be given daily orally to reduce the frequency of decompensations. A mitochondrial cocktail consisting of vitamins and supplements such as coenzyme Q10, riboflavin, L-carnitine, and creatine may be prescribed to aid mitochondrial function. Additionally, intercurrent illness and dehydration are treated quickly and aggressively with antipyretics and IV fluid rehydration.

Neurologic manifestations include stroke-like episodes, dementia, epilepsy, migraine headaches, sensorineural hearing impairment, peripheral neuropathy, developmental delays, psychiatric illness, and myopathy. Seizures generally respond to symptomatic treatment with antiepileptic medications, but valproic acid should be avoided, as it has been reported to cause mitochondrial toxicity and hepatopathy and/or encephalopathy in mitochondrial disease patients. Sensorineural hearing loss has been treated with cochlear implants in some cases.

Other multisystem manifestations include cardiac manifestations (dilated, hypertrophic cardiomyopathy, cardiac conduction defects), gastrointestinal manifestations (cyclic vomiting, diarrhea, constipation, gastric dysmotility, intestinal pseudoobstruction, pancreatitis), endocrine manifestations (diabetes, short stature, hypothyroidism, hypogonadotropic hypogonadism, hypoparathyroidism), renal disease, pulmonary hypertension, dermatologic disease (vitiligo, diffuse erythema with reticular pigmentation, hypertrichosis), and chronic anemia. Treatment is symptomatic.

Annual ophthalmology, cardiology (echocardiogram and EKG), and audiology examinations as well as urinalysis and fasting blood glucose are recommended.

- **Caveats**: MELAS is a multisystem disorder with variable expressivity (Section 2.8) in affected individuals. In general, individuals with higher heteroplasmy levels of the mutation have earlier and more severe symptoms; however, this is not always the case. Additionally, heteroplasmy levels can differ from tissue to tissue. Therefore, clinical recommendations may change over time based on emerging symptoms. Furthermore, protocols for "asymptomatic carriers" of the MELAS mutation are not straightforward and are managed on a case-by-case basis.
- **Differential Diagnosis**: Clinical features seen in MELAS can be seen in a wide variety of other mitochondrial diseases including myoclonic epilepsy with ragged red fibers (MERRF) and POLG-associated disorders.
- **Implications for Other Family Members**: It is very difficult to predict clinical features of MELAS in individuals who are heteroplasmic for the mtDNA mutation. Juan's maternal family members are given appropriate clinical workup for clinical features that can be associated with the m.3243A > G mutation and are being followed for management of their symptoms.
- **Outcome**: Juan clinically recovers from his stroke-like event with minimal residual hemiparesis. His family is provided with an emergency plan for medical management in the setting of future changes in his medical or clinical status. He regularly follows up with pediatric neurology and metabolic genetics.

The Diagnosis

Clinical Diagnosis: Patient—mitochondrial encephalomyopathy, lactic acidosis, and stroke-like episodes (MELAS). Mother—maternally inherited diabetes and deafness (MIDD). Sister—migraines.

Genetic Diagnosis: Patient, Mother, and Sister—*MT-TL1* gene, heteroplasmic pathogenic variant.

Further Reading

El-Hattab, A.W., Adesina, A.M., Jones, J., Scaglia, F., 2015. MELAS syndrome: clinical manifestations, pathogenesis, and treatment options. Mol. Genet. Metabol. 116, 4–12. PMID 26095523.

Koenig, M.K., Emrick, L., Karaa, A., Korson, M., Scaglia, F., Parikh, S., et al., 2016. Recommendations for the management of strokelike episodes in patients with mitochondrial encephalomyopathy, lactic acidosis, and strokelike episodes. JAMA Neurol. 73, 591–594. PMID 26954033.

Case 11.6.6

Metabolic Cases: Hunger and Protein Intake in a Metabolic Condition

George A. Diaz

Department of Genetics and Genomic Sciences, Icahn School of Medicine at Mount Sinai, New York, NY, United States

Abstract

A 12-year-old male with phenylketonuria undergoes adjustment of diet.

The Case

Harry is a new patient to your pediatric practice who has moved with his family from another state. Harry is 12 years old and was identified as having phenylketonuria (PKU) by newborn screening. His phenylalanine levels have generally been in the goal range (<360 uM) but he has been having increasing difficulty complying with protein restriction (Section 5.4.3) as his physical activity level has increased with involvement with school sports and he frequently feels hungry. The family asks you if you know about any new developments in treatment for PKU. You assess Harry's growth and note that he is slightly overweight but his height is appropriate. You order screening blood work including hemoglobin, hematocrit, and dyslipidemia screening.

The Issues

- **Will Harry always need to be protein restricted or can he eat more natural protein now that he is becoming an adolescent?** PKU is a lifelong disorder. Longitudinal data have supported treatment for life based on neurocognitive deficits evident in treated individuals who are nonadherent to diet. Protein tolerance can vary based on rate of growth, but dietary restrictions in classical PKU are always stringent and require a metabolic specialist for management.
- **What are the long-term risks of PKU treatment for Harry's health?** The use of medical formulas and protein-restricted diets incur the risk of iatrogenic malnutrition, which can have consequences on growth and intellectual development. Bone health in adulthood is a significant concern in patients who are unable to take in adequate dietary calcium because of dietary protein (including dairy) restriction.
- **What is the next step?** Referral to a metabolic specialist is indicated for ongoing management of dietary phenylalanine restriction and consideration for new therapeutic options.

Available diet history, other laboratory studies, and a record of recent phenylalanine levels will be useful for the metabolic dietician to assess and make treatment recommendations.

Genomics in the Clinic. https://doi.org/10.1016/B978-0-12-816478-5.00052-2

The Referral

- **Genetic Evaluation**: Harry is evaluated by the metabolic geneticist who reviews his dietary control and reminds the family that PKU is a lifelong autosomal recessive condition requiring the dietary restrictions. Although dietary therapy has been the mainstay of treatment for decades, therapeutic options have been expanding in recent years (Section 5.4.3). An important recent addition has been sapropterin dihydrochloride, an analogue of tetrahydrobiopterin. This molecule is an essential cofactor for the phenylalanine hydroxylase (PAH) enzyme, which is deficient in PKU, and serves as a chaperone to rescue protein function for certain types of *PAH* variants. A genotype-independent enzyme substitution therapy has recently been approved which utilizes the bacterial enzyme phenylalanine ammonia lyase to reduce phenylalanine levels as a subcutaneous injectable formulation. While approved to treat PKU in adults, the side effect profile is significant and includes risk of anaphylaxis and the drug is not approved for pediatric use. Overall nutritional status is assessed including micronutrient testing.

  The family meets with the metabolic dietician and reviews Harry's protein intake, calorie intake, and micronutrient status. Based on Harry's growth and laboratory data, the family is reassured that he is growing well on the currently prescribed protein intake but that he will need ongoing monitoring, particularly as he will be undergoing rapid growth during adolescence.
- **Management and Recommendations**: The option of sapropterin dihydrochloride treatment is discussed with the family, who is interested. While long-term outcome data are limited, the safety profile of the drug is favorable. Unfortunately, a beneficial effect on increased dietary protein tolerance is only seen in about half of patients trialed on the drug. There are some genotype–phenotype associations with expected responsiveness to sapropterin dihydrochloride, so sequencing of the *PAH* gene (Section 6.2.2) is recommended prior to starting a therapeutic trial. The identification of *PAH* pathogenic variants is not essential prior to initiating a therapeutic trial but given the increased amount of testing that will be required to monitor for a response to therapy; this information can be helpful for families or patients weighing whether to try the drug.
- **Results**: Five weeks after ordering the molecular testing, Harry's *PAH* results are returned. He is heterozygous (Section 2.3.4) for two pathogenic variants in this gene (Section 2.7.1). He carries one variant (Section 7.1) inherited from his father that has not been described in other patients in the medical literature but is predicted to be deleterious by bioinformatics analysis, and one known variant inherited from his mother that has been reported to respond well to sapropterin dihydrochloride.
- **Caveats**: Despite the introduction of new therapeutic agents to increase natural protein tolerance, it is not unusual for an adolescent patient with PKU to assert autonomy by ingesting increased amounts of natural protein in social settings, resulting in persistently elevated phenylalanine levels. A subset of patients may manifest neurological and physical symptoms of phenylalanine excess such as lighter hair pigmentation, tremor, anxiety, and executive dysfunction.
- **Outcome**: Harry is treated with 1 mg/kg of sapropterin dihydrochloride daily for 4 weeks with weekly phenylalanine monitoring. His phenylalanine levels decrease on his prescribed diet by more than 30% and his diet is liberalized to introduce additional protein in a stepwise manner. He is happy that he can eat additional protein items in his diet, though he still has to be very conscious of his daily protein intake. He will continue to be monitored by the metabolic team and adjustments to his dietary protein intake made as he progresses through adolescence.

The Diagnosis

Clinical diagnosis: Phenylketonuria, classic form.
Genetic diagnosis: *PAH* gene, compound heterozygous pathogenic variants.

Further Reading

Longo, N., Dimmock, D., Levy, H., Viau, K., Bausell, H., Bilder, D.A., et al., 2019. Evidence- and consensus-based recommendations for the use of pegvaliase in adults with phenylketonuria. Genet. Med. 21, 1851–1867. PMID 30546086.

Muntau, A.C., Adams, D.J., Belanger-Quintana, A., Bushueva, T.V., Cerone, R., Chien, Y.H., et al., 2019. International best practice for the evaluation of responsiveness to sapropterin dihydrochloride in patients with phenylketonuria. Mol. Genet. Metabol. 127, 1–11. PMID 31103398.

Newborn Screening, 2012. Newborn Screening ACT Sheet [Increased Phenylalanine] Phenylketonuria (PKU) [Online]. American College of Medical Genetics and Genomics. Available: https://www.acmg.net/PDFLibrary/Phenylalanine.pdf [Accessed January 2022].

Vockley, J., Andersson, H.C., Antshel, K.M., Braverman, N.E., Burton, B.K., Frazier, D.M., et al., 2014. Phenylalanine hydroxylase deficiency: diagnosis and management guideline. Genet. Med. 16, 188–200. PMID 24385074.

Section 11.7

Endocrine and Growth Cases

Case 11.7.1

Endocrinologic and Growth Cases: Overgrowth

Rosanna Weksberg[1,2] and Cheryl Cytrynbaum[3,4]

[1]Division of Clinical and Metabolic Genetics, Genetics and Genome Biology, The Hospital for Sick Children, University of Toronto, Toronto, ON, Canada; [2]Departments of Pediatrics and Molecular Genetics, University of Toronto, ON, Canada; [3]Department of Genetic Counselling, Genetics and Genome Biology, The Hospital for Sick Children, Univeristy of Toronto, ON, Canada; [4]Department of Molecular Genetics, Univeristy of Toronto, ON, Canada

Abstract

A 1-week-old girl is noted to have large size for gestational age, hypoglycemia, and a large tongue and is found to have Beckwith–Wiedemann syndrome due to paternal uniparental disomy of 11p15.5.

The Case

Eliza is seen at 1 week of age for her first assessment, by you, a pediatrician. She has been referred to you by her family physician who has requested your opinion regarding her advanced growth parameters, specifically height and weight measurements which plot above the 97th percentile. Her birth weight at term was 5.2 kg and her birth length was 56 cm. Review of the neonatal history is remarkable for hypoglycemia which responded well to dextrose infusion. There were no other neonatal concerns reported. On examination you note that Eliza has a small umbilical hernia and that her tongue protrudes from her mouth at rest. The remainder of your assessment is unremarkable.

The Issues

- **How concerned are you by her clinical findings?** There are several different reasons for a baby to be large for gestational age (LGA). Maternal diabetes is the most common cause of large for gestational age. Review of the prenatal records indicates that there was no history of gestational diabetes and the maternal weight gain during pregnancy was 15 kg (33 lb). Some babies are large because their parents are large. Eliza's growth pattern is out of keeping with the family history as both parents had average birth weights and their adult heights both plot at the 50th percentile. You decide that Eliza has macrosomia. In addition, you suspect that she has macroglossia as the etiology of her protruding tongue.
- **What is the significance of her large size?** Given the combination of macrosomia, macroglossia, and neonatal hypoglycemia (in the absence of gestational diabetes), you suspect that Eliza may have an overgrowth syndrome. Overgrowth syndromes comprise a heterogeneous group of disorders with overlapping clinical manifestations, including a variety of systemic anomalies. Transient neonatal hypoglycemia, although common in large for gestational age babies, also occurs with increased frequency in several overgrowth syndromes. You recall that early diagnosis is

Genomics in the Clinic. https://doi.org/10.1016/B978-0-12-816478-5.00053-4

important because several overgrowth syndromes are associated with an increased risk for embryonal tumors, such as Wilms tumor and hepatoblastoma.

- **Will there be any implications for her medical management?** You explain that if Eliza has an overgrowth syndrome, she may be at risk for certain tumors. The identification of a specific syndromic diagnosis is of critical importance to clarify whether this risk applies to Eliza and to ensure appropriate surveillance for such tumors is initiated, if indicated. A genetic evaluation can help establish a clinical diagnosis and guide genetic testing. You make a referral to pediatric genetics.

The Referral

- **Genetic Evaluation**: Eliza and her parents are seen in the pediatric genetic clinic by the clinical geneticist and genetic counselor 3 weeks later. The pregnancy, medical, and family histories are reviewed and a detailed physical examination is performed by the geneticist. Eliza's weight and length continue to plot above the 97th percentile. Her head circumference plots at the 75th percentile. In addition to the previously noted umbilical hernia and macroglossia (Fig. 11.7.1.1), the geneticist also notes anterior linear ear lobe creases (Fig. 11.7.1.2). Her limbs appear symmetric.
- **Genetic Counseling**: The geneticist and genetic counselor review the findings and return to discuss their impression with Eliza's parents. They strongly suspect that Eliza has Beckwith–Wiedemann syndrome (BWS). BWS is a growth disorder variably characterized by neonatal hypoglycemia, macrosomia, macroglossia (Fig. 11.7.1.1), hemihyperplasia, omphalocele, visceromegaly, adrenocortical cytomegaly, renal abnormalities (e.g., medullary dysplasia, nephrocalcinosis, medullary sponge kidney, and nephromegaly), and ear creases/pits (Fig. 11.7.1.2). There is an increased risk for embryonal tumors (e.g., Wilms tumor, hepatoblastoma, neuroblastoma, and rhabdomyosarcoma). BWS is considered a clinical spectrum, in which affected individuals may have a variable number of the characteristic clinical features. Development is usually normal in children with BWS unless there is an associated chromosome abnormality, brain malformation, or history of hypoxia or significant untreated hypoglycemia. Although consensus clinical diagnostic criteria for BWS do not exist, a clinical diagnosis is typically established when an individual has at least three of the major criteria, including Eliza's features of macrosomia, macroglossia, umbilical hernia, and anterior ear lobe creases. The increased risk for embryonal tumors is 7.5% particularly in the first 8 years of life. Tumor development in affected individuals older than age 8 years is uncommon. Screening recommendations are discussed below. Other types of tumors can also occur in these children and must be considered if there are relevant presenting findings, e.g., rhabdomyosarcoma.

  BWS is caused by molecular alterations that impact the regulation of imprinted genes (Sections 2.7.5, 6.4.3) in the chromosome 11p15.5 region. There are several different genetic mechanisms that can cause BWS, including DNA methylation changes, paternal uniparental disomy, pathogenic variants (Section 7.1.3) in the *CDKN1C* gene (on chromosome 11p15.5), and genomic alterations of chromosome 11p15.5. A molecular cause is identified in about 80% of individuals. Most of the time, it is a sporadic condition (85%) in which case the recurrence risk is less than 1%, but for some families the recurrence risk may be as high as 50% (e.g., female with BWS associated with pathogenic variant in the *CDKN1C* gene). A molecular testing strategy is discussed below.

FIGURE 11.7.1.1 4-month-old female with Beckwith–Wiedemann syndrome presenting with macroglossia.

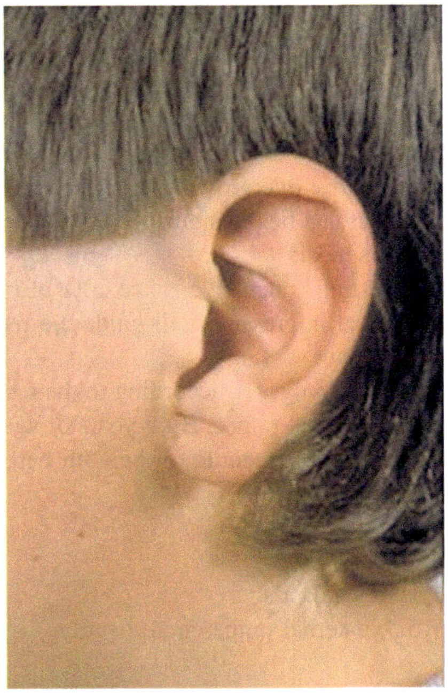

FIGURE 11.7.1.2 Anterior linear ear lobe crease in a child with Beckwith–Wiedemann syndrome.

- **Management and Recommendations**: Although most individuals with BWS show rapid growth in early childhood, height typically remains at the upper range of normal. Growth rate usually appears to slow around ages 7–8 years. Children with macroglossia may benefit from tongue reduction surgery in infancy or early childhood (if indicated) and from speech therapy. Recommendations for screening for embryonal tumors include abdominal ultrasound examination every 3 months until age 8 years for early detection of Wilms tumor and hepatoblastoma, and measurement of serum alpha-fetoprotein (AFP) concentration every 3 months in the first 4 years of life for early detection of hepatoblastoma. Notably, there are currently disparities in recommendations across different publications and in their implementation across different centers. Between age 8 years and midadolescence annual renal ultrasound examinations are recommended to identify findings such as nephrocalcinosis and medullary sponge kidney disease requiring further evaluation. Eliza is referred for abdominal ultrasound, which should be ordered every 3 months by her pediatrician. See references below for additional information regarding medical management.

  Alpha-fetoprotein (AFP) serum concentration may be elevated in children with BWS in the first year of life. If the AFP is elevated and imaging reveals no suspicious lesion, follow-up measurement of serum AFP concentration plus baseline liver function tests 1 month later can be used to determine the trend in serum AFP concentrations over time. If the concentration is not decreasing, it is appropriate to refer to oncology for an extensive search for an underlying tumor. Eliza also is referred for an AFP level, which should be ordered every 3 months by her pediatrician.

  Hemihyperplasia (Section 11.10.1) that may involve the leg, arm, face, or tongue can generally be appreciated at birth but may become more or less evident as the child grows. If hemihyperplasia results in a significant difference in leg length, surgery may be necessary during early puberty to equalize the final leg lengths.
- **Genetic Testing**: Genetic testing should begin with DNA methylation analysis (by MS-MLPA—Section 6.4.3) which can identify a methylation abnormality in 75% of individuals. Further genetic studies may need to be undertaken to define the specific mechanism that caused the methylation abnormality (i.e., testing for uniparental disomy or a chromosome abnormality—Sections 2.2.1, 2.7.5). If the DNA methylation analysis is normal, and the patient has a cleft palate or a positive family history, then sequencing (Section 6.2.2) of the *CDKN1C* gene can be done. Genetic testing is typically done on a blood sample, but in some cases the genetic change may be present in a mosaic state (present in some cells and not others—Section 2.2.2); therefore testing of other tissues (i.e., buccal, skin) can be considered if indicated. Eliza is found to have paternal uniparental disomy of chromosome 11p15.5.
- **Differential Diagnosis**: There are several overgrowth syndromes which have some clinical overlap with BWS including Simpson–Golabi–Behmel syndrome (characterized by pre- and postnatal overgrowth, congenital anomalies,

organomegaly, distinctive facial features, and increased tumor risk), Perlman syndrome (characterized by neonatal macrosomia, organomegaly, distinctive facial features, and increased risk for Wilms tumor), Costello syndrome (characterized by cardiac anomalies, coarse facial features, hypotonia, and neonatal edema that may be deemed "macrosomia" followed by onset of failure to thrive and loose skin as well as increased tumor risk), Sotos syndrome (characterized by overgrowth, distinctive facial features, and intellectual disability—Section 11.7.2), and mosaic genome-wide paternal uniparental isodisomy (characterized by overgrowth, hemihyperplasia, hyperinsulinism, and increased tumor risk—Section 2.7.5). Most of these conditions can be distinguished on a clinical basis, but in some cases genetic testing may be required to establish a definitive diagnosis.

- **Caveats**: In the event that Eliza's genetic testing is negative, since 20% of individuals with BWS do not have an identified genetic etiology, the geneticist's clinical impression will guide the recommendations for medical management, specifically the indication for tumor surveillance.
- **Outcome**: Eliza follows the tumor surveillance program, according to the recommended protocol, until completing it at 8 years of age, and does not develop an embryonal tumor. By 5 years of age her growth has slowed, and at 8 years of age her height has dropped from above the 97th percentile to the 75th percentile. She does well in school.

The Diagnosis

Clinical Diagnosis: Beckwith–Wiedemann syndrome.
Genetic Diagnosis: Chromosome 11p15.5, paternal uniparental disomy.

Further Reading

Kalish, J.M., Doros, L., Helman, L.J., Hennekam, R.C., Kuiper, R.P., Maas, S.M., et al., 2017. Surveillance recommendations for children with overgrowth syndromes and predisposition to Wilms tumors and hepatoblastoma. Clin. Cancer Res. 23, e115–e122. PMID 28674120.

Shuman, C., Beckwith, J.B., Weksberg, R., March 3, 2000. Beckwith-Wiedemann syndrome. Adam, M.P., Ardinger, H.H., Pagon, R.A., Wallace, S.E., Bean, L.J.H., Stephens, K., Amemiya, A., (editors). In GeneReviews. University of Washington, Seattle (WA), 1993–2022 [updated August 11, 2016]. PMID 20301568.

Case 11.7.2

Endocrinologic and Growth Cases: Macrocephaly and Tall Stature in a Toddler

Antonie D. Kline

Harvey Institute for Human Genetics, Greater Baltimore Medical Center, Baltimore, MD, United States

Abstract

A 15-month-old male with macrocephaly, tall stature, and delayed speech is found to have Sotos syndrome.

The Case

Jaden comes in for his check-up with you, his family physician, at 15 months of age. You recall that he has always been healthy, although initially had feeding difficulty and jaundice, and prenatally there were no complications. His parents report that he is only using one or two specific words such as "mama" and "no", but he seems to understand more than that. He is not yet walking independently and has just started feeding himself with a spoon. He has always been tall for his age, even at birth, and he has had macrocephaly since the prenatal period. You note that he weighs 31 pounds, he is 34 inches long, and his head circumference is 50 cm, all greater than the 95th percentiles.

The Issues

- **How concerned are you about his growth?** You feel somewhat concerned, especially because his parents are of average size, but also reassured because he has always been above but parallel to the 95th percentile growth curves, following his own curve, and overall physically he looks normal. You are aware that macrocephaly can be associated with autism spectrum disorder, but behaviorally he seems interactive.

You could obtain a bone age X-ray to assess bone age compared to chronologic age. If advanced, it is more likely due to an underlying syndrome. You also could consider hormonal studies.

- **How concerned are you about his development**? You discuss that he is developmentally slightly behind his expected milestones and that you want to refer him for an evaluation to the state program for early intervention for infants and toddlers, and likely for early intervention therapy. You note that he has both growth and development affected.
- **Would you refer to a specialist at this point**? You tell his parents that you can wait and see how he is doing at his 18-month-old check-up, and consider a genetic evaluation if he has not made progress. His parents are starting to think about having another baby and say they would prefer to be referred at this time. You refer him for a genetic evaluation.

Genomics in the Clinic. https://doi.org/10.1016/B978-0-12-816478-5.00054-6

It is helpful for you to send a copy of the growth charts and records prior to the child being seen by genetics, as well as results of any studies that you have done. It may be helpful for the family to know ahead of time that this will be a long visit, with very thorough discussion of history and family history of three generations, as well as a very detailed examination with lots of measurements, and findings documented by photography.

The Referral

- **Genetic Evaluation**: Jaden and his parents are seen in the pediatric genetic clinic by the pediatric geneticist and genetic counselor 6 weeks later. History is reviewed and a family history taken. His mother had preeclampsia but no other complications and he was born by Cesarean section for failure to progress. No history of abnormal growth or development is noted on family history; the mother grew slowly as a child, she is 5′4″ in height, and the father is 5′11″ in height and both of their head circumferences are in the normal range. A physical exam is performed, and the patient is photographed. On exam, growth parameters are noted to be greater than 95th percentile with head greater than 98th percentile (nearly 4 SD above the mean). Facies is long and minimally dysmorphic with tall forehead, sparse frontotemporal hair, slightly downslanting palpebral fissures, and small pointed chin (Fig. 11.7.2.1). Hands are normal and feet appear flat. Neurologically, he has mild hypotonia.

Baby photographs of the parents and other family members can be helpful for the parents to bring to the visit, especially if they perceive a strong resemblance to the child.

- **Genetic Counseling**: The geneticist and genetic counselor confer and then discuss with Jaden's parents that he may have an underlying genetic syndrome causing his tall stature, and macrocephaly, and they think that he most likely has a condition called Sotos syndrome. It includes tall stature and macrocephaly, and typical facial features like Jaden's. Hypotonia is often present and can lead to delayed motor skills. There also is often some degree of learning disability, including some children with intellectual disability and/or autism spectrum disorder. This is an autosomal dominant (Section 2.7.1) disorder that is most often sporadic but can be inherited in about 5% of the time. Genetic testing is available, and most patients will have pathogenic variants or mutations in, or larger deletions (Section 2.4.1) including, the *NSD1* gene. There is always a small possibility that one of his parents carries a germline mutation and thus is mosaic (Section 2.4.2), giving a slightly increased risk for recurrence in future children. It is important to do genetic testing, not only because of this but also because in some overgrowth disorders (Section 5.3) there is high risk for embryonal tumors that are treatable when detected early.

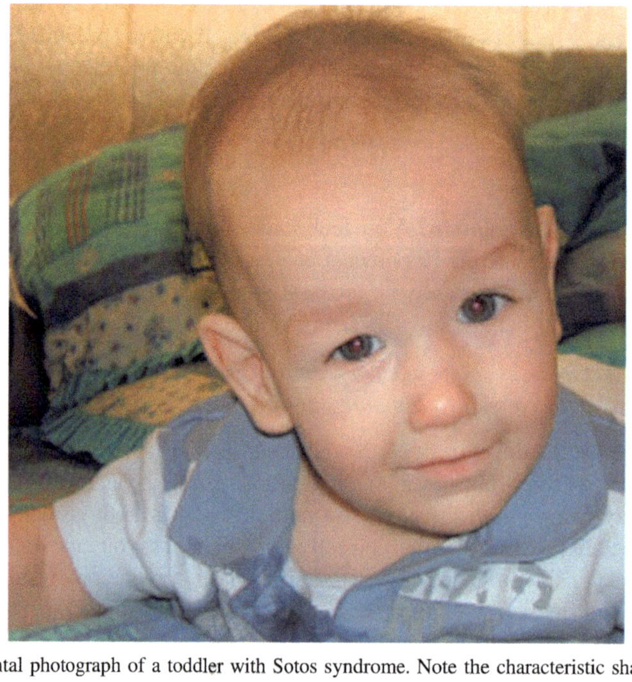

FIGURE 11.7.2.1 Frontal photograph of a toddler with Sotos syndrome. Note the characteristic shape of the face and head.

- **Management and Recommendations**: The geneticist concurs with the referral for early intervention evaluation and therapy. Bone age X-ray is recommended to confirm advanced bone age, typically seen in prepubertal Sotos syndrome. Echocardiogram is recommended to rule out congenital heart disease, found in up to 20% of individuals with this condition. Renal ultrasound is recommended to rule out renal anomaly. He will need to be clinically followed for scoliosis with aging. In some cases of overgrowth there is an increased risk for embryonal tumor, with need for abdominal ultrasounds every 3 months at least until age 8 years, and alpha-fetoprotein measurement in blood every 3 months until age 4 years, but this is generally not recommended in Sotos syndrome because the overall incidence of this type of tumor is low. Referral for physical therapy to help with the hypotonia is recommended. MRI of the brain can be done to assess for anomalies but would not be recommended at this time. There is a parent support organization (Section 3.5) to which the parents are referred if interested, for further information.
- **Genetic Testing**: Gene sequencing (Section 6.2.2) for the *NSD1* gene will detect up to 90% of cases. Insurance coverage approval is recommended before sending. Jaden's parents agree to this, approval is received, and testing is sent. If negative, a macrocephaly/overgrowth panel (Section 6.2.3) can be sent, available from various genetic laboratories.
- **Results**: Jaden is found to have a pathogenic variant (Section 7.1.3) in the *NSD1* gene, confirming that he has Sotos syndrome. This information can be used for prenatal testing if the parents are interested.
- **Differential Diagnosis**: There are other overgrowth conditions, although most also exhibit increased weight; these include Beckwith–Wiedemann syndrome, an imprinting disorder with organomegaly, omphalocele, and risk for embryonal tumors (Section 11.7.1); Weaver syndrome, which overlaps with Sotos syndrome but has different facies and is autosomal recessive (Section 2.7.1); and Simpson–Golabi–Behmel syndrome, with coarse facies and more internal organ malformations, and which is X-linked (Section 2.7.2). Individuals with *PTEN* variants present with macrocephaly and also may have autism spectrum disorder, and are at risk for specific cancers (Section 11.17.1). Mutations in additional genes (e.g., *SETD2*) can present with similar features to Sotos syndrome.
- **Caveats**: When a parent of a child with overgrowth is tall with macrocephaly, the child could still have an underlying genetic condition with more than just these familial traits. In this instance, 5% of the time a parent will be affected.
- **Outcome**: Jaden enrolls in early intervention and makes good progress with speech and language. He continues to have very mild hypotonia, and is walking by 18 months. He does not develop signs of autism spectrum disorder, but he remains somewhat immature compared to his peers. At age 3 years he develops seizures. A brain MRI reveals dilated ventricles and absence of the corpus callosum. In elementary school he is taller than his peers (Fig. 11.7.2.2).

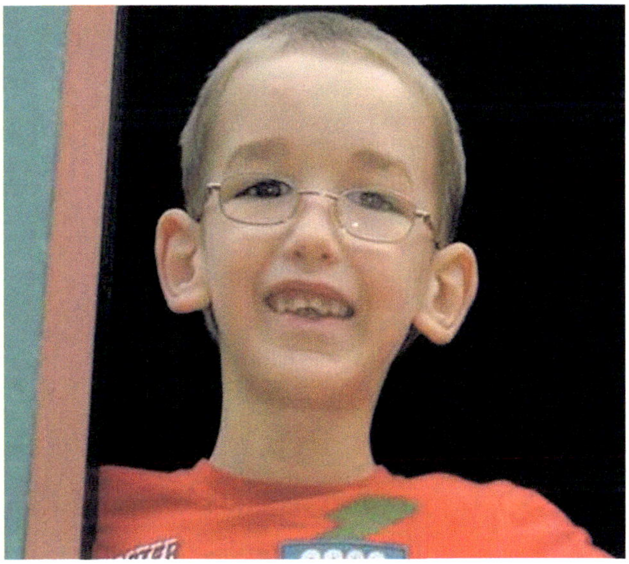

FIGURE 11.7.2.2 An older child with Sotos syndrome still has sparse frontotemporal hair, broad forehead, and downslanting palpebral fissures.

The Diagnosis

Clinical Diagnosis: Sotos syndrome.
Genetic Diagnosis: *NSD1* gene, heterozygous pathogenic variant.

Further Reading

Brioude, F., Toutain, A., Giabicani, E., Cottereau, E., Cormier-Daire, V., Netchine, I., 2019. Overgrowth syndromes—clinical and molecular aspects and tumour risk. Nat. Rev. Endocrinol. 15, 299–311. PMID 30842651.

Nicita, F., Ruggieri, M., Polizzi, A., Mauceri, L., Salpietro, V., Briuglia, S., et al., 2012. Seizures and epilepsy in Sotos syndrome: analysis of 19 Caucasian patients with long-term follow-up. Epilepsia 53, e102–e105. PMID 22352460.

Tatton-Brown, K., Cole, T.R.P., Rahman, N., December 17, 2004. Sotos syndrome. Adam, M.P., Ardinger, H.H., Pagon, R.A., Wallace, S.E., Bean, L.J.H., Stephens, K., Amemiya, A., (editors). In GeneReviews. University of Washington, Seattle (WA), 1993–2022 [updated August 1, 2019]. PMID 20301652.

Case 11.7.3

Endocrinologic and Growth Cases: Diabetes

Toni I. Pollin[1,2]

[1]*Medicine and Epidemiology and Public Health, University of Maryland School of Medicine, Baltimore, MD, United States;* [2]*Human Genetics, University of Maryland School of Medicine, Baltimore, MD, United States*

Abstract

A preadolescent is found to have hyperglycemia incidentally, and on genetic testing has heterozygous glucokinase deficiency, known as *GCK*-MODY (maturity onset diabetes of the young), or MODY2, maternally inherited.

The Case

Cory, the 12-year-old son of a woman with type 2 diabetes (diabetes mellitus, type 2), presents to you, his pediatrician, with elevated blood sugar. Both he and his mother are lean and physically active. His mother was first found to have diabetes at the age of 28 during her first pregnancy when she was diagnosed with gestational diabetes with an elevated fasting glucose. She was treated with insulin throughout the pregnancy. Following the birth of her first child, a daughter (now 14 years and healthy) who weighed 10 pounds 4 ounces, the mother's hyperglycemia persisted, and she was put on metformin, which did not make her glucose levels normal, but now her fasting glucose is stable at about 105 mg/dL. During her pregnancy with her son, she went back on insulin and attempted to control her glucose level strictly, though it was still hard to get it with the normal range. Cory weighed 6 pounds 1 ounce at birth. He has been healthy with normal growth and development. Recently, however, he was curious about what his blood sugar was and tested himself and found it was elevated, about 110 mg/dL.

The Issues

- **Is Cory on his way to developing type 1 diabetes?** Since home glucose monitoring devices could be inaccurate, you order a laboratory fasting glucose and a hemoglobin A1C (HBA1c) to confirm and find that he has a fasting glucose level of 108 mg/dL and a HBA1c of 6.1%. He seems too young for type 2 diabetes, especially given that he is lean and active. Type 1 diabetes requires treatment with insulin. You wonder if this was a fortuitous finding of elevated fasting glucose that might prompt treatment with insulin.
- **Why was a lean young woman diagnosed with type 2 diabetes?** You learn that when her diabetes persisted after pregnancy, the mother was suspected to have type 1 diabetes and tested for several diabetes autoantibodies, and all were negative. She also had a positive C-peptide 1 year after delivery. Even though she is lean, this is consistent with type 2 rather than type 1 diabetes, hence this diagnosis.

Genomics in the Clinic. https://doi.org/10.1016/B978-0-12-816478-5.00055-8

- **Is there anyone else in the family with diabetes?** The mother says that her mother and "lots of people" in her family on her mother's side have diabetes, including a great-uncle who is 100 years old!
- **Should this family be referred to other clinical specialists?** You refer Cory to pediatric endocrinology. He is tested for a number of diabetes autoantibodies and is found to be negative for GAD65, IA-2, and ZnT8. His C-peptide is found to be normal. He also undergoes a 75-g oral glucose tolerance test as an additional measure of his glycemia. Consistent with previous measurements, his fasting glucose is in the prediabetic range (108 mg/dL), but his 2-hour glucose level is only 135 mg/dL.

The American Diabetes Association recommends consultation with a center specializing in diabetes genetics for individuals suspected of having monogenic diabetes. You refer Cory to such a center at the local university.

*The number of such centers for diabetes genetics is limited. The Monogenic Diabetes Research and Advocacy Consortium (*http://www.mdrac.org*) is developing a network of resources in this area. If unable to obtain a consultation for the patient with a diabetes genetic specialist, consulting with a pediatric endocrinologist as well as a genetic counselor in a collaborative manner is likely to be helpful. Endocrinology appointments are likely to have shorter wait times than specialized genetic appointments.*

The Referral

- **Genetic Evaluation**: Cory and his mother are evaluated by the endocrine genetic specialist and genetic counselor. The history is reviewed, and a physical exam is performed on the boy. The genetic counselor takes a detailed three-generation family history and learns that the mother's mother (diagnosed at 38), two maternal uncles (diagnosed at 40 and 42), mother's maternal grandmother (diagnosed at 25), and one maternal great-aunt (diagnosed at 45 and now deceased) and one great-uncle (diagnosed at 27 and now 100 years old) all have a diagnosis of type 2 diabetes. All are lean except the great-aunt, who was reportedly very overweight. The mother's paternal great-grandfather also had type 2 diabetes, was overweight, and died of a heart attack at age 75. Some of these relatives are taking various oral medications, one is on insulin, and two are watching their diet and activity levels. Other than the great-grandfather, none have experienced any complications of diabetes.
- **Genetic Counseling**: The endocrine genetic specialist and genetic counselor discuss that several features suggest that the patients in this family may be among about 0.4% of diabetic individuals who have monogenic diabetes, typically autosomal dominant (Section 2.7.1). The most common forms of these are due to pathogenic variants in genes encoding pancreatic beta cell transcription factors and the enzyme glucokinase. These features include a diagnosis of type 2 diabetes as in the mother without usual risk factors for type 2 (age, adiposity), incidental mild hyperglycemia as in Cory, and strong suggestion of autosomal dominant inheritance pattern of often young onset, non-insulin-requiring diabetes.

  The lack of complications and stable mild hyperglycemia particularly suggest heterozygosity for a mutation in the glucokinase gene, causing *GCK* diabetes, or maturity onset diabetes of the young, type 2 (MODY2), which is distinct and should not be confused with diabetes type 2. Glucokinase is an enzyme that catalyzes the phosphorylation of glucose in the pancreatic beta cell (as well as hepatic cells), which is required for the initiation of glycolysis and ultimately the secretion of insulin. Individuals with MODY2 or heterozygous glucokinase deficiency are generally not at risk for diabetes complications, with the possible exception of mild retinopathy which has not been shown to be clinically significant. The family is counseled about the inheritance pattern, and 50% chance of an affected child if the result is positive.
- **Management and Recommendations**: *GCK* diabetes rarely requires treatment to prevent complications, so a molecular diagnosis is important to determine if there is a need for management. Although *GCK* diabetes is most likely, panel testing (Section 6.2.3) for multiple monogenic diabetes genes is probably the best course of action because panel testing is now usually not more expensive than single gene testing and this could be an atypical presentation of another form of monogenic diabetes. The most common other forms of monogenic diabetes are *HNF1A* or *HNF4A* diabetes, which are progressive and usually require treatment with low dose sulfonylureas and sometimes insulin.
- **Results**: The genetic counselor chooses to order testing on Cory first because of the utility of a diagnosis for management. A multigene sequencing panel uncovers a pathogenic nonsense variant (Sections 2.4.1, 7.1.3) in the *GCK* gene, leading to a diagnosis of glucokinase diabetes, or MODY2. The mother undergoes targeted single site testing (Section 6.2.1) for Cory's variant and tests positive.

  The family is counseled that, barring any other events or risk factors, Cory's and his mother's hyperglycemia resulting from the *GCK* mutation is a benign condition that does not require treatment. It is important that the family and all of his care providers are aware of this so that a future detection of hyperglycemia, such as during an emergency room admission, does not incorrectly trigger a diagnosis of type 1 diabetes.

- **Caveats**: The finding that Cory and his mother have *GCK* diabetes is very good news because it indicates that the explanation for their hyperglycemia is heterozygous glucokinase deficiency. It is important for individuals with this condition to understand, however, that they are at least at population risk for type 1 and type 2 diabetes and their complications. They are not "protected" from type 2 diabetes. It is still important to maintain a healthy lifestyle, since obesity and a sedentary lifestyle are risk factors for type 2 diabetes, which is fairly common in the general population in the United States.

  For future pregnancies for the mother or other affected individuals in the family, the obstetricians should be made aware of the *GCK* mutation. Maternal mutations raise the risk of macrosomia and associated perinatal complications when the fetus does not inherit the mutation. *Because of the impact of a fetal variant on reducing ability to take up glucose, affected babies of affected mothers tend to have normal birthweight or even low birthweight if the maternal glucose is very strictly controlled.* It is important to be aware of the potential danger of overtreating hyperglycemia in affected women carrying affected fetuses for decreasing birthweight. Also, fetal variants raise the risk of low birthweight, even when the mother does not carry the mutation.

  Of note, individuals who are homozygous or compound heterozygous (Section 2.7.1) for *GCK* loss of function variants (Section 7.1.3) usually have neonatal diabetes requiring insulin from birth. Thus couples where both are heterozygous with MODY2 have a 25% risk for having a child with neonatal diabetes.
- **Implications for Other Family Members**: It also is recommended that the maternal relatives, particularly those who are diabetic, undergo single site testing (Section 6.2.1) to determine if the *GCK* mutation is the cause of their diabetes. Relatives who have not been diagnosed with diabetes or prediabetes and have never had their blood sugar tested should undergo glucose and/or HbA1c testing before having testing for the *GCK* variant if they want to know their status. This is also relevant for Cory's 14-year-old sister, who should be made aware of her risk of having the mutation so it can be considered as an explanation if her blood sugar tests high.

Of note, the sister's high birthweight does reduce the likelihood of having the monogenic diabetes, since unaffected offspring of mothers with GCK mutations tend to have high birthweight because of the usual impact of a hyperglycemic environment on a genotypically normal fetus.

- **Differential Diagnosis**: The differential diagnosis includes: type 1 diabetes (T1D), type 2 diabetes (T2D), other forms of monogenic diabetes including other forms of MODY, syndromic diabetes, recurrence of missed transient neonatal diabetes, transient hyperglycemia, and lipodystrophy (abnormal fat distribution, which can be genetic or acquired and can cause insulin resistance/diabetes).
- **Outcome**: The family understands the implications of the *GCK* variant for Cory and his mother, and potentially other relatives. You note in Cory's chart that his hyperglycemia is due to a *GCK* variant and he does not require treatment at his present glycemic levels, nor does he need to be seen for medical follow-up more than typical for a child his age. This may apply to his sister as well, depending on if she is tested and the outcome. You refer the mother back to her own PCP, who is able to take her off of metformin on the recommendation of the genetic and endocrine specialists. Cory does well, without further medical involvement.

The Diagnosis

Clinical Diagnosis: Patient and Mother—maturity onset diabetes of the young type 2.
Genetic Diagnosis: Patient and Mother—*GCK* gene, heterozygous pathogenic variant.

Further Reading

American Diabetes, A., 2020. 2. Classification and diagnosis of diabetes: standards of medical care in diabetes-2020. Diabetes Care 43, S14–S31. PMID 31862745.

Chakera, A.J., Steele, A.M., Gloyn, A.L., Shepherd, M.H., Shields, B., Ellard, S., et al., 2015. Recognition and management of individuals with hyperglycemia because of a heterozygous glucokinase mutation. Diabetes Care 38, 1383–1392. PMID 26106223.

Hattersley, A.T., Greeley, S.A.W., Polak, M., Rubio-Cabezas, O., Njolstad, P.R., Mlynarski, W., et al., 2018. ISPAD Clinical Practice Consensus Guidelines 2018: the diagnosis and management of monogenic diabetes in children and adolescents. Pediatr. Diabetes 19 (Suppl. 27), 47–63. PMID 30225972.

Stein, S.A., Maloney, K.L., Pollin, T.I., 2014. Genetic counseling for diabetes mellitus. Curr. Genet. Med. Rep. 2, 56–67. PMID 25045596.

Case 11.7.4

Endocrinologic and Growth Cases: Short Stature

Rosanna Weksberg[1,2] and Cheryl Cytrynbaum[3,4]

[1]*Division of Clinical and Metabolic Genetics, Genetics and Genome Biology, The Hospital for Sick Children, University of Toronto, Toronto, ON, Canada;* [2]*Departments of Pediatrics and Molecular Genetics, University of Toronto, ON, Canada;* [3]*Department of Genetic Counselling, Genetics and Genome Biology, The Hospital for Sick Children, Univeristy of Toronto, ON, Canada;* [4]*Department of Molecular Genetics, Univeristy of Toronto, ON, Canada*

Abstract

A 2-year-old male is referred for short stature with a history of small size for gestational age and feeding difficulties and is found to have Russell—Silver syndrome due to uniparental disomy of chromosome 7.

The Case

David is a 2-year-old male who has been referred to you for a pediatric assessment by his family physician because of concern regarding growth delay. He has a history of being small for gestational age with a birth weight of 2.5 kg and a birth length of 47 cm, at term. Postnatally he demonstrated growth deficiency, with height and weight dropping below the second percentile in the first 2 months of life. Parents report that David has had a small appetite since birth, with little interest in eating. They have seen a dietician and are fortifying his diet with calorie dense foods. His growth velocity over the last year has been stable, but his height and weight remain below the second percentile. On examination you note that David has relative macrocephaly with head circumference at the 50th percentile. There are no developmental concerns.

The Issues

- **Why is David's growth behind?** There are a wide variety of causes for delayed growth. David's growth trajectory will depend on the underlying cause for his growth delay. Some of the more common causes of growth delay include a family history of short stature, constitutional growth delay, hormonal abnormalities (i.e., growth hormone deficiency, hypothyroidism), or a genetic syndromic etiology. David's growth pattern is out of keeping with the projected midparental height, which is at the 50th percentile, and there is no family history of short stature. You send endocrine investigations, including thyroid function testing and growth hormone, as well as a complete metabolic profile, and all are normal. The fact that David' growth parameters were impacted before birth and that further decline in growth occurred in the first 2 months of life makes constitutional growth delay unlikely.
- **Is there any treatment for short stature?** Options for treatment depend on the cause of the short stature. The primary treatment for many types of significant short stature is growth hormone therapy, which has been shown to be beneficial

Genomics in the Clinic. https://doi.org/10.1016/B978-0-12-816478-5.00056-X

even in the absence of growth hormone deficiency. Eligibility for growth hormone treatment can be discussed with an endocrinologist once more information is known about the cause of short stature.

- **Is there a genetic cause for his short stature?** It is unclear at this time. The absence of a family history of short stature does not mean there is not a genetic cause. David's growth delay could be the result of a new genetic change in him, not necessarily something that was inherited. You refer him for a pediatric genetic evaluation.

The Referral

- **Genetic Evaluation**: David and his parents are seen in the pediatric genetic clinic by the clinical geneticist and the genetic counselor a few months later. The pregnancy, medical, and family histories are reviewed and a detailed physical examination is performed by the geneticist. In addition to the previous findings of growth delay (height and weight below the second percentile) with head sparing, the geneticist notes that David has a prominent forehead and a narrow chin, giving the impression of a "triangular facies" (Fig. 11.7.4.1). He also has bilateral fifth finger clinodactyly. There are symmetric limbs and no skin lesions. No developmental or cognitive concerns are identified.
- **Genetic Counseling**: The geneticist and genetic counselor review the findings and return to discuss their impression with David's parents. They suspect that David has Russell–Silver syndrome, which is a genetic condition characterized by prenatal and postnatal growth deficiency. Individuals with Russell–Silver syndrome are typically born small for gestational age with continued postnatal growth delay. Other characteristic features include relative macrocephaly, typical facial features including a prominent forehead and narrow chin ("triangular facies"), somatic asymmetry (with one limb being smaller due to hemihypoplasia), and feeding difficulties. Individuals with Russell–Silver syndrome also are at risk for hypoglycemia, particularly with any prolonged fast, as they have little subcutaneous fat and poor appetites. There is also an increased risk for developmental delay and learning difficulties.

  Russell–Silver syndrome is currently a clinical diagnosis based on a combination of characteristic features. There is extensive variability in the clinical presentation of Russell–Silver syndrome, and consensus clinical diagnostic criteria have not yet been established, although several clinical scoring systems have been proposed. The Netchine–Harbison scoring system proposed in 2015 is commonly used, which suggests that a clinical diagnosis of Russell–Silver syndrome is considered if an individual has at least four of six criteria, and for David these include his features of small for gestational age, postnatal growth failure, relative macrocephaly, protruding forehead, and feeding difficulties.

  There are two genetic mechanisms which are known to cause Russell–Silver syndrome. The more common mechanism is the impact from molecular alterations on the regulation of imprinted genes (Sections 2.7.5, 6.4.3) in the chromosome 11p15.5 region (40%–60% of patients); this includes loss of paternal or gain of maternal DNA methylation (Section 2.6) associated in some cases with genomic alterations (duplications or deletion of the chromosome 11p15 region). Of note, the DNA methylation changes in the chromosome 11p15 region that cause Russell–Silver syndrome are opposite to those that occur in Beckwith–Wiedemann syndrome (an overgrowth condition) (Section 11.7.1). The second mechanism known to cause Russell–Silver syndrome is maternal uniparental disomy (Section 2.7.5) of chromosome 7 (5%–10% of patients). In 30%–55% of patients with a clinical diagnosis of Russell–Silver syndrome, the genetic mechanism is not known. Most of the time it is a sporadic condition in which case the recurrence risk is less than 1%, but for families with a documented genomic alteration, the risk may be as high as 50%.

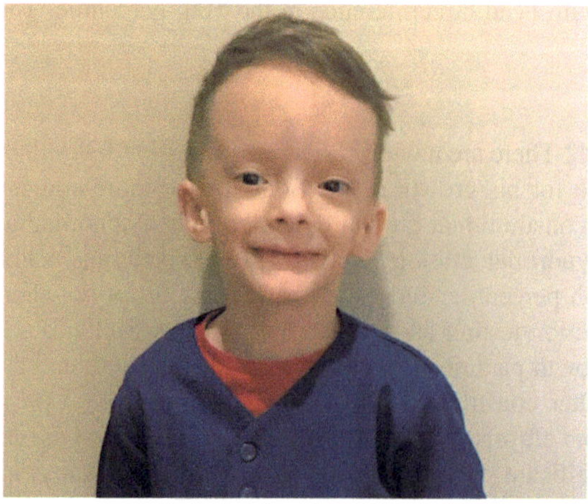

FIGURE 11.7.4.1 6-year-old male with Russell–Silver syndrome. Note relative macrocephaly with a prominent forehead and triangular-shaped face.

- **Management and Recommendations**: Infants or children with Russell—Silver syndrome with poor appetites and feeding difficulties benefit from nutritional support to ensure appropriate caloric intake, to prevent calorie-related length deficit, and to prevent hypoglycemia. David should continue working with the dietician to ensure adequate caloric intake. Young children with Russell—Silver syndrome, particularly under 5 years, are at increased risk of fasting hypoglycemia. If preoperative fasting is required, they may require intravenous glucose (10% dextrose). David's parents are advised of the risk for hypoglycemia and appropriate precautions. Treatment with growth hormone has been shown to be of benefit for individuals with Russell—Silver syndrome resulting in significant growth acceleration and improved final height. The growth response rate depends on several factors including the patient's age, growth hormone dose, height deficit, and rate of weight gain. A referral to endocrinology, to discuss growth hormone treatment, is initiated at the request of his parents. If leg length asymmetry is present, it should be monitored regularly and orthopedic management should be considered for significant leg length differences. Given the risk of developmental delay and learning difficulties, development should be monitored so that appropriate intervention and supports can be put in place if indicated. See Wakeling reference for additional information regarding medical management.
- **Genetic Testing**: Genetic testing should begin with DNA methylation analysis of chromosome region 11p15.5 and chromosome 7. This is recommended for David. Further genetic studies may need to be undertaken to define the specific mechanism that caused the methylation abnormality (i.e., testing for a chromosome abnormality). If the DNA methylation analysis is normal, then consider chromosome analysis (microarray—Section 6.1.3) to look more extensively for a chromosome deletion/duplication. It is important to note that in some cases DNA methylation changes may be present in a mosaic state (present in some cells and not others—Sections 2.2.2, 2.4.2), therefore testing of other tissues (i.e., buccal, skin) can be considered if indicated. David is found to have maternal uniparental disomy of chromosome 7.
- **Differential Diagnosis**: The are several other conditions that can cause intrauterine growth retardation and short stature which have some clinical overlap with Russell—Silver syndrome, including 3M syndrome (characterized by short stature, skeletal anomalies, and characteristic facial features), Mulibrey nanism (characterized by profound growth delay and distinctive abnormalities of the muscles, liver, brain, and eyes), SHORT syndrome (characterized by short stature, hyperextensibility, hernia, ocular anomalies, teething delay, and characteristic facial features), and IMAGE syndrome (characterized by intrauterine growth restriction, metaphyseal dysplasia, adrenal hypoplasia, and genitourinary abnormalities). Individuals with Russell—Silver syndrome have a normal head circumference; therefore if the head circumference is more than two standard deviations below the mean, another etiology should be considered. In addition to microcephaly, significant developmental delay/intellectual disability or the presence of additional congenital anomalies is suggestive of another diagnosis.
- **Caveats**: In the event that David's genetic testing is negative, since 30%—55% of individuals with Russell—Silver syndrome do not have an identified genetic etiology, the geneticist's clinical impression will guide the recommendations for medical management. Although there is relative macrocephaly, the head size is more normal or spared in comparison to the other growth parameters, so this does not warrant head imaging.
- **Outcome**: David does well on growth hormone therapy and by first grade his height is plotting at the 10th percentile. His appetite also has improved on treatment and is plotting at the fifth percentile. He did experience some speech delay and received speech therapy between 3 and 6 years of age.

The Diagnosis

Clinical Diagnosis: Russell—Silver syndrome.
Genetic Diagnosis: Chromosome 7, maternal uniparental disomy.

Further Reading

Azzi, S., Salem, J., Thibaud, N., Chantot-Bastaraud, S., Lieber, E., Netchine, I., et al., 2015. A prospective study validating a clinical scoring system and demonstrating phenotypical-genotypical correlations in Silver-Russell syndrome. J. Med. Genet. 52 (7), 446—453. https://doi.org/10.1136/jmedgenet-2014-102979. https://www.ncbi.nlm.nih.gov/pmc/articles/PMC4501172/pdf/jmedgenet-2014-102979.pdf.

Eggermann, K., Bliek, J., Brioude, F., Algar, E., Buiting, K., Russo, S., et al., 2016. EMQN best practice guidelines for the molecular genetic testing and reporting of chromosome 11p15 imprinting disorders: Silver-Russell and Beckwith-Wiedemann syndrome. Eur. J. Hum. Genet. 24, 1377—1387. PMID 27165005.

Wakeling, E.L., Brioude, F., Lokulo-Sodipe, O., O'Connell, S.M., Salem, J., Bliek, J., et al., 2017. Diagnosis and management of Silver-Russell syndrome: first international consensus statement. Nat. Rev. Endocrinol. 13, 105—124. PMID 27585961.

Case 11.7.5

Endocrine and Growth Cases: Small Stature and Delayed Puberty

Natalie Blagowidow

Harvey Institute for Human Genetics, Greater Baltimore Medical Center, Baltimore, MD, United States

Abstract

A 13-year-old girl presents with short stature, delayed puberty, and mild cognitive delay and is found to have Turner syndrome.

The Case

Tammy, a 13-year-old girl, comes in for a new pediatric appointment. You learn she weighed 2600 g at term birth, after an unremarkable pregnancy. She has had no significant medical issues, although she was always on the small side. Her developmental milestones were on schedule. Tammy requires special assistance with mathematics and previously had speech and reading assistance. On physical exam you note her height is less than the first percentile. You do not notice any unusual physical findings, but there is no evidence of pubertal development. Her mother reports her own menarche at 10 years of age.

Some breast development (thelarche) should be present in a 13-year-old. This is an indication of delayed puberty and should now be addressed. The delay is more striking given the maternal history of menstruation at a younger age.

The Issues

- **Why is Tammy so short?** Short stature can be the result of a variety of causes. These include constitutional delay, endocrine etiologies such as hypothyroidism and panhypopituitarism, and systemic disease. Some genetic disorders with short stature are Down syndrome (Section 11.13.2), Russell–Silver syndrome (Section 11.7.4), Prader–Willi syndrome (Section 11.13.1), Noonan syndrome (Section 11.4.3), and Turner syndrome. You order a bone age, which shows delay.
- **What is the next step?** You do not note any atypical physical features, and thus do not suspect any of the above genetic syndromes including Turner syndrome. You refer her to a pediatric endocrinologist, who finds a normal thyroid-stimulating hormone (TSH), elevated follicle-stimulating hormone (FSH), low estradiol, and 45,X, Turner syndrome on chromosomal analysis. The endocrinologist starts growth hormone and oxandrolone treatment and plans to follow with estrogen therapy once she has achieved a desired height. The patient and her mother still have many questions about Turner syndrome.

The physical findings associated with Turner syndrome are variably expressed and can be subtle. Obvious features such as a webbed neck are not seen in all affected individuals. The delayed puberty is consistent with Turner syndrome.

Genomics in the Clinic. https://doi.org/10.1016/B978-0-12-816478-5.00057-1

- **Are there medical issues that need immediate attention**? You know that there are cardiac issues that can be associated with Turner syndrome, and so you refer her for cardiac evaluation. The cardiac evaluation is normal. And you are aware that other systems can be affected in this disorder, but are unsure which should be studied, and so you refer to genetics. The geneticist will be familiar with what further evaluations to recommend and can provide the family with more information about the syndrome.
- **Are Tammy's academic issues related to Turner Syndrome**? You recall that there can be learning difficulties associated with Turner syndrome but do not know if there are specific cognitive deficits associated with the disorder. The genetic consultation will also address this concern.

The Referral

- **Genetic Evaluation:** Tammy and her mother are seen by the pediatric geneticist and genetic counselor. The family, medical, and developmental history is reviewed. In addition to the learning issues already noted, difficulty with short- and long-term memory is reported. Tammy's height and weight are less than the third percentile. She is found to have epicanthal folds, a broad neck with a low posterior hairline, and convex nails on exam (Fig. 11.7.5.1).
- **Genetic Counseling:** Turner syndrome is a chromosomal disorder defined by the absence of a second X chromosome in some or all cells (Sections 2.2.1 and 2.2.2). In addition to hypogonadism, Turner syndrome can also feature cardiac and renal anomalies. Tammy's subtle physical findings identified are explained as typical of Turner syndrome. The sporadic nature of the syndrome is reviewed. Most pregnancies with a 45,X chromosomal makeup are lost in the early first trimester, with 99% resulting in spontaneous abortion (Table 2.1). There is an incidence of about 1/3000 liveborn females. Only 10% will have significant developmental delay requiring special education. Specific learning deficits include visual spatial difficulties and right-left discrimination issues. Mathematics may be particularly challenging, and memory problems, such as Tammy's, can be seen. Although a normal uterus is typically present, ovarian function is rare, and assisted reproduction may be considered in the future.
- **Management and Recommendations:** Treatment is recommended to achieve puberty once final adult height has been reached, and therapy has already been initiated to improve height. Hypothyroidism occurs more often in Turner syndrome and an annual evaluation of thyroid status should be performed. Other autoimmune disorders may occur, such as type 1 diabetes, and juvenile rheumatoid arthritis. There is an increased incidence of celiac disease as well, and screening even through adulthood is recommended with symptoms. She should have regular evaluations for scoliosis while on growth hormone, due to an increased risk for this to develop.

  Tammy has had a normal echocardiogram, which ruled out coarctation of the aorta, and bicuspid aortic valve. The ascending aorta was reported as normal, but there is an increased risk for aortic dilatation and dissection or potential rupture after age 16 years, with Z-scores specific for Turner syndrome. Women with Turner syndrome also have a higher incidence of hypertension, starting in their mid-20s. Cardiac MRI is recommended by some at diagnosis if it

FIGURE 11.7.5.1 Features characteristic of Turner syndrome include prominent, posteriorly rotated ears and a broad neck [*permission obtained*].

can be tolerated, especially if there is need to see more of the chest vessels, including the pulmonary venous return. With a normal echocardiogram and MRI, Tammy can have an echocardiogram (or MRI) repeated every 5 years if stable, and every 10 years as an adult or before pregnancy.

There are additional clinical recommendations. A renal ultrasound should be done, since renal anomalies are seen in more than 30% of affected individuals. Horseshoe kidney is the most common, but also duplicated kidney or collecting system anomalies can be seen. An ophthalmologic exam will be of benefit since 40% will have refractive errors. Annual liver function testing is recommended after age 10 years to screen for elevations, and if elevated should prompt referral for evaluation of liver disease. Cognitive deficiencies should be addressed. Tammy has good academic support, but the learning issues that are seen more frequently in Turner syndrome should be shared with the school to make sure they are not overlooked. If there is any question in the future, a developmental evaluation could be done.

- **Caveat:** Approximately half of all individuals with Turner syndrome are mosaic for the disorder or have a structurally abnormal X or Y chromosome (Section 2.2.2). If a 46,XX cell line also is present, findings may be milder and there can be ovarian function. If a 46,XY cell line is present, there is an increased risk for gonadal tumors, and gonadectomy is recommended. Both conventional cytogenetics and microarray can detect mosaicism, but the latter may not reliably detect low level ($<10\%$) mosaicism (Section 6.1).
- **Outcome:** A horseshoe kidney is identified by ultrasound requiring no additional treatment. Tammy gains a centimeter of height attributed to the growth hormone therapy in the first year of treatment. She continues to progress well in school.

The Diagnosis

Clinical Diagnosis: Turner syndrome.
Genetic Diagnosis: 45,X.

Further Reading

Gravholt, C.H., Andersen, N.H., Conway, G.S., Dekkers, O.M., Geffner, M.E., Klein, K.O., et al., 2017. Clinical practice guidelines for the care of girls and women with Turner syndrome: proceedings from the 2016 Cincinnati International Turner Syndrome Meeting. Eur. J. Endocrinol. 177, G1–G70. PMID 28705803.

Silberbach, M., Roos-Hesselink, J.W., Andersen, N.H., Braverman, A.C., Brown, N., Collins, R.T., et al., 2018. Cardiovascular health in Turner syndrome. Circ. Genom. Precis. Med. 11, e000048. PMID 30354031.

Stephure, D.K., Canadian Growth Hormone Advisory Committee, 2005. Impact of growth hormone supplementation on adult height in Turner syndrome: results of the Canadian randomized controlled trial. J. Clin. Endocrinol. Metab. 90, 3360–3366. PMID 15784709.

Case 11.7.6

Endocrine and Growth Cases: Pubertal Delay, Hearing Loss, and Lack of Smell

Dena R. Matalon and Louanne Hudgins

Department of Pediatrics, Division of Medical Genetics, Stanford University, Stanford, CA, United States; Division of Medical Genetics, Stanford University, Perinatal Genetics, Lucile Packard Children's Hospital, Stanford, CA, United States

Abstract

A 16-year-old female presents with no menarche, also with sensorineural hearing loss, anosmia, and bimanual synkinesis on exam and is found to have Kallmann syndrome.

The Case

A 16-year-old female, Nisha, comes to see you, her pediatrician, because she has not had her first menstrual period. You know her well and her only other medical concern is bilateral sensorineural hearing loss. On exam she is noted to have sexual maturity rating I for breast development and sexual maturity rating III for pubic hair. Her family history is notable for parents with normal pubertal development.

The Issues

- **What testing would you perform to assess pubertal status and aid in your differential diagnosis?** You request a serum estradiol, luteinizing hormone (LH), and follicle-stimulating hormone (FSH). Estradiol and LH return low and FSH returns low normal. You are now thinking that her delayed menarche is due to hypogonadotropic hypogonadism.
- **Should she be referred to other clinical specialists?** You refer Nisha to endocrinology who will not only evaluate anterior pituitary function, prolactin level, and likely perform a brain MRI with specific pituitary views but also be able to manage any hormone replacement needed. Based on her findings of sensorineural hearing loss and hypogonadotropic hypogonadism, you also refer Nisha to a clinical geneticist to determine if she may have a syndrome.

The Referral

- **Genetic Evaluation:** Nisha is seen by the clinical geneticist and the genetic counselor. A full history of present illness, past medical and surgical history, review of systems, developmental history, social history, and physical examination are obtained by the geneticist. Nisha had an unremarkable prenatal history with normal ultrasounds. She was the product of a normal full-term spontaneous vaginal delivery. The genetic counselor obtains a detailed three-generation pedigree (Sections 2.7 and 3.2) and no one else in the family is found to have hearing loss or delayed puberty. On

Genomics in the Clinic. https://doi.org/10.1016/B978-0-12-816478-5.00058-3

examination, Nisha is found to have bimanual synkinesis (mirror movements). The remainder of her exam is normal with no dysmorphic features (Section 5.1) noted.

The geneticist considers Kallmann syndrome as the likely diagnosis and recommends the University of Pennsylvania smell identification test (UPSIT), which reveals Nisha has anosmia. She has a history of bilateral sensorineural hearing loss and delayed puberty, as well as bimanual synkinesis on exam, which are all findings associated with Kallmann syndrome.

Kallmann syndrome is defined as hypogonadotropic hypogonadism with anosmia. It is clinically and genetically heterogeneous (Section 2.8). In Kallmann syndrome, the remainder of the hypothalamic and pituitary hormones as well as the radiographic appearance of the pituitary tend to be normal. Thus, Nisha's clinical presentation is consistent with a diagnosis of Kallmann syndrome. Most cases are diagnosed at the time of puberty due to the lack of sexual development, which includes small testes without virilization in males and lack of breast development with primary amenorrhea in females.

Findings on physical examination can be normal. They can also include bimanual synkinesis. This means when one hand does a specific movement, the other hand also does this movement involuntarily. Additional findings associated with Kallmann syndrome include hearing loss, abnormal eye movements, chorioretinal colobomas, renal agenesis, agenesis of the corpus callosum, cleft lip or palate, limb/spine bony abnormalities, and hypodontia.

- **Genetic Counseling:** The geneticist discusses with Nisha's parents that Kallmann syndrome can be transmitted in a variety of ways including: X-linked recessive, autosomal recessive, autosomal dominant (with incomplete penetrance) (Sections 2.7.1, 2.7.2), and oligogenic inheritance (Section 2.7.3).

  Multiple genes are associated with Kallmann syndrome. Nisha is a female with no other significant family history. Thus, you suspect autosomal recessive inheritance (Section 2.7.1) or a de novo variant to be the genetic cause for Kallmann syndrome in Nisha.
- **Genetic Testing:** Genetic testing strategy is based on the patient's sex, family history, and presence of any other pertinent clinical features. Genomic sequencing via next-generation sequencing allows multiple genes to be analyzed at one time through multigene panel testing (Section 6.2.3). A pathogenic variant (Section 7.1.3) is only identified in around 40% of cases of Kallmann syndrome. If a pathogenic variant is identified, recurrence risk counseling (Section 4.4.3) can be provided, carrier testing (Section 4.1) can be offered for at-risk relatives, and prenatal diagnostic testing (Section 4.4.2) for pregnancies at increased risk is possible. Panel testing for Kallmann syndrome is recommended for Nisha.
- **Results:** The result of Nisha's genetic sequencing reveals no pathogenic variants and is nondiagnostic. Unfortunately, Nisha is in the majority of patients with Kallmann syndrome in which no genetic cause can be identified.
- **Management and Recommendations:** Treatment for Kallmann syndrome is centered around addressing the hypogonadism. Patients should be followed closely by endocrinology. First, hormonal treatment is initiated to stimulate breast development in females and virilization in males. Subsequently, the focus will be on fertility. Typically, hormone replacement therapy is used including estrogen and progesterone for females and testosterone for males. In terms of fertility, pulsatile gonadotropin-releasing hormone (GnRH) stimulation and exogenous gonadotropins can be used. Management also includes evaluating for any other features that can be associated with Kallmann syndrome. Thus, one can consider a renal bladder ultrasound to rule out renal agenesis, audiology evaluation for hearing loss, dental exam for hypodontia, ophthalmology exam for any chorioretinal changes, and skeletal survey or X-rays for any bony abnormalities. Additionally, patients are at risk for osteoporosis and will need to be monitored for this complication over time.
- **Differential Diagnosis:** Anosmia is an uncommon finding and could be due to anatomic causes. Few other genetic conditions present with anosmia, but they include some cases of Waardenburg syndrome (Section 11.3.7), some cases of peroxisomal disorders, and some forms of ichthyosis.
- **Caveats:** While a pathogenic variant was not identified in this case, it does not mean that there is not a genetic cause of Kallmann syndrome for Nisha. A genetic cause may be elucidated in the future by additional testing or as technology advances and information accumulates.
- **Outcome:** Nisha and her family are grateful to know and understand the underlying cause of her findings. They also know that new genetic information is regularly available. Nisha plans to check back in with prenatal genetics prior to becoming pregnant to learn if there are any new genetic sequencing options at that time. Additionally, she is referred for a renal bladder ultrasound, which is normal, and an ophthalmology exam, which also is normal. She has established care with endocrinology. The geneticist will continue to follow Nisha to be sure all of her specialty care is coordinated and monitor for any other symptoms that may be related to her condition.

The Diagnosis

Clinical Diagnosis: Kallmann syndrome.
Genetic Diagnosis: Pathogenic variant not detected.

Further Reading

Ahmadzadeh, A., Ghods, E., Mojarrad, M., Aboutorabi, R., Afkhamizadeh, M., Bonakdaran, S., et al., 2015. Study on KAL1 gene mutations in idiopathic hypogonadotropic hypogonadism patients with X-Linked recessive inheritance. Int. J. Mol. Cell Med. 4, 152–159. PMID 26629483.

Balasubramanian, R., Crowley Jr., W.F., May 23, 2007. Isolated gonadotropin-releasing hormone (GnRH) deficiency. Adam, M.P., Ardinger, H.H., Pagon, R.A., Wallace, S.E., Bean, L.J.H., Stephens, K., Amemiya, A., (editors). In GeneReviews. University of Washington, Seattle (WA), 1993–2022 [updated March 2, 2017]. PMID 20301509.

Dode, C., Hardelin, J.P., 2009. Kallmann syndrome. Eur. J. Hum. Genet. 17, 139–146. PMID 18985070.

Stamou, M.I., Georgopoulos, N.A., 2018. Kallmann syndrome: phenotype and genotype of hypogonadotropic hypogonadism. Metabolism 86, 124–134. PMID 29108899.

Section 11.8

Genitourinary Cases

Case 11.8.1

Genitourinary Cases: Ambiguous Genitalia

Antonie D. Kline

Harvey Institute for Human Genetics, Greater Baltimore Medical Center, Baltimore, MD, United States

Abstract

A baby is born with ambiguous genitalia and is subsequently found to have partial androgen insensitivity syndrome.

The Case

You are called to the hospital to see a new baby born to a family in your practice; you already follow a healthy older daughter. The hospital staff cannot tell if the new baby is a boy or a girl. You arrive at the nursery to see Kamil, the new baby. This was an uneventful term pregnancy, born to a 35-year-old healthy mother following a spontaneous vaginal delivery. When you examine the baby, you note normal birth weight and length, genitalia with a rudimentary phallus, fused folds, one palpable testis, and no additional anomalies (Fig. 11.8.1.1). You confirm ambiguous genitalia and meet with the parents.

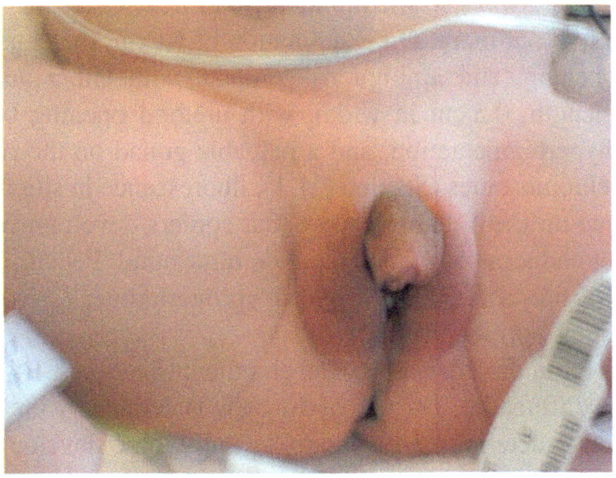

FIGURE 11.8.1.1 Note ambiguous genitalia with fused pigmented rugated labioscrotal folds and rudimentary phallus.

Genomics in the Clinic. https://doi.org/10.1016/B978-0-12-816478-5.00059-5

The Issues

- **How should the baby's sex be assigned?** You discuss with the parents that the baby's genitalia appear ambiguous, and that you are unable to tell whether or not the baby appears male or female. You discuss that you cannot determine the cause of the baby's ambiguity, or decide on the sex assignment of the baby at this time, but through laboratory studies sent, these will be determined as soon as possible. You assure them that you will make appropriate referrals, including pediatric endocrinology and pediatric genetics. There can be life-threatening causes, so there is an urgency to determining etiology. Because one of the first questions asked of parents is whether they had a boy or girl, it is helpful for the parents to have emotional support from your team. They can tell family members and those close to them that they are not certain at this time if they have a boy or a girl.

It is appropriate to refer to both pediatric endocrinology and pediatric genetics, as well as pediatric urology. Ambiguous genitalia are considered a neonatal emergency, both medically and psychosocially. Although sex identity can be determined through anatomical, genetic, and hormonal results, gender identity is a complex topic, and other than the major issues including genetics discussed below, others are beyond the scope of this case.

- **Were there any prenatal clues to the situation?** The mother says that early in pregnancy she had a noninvasive prenatal screening test (Section 4.4.1), and this indicated Y chromosome material. When she had the mid-gestation anatomy ultrasound (Section 4.3), the genitalia did not look clearly male or female. You ask about additional definitive testing done, and the couple says that after discussing it with the obstetrician, they elected not to do prenatal diagnostic testing (Section 4.4.2) because they were not worried and would not be interested in termination of a baby with abnormal or ambiguous genitalia. No genetic counseling was done.

Important to note that even though a couple whose fetus has anomaly(ies) says they would not be interested in terminating a pregnancy, it is still appropriate to offer genetic counseling to discuss the scope and implications of the anomaly(ies) as well as the options of prenatal diagnostic testing so that they are fully informed. The genetic counselor also can help coordinate appropriate studies immediately after birth, including on cord blood in the delivery room.

- **Are there any other associated factors?** You ask if there is any known family history of other children with abnormal genitalia. The mother says that her brother may have been born with something, and he needed surgery when he was a baby, but her family did not discuss it and she is not in communication with him currently. She thinks he had breast development at puberty. The family awaits the endocrine and genetic consultations.

The Referral

- **Genetic Evaluation:** A full pregnancy history, medical history, and family history are obtained, and a detailed physical examination is performed by the pediatric geneticist and genetic counselor. Kamil's mother underwent normal puberty and had menarche around age 11 years, as did her own mother. In addition to the family history above, the geneticist finds out that the maternal uncle has normal intelligence, never married or had children, always urinated sitting down, had gynecomastia as a teenager, and always seemed to identify as a male. He may have had surgery, but the mother is unsure and has not been able to find out more. On examination of Kamil, the geneticist notes normal growth, nondysmorphic facies, normal anal opening, and ambiguous genitalia with either an enlarged clitoris or small phallus measuring 1.5 cm in stretched length, 0.5 cm in width, with urethral opening below the phallus, labioscrotal folds with minimal rugae and some hyperpigmentation, and a palpable gonad on the right but not on the left.

  The geneticist recommends chromosomes (Section 6.1.1), fluorescence in situ hybridization (FISH) testing (Section 6.1.2) for the *SRY* gene (the gene on the Y chromosome that confers development of the Wolffian system), baseline electrolytes, hormone studies per endocrinology, and a pelvic ultrasound. Pediatric endocrinology recommends testosterone to be drawn at 72 hours of life, as well as 17-hydroxyprogesterone, 17-hydroxypregnenolone, luteinizing hormone (LH), and follicle-stimulating hormone (FSH).
- **Genetic Counseling:** This baby has a disorder of sex development (Section 5.1.5). Androgen insensitivity syndromes are rare but important within this differential. The sex of the baby is defined by the chromosomes present at conception, and the hormonal balance, which is determined very early in embryonic development. If testosterone is made by the embryo following a signal originating from the *SRY* gene on the Y chromosome, the fetal genitalia develop into a defined male (Wolffian) system, whereas if no testosterone is detected by the embryo, development of the fetal genitalia proceeds to a defined female (Mullerian) system. Palpable gonads in a newborn baby are most likely testes, and the chromosomes (Section 6.1.1) are most often 46,XY. There are disorders involving poor or no detection of testosterone by the androgen receptors. If a uterus is present as detected by the pelvic ultrasound, the chromosomes are likely 46,XX, but there can be complete gonadal dysgenesis involving pathogenic variants or mutations in the *SRY* or related genes.

The geneticist and genetic counselor discuss with the parents that there are many causes of disorders of sex development. Studies need to be done to determine the chromosomes and hormone levels of Kamil. Although the sex of the baby cannot be readily identified, since there are palpable gonads it is more likely that a Y chromosome is present. If baby is chromosomally male, there has been undervirilization leading to a bifid scrotum and severe hypospadias. If baby is chromosomally female, there has been virilization with fused labial folds. The pelvic ultrasound will determine if a uterus is present and it can sometimes visualize gonads. There are a number of potential causes, most of them genetic in etiology.

The gender identity of any individual, however, is less biologically determined and is related to fetal brain and psychosexual development, induced by sex hormones early in development, as well as many other factors. The decision about which gender to assign the baby will be better able to be made once the recommended studies are completed, and there has been discussion with pediatric genetics, pediatric urology, and pediatric endocrinology. It is important to determine the most appropriate gender assignment in the neonatal period, but it is very possible that this could be reassigned subsequently.

Determination of the baby's sex should be a multidisciplinary decision with parental input. Not only the neonatal examination but also future pubertal changes and outcomes play a role. In addition to the subspecialists, social work and/or behavioral psychology can have important input, especially at puberty or if dissatisfaction with assigned gender identity arises.

- **Results:** This patient's pelvic ultrasound shows no uterus and identifies two testicles, one in the canal and one high in the pelvis. The electrolytes are normal. Chromosomes show a normal male karyotype, 46,XY, with no rearrangement involving X or Y. In addition, LH and FSH are elevated, testosterone is very low, and 17-hydroxyprogesterone and 17-pregnenolone are normal. Congenital adrenal hyperplasia is ruled out.

Congenital adrenal hyperplasia can lead to electrolyte imbalance and death if not treated. Many states have newborn screening for this reason.

With a male karyotype, testes present, and a family history consistent with an X-linked (Section 2.7.2) condition, the suspicion for partial androgen insensitivity syndrome (PAIS) seems likely. This is an X-linked disorder of sex development due to defects in androgen action leading to a spectrum of effects. Androgens are formed but not fully detected by the body. In PAIS, there is ambiguity of the genitalia in males, with typical findings including microphallus with hypospadias, fused scrotum, and gynecomastia in puberty. There is impaired spermatogenesis and decreased body hair. Typically, there is identification with male gender, and also interest in female relationships. Females, 46,XX, carrying the variant for this condition have normal genitalia and puberty.

- **Management and Recommendations:** Pediatric urologic surgery for PAIS includes repair of the severe hypospadias during infancy and bilateral orchiopexy. Gonadectomy is recommended prior to puberty if the child is to be raised female to prevent (further) masculinization and does not have to be done in the neonatal period. Testing includes sequencing (Section 6.2.2) of the androgen receptor gene on the X chromosome. Gynecomastia may lead to a need for reduction mammoplasty during puberty.

This is distinguished from complete androgen insensitivity syndrome, in which there is no detection of androgens by the body, and fully female genitalia result with subsequent primary amenorrhea, although inguinal hernias with gonad(s) are often noted earlier. There is also mild androgen insensitivity syndrome, in which males present only with gynecomastia and/or infertility.

- **Caveats:** About 50% of individuals with PAIS will not have a detectable pathogenic variant on testing. There are many potential causes of disorders of sex development in males with undervirilization.
- **Outcome:** Sequencing of the androgen receptor, *AR,* gene detects a pathogenic variant, confirming PAIS in Kamil. The family chooses to assign Kamil as male.

The Diagnosis

Clinical Diagnosis: Partial androgen insensitivity syndrome.
Genetic Diagnosis: *AR* gene, hemizygous variant.

Further Reading

Fisher, A.D., Ristori, J., Fanni, E., Castellini, G., Forti, G., Maggi, M., 2016. Gender identity, gender assignment and reassignment in individuals with disorders of sex development: a major of dilemma. J. Endocrinol. Invest. 39, 1207–1224. PMID 27287420.

Hughes, I.A., Werner, R., Bunch, T., Hiort, O., 2012. Androgen insensitivity syndrome. Semin. Reprod. Med. 30, 432–442. PMID 23044881.

Lucas-Herald, A., Bertelloni, S., Juul, A., Bryce, J., Jiang, J., Rodie, M., et al., 2016. The long-term outcome of boys with partial androgen insensitivity syndrome and a mutation in the androgen receptor gene. J. Clin. Endocrinol. Metab. 101, 3959–3967. PMID 27403927.

Case 11.8.2

Genitourinary Cases: Familial Renal Disease

Kristin Meliambro and Kirk Campbell
Icahn School of Medicine at Mount Sinai, Department of Medicine, Icahn School of Medicine at Mount Sinai, New York, NY, United States

Abstract
A woman considers donating her healthy kidney to her 33-year-old sister, who has focal segmental glomerulosclerosis that is found to have a genetic basis on testing.

The Case

Ruby, a 26-year-old woman, presents for a routine physical exam. She is in good health and mentions that she is considering donating a kidney to her older sister, Pearl. Pearl, who is also a patient in the same primary care practice, is 33 years old and has a 20-year history of biopsy-proven focal segmental glomerulosclerosis (FSGS). Since being diagnosed, she has failed to respond sufficiently to treatment with prednisone and cyclosporine. A repeat kidney biopsy 5 years ago confirmed FSGS progression with severe renal fibrosis and scarring. Her creatinine is now high at 4.2 mg/dL with a low estimated glomerular filtration rate (eGFR) of 13 cc/min. There are six total siblings: three with glomerular disease and three without.

The Issues

- **Is there a strong genetic basis for focal segmental glomerulosclerosis?** To date, more than 60 genes have been associated with FSGS and steroid-resistant nephrotic syndrome. The genes are expressed in kidney podocytes, visceral epithelial cells that line the urinary space, that are essential to maintenance of an intact glomerular filtration barrier. Injury to podocytes due to genetic or other etiologies can cause proteinuria and progressive renal disease. Recent estimates suggest that genetic etiologies account for 20%–30% of steroid-resistant FSGS cases. Because of the family history of two other affected siblings, it seems likely that there is a genetic cause in this family.
- **What are the prognostic implications for Pearl of having a genetic basis for focal segmental glomerulosclerosis?** Pearl's clinical course is also consistent with a genetic etiology for FSGS. These individuals tend to be resistant to steroids and immunosuppressive agents like cyclosporine. They also have faster progression to end-stage kidney disease than patients with nongenetic disease. From an encouraging perspective, while FSGS has a 30%–40% recurrence rate posttransplant, the likelihood of recurrence is much lower for patients with genetic etiologies.

Genomics in the Clinic. https://doi.org/10.1016/B978-0-12-816478-5.00060-1

- **Is it safe for Ruby to donate a kidney to Pearl?** To answer this question, it will be important for Ruby to undergo a complete general health assessment with a focus on any existing kidney disease (including renal imaging, eGFR, albuminuria) or associated risk factors. Even if these are ruled out, her risk of having familial FSGS can only be confirmed with genetic testing. Ruby undergoes lab testing, which is unrevealing for existing renal disease. She is referred for a genetic evaluation for genetic testing, appropriate to fully define her risk given the intention to be a kidney donor. The primary goal is to ensure that Ruby is not a silent carrier of the same causative mutation in Pearl.

The Referral

- **Genetic Evaluation:** Ruby and Pearl both attend the genetic appointment. A full medical history and family history are obtained and a detailed physical examination performed by the geneticist and genetic counselor. It is verified that three of six siblings were diagnosed with biopsy-proven FSGS in the 12–15 year age range. No other family members have kidney disease and nobody in the family has ever had genetic testing. Both sides of the family are of African American ancestry.
- **Genetic Counseling:** The geneticist and genetic counselor explain that both autosomal dominant and recessive inheritance patterns (Section 2.7.1) have been observed with disease-causing FSGS genes. Autosomal recessive traits have been more commonly identified in children and adolescents. An autosomal dominant inheritance for FSGS genes has been observed more frequently in adult onset disease. These familial mutations are highly penetrant (Section 2.8) and to be distinguished from more common variants that act as disease modifiers. Being of African descent, both women have more than a 30% chance of carrying high-risk apolipoprotein L1 (*ApoL1*) alleles which, though not disease causing, markedly increase the risk of nondiabetic kidney disease in general and FSGS in particular.
- **Genetic Testing:** Ruby and Pearl opt for exome sequencing (Section 6.3.1) with the American College of Medical Genetics and Genomics (ACMG) recommendations followed to categorize secondary findings (Section 7.2). Variants in genes associated with FSGS and nephrotic syndrome are examined. Using in silico algorithms, no disease-causing FSGS genes are identified in Ruby. She is found to carry one high-risk G1 *ApoL1* allele not independently thought to increase the risk of kidney disease. Pearl was found to have a homozygous (Section 2.7.1) pathogenic (Section 7.1.3) R138Q variant in the *NPHS2* gene that encodes podocin. This is a known recurrent mutation in this gene associated with FSGS. She is not found to carry any copies of high-risk G1 or G2 *ApoL1* alleles.
- **Management and Recommendations:** Ruby is notified that she has a low genetic risk for glomerular disease, thereby providing reassurance as she considers becoming a kidney donor. Pearl is equally reassured that, based on available evidence, although the genetic disease is aggressive and treatment-resistant in her native kidneys, she is a lower risk of FSGS recurrence posttransplant. Because she does not carry any high-risk *ApoL1* alleles, she does not have additional risk for development of hypertension and nondiabetic renal disease that is present in carriers of two high-risk *ApoL1* alleles.
- **Caveats:** Recent evidence suggests that up to 20% of patients with steroid-resistant FSGS have a genetic etiology. This only accounts for known disease-causing variants and likely still underestimates the true genetic contribution. The expansion of exome (Section 6.3.1) and genome (Section 6.3.2) sequencing in affected patients will likely increase the genetic diagnostic rate.
- **Outcome:** Ruby successfully donates a kidney to Pearl. They both recover well after surgery. Pearl is able to avoid dialysis, and 3 years posttransplant, has no evidence of FSGS recurrence. Both patients have encouraged their immediate and extended family to be screened for kidney disease and to obtain genetic testing if recommended.

The Diagnosis

Clinical Diagnosis: Patient—low risk for glomerular disease. Sister—Focal segmental glomerulosclerosis.
Genetic Diagnosis: Patient—*APOL1*, heterozygous G1 allele. Sister—*NPSH2* gene, homozygous pathogenic variants.

Further Reading

Buscher, A.K., Kranz, B., Buscher, R., Hildebrandt, F., Dworniczak, B., Pennekamp, P., et al., 2010. Immunosuppression and renal outcome in congenital and pediatric steroid-resistant nephrotic syndrome. Clin. J. Am. Soc. Nephrol. 5, 2075–2084. PMID 20798252.

Greka, A., Mundel, P., 2012. Cell biology and pathology of podocytes. Annu. Rev. Physiol. 74, 299–323. PMID 22054238.

Jungraithmayr, T.C., Hofer, K., Cochat, P., Chernin, G., Cortina, G., Fargue, S., et al., 2011. Screening for NPHS2 mutations may help predict FSGS recurrence after transplantation. J. Am. Soc. Nephrol. 22, 579–585. PMID 21355056.

Yao, T., Udwan, K., John, R., Rana, A., Haghighi, A., Xu, L., et al., 2019. Integration of genetic testing and pathology for the diagnosis of adults with FSGS. Clin. J. Am. Soc. Nephrol. 14, 213–223. PMID 30647093.

Case 11.8.3

Genitourinary Cases: Renal Cysts

Kristin Meliambro and Kirk Campbell

Icahn School of Medicine at Mount Sinai, Department of Medicine, Icahn School of Medicine at Mount Sinai, New York, NY, United States

Abstract

A 32-year-old man with hypertension, microalbuminuria, and slightly elevated creatinine is found to have autosomal dominant polycystic kidney disease.

The Case

Paul, 32-year-old male, presents to you, his primary care physician, for a routine physical exam. The patient has hypertension that has been treated with amlodipine 10 mg daily for the past 5 years. His blood pressure is well controlled at 119/68. Lab studies you send reveal that he has microalbuminuria (urinary albumin is 100 mg/g on a spot sample) with a slightly elevated creatinine of 1.2 mg/dL and a low, estimated glomerular filtration rate (eGFR) of 56 mL/min per 1.73 m^2 using the Chronic Kidney Disease Epidemiology Collaboration (CKD-EPI) equation. You obtain a renal ultrasound, which reveals three bilateral simple cysts in 12 cm kidneys. The patient was adopted and has no available biological family history. He is otherwise well with a normal body mass index (BMI).

The Issues

- **Does this patient meet the diagnostic criteria for polycystic kidney disease (PKD)?** For individuals aged 15−39 years with a family history of polycystic kidney disease, the presence of three renal cysts (bilateral or unilateral) on ultrasound imaging is sufficient for the diagnosis of autosomal dominant (Section 2.7.1) polycystic kidney disease (ADPKD). Given the lack of family history, this patient's ultrasonographic findings are equivocal and further investigation will be required.
- **If there is polycystic kidney disease, what other symptoms or complications should the patient be concerned about?** If this patient has polycystic kidney disease, then his male gender, early age of onset of hypertension, and proteinuria are risk factors for progression to end-stage kidney disease. More than 80% of patients with polycystic kidney disease have detectable liver cysts by age 30. Intracranial aneurysms have a 9%−12% prevalence in these patients compared to 2%−3% in the general population. Other cardiovascular manifestations such as left ventricular hypertrophy, coronary aneurysms, and valvular heart disease (especially mitral valve prolapse and mitral regurgitation) are also prevalent in patients with polycystic kidney disease.
- **Is there treatment available?** Care for patients with polycystic kidney disease had been largely supportive until April 2018 when the US Food and Drug Administration (FDA) approved the use of tolvaptan for patients with polycystic kidney disease at high risk for chronic kidney disease progression. Because of all of these implications, you refer Paul for genetic evaluation and testing.

Genomics in the Clinic. https://doi.org/10.1016/B978-0-12-816478-5.00061-3

The Referral

- **Genetic Evaluation:** Paul undergoes a complete past medical and surgical history, review of systems, development history, social history, and physical examination by the clinical geneticist. No family history is available. His physical exam is normal except for an S4 on his cardiac exam. The ultrasound studies previously obtained are reviewed.
- **Genetic Counseling:** The geneticist and genetic counselor explain to Paul that his physical, lab, and imaging findings are suggestive, though not diagnostic, of autosomal dominant polycystic kidney disease. The differential diagnosis would include syndromic conditions such as tuberous sclerosis, Von Hippel–Lindau syndrome, glomerulocystic disease, medullary cystic disease, and medullary sponge kidney. Nonsyndromic conditions like simple renal cysts and acquired renal cystic disease also have to be considered. Additional imaging with CT scan or MRI with heavy-weighted unenhanced T2 images are considered, but deemed unlikely to significantly improve the diagnostic yield. Genetic testing is offered.
- **Genetic Testing:** The absence of a family history renders the imaging findings equivocal and makes linkage analysis, one approach to identifying variants in the polycystic kidney disease genes, impossible. Sanger sequencing (Section 6.2.2) of the *PKD1* and *PKD2* exons is the method of choice for most patients, and in sequence-negative cases multiplex ligation-dependent probe amplification (Section 6.2.1) approaches can be used to detect large gene rearrangements that rarely occur. Given the uncertainly with this patient's clinical diagnosis and genetic underpinnings, a sequence capture-based next-generation sequencing (NGS) panel (Section 6.2.3) approach for all genes known for cystic and polycystic kidney disease is recommended. This provides an additional advantage over Sanger sequencing due to its ability to detect copy number variations (Section 2.4.1).
- **Results:** Paul is found to have a 109delT frameshift variant (Section 2.4.1) in exon 1 of the *PKD1* gene (Cys37fs36X) defined as definitely pathogenic (Section 7.1.3) on the Mayo Clinic ADPKD Mutation Database (pkdb.mayo.edu). Disease-causing variants are not identified in other known autosomal dominant polycystic kidney disease genes such as *PKD2*, *GANAB*, and *DNAJB11*.
- **Management and Recommendations:** With the autosomal dominant polycystic kidney disease diagnosis now confirmed, Paul can be treated according to established algorithms to slow progression of chronic kidney disease. His blood pressure medication is switched from amlodipine to losartan since angiotensin-converting enzyme inhibitors and angiotensin receptor blocker have been shown to reduce the rate of kidney growth and may slow eGFR decline. Paul is also advised to limit daily sodium intake to 2 gm and increase fluid intake to more than 3 L per day.

  The risk of developing a cerebral aneurysm is elevated in patients with polycystic kidney disease, and higher if there is a family history, unknown in this case. Paul is scheduled for an MRA of the brain to screen for cerebral aneurysms. He is also scheduled for an echocardiogram given physical exam findings suggestive of left ventricular hypertrophy.
- **Caveats:** Up to 8%–10% of patients with suspected autosomal dominant polycystic kidney disease will have no detectable variants on genetic testing.
- **Outcome:** The patient continues to follow up with his nephrologist with regular assessments of risk for chronic kidney disease progression. During follow up he is noted to have a PROPKD score of 9, suggesting high risk for eGFR decline. He also has Mayo Class 1C with a height adjusted kidney volume of 600 mL/m. The patient is started on tolvaptan and continues regular follow up. His magnetic resonance angiogram (MRA) is negative. He is found to have left ventricular hypertrophy with normal ejection fraction on echocardiogram. Blood pressure control and in particular angiotensin receptor blockade, which has already been initiated in Paul, are also the mainstays of therapy to promote regression of left ventricular hypertrophy.

The Diagnosis

Clinical Diagnosis: Polycystic kidney disease.
Genetic Diagnosis: *PKD1* gene, heterozygous pathogenic variant.

Further Reading

Chebib, F.T., Torres, V.E., 2018. Recent advances in the management of autosomal dominant polycystic kidney disease. Clin. J. Am. Soc. Nephrol. 13, 1765–1776. PMID 30049849.

Cornec-Le Gall, E., Audrezet, M.P., Rousseau, A., Hourmant, M., Renaudineau, E., Charasse, C., et al., 2016. The PROPKD score: a new algorithm to predict renal survival in autosomal dominant polycystic kidney disease. J. Am. Soc. Nephrol. 27, 942–951. PMID 26150605.

Irazabal, M.V., Abebe, K.Z., Bae, K.T., Perrone, R.D., Chapman, A.B., Schrier, R.W., et al., 2017. Prognostic enrichment design in clinical trials for autosomal dominant polycystic kidney disease: the HALT-PKD clinical trial. Nephrol. Dial. Transplant. 32, 1857–1865. PMID 27484667.

Pei, Y., Obaji, J., Dupuis, A., Paterson, A.D., Magistroni, R., Dicks, E., et al., 2009. Unified criteria for ultrasonographic diagnosis of ADPKD. J. Am. Soc. Nephrol. 20, 205–212. PMID 18945943.

Section 11.9

Hematologic and Immunologic Cases

Case 11.9.1

Hematologic and Immunologic Cases: Bleeding Disorder

Michele P. Lambert
Perelman School of Medicine at the University of Pennsylvania, Division of Hematology, Children's Hospital of Philadelphia, Philadelphia, PA, United States

Abstract
A 2-day-old male with blood oozing from a circumcision is found to have factor VIII deficiency and hemophilia A.

The Case

A 2-day-old male, Jason, is seen by you, the pediatrician in the newborn nursery. He is a full-term product of an uncomplicated pregnancy and he is the first baby born to this couple. He was born by an uncomplicated vaginal delivery and received his vitamin K injection in the delivery room as well as hepatitis B immunization. He was doing well and his parents asked for circumcision, which was performed just this morning, and now he continues to have blood oozing from the circumcision site. He has a palpable knot at the site of his vitamin K injection with an overlying bruise. He has multiple other small bruises and docs have a mild cephalohematoma. He is feeding well and otherwise appears to be doing well.

The Issues

- **Why does this baby have blood oozing from his circumcision site?** It is possible that there are technical issues with the circumcision, but the more likely explanation, given the other symptoms, is that this baby has a bleeding disorder.
- **Are there any other clues?** The other clues to the diagnosis of a bleeding disorder are the palpable hematoma at the site of the vitamin K administration, and the other small bruises (unusual in a newborn), as well as the cephalohematoma (not uncommon but, in combination with the other bleeding, is concerning).
- **What testing should be done?** Referral to hematology (non-oncology) is made to help guide further evaluation and management. The hematologist recommends testing with coagulation screening (PT/PTT) and factor levels, in order of priority, a factor 8 (FVIII), factor 9 (FIX), and factor eleven (FXI) levels—assuming only the PTT is prolonged. While awaiting the results, empiric treatment is recommended with recombinant or plasma-derived factor VIII product for just one dose (since the results of the factor levels may take >24 h to return), and then rechecking the PTT and looking at hemostatic effects. These labs cannot be drawn by arterial puncture and pressure should be held at the venipuncture site after phlebotomy.
- **What is the next step?** It is important to try to get control of the bleeding quickly and topical measures can be helpful but in the setting of hemophilia, it is often necessary to replace the missing factor. Since factor VIII (FVIII) deficiency is much more common than factor IX (FIX) deficiency, and giving factor is better than giving fresh frozen

Genomics in the Clinic. https://doi.org/10.1016/B978-0-12-816478-5.00062-5

plasma for replacement (15 cc/kg gives at best a 30% correction), it may be better to give FVIII followed by FIX if there is no effect on the PTT in this suspected new case of hemophilia. The diagnosis is established by sending factor levels. Genetic testing is recommended, especially to identify the carrier status of the mother (for counseling for risk for additional affected children) and for later in childhood because of the risk of low factor levels in female carriers, e.g., female siblings. Jason's PTT was 67 s (normal is generally below 48 s in a neonate). Factor VIII level is below 1% (normal is 50%−150%), confirming severe factor VIII deficiency. He is referred for genetic evaluation.

The Referral

- **Genetic Evaluation:** A full pregnancy history, a medical history, and a family history are obtained, and a detailed physical examination is performed by the pediatric geneticist and genetic counselor. There is no family history of any bleeding disorders, although the mother does have some bleeding symptoms herself. She has a history of heavy menstrual bleeding that improved with hormonal contraceptives, and then got worse again when she stopped the combination oral contraceptive when getting ready to have children. She also has a history of iron deficiency anemia from the heavy menstrual bleeding.
- **Genetic Counseling:** Hemophilia is inherited in an X-linked fashion (Section 2.7.2). The type of hemophilia that Jason has, hemophilia A (factor VIII deficiency) is more common than hemophilia B (factor IX deficiency). The risks to the siblings of a proband depend on the carrier status of the mother. If she is a carrier, there is a 50% chance any male offspring will be affected. Daughters that inherit the variant will be carriers and may be symptomatic depending on the FVIII expression level. Genetic testing is able to identify a pathologic variant (Section 7.1.3) in 98% of patients with abnormal FVIII levels. The most common causes of FVIII deficiency are inversions (Section 2.2.1) in intron 22 or intron 1 and result in a severe deficiency (FVIII levels of below 1%), so many labs will test for these two changes first and perform sequencing (Section 6.2.2) only if the inversions are not present. Deletions, splice site mutations, as well as missense mutations and nonsense mutations (Section 2.4.1) have all been described. Prenatal counseling will be recommended for Jason's parents, and prenatal diagnostic testing (Section 4.4.2) to be considered for future pregnancies.
- **Management and Recommendations:** Many clinical laboratories offer testing for hemophilia. Jason is recommended to undergo testing. Jason and his mother are referred to a hemophilia treatment center. His circumcision bleeding is controlled with replacement FVIII and he is discharged home after education on the new diagnosis is given to his family. He is seen shortly after discharge and discussions about ongoing management including questions about when to start prophylaxis and with which product, as well as the results of the genetic testing.
- **Results:** Jason's testing is positive for a hemizygous intron 22 inversion in *F8*, which he inherited from his mother, confirming his diagnosis of hemophilia A (with the most common variant seen in 45% of patients with severe hemophilia A). Thus his mother is a carrier but with some symptoms. This change is associated with a 20%−30% risk of inhibitor development when patients are treated with recombinant FVIII products for prophylaxis to prevent life-threatening spontaneous hemorrhage.
- **Implications for Other Family Members:** In addition to testing Jason, it was important to perform carrier testing (Section 4.1) on his mother in order to better counsel his parents about risk to future pregnancies. Severe deficiency in carrier females is reported in cases of abnormally skewed lyonization (Section 2.7.2). In 30% of families, there is no family history but there is mosaicism (Section 2.4.2) with the new variant arising in the maternal grandfather's gametes making the mother a carrier. This is particularly true for the inversion in *F8* found in Jason. In addition, it is becoming apparent that many female carriers of hemophilia can have abnormally low FVIII levels and are symptomatic carriers with mild hemophilia. These women may have abnormal uterine bleeding with menses or delivery and can have significant clinical benefit from strategies to increase FVIII levels. Because Jason's mother is a carrier, for future children, 50% of males will be affected and 50% of females will be carriers.
- **Caveats:** Treatment paradigms for hemophilia may be shifting in the next few years as emerging data and novel therapies progress. Bipotent chimeric antibodies mimicking FVIII activity that can be administered subcutaneously may change the need for IV access, gene therapy trials have demonstrated efficacy in adult patients, especially in FIX deficiency, and data about use of recombinant versus plasma-derived products in prophylaxis with regard to risk of inhibitor development are emerging. In addition, some patients with severe deficiency (about 2%) do not have detectable variants in FVIII or FIX. Understanding these mechanisms may lead to novel therapies.
- **Outcome:** Jason has hemophilia A, but his bleeding has stabilized. Jason's mother is a carrier of the inversion, but her FVIII level is only 28% when she is tested about 4 months after delivery. The family is followed at a local hemophilia treatment center, and Jason is doing well currently on intermittent treatment.

The Diagnosis

Clinical Diagnosis: Patient—hemophilia A. Mother—symptomatic carrier.

Genetic Diagnosis: Patient—*F8* gene, hemizygous pathogenic variant. Mother—*F8* gene, heterozygous pathogenic variant.

Further Reading

Arruda, V.R., Doshi, B.S., Samelson-Jones, B.J., 2017. Novel approaches to hemophilia therapy: successes and challenges. Blood 130, 2251–2256. PMID 29018078.

Konkle, B.A., Johnsen, J.M., Wheeler, M., Watson, C., Skinner, M., Pierce, G.F., et al., 2018. Genotypes, phenotypes and whole genome sequence: approaches from the My Life Our Future haemophilia project. Haemophilia 24 (Suppl. 6), 87–94. PMID 29878652.

Von Der Lippe, C., Frich, J.C., Harris, A., Solbraekke, K.N., 2017. "It was a lot tougher than I thought it would be". A qualitative study on the changing nature of being a hemophilia carrier. J. Genet. Counsel. 26, 1324–1332. PMID 28547664.

Case 11.9.2

Hematologic and Immunologic Cases: Recurrent Infections in Childhood

Jessica C. Barry[1] and Donna M. McDonald-McGinn[2,3]

[1]*LVPG Maternal Fetal Medicine, Lehigh Valley Health Network, Allentown, PA, United States;* [2]*Section of Genetic Counseling, Clinical Genetics Center, 22q and You Center, the Children's Hospital of Philadelphia, Department of Pediatrics, Perelman School of Medicine of the University of Pennsylvania, Philadelphia, PA, United States;* [3]*Department of Human Biology and Medical Genetics, Sapienza University, Rome, Italy*

Abstract

A 5-year-old boy has had lifelong recurrent infections, gastroesophageal reflux disease, expressive language delays, and growth parameters tracking along the lower percentiles. He is found to have the 22q11.2 deletion syndrome.

The Case

Wyatt is a 5-year-old boy who is presenting with his fifth upper respiratory infection this year. Upon reviewing his history, you note that this is not an uncommon trend. Wyatt has a longstanding history of nasal regurgitation in infancy followed by recurrent upper respiratory infection including chronic otitis media (status post four myringotomy tube placements) with mild conductive hearing loss and more recently recurrent sinusitis. He also has been hospitalized at least three times due to respiratory syncytial virus pneumonia. Parents report that he was a fussy baby requiring multiple formula changes and ultimately medical management to treat his gastroesophageal reflux disease (GERD), as well as constipation occasionally necessitating enemas. During your conversation with his family, his mother notes that they are considering having him repeat kindergarten as he does not seem ready to move on to first grade. Wyatt's motor milestones were generally on time (sitting unsupported at 8 months and walking at 16 months); however, he had significant delays in emergence of language (single words at 22 months and short sentences at 32 months) which qualified him for early intervention services. The speech pathologist noted Wyatt's receptive language skills to be relatively preserved in comparison with expressive language and she raised concern over speech intelligibility due hypernasality. Parents also report apprehension about Wyatt moving on to first grade given his struggles keeping up with his peers cognitively and emotionally. Moreover, Wyatt has always tracked at the lower growth curves (10th percentile for height and weight), thus his parents feel that his size also impacts his ability to keep up with his peers.

The Issues

- **How concerned are you by his recurrent infections?** Some children seem to have weaker immune systems than their peers and are more prone to infections and illnesses, but the frequency and severity of Wyatt's infections has you concerned for the possibility of a primary immunodeficiency. Concurrently, you are considering Wyatt's history of GERD as possibly contributing to his upper respiratory infections. Wyatt's mother mentions that he previously received

Genomics in the Clinic. https://doi.org/10.1016/B978-0-12-816478-5.00063-7

childcare in his grandparents' home (from infancy until preschool) so he was not exposed to many sick children until recently.

- **What should the workup be for these symptoms?** There are a number of causes for recurrent infections in the pediatric population, including general risk factors (day-care, school attendance, older siblings, second-hand smoke). Recurrent infection could also be a sign of a primary or secondary immunodeficiency. Recommended preliminary bloodwork includes a CBC with differential, peripheral blood smear, electrolytes, glucose, urinalysis, renal and liver function studies, and HIV testing. You order some preliminary blood studies to probe the question of immunodeficiency. The results of these studies in Wyatt are all within normal limits, although his platelets are on the lower side of normal.
- **How concerned are you by his growth and development?** While Wyatt is on the smaller side, he has always tracked along his growth curve and has made slow but steady progress. Similarly, he has always been slightly behind in meeting his milestones but is always successful in reaching them. It is possible that Wyatt is just a late bloomer? But the combination of his growth, developmental delay, and infection history has you concerned that there could be an underlying and perhaps unifying etiology.
- **Could Wyatt have a genetic condition?** It is not clear if Wyatt's medical issues are due to an underlying genetic diagnosis or are just isolated, unrelated issues. The combination of these findings has raised enough of a red flag for you that you tell Wyatt's family you would like to refer him to Genetics for a full evaluation.

You should provide the genetics office evaluating Wyatt with a copy of his most recent pediatric records, growth chart, and the results of any laboratory studies he has had to date (both normal and abnormal).

The Referral

- **Genetic Evaluation**: A full pregnancy, perinatal, medical, and family history are reviewed during Wyatt's evaluation with the pediatric geneticist and genetic counselor. The pregnancy was remarkable only for polyhydramnios. Wyatt was born at term with normal growth parameters and Apgar scores. He had poor feeding in infancy with nasal regurgitation without evidence of an overt cleft palate. He was discharged on time. Following considerable weight loss as a neonate, supplemental nasogastric tube feeding was considered. However, Wyatt's feeding improved and the discussion was abated. He later benefited from ranitidine therapy for GERD outgrowing his dose by 18 months of age. He continues to have constipation, at times requiring enemas. In fact, he has been evaluated in the emergency department twice for stool impaction presenting as severe abdominal pain. His motor milestones were on the late side of normal, sitting unsupported at 8 months and walking at 16 months. His speech was delayed. Wyatt did not babble. He did not speak his first words until 22 months and he didn't string words together until 32 months. It is notable that he continues to struggle with articulation and it is difficult for him to be understood by anyone outside his immediate family. This often results in significant frustration manifesting as temper tantrums. A review of past medical records indicates that he has had four sets of myringotomy tubes placed following chronic otitis media with mild conductive hearing loss. Otolaryngology also noted tonsillar and adenoid hypertrophy.

  The family history reveals Wyatt to be the only child born to his parents, whose medical and family histories are unremarkable. Both parents are of Western European descent and deny consanguinity (Section 3.2.4). Neither parent reports any history of learning differences or behavioral problems. A four-generation pedigree (Section 3.2.1) reveals the remainder of the family history to be unremarkable. On physical examination, Wyatt's growth parameters are proportionate at the 10th percentile including his head circumference. He has mildly dysmorphic facial features including malar flatness, hooded eyelids, hypertelorism, protuberant ears with thick overfolded helices and attached lobes, a bulbous nasal tip with hypoplastic alae nasi, and a small mouth with an asymmetric smile (Fig. 11.9.2.1). His fingers are long and slender and he has slight hallux valgus with mild two to three syndactyly of his toes. He also has a small umbilical hernia. Hypernasal speech is noted.
- **Genetic Counseling**: The geneticist and genetic counselor review Wyatt's history and findings on physical exam. They inform his family that he may have a genetic condition known as chromosome 22q11.2 deletion syndrome (22q11.2DS). 22q11.2 deletion syndrome is the most common chromosome microdeletion syndrome (Sections 2.4, 6.1.3) resulting in a multisystem disorder, including but not limited to immunodeficiency (77%), congenital heart disease (69%) (most often conotruncal defects), palatal abnormalities (68%) (most often submucosal cleft palate or velopharyngeal insufficiency/incompetence), gastrointestinal differences (66%) (GERD, dysphagia, intestinal malrotation, umbilical hernia, constipation, imperforate anus, Hirschsprung's disease), endocrine abnormalities (55%) (hypocalcemia, thyroid disease, growth hormone deficiency), skeletal anomalies (50%) (vertebral anomalies, scoliosis, club foot,

(A) (B)

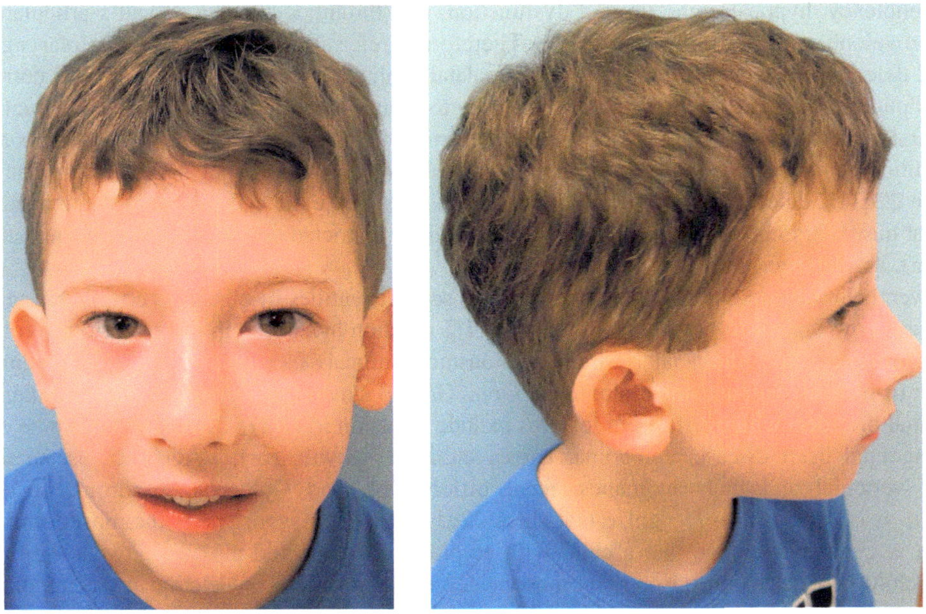

FIGURE 11.9.2.1 (A) and (B) Male patient with a standard LCR22A-LCR22D chromosome 22q11.2 deletion syndrome demonstrating typical craniofacial features including: malar flatness, hooded eyelids, mildly upslanting palpebral fissures, allergic shiners, prominent nasal root, bulbous nasal tip with hypoplastic alae nasi, protuberant ears, and mild micrognathia.

polydactyly), and genitourinary anomalies (33%) (absent kidney, multicystic dysplastic kidney, cryptorchidism, hypospadias, absent uterus). Developmental delay, particularly in speech, is common. Most individuals have cognitive deficits (more than 95%) (mean full scale IQ is ~75), although 65% of affected individuals have a nonverbal learning disability with a more than 10 point split between their verbal and performance IQ, with the verbal IQ being higher. This type of learning difference is uncommon in the general population, seen in only ~1%. Neuropsychiatric comorbidities are also frequent (60%) including an increased risk for autism spectrum disorder, ADHD, anxiety, obsessive compulsive disorder, and psychosis. In fact, 22q11.2 deletion syndrome is the most common cause of schizophrenia (25%) (Section 11.14.2). Facial features are mildly dysmorphic (Section 5.1.2, Fig. 5.2B, Fig. 11.9.2.1).

- **Results**: SNP microarray analysis (Section 6.1.3) is ordered and is positive for the common 2.5 Mb deletion of chromosome 22q11.2. This deletion results in loss of ~50 genes, including the important developmental gene *TBX1*, known to be associated with the triad of DiGeorge syndrome (conotruncal cardiac anomalies, T cell deficiency, and hypocalcemia due to hypoparathyroidism). Given the level of intrafamilial variability and that approximately 10% of patients inherit the deletion from a more mildly affected parent, parental testing is recommended (Section 7.4.3) and is negative. Thus, the 22q11.2 deletion is de novo, or new, in Wyatt.

  Wyatt will have a 50% (Sections 2.7.1, 4.4.3) chance of passing on the 22q11.2 deletion to each of his future children. When he reaches childbearing age, he can discuss reproductive options with his partner and healthcare provider including preimplantation genetic testing (Section 4.5.1) using in vitro fertilization, prenatal diagnostic genetic testing (Section 4.4.2), adoption, and the general variability of the 22q11.2 deletion syndrome among family members, including the risk to have a more severely affected child.

Both SNP microarray (Section 6.1.3) and multiplex ligation−dependent probe amplification (Section 6.2.3) studies can detect the common 2.5 Mb deletion as well as atypical nested deletions (1.5−0.7 Mb) on chromosome 22. FISH (fluorescence in situ hybridization) studies (Section 6.1.2) were developed in the early 1990s using probes to one specific area of the deletion located between low copy repeats (LCRs) A to B (LCR22A−LCR22B). These studies will miss any patient whose deletion does not include the "A-B" region of the chromosome and are therefore no longer the preferred testing method.

- **Management and Recommendations**: Individuals with a new diagnosis of 22q11.2 deletion syndrome require a thorough clinical evaluation at diagnosis and periodically thereafter as outlined in the established international healthcare guidelines. These include a cardiac evaluation with an echocardiogram and electrocardiogram, and possibly cardiac MRI to rule out an aortic arch anomaly/vascular ring. A renal ultrasound is recommended to rule out renal anomalies including a solitary kidney. Laboratory studies include a CBC with differential, calcium, thyroid, and an immune assessment (including assessment of T cell number and function, immunoglobulins, and antibody studies) due to the association with immunodeficiency, hypocalcemia, thyroid dysfunction, and thrombocytopenia. Wyatt's articulation deficits may be secondary to velopharyngeal dysfunction (VPD). Therefore, a referral to a plastic surgeon/otolaryngologist familiar with 22q11.2 deletion syndrome will allow for a full palatal evaluation with options for surgical correction as needed, with or without a primary adenoidectomy/tonsillectomy to reduce the risk of postoperative obstructive sleep apnea. A developmental pediatric/neurology evaluation is recommended to address Wyatt's cognitive deficits and behavioral differences. Additional initial evaluations include an ophthalmology exam, audiogram, speech evaluation by a therapist familiar with the 22q11.2 deletion syndrome, and orthopedics to examine the cervical spine for anomalies (seen in 50% of patients that may lead to instability with the resultant need for restricted activities, such as gymnastics, downhill skiing, and diving), and later to monitor for scoliosis.

- **Differential Diagnosis**: Several clinical conditions initially described as separate entities by clinicians concentrating on specific areas of expertise are now known to be caused by the 22q11.2 deletion syndrome in the majority of cases. These include: DiGeorge syndrome, velocardiofacial syndrome, conotruncal anomaly face syndrome, Cayler cardiofacial syndrome, and a subset of patients with autosomal dominant Opitz G/BBB syndrome. Depending on the patient's presenting findings, additional disorders with overlapping features include mutation in *TBX1*, CHARGE syndrome due to a *CHD7* mutation (congenital heart disease, coloboma, atresia choanae, growth, ear), VACTERL association (congenital heart disease, vertebral and limb anomalies, renal differences), and Goldenhar syndrome (oculo-auriculo-vertebral spectrum). It is important to recognize that since 22q11.2 deletion syndrome is the most common microdeletion syndrome, it is certainly possible that a child with atypical features could also have two conditions by chance alone or that a child has a deletion on one chromosome 22 and a mutation or deletion in an important gene on the intact chromosome 22q11.2 region, resulting in unmasking of an autosomal recessive condition (e.g., Bernard–Soulier syndrome resulting in thrombocytopenia and giant platelets due to mutations in the *GP1BB* gene; CEDNIK syndrome resulting in polymicrogyria and keratoderma due to mutations in *SNAP29*; and CGS syndrome, resulting in craniosynostosis, short stature, and limb differences as a result of mutations in *CDC45*, and *TANGO2* related diseases).
- **Caveats**: As a cautionary note, adenoidectomy in children with a submucous cleft palate due to 22q11.2 deletion syndrome may result in worsening hypernasality. Likewise, overtreatment of hypocalcemia may cause nephrocalcinosis.
- **Implications for Other Family Members**: Parental testing for the 22q11.2 deletion is negative. However, as there are a handful of reports of germline mosaicism (Section 2.4.2), the family is counseled that the recurrence risk is close to zero, but not zero. It is therefore recommended that they discuss future pregnancy monitoring with their obstetric provider, including noninvasive prenatal screening (Section 4.4.1), fetal ultrasound with fetal echocardiography, and prenatal diagnostic testing (chorionic villus sampling/amniocentesis) (Section 4.4.2).
- **Outcome**: In light of his diagnosis, Wyatt begins receiving additional therapeutic services through school with the support of recommendations set forth by the developmental pediatrician, including pull-outs (small group strategies) for academic subjects like beginning reading and math, as well as intensive one-on-one speech therapy to help overcome his compensatory articulation deficits. His parents elect to have him continue on to first grade and with the additional supports, he has been able to do well enough. Following presurgical studies of his cervical spine, nasendoscopy, magnetic resonance angiogram (MRA) of the carotid arteries, and tonsillectomy, he undergoes VPD repair which has greatly improved his speech resonance and has improved his level of frustration, concomitantly reducing his behavioral outbursts. He continues to have speech therapy to address his language deficits. A vascular ring is confirmed by chest MRI following a suspected right aortic arch with aberrant left subclavian artery on the echocardiogram. Wyatt is monitored annually by cardiology. Surgical intervention will only be considered should he become symptomatic.

The Diagnosis

Clinical Diagnosis: 22q11.2 deletion syndrome.
Genetic Diagnosis: Chromosome 22q11.2 microdeletion.

Further Reading

Boot, E., Óskarsdóttir, S., Loo, J.C.Y., Crowley, T.B., Orchanian-Cheff, A., Andrade, D.M., et al., 2023. Updated clinical practice recommendations for managing adults with 22q11.2 deletion syndrome. Genet. Med. 25, 100344. PMID 36729052.

Campbell, I.M., Sheppard, S.E., Crowley, T.B., McGinn, D.E., Bailey, A., Mcginn, M.J., et al., 2018. What is new with 22q? An update from the 22q and You Center at the Children's Hospital of Philadelphia. Am. J. Med. Genet. 176, 2058–2069. PMID 30380191.

Cohen, J.L., Crowley, T.B., McGinn, D.E., McDougall, C., Unolt, M., Lambert, M.P., et al., 2018. 22q and two: 22q11.2 deletion syndrome and coexisting conditions. Am. J. Med. Genet. 176, 2203–2214. PMID 30244528.

McDonald-McGinn, D.M., Hain, H.S., Emanuel, B.S., Zackai, E.H., September 23, 1999. 22q11.2 deletion syndrome. Adam, M.P., Ardinger, H.H., Pagon, R.A., Wallace, S.E., Bean, L.J.H., Gripp, K.W., Mirzaa, G.M., Amemiya, A., (editors). In GeneReviews. University of Washington, Seattle (WA), 1993–2022 [updated February 27, 2020]. PMID 20301696.

Mcdonald-McGinn, D.M., Sullivan, K.E., Marino, B., Philip, N., Swillen, A., Vorstman, J.A., et al., 2015. 22q11.2 deletion syndrome. Nat. Rev. Dis. Prim. 1, 15071. PMID 27189754.

Óskarsdóttir, S., Boot, E., Crowley, T.B., Loo, J.C.Y., Arganbright, J.M., Armando, M., et al., 2023. Updated clinical practice recommendations for managing children with 22q11.2 deletion syndrome. Genet. Med. 25, 100338. PMID 36729053.

Unolt, M., Kammoun, M., Nowakowska, B., Graham, G.E., Crowley, T.B., Hestand, M.S., et al., 2019. Pathogenic variants in CDC45 on the remaining allele in patients with a chromosome 22q11.2 deletion result in a novel autosomal recessive condition. Genet. Med. 22, 326–335. PMID: 31474763.

Case 11.9.3

Hematologic and Immunologic Cases: Chronic Thrombocytopenia in a Child

Michele P. Lambert

Perelman School of Medicine, University of Pennsylvania, Division of Hematology, Children's Hospital of Philadelphia, Philadelphia, PA, United States

Abstract

An 8-year-old girl with chronic thrombocytopenia since infancy and positive family history is found to have *MYH9*-related thrombocytopenia inherited from the paternal side of the family.

The Case

An 8-year-old female, Nia, with a history of chronic immune thrombocytopenia (ITP, idiopathic or immune thrombocytopenic purpura) is seen for her well-child follow-up as a new patient to the practice after relocating to the area. Her low platelet count was initially diagnosed at 9 months of age, after presenting with an incidental finding of thrombocytopenia during a screening cell blood count (CBC) for iron deficiency. She has had multiple CBCs since then with platelet counts routinely between 40,000 and 60,000, and she has never responded to any therapy for her low platelet count. She has been otherwise completely well and has not had any other major problems although she is mildly overweight, in part because her parents and previous providers would not allow her to participate in any physical activities. She has not had any other significant medical issues. Recent bloodwork at your office demonstrated a macrothrombocytopenia with a platelet count of 45×10^9/L with an elevated MPV of 18.2 fL. She had an otherwise normal CBC and has had no bleeding symptoms. Her family history is significant for a father with hearing loss diagnosed in his mid-30s and a paternal uncle with cataracts and renal disease.. Nia has a paternal grandmother with thrombocytopenia that was noted during her pregnancies that never fully resolved but she never had bleeding. The family thinks that Nia's father may have had a low platelet count at some point. Her physical examination is normal and she has no evidence of bleeding on her exam (i.e., abnormal bruising, petechiae).

The Issues

- **Does this patient have chronic immune thrombocytopenia?** While chronic ITP, an acquired, autoimmune disease that causes low platelet counts in 1:20,000 children per year, has been classically considered the most common cause of low platelet counts in pediatric patients, certain clues can help clinicians identify those children who are more likely to have congenital thrombocytopenia. Family history of thrombocytopenia, especially if there is more than one family member who may have low platelet counts, suggests congenital thrombocytopenia. But it can be difficult to assess if the

Genomics in the Clinic. https://doi.org/10.1016/B978-0-12-816478-5.00064-9

count dropped from a previously normal level or if there has never been a prior platelet count. A platelet count that does not respond to treatment also makes you wonder if the diagnosis of chronic ITP is correct.

- **Are there any other clues?** Associated features in some platelet disorders such as large platelets, hearing loss, renal disease, and cataracts or in others dysmorphic features, developmental delay, and cardiac abnormalities may suggest syndromic platelet disorders. This patient has none of these.
- **Why is the correct diagnosis important?** Nia is mildly overweight and has been prohibited from participation from physical activities by her parents and previous providers due to concerns for bleeding. This can have serious health consequences in terms of health-related quality of life, mental health, and long-term health outcomes. With the fresh eyes of a new evaluation, you wonder if cognitive bias of the established diagnosis played a role in the patient's prior evaluations and suspect that the diagnosis of chronic ITP may not be the best explanation for her thrombocytopenia
- **What's the best way to establish the diagnosis of congenital/familial thrombocytopenia?** Referral to a pediatric hematologist with expertise in benign hematology can help guide further evaluation. Often, it is helpful to have records from family members suspected of having thrombocytopenia available for the hematology evaluation in addition to the pediatric records including laboratory studies and growth charts. In addition, the family will benefit from evaluation by clinical genetics to aid in establishing the diagnosis and counseling about risk. You refer to both specialists.

The Referral

- **Genetic Evaluation**: A full pregnancy history, a medical history, and a family history are obtained, and a detailed physical examination is performed by the pediatric geneticist. Family history reveals that both the father and paternal uncle may have been told about low platelets in the past, but because they have never had bleeding issues, the father is not sure about whether this is still an issue for his brother. During Nia's hematology evaluation, the father's platelet count is found to be low (60,000). Nia's father remembers that his mother had a low platelet count because there was some concern when she was older about leukemia, which she did not ultimately have. Nia's paternal great-grandmother also had hearing loss (like Nia's father). Only the paternal uncle has renal disease and cataracts. Nia's physical exam is normal.
- **Genetic Counseling**: The geneticist and genetic counselor discuss with Nia's parents that there are several platelet disorders that are autosomal dominant (Section 2.7.1), which seems to be the situation in this family, with a 50% risk of inheritance from affected parent to child. Unfortunately, molecular diagnosis of some of the inherited platelet disorders is still not possible, but for about 75% of patients, the cause can be identified with testing, allowing for more precise counseling and screening for associated problems.
- **Genetic Testing**: There are multiple inherited platelet disorder gene panels (Section 6.2.3) available for clinical diagnosis. Most of these panels also provide some variant interpretation using available databases. ClinGen (https://clinicalgenome.org/) developed several expert panels to help curate genes and variants for the inherited platelet disorders (Section 7.1). A platelet panel is recommended for Nia.
- **Results**: Nia's genetic evaluation demonstrates a known pathogenic variant (Section 7.1.3) (p.R702H) in the gene *MYH9* (myosin heavy chain 9), which encodes for the nonmuscle myosin heavy chain IIA (myosin IIA). Clinically, patients may present at birth with macrothrombocytopenia and spindle-shaped Döhle leukocyte inclusions. These inclusions stain positively for myosin II when immunostaining is performed, suggesting that the abnormal protein precipitates and forms these inclusions. In childhood or adult life, patients may develop sensorineural hearing loss, cataracts, and glomerulonephritis that may progress to severe renal failure. Thrombocytopenia is usually noted incidentally or on routine screening in patients with *MYH9*-related disease and does not usually cause a symptomatic bleeding diathesis, although a few patients may have severe bleeding. Nia's father undergoes targeted variant analysis (Section 6.2.1) for the specific mutation and is found to have it.
- **Management and Recommendations**: Further information is shared with Nia's parents. *MYH9*-related disease is the most common inherited thrombocytopenia, and its earliest manifestation is macrothrombocytopenia with additional manifestations developing throughout the lifetime. Based on the particular variant, there are some genotype–phenotype correlations that can be predicted, but some variability exists in phenotype even within the same family (Section 2.8). Therefore, recommendations currently are for all affected individuals to have screening on a regular basis for hearing loss, urinalysis for early evidence of glomerulonephritis, and eye screening for cataracts in addition to discussion of the potential for increased risk of bleeding particularly with trauma and surgical intervention. Some families also have mild elevation of liver enzymes.
- **Caveats**: Just because someone has negative genetic testing for an inherited platelet disorder, with a strong family history of thrombocytopenia, there could still be an inherited disease, with the molecular basis for that family's

thrombocytopenia not yet determined. New platelet disorders are being described at the rate of three to four new disorders (new genes) per year, so it is important to revisit the diagnosis and the recommendations every few years. Additionally, it is extremely important to discuss risks and benefits of genetic testing for establishing the diagnosis in families prior to undergoing the testing. In some cases, where an inherited platelet disorder is suspected but the only additional manifestation may be a predisposition to cancer, the molecular testing may be deferred until the child is old enough to participate in the decision-making process (Section 10.1.2). In this case, a syndromic disorder with multisystem disease was more likely, so earlier testing is appropriate to establish the correct diagnosis and ensure proper monitoring.

- **Differential Diagnosis**: There are several other familial macrothrombocytopenias that have been described to be autosomal dominant associated with pathogenic variants in *ACTN1*, *TUBB1*, and *DIAPH1*. Both *MYH9*- and *DIAPH1*-related thrombocytopenias have associated hearing loss. In addition, with the family history of renal disease and cataracts, it is possible that there is some familial immune dysregulation that is autosomal dominant leading to this type of presentation such as can be seen in *CTLA4* deficiency or *LRBA* deficiency although the latter is autosomal recessive inheritance.
- **Outcome**: Nia is found to have an inherited platelet disorder instead of chronic ITP and returns to hematology for further follow-up. Management is not aimed at raising the platelet count by measures that are not effective (such as immunomodulation or splenectomy) but instead focuses on education and monitoring for associated syndromic features, health maintenance, and use of effective therapies to modulate platelet counts if needed (such as transfusion or thrombopoietin receptor agonists). Nia is encouraged to resume exercise and participate in physical education at school. She is referred to nephrology for evaluation of her renal function, as early treatment with ACE inhibitors has been shown to be beneficial in preventing progression to renal failure; she is referred to ophthalmology for evaluation for cataracts and to audiology for hearing evaluation.

The Diagnosis

Clinical Diagnosis: Patient and Father—myosin heavy chain 9—related inherited thrombocytopenia.
Genetic Diagnosis: Patient and Father—*MYH9* gene, heterozygous pathogenic variant.

Further Reading

Balduini, A., Raslova, H., Di Buduo, C.A., Donada, A., Ballmaier, M., Germeshausen, M., et al., 2018. Clinic, pathogenic mechanisms and drug testing of two inherited thrombocytopenias, ANKRD26-related thrombocytopenia and MYH9-related diseases. Eur. J. Med. Genet. 61, 715–722. PMID 29545013.

Melazzini, F., Zaninetti, C., Balduini, C.L., 2017. Bleeding is not the main clinical issue in many patients with inherited thrombocytopaenias. Haemophilia 23, 673–681. PMID 28594466.

Nava, T., Rivard, G.E., Bonnefoy, A., 2018. Challenges on the diagnostic approach of inherited platelet function disorders: is a paradigm change necessary? Platelets 29, 148–155. PMID 29090587.

Pecci, A., Klersy, C., Gresele, P., Lee, K.J., De Rocco, D., Bozzi, V., et al., 2014. MYH9-related disease: a novel prognostic model to predict the clinical evolution of the disease based on genotype-phenotype correlations. Hum. Mutat. 35, 236–247. PMID 24186861.

Case 11.9.4

Hematologic and Immunologic Cases: Thrombocytopenia in an Adult

Ayuko Iverson and Manisha Balwani

Departments of Genetics and Genomic Sciences and Medicine, Icahn School of Medicine at Mount Sinai, New York, NY, United States

Abstract

A 30-year-old woman with history of idiopathic thrombocytopenia and splenomegaly is pregnant and is found to have Gaucher disease type 1 on expanded carrier screening.

The Case

Miriam, a 30-year-old Ashkenazi Jewish female with a history of idiopathic thrombocytopenia, is presenting to you, her obstetrician, for a preconception visit after an expanded carrier screening (Section 4.1) found two pathogenic variants in the gene for Gaucher disease type 1, thus diagnosing her with this condition. The rest of her expanded carrier screening was negative. Her partner Michael, who is also Ashkenazi Jewish, is found to be a carrier for Gaucher disease type 1. Miriam is concerned about what this diagnosis means for her health, carrying a pregnancy, and passing the disease on to her future children. You recommend basic lab work, including a complete blood count, which shows stable thrombocytopenia.

The Issues

- **What is the risk of a future pregnancy to be affected with Gaucher disease?** Gaucher disease is an autosomal recessive (Section 2.7.1) condition. Since Miriam is affected and Michael is a carrier, there is 50% chance their child will be affected with Gaucher disease and a 50% chance their child will be a carrier.
- **How accurate are these carrier testing results?** Carrier screening for Gaucher disease is performed by genotyping. A negative test does not completely eliminate the possibility of being a carrier; however, a positive result is considered definitive. Carrier screening for Gaucher disease detects greater than 95% of cases in Ashkenazi Jews. In other ethnic populations (Section 3.2.3), the detection rate is lower. In rare cases, such as this, individuals who are homozygous (Section 2.3.4) for an autosomal recessive condition (Section 2.7.1) are picked up on carrier screening (Section 4.1). You refer Miriam and Michael to a clinical geneticist to review their genetic testing results and to discuss further management of Gaucher disease.

The Referral

- **Genetic Evaluation**: Miriam and Michael meet with the clinical geneticist, who obtains a medical history and completes a physical exam. Miriam notes that she bruises easily and bleeds excessively with minor cuts but does not

Genomics in the Clinic. https://doi.org/10.1016/B978-0-12-816478-5.00065-0

have a history of menorrhagia. She denies bone pain, fractures, early satiety, gallstones, or fatigue. Her physical exam is significant for splenomegaly.

A three-generation pedigree (Section 3.2) is obtained of Miriam's family. She has a younger brother who is healthy and otherwise has no family history of Gaucher disease, or any other genetic condition.

- **Genetic Counseling**: The clinical geneticist reviews the diagnosis of Gaucher disease with Miriam. The disease, a lysosomal storage disorder, is caused by the accumulation of glucosylceramide (GL1) due to the deficient activity of glucocerebrosidase. The three clinical types of Gaucher disease are determined by the presence (types 2 and 3) or absence (type 1) of primary neurologic disease. Miriam was homozygous (Section 2.7.1) for the p.Asn409Ser pathogenic variant (formerly known as p.N370S) within *GBA*, the gene associated with Gaucher disease, on her carrier screening (Section 4.1). As individuals with at least one p.Asn409Ser variant do not develop primary neurologic disease, Miriam has Gaucher disease type 1. Michael also carries the p.Asn409Ser variant and thus an affected child would have Gaucher disease type 1.

  Gaucher disease type 1 is characterized by bone disease (osteopenia, lytic lesions, and osteonecrosis), hepatosplenomegaly, and cytopenias (Section 5.4.1). There is also an increased risk for Parkinson disease and multiple myeloma. Carriers for Gaucher disease also have an increased risk of Parkinson disease (Section 11.17.4). Miriam is surprised to hear that her thrombocytopenia is likely secondary to Gaucher disease. See also Section 11.17.4 for further discussion of Gaucher disease and implications for carriers.

  Gaucher disease type 1 is prevalent in the Ashkenazi Jewish population, with four variants, including p.Asn409Ser and p.Leu483Pro, accounting for approximately 90% of the pathogenic variants (Section 7.1.3) found. The estimated carrier frequency in Ashkenazi Jews is 1:18, with a disease prevalence of 1:855.
- **Management and Recommendations**: In addition to the expanded carrier screen and complete blood count that were obtained previously, the clinical geneticist recommends baseline labs and imaging as part of Miriam's evaluation. Liver function testing, a coagulation profile, serum protein electrophoresis (SPEP) with immunofixation and quantification of immunoglobulins, ferritin, and an iron panel are drawn. Glucocerebrosidase activity is measured. Biomarkers of disease burden, plasma chitotriosidase activity, and glucosylsphingosine (lysoGL1), are obtained.

  For imaging, an abdominal MRI to evaluate liver and spleen volumes, an MRI of the femurs and lumbar spine to evaluate bone marrow infiltration for Gaucher cells, and/or infarcts, and a DEXA scan are ordered.

  Miriam asks about treatment. Enzyme replacement therapy (ERT) and substrate reduction therapy (SRT) (Section 5.4.3), which provide exogenous enzyme to clear GL1 or limit the amount of GL1 that is produced, respectively, are available treatment options to treat symptoms and prevent complications. Increase in platelet counts and decrease in liver and spleen size generally occur within a year of initiation of therapy, and bone density improves 3–5 years after initiating therapy.

  Pregnancy may worsen existing symptoms as well as cause new symptoms, such as bone pain. Postpartum bleeding is the most common serious complication. Substrate reduction therapy is contraindicated in pregnancy, thus enzyme replacement therapy is the best option for Miriam. The clinical geneticist recommends starting treatment now due to her thrombocytopenia and continuing treatment throughout pregnancy.
- **Results**: Miriam's liver function profile, coagulation profile, SPEP, ferritin, and iron panel are normal. Her baseline chitotriosidase activity and lyso-GL1 levels are elevated and her glucocerebrosidase activity is significantly decreased, as expected. Her MRI abdomen reveals a normal liver size and an enlarged spleen about six times normal. Her MRI of her femurs and spine shows infiltration by Gaucher cells throughout the spinal bone marrow and medullary cavity of both femurs and proximal third of both tibias. Her DEXA scan shows osteopenia.
- **Implications for Other Family Members**: As Miriam and Michael's family members are also at risk for being either carriers or affected with Gaucher disease, the clinical geneticist recommends they receive genetic counseling and testing. Miriam's younger brother in particular should be tested, as he also may be affected.
- **Differential Diagnosis**: The clinical features of Niemann–Pick disease type B (acid sphingomyelinase deficiency) including hepatosplenomegaly overlap with Gaucher disease type 1. Features that differentiate Niemann–Pick from Gaucher disease are interstitial lung disease and low HDL. Hepatosplenomegaly and thrombocytopenia can also be seen in other storage diseases such as lysosomal acid lipase deficiency which also presents with serum lipid abnormalities, xanthelasma, and elevated liver enzymes in later onset forms.
- **Caveats**: In non-Jewish individuals, *GBA* genotyping of the four most common variants (Section 6.2.1) will only detect approximately 50%–60% of cases, thus sequence analysis of the entire gene is recommended (Section 6.2.2). While individuals with at least one p.Asn409Ser variant do not develop primary neurologic disease, genotype–phenotype correlations are otherwise limited due to a wide range of phenotypic variability (Section 2.8).

- **Outcome**: Miriam starts enzyme replacement therapy every 2 weeks under the care of her clinical geneticist. She becomes pregnant a few months after her initial visit and she decides to do prenatal genetic testing (Section 4.4.2) for Gaucher disease via chorionic villous sampling at week 11 of gestation. Michael and Miriam are told their child is a carrier for p.Asn409Ser and thus is not affected with Gaucher disease. Miriam's pregnancy and delivery are uncomplicated and their child is healthy. Miriam's platelet count normalizes a year after starting treatment and her chitotriosidase and lyso-GL1 levels are significantly reduced, indicating decreased disease burden. The clinical geneticist recommends continued yearly follow-up.

The Diagnosis

Clinical Diagnosis: Patient—Gaucher disease, type 1. Partner—carrier.
Genetic Diagnosis: Patient—*GBA* gene, homozygous pathogenic variants. Partner—*GBA* gene, heterozygous pathogenic variant.

Further Reading

Biegstraaten, M., Cox, T.M., Belmatoug, N., Berger, M.G., Collin-Histed, T., Vom Dahl, S., et al., 2018. Management goals for type 1 Gaucher disease: an expert consensus document from the European working group on Gaucher disease. Blood Cells Mol. Dis. 68, 203–208. PMID 28274788.

Cox, T.M., Aerts, J.M., Belmatoug, N., Cappellini, M.D., Vom Dahl, S., Goldblatt, J., et al., 2008. Management of non-neuronopathic Gaucher disease with special reference to pregnancy, splenectomy, bisphosphonate therapy, use of biomarkers and bone disease monitoring. J. Inherit. Metab. Dis. 31, 319–336. PMID 18509745.

Elstein, Y., Eisenberg, V., Granovsky-Grisaru, S., Rabinowitz, R., Samueloff, A., Zimran, A., et al., 2004. Pregnancies in Gaucher disease: a 5-year study. Am. J. Obstet. Gynecol. 190, 435–441. PMID 14981386.

Pastores, G.M., Weinreb, N.J., Aerts, H., Andria, G., Cox, T.M., Giralt, M., et al., 2004. Therapeutic goals in the treatment of Gaucher disease. Semin. Hematol. 41, 4–14. PMID 15468045.

Zuckerman, S., Lahad, A., Shmueli, A., Zimran, A., Peleg, L., Orr-Urtreger, A., et al., 2007. Carrier screening for Gaucher disease: lessons for low-penetrance, treatable diseases. JAMA 298, 1281–1290. PMID 17878420.

Case 11.9.5

Hematologic and Immunologic Cases: Bone Marrow Failure

Michele P. Lambert

Perelman School of Medicine at the University of Pennsylvania, Division of Hematology, Children's Hospital of Philadelphia, Philadelphia, PA, United States

Abstract

A 14-year-old male with progressive fatigue, exercise intolerance, a history of familial short stature, and bone marrow failure is found to have Fanconi anemia.

The Case

Nathan is a 14-year-old male, previously healthy except for short stature, who presents to your office with progressive fatigue and decreasing exercise tolerance over the past several months. In the past week, he and his family have noted that he has some increased bruising, and yesterday he had his first episode of epistaxis. He has had some low-grade fever for the past 2–3 days which, combined with the new bleeding symptoms, prompted the family to make the appointment with you, his pediatrician. He has not had contact with anyone sick recently. He does not have any lymphadenopathy or organomegaly on exam, but he is pale, with increased bruises and petechiae. His vital signs are normal at the time of the visit and he is afebrile. You note that his height is at the third centile, while his parents appear to be more average height.

The Issues

- **What is the cause of the pallor, increased bruising?** Given these symptoms, cytopenia (anemia, thrombocytopenia) is likely and a cell blood count (CBC) is a critical diagnostic test and should be performed STAT. You appreciate that Nathan probably should be referred to tertiary care center able to perform some further diagnostic evaluation because the differential diagnosis at this point includes bone marrow failure, acquired aplastic anemia, and leukemia.
- **Are there any other clues?** The long latency of symptoms, lack of organomegaly along with the short stature make inherited bone marrow failure syndrome (IBMFS) a more likely diagnosis, but the IBMFSs are more likely to develop leukemia.
- **What does the workup show?** The patient is referred to the emergency room where workup demonstrates normal chemistry panel, but a complete blood count shows pancytopenia with profound anemia, thrombocytopenia, and leukopenia. The patient is admitted for further diagnostic evaluation and a bone marrow demonstrates a hypocellular marrow without evidence of malignancy. Therefore, the differential diagnosis is narrowed to acquired aplastic anemia versus IBMFSs.

Genomics in the Clinic. https://doi.org/10.1016/B978-0-12-816478-5.00066-2

- **Is there a role for genetics in evaluation of the patient with apparent aplastic anemia?** Although the clinical features of the IBMFSs can be diagnostic, incomplete disease penetrance and variable expressivity (Section 2.8) result in sometimes subtle phenotypes and delayed presentation, and these conditions may present as diagnostic dilemmas. Next-generation sequencing (Section 6.2.2) has contributed greatly to our understanding of the genetic etiology of the IBMFSs and has allowed for the expansion of phenotypes and identification and novel molecular etiologies. However, this increased complexity has made genetic evaluation of all patients presenting with pancytopenia with possible bone marrow aplasia imperative prior to definitive therapy. The patient is referred for a genetic evaluation.

The Referral

- **Genetic Evaluation**: A full family history, physical examination, and medical history are performed by the clinical geneticist and genetic counselor. There is no family history (Section 3.2) of any bleeding disorders, nor is there family history of cancer or bone marrow failure. The growth charts are reviewed for Nathan and it is noted that he has always been small with growth well below the 10th percentile for all of his life. Physical examination also reveals some subtle abnormalities that the pediatrician did not appreciate. Nathan's face is more triangular in shape than his parents or his 12-year-old sibling (who has normal height and growth velocity). Nathan has a few café au lait spots (Section 5.1.7) on his trunk and lower extremities and he has a flattened thenar eminence with shortened thumbs. His parents and sibling are phenotypically normal.

  The geneticist and genetic counselor discuss that Nathan's physical examination of short stature, abnormal thumbs, café au lait macules, and his bone marrow failure suggest Fanconi anemia (FA). Patients with FA develop variable degrees of pancytopenia in childhood with 40% developing bone marrow failure by age 20 years. They generally have short stature and skin abnormalities, such as café au lait macules and/or hyper- or hypopigmented spots. Upper limb abnormalities (Section 5.1.6) (absent, hypoplastic, bifid, duplicated or rudimentary thumb; absent or hypoplastic radius, flat thenar eminence, clinodactyly or polydactyly) are common, occurring in about 30% of patients. While one-third of patients with FA may have no congenital abnormalities and present only with the hematologic manifestations, others can be quite severely affected from the neonatal period with microcephaly, a triangular "Fanconi" face, Vertebral defects, Anal atresia, Cardiac defects, Tracheo-Esophageal fistula or esophageal atresia, Renal anomalies, and Limb abnormalities (features of the VACTERL association). There is also an increased risk for solid tumors and acute myelogenous leukemia (AML).
- **Genetic Counseling**: There are at least 20 genes in the FA/BRCA pathway that have been implicated in FA. Most FA is inherited in an autosomal recessive form (Section 2.7.1) through pathogenic variants (Section 7.1.3) in one of the key DNA repair genes (*FANCA, FANCC, BRCA2/FANCD1, FANCD2, FANCE, FANCF, FANCG, FANCI, BRIP1/FANCJ/BACH1, FANCL, FANCM, PALB2/FANCN, RAD51C/FANCO, SLX4/FANCP, ERCC4/FANCQ/XPF, RAD51/FANCR, BRCA1/FANCS/, UBE2T/FANCT, XRCC2/FANCU*), and there is one X-linked gene (*FANCB*) (Section 2.7.2). The carrier frequency in the United States is about 1 in 180, generally, but much more common in some populations (Section 3.2) (such as the Ashkenazi Jewish population with carrier frequency estimated at 1 in 77–108). Identification of pathogenic variants in FA genes in Nathan along with the mode of inheritance allows for appropriate management of not only his bone marrow failure but also counseling for his family about their risk, identification of potentially affected family members with decreased penetrance, and selection of suitable bone marrow donors.

Of note, FANCD1 is also called BRCA2 and some families have history of solid tumors such as pancreas and prostate cancer, melanoma, and breast cancer (particularly in men) but have less bone marrow failure.

- **Management and Recommendations**: Laboratory diagnosis generally proceeds with chromosomal breakage studies (Section 2.2.1) to demonstrate the functional defect in FA while pursuing genetic diagnosis. Generally, patients are tested with a combination of panel testing (Section 6.2.3), exome sequencing (Section 6.3.1), array CGH (Section 6.1.3), and even RNA sequencing (Section 6.3.3) depending on the results of functional testing and degree of suspicion, coverage on the panel, exome sequencing, and other factors of individual tests. Bone marrow transplant is recommended, depending on the molecular results.
- **Results**: Nathan is found to harbor heterozygous pathologic variants in *FANCA*, one of which was de novo and the other inherited from his father. His sibling has normal chromosomal breakage studies and did not inherit the *FANCA* variant, and is a 9/10 HLA match for bone marrow transplant. Therefore, Nathan is prepped for a modified bone marrow transplant using a reduced intensity regimen (because patients with FA have difficulty tolerating chemotherapy and may have increased late effects) from his matched sibling. Because Nathan has been diagnosed with FA, he will

now require additional monitoring and testing including: assessment of his liver and renal structure and function, endocrine assessment (thyroid disease, diabetes, and growth hormone abnormalities are common), hearing assessment, eye examination, and careful head and neck examination and screening for head and neck cancers.

- **Differential Diagnosis**: Other conditions that can present with a hypocellular appearing bone marrow or pancytopenia include myelodysplastic syndromes (which are rare in childhood but can be associated with variants in some genes such as *ETV6*, *RUNX1*, and *MECOM*). Other canonical IBMFSs include dyskeratosis congenita, Pearson syndrome, amegakaryocytic thrombocytopenia, Shwachman–Diamond syndrome, severe congenital neutropenia, Blackfan–Diamond anemia, and thrombocytopenia absent radii. Novel bone marrow failure syndromes are being described due to variants in many different genes involved in hematopoiesis.
- **Caveats**: The pathway governing DNA repair is still not completely understood, nor are all of the disease mechanisms for FA. Patients are still identified as having abnormal chromosome breakage, or clinical phenotype consistent with FA, but without definitive molecular testing. These patients present a challenge for management, especially if considering transplant from matched related family members.
- **Outcome**: Nathan has full bone marrow recovery and is now undergoing routine screening and surveillance for the associated increased solid tumor risks.

The Diagnosis

Clinical Diagnosis: Fanconi anemia.
Genetic Diagnosis: *FANCA* gene, compound heterozygous pathogenic variants.

Further Reading

Kallen, M.E., Dulau-Florea, A., Wang, W., Calvo, K.R., 2019. Acquired and germline predisposition to bone marrow failure: diagnostic features and clinical implications. Semin. Hematol. 56, 69–82. PMID 30573048.

Wegman-Ostrosky, T., Savage, S.A., 2017. The genomics of inherited bone marrow failure: from mechanism to the clinic. Br. J. Haematol. 177, 526–542. PMID 28211564.

Section 11.10

Skeletal and Connective Tissue Cases

Case 11.10.1

Skeletal and Connective Tissue Cases: Asymmetric Leg Size in an Infant

Rosanna Weksberg[1,2] and Cheryl Cytrynbaum[3,4]

[1]*Division of Clinical and Metabolic Genetics, Genetics and Genome Biology, The Hospital for Sick Children, University of Toronto, Toronto, ON, Canada;* [2]*Departments of Pediatrics and Molecular Genetics, University of Toronto, ON, Canada;* [3]*Department of Genetic Counseling, Genetics and Genome Biology, The Hospital for Sick Children, University of Toronto, ON, Canada;* [4]*Department of Molecular Genetics, University of Toronto, ON, Canada*

Abstract

A 6-month-old male is found to have a larger right leg than left, which is diagnosed to be an isolated hemihyperplasia.

The Case

Colton is a 6-month-old male, whom you have been following in your family practice. He is being seen for a routine well-child visit. During the physical exam you note that his right thigh and calf are larger in girth than his left. The remainder of your assessment is unremarkable. Colton is growing well, with all growth parameters at the 50th percentile. There is no other evidence of somatic asymmetry. He has been meeting all of his developmental milestones. His parents had not noted the asymmetry prior to your observation.

The Issues

- **Why is one leg larger than the other?** When one part or side of the body is larger than the other, this is called hemihyperplasia. This occurs because of an abnormality of cell growth in which the cells on the larger side of the body are growing faster; this may involve differences in growth of the soft tissue, bones, or both. This is likely something that was present from birth but has become more evident as Colton is growing.
- **Could he have other medical problems?** In most cases, hemihyperplasia (also known as lateralized overgrowth) is isolated, meaning that it occurs without signs of other medical problems; however, it may be one symptom of an underlying genetic syndrome. This is important to clarify for Colton's ongoing medical management. You reassure the parents that his growth otherwise and his development appear normal.
- **What is the next step?** You recall that hemihyperplasia can be associated with an increased risk for tumors in the kidney and liver. Therefore, you recommend that Colton undergo a genetic assessment to try to determine the cause of his hemihyperplasia (isolated or part of a genetic condition) which will provide them with more information about any associated tumor risks and screening recommendations. You will make a referral and in the interim you will arrange for Colton to have an abdominal ultrasound and serum alpha-fetoprotein (AFP).

Genomics in the Clinic. https://doi.org/10.1016/B978-0-12-816478-5.00067-4

The Referral

- **Genetic Evaluation:** Colton and his parents are seen in the pediatric genetic clinic by the clinical geneticist and genetic counselor 4 weeks later. The pregnancy, medical, and family histories are reviewed and a detailed physical examination is performed by the geneticist. The geneticist confirms the referring doctor's impression of somatic asymmetry involving Colton's legs. Clinical measurements are taken where visually the difference between the right and left limbs appears to be maximal. These indicate that the difference in the girth of his thighs (measured 12 cm above the top of the patella) is 2.0 cm and the difference between his calves (measured 10 cm below the bottom of the patella) is 1.5 cm. At this time the difference between his leg lengths (measured from the pubic symphysis) measures 1.2 cm, just above what is considered within the normal range of discrepancy (1.0 cm). There is no other evidence of asymmetry involving his arms, torso, face, or tongue. There are no pigmentary findings, specifically no hyperpigmented macules and no vascular lesions. There are no ear pits or creases (Section 11.7.1, Fig. 11.7.1.2). Colton's growth continues to plot at the 50th percentile and there are no concerns regarding his developmental progress.
- **Genetic Counseling:** The geneticist and genetic counselor review the findings and return to discuss their impression with the Colton's parents. Their assessment suggests that Colton has isolated hemihyperplasia (Fig. 11.10.1.1). There are no findings at this time to suggest he has a syndromic cause for his hemihyperplasia. As mentioned by their family doctor, there is an increased risk for embryonal tumors in individuals with isolated hemihyperplasia. This risk is about 6% and appears to be higher in the first 8 years of life. Tumor development in affected individuals older than age 8 years is uncommon. Screening recommendations are discussed below.

  The majority of patients with isolated hemihyperplasia have no identified cause. Approximately 20% of patients have a molecular alteration that impacts the regulation of imprinted genes (Sections 2.7.5 and 6.4.3) in the chromosome 11p15.5 region, specifically DNA methylation changes in this region or paternal uniparental disomy. These are the same changes that have been described in individuals with Beckwith–Wiedemann syndrome (BWS) (Section 11.7.1). It is not clear at this time why individuals with the same molecular finding can present with either isolated hemihyperplasia or BWS. These two conditions may be part of a spectrum where the phenotypic expression may depend, in part, on the level of somatic mosaicism (Section 2.4.2) (how many and which cells carry the genetic change). Genetic testing is discussed below.
- **Management and Recommendations:** Recommendations for screening for embryonal tumors include abdominal ultrasound examination every 3 months until age 8 years to screen for Wilms tumor and hepatoblastoma, and measurement of serum AFP concentration every 3 months in the first 4 years of life for early detection of hepatoblastoma. Other types of tumors can also occur in these children and must be considered if there are relevant presenting findings, e.g., rhabdomyosarcoma. Between age 8 years and mid-adolescence annual renal ultrasound examinations are

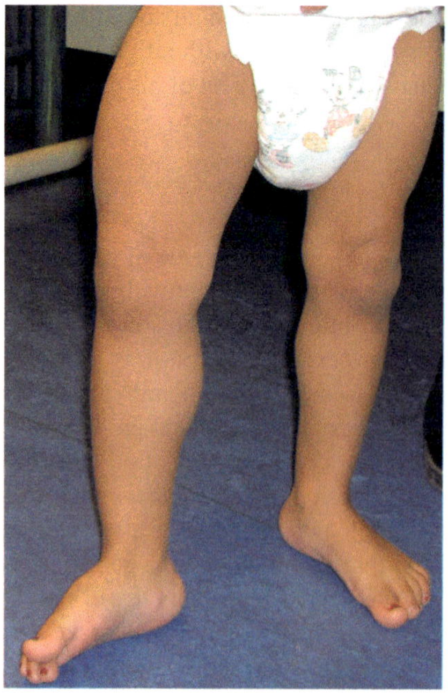

FIGURE 11.10.1.1 Hemihyperplasia of the lower limbs with the right leg larger than the left.

recommended to identify those requiring further evaluation for findings such as nephrocalcinosis and medullary sponge kidney disease.

Hemihyperplasia may become more or less evident as the child grows. Measurement of girth differences can be difficult and may not accurately capture a visually apparent difference.

If hemihyperplasia results in a clinically significant difference in leg length (greater than or equal to 1.5 cm), more accurate measurements can be obtained radiologically. In this situation, Colton's leg length should continue to be monitored annually by orthopedics in order to predict from serial radiologic measurements whether surgical intervention during early puberty is indicated. In some cases, it may be beneficial to close the growth plate of the longer leg in order to equalize the final leg lengths.

- **Genetic Testing:** Genetic testing includes DNA methylation analysis of chromosome region 11p15.5 which can identify a methylation abnormality (Sections 2.6, 2.7.5, and 6.4.3) in about 20% of individuals with isolated hemihyperplasia (based on findings in the literature). Genetic testing is typically done on a blood sample, but in some cases the genetic change may be present in a mosaic state (present in some cells and not others) (Section 2.4.2), therefore testing of other tissues (i.e., buccal, skin) can be considered if indicated. In the event that Colton's genetic testing is negative, the geneticist's clinical impression will guide the recommendations for medical management, specifically the indication for tumor surveillance.
- **Differential Diagnosis:** Hemihyperplasia can occur in isolation but can also occur as a feature of several overgrowth syndromes including BWS (Section 11.7.1), Proteus syndrome (characterized by disproportionate, asymmetric overgrowth that can affect bone, connective tissue, fatty tissue, skin, and internal organs), neurofibromatosis type 1 (characterized by pigmentary changes and growth of tumors in the nervous system and other parts of the body—Section 11.11.2), mosaic trisomy 8 (a chromosome disorder characterized by multiple congenital anomalies, advanced growth, and intellectual disability), and disorders associated with vascular malformations including Klippel–Trenaunay syndrome (characterized by vascular malformations, lymphatic anomalies, and overgrowth of soft tissues and bone) and megalencephaly-capillary malformation syndrome (and other *PIK3CA*-related overgrowth disorders associated with segmental body overgrowth). Most of these conditions can be distinguished on a clinical basis. Also, hemihyperplasia should be distinguished from hemihypoplasia (see below).
- **Caveats:** It is important to distinguish hemihyperplasia from hemihypoplasia (one side of the body smaller than the other) as hemihypoplasia is not associated with an increased tumor risk. However, this distinction can be subjective and difficult to make. Hemihypoplasia can be seen in isolation or as part of Russell–Silver syndrome (RSS), a condition associated with delayed growth, which has variable presentation (Section 11.7.4). Of note, isolated hemihypoplasia and RSS can be associated with DNA methylation changes at the chromosome 11p15.5 region which are opposite to those seen in BWS. Therefore, genetic testing may be valuable in distinguishing between hemihyperplasia and hemihypoplasia.
- **Outcome:** Colton undergoes genetic testing on a blood sample but is not found to have a molecular alteration at chromosome region 11p15.5. The recommendation is to continue to undergo screening for tumor surveillance, according to the recommended protocol, until 8 years of age, given the clinical impression of isolated hemihyperplasia. The option of performing genetic testing on another tissue sample is discussed but declined, as negative results would not alter the recommendation for continued tumor surveillance. As Colton grows, he continues to have normal development and do well.

The Diagnosis

Clinical Diagnosis: Hemihyperplasia (also known as isolated lateralized overgrowth).
Genetic Diagnosis: Unknown genetic cause.

Further Reading

Clericuzio, C.L., Martin, R.A., 2009. Diagnostic criteria and tumor screening for individuals with isolated hemihyperplasia. Genet. Med. 11, 220–222. PMID 19367194.

Mussa, A., Carli, D., Cardaropoli, S., Ferrero, G.B., Resta, N., 2021. Lateralized and segmental overgrowth in children. Cancers (Basil). 13, 6166. PMID 34944785.

Shin, C.H., Lim, C., Kim, H.Y., Yoo, W.J., Cho, T.-J., Choi, I.H., et al., 2021. Prospective study of epigenetic alterations responsible for isolated hemihyperplasia/hemihypoplasia and their association with leg length discrepancy. Orphanet. J. Rare Dis. 2021. 16, 418. PMID 34627330.

Case 11.10.2

Skeletal and Connective Tissue Cases: Disproportionate Short Stature

Julie Hoover-Fong

Greenberg Center for Skeletal Dysplasias, Department of Genetic Medicine, Johns Hopkins University, Baltimore, MD, United States

Abstract

A 3-year-old male whose height has fallen below the 2nd percentile and who has disproportionately short limbs is found to have hypochondroplasia.

The Case

Ben is a 3-year-old male patient who presents for his annual well-child checkup. He was born at term to a 31-year-old mother and 49-year-old father after an uncomplicated pregnancy. At birth, his length was at the 50th percentile, his weight was at the 25th percentile, and his head was at the 90th percentile. Ben has been a healthy child, having only typical childhood infections, which he resolved easily. However, while plotting his growth parameters today and reviewing them in the context of his longitudinal length/height, weight, and head circumference since birth, discrete trends are now apparent. Ben is growing in height, but his percentile position has dropped steadily from birth at the 50th to 25th percentiles by 6 months of age, 5th percentile at 12 months, and below the 2nd percentile by 18 months of age. His height has remained at minus 3 standard deviations (SD) below the mean for the last 18 months. His weight has remained on the WHO normative male chart, consistently between the 10th and 30th percentiles since birth, while his head circumference was always at the upper limit of normal (90th to 97th percentiles) and now crossed to +2.5 SD above the mean. You share the image of Ben's growth curves with his mother who responds that it looks normal to her because she herself was small when she was young and also a 'late bloomer.' You carry on with your history and exam, noting that Ben is wearing 15-month pants and a 3T shirt with the sleeves rolled up. Once undressed, his limbs appear shorter than expected for the length of his trunk. He has no striking facial dysmorphisms or known birth defects. You confirm he is meeting all developmental milestones appropriately and has no behaviors or signs of increased intracranial pressure associated with his macrocephaly.

The Issues

- **Are you concerned by his growth?** There are several aspects of Ben's growth parameters that are concerning and suggestive of an underlying diagnosis including: (1) his 'asymmetric' growth pattern in length/height (short) relative to weight (average) and head circumference (large), (2) the position of two of the three parameters outside 2 SD below (height) and above (head circumference) the normal values for age, and (3) the appearance of disproportionate limbs relative to his trunk.

Genomics in the Clinic. https://doi.org/10.1016/B978-0-12-816478-5.00068-6

In general, we expect that about 95% of children will have growth parameters (height, weight, and head circumference) between ± 2 SD from the mean (approximately the 2.5th to 97.5th percentiles). Usually the percentiles of these three growth parameters also are similar (e.g., all three on the upper quartile or all three between the 1st and 50th centiles). Deviations from these general guidelines deserve further consideration.

- **Are there any patterns consistent with Ben's short stature, his average weight, and his macrocephaly?** Different longitudinal patterns of growth also may point toward a specific etiology. On the lower end of the size spectrum, a child with all three growth parameters below the curve, especially if accompanied by cognitive impairment/intellectual disability and/or birth defects (e.g., heart anomaly, cleft palate), warrants a comprehensive workup for a unifying diagnosis (e.g., chromosome microdeletion syndrome). If only one parameter is below the curve, such as weight, this may indicate an overall caloric deficit. This may be caused by an inability to eat (e.g., hypotonia, discoordination), malabsorption (e.g., celiac disease), or excessive caloric utilization (e.g., tachypnea with heart failure). In these medical conditions, the caloric deficit typically causes the weight percentile to be compromised first, then over time linear growth, and last the head circumference in order to preserve brain growth. For Ben, his length is below the curve (not his weight or head circumference); he has normal development and no recognized birth defects.

  On the upper end of the growth spectrum, children who are overweight or obese from excessive caloric intake and/or low physical activity are usually concurrently tall. If longitudinal data are available, the weight is usually first to increase in this situation, then the height, and last the head circumference, which may become generous for age but proportionate to the increased weight and height. An obese child who is short does not fit this pattern and deserves further evaluation specifically targeting hormonal conditions (e.g., hypothyroidism, growth hormone deficiency). If there is cognitive impairment with obesity, then further workup for genetic obesity or overgrowth syndromes (e.g., Prader–Willi syndrome, Section 11.13.1) is indicated. Ben does not fit any of these patterns because only his head circumference is above the curve. Overall you are convinced Ben's growth is not normal and warrants further evaluation.
- **How do you address mother's comment about her own growth?** You acknowledge that there are two variations on normal longitudinal growth: familial short stature and familial slow maturation. The former is a child born to shorter parents, therefore the genetic potential of final adult height is lower than average. A child with familial short stature will typically have normal birth parameters and have a slower pace of linear growth in childhood but reach adolescence at a 'normal' age with otherwise 'normal' development. He/she will be a shorter than average adult but within normal limits and will be proportionate in limb length compared to the trunk. A person with familial slow maturation is also known a 'late bloomer.' It is relatively common to elicit a history of 'late bloomers' in their family. A child with familial slow maturation will be short for his/her chronologic age, but normal for his/her delayed bone age. For example, a girl's height may be −2.1 SD below the mean at her chronologic age of 8 years 6 months. But when plotted at her bone age of 7 years, her height is between the 25th to 50th percentiles. Longitudinal follow-up is recommended to ensure the bone age advances closer to the chronologic age as the child crosses to higher height percentiles for age. Puberty usually is later in onset (but otherwise normal), and after a generous adolescent growth spurt, the final adult height is average.

  You do not have mother's childhood growth curves, but she confirms she was smaller than most classmates in elementary school, menses began later than average (e.g., 14 or 15 years), and she is now an average statured female (5′ 7″); familial slow maturation is a reasonable diagnosis for the mother. You recall this is a normal variation of growth and you suspect that Ben's growth deficit is not normal. Ben's father had average growth in childhood, normal pubertal onset, and is 5′11″ tall. To further support that Ben's current height is outside the range of normal, you calculate his midparental height range to be 66.4″ + 3.3″ (168.7 cm) (Fig. 11.10.2.1). Using the recommended Center for Disease

Male child:	In inches (father's height + mother's height + 5) / 2 In cm (father's height + mother's height + 13) / 2
Female child:	In inches (father's height – 5 + mother's height) / 2 In cm (father's height -13 + mother's height) / 2
± 8.5 cm to calculated value is 3-97th %ile range	

FIGURE 11.10.2.1 Adult height prediction from midparental height. Reference: Tanner, J.M., Goldstein, H., Whitehouse, R.H., 1970. Standards for children's height at ages 2–9 years allowing for heights of parents. Arch Dis Child 45 (244), 755–62. PubMed ID: 5491878.

Control and Prevention (CDC) height curve for 2–18 years of age, you demonstrate that Ben's trajectory from his current position at -3 SD below the mean will be far below his expected genetic potential at 59″ (~150 cm) or less.

- **Do you think Ben's short stature could be 'genetic'?** You are beginning to suspect that he could have an underlying genetic condition. You have a teenage patient in your practice with achondroplasia, and Ben's shortened limbs with his longer trunk remind you of the teen, with less pronounced facial features. The teen's head is large for the size of his body but also quite large compared to average stature adult males, and his midface and nasal bridge are recessed. He has also had chronic middle ear fluid and had myringotomy tubes placed as well as chronic back pain; Ben doesn't have these additional features.

  While you are contemplating which tests and studies you are going to pursue, along with a genetic referral, his mother adds the following: "There aren't any 'genetic diseases' on my side of the family or on his father's side. His side only has some cancer and heart attacks. Ben being short didn't come from either of us, so it's not genetic, right?"

  This is one of the most difficult myths in genetics to dispel when talking with families about possible genetic diagnoses. Inheriting a dominant (Section 2.7.1) genetic condition from a similarly affected parent or inheriting a recessive (Section 2.7.1) condition from two carrier parents is often easier to recognize as a 'genetic' condition because a gene pathogenic variant or mutation(s) can be inherited. However, a new pathogenic variant that causes a condition or syndrome for the first time in an individual is also 'genetic' even though it was not inherited (Section 3.4). That affected individual may pass their variant on to their offspring. Every genetic condition had to start with one individual whether it was inherited or not. You explain this to Ben's mother.

For completeness, there is a rare third situation genetic inheritance pattern called gonadal mosaicism (Section 2.4.2). *This occurs when a parent has a pathogenic variant in a gene isolated to their gonadal tissue and nowhere else in their body. They may pass on this pathogenic variant to more than one of their offspring even though the parent does not manifest any of the features of that condition since the mutation is NOT in the other cells of their body.*

- **What is the next step?** You recommend a genetic clinic consultation to Ben's mother to start a workup of Ben's asymmetric and disproportionate small size. You are reassuring that Ben's development is appropriate for age and he does not have any major malformations or complicating medical conditions, so a complex genetic syndrome or large chromosomal deletion or duplication seems less likely. In preparation for the genetic clinic appointment, you also recommend the following first-line tests and studies to start the process: CBC, comprehensive metabolic panel (CMP), phosphorus, free T4, TSH, IGF-1, C-reactive protein (CRP), celiac screening, urinalysis, and a bone age.

It is important to share the results of the preliminary workup and the longitudinal growth curves with the geneticist for the scheduled appointment.

The Referral

- **Genetic Evaluation:** The genetic evaluation includes a complete history and a detailed physical exam with a three-generation pedigree (Sections 2.7 and 3.2.1) by the geneticist and genetic counselor. In the history, particular attention is paid to factors that could influence postnatal growth, such as modality of conception, prenatal exposures, and concurrent medical complications. During the exam, close examination and measurements of body proportions (e.g., arm span to height, upper-to-lower segment ratios, and limb segments) are completed. In his medical and family history, there are no contributing factors to Ben's asymmetric and disproportionate short stature.

  However, in addition to the discordance in his height, weight, and head circumference, his body proportion measurements are abnormal. For example, his upper to lower segment ratio is much higher than the expected for his age (Fig. 11.10.2.2) and the distal segments of his arms and legs are relatively shorter than the proximal segments. The laboratory evaluations and bone age performed before Ben's appointment are all normal.
- **Genetic Counseling:** The geneticist and genetic counselor confer following the evaluation. They both suspect that Ben could have an underlying skeletal dysplasia (Section 5.1.6). Although proportionately he resembles patients with achondroplasia, the most common short stature dysplasia, his findings are milder. They discuss with the parents that he may have a milder but related dysplasia known as hypochondroplasia. Both of these conditions are autosomal dominant (Section 2.7.1), but likely sporadic (occurred at conception rather than inherited), since neither parent is affected.

Advanced paternal age at the time of conception is more commonly associated with children with achondroplasia, hypochondroplasia, and other spontaneous dominant conditions.

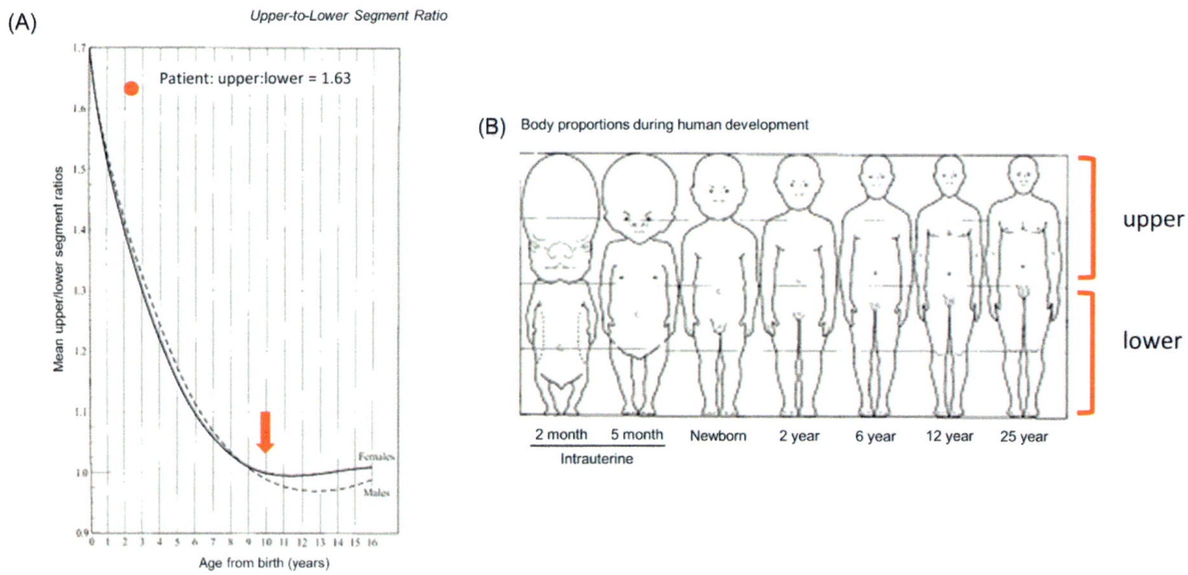

FIGURE 11.10.2.2 (A) Upper-to-lower segment ratio, both sexes, birth to 16 years. Note ratio of 1:1 achieved ~ 9–10 years of age (red arrow). Upper segment and lower segment indicated at 25 years (red bracket). (B) Normal body proportions over development, fetal to adulthood. Reference: Headings, D.L. (Ed.), 1975. The Harriet Lane Handbook, seventh ed. Year Book Medical Publishers, Chicago.

- **Genetic Testing**: If your patient appears to be female, a karyotype (Section 6.1.1) is recommended with specific instructions to the lab to count the chromosomes in at least 50 cells to assess mosaicism (Section 2.2.2) for Turner syndrome, another relatively common cause of short stature in which one of the X chromosomes is missing in all cells (45,X—complete Turner syndrome) or some proportion of the cells (mosaic Turner syndrome) (Section 11.7.5). Because Ben appears male, mosaicism with a full male karyotype and one with a missing Y chromosome, or 45,X/46,XY, would be a possibility. Therefore, a karyotype is recommended for him. The gene involved in Turner syndrome (including mosaicism) which specifically causes short stature is the *SHOX* (short stature homeobox) gene. A single-nucleotide variant, deletion or duplication (Section 2.4.1) (of any size including the entire gene) in *SHOX* and even some alterations in the surrounding DNA sequence that influences *SHOX*, can also cause short stature. A large deletion or duplication may be visible on the karyotype. If not and a *SHOX* gene condition is still suspected, then targeted genetic testing (Section 6.2.1) of that gene is indicated.

  The main gene that should be tested in Ben's case is *FGFR3*, which causes both achondroplasia and hypochondroplasia. Targeted single site testing (Section 6.2.1) for only the common *FGFR3* mutation associated with 99% of all cases of achondroplasia will not detect mutations throughout the rest of the gene that are associated with hypochondroplasia. Single gene full sequencing (Section 6.2.2) is indicated if hypochondroplasia is suspected. A blood sample from Ben for full *FGFR3* sequencing and deletion/duplication testing is sent to the lab.
- **Management and Recommendations:** A skeletal survey is recommended to assess bone maturity (bone age as well as knee epiphyses) and for other structural anomalies that may point to a skeletal dysplasia (Section 5.1.6). The latter could include diffuse abnormalities in the epiphyses, metaphyses, vertebrae, or other structural differences (e.g., distal radial metaphyseal angulation consistent with a Madelung deformation which is seen in *SHOX* gene conditions). There are specific management recommendations depending on whether the diagnosis of achondroplasia or hypochondroplasia is correct.

  The skeletal survey reveals generalized metaphyseal widening, shortened long bones, overgrowth of the fibulae, broad iliac bones, and brachydactyly. These features are suggestive of a diagnosis in the *FGFR3* disease spectrum including hypochondroplasia or achondroplasia. There is no 'V-shaped' deformation at the distal radius and ulna which is seen in patients with *SHOX* gene abnormalities with a Madelung deformity.

  In terms of treatment, there is some evidence that growth hormone therapy may increase final adult height in patients with hypochondroplasia. There are also several new medications in clinical trials for the treatment of patients with achondroplasia which may be utilized in patients with hypochondroplasia if approved by the Food and Drug Administration (FDA). The effect of these medications on the long-term skeletal manifestations of any of the *FGFR3*-related

FIGURE 11.10.2.3 Boy with hypochondroplasia at age 2 years old on the left and 4 years old on the right. Note rhizomelic shortening and short chest at both ages, as well as relative macrocephaly and normal facies. Broad hands and feet and mild joint laxity can also be found in hypochondroplasia. Publishing permission obtained.

conditions is unknown at this time. There is also controversy within the short stature community as to the need and/or desire of increasing height through the use of these medications.

Surgical limb lengthening may also be performed to increase height. This is an involved surgical process with extensive physical therapy and close medical follow-up required in the postoperative period.

Ben and his family are also referred to the Little People of America (LPA) the largest patient support organization (Sections 3.5 and 7.4) for those with short stature. The LPA can be contacted via their website lpaonline.org, email info@lpaonline.org, or by calling their national office at 888-LPA-2001 or 714-368-3689.

- **Results:** The sequencing reveals a previously described mutation in the *FGFR3* gene that was found in a patient with mild adult short stature (4′7″), mild lordosis, and macrocephaly consistent with hypochondroplasia. Ben's diagnosis is confirmed (Fig. 11.10.2.3). The karyotype is consistent with a normal male, 46,XY.
- **Differential Diagnosis:** Besides the possibilities of hypochondroplasia or achondroplasia, there could also be Turner syndrome, or a *SHOX* gene abnormality, as mentioned above. All skeletal dysplasias will have some differences on X-rays.
- **Caveats:** Failure to notify a genetic laboratory that you are considering a diagnosis of Turner syndrome (or other variant of this condition as discussed previously) will result in a standard karyotype report based on the analysis of only a few cells. This is not sufficient to find cells that may be missing an entire sex chromosome such that a diagnosis of mosaic Turner syndrome may be overlooked.
- **Outcome:** Ben is referred to a skeletal dysplasia clinic to monitor his growth and weight, advise on physical adaptations of his environment in school and at home, and coordinate multiple medical specialists' care. An orthopedist will monitor his lower extremities for genu varus, an otolaryngologist will monitor his hearing and middle ear fluid/chronic infections, a pulmonologist will evaluate for sleep disordered breathing, and a neurosurgeon will monitor the spine. Overall Ben is expected to live a long, healthy life with normal cognitive development and fertility.

The Diagnosis

Clinical Diagnosis: Hypochondroplasia.

Genetic Diagnosis: *FGFR3* gene, heterozygous pathogenic variant.

Further Reading

Arenas, M.A., Del Pino, M., Fano, V., 2018. FGFR3-related hypochondroplasia: longitudinal growth in 57 children with the p.Asn540Lys mutation. J. Pediatr. Endocrinol. Metab. 31, 1279–1284. PMID 30335613.

Brandt, J.S., Cruz Ithier, M.A., Rosen, T., Ashkinadze, E., 2019. Advanced paternal age, infertility, and reproductive risks: a review of the literature. Prenat. Diagn. 39, 81–87. PMID 30520056.

Malaquias, A.C., Scalco, R.C., Fontenele, E.G., Costalonga, E.F., Baldin, A.D., Braz, A.F., et al., 2013. The sitting height/height ratio for age in healthy and short individuals and its potential role in selecting short children for SHOX analysis. Horm. Res. Paediatr. 80, 449–456. PMID 24296787.

Case 11.10.3

Skeletal and Connective Tissue Cases: Multiple Fractures

Julie Hoover-Fong

Greenberg Center for Skeletal Dysplasias, Department of Genetic Medicine, Johns Hopkins University, Baltimore, MD, United States

Abstract

A 12-year-old male suffers a radius and ulna fracture, with a history of multiple other fractures when younger, and is diagnosed as having osteogenesis imperfecta (OI). With a paternal family history of similar findings, he is found to have the type 1 autosomal dominant form of OI.

The Case

Ricardo is a 12-year-old male presenting to your office for follow-up after an Emergency Department visit for a right radius and ulna fracture. He fell on his outstretched arm while playing soccer in gym class about a week ago. Ricardo was casted after a nonsurgical reduction in the Emergency Department. He and his parents are pleased the pain around the fracture has largely resolved, but they were told to follow up with you because the X-ray of his fractured arm, "looked like an 80-year-old man."

You know Ricardo to be a healthy child overall with no hospitalizations, up-to-date immunizations, mild exercise-induced reactive airway disease, and normal development. You check his growth parameters today; he is 25th percentile for weight and 10th percentile for height, which is the lower limit of his expected midparental height percentile. Ricardo and his two healthy siblings have been followed in your group practice for several years by you and your three partners. While reviewing his medical history, you notice that each of your partners has recorded at least one fracture in Ricardo's medical history: a tibia fracture at 18 months old falling off the couch and hitting the coffee table, a left radius fracture falling while running on a playground at 5 years, digit fractures at 4, 8, and 10 years with minimal trauma, and now this right radius and ulna fracture.

You ask if anyone else in the family has had fractures like Ricardo. Ricardo and his mother respond that Ricardo takes after his father who has had at least a dozen fractures in his life including a recent diagnosis of a compression fracture of two lumbar vertebrae. But they quickly attribute their similarities to their physical activity and 'being clumsy.' With further questioning, you also learn that Ricardo's paternal aunt has osteoporosis and some fractures (unknown how many) and her mother (paternal grandmother to Ricardo) had them as well.

The Issues

- **What is wrong with Ricardo's bones?** Considering Ricardo's full fracture history, you are concerned this is too many for his age especially since several were sustained with minimal trauma. His father's fracture history is also concerning

Genomics in the Clinic. https://doi.org/10.1016/B978-0-12-816478-5.00069-8

and you are not willing to accept that sharing his father's clumsiness is a sufficient explanation for Ricardo's fractures. You explain that the emergency doctor likely meant Ricardo's bones in the X-rays appeared more transparent and thinner than expected for his age. This appearance could be a sign of bone demineralization, allowing the skeletal to be susceptible to fracture.

- **What can help direct your workup?** You collect more medical history and examine Ricardo closely for abnormalities in other body systems related to the skeleton that also may be affected (e.g., teeth, eyes, hearing). You explain that the fractures, osteoporosis and the additional history in Ricardo, his father, and his paternal aunt and paternal grandmother are strong evidence of a common, inherited bone fragility condition that deserves further workup.
- **What tests and studies can be done to figure out what is wrong with Ricardo's bones?** There are many causes of poor bone quality predisposing to fractures. However, Ricardo's general bone health has not been assessed and you decide to do this in preparation for a referral to genetic clinic for testing. General bone health screening could include a comprehensive metabolic profile including alkaline phosphatase, calcium, phosphorus, 25-hydroxy vitamin D level, and parathyroid hormone levels. If possible, you and his family gather copies of radiographs from his prior fractures (i.e., at the acute event and after healing) to examine the gross integrity of the bone at the time of fracture and the type of callus that formed during healing. A copy of his growth curve also is helpful for the genetic referral.

If additional history suggests an underlying, primary medical problem that could negatively impact calcium and phosphorus availability to the bones and/or absorption of vitamin D (e.g., malabsorption, anorexia, renal disease) you would consider adding labs for celiac disease, screening questions of an eating disorder and/or depression, and urinalysis with spot quantification of calcium, phosphorus, creatinine, in addition to other indicated labs.

The Referral

- **Genetic Evaluation:** The genetic clinic evaluation includes a complete history and a dysmorphology exam with a three-generation pedigree (Sections 2.7 and 3.2). In the history, particular attention is paid to factors that could compromise bone integrity including risk factors for osteopenia of prematurity (e.g., inability to utilize the enteral system for feeding, prolonged diuretic use), malabsorption, eating disorder, or underlying chronic inflammatory or infectious conditions. In the family history, careful inquiry about fractures (i.e., age, mechanism, healing), teeth, hearing, and scleral color is performed. The geneticist and genetic counselor learn that Ricardo's father and aunt have experienced cracking of their teeth and their mother had dentures by the time she was 40 years old. Ricardo has a bluish coloration to his sclerae but those of his father are normal (i.e., white). Ricardo's mother is convinced her husband has some hearing deficit which he denies; his mother (Ricardo's paternal grandmother) wears hearing aids.

  The screening labs collected by the primary care provider are all normal. Therefore, other bone fragility conditions such as hypophosphatasia (would have low alkaline phosphatase), X-linked hypophosphatemic rickets (would have low phosphorus), and vitamin D–deficient rickets (would have low vitamin D) clinically are ruled out.
- **Genetic Counseling:** Based on the family history and involvement of the skeleton plus the teeth, hearing, and eyes as outlined above, the geneticist and genetic counselor suspect a unifying diagnosis affecting the type 1 collagen 'backbone' structure of these tissues. Osteogenesis imperfecta (OI) is at the top of the differential diagnosis.

Multiple fractures with minimal or moderate trauma in a child should also raise a concern of nonaccidental trauma (NAT). This is especially important to consider in children less than 3 years of age when accompanied by intracranial bleeding, retinal damage, and certain fracture locations (e.g., rib, digit, bucket-handle metaphyseal fractures). NAT is far more likely than osteogenesis imperfecta to be the source of fractures in a young child. The incidence of osteogenesis imperfecta is estimated at 1:10,000, meaning about 400 new affected individuals are born each year in the United States. Though this estimate includes the entire severity spectrum of osteogenesis imperfecta, over 95% have a nonlethal form of this condition. Pepin and Byers published an important article in 2015 outlining the relative risk between the two etiologies of fractures in children (highly recommended to review). Concurrently, it is estimated that 15,000 children between birth and 3 years of age are hospitalized in the United States each year due to fractures. If all 400 new patients with osteogenesis imperfecta from birth to 1 year, 1–2 years, and 2–3 years were hospitalized (extreme overestimate), these 1200 children with osteogenesis imperfecta would still only represent about 12% (1200/15,000) of the total population of young children hospitalized with fractures.

For Ricardo, his fractures began at a relatively young age (18 months) but they were not concurrent with these other NAT-associated injuries. His fractures have continued after the age of greatest vulnerability to abuse (below 3 years) and he has a family history highly suspicious of an autosomal dominant (Section 2.7.1) bone fragility condition. For all of these reasons as well as his normal screening labs, osteogenesis imperfecta is the leading diagnostic consideration for Ricardo, and the genetic team is not considering highly NAT in this case.

The different forms of osteogenesis imperfecta are reviewed with the family including the four main autosomal dominant (Section 2.7.1) forms (i.e., the Sillence groups of types I, II, III, and IV) which are differentiated by phenotype:

- Type I: Type I OI is the mildest form, usually marked by fractures after children begin walking and may also include blue sclerae and low-normal length/height for age and genetic potential.
- Type II: Type II is the most severe form and considered lethal. It is marked by short limbs, small chest, diffuse fractures of varying ages at birth, and poorly developed lungs.
- Type III: Type III OI is the most severe form in children who survive the newborn period. Frequent fractures, pulmonary insufficiency, feeding problems, dentinogenesis imperfecta, and blue sclerae are common.
- Type IV: Type IV is intermediate in severity and marked by fractures, shorter stature, long bone bowing are common with variable dentinogenesis imperfect and scleral discoloration.

Based on Ricardo's relatively mild presentation, the geneticist and genetic counselor are only considering these common autosomal dominant forms of osteogenesis imperfecta for him and his family. His comparatively mild phenotype suggests type I osteogenesis imperfecta as the leading subtype.

- **Management and Recommendations:** If possible, referral to the closest Osteogenesis Imperfecta Clinic is made. Standardization of care for patients with osteogenesis imperfecta is optimized at specialized treatment centers as well as research opportunities. Bisphosphonate therapy may be considered, particularly if fractures occur more frequently, with less trauma and/or at vulnerable sites (e.g., vertebral compression). Hearing screening is recommended, and this should be done regularly. Regular visits to a pediatric dentist should occur for evaluation for dentinogenesis imperfecta and oral hygiene management. A DXA scan should be considered to follow osteopenia or osteoporosis, if present. Physical therapy consult should be made to assess for occult physical function compromise. The family also is referred to a family support group (Sections 3.5 and 7.4.4).
- **Genetic Testing:** All four types of osteogenesis imperfecta above are the result of a pathogenic variant (Sections 2.4.1, 7.1.3) in either the *COL1A1* or *COL1A2* gene. These are the initial genes to assess. Two strands of the protein product of the *COL1A1* gene and one strand of the protein product of the *COL1A2* gene form a triple helix to make type 1 collagen. Type 1 collagen is found throughout the body including the skeleton, eye, inner ear, and skin, thus explaining the somatic distribution of the manifestations of osteogenesis imperfecta. In recent years, many more genes have been discovered and associated with OI. These genes collectively constitute the remaining 5% of all cases of osteogenesis imperfecta. One additional rare autosomal dominant form is caused by mutations in the *IFITM5* gene. It confers a relatively severe phenotype but is also uniquely marked by hypertrophic callus of the interosseous membrane of the forearm. The rest of the currently recognized forms of OI are autosomal recessive (Section 2.7.1) in their inheritance pattern with severe manifestations at a young age. The function of the proteins that are produced by these recessive genes is to process, modify, and refine the type 1 collagen that is encoded by the *COL1A1* and *COL1A2* genes.

It should be noted that there is some controversy in the medical literature about how many distinct osteogenesis imperfecta types there are. Some argue that each disease-causing gene should have its own 'type' (i.e., types V, VI, etc.) while others prefer to maintain the original Sillence groups (i.e., I, II, III, IV) and 'fit' the new genes into these groups based on phenotype severity.

- **Results:** Sequencing of the *COL1A1* and *COL1A2* genes (Section 6.2.2) is ordered on Ricardo, and a previously described pathogenic variant in the *COL1A1* gene is detected. This mutation produces a protein product that can combine with the protein product of the *COL1A2* gene in the correct ratio (i.e., resulting in the correct type 1 collagen structure), but in an overall decreased amount. This confirms a diagnosis of type 1 osteogenesis imperfecta in Ricardo.
- **Implications for Other Family Members:** Since a disease-causing mutation has been detected in Ricardo, targeted testing (Section 6.2.1) of his relatives with features suggestive of osteogenesis imperfecta is pursued. Ideally relatives *without* the osteogenesis imperfecta phenotype would also be tested to ensure the variant is segregating with the disease in this family. Since this is a dominant (Section 2.7.1) condition, Ricardo, his father, his paternal aunt, and grandmother all have the same variant while Ricardo's sister and mother do not. Thus the variant is segregating in a dominant fashion in the way expected based on who has the osteogenesis imperfecta phenotype.
- **Differential Diagnosis:** The differential diagnosis for osteogenesis imperfecta includes NAT (see above), osteopenia of prematurity, hypophosphatasia, vitamin D–deficient rickets and X-linked (Section 2.7.2) hypophosphatemic rickets. Lab results and X-rays should be helpful to distinguish these conditions.

- **Caveats:** Unfortunately there is no 'rule' that a child with osteogenesis imperfecta cannot also be abused and suffer NAT.
- **Outcome:** Ricardo is followed at the Osteogenesis Imperfecta Clinic in the nearby city along with his father. They have regular surveillance of their hearing, pain, function, and skeletal health. As he goes into and through puberty, the frequency of his fractures decreases and overall pain is minimal so he is not treated with bisphosphonates. However, Ricardo's father does start treatment with bisphosphonates because of his lumbar compression fractures and associated pain.

The Diagnosis

Clinical Diagnosis: Patient, Father, Paternal Grandmother, and Paternal Aunt - osteogenesis imperfecta type 1.
Genetic Diagnosis: Patient, Father, Paternal Grandmother, and Paternal Aunt - *COL1A1* gene, heterozygous pathogenic variant.

Further Reading

Osteogenesis Imperfecta Foundation, 2022. Osteogenesis Imperfecta Foundation [Online]. Osteogenesis Imperfecta Foundation. Available: https://oif.org/

Pepin, M.G., Byers, P.H., 2015. What every clinical geneticist should know about testing for osteogenesis imperfecta in suspected child abuse cases. Am. J. Med. Genet. C Semin. Med. Genet. 169, 307–313. PMID 26566591.

Steiner, R.D., Basel, D., January 28 2005. COL1A1/2 osteogenesis imperfecta. Adam, M.P., Ardinger, H.H., Pagon, R.A., Wallace, S.E., Bean, L.J.H., Gripp, K.W., Mirzaa, G.M., Amemiya, A., (editors). In GeneReviews. University of Washington, Seattle (WA), 1993–2022 [updated May 6, 2021]. PMID 20301472.

Case 11.10.4

Musculoskeletal Cases: Tall and Lean with Nonspecific Hip Pain

Houriya Ayoubieh[1], Gretchen MacCarrick[2] and Harry C. Dietz[3]

[1]Department of Medical Education, Texas Tech University Health Sciences Center El Paso, Paul L. Foster School of Medicine, El Paso, TX, United States; [2]McKusick-Nathans Department of Genetic Medicine, Department of Genetic Medicine, Baltimore, MD, United States; [3]Howard Hughes Medical Institute, William S. Smilow Center for Marfan Syndrome Research, McKusick-Nathans Institute of Genetic Medicine, Departments of Pediatrics, Medicine, Molecular Biology & Genetics, Johns Hopkins University School of Medicine, Baltimore, MD, United States

Abstract

A 12-year-old boy, who is tall and thin with scoliosis and myopia, develops hip pain and is found to have Marfan syndrome.

The Case

Todd and his mother just moved to the area and are presenting to establish care at your office. Todd is a 12-year-old boy, former product of a term pregnancy and uncomplicated vaginal delivery to a 20-year-old mother. He had normal growth and developmental parameters, but his weight lagged behind with age. He was evaluated by gastroenterology in their hometown for failure to thrive, which was unrevealing. The mother describes him as a picky eater and remarks that he is just tall and thin like the rest of her family. He has history of scoliosis that was diagnosed at 5 years old and was initially managed with bracing, but later required rod placement at 8 years old due to progression. Past surgical history includes inguinal hernia repair at one year of life. He wears glasses for myopia. He is a good student, loves sports, and is on his school basketball team. Lately he has been complaining of hip pain after practice. The mother is similarly lean and tall, wears thick glasses, and has similar facial features with a long face and flat cheeks. She shares with you that she is 6 weeks pregnant. There is a family history of a collapsed lung in the maternal uncle at age 18 years. The maternal grandfather died young after collapsing at work of an unknown cause. On exam, you note a tall and lean boy. He is at the 5th and 75th percentiles for weight and height. His vital signs are within the average range. He wears thick eye glasses. He has a pectus excavatum and mild residual scoliosis with a well-healed surgical scar. His limbs appear longer than his trunk, and he has flat feet. There is no hip tilt on Trendelenburg testing. His right hip exam reveals a click but otherwise full active and passive range of motion with mild pain on internal rotation. His pulmonary, cardiac, and abdominal exams are within normal limits.

The Issues

- **Are Todd's hip pain and other physical traits of concern**? You decide to measure Todd's arm span to height ratio, which is found to be greater than 1.05. Typically, the arm span should be less than the height. You are also concerned

Genomics in the Clinic. https://doi.org/10.1016/B978-0-12-816478-5.00070-4

about the other musculoskeletal findings on his skeletal exam, such as his flat feet and the pectus excavatum. You recommend that Todd have X-ray imaging of his hip given his physical traits.

- **Is the family history significant**? The mother says that her brother was perfectly healthy and the collapsed lung happened spontaneously and recurred a few years later. He also has a pectus excavatum. Although the circumstances of the grandfather's sudden death are unclear, the history of pneumothorax in the maternal uncle raises your suspicion for a disorder of connective tissue (Section 5.1.6) given Todd's skeletal features and history. You explain that when a pneumothorax occurs at a young age without history of prior lung disease or trauma, it could be because of a genetic condition affecting the connective tissue.

Genetic conditions usually run in families, but different family members with the same genetic condition can have different manifestations that may be more or less severe than other members (variable expressivity—Section 2.8).

- **What is the significance of the maternal grandfather's history?** The mother asks if it is possible that her father died from a collapsed lung? You explain that there are many causes of sudden death that may or may not be related to a genetic condition affecting connective tissue, most of which would be related to cardiovascular status. You refer them to the geneticist for further evaluation.

The Referral

- **Genetic Evaluation:** A full pregnancy, medical, and family histories are obtained, and a detailed physical examination is performed by the geneticist. Additional features on Todd's physical exam include: normocephaly with normal cranial sutures, long face, malar hypoplasia, high arched palate with singular uvula, crowded dentition, disproportionately long limbs for the size of the trunk (dolichostenomelia), limited elbow extension, and positive wrist and thumb signs (Fig. 11.10.4.1). His hip X-ray shows protrosio acetabulae, which is intrapelvic displacement of the acetabulum and femoral head, so that the femoral head projects medial to the ischioilial line. The mother and maternal uncle have similar features and body habitus on their exams. Additional family history is significant for myopia in the mother who is currently 8 weeks pregnant. The maternal great uncle had retinal detachment and died from a ruptured aortic aneurysm at age 60.
- **Genetic Counseling:** Todd's constellation of physical features with a systemic score of 12 (see Table 11.10.4.1) and family history of pneumothorax, sudden death, retinal detachment, and ruptured aorta raises the concern for Marfan syndrome. Marfan syndrome is an autosomal dominant condition (Section 2.7.1) caused by heterozygous pathogenic

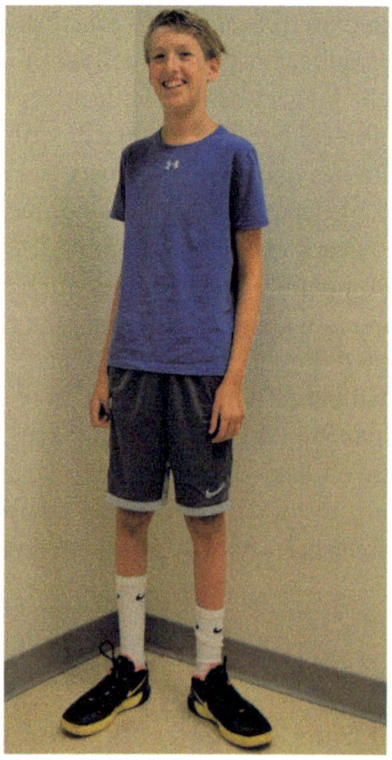

FIGURE 11.10.4.1 Typical features seen in Marfan syndrome. Note tall lean body habitus, long facies with malar flattening, disproportionately long arms with long fingers, and long legs.

TABLE 11.10.4.1 Systemic score in Marfan syndrome, value for each feature and patient's score. A systemic score of ≥7 is suggestive for Marfan syndrome.

Clinical feature	Value	Value in patient
Wrist AND thumb sign	3	3
Wrist OR thumb sign	1	
Pectus carinatum deformity	2	
Pectus excavatum or chest asymmetry	1	1
Hindfoot deformity	2	
Plain flat foot (pes planus)	1	1
Pneumothorax	2	
Dural ectasia	2	
Protrusio acetabulae	2	2
Reduced upper segment/lower segment AND increased arm span/height ratios	1	1
Scoliosis or thoracolumbar kyphosis	1	1
Reduced elbow extension	1	1
3 of 5 facial features	1	1
Skin striae	1	
Myopia	1	1
Mitral valve prolapse	1	
Total		12

Information from: https://www.marfan.org/dx/score.

variants in the *FBN1* gene, encoding fibrillin 1, the main glycoprotein found in the microfibrils of the extracellular matrix. Affected individuals can have variable involvement of the ocular, skeletal, and cardiovascular systems and can be at risk for life-threatening complications, which may be prevented by early diagnoses and surveillance. The most common ocular pathology is myopia. Other ocular features include ectopia lentis which is best seen at maximal pupillary dilation with a slit lamp, retinal detachment, cataract, and glaucoma. Skeletal features include dolichostenomelia, long slender fingers, pectus deformity of the chest, scoliosis, protrosio acetabulae, and flat feet. Cardiovascular involvement includes dilation of the aorta at the aortic root (Ao) which can predispose to tear and rupture and mitral valve prolapse. Other manifestations include pneumothorax, skin striae, and stretching of the dural sac in the lumbosacral region (dural ectasia) which can cause pain, weakness, and numbness from nerve entrapment. The diagnosis of Marfan syndrome in an index patient is made by the following criteria:

- Without family history (FH)
  Ao ($Z \geq 2$) AND ectopia lentis
  OR
  Ao ($Z \geq 2$) AND *FBN1* pathogenic variant
  OR
  Ao ($Z \geq 2$) AND Systemic score ($\geq$7 pts)
  OR
  Ectopia lentis AND *FBN1* pathogenic variant
- In presence of an affected first-degree family member:
  Ectopia lentis
  OR
  Systemic score ($\geq$7 pts)
  OR
  Ao ($Z \geq 2$ above 20 years old, $\geq$3 below 20 years)

Key: Ao: aortic diameter at the sinuses of valsalva above indicated Z-score or aortic root dissection. Z: Z-score describes how many standard deviations the size of the aorta lies above or below a size- or age-specific population mean.

The *FBN1* pathogenic variant (Section 7.1.3) can be inherited in an autosomal dominant manner or can occur de novo in up to 30% of probands. In an affected individual, there is a 50% risk of recurrence in their offspring. An aneurysm gene panel (Section 6.2.3) is recommended.

- **Management and Recommendations:** The geneticist recommends that Todd, his mother, and maternal uncle get baseline echocardiograms. Given the gravity of the cardiovascular manifestations, regular echocardiogram imaging is recommended to monitor the aortic root size. There are online Marfan Z-score calculators available to clinicians to ensure that the aortic measurement is standardized for age and body size (https://www.marfan.org/dx/zscore). The echocardiogram shows an aortic root diameter at the upper limits of normal in Todd, and a yearly echocardiogram is recommended. Baseline computed topography (CT) or magnetic resonance angiography (MRA) of the entire aorta will be pursued in early adulthood. Initiation of beta blockers and/or angiotensin receptor blockers should be considered at diagnosis or if aortic Z-score crosses the 95th–99th percentiles. It is important to address activity recommendations, including avoidance of: contact sports, exercising to exhaustion, and isometric exercises (sit-ups, push-ups, pull-ups, and weight lifting). He is encouraged to participate in low impact sports like swimming. Todd also is referred to ophthalmology for a dilated eye exam and he should continue to follow with orthopedics.
- **Results:** An aneurysm gene panel sent on Todd shows a pathogenic variant (Section 7.1.3) in the *FBN1* gene, confirming Marfan syndrome. Familial testing for the identified *FBN1* gene variant (Section 6.2.1) in maternal relatives is recommended to guide management by a multidisciplinary team that includes: genetics, ophthalmology, cardiology, cardiovascular surgery, and orthopedics.

In light of a family history of aortic aneurysm and rupture, if molecular testing comes back negative, given the family history, it is important to manage this family with a presumed diagnosis of a hereditary disorder of connective tissue (Fig. 11.10.4.2).

- **Implications for Other Family Members:** Todd's mother has a mildly dilated aortic root on echocardiogram. She also tests positive for the *FBN1* variant. She is referred to prenatal genetics at 16 weeks of pregnancy. She declines amniocentesis (Section 4.4.2) for genetic testing and opts for site-specific, targeted genetic testing (Section 6.2.1) when the baby is born. She is started on atenolol and referred to a high-risk obstetrician. She has cardiovascular imaging with echocardiography performed every 2–3 months during the pregnancy and the postpartum period because of the increased risk for dissection. The maternal uncle has an enlarged aortic root at 4.8 cm. He undergoes CT and MRA of the entire aorta and is referred to cardiothoracic surgery as surgical repair of the aorta is recommended when the aortic root approaches 5 cm in adults with Marfan syndrome or there is rapid increase in the size of the aorta of 0.5 cm per year. Angiotensin receptor blocker therapy is initiated.
- **Differential Diagnosis:** Other potential genetic conditions for this patient's skeletal and cardiovascular features include: vascular Ehlers–Danlos syndrome which is associated with spontaneous aortic or arterial dissection, pneumothorax, and hollow organ rupture; Shprintzen–Goldberg syndrome which is associated with a Marfanoid body habitus and intellectual disability; and Loeys–Dietz syndrome which is associated with craniosynostosis, bifid uvula or cleft palate, clubfoot, allergic and inflammatory gastrointestinal disease, and arterial tortuosity with risk for aneurysms/dissections throughout the body.
- **Caveats:** Nonspecific connective tissue features (hypermobility, mild scoliosis, striae, flat feet, etc.) are observed in about 15% of the population, and not everyone in this group is at risk for cardiovascular complications (Fig. 11.10.4.2). Obtaining an echocardiogram can be a good first screening test to triage patients. Genetic testing in Marfan syndrome is approximately 90%–95% sensitive. As well, panel testing (Section 6.2.3) in aortopathies will identify a genetic diagnosis in about 30%–40% of patients, so much of the genetic basis of aneurysm and connective tissue disorders is unknown.
- **Outcome:** Todd reluctantly gives up basketball when the season is over and takes up recreational swimming. His mother does well with her pregnancy and continues on blood pressure medication after delivery. His new baby brother is found to be negative on testing after birth.

The Diagnosis

Clinical Diagnosis: Patient, Mother, Maternal Uncle, and Maternal Grandfather—Marfan syndrome.
Genetic Diagnosis: Patient and Mother - *FBN1* gene, heterozygous pathogenic variant.

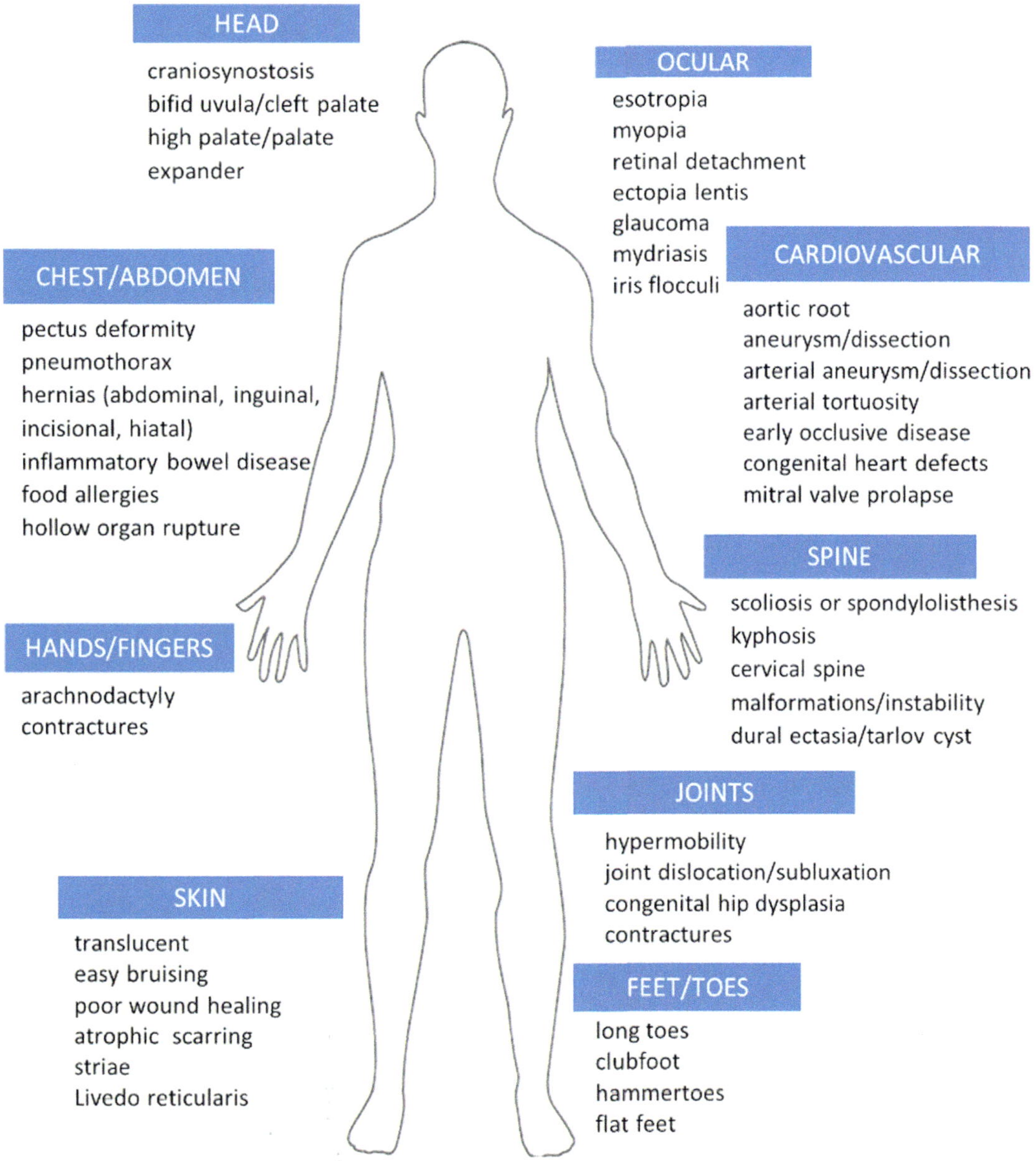

FIGURE 11.10.4.2 Systemic manifestations suggestive of syndromic heritable thoracic aortic disease (HTAD). *Reprinted from Demo, E., Rigelsky, C., Rideout, A.L., Graf, M., Pariani, M., Regalado, E., MacCarrick, G., Genetics and precision medicine: heritable thoracic aortic disease. Med. Clin. North Am. 103 (6), 1005–1019, Page 1008, Figure 2, 2019 with permission from Elsevier.*

Further Reading

Dietz, H., 2001 Apr 18. FBN1-related Marfan syndrome. In: Adam, M.P., Ardinger, H.H., Pagon, R.A., Wallace, S.E., Bean, L.J.H., Stephens, K., Amemiya, A. (Eds.), GeneReviews [Internet]. University of Washington, Seattle, Seattle (WA); 1993–2022 [updated 2022 Feb 17]. PMID 20301510.

Lacro, R.V., Dietz, H.C., Sleeper, L.A., Yetman, A.T., Bradley, T.J., Colan, S.D., et al., 2014. Atenolol versus losartan in children and young adults with Marfan's syndrome. N. Engl. J. Med. 371, 2061–2071. PMID 25405392.

Loeys, B.L., Dietz, H.C., Braverman, A.C., Callewaert, B.L., De Backer, J., Devereux, R.B., et al., 2010. The revised Ghent nosology for the Marfan syndrome. J. Med. Genet. 47, 476–485. PMID 20591885.

The Marfan Foundation, 2014. The Marfan Foundation [Online]. Marfan Foundation, New York. Available: https://www.marfan.org/. Accessed April 2022.

Case 11.10.5

Musculoskeletal Cases: Chronic Musculoskeletal Pain and Fatigue

Clair A. Francomano

Department of Medical and Molecular Genetics, Indiana University School of Medicine, Indianapolis, IN, United States

Abstract

A 32-year-old woman who presents with chronic muscle pain since childhood, as well as flexible joints, chronic fatigue, and dizziness on standing, is found to have hypermobile Ehlers–Danlos syndrome and dysautonomia.

The Case

A 32-year-old woman, Helen, presents to your internal medicine office as a new patient with a complaint of chronic musculoskeletal pain and fatigue. She reports that she began having muscle aches and pains around age 12–13 years, when she went through puberty, and that the pain has been accelerating and is causing her to have absences from work and difficulty performing her job as a librarian. She reports that she has had chronic pain since childhood, including "growing pains" in her legs, and that neck and back pain have been a constant feature in her life. She has hand pain from writing and reports that her fingers will sometimes "lock" when she is knitting. She has always been "double jointed." Her primary concerns on presentation are body aches and muscle pains around her joints, chronic fatigue, and ankle edema. Early childhood was notable for in-toeing that required bracing. Her past medical history is notable for chondromalacia, Crohn's disease, irritable bowel syndrome (diarrhea-dominant), mitral valve prolapse, and venous insufficiency. Her surgical history includes extraction of six teeth for dental crowding and two procedures for venous ablation at age 29 years. Review of systems is positive for chronic fatigue, tinnitus, myopia, racing heart beat and dizziness on standing, diarrhea, non-migraine headaches approximately once a month that are sometimes severe enough to induce vomiting, and cold intolerance.

Her physical examination is notable for a well-appearing young woman who appears younger than her stated age. Vital signs are notable for low blood pressure (102/60) and elevated heart rate (110). She has a high, narrow palate, a midsystolic click on cardiac auscultation, genu recurvatum, and very flexible fingers and wrists. The skin is soft and appears translucent. Neurological exam is within normal limits.

The Issues

- **What is the reason for the chronic pain in this patient?** You recognize that Helen has evidence of a hereditary connective tissue disorder, with joint hypermobility and soft, translucent skin. The history of tooth extractions and finding of a high narrow palate are also in keeping with this diagnosis.

Genomics in the Clinic. https://doi.org/10.1016/B978-0-12-816478-5.00071-6

- **Is there a unifying diagnosis for this young woman's pain, fatigue, and dizziness?** You suspect dysautonomia based on Helen's history of dizziness upon standing. A quick literature review reveals that dysautonomia can be a co-morbid condition with Ehlers–Danlos syndrome (EDS), and you suspect that your patient may have both the hypermobile type of EDS and postural orthostatic tachycardia syndrome (POTS), a form of postural orthostatic intolerance. Helen's fatigue could be secondary to the POTS.
- **What are the best next steps?** You refer Helen to a clinical geneticist for further evaluation to establish a definitive diagnosis. In the meantime, ensuring adequate hydration and salt intake may help improve her dizziness on standing. Compression stockings or other compression garments may also help with this symptom. You discuss a referral to physical therapy for her musculoskeletal pain and to hand therapy for the "locking" and pain in her fingers.

The Referral

- **Genetic Evaluation:** Full pregnancy, medical, and family histories are obtained, and a detailed physical examination is performed by the geneticist and genetic counselor. The following historical information is obtained. Her Karnofsky score, a measure of how well she is able to function, is 80 (range 0–100): "normal activity with effort, some signs of symptoms or disease." On the Brief Pain Inventory, the patient reports the following:
    - Pain location: ankles, hips, lower back
    - Pain level at worst, on a scale from 0 to 10: 7
    - Pain level at least, on a scale from 0 to 10: 3
    - Pain level on average, on a scale from 0 to 10: 5
    - Pain management modalities: ice, elevating her legs

  She has never had physical therapy and does not take anything for pain other than ibuprofen or acetaminophen.

  The genetic counselor obtains the following additional information on the family history: Helen has a 25-year-old sister with myopia, dental crowding, and a high, narrow palate; one maternal first cousin with a narrow palate that required expansion; and another maternal first cousin who has been diagnosed with EDS, unknown type. There is no family history of lens dislocations, retinal detachment, aneurysm, cleft palate, early hearing loss, early osteoarthritis, or early unexpected deaths.

  The geneticist notes the following additional features on physical examination: normal body proportions including arm span to height ratio and upper/lower segment ratio, blue/gray sclerae, a high, narrow palate, singular uvula, moderate dental crowding, midsystolic click on cardiac auscultation, marked joint hypermobility with a Beighton score of 8/9 (Fig. 11.10.5.1), soft, mildly stretchy skin with striae over the hips and thighs, two atrophic scars over her anterior shins, pes planus with bilateral ankle pronation, and bilateral piezogenic papules at the heels.
- **Genetic Counseling**: Helen meets the 2017 diagnostic criteria for hypermobile EDS (Fig. 11.10.5.1). Dysautonomia can be a comorbid condition with EDS, and it is likely that Helen has both the hypermobile type of EDS and postural orthostatic intolerance or POTS. Her fatigue could be secondary to the POTS; dysautonomia can also contribute to poor sleep architecture which may exacerbate fatigue as well.

  The hypermobile type of EDS (hEDS) is inherited as an autosomal dominant condition (Section 2.7.1). Clinical experience suggests that women are more likely to suffer from chronic pain and other comorbid conditions associated with this condition. However, the risk to children of affected persons is 50%, in keeping with the autosomal dominant pattern of inheritance.

  Patients who do not meet the diagnostic criteria for hypermobile EDS but have generalized joint hypermobility and pain likely fall into one of the Hypermobile Spectrum Disorders. These patients may present with comorbid conditions similar to those seen in hypermobile EDS (see below). A careful history and physical examination should lead to a suspicion of a hereditary disorder of connective tissue and prompt a referral for a genetic evaluation.
- **Genetic Testing:** No genetic testing is recommended since the gene(s) underlying hypermobile EDS are not yet known. However, if the type of EDS is unclear or seems more consistent with one of the other types, such as classical EDS with very stretchy friable skin and similar joint complications, or vascular EDS with personal or family history of clubfoot and/or hip dislocation (Section 5.1.6) and increased risk for arterial aneurysms (Section 5.1.3), then panel testing (Section 6.2.3) would be indicated.
- **Management and Recommendations:** Helen is referred for an echocardiogram to evaluate her aortic root and valvular anatomy. The echocardiogram reveals only mitral valve prolapse. She is sent for a sleep study to evaluate the sleep architecture. She is found to spend very little time in stage 3 and REM sleep and is placed on a beta blocker to minimize the effect of the sympathetic nervous system during sleep. She is encouraged to see a physical therapist for manual therapy including dry needling if necessary and to get a minimum of 30–40 min of low impact exercise 4–5 days

The International Consortium on Ehlers-Danlos Syndromes & Related Disorders
In Association with The Ehlers-Danlos Society

Diagnostic Criteria for Hypermobile Ehlers-Danlos Syndrome (hEDS)

This diagnostic checklist is for doctors across all disciplines to be able to diagnose EDS

Distributed by The Ehlers Danlos Society

Patient name: ______________________ DOB: ____________ DOV: ____________ Evaluator: ______________

The clinical diagnosis of hypermobile EDS needs the simultaneous presence of all criteria, 1 **and** 2 **and** 3.

CRITERION 1 – Generalized Joint Hypermobility

One of the following selected:

- ☐ ≥6 pre-pubertal children and adolescents
- ☐ ≥5 pubertal men and woman to age 50
- ☐ ≥4 men and women over the age of 50

Beighton Score: _____/9

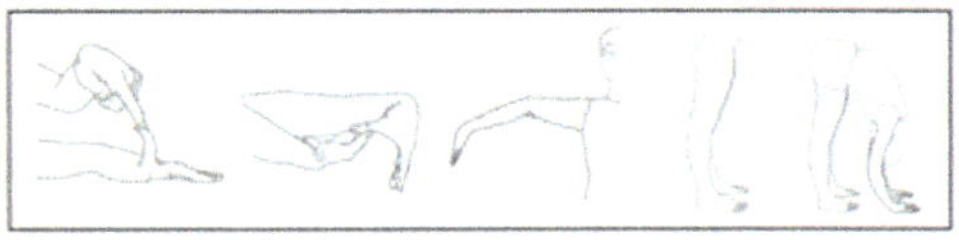

If Beighton Score is one point below age- and sex-specific cut off, two or more of the following must also be selected to meet criterion:

- ☐ Can you now (or could you ever) place your hands flat on the floor without bending your knees?
- ☐ Can you now (or could you ever) bend your thumb to touch your forearm?
- ☐ As a child, did you amuse your friends by contorting your body into strange shapes or could you do the splits?
- ☐ As a child or teenager, did your shoulder or kneecap dislocate on more than one occasion?
- ☐ Do you consider yourself "double jointed"?

CRITERION 2 – Two or more of the following features (A, B, or C) must be present

Feature A (five must be present)

- ☐ Unusually soft or velvety skin
- ☐ Mild skin hyperextensibility
- ☐ Unexplained striae distensae or rubae at the back, groins, thighs, breasts and/or abdomen in adolescents, men or pre-pubertal women without a history of significant gain or loss of body fat or weight
- ☐ Bilateral piezogenic papules of the heel
- ☐ Recurrent or multiple abdominal hernia(s)
- ☐ Atrophic scarring involving at least two sites and without the formation of truly papyraceous and/or hemosideric scars as seen in classical EDS
- ☐ Pelvic floor, rectal, and/or uterine prolapse in children, men or nulliparous women without a history of morbid obesity or other known predisposing medical condition
- ☐ Dental crowding and high or narrow palate
- ☐ Arachnodactyly, as defined in one or more of the following: (i) positive wrist sign (Walker sign) on both sides, (ii) positive thumb sign (Steinberg sign) on both sides
- ☐ Arm span-to-height ratio ≥1.05
- ☐ Mitral valve prolapse (MVP) mild or greater based on strict echocardiographic criteria
- ☐ Aortic root dilatation with Z-score >+2

Feature A total: _____/12

Feature B

- ☐ Positive family history; one or more first-degree relatives independently meeting the current criteria for hEDS

Feature C (must have at least one)

- ☐ Musculoskeletal pain in two or more limbs, recurring daily for at least 3 months
- ☐ Chronic, widespread pain for ≥3 months
- ☐ Recurrent joint dislocations or frank joint instability, in the absence of trauma

CRITERION 3 - All of the following prerequisites MUST be met

1. Absence of unusual skin fragility, which should prompt consideration of other types of EDS
2. Exclusion of other heritable and acquired connective tissue disorders, including autoimmune rheumatologic conditions. In patients with an acquired CTD (e.g. Lupus, Rheumatoid Arthritis, etc.), additional diagnosis of hEDS requires meeting both Features A and B of Criterion 2. Feature C of Criterion 2 (chronic pain and/or instability) cannot be counted toward a diagnosis of hEDS in this situation.
3. Exclusion of alternative diagnoses that may also include joint hypermobility by means of hypotonia and/or connective tissue laxity. Alternative diagnoses and diagnostic categories include, but are not limited to, neuromuscular disorders (e.g. Bethlem myopathy), other hereditary disorders of the connective tissue (e.g. other types of EDS, Loeys-Dietz syndrome, Marfan syndrome), and skeletal dysplasias (e.g. osteogenesis imperfecta). Exclusion of these considerations may be based upon history, physical examination, and/or molecular genetic testing, as indicated.

Diagnosis: ____

FIGURE 11.10.5.1 2017 diagnostic criteria for hypermobile Ehlers–Danlos syndrome. *Source: https://www.ehlers-danlos.com/heds-diagnostic-checklist/.*

weekly to optimize her pain management and dysautonomia symptoms. Salt supplementation and improved hydration may help with her dizziness on standing, and she is encouraged to wear her compression hose and/or other compression garments regularly. Pain management can also include topical products such as patches with lidocaine, antiinflammatory cream, or cream and/or patches made from antiinflammatory and analgesic Chinese herbs. Epsom salts baths or footbaths may also help with muscle spasm.

In addition, other possible comorbid conditions may be present and require evaluation. Other comorbid conditions often seen with hypermobile EDS include mast cell activation syndrome, gastrointestinal dysmotility, which may be severe, and neurologic findings secondary to instability at the craniocervical junction or cervical spine. Workup for these may be indicated depending on the specific presenting symptoms and signs.

- **Differential Diagnosis:** The differential diagnosis includes other hereditary disorders of connective tissue characterized by joint hypermobility, including Marfan syndrome (Section 11.10.4), Loeys–Dietz syndrome, and Stickler syndrome (Section 11.3.5). Cardinal features for Marfan syndrome and Loeys–Dietz syndrome include ascending thoracic aneurysm and risk of dissection, so it is important to ask about a family history of aortic aneurysm or sudden unexpected deaths. Stickler syndrome is characterized by early-onset osteoarthritis, cleft palate, hearing loss, and retinal detachment. Helen has had a normal echocardiogram and does not have a family history of sudden, unexpected deaths, cleft palate, or early arthritis, so the hereditary disorders of connective tissue causing aortic aneurysm and Stickler syndrome are unlikely.
- **Caveats:** It is common for young women who present with multiple complaints such as these to be dismissed as malingering or that the symptoms are "all in their head." This can lead to a long delay in diagnosis and a failure of the medical system to properly address the multiple comorbid conditions that may be seen with hypermobile EDS.
- **Outcomes:** On a follow-up visit, Helen reports that supplemental salt and hydration as well as regular use of the compression hose have dramatically improved her ability to stand for long periods of time. Physical therapy and dry needling have substantially improved her pain. A low-dose beta blocker at night has improved the quality of her sleep. She reports less fatigue and she is experiencing less absenteeism and better performance at her job. She is profoundly grateful to have a diagnosis and to understand the physical and physiological factors that have disrupted her quality of life for so long.

The Diagnosis

Clinical Diagnosis: Patient—hypermobile Ehlers–Danlos syndrome, dysautonomia, with maternal family history.
Genetic Diagnosis: Not tested.

Further Reading

Castori, M., Tinkle, B., Levy, H., Malfait, F., Hakim, A., 2017. A framework for the classification of joint hypermobility and related conditions. Am. J. Med. Genet. C Semin. Med. Genet. 175 (1), 148–157. PMID 28145606.

Chopra, P., Tinkle, B., Hamonet, C., Brock, I., Gompel, A., Bulbena, A., et al., 2017. Pain management in the Ehlers–Danlos syndromes. Am. J. Med. Genet. C Semin. Med. Genet. 175, 212–219. PMID 28186390.

Engelbert, R.H., Juul-Kristensen, B., Pacey, V., De Wandele, I., Smeenk, S., Woinarosky, N., Sabo, S., Scheper, M.C., Russek, L., Simmonds, J.V., 2017. The evidence-based rationale for physical therapy treatment of children, adolescents, and adults diagnosed with joint hypermobility syndrome/hypermobile Ehlers Danlos syndrome. Am. J. Med. Genet. C Semin. Med. Genet. 175, 158–167. PMID 28306230.

Hakim, A., Tinkle, B.T., Francomano, C.A., 2021. Ehlers-Danlos syndromes, hypermobility spectrum disorders, and associated co-morbidities: Reports from EDS ECHO. Am. J. Med. Genet. C Semin. Med. Genet. 187 (4), 413–415. PMID 34793630.

Malfait, F., Francomano, C., Byers, P., Belmont, J., Berglund, B., Black, J., Bloom, L., Bowen, J.M., Brady, A.F., Burrows, N.P., Castori, M., Cohen, H., Colombi, M., Demirdas, S., De Backer, J., De Paepe, A., Fournel-Gigleux, S., Frank, M., Ghali, N., Giunta, C., Grahame, R., Hakim, A., Jeunemaitre, X., Johnson, D., Juul-Kristensen, B., Kapferer-Seebacher, I., Kazkaz, H., Kosho, T., Lavallee, M.E., Levy, H., Mendoza-Londono, R., Pepin, M., Pope, F.M., Reinstein, E., Robert, L., Roherbach, M., Sanders, L., Sobey, G.J., Van Damme, T., Vandersteen, A., van Mourik, C., Voermans, N., Wheeldon, N., Zschocke, J., Tinkle, B., 2017. The 2017 international classification of the Ehlers–Danlos syndromes. Am. J. Med. Genet. C Semin. Med. Genet. 175, 8–26. PMID 28306229.

Mathias, C.J., Owens, A., Iodice, V., Hakim, J., 2021. Dysautonomia in the Ehlers-Danlos syndromes and hypermobility spectrum disorders-With a focus on the postural tachycardia syndrome. Am. J. Med. Genet. C Semin. Med. Genet. 187 (4), 510–519. PMID 34766441.

Russek, L., Stott, P., Simmonds, J., 2019. Recognizing and Effectively Managing Hypermobility-Related Conditions. Physical Therapy 99 (9), 1189–1200. PMID 31158283.

Section 11.11

Dermatologic Cases

Case 11.11.1

Dermatologic Cases: Infant with Nystagmus and Fair Skin

Mary Beth Palko Dinulos

Division of Genetics and Child Development, Department of Pediatrics, and Department of Pathology, Dartmouth-Hitchcock Medical Center, The Geisel School of Medicine at Dartmouth, Lebanon, NH, United States

Abstract

A 4-month-old male has horizontal nystagmus, fair skin, and blonde hair and is diagnosed with albinism.

The Case

Matthew, a 4-month-old patient of yours, comes into the office for his well-child check. You review his medical history before you see him, and note that he has been healthy overall, with normal growth and development. He passed all of his newborn screening tests. When you ask his mother how he has been since his last visit, she mentions that he is breastfeeding well, almost sleeping through the night, smiling, and vocalizing. She does have one concern, however. Matthew has constant rapid back-and-forth movements of his eyes. She has noticed these jerky eye movements within the past month. She states that she believes that he can see, though, as he will follow objects and recognizes her. Mother also notes that he does not enjoy being outside in bright light. On examination, you note a well-developed baby boy with continual horizontal nystagmus. His exam is otherwise normal, including normal red reflexes bilaterally. You note that he has blue eyes, blonde hair, and fair skin, similar to his mother.

The Issues

- **How concerned are you about his nystagmus?** Matthew's infantile nystagmus is quite concerning to you due to its constant nature. Sometimes intermittent nystagmus in the newborn period can be normal; but mother reports she has noticed continual nystagmus for the past several weeks. The differential diagnosis for infantile nystagmus is quite broad.

Infantile nystagmus typically presents within the first few months of life. Infantile nystagmus can be an isolated abnormality or associated with a wide variety of underlying disorders. The majority of infants with infantile nystagmus have a visual sensory disorder of the eye that is either anatomical or functional: corneal opacities, aniridia, albinism, retinopathy of prematurity, achromatopsia (nonprogressive hereditary visual disorder characterized by decreased vision, light sensitivity, and the absence of color vision), Leber's congenital amaurosis, and optic nerve abnormalities. Less commonly, infants may have an underlying metabolic or neurological disorder, a central nervous system (CNS) malformation of the cerebellum or brainstem, hydrocephalus, or an underlying visual pathway tumor. As a result, the infant is usually referred either to a pediatric ophthalmologist, a pediatric neurologist, or directly for neuroimaging.

Genomics in the Clinic. https://doi.org/10.1016/B978-0-12-816478-5.00072-8

- **Should you refer to ophthalmology and neurology or do immediate neuroimaging?** Infantile nystagmus with onset within the first 2–6 months of life is less worrisome than onset after 6 months of age. Late infantile onset nystagmus (>6 months of age) is more likely to have an underlying CNS etiology. You note no other neurologic findings in Matthew. He has no signs or symptoms of hydrocephalus or increased intracranial pressure. You refer Matthew to a pediatric ophthalmologist for further evaluation.
- **What does the ophthalmologic evaluation show?** Matthew is seen at 5 months of age by a pediatric ophthalmologist. On examination, Matthew is visually alert and fixes and follows well with both eyes. He has a harsh horizontal nystagmus in all positions of gaze. At times he appears to have a right exotropia but this is difficult to assess because of the nystagmus. His ocular exam reveals iris transillumination. There is slightly decreased pigment in the retina. Dilated fundus exam shows a decreased foveal reflex and blonde fundus. Mother also is noted to have iris transillumination on cursory exam in the office. She is not aware of any family history of nystagmus but reports that her niece and nephew have poor vision. Ophthalmology evaluation is usually sufficient to make a diagnosis of albinism in an infant, but it is helpful to refer to genetics for counseling and underlying diagnosis. Matthew is referred to genetics with a clinical diagnosis of albinism based on his ophthalmologic findings.

The Referral

- **Genetic Evaluation:** Matthew is seen at 6 months of age by a pediatric geneticist and a genetic counselor. A complete pregnancy and birth history, medical history, and family history (Sections 2.7 and 3.2) are obtained. His mother relates that following Matthew's ophthalmology appointment she contacted her mother and obtained additional information regarding the eye findings in the family. A three-generation pedigree is drawn. The mother has six sisters, one of whom has a son and daughter with vision problems. Those children are otherwise normal. The maternal grandmother has four sisters, two of whom have a son with vision problems but no other issues. Matthew's mother has normal vision but has iris transillumination noted by the ophthalmologist. On physical examination, Matthew has normal growth. He has rapid horizontal nystagmus and bilateral red reflexes. He has blue eyes. He has blonde eyelashes, eyebrows, and scalp hair. He has fair skin.
- **Genetic Counseling:** The geneticist and genetic counselor spend time discussing Matthew's history, family history, and physical examination findings with the family. They review that the ophthalmologist noted physical findings in Matthew consistent with a diagnosis of albinism (Section 5.1.7). Albinism typically presents in the infant as nystagmus. The specific type of albinism may not be readily apparent by history or physical examination and usually requires molecular diagnosis. They discuss that albinism may include only ocular findings (Section 5.1.2) (ocular albinism) or eye and skin findings (oculocutaneous albinism). Some forms of albinism may have other associated findings as well, such as hearing loss or bleeding disorders. Different forms of albinism may have differences in long-term vision prognosis as well.

  In detailed review of the family history, they note that multiple individuals have vision loss. This appears to be following an X-linked inheritance pattern (Section 2.7.2), as there have been three affected males through unaffected females. In addition, Matthew's mother was noted to have findings of possible carrier status for albinism detected on cursory examination by the ophthalmologist. No one is reported to have cutaneous involvement. Interestingly, there is also one maternal female cousin to Matthew who has been diagnosed with vision loss, although her visual acuity is less impaired than that of her brother. They discuss that X-linked conditions may present in females with Turner syndrome (Section 11.7.5) (girls/women with 45, X chromosome complement will manifest X-linked conditions due to the presence of only one X chromosome), a structural X chromosome anomaly that disrupts a gene for ocular albinism (e.g. translocation – Section 2.2.1 – with a breakpoint in a gene responsible for albinism), or due to skewed X-inactivation (where one X chromosome is preferentially used more than the other) (Section 2.7.2). Due to her unusual presentation, genetic testing had been advised in Matthew's cousin, but her family declined testing and follow-up.

  The albinism appears to be X-linked in this family; however, Matthew's female cousin has significant vision loss as well. In addition, some forms of albinism have associated findings, including hearing loss and bleeding disorders. Therefore, genetic testing may help determine appropriate management for affected individuals in this family, as well as provide more precise reproductive risks for Matthew's family.
- **Management and Recommendations:** The geneticist recommends that Matthew have annual ophthalmologic examinations until 16 years of age, and thereafter when needed. Refractive errors should be treated as early as possible. If Matthew develops abnormal head posturing he may need prismatic spectacle correction. Strabismus surgery usually is not required. Photodysphoria may be relieved by sunglasses, special filter glasses, or transition lenses. Dermatology referral is important for counseling regarding age-appropriate sun-protective lotions and clothing. Educational supports are also recommended for children with vision impairment.

- **Genetic Testing:** The geneticist advises genetic testing for albinism in Matthew to determine the molecular basis of albinism in this family. A next-generation sequencing panel (Section 6.2.3) including 28 genes known to cause syndromic and nonsyndromic forms of albinism is recommended by the geneticist. The genetic counselor reviews molecular genetic testing with the family and a genetic testing consent form is signed. Insurance prior authorization is obtained and blood is drawn and sent to the laboratory (Section 6.6).
- **Results:** Four weeks later, the genetic counselor receives the results of Matthew's genetic testing. The lab report notes that no clinically significant sequence variants or copy number changes (Section 2.4.1) were identified which could explain the primary clinical concerns for this patient. However, one variant of unknown significance (VUS) (Section 7.1.5) was identified in the *HPS4* gene and clinical correlation would be necessary to determine if this would be important in terms of Matthews's management.

  The genetic counselor reviews the report with the geneticist and they make a plan for Matthew and his family based on his clinical findings and the genetic testing results. The genetic counselor contacts the family and discusses the results by telephone. She explains that there is no immediate diagnosis made based on the results of the testing. However, the lab did report a single variant in 1 of the 28 genes tested, *HPS4*. She discusses that when genetic testing is completed, identification of a unique gene change is not uncommon. She reviews the various types of variants including those of uncertain significance (VUS). She explains that mutations within the *HPS4* gene cause a condition called Hermansky–Pudlak syndrome (HPS). HPS is a syndromic form of albinism associated with a significant bleeding disorder due to abnormal platelets. HPS is inherited in an autosomal recessive manner (Section 2.7.1) and therefore pathogenic variants (Section 7.1.3) in both copies of the *HPS4* gene would be needed to cause symptoms. Matthew was found to have only one variant of uncertain significance in the *HPS4* gene. However, if the VUS was eventually found to be pathogenic, and another variant was missed, he could have HPS. As there is a functional platelet study available to assess for HPS, the geneticist recommends that Matthew have that study performed for definitive diagnostic purposes. Blood on Matthew is then sent for platelet transmission electron microscopy (PTEM) testing looking for the characteristic dense bodies/delta granules in platelets that are characteristic of HPS. Matthew's PTEM testing is normal. Therefore, he has at least one functioning copy of the *HPS4* gene and does not have HPS. He is merely a carrier for a variant in the *HPS4* gene.

  Overall, the geneticist is not able to find the underlying genetic cause of Matthew's albinism. However, based upon his clinical findings and his family history, he most likely has X-linked ocular albinism and is counseled accordingly.
- **Differential Diagnosis:** Congenital or infantile nystagmus is not specific to X-linked ocular albinism and can appear as an isolated finding (idiopathic infantile nystagmus) or part of a broader syndrome, such as ocular albinism with sensorineural deafness, Leber congenital amaurosis (Section 11.3.3), the oculocutaneous albinisms (e.g., OCA1, OCA2, OCA3, OCA4), and HPS (OCA, bleeding diathesis, and possible pulmonary fibrosis).
- **Caveats:** It is important to note that infants with various forms of oculocutaneous albinism may have pigmented hair and irides, with lighter skin tone than their families. Their findings are indistinguishable from infants with ocular albinism. However, infants with oculocutaneous albinism type 1 (OCA1) typically have white skin, white hair, and blue irides, although they may develop pigment over time. Matthew and his mother are both fair-skinned and blue-eyed, but this is much more likely due to heredity factors than oculocutaneous albinism.
- **Outcome:** Matthew has significant refractive errors, strabismus, photodysphoria, and reduced binocular vision. However, because X-linked ocular albinism is not typically a progressive disorder, we anticipate that Matthew's visual acuity will remain stable throughout his life and may actually slowly improve into his mid-teens.

The Diagnosis

Clinical Diagnosis: Patient—ocular albinism. Maternal family history consistent with X-linked inheritance.
Genetic Diagnosis: Patient—unknown genetic cause. *HPS4* gene, heterozygous variant of uncertain significance.

Further Reading

Lewis, R.A., March 12, 2004. Ocular albinism: X-linked. Adam, M.P., Ardinger, H.H., Pagon, R.A., Wallace, S.E., Bean, L.J.H., Stephens, K., Amemiya, A., (editors). In GeneReviews. University of Washington, Seattle (WA). 1993–2022 [updated November 19, 2015]. PMID 20301517.

Weiss, A.H., 2017. *Infantile Nystagmus* [Online]. American Academy of Ophthalmology. Available from: https://www.aao.org/disease-review/infantile-nystagmus. (Accessed April 2022).

Case 11.11.2

Dermatologic Cases: Multiple Café au Lait Spots

Ethylin Wang Jabs*

Department of Genetics and Genomic Sciences, Department of Pediatrics, Department of Cell, Developmental and Regenerative Biology, Icahn School of Medicine at Mount Sinai, New York, NY, United States

Abstract

An infant presents with four café au lait spots and no other health issues or findings on physical exam and is found to have a de novo, pathogenic variant in the *NF1* gene.

The Case

A 10-month-old female, Jane, comes to see you, her pediatrician, for her well-child care visit. She is noted to have four café au lait spots and no other skin lesions. Her parents mention that they believe three of these spots were present at birth and one developed since then. They say these spots have not grown in size or changed in texture in the past 6 months. Her head circumference is normal and her development is on track as she pulls from a seated position to a stand, cruises, has a pincer grasp, babbles, and follows one-step commands. You consider neurofibromatosis type 1 in your differential diagnosis.

The Issues

- **Does she have a clinical diagnosis of neurofibromatosis type 1?** You review Jane's family history to determine if anyone else has symptoms or the clinical diagnosis of neurofibromatosis type 1 (NF1). You learn that no other relatives have café au lait spots, neurofibromas, or other skin, neurologic, ophthalmologic, or bony lesions.

  You are not sure if she has NF1 because, although she may not meet the clinical criteria for NF1 at this time, she may fulfill the clinical criteria when she is older and phenotypic variability and incomplete penetrance (Section 2.8) can be associated with autosomal dominant (Section 2.7.1) conditions.
- **Should she be referred to other clinical specialists to establish the diagnosis of neurofibromatosis type 1?** You can refer Jane to neurology and ophthalmology for a slit lamp exam to detect additional clinical findings consistent with neurofibromatosis type 1, but at this time she does not have any obvious neurologic or visual abnormalities on your examination. Based on her findings, you refer Jane to a clinical geneticist to determine if she has neurofibromatosis type 1.

* Author EWJ was employed by Icahn School of Medicine at Mount Sinai at the time of writing the manuscript, but is currently employed by Mayo Clinic

Genomics in the Clinic. https://doi.org/10.1016/B978-0-12-816478-5.00073-X

The Referral

- **Genetic Evaluation**: The clinical geneticist obtains a full present illness, past medical and surgical history, review of systems, developmental history, social history, and physical examination. Her mother reports that she was 37 years old, as was the father, at Jane's birth. She had routine ultrasounds and prenatal genetic bloodwork for advanced maternal age (Section 4.2.2) during the pregnancy that were normal. Jane was a product of a normal full-term spontaneous vaginal delivery. The genetic counselor obtains a detailed three-generation pedigree (Section 3.2) and no one else in the family is found to have symptoms of neurofibromatosis. On examination, Jane is found to have four café au lait spots of at least 5 × 5 mm in size (Fig. 11.11.2.1). The geneticist draws a body diagram with the location and size of each café au lait spot and takes photos of the skin lesions for documentation in Jane's medical records.

  Jane does not have axillary or inguinal freckling, neurofibromas, or a plexiform neurofibroma. She does not have osseous lesions.

  The National Institute of Health (NIH) diagnostic criteria was revised by the International Consensus Group on Neurofibromatosis Diagnostic Criteria for neurofibromatosis type 1 in an individual who does not have a parent diagnosed with NF1 to include two or more of the following: (1) six or more café au lait spots over 5 mm in greatest diameter in prepubertal individuals and over 15 mm in greatest diameter in postpubertal individuals, (2) two or more neurofibromas of any type or one plexiform neurofibroma, (3) freckling in the axillary or inguinal region, (4) optic pathway glioma, (5) two or more iris Lisch nodules identified by slit lamp examination or two or more choroidal abnormalities defined as bright, patchy nodules imaged by optical coherence tomography (OCT)/near-infrared reflectance (NIR) by imaging, (6) distinctive osseous lesion such as sphenoid dysplasia, anterolateral bowing of the tibia, or pseudarthorsis of the long bone, and (7) heterozygous pathogenic *NF1* variant with a variant alelle fraction of 50% in apparently normal tissue such as white blood cells. A child of a parent who meets the diagnostic criteria merits a diagnosis of NF1 if one or more of the criteria are met.

  Jane does not currently meet any of the clinical criteria for NF1. She only has four café au lait spots, and she has no other findings nor family members with symptoms of NF1. Only about half of the children with NF1 and no known family history meet clinical criteria by age 1 year, but almost all do by age 8 years and all do by 20 years old. Penetrance is usually complete after childhood.

  The usual order of appearance of the clinical features listed on the NIH criteria is: café au lait macules, axillary freckling, Lisch nodules, and neurofibromas. Symptomatic optic glioma is usually diagnosed by age 3 years, and 90% of patients will have Lisch nodules by age 6 years. Characteristic osseous lesions are usually apparent within the first year of life but can be present at birth. About half of the individuals with NF1 develop plexiform neurofibromas, also typically present at birth, but they may be missed clinically because most are internal. In addition, at least half of

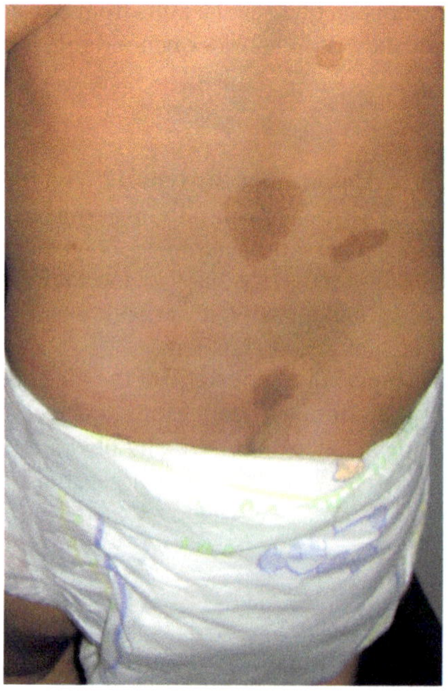

FIGURE 11.11.2.1 Café au lait spots, four of more than 5 mm in greatest diameter on the back.

individuals with NF1 have learning disabilities and/or behavioral involvement. Less common features include central nervous system gliomas, malignant peripheral nerve sheath tumors, scoliosis, and vasculopathy. Therefore, it will be important for Jane to continue to be followed by her pediatrician for the development of additional findings of NF1.

- **Genetic Counseling**: The genetic counselor discusses with Jane's parents that NF1 is autosomal dominant (Section 2.7.1) and can be new in a child or be inherited through an affected parent 50% of the time. The clinical geneticist explains there are genotype−phenotype correlations so that determining the exact gene variant in some cases may predict the severity of the condition. A deletion of the entire *NF1* gene results in a severe phenotype of cognitive disabilities, more neurofibromas, and increased risk of malignant peripheral nerve sheath tumors and cardiovascular malformations. A specific 3 base deletion in exon 22 (c.2970-2972 delAAT) is associated with a mild phenotype characterized by café au lait macules and skinfold freckles with no neurofibromas. An amino acid substitution of Arg1809 leads to a Noonan-like syndrome with pulmonic stenosis and short stature (Sections 11.4.3, 11.12.2). Some missense and splice variants are found in patients with large number of internal tumors of the spinal nerve roots and deep peripheral nerves with few pigmentary findings or few or no neurofibromas, and normal cognitive function.

  The geneticist also discusses autosomal dominant Legius syndrome with Jane's family. This is a genetic syndrome that presents with 6 or more café au lait spots bilaterally distributed and freckling, without neurofibromas or the other tumor manifestations associated with NF1. There is phenotypic variability (Section 2.8) associated with Legius syndrome with additional findings of macrocephaly, learning disabilities, attention-deficit hyperactivity disorder (ADHD), and/or developmental delays. Single gene sequencing (Section 6.2.2) for *SPRED1* variants in Legius syndrome is available. In addition, there are more than 100 genetic conditions and multiple congenital anomaly syndromes that include café au lait spots or other individual features of NF1.
- **Genetic Testing**: Molecular genetic testing is indicated for individuals in whom NF1 is suspected, but who do not fulfill the diagnostic criteria. The geneticist discusses single gene sequencing (Section 6.2.2) for *NF1* that identifies more than 95% of patients, who fulfill the diagnostic criteria. If the *NF1* gene sequencing comes back negative, there is still a small possibility that the patient has NF1, as the yield of testing is less than 100% because gene sequencing, particularly for large genes like *NF1*, only examines the coding regions and some of the flanking regions of the exons of the gene. Deletion analysis (Sections 2.4.1, 6.2.1) for the *NF1* gene can be performed that identifies about 5% of NF1 cases. The geneticist discusses the different test results that can come back, including a positive result with a known disease-causing variant, a negative result, and a variant of uncertain significance (VUS) (Section 7.1.5). If a VUS is found, parental samples can aid in the interpretation because if one of the parents has the variant and by examination does not have symptoms of NF1, then the variant is less likely to be pathogenic. Jane's parents express understanding and decide that they would like to pursue gene sequencing for the *NF1* gene first, and if this test is negative, then reflex to deletion analysis. If the deletion analysis is negative, then they will discuss gene sequencing of *SPRED1*.
- **Results**: The sequencing test for the *NF1* gene is ordered and Jane's saliva sample rather than a blood sample is sent because it is difficult to obtain her blood sample. Jane is found to have a pathogenic missense variant (Section 7.1.3). Both parents undergo site specific, targeted testing (Section 6.2.1) for the missense variant and are not found to have the change.

  Jane is a de novo case or the first in her family with NF1. Half of affected individuals have NF1 as the result of a new variant. Her parent's recurrence risk for having another child with NF1 is slightly higher than the general population. If Jane's parents are considering having more children, they should seek prenatal counseling.

  Jane's offspring, however, will have a 50% risk of inheriting the variant and the disease manifestations can be variable, even between parents and offspring within a family. The optimal time to discuss risks to offspring and reproductive options is when Jane is of childbearing age.
- **Management and Recommendations**: Jane needs to be monitored for development of additional symptoms of NF1. Annual physical examination includes neurologic exam by a physician familiar with the disorder; twice annual ophthalmologic examination to regularly assess vision and rule out an optic glioma and Lisch nodules in children, less frequently in adults; regular developmental assessment of children; regular blood pressure monitoring; MRI for follow-up of clinically suspected intracranial tumors; and other internal tumors. As an adult, malignant peripheral nerve sheath tumor, breast cancer, cutaneous neurofibromas, and both psychiatric and neurologic diagnoses are common problems in patients with NF1. Although many issues can be managed by a pediatrician or family physician, monitoring and care coordination with a specialized NF1 clinic or clinical geneticist is encouraged. The American College of Medical Genetics and Genomics has developed guidelines and recommendations for health supervision from infancy

through adulthood. The FDA has approved an MEK1/2 protein kinase inhibitor, for neurofibromatosis type 1 with symptomatic, inoperable plexiform neurofibromas.

- **Caveats**: There is a low empiric risk of 1%–5% that Jane has NF1 as a result of germline mosaicism (Section 2.4.2) in one of her parents leading to a recurrence risk for her parents of having another child with NF1 of greater than the general population. It should be noted that single-nucleotide changes often occur in the father rather than the mother with the association of advanced paternal age and Jane's father is 37 years old (Section 4.2.2). NF1 should not be confused with neurofibromatosis type 2 (NF2), another autosomal dominant condition. NF2 is characterized by bilateral vestibular schwannomas, also known as acoustic neuromas, which result in hearing loss and imbalance. There can be noncancerous tumors including meningiomas, schwannomas, and gliomas of the brain or spinal cord. NF2 presents as early as late teens with full penetrance by 30 years of age. The condition is associated with variants in *NF2*, a different gene from *NF1*.
- **Outcome**: The family understands the genetic risk for NF1 in their family and will follow up to monitor for symptoms related to NF1 that may arise in Jane. The parents are given information how to contact a comprehensive, multidisciplinary neurofibromatosis clinic as well as a support group (Section 3.5) should they wish to learn more from other families with this condition.

The Diagnosis

Clinical Diagnosis: Neurofibromatosis type 1.
Genetic Diagnosis: *NF1* gene, heterozygous pathogenic variant.

Further Reading

DeBella, K., Szudek, J., Friedman, J.M., 2000. Use of the National Institutes of Health criteria for diagnosis of neurofibromatosis 1 in children. Pediatrics 105 (3 Pt 1), 608–614. PMID 10699117.

Legius, E., Messiaen, L., Wolkenstein, P., Pancza, P., Avery, R.A., Berman, Y., et al., 2021. Revised diagnostic criteria for neurofibromatosis type 1 and Legius syndrome: an international consensus recommendation. Genet. Med. 23 (8), 1506–1513. PMID 34012067.

Miller, D.T., Freedenberg, D., Schorry, E., Ullrich, N.J., Viskochil, D., Korf, B.R., et al., 2019. Health supervision for children with neurofibromatosis type 1. Pediatrics 143 (5), e20190660. PMID 31010905.

Stewart, D.R., Korf, B.R., Nathanson, K.L., Stevenson, D.A., Yohay, K., 2018. Care of adults with neurofibromatosis type 1: a clinical practice resource of the American College of Medical Genetics and Genomics (ACMG). Genet. Med. 20 (7), 671–682. PMID 30006586.

Case 11.11.3

Dermatologic Cases: Abnormal Teeth in a Toddler

Douglas K. Clemens[1] and Antonie D. Kline[2]

[1]*Cross Keys Dental Associates, Baltimore, MD, United States;* [2]*Harvey Institute for Human Genetics, Greater Baltimore Medical Center, Baltimore, MD, United States*

Abstract

A 3-year-old with conical teeth, missing teeth, sparse hair, and decreased sweating is diagnosed with ectodermal dysplasia, paternally inherited.

The Case

Derek, a 3-year-old male, is seeing his pediatric dentist, for the first time. His mother says that his teeth have been coming in pointy, and there does not seem to be as many of them as there were for his older sister. The pediatric dentist notes that many of his teeth including upper canine teeth look conical (Fig. 11.11.3.1), his teeth are small, and he only has 13 primary teeth

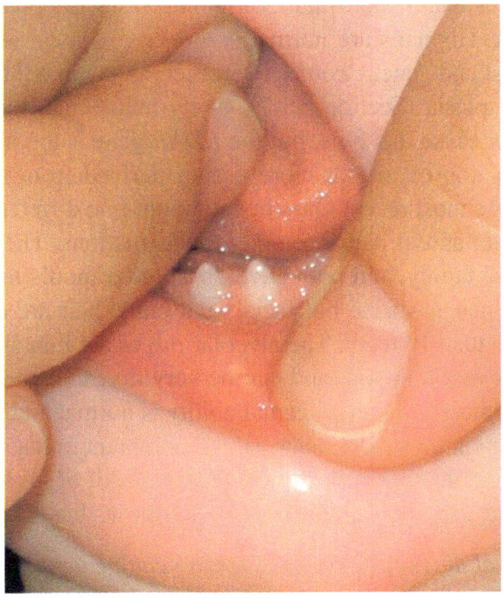

FIGURE 11.11.3.1 Conical teeth erupted in the mandible of a toddler. Please note that the parent's fingers are holding open the child's mouth.

Genomics in the Clinic. https://doi.org/10.1016/B978-0-12-816478-5.00074-1

instead of the usual 20. His blonde hair looks sparse, and his mother says that it tends to break off and it has barely grown since infancy. His mother says that he takes after his father, Solomon, who is bald, and who had all his teeth pulled when he was younger. The pediatric dentist asks about sweating, and his mother has not noticed anything unusual about this.

The Issues

- **What are the implications of the abnormal hair and teeth?** The pediatric dentist suspects a form of ectodermal dysplasia. There are more than 200 types of ectodermal dysplasias, and it is likely that Derek has this. The hair is often brittle and can indicate a problem with the teeth. The teeth tend to be smaller and often many are missing. There is the potential to have dental implants supporting dentures if there are too many missing teeth, and/or removable partial prostheses if needed. The bones tend to be thin, and a bone graft might be needed. Derek is too young to undergo a panorex X-ray, but the dentist recommends that he have this when he is 4–6 years of age.
- **Could his father also be affected?** Although the dentist knows that the more common type of ectodermal dysplasia is inherited from the mother, it sounds possible that Solomon could also be affected.
- **How serious are these findings?** The pediatric dentist knows that there can be a concern about sweating, which could be of concern in warm weather or during sports activities. The dentist reassures the mother that Derek should be fine but recommends referral to the pediatric geneticist for the family to find out more information. The family relays this to you, Derek's pediatrician. You refer him to the pediatric geneticist at the local hospital.

The Referral

- **Genetic Evaluation**: Prior to the appointment, after the previsit forms are completed, it is recommended that Solomon, Derek's father, also come on the day of the evaluation. The family is seen by the pediatric geneticist and genetic counselor. Medical history is reviewed and a three-generation family history (Sections 2.7.1, 3.2.1) is obtained. On detailed questioning, the parents say that they have never felt Derek's head to be sweaty even on the warmest nights, and he often becomes flushed with slight exertion.

  Solomon reveals that he had "horrible tiny teeth" growing up, and he had a lot of crowns placed, but got cavities underneath them, and when he was a teenager lost all his teeth. He has since had dentures but he didn't wear them properly and they never fit correctly. He also never sweats, and when he was younger had several episodes of heat exhaustion before stopping playing sports. His family history reveals that Solomon's mother always had dentures and has very thin hair. She "was never a sports person" and Derek's parents do not know if she ever sweats. No one else in the family seems to have similar findings.

  On exam, Derek is on the 10th percentile for all parameters. His blonde hair is sparse and thin. His facies appears slightly pinched, with shallow orbits and midfacial hypoplasia. His teeth are small and both the upper and lower canines are conical (Fig. 11.11.3.1), and he has quite a few missing teeth. His skin feels normally formed. His fingernails and toenails look normal. The rest of his exam is normal. Solomon refuses to open his mouth. He is bald but has a beard, and says he has normal pubic hair. His nails are normal.
- **Genetic Counseling**: The geneticist and genetic counselor confer, and then tell the family that both Derek and Solomon have hypohydrotic ectodermal dysplasia (Section 5.1.7), likely autosomal dominant type (Section 2.7.1). There are several genes with variants that can cause this type, but the *EDAR* gene is the most common one involved. Ectodermal dysplasia involves abnormal formation of two or more tissues derived from the ectoderm, namely hair, skin, sweat glands, teeth, and nails, based on the most recent classification with five different classifications based on genetic pathways. Some forms will have cleft lip and/or palate or other malformations. There is variable presentation, with incomplete penetrance (Section 2.8) in the family, and both Derek's and Solomon's findings are typical. The hair on the head is usually sparse and brittle, body hair is sparse but facial, and pubic hair are normal. Most adults with hypohydrotic ectodermal dysplasia have no more than 10 teeth. The alveolar ridges are thin and can be weak. The skin has decreased or absent sweat glands, and poor sweat function, and can be very dry. In some forms, nails are dysplastic. Both nasal and ear concretions may form. For most types, intellectual ability is normal and overall growth and other organ systems are unaffected. In this family, Solomon may have inherited the dominant mutation (Section 2.7.1) from his mother, or it could have been de novo in him, and then he passed it on to Derek. All of his children will have a 50% risk for inheriting it. Derek's parents decide that genetic testing does not matter at this point and decline it.

Ectodermal dysplasia remains a clinical diagnosis, but gene testing can be done, including gene panels (Section 6.2.3) for ectodermal dysplasias, especially for families who might be interested in pursuing prenatal diagnosis (Section 4.4.2) or preimplantation genetic testing (Section 4.5.1).

- **Management and Treatment**: The most important medical implication is the decreased ability to sweat. There is a risk for hyperthermia as early as infancy, and both hydration beforehand and during activities and rehydration afterward are extremely important. Remaining in the cool indoors is recommended during very hot days. All affected children should be referred to the pediatric dentist after the first teeth erupt. Good oral hygiene is crucial, including brushing, flossing, and fluoridated water or oral fluoride. Orthodontics may be helpful, including palate expansion. Dental implants and dentures attached to the implants can be used, but because the alveolar ridges are thin may be less successful. If possible, this can improve chewing, speech, and appearance. Partial prostheses in both the maxilla and mandible may be recommended when older. Topical medications and hair care treatments have been used successfully on the sparse hair.
- **Differential Diagnosis**: Brittle hair is seen in Menkes disease, due to copper deficiency, and trichothiodystrophy, due to sulfur deficiency, as well as some forms of ichthyosis, with thick scaly skin, easily broken hair, and ectropion. Sparse hair can be seen in trichorhinophalangeal syndrome, also with bone malformations, dysmorphic facies, and due to deletions on the long arm of chromosome 8, and cardio-facial-cutaneous syndrome, a rasopathy disorder like Noonan syndrome with cardiac defects and dysmorphic features. Missing teeth can be present in cleidocranial dysostosis, a skeletal dysplasia with absent clavicles and wide fontanelles with delayed closure. In addition, both Down syndrome (Section 11.13.2) and some of the oral-facial-digital syndromes can have missing teeth, but many people without underlying syndromes may have several missing teeth. Decreased sweating can occur in some ichthyoses and a form of frontonasal dysplasia, with widened facies and some midline defects.
- **Caveats**: Hyperthermia can occur even in infancy in ectodermal dysplasia. Adults who have had dental problems throughout childhood can be very sensitive to a discussion about their teeth, and oral health quality of life is reported as poor.
- **Outcome**: Derek does well overall; his parents are careful not to let him get overheated and to always make sure he is hydrated. Solomon obtains dentures that fit better.

The Diagnosis

Clinical Diagnosis: Patient and Father—ectodermal dysplasia, hypohydrotic type.
Genetic Diagnosis: Unknown genetic cause.

Further Reading

Hanisch, M., Sielker, S., Jung, S., Kleinheinz, J., Bohner, L., 2019. Self-assessment of oral health-related quality of life in people with ectodermal dysplasia in Germany. Int. J. Environ. Res. Publ. Health 16, 1933. PMID 31159198.

Wright, J.T., Fete, M., Schneider, H., Zinser, M., Koster, M.I., Clarke, A.J., et al., 2019. Ectodermal dysplasias: classification and organization by phenotype, genotype and molecular pathway. Am. J. Med. Genet. 179, 442–447. PMID 30703280.

Wright, J.T., Grange, D.K., Fete, M., Hypohidrotic Ectodermal Dysplasia, April 28, 2003. Adam, M.P., Ardinger, H.H., Pagon, R.A., Wallace, S.E., Bean, L.J.H., Gripp, K.W., Mirzaa, G.M., Amemiya, A., (editors). In GeneReviews. University of Washington, Seattle (WA), 1993–2022 [updated June 1, 2017]. PMID 20301291.

Section 11.12

Neurodevelopmental Cases

Case 11.12.1

Neurodevelopmental Cases: Developmental Stagnation and Regression

Constance L. Smith-Hicks

Department of Neurology and Developmental Medicine, Center for Synaptic Disorders, Kennedy Krieger Institute, Department of Neurology, Johns Hopkins University School of Medicine, Baltimore, MD, United States

Abstract

A 24-month-old female presents with developmental delay and regression, irregular respirations, axial hypotonia and focal epilepsy and is found to have Rett syndrome.

The Case

JoAnn is a 24-month-old female who comes today for her check up with you, her pediatrician. At her previous visit you noted that she was slow to make developmental gains and today you note that she is delayed in several domains of development and has regressed in her social interaction, communication, and hand use. She is not babbling, she does not wave or point, and she does not yet pull to stand, but she has begun to sit without support. In addition, she has begun to have eye blinking and eye rolling spells, grinding of her teeth, inconsistent alertness, and inconsistent social smile. On her exam you note irregular respiration, a flat affect, a wide face, sitting without support but with axial hypotonia, and she intermittently rubs her hands together.

The Issues

- **How delayed is she?** She is delayed across several areas of development with language and hand use being most affected. She should be referred for developmental assessment and early intervention services. This could be done through the state or privately. You discuss this with her parents and refer her to the state's early intervention program for infants and toddlers.
- **What is the significance of the eye movements?** You witness the eye blinking and eye rolling during the visit. You suspect seizures and consult with your local pediatric neurologist. You order an EEG and refer her for a neurologic evaluation.
- **What is the cause of her developmental delay?** We cannot be sure at this time but there are several possible reasons for her delay. This child's delay has both a progressive and a regressive quality to her clinical symptoms, since she has both lost skills and gained atypical behaviors. The differential diagnosis is large and warrants a referral to the geneticist.

Genomics in the Clinic. https://doi.org/10.1016/B978-0-12-816478-5.00075-3

It is helpful to send the results of the neurologic and developmental assessments to the geneticist prior to the child being seen. You refer her to a pediatric geneticist.

The Referral

- **Genetic Evaluation:** JoAnn is brought by her parents to the genetic clinic where she is seen by a geneticist and genetic counselor. Detailed pregnancy, medical, developmental, and family histories are obtained. Previous test results and evaluations are reviewed. The pregnancy, perinatal, and family histories are unremarkable and her parents and siblings are in good health. Her physical exam is notable for a decline in her rate of head growth, hand tapping, bruxism, hyperventilation, abdominal distention, and axial hypotonia. Her EEG is consistent with focal epilepsy.
- **Genetic Counseling:** The geneticist and genetic counselor suspect that she has Rett syndrome (Section 5.2) and discuss this possibility with the parents. Rett syndrome is an X-linked (Section 2.7.2) neurodevelopmental disorder that most commonly affects girls. Affected girls may present with loss of previously acquired developmental skills usually affecting language, social interaction, and gait. Stereotypic hand movements (e.g., hand wringing or washing) replace functional hand use. After the period of regression, patients enter a period of stability and may gain some skills.

Hemizygous males (Section 2.7.2) *with Rett syndrome will typically present with encephalopathy and severe seizures during infancy, with abnormal breathing, and early death.*

- **Genetic Testing:** The majority of individuals with Rett syndrome have pathogenic variants (Section 7.1.3) in the *MECP2* gene but genetic testing can be negative in 5% of individuals with classic features of Rett syndrome. The variant is generally de novo but can be inherited from a carrier mother or result from gonadal mosaicism (Section 2.4.2), in mother or father, thus increasing the risk for the parents to have another child with Rett syndrome. *MECP2* gene sequencing (Section 6.2.2) and deletion/duplication analysis by multiplex ligation–dependent probe amplification test are recommended (Sections 2.4.1 and 6.2.1). The parents agree to have JoAnn tested, and the test reveals an R168X variant which is commonly associated with Rett syndrome. Maternal testing is recommended to be considered if contemplating further children.
- **Management and Recommendations:** Each child with Rett syndrome has a varied developmental outcome with some being more affected than others. Some have preserved speech and gait while others are nonverbal and nonambulatory. The phenotypic variation is due in part to X-inactivation (Section 2.6) and genetic background effects. Each child makes progress along their own path and referral to a multidisciplinary Rett syndrome clinic for expert care is recommended. If not available, the child should be followed by pediatric neurology, orthopedics, gastroenterology, nutrition, and behavioral psychology and receive physical, occupational, and speech therapy.

  Recommended evaluations include: monitor oromotor skills and weight, consider gastrostomy tube if having feeding difficulties, manage constipation, perform annual EKG to evaluate for prolonged QT interval, monitor for scoliosis, and evaluate bone density. Consider sleep study if indicated. Referral to the parent support group is helpful for the family (Sections 3.5 and 7.4).
- **Differential Diagnosis:** Rett syndrome is defined by a set of clinical criteria that delineates a classic form of the disorder from an atypical form of the disorder. Pathogenic variants (Sections 7.1.3) in *MECP2* account for the majority of cases of Rett (95% in classic and 75% in atypical Rett). Some individuals with pathogenic variants in cyclin-dependent kinase-like 5 (*CDKL5*), forkhead box protein G1 (*FOXG1*), myocyte-specific enhancer factor 2C (*MEF2C*), and transcription factor 4 (*TCF4*) genes have some features that are shared with Rett syndrome. Girls with loss-of-function mutations in the *SMC1A* gene present with intractable seizures and significant developmental delay; they often develop similar hand stereotypies, with hands constantly in their mouths. Angelman syndrome is also in the differential and presents with speech delay, gait ataxia, epilepsy, and a happy disposition.
- **Caveats:** Males with Klinefelter syndrome, 47,XXY (Section 11.12.5), and pathogenic variants in *MECP2* can present similarly to females with Rett syndrome rather than with the severe encephalopathic form because of X-inactivation mosaicism (Section 2.6). Postzygotic *MECP2* variants also may present in females as Rett syndrome because of somatic mosaicism (Section 2.4.2).
- **Outcome:** JoAnn makes slow progress with therapy, and her seizures are well controlled. She is seen in a Rett syndrome specialty clinic. A gastrostomy tube is placed due to poor weight gain. She subsequently develops scoliosis.

The Diagnosis

Clinical Diagnosis: Rett syndrome.
Genetic Diagnosis: *MECP2* gene, heterozygous pathogenic variant.

Further Reading

Naidu, S., Johnston, M.V., 2011. Neurodevelopmental disorders: clinical criteria for Rett syndrome. Nat. Rev. Neurol. 7, 312–314. PMID 21587242.

Neul, J.L., Kaufmann, W.E., Glaze, D.G., Christodoulou, J., Clarke, A.J., Bahi-Buisson, N., Leonard, H., Bailey, M.E.S., Schanen, N.C., Zappella, M., Renieri, A., Huppke, P., Percy, A.K.; RettSearch Consortium, 2010. Rett syndrome: revised diagnostic criteria and nomenclature. Ann. Neurol. 68, 944–950. PMID 21154482.

Percy, A.K., Neul, J.L., Glaze, D.G., Motil, K.J., Skinner, S.A., Khwaja, O., Lee, H.-S., Lane, J.G., Barrish, J.O., Annese, F., McNair, L., Graham, J., Barnes, K., 2010. Rett syndrome diagnostic criteria: lessons from the Natural History Study. Ann. Neurol. 68, 951–955. PMID 21104896.

Case 11.12.2

Neurodevelopmental Cases: Delayed Speech, Polydactyly, and Short Stature

Anne Slavotinek

Division of Medical Genetics, Department of Pediatrics, Cincinnati Children's Hospital Medical Center, Cincinnati, United States

Abstract

A 3.5-year-old girl with history of postaxial polydactyly and delayed development, and short stature, ptosis, and hypotonia on exam is found to have Noonan syndrome, as well as Lynch syndrome.

The Case

Jaclyn is a 3 year-6 month-old female who has recently transferred to your care. She was delivered vaginally at term following a typical pregnancy; birth weight was 3080 g and length was 52 cm. However, she required oxygen after delivery and was noted to have low tone, with Apgar scores of 6 and 8 at 1 and 5 min, respectively. Extra digits on the outer aspect of each hand were also noted at birth; these additional digits were small and have been surgically removed. Jaclyn was slow to latch and required support with feeding. Her parents noted that she was less responsive than her 6-year-old brother from an early age. She rolled over at 6 months of age, sat with support at 10 months, and crawled at 18 months before walking independently at 22 months of age. Her speech was delayed and she first uttered single words at 2 years of age and has only 10–20 single words with unclear pronunciation at this visit. She has no allergies and takes a daily multivitamin. Your examination notes that her weight is on the 50th percentile, but that her height is on the 3rd percentile. She looks different from her parents and has widely spaced eyes, droopy eyelids, anomalous ears, and a chest deformity. Her tone is still low, but there are no other neurological signs. Both parents and older brother are healthy and her mother mentions to you that she would like to plan another pregnancy.

The Issues

- **You are concerned by her development. Do you order any testing or place any referrals?** Jaclyn's milestones have been delayed since birth and she has global developmental delays, meaning that gross motor, fine motor, expressive and receptive speech, and social skills are affected. Global developmental delays can be caused by many factors, including prenatal exposures, perinatal events, and chronic disease, that should be considered and excluded as appropriate. In addition, hearing loss can contribute to speech delay and vision loss can influence motor progression, but since Jaclyn's parents have no concerns about either, you do not refer to ophthalmology at this time. You send a complete blood count (CBC), a comprehensive metabolic panel, and thyroid function testing, and all of these tests return with unremarkable results. A hearing screen is difficult to perform, but the audiologist concludes that her hearing is within the normal range. You refer to a child development specialist for evaluation for services.

Genomics in the Clinic. https://doi.org/10.1016/B978-0-12-816478-5.00076-5

- **Should you refer her for a genetic evaluation?** It is often useful to determine if there are any specific clinical features that might indicate a genetic etiology for which targeted testing (Section 6.2.1) can be performed. Jaclyn's findings could be caused by a genetic condition; in this case, it appears likely that she will have a form of syndromic intellectual disability (Section 5.2), rather than a nonsyndromic form, due to having more than one clinical feature, including her facial anomalies, growth delays, and hypotonia. It would be very reasonable to refer for genetic testing, in particular because the parents would like to have another baby. In addition to determining recurrence risk (Section 3.4), a genetic etiology can be helpful for guiding patient management. You place an order for a genetic evaluation at your local hospital.

  A few days later, a genetics provider calls back and asks you to order a chromosome analysis (Section 6.1.1) and single-nucleotide polymorphism (SNP) microarray (Section 6.1.3) prior to the genetic appointment. You order the tests, and the results return normal.

In many centers, the geneticist would see the patient first, and then discuss and order the SNP microarray, which will detect microdeletions and microduplications (copy number variants) along the chromosomes, often containing several to many genes. Either way, appropriate counseling regarding limitations of this testing and possible unexpected results should be discussed prior to sending it. This test can result in a definitive answer, but also can show a finding that is unclear. Parental testing may be recommended. Limitations include that if this test is negative, there could still be a genetic cause, and a detected finding could merely be familial variation and unrelated. Potential unexpected findings include the detection of consanguinity (Section 3.2.4), i.e. that the parents have a common ancestor, or detection of a finding unrelated to the reason testing was sent, but which could have implications in the future both for the health of the child and the parents.

The Referral

- **Genetic Evaluation:** The clinical geneticist and genetic counselor meet with the family. The medical history is reviewed and a physical examination performed. A three-generation pedigree (Section 3.2.1) is obtained to collect information about more distant relatives who may share similar health problems. In this case, Jaclyn's maternal grandmother had an additional toe that was 'tied off' at birth and her brother was born with extra digits on all four limbs. Her maternal grandfather died of colon cancer in his 60s. Jaclyn's father had support for reading in elementary school, but graduated from high school. Both parents are of mixed European and Hispanic ethnicity and there is no known consanguinity. Jaclyn's physical exam shows height at the 3rd percentile, weight at the 50th percentile, and head circumference at the 90th percentile. Facies is significant for widely spaced eyes (hypertelorism), bilateral ptosis, and low-set, anomalous, and short ears. Her neck is broad and she has a pectus deformity. Hands and feet are normal, and she has hypotonia.
- **Genetic Counseling:** The clinical geneticist and genetic counselor confer and return to tell the parents that they suspect an underlying genetic condition. Although Jaclyn's SNP microarray did not reveal any chromosome aberrations or copy number variants, further genetic testing could be performed with a panel for intellectual disability, or with exome sequencing or genome sequencing (Sections 6.3.1 and 6.3.2) to look for sequence variants in genes that cause syndromic intellectual disability. Because Jaclyn has other findings in addition to the delays, panel testing for intellectual disability may not include the relevant disease genes. After consideration, exome sequencing is ordered and written consent is obtained for testing in Jaclyn and both parents. The parents would like to know the etiology of her delays, low tone, and extra fingers, but would also like to receive secondary findings (Section 7.2), if any are identified.

  Although Jaclyn's parents are satisfied with the genetic appointment, they complain that the test will take 4 months to run. Jaclyn's health is stable and her parents now reveal that they have decided not to have another child. Exome sequencing can take several months to perform in view of the complexity of the test and its interpretation. You reassure the parents that there does not seem to be a need to expedite the test at this stage.

Rapid exome testing, with results available in weeks rather than months, can be ordered in some laboratories, but this should be reserved for acute situations, hospitalized patients, or other emergency-related causes.

- **Results:** In due course, the results of the exome sequencing test return and show that Jaclyn has a de novo variant in the *SOS1* gene that the laboratory has interpreted as 'likely pathogenic' (Section 7.1.3). In addition, she has a secondary finding, a pathogenic variant in the *PMS2* gene that has been inherited from her mother. The family receives genetic counseling regarding these results, and Jaclyn is diagnosed with a mild form of Noonan syndrome (Section 11.4.3) as well as Lynch syndrome (Sections 8.1.3 and 8.3.3). Her mother also receives a diagnosis of Lynch syndrome.

- **Management and Recommendations:** Noonan syndrome is characterized by developmental delays, including speech delays, with additional syndromic features. Facial anomalies can include wide-set eyes, droopy eyelids (ptosis), epicanthic folds, low-set and posteriorly rotated ears with thickened helices, and a deeply grooved philtrum. Birth length is typically within the normal range, but frequently there is slowing of postnatal growth, with a final height that reaches the lower end of the normal range. Growth hormone testing and treatment is recommended for those individuals whose height is smaller than 2 standard deviations below the mean. Heart defects are found in 50%−80% of affected individuals and pulmonary valve stenosis and hypertrophic cardiomyopathy are characteristic; echocardiogram is recommended for all upon diagnosis. Other physical findings include a broad or webbed beck, superior pectus carinatum and inferior pectus excavatum (which can be noted on chest X-ray clinically), widely spaced nipples, cryptorchidism, coagulation defects, and lymphatic dysplasia. Noonan syndrome demonstrates genetic heterogeneity (Sections 2.8 and 11.4.3), and affected individuals may have causative sequence variants in the *PTPN11*, *SOS1*, *RAF1*, *RIT1*, and *KRAS* genes. Those with pathogenic variants in the *SOS1* gene tend not to be as short and to appear milder. Jaclyn has many features noted in Noonan syndrome, and on closer review of the report, the variant found is not listed in ClinVar, the database repository of variants (https://www.ncbi.nlm.nih.gov/clinvar/), nor has it been reported in a large panel of unaffected individuals and also the protein change is predicted to be deleterious. The geneticist and genetic counselor conclude that this should be considered pathogenic.

  Jaclyn and her mother also have a pathogenic variant in the *PSM2* gene. Single, or heterozygous (Section 2.4.1), pathogenic sequence variants in *PSM2* confer a diagnosis of Lynch syndrome, a cancer predisposition syndrome that is associated with a lifetime risk of 50%−80% for developing colorectal cancer. Risks for a variety of other tumor types are also elevated, including endometrial cancer, cancer of the stomach, ovary, small intestine, hepatobiliary tract, brain, and skin. Although Jaclyn is still too young for screening, her mother should be counseled about her higher risk of developing bowel and endometrial cancer and referred to a cancer genetic clinic to discuss surveillance with clinical screening and testing of other relatives at risk.
- **Caveats:** Polydactyly, or extra digits, is not a typical finding in Noonan or Lynch syndrome. However, the geneticist notes that several other relatives in your patient's family have had polydactyly without other medical concerns and that this can be an isolated trait. The geneticist and genetic counselor reassure the family that this finding is not likely to be related to either of the reported gene variants, and that sometimes exome sequencing does not explain all of the findings (Section 7.3). There are also times when exome sequencing returns with no variants. This is likely because the significance of certain sequence variants is not yet known (Section 7.3.3). Variant return can also be influenced by the clinical information that is given to the laboratory.
- **Outcome:** Jaclyn's parents also want more information about Noonan syndrome and bowel cancer, so they are provided with information regarding a support group (Sections 3.5 and 7.4) for Noonan syndrome, information on Lynch syndrome, and recommendations for screening for cancer in this condition. Reviews for genetic conditions can be found on Pubmed and include GeneReviews articles and Orphanet reviews. Genetics Home Reference can also provide summaries of genetic conditions. The geneticist plans to follow Jaclyn regularly, but she will also continue to be followed in your clinic. Jaclyn's heart is found to be normal and she receives an IEP in school and starts therapy.

The Diagnosis

Clinical Diagnosis: Patient—Noonan syndrome. Lynch syndrome. Familial polydactyly. Mother—Lynch syndrome.
Genetic Diagnosis: Patient—*SOS1* gene, heterozygous likely pathogenic variant. *PMS2* gene, heterozygous pathogenic variant. Mother—*PMS2* gene, heterozygous pathogenic variant.

Further Reading

Idos, G., Valle, L., Lynch syndrome. Adam, M.P., Ardinger, H.H., Pagon, R.A., Wallace, S.E., Bean, L.J.H., Gripp, K.W., Mirzaa, G.M., Amemiya, A., (editors). In GeneReviews. University of Washington, Seattle (WA), 1993−2022 [updated February 4, 2021]. PMID 20301390.

Roberts, A.E., November 15, 2001. Noonan syndrome. Adam, M.P., Ardinger, H.H., Pagon, R.A., Wallace, S.E., Bean, L.J.H., Gripp, K.W., Mirzaa, G.M., Amemiya, A., (editors). In GeneReviews. University of Washington, Seattle (WA), 1993−2022 [updated February 27, 2022]. PMID 20301303.

Roberts, A.E., Allanson, J.E., Tartaglia, M., Gelb, B.D., 2013. Noonan syndrome. Lancet 381, 333−342. PMID 23312968.

Case 11.12.3

Neurodevelopmental Cases: Delayed Development in Early Childhood

Randi J. Hagerman

MIND Institute, Department of Pediatrics, University of California, Davis, Sacramento, CA, United States

Abstract

A 3-year-old boy with global developmental delay, macrocephaly, and hypotonia has perseverative behavior, aggression, and tactile defensiveness and meets the criteria for autism spectrum disorder. He is found to have fragile X syndrome and there is a strong maternal family history of complications related to this condition.

The Case

You follow John, who has continued to have developmental delays. He is now a 3.5-year-old boy with a history of a normal prenatal and birth history, with birth weight of 7 lb 12 oz, but because he had difficulty with latching to the breast after birth, his mother bottle fed him. He had frequent vomiting in the first year due to gastroesophageal reflux and had recurrent otitis media in the first 2 years of life, necessitating the placement of PE tubes by 2.5 years. All developmental milestones are delayed. You referred for occupational and physical therapy at 15 months for hypotonia, which helped initiate walking at 16 months. He was referred for early language intervention at 2 years old, and said his first words at 30 months after the ear tube surgery. His therapists recommend a developmental preschool at age 3 years.

Behaviorally, he has poor eye contact and stereotypic behavior including hand flapping when excited. He has daily tantrums if he does not get what he wants, and at home his tantrums include screaming and falling onto the floor with occasional biting and hitting his mother but never his father. His teacher is concerned about his perseverative behavior and aggression, with frequent impulsive hitting of another boy in the class and occasionally the teacher.

Your examination reveals a boy difficult to examine because of tactile defensiveness, poor eye contact, and lack of social interaction, but when you sing to him he smiles. His longest phrase is "all done, go bye-bye," which he says over and over again throughout the exam. His vital signs are normal but head circumference is above the 99th percentile and his height and weight are at the 80th and 60th percentiles, respectively. The rest of his examination is normal except his forehead is broad, ears are slightly prominent, his feet are flat with a mild degree of pronation, and he has hypotonia.

The Issues

- **Is there any significant family history?** The family history is significant for his mother who is 35 years old, with a history of recurrent depression and anxiety, and hypothyroidism at age 32. His maternal grandfather was diagnosed with atypical Parkinson disease, but there is no family history of autism or intellectual disability.

Genomics in the Clinic. https://doi.org/10.1016/B978-0-12-816478-5.00077-7

- **How should he be formally evaluated?** You are concerned for autism and the fact that John's head is large and his development is delayed. You refer him for developmental testing and an autism evaluation. Results of the Mullen Scales of Early Development at 41 months demonstrate: gross motor at 21 months, visual reception at 20 months, receptive language at 22 months, and expressive language at 18 months with an Early Learning Composite of 49. He thus has intellectual disabilities with delay in all areas of development. On the ADOS2 testing he meets diagnostic criteria of autism spectrum disorder (ASD) with a total score of 18.

You most likely would consider sending a SNP microarray (Section 6.1.3) to rule out any unbalanced chromosomal rearrangement that could predispose to similar findings. Metabolic testing can be second tier testing, once the more common entities are ruled out. The presence of autism with macrocephaly can also be due to PTEN mutations, another second tier test.

- **Is there any specific testing to consider?** You remember that the most common inherited cause of intellectual disability or autism is fragile X syndrome (FXS) and he does have some physical features that are commonly seen in this disorder, including hypotonia, perseverative speech, hand flapping, prominent ears, high forehead, and flat feet, although there is no family history of intellectual disability or autism. You decide to order the fragile X DNA test, also called the *FMR1* DNA test. This test comes back positive with a smear from 450 to 800 CGG triplet repeats (Section 6.4.1) with abnormal methylation on the front end of the *FMR1* gene.

Normal alleles typically have the number of CGG triplet repeats at the 5′ end of the FMR1 gene of 44 or less. From 45 to 54 repeats is considered the intermediate range, in which there can be expansion but not into the full mutation range and does not cause a phenotype. A premutation means having between 55 and 200 CGG repeats in length.

- **Should he have a neurologic workup?** You note his global developmental delay and presence of ASD. You remember that he also has an enlarged head, not uncommon in ASD, but this is typical for fragile X syndrome. You wonder whether you should order an MRI of the brain to look for brain abnormalities; however, this would need sedation in a young child. You also wonder if you should send metabolic testing for the hypotonia. You decide to refer John to genetics.

The Referral

- **Genetic Evaluation**: A full medical history, family history, and detailed physical examination are performed by the pediatric geneticist. John has fragile X syndrome, with a full mutation that is completely methylated (Sections 2.4.1, 6.4.1). His history of a poor suck after birth, frequent vomiting in the first year secondary to reflux, and his diagnosis of ASD are also consistent with his diagnosis of fragile X syndrome. John's examination by the geneticist demonstrates loose connective tissue, including hyperextensible finger joints, large prominent ears (Fig. 11.12.3.1), and

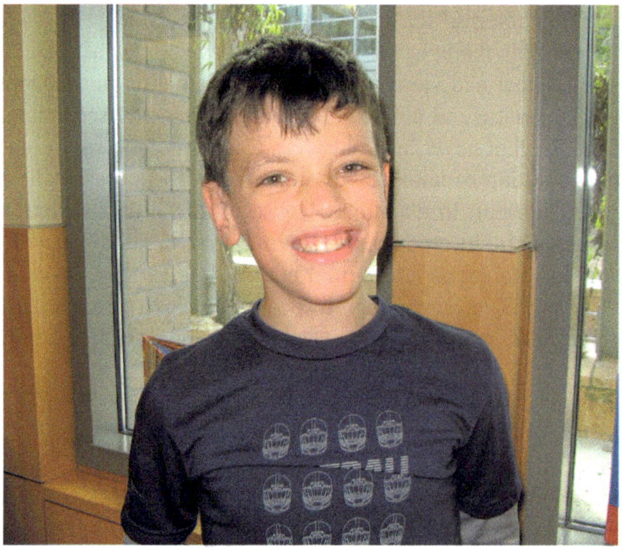

FIGURE 11.12.3.1 A 12-year-old boy with fragile X syndrome, with mildly prominent ears.

significant pronation accompanying his flat feet, all commonly seen in fragile X syndrome. Other findings could include long face, high palate, flat feet, and macroorchidism, which begins at about age 10 years and reaches maximal size at about 16 years and then stabilizes. His behavior findings are typical.

A more detailed family history is obtained. John's mother is an obligate fragile X carrier; although she could either have a premutation (55−200 CGG repeats) or a full mutation (more than 200 CGG repeats), it is likely the former because of her father's history. He has a tremor and ataxia, which can be found in older premutation carriers with the fragile X−associated tremor/ataxia syndrome (FXTAS) (Section 11.13.11), although he has been diagnosed with atypical Parkinson disease. Further family history reveals that the maternal grandfather has a 70-year-old sister who also has demonstrated a tremor and balance problems, in addition to a history of menopause when she was 35 years old, and a diagnosis of fibromyalgia that she has suffered from for many years.

- **Genetic Counseling**: Genetic counseling is given to John's family, and all family members at risk for a fragile X mutation. Typical physical and behavioral features are explained for the children with full mutations, and the premutation carriers (Section 11.17.3). Those with fragile X syndrome (full mutations) typically demonstrate hyperactivity and inattentive and impulsive behavior consistent with attention deficit hyperactivity disorders, in addition to overreactivity to sensory stimuli leading to tantrums and sometimes aggressive behavior. Approximately 60% of boys with fragile X syndrome are also diagnosed with ASD.

  Those with a premutation often suffer from anxiety and/or depression along with other psychiatric conditions known as the fragile X−associated neuropsychiatric disorders (FXANDs). They also may have medical problems, such as: chronic fatigue, hypertension, fibromyalgia, hypothyroidism, early menopause before age 40 years called fragile X−associated primary ovarian insufficiency (FXPOI), or FXTAS (Section 11.13.11). FXPOI occurs in approximately 20% of female carriers, FXTAS in 40% of male carriers and 16% of female carriers after age 50 years, and FXAND in approximately 50% of carriers at any age including childhood.
- **Genetic Testing**: DNA for fragile X syndrome is offered to all family members who could be a fragile X carrier, either affected by a premutation or the full mutation.
- **Management and Recommendations**: John is referred for a pediatric ophthalmology evaluation to assess for strabismus and will have ongoing assessments. Since his cardiac exam is normal, with no murmur or click, he is not referred for echocardiogram or EKG.

  Because John has significant behavioral problems and ASD, he is referred for an applied behavioral analysis (ABA) program for ASD, and his work with his therapist will also include ways to decrease his aggression. He needs ongoing intensive speech and language therapy and occupational and physical therapy intervention in his developmental preschool. He should be referred for pediatric ophthalmology evaluation by the age of 4 years, or sooner if has strabismus or nystagmus.

  The family is also referred to their nearest family support center listed on the National Fragile X Foundation (NFXF, www.fragileX.org) and also to the nearest Fragile X Clinical Research Center (FXCRC, listing on the NFXF website). FXCRCs offer the latest in interventions for those with FXS and support to the families.
- **Implications for Other Family Members:** John's mother has 105 CGG repeats in *FMR1*, so she is a premutation carrier, and her anxiety and depression are consistent with a diagnosis of FXAND. She is at 50% risk for having another baby with fragile X syndrome in subsequent pregnancies, so reproductive options, including prenatal diagnosis with subsequent pregnancies, egg donation, IVF, and adoption, are all discussed at length. She is also at a 20% risk for fragile X-associated primary ovarian insufficiency. She is referred to a therapist to help with anxiety and depression, and it is recommended that she discuss possible premature ovarian failure with her gynecologist. She also is offered testing for the number of AGG interruptions in the CGG repeat area to assess stability and likelihood of expansion; the more AGG interruptions, the less likely for expansion in meiosis.

Every 10 or so CGG repeats usually have an AGG anchor before the next 10 CGG repeats. Some premutation carriers have no anchors instead of the usual one or two anchors and if the anchors are lost, then the chance that the CGG repeat will expand to a full mutation when it passes through a carrier female to the next generation is much higher.

John's maternal grandfather is found to have a CGG repeat number of 70 and his examination is consistent with the tremor and ataxia associated with fragile X-associated tremor/ataxia syndrome (Section 11.13.11). John's maternal great-aunt is found to have a premutation also, with 82 CGG repeats, and she is diagnosed with FXTAS and FXPOI. An MRI of the brain is recommended for John's grandfather and demonstrates white matter disease in the middle cerebellar peduncles (MCP sign present in various conditions, particularly neurodegenerative diseases) in addition to brain atrophy, consistent with a diagnosis of Definite FXTAS, instead of atypical Parkinson disease. He and John's great-aunt are referred to a neurologist who specializes in movement disorders for their FXTAS.

- **Caveats**: The size of the CGG repeat length determines the likelihood of expansion in the offspring into the full mutation range in female meiosis, and this increases exponentially as the size increases.
- **Outcomes**: After one year of ABA therapy, John is doing much better with behavior. He has ongoing early intervention therapy through the school system and is starting to communicate better. His mother's anxiety and depression have largely resolved.

The Diagnosis

Clinical Diagnosis: Patient—fragile X syndrome. Mother—fragile X–associated neuropsychiatric disorder. Maternal Grandfather and Great-aunt—fragile X–associated tremor/ataxia syndrome. Great-aunt—fragile X–associated primary ovarian insufficiency.

Genetic Diagnosis: Patient—*FMR1* gene, hemizygous full mutation triplet repeats, abnormal methylation. Mother and Maternal Great-aunt—*FMR1* gene, heterozygous premutation triplet repeats. Maternal Grandfather—*FMR1* gene, hemizygous premutation triplet repeats.

Further Reading

Erickson, C.A., Kaufmann, W.E., Budimirovic, D.B., Lachiewicz, A., Haas-Givler, B., Miller, R.M., et al., 2018. Best practices in fragile X syndrome treatment development. Brain Sci. 8, 224. PMID 30558274.

Hagerman, R.J., Berry-Kravis, E., Hazlett, H.C., Bailey Jr., D.B., Moine, H., Kooy, R.F., et al., 2017. Fragile X syndrome. Nat. Rev. Dis. Prim. 3, 17065. PMID 28960184.

Hagerman, R.J., Protic, D., Rajaratnam, A., Salcedo-Arellano, M.J., Aydin, E.Y., Schneider, A., 2018. Fragile X-associated neuropsychiatric disorders (FXAND). Front. Psychiatr. 9, 564. PMID 30483160.

Case 11.12.4

Neurodevelopmental Cases: Autism Spectrum Disorder

Vivien Narcisa Triano[1] and Suma Shankar[2]

[1]*Kaiser Permanente Northern California, Sacramento, CA, United States;* [2]*Division of Genomic Medicine, Departments of Pediatrics and Ophthalmology, University of California Davis, MIND Institute, Sacramento, CA, United States*

Abstract

A 5-year-old girl is referred for nonsyndromic autism spectrum disorder, which is found to be due to supernumerary isodicentric chromosome 15.

The Case

Lily is a 5-year-10-month-old girl presenting to your pediatric primary care practice to establish care. She comes from a military family and they have just moved to your town from out of state. Per parental report, her previous pediatrician had raised the question of possible autism spectrum disorder (ASD), but because of the family's relocation and prior issues with travel, Lily has not yet had a formal evaluation for this. She was born at term by normal spontaneous vaginal delivery, with prenatal care history of maternal diabetes treated with Glyburide. There were no immediate neonatal complications and by report, she passed the newborn screening in her state of birth.

The family first became concerned about her development when her milestones were delayed. Although she sat unassisted at 7 months, she started walking at 18 months of age. Her first words were at 2 years of age and phrase speech was at 3–4 years. Her language continues to be delayed. Lily did not make eye contact or engage with her peers until approximately 4.5 years of age; her family describes her as "being in her own world."

Behaviorally, she has a history of repetitive behaviors such as hand flapping and self-injurious habits like finger and toenail biting to the bottom. Her mother reports that she cannot take care of any personal hygiene and is unable to dress herself. She still prefers a sippy cup for drinking and likes to be swaddled, particularly when she becomes upset and needs to be calmed. Toilet training has been a significant challenge and she is still in diapers, though she has started to show some recent interest. In addition to the behavioral challenges, her family reports that she has difficulty remembering academic information. She has not had any seizures.

Lily is not enrolled yet in school and is not receiving any behavioral interventions. The family is now more permanently established in your town, so they are interested in pursuing a more formal diagnosis with the goal of identifying appropriate educational and behavioral interventions.

Genomics in the Clinic. https://doi.org/10.1016/B978-0-12-816478-5.00078-9

The Issues

- **Does Lily have autism?** ASD is a diagnosis which impacts an individual's social interaction and communication (Section 5.2). Sometimes, repetitive, restricted behaviors and/or interests also are present. The concerns reported by Lily's parents are appropriate for ASD, therefore further evaluation by a developmental pediatrician and/or a licensed psychologist familiar with this diagnosis is recommended to confirm her diagnosis and assess her developmental abilities. Ideally, diagnoses are made by standardized, play-based evaluations such as the Autism Diagnostic Observation Schedule (ADOS) or by comprehensive interview of the caregiver and/or educator. You refer Lily to a clinical psychologist for behavioral evaluation, given that her main concerns appeared to be primarily behavioral in nature, and the developmental pediatrician, both at the local university hospital, for further evaluation.
- **What does the evaluation show?** Formal developmental evaluation demonstrates that Lily does indeed meet DSM-5 diagnostic criteria for ASD. Specifically, she demonstrates deficits in social communication and interaction across multiple contexts; restricted, repetitive patterns of behavior, interests, and activities; and these deficits were present in her early development. In addition, she is found to have borderline intellectual ability, particularly in nonverbal cognition.
- **What are the subsequent recommendations?** As part of the developmental pediatrician's and psychologist's recommendations for new diagnoses of autism and borderline intellectual abilities, applied behavioral analysis (ABA) therapy, speech, physical, and occupational therapies are recommended. To address social communication deficits, participation in a social skills group and individualized social communication therapy are recommended as well. Finally, a referral for a genetic evaluation is also recommended to assess for the possibility of an underlying syndrome as the cause for her developmental and cognitive impairment.

The Referral

- **Genetic Evaluation**: A combined detailed evaluation of Lily is performed by the medical geneticist and genetic counselor with a review of the personal and family medical history and a physical exam to evaluate for syndromic or dysmorphic features. Aside from the new diagnosis of ASD, Lily appears to be healthy without additional medical complications.

  Lily's family history is significant for two older sisters, one of whom was previously diagnosed with "borderline" high-functioning autism, attention-deficit hyperactivity disorder, and bipolar disorder. Her other sister is healthy and doing well in school. Neither of her parents reports developmental or medical concerns. There is also a male first cousin, through a paternal aunt who has a diagnosis of autism. Both her parents are of mixed European ancestry and there is no known consanguinity.

  Her physical exam reveals her to be at the 70th percentile for length, 97th percentile for weight, and 93rd percentile for head circumference. She has no other distinct physical features indicating a specific genetic syndrome, and she is nondysmorphic. Lily's evaluation suggests that she has nonsyndromic ASD, in the absence of additional medical diagnoses.
- **Genetic Counseling**: Studies of twins and families with multiple affected siblings with ASD have shown that there is a strong genetic link for autism (Section 5.7.2). Approximately 25% of individuals referred for clinical genetics evaluation for the diagnosis of ASD are syndromic (with additional clinical findings) while the remaining 75% of individuals have "essential" or nonsyndromic autism. Because of this, evaluation by a clinical geneticist and/or genetic counselor is recommended following a new behavioral diagnosis of ASD.

  Appropriate genetic or genomic testing options are reviewed with the family, to include discussion of the benefits and limitations of testing, results (including possible secondary or unanticipated findings—Section 7.2), test costs, insurance coverage, and estimated turnaround time (Sections 6.6, 7.3, 10.2, 10.3). Clarifying the presence of an underlying genetic etiology for a patient's neurodevelopmental diagnosis may provide anticipatory guidance for screening and management of additional comorbidities such as her borderline intellectual disability, risk for a seizure disorder, or other systemic involvement. If an identifiable molecular diagnosis is revealed, more accurate recurrence risk information can also be provided to the family. It should be emphasized that as ASD is a behavioral diagnosis, a molecular genetic test is not intended to diagnose an individual with ASD, but rather to clarify the underlying etiology. In other words, a positive genetic test will not diagnose an individual with autism, in the same way that a negative result does not exclude the diagnosis.
- **Genetic Testing**: In light of this, rather than recommended targeted single gene or panel testing in Lily, broader chromosome microarray is sent to evaluate for copy number variants (chromosomal deletions and/or duplications) (Section 6.1.3). Depending on the technique, of the individuals with ASD who have an underlying genetic etiology, between 2%

and 10% of individuals have gross or submicroscopic chromosome abnormalities. Because of this, the American College of Medical Genetics and Genomics (ACMG) recommends chromosomal microarray testing as a first-line molecular test in the initial evaluation of individuals with ASD, in apparently nonsyndromic developmental delay/intellectual disability, or multiple anomalies not specified to a well-delineated genetic syndrome. Similarly, revised practice guidelines from the American Academy of Pediatrics (AAP) recommend chromosomal microarray for all individuals with a diagnosis of global developmental delay as well as fragile X DNA analysis (Sections 5.2, 11.12.3, 11.13.11, 11.173) for all. Clinical chromosome microarray test results typically return within 2–4 weeks.

- **Results**: After about 2–3 weeks, Lily's chromosome microarray results return, which demonstrate that she has a partial tetrasomy of 15q (4 copies) of the Prader–Willi/Angelman syndrome critical region located at chromosome 15q11.2 (Section 11.13.1). This region contains the *SNRPN* gene. This microarray result is suggestive of either a tandem triplication (interstitial, within a single chromosome) or a supernumerary isodicentric 15 (an extra chromosome, made up of two attached copies of the long arm). As chromosome microarray only evaluates for copy number, additional testing by fluorescent in situ hybridization (FISH) is recommended to clarify the orientation of the extra chromosomal material (Section 6.1.2). Follow-up FISH in Lily demonstrates that she has the supernumerary isodicentric chromosome 15. This finding is consistent with the idic(15q) syndrome, in which 15q refers to the long arm of chromosome 15.

  Idic(15q) is the most frequent chromosomal finding on G-banded karyotypes (Section 2.2) in patients with ASD. It is typically derived from the maternal homolog when associated with ASD and/or intellectual disability. It also appears in approximately 1% of microarrays in individuals with ASD who have had a previously normal karyotype. Idic(15q) is characterized by hypotonia, motor delay, intellectual disability, and epilepsy including infantile spasms. Rarely, dup15q can be associated with psychosis or sudden unexplained death. There have been individuals described with cleft lip and/or palate, cardiac, renal, or other systemic abnormalities, small stature, hypogonadism, and pubertal disorders.
- **Management and Recommendations**: As there is an increased risk for possible seizures associated with Lily's new molecular diagnosis of idic(15q), baseline neurologic evaluation is recommended, including assessment for seizure activity and an EEG. Furthermore, continued supportive care to enhance Lily's developmental and cognitive potential is recommended. This may include occupational, physical, speech, and behavioral therapies, as well as pharmacologic interventions to manage behaviors and/or seizures.
- **Implications for Other Family Members**: For idic(15q), all reported cases have been de novo, and the presence of the supernumerary chromosome 15 has not been inherited from a parent. Because of this, risk of recurrence to siblings is low, but not completely zero, as there is minimal risk above the general population due to possible maternal germline mosaicism (Section 2.2.2). We would not anticipate either of Lily's parents to carry the isodicentric chromosome since they are unaffected. To provide reassurance of low recurrence for future pregnancies, chromosome analysis can be considered in both parents if desired.

It should be noted that the recurrence risk (Section 4.4.3) for interstitial (not terminal) copy number gains of 15q is distinct from idic(15q). This is important to clarify as the recurrence risk is based on the orientation of the duplicated material (isodicentric vs. interstitial). Had Lily's FISH studies returned indicating that the copy number gain was interstitial, the recurrence risk for her parents would not be minimal. ***Occurrences of the interstitial copy number gains have been reported to be*** *de novo* ***in 85% of cases, whereas 15% have been reported to be inherited.*** *If the mother of a proband with an interstitial copy number gain is found to carry the interstitial triplication, the risk for each sibling or future pregnancy is 50%. To date, 15q copy number gains have almost exclusively been reported to be maternally derived (arising from chromosome errors originating in the ovum).*

Genetic evaluation and testing is offered for Lily's sister with mild autism. In light of Lily's diagnosis, presence or absence of the isodicentric 15 in her sister could help clarify possible inheritance. If her affected sister also carried the same genetic diagnosis, this would imply that recurrence of isodicentric 15 in these sisters was due to germline mosaicism.

Her sister's genetic evaluation and test results, however, return negative; the chromosome abnormality is not present in her mildly affected sister. Remaining explanations for her sister's milder presentation include a separate, distinct genetic etiology from the idic(15q) finding, presence of a genetic variant in an unknown gene not detected by available technology, or a combination of both genetic and nongenetic factors (i.e., multifactorial causes - Section 2.7.3) which resulted in a milder presentation.

- **Caveats**: Lily's case is an example of capturing the underlying cause of ASD using the first-line test. In some cases of suspected syndromic autism, chromosome microarray may not detect an identifiable molecular finding. For example, an individual with the *PTEN* hamartoma tumor syndrome (PHTS) may present with macrocephaly, dermatologic findings such as lipomas, or penile freckling in addition to the developmental diagnosis of autism (Section 11.7.1). In this case, chromosome microarray would not be the optimal molecular test to recommend initially, but rather single gene testing (Section 6.2.2) or panel testing (Section 6.2.3) including the *PTEN* gene would be indicated, in order to evaluate for single gene sequence variants rather than chromosomal copy number changes.

For a proportion of syndromic cases of ASD and the vast majority of nonsyndromic cases, an identifiable genetic cause is not yet available. Because behavioral phenotypes such as ASDs are complex, even more comprehensive molecular techniques such as exome or genome sequencing (Sections 6.3.1, 6.3.2) may be unrevealing. This suggests that additional means of evaluation including assessing oligogenic effects and combined environmental and genetic background effects may be required to understand the underlying etiology of all ASDs.

- **Outcome**: Following her diagnosis of idic(15q), Lily is enrolled in ABA and early intervention therapy and makes some improvements in communication. Her baseline EEG is negative, and she does not develop seizures. While she has made developmental progress with intensive therapies, when Lily ages out of her early intervention program, she enrolls in a special education classroom with a 1:1 aide. She enjoys water therapy and music and will continue these.

The Diagnosis

Clinical Diagnosis: Autism spectrum disorder, supernumerary isodicentric chromosome 15 syndrome.
Genetic Diagnosis: 47,XX,+idic(15)(q12).

Further Reading

Carter, M.T., Scherer, S.W., 2013. Autism spectrum disorder in the genetics clinic: a review. Clin. Genet. 83, 399–407. PMID 23425232.

Finucane, B.M., Lusk, L., Arkilo, D., Chamberlain, S., Evinsky, O., Dindot, S., et al., June 16, 2016. Maternal 15q duplication syndrome and related disordersAdam, M.P., Ardinger, H.H., Pagon, R.A., Wallace, S.E., Bean, L.J.H., Stephens, K., Amemiya, A., (editors). In GeneReviews. University of Washington, Seattle (WA), 1993–2020 [updated July 15, 2021]. PMID 27308687.

Schaefer, G.B., Mendelsohn, N.J., Professional Practice & Guidelines Committee, 2013. Clinical genetics evaluation in identifying the etiology of autism spectrum disorders: 2013 guideline revisions. Genet. Med. 15, 399–407. PMID 23519317.

Yehia, L., Eng, C., November 29, 2001. PTEN hamartoma tumor syndrome. Adam, M.P., Ardinger, H.H., Pagon, R.A., Wallace, S.E., Bean, L.J.H., Gripp, K.W., Mirzaa, G.M., Amemiya, A., (editors). In GeneReviews. University of Washington, Seattle, Seattle (WA), 1993–2022. [updated February 11, 2021]. PMID 20301661.

Case 11.12.5

Neurodevelopmental Cases: Cognitive Issues and Tall Stature

Natalie Blagowidow

Harvey Institute for Human Genetics, Greater Baltimore Medical Center, Baltimore, MD, United States

Abstract

A 9-year-old boy previously diagnosed with Klinefelter syndrome presents with mild cognitive issues and tall stature.

The Case

Philip is a 9-year-old boy who comes in with his parents to see you for a pediatric appointment. He was found to have Klinefelter syndrome, 47,XXY, by prenatal genetic amniocentesis performed for advanced maternal age. Philip had hypotonia as an infant and toddler. At 5 years of age he was noted to have some fine motor delay and required speech and occupational therapy. He is now in fourth grade. He has tutoring for language arts and continues speech therapy for speech delay. On physical exam his height is at the 90th percentile, and there is a soft systolic murmur. External genitalia are normal for a prepubertal male. Philip's parents are now raising questions about how their son's cognitive issues are related to having Klinefelter syndrome and how it will affect his health and development going forward.

Klinefelter syndrome is diagnosed prenatally in a small percent of individuals. It is infrequently diagnosed in childhood, usually part of an evaluation for developmental delay or hypogonadism. The majority of diagnoses are made as part of an adult infertility evaluation. It is estimated that over 50% of affected individuals are never diagnosed.

The Issues

- **How is having Klinefelter syndrome affecting Philip's health?** You tell the parents that he is generally in good health, but that you want to refer him for an echocardiogram since you hear a heart murmur today. Mitral valve prolapse is found.

Klinefelter syndrome is associated with a small increased risk for cardiac structural defects, including mitral and aortic valve anomalies and left diastolic dysfunction. A baseline echocardiogram may be of benefit in this population.

- **Does Philip need to take hormonal treatment to go into puberty?** You are not sure, so you order LH, FSH, and testosterone levels to address this concern. The results are normal for his age. You consider sending him for an endocrine evaluation.

These hormonal studies should have been monitored annually. If hypogonadism was detected, testosterone treatment could be offered. Although significant effects on cognition have not been seen, treatment may benefit psychosocial function.

Genomics in the Clinic. https://doi.org/10.1016/B978-0-12-816478-5.00079-0

There has been growing evidence that some androgen treatment before puberty can be of benefit but this is still being investigated.

- **Are Philip's cognitive issues due to Klinefelter syndrome?** You know that there are some individuals with learning problems related to the syndrome but are not sure what developmental and behavioral issues to anticipate in the future. You refer the patient for a genetic consultation.

The geneticist will have comprehensive information about the particular intellectual and physical developmental issues that occur in Klinefelter syndrome at different ages. The genetic consultation will provide the family with more recommendations for evaluation and treatment of problems seen in this syndrome. It is helpful to send your recent note and prior lab studies prior to the genetic evaluation.

The Referral

- **Genetic Evaluation**: Philip and his parents are seen by the pediatric geneticist and genetic counselor. A complete review of the family history, as well as a medical and developmental history, is performed. Overall, Philip has been healthy, and both vision and hearing have been normal. Although his early motor milestones were delayed due to the hypotonia, he did well subsequently. His speech delay always exhibited receptive language higher than expressive language. He had frustration related to communication when younger but has improved quite a bit with the speech therapy. He has a reading disability, which is being addressed by the tutoring. In addition to the academic help Philip receives, there have been some social issues. He tends to play with younger children, and less with his peers. He seems somewhat introverted and tends to be immature. He was said to be easily distracted when younger but never met criteria for ADHD. The pediatrician's physical exam is confirmed. Tall stature is noted and a low upper to lower segment ratio is seen. Hypertelorism and a high arched palate are noted. Gynecomastia is not appreciated. The genitalia appear prepubertal, with a stretched penile length of 6 cm and small testes.

Gynecomastia occurs in some individuals with Klinefelter syndrome, with onset after puberty. Antiestrogen therapy may help; cosmetic mastectomy is also an option. There is a 20- to 30-fold increased risk for breast cancer, so regular breast exams are recommended in adulthood. Mammogram is not, as the absolute incidence of breast cancer is still low.

- **Genetic Counseling**: The physical findings including height and weight identified on the exam are explained as being consistent with Klinefelter syndrome. The sporadic nature of the syndrome is reviewed, with an incidence of about 1/600 liveborn males. Most have normal intelligence, but 70%, like Philip, will have minor developmental issues with normal cognition. Delayed speech and expressive language is seen, and dyslexia and attention-deficit disorder occur more often than unaffected males. Social skills may be poor with lower adaptive functioning. Normal puberty is expected, but testicular growth arrests in mid-adolescence resulting in low testosterone levels and infertility due to absent to low sperm production. Replacement testosterone is started when the natural production plateaus and leads to virilization, increased muscle mass and bone density, and ultimately improved sexual function. Gender identity is typically male but gender dysphoria has been reported.
- **Management and Recommendations**: Monitor weight, with appropriate attention to nutrition and exercise. Testosterone, LH, and FSH screening should continue annually. After puberty, screening for metabolic syndrome, dyslipidemia, and hypothyroidism is recommended annually. A DEXA scan for bone mineral density is recommended every 3 years after puberty due to an increased risk to develop osteopenia/osteoporosis. Regular dental exams are important, including X-ray screening for taurodontism, a condition characterized by an abnormal pulp to enamel ratio and associated with increased dental decay. The learning disabilities should continue to be monitored, with appropriate support provided, including intervention for behavior issues if necessary. Testicular sperm extraction for sperm cryopreservation may be considered in late adolescence or adulthood. In adulthood, there is a 5-fold increased risk for developing metabolic syndrome and a small increased risk for elevated lipids. There is a slightly increased risk for autoimmune disorders such as hypothyroidism.
- **Caveat**: Mosaicism is seen in 10%–20%, mainly 46,XY/47, XXY (Section 2.2.2, Table 2.1). This may be associated with a milder phenotype. An additional X chromosome (48,XXXY) is seen in a small percentage of individuals and is associated with more significant developmental delays.
- **Outcome**: Philip goes through a normal puberty, with appropriate sexual development and growth; testicular size remains small. Testosterone levels are being monitored, and supplementation is planned once levels drop. Consultation for sperm cryopreservation is planned. Philip develops a small group of good friends. He is not involved in competitive sports but enjoys swimming. He is getting average grades in school.

The Diagnosis

Clinical Diagnosis: Klinefelter syndrome.
Genetic Diagnosis: 47,XXY.

Further Reading

Aksglaede, L., Link, K., Giwercman, A., Jorgensen, N., Skakkebaek, N.E., Juul, A., 2013. 47,XXY Klinefelter syndrome: clinical characteristics and age-specific recommendations for medical management. Am. J. Med. Genet. C Semin. Med. Genet. 163C, 55–63. PMID 23345262.

Davis, S., Howell, S., Wilson, R., Tanda, T., Ross, J., Zeitler, P., et al., 2016. Advances in the interdisciplinary care of children with Klinefelter syndrome. Adv. Pediatr. 63, 15–46. PMID 27426894.

Gies, I., Unuane, D., Velkeniers, B., De Schepper, J., 2014. Management of Klinefelter syndrome during transition. Eur. J. Endocrinol. 171, R67–R77. PMID 24801585.

Ross, J.L., Kushner, H., Kowal, K., Bardsley, M., Davis, S., Reiss, A.L., et al., 2017. Androgen treatment effects on motor function, cognition, and behavior in boys with Klinefelter syndrome. J. Pediatr. 185, 193–199 e4. PMID 28285751.

Section 11.13

Neurologic and Muscular Cases

Case 11.13.1

Neurologic and Muscular Cases: Congenital Hypotonia and Feeding Difficulties

Antonie D. Kline
Harvey Institute for Human Genetics, Greater Baltimore Medical Center, Baltimore, MD, United States

Abstract
A newborn with bilateral cryptorchidism and mild hypotonia has difficulty feeding and is found to have a microdeletion at chromosome 15q11.2 and abnormal methylation consistent with Prader–Willi syndrome.

The Case

A newborn male infant, Malik, has had feeding difficulties due to poor suck. You, the pediatrician, are called on day 3 of life by the nursery to discuss transfer to the NICU for significant weight loss and difficulty feeding. On your partner's initial physical exam, she noted average size at birth, no unusual features, bilateral cryptorchidism, and mild hypotonia. This is the first baby born to parents in their 30s with uneventful pregnancy, delivery at 37 weeks gestation. You head to the hospital when your afternoon patient session is over. You note current weight more than 10% down from birth weight, bilateral cryptorchidism, moderate to marked hypotonia, and a weak cry.

The Issues

- **Is there an underlying cause for the poor feeding and hypotonia?** You study the infant's facial features. There may be subtle differences, but you see nothing specific. The baby could be slightly more premature than thought, especially if the mother's menstrual periods were farther apart than average. You concur with the recommendation to transfer the baby and discuss this with his parents. You discuss hypotonia and the many potential causes. You mention metabolic conditions and the newborn screening that Malik has already had. You also say that if there is a concern for an underlying condition, genetics will be consulted.

You are not sure if he has an underlying genetic condition, but you and the neonatologist should have a low threshold for consulting genetics if there is any question. It is preferable to consult genetics prior to sending any workup other than the standard newborn screening.

- **Should he be seen by pediatric neurology for the hypotonia?** For the moment, Malik is breathing well and does not appear to show any muscle weakness. The neonatologist decides not to consult neurology but does consult occupational

Genomics in the Clinic. https://doi.org/10.1016/B978-0-12-816478-5.00080-7

therapy to help with feeding, and the baby is observed for several days. Slowly, the infant starts to gain some weight, but the hypotonia and poor feeding persist.

- **What is the next step?** Based on the findings of poor feeding, bilateral cryptorchidism, and hypotonia, pediatric genetics is consulted on day 5 of life.

The Referral

- **Genetic Evaluation:** A full pregnancy history, a medical history, and a family history are obtained, and a detailed physical examination performed by the pediatric geneticist and genetic counselor, with photographs taken with permission from parents. His mother reports routine ultrasounds were normal, as was the pregnancy, but on further questioning, fetal movement was late and may have been decreased. Malik remained in breech position, and because of this, delivery was by Cesarean section.

  Family history is unremarkable. His mother reports no neurologic deficits, she has no signs of a neuropathic condition, and her handshake is normal. She had a negative carrier screening (Section 4.1) in early pregnancy. There is no consanguinity.

  On exam, Malik has low weight, but normal growth parameters at birth. He is normocephalic, with facies slightly long, and he has mildly dysmorphic features (Section 5.1) with shallow orbits, almond-shaped eyes, epicanthal folds, normally formed ears in normal position, small mouth with normal intact palate, no tongue fasciculations, normal nuchal region, symmetric chest with slightly widely spaced nipples, no organomegaly, normally formed extremities with full range of motion of joints and no contractures, normal hands with normally formed creases and normal nails, bilateral metatarsus adductus deformities of feet, bilateral cryptorchidism with relatively short penis and small scrotum, marked hypotonia with head lag, and normal deep tendon reflexes.
- **Genetic Counseling:** The geneticist and genetic counselor discuss with the NICU staff and parents that there are a number of findings of concern, and it is likely there is an underlying genetic condition. A diagnosis of Prader–Willi syndrome is strongly considered in this baby, given the history of poor feeding, failure to thrive, bilateral cryptorchidism with relatively short penis, and marked hypotonia of prenatal onset. This condition is a microdeletion syndrome (Sections 2.2.1, 2.4.1, Table 6.1) involving genes on the long arm of chromosome 15 within the region of 15q11.2 to 15q13, and these genes undergo imprinting (Sections 2.7.5, 6.4.3), with methylation inactivation depending on the parent of origin of the deletion. Prader–Willi syndrome is due to abnormal parental imprinting, specifically a deletion or missing component of the paternal allele 70% of the time, and maternal uniparental disomy (Section 2.7.5) 25%–30% of the time. Methylation testing is the first line of analysis and will detect 99% of cases. The diagnosis of Angelman syndrome, due to a deletion or missing component of the maternal allele or paternal uniparental disomy of the same region, is not discussed.
- **Genetic Testing:** Methylation testing for the Prader–Willi syndrome region is recommended. Because Malik has so many findings, a chromosome microarray analysis (Section 6.1.3) is also recommended to rule out microdeletions and/or microduplications of chromosomal material; if a deletion is present at 15q11.2, this will be detected by this method as well. A creatine kinase level is also recommended because of the hypotonia. The NICU staff will check on results of the newborn screening, since two samples have already been sent to the state lab per protocol, but additional metabolic testing will be done if the above tests are normal.
- **Management and Recommendations:** Physical therapy is recommended for the hypotonia, which mostly resolves by adulthood. An abdominal ultrasound can be obtained to confirm the presence of the testes. Further information is shared with the parents. There is hypogonadotropic hypogonadism, with lack of satiety and marked truncal obesity, if growth hormone treatment is not initiated during the first 2 years of life, and referral to pediatric endocrinology is recommended during infancy. Growth hormone treatment is indicated through adulthood. Hyperphagia can persist into adulthood, but with education, exercise, and continued growth hormone, morbid obesity can be avoided. At adolescence, sex hormone replacement is recommended. With aging, type 2 diabetes, hypothyroidism, and adrenal insufficiency can develop. Infertility is typical in adult males. There also may be eye findings (decreased pigment, strabismus), scoliosis, and sleep dysfunction with hypersomnolence. Individuals with Prader–Willi syndrome will have mild intellectual disability on average, and early intervention therapy should continue, particularly physical therapy, with monitoring of developmental milestones. There are behavioral outbursts and other behavioral issues, with increased risk for skin picking, ADHD, and psychiatric illness. Autism is reported. Referral for behavioral psychology may be indicated. There is a parent support organization (Section 3.5), to which the family is referred. Future prenatal testing (Section 4.4.2) is briefly discussed, with methylation testing available on invasive prenatal diagnostic testing, and the option of preimplantation genetic testing (Section 4.5.1) is also possible. Follow-up is arranged for 3–4 months in genetics.

- **Results:** Malik continues to have occupational therapy, and the feeding continues to improve. The newborn screening has all been negative. The CK level is normal. He starts physical therapy and the parents are given exercises to do at home after discharge. He gains enough weight to be able to be discharged.
  Three weeks later he and his parents return for a follow-up in genetics. The methylation testing is positive, with abnormal methylation on the allele inherited from his father, confirming a diagnosis of Prader–Willi syndrome. The chromosome microarray shows a microdeletion at 15q11.2 from breakpoint BP3 to BP1, also confirming the diagnosis.
- **Differential Diagnosis:** Babies can potentially develop hypotonia from involvement of different aspects of the central and peripheral nervous system. These could include: brain abnormalities such as holoprosencephaly or leukodystrophies, and other brain lesions including neonatal or prenatal stroke, spinal cord anomalies such as vertebral malformation compressing a nerve, neuromuscular junction disruption such as the congenital myasthenia gravis syndromes, or primary muscle disorders such as muscular dystrophy (Section 11.13.5). Other potential causes include chromosomal disorders such as Down syndrome (Section 11.13.2), additional single gene disorders such as the tubulinopathy disorders which usually also have eye findings, or inborn errors of metabolism (Section 5.4) such as congenital disorders of glycosylation, and conditions secondary to maternal pathology such as congenital myotonic dystrophy (Section 11.13.4).
- **Caveats:** Malik's mother could have had a limited panel of carrier screening (Section 4.1), which could have been falsely reassuring about no newborn issues, but it would not have detected this. Prader–Willi syndrome is more severe clinically if there is uniparental disomy (Section 2.7.5) than if there is a detectable deletion on chromosomes.
- **Outcome:** The family provides intensive therapy and works with Malik at home. As he grows, the hypotonia improves. He starts growth hormone therapy and his weight remains appropriate, and his appetite is stable. He has some mild developmental delays, but overall does well.

The Diagnosis

Clinical Diagnosis: Prader–Willi syndrome.
Genetic Diagnosis: Chromosome 15q11.2 microdeletion, abnormal methylation of paternal allele.

Further Reading

Harris, S.R., 2008. Congenital hypotonia: clinical and developmental assessment. Dev. Med. Child Neurol. 50, 889–892. PMID 19046184.

Driscoll, D.J., Miller, J.L., Schwartz, S., Cassidy, S.B., 1998. Prader-Willi syndrome. Adam, M.P., Ardinger, H.H., Pagon, R.A., Wallace, S.E., Bean, L.J.H., Stephens, K., Amemiya, A., (editors). In GeneReviews. University of Washington, Seattle (WA), 1993–2022 [updated December 14, 2017]. PMID 20301505.

Ramsden, S.C., Clayton-Smith, J., Birch, R., Buiting, K., 2010. Practice guidelines for the molecular analysis of Prader-Willi and Angelman syndromes. BMC Med. Genet. 11, 70. PMID 20459762.

Sparks, S.E., 2015. Neonatal hypotonia. Clin. Perinatol. 42, 363–371 ix. PMID 26042909.

Whittington, J., Holland, A., Webb, T., Butler, J., Clarke, D., Boer, H., 2004. Cognitive abilities and genotype in a population-based sample of people with Prader-Willi syndrome. J. Intellect. Disabil. Res. 48, 172–187. PMID 14723659.

Case 11.13.2

Neurologic and Muscular Cases: Dysmorphic Features and Hypotonia

Antonie D. Kline
Harvey Institute for Human Genetics, Greater Baltimore Medical Center, Baltimore, MD, United States

Abstract
A 2-month-old with mild hypotonia and dysmorphic features is diagnosed with Down syndrome.

The Case

A newborn male infant, Sonam, is born at term to a 39-year-old Gravida 4 Para 2012 mother, with an uneventful prenatal course, including normal ultrasounds and first trimester screening, and uncomplicated vaginal delivery. On his initial physical exam you, his pediatrician, note that the baby is average size for gestational age at birth, with features consistent with parental country of origin, normal cardiac exam, and mild hypotonia. He feeds well, stools, and urinates, and by the end of the second day is ready for discharge. He has failed the newborn hearing screening twice on the right side, but passes prior to discharge. The newborn nurse asks you if you think his features look unusual. You say that he resembles his parents and approve the discharge. You see him 2 days later and he continues to feed without difficulty and do well. At the 2-week visit you repeat the newborn screening. He is nearly back to birth weight and continuing to do well, still with mild hypotonia. At the 2-month visit you note features that could be consistent with Down syndrome, as well as persistent hypotonia.

The Issues

- **Why does the baby still have hypotonia?** You know that there are multiple causes for hypotonia in an infant. The baby could have a brain abnormality, such as neonatal stroke, that could lead to hypotonia. He could have an underlying genetic disorder, including a chromosomal abnormality, e.g., trisomy or Prader–Willi syndrome (Section 11.13.1); an inborn error of metabolism such as the lysosomal storage diseases; or a primary neurologic cause such as a neuropathy. There also could be a maternal cause, such as myotonic dystrophy in the mother (Section 11.13.4). You consider a neurologic evaluation and/or brain study.
- **Could the baby have Down syndrome?** When you recall the newborn period for this baby, you remember mild hypotonia since birth and the newborn nurse raising the question of dysmorphic features. You know that the family is of Asian background and felt that the baby resembled his parents, but now you are not sure that this is the sole explanation.

Genomics in the Clinic. https://doi.org/10.1016/B978-0-12-816478-5.00081-9

It is strongly recommended to consider carefully any unusual feature or presentation mentioned by the neonatal nurses. They are typically very experienced, excellent nurses who manage unaffected babies all day. If they feel that a baby stands out for any reason, it is prudent to reevaluate the baby with that reason in mind.

- **Should the baby undergo any studies at this time?** You discuss the case with the pediatric geneticist, who is happy to fit the baby into the schedule in the next week. She recommends obtaining an echocardiogram at once and to send chromosomes with fluorescence in situ hybridization (FISH) testing (Section 6.1.2) for chromosome 21. You order these, and refer the baby to the pediatric geneticist.

The Referral

- **Genetic Evaluation:** A full pregnancy history and medical history are obtained when Sonam and his parents are seen by the pediatric geneticist 3 days later. The parents report that the prenatal first trimester screening (Section 4.4.1) was normal. They thought that this screening ruled out chromosome abnormalities and declined any invasive testing. Family history is obtained by the genetic counselor. There are two healthy older sisters and his parents had a first trimester spontaneous abortion. Both sides of the families are originally from Tibet. There is no consanguinity.

  A detailed physical examination is performed by the geneticist. Sonam is found to have length and weight at the 10th percentile for 2-month-olds, and head circumference below the 2nd percentile. Both his anterior and posterior fontanelles are open with the sagittal suture widened in between. His facies shows upslanting palpebral fissures, bilateral epicanthal folds, short dysplastic ears with overfolded helices, flat midfacial region, and downturned corners of the mouth (Fig. 5.2A). There is excess nuchal skin. Cardiac exam is normal. Genitalia are normal. The extremities reveal slightly short upper arms, hands with brachydactyly (Fig. 5.3A), with bilateral clinodactyly of the fifth fingers, bilateral single palmar creases, shallow acetabular fossae on hip exam, and wide space between first and second toes (Fig. 5.3B). He is hypotonic with normal reflexes. The echocardiogram ordered by the pediatrician was normal.
- **Genetic Counseling:** The geneticist and genetic counselor discuss with the parents that they think that clinically Sonam has Down syndrome. Of the 10 cardinal features (flat facial profile, anomalous ears, upslanting palpebral fissures, excess skin on back of neck, hyperflexibility of joints, dysplasia of pelvis, dysplasia of mid-phalanx, single palmar crease, hypotonia and poor Moro reflex) seen in newborns with Down syndrome, and 90% of babies with Down syndrome will have 6 or more of these features, he has 8 (does not have hyperflexible joints or poor Moro reflex). They discuss that the most likely chromosome result will be trisomy 21, with an extra chromosome 21 in all cells (Section 2.2.1). The parental origin of the additional chromosome 21 cannot be determined in any baby but is typically due to nondisjunction (Section 2.2.1) in formation of the gametes. This is more likely to happen in the mother's egg as the mother gets older but can occur in either. It is due to chance 95% of the time, but 5% of the time there is a chromosome 21 attached to a different chromosome due to a translocation (Section 2.2.1), which could be carried in a balanced formation by one of the parents. In addition, there could be a mosaic pattern (Section 2.2.2), in which some of the cells have the normal male chromosome count, i.e., 46,XY, and others have the extra chromosome 21, i.e., 47,XY,+21. The FISH test (Section 6.1.2) and the karyotype (Section 6.1.1) will determine all of this.
- **Results:** The FISH test confirms an extra chromosome 21 in half of the cells the next day. When the karyotype is completed 2 days later, Sonam is found to be mosaic for trisomy 21, namely 46,XY/47,XY,+21. The geneticist discusses that in any case with mosaicism, it is possible that all organs will have a mixed pattern of chromosomes. This could include the brain, and thus it is possible that those with mosaic Down syndrome could perform slightly better neurodevelopmentally than those with full trisomy 21 and generally could have a milder involvement. Also, because this is a trisomy 21 variant, there is no need to check the parents for a balanced translocation (Section 2.2.1).
- **Management and Recommendations:** Further information about Down syndrome is shared with the parents. There are published guidelines. It is best to follow growth using Down syndrome–specific growth charts, and if there is any fall-off in growth to consider hypothyroidism initially. The newborn screening results should be confirmed to be normal for thyroid function testing, which should be checked annually; he should be referred to pediatric endocrinology to be treated for likely hypothyroidism if abnormal. His hemoglobin should be followed annually, initially for a possible congenital leukemoid reaction, and later to screen for leukemia, particularly acute lymphoblastic leukemia. It was important to assess for any congenital heart defect, since 45% of all children with Down syndrome will have congenital heart disease, particularly ventricular septal defect or atrioventricular canal defect. Other congenital malformations can occur but Sonam does not appear to have any feeding difficulty, thus duodenal stenosis is unlikely, or constipation, thus Hirschsprung disease also is unlikely. Sonam will be at risk for myopia and strabismus and should be referred for pediatric ophthalmology evaluation by 6 months and regularly thereafter. His hearing should be reassessed at 12 months

of age and then regularly as he gets older especially if having difficulty with speech. The hypotonia will delay both gross and fine motor development, as well as speech because of the low muscle in the tongue. Physical therapy is indicated during infancy and will benefit Sonam greatly. Occupational therapy likely will be needed as he gets older and starts using his hands more. Intellectual disability is likely to occur, but there is a broad range of abilities, and family background plays a role. Speech therapy will be needed, initially for the hypotonic tongue, but also to help developmentally as he gets older. With aging, there is also an increased risk for early Alzheimer disease. Finally, support from other families is helpful and there are several parent organizations (Section 3.5) for Down syndrome.

- **Caveats:** Just because an individual is of a certain ethnic background, do not discount the findings on exam. If his chromosomes had been normal, he still would have needed a workup for hypotonia.
- **Outcome:** Sonam is a happy, outgoing child. He is enrolled in the state early intervention program and receives physical therapy weekly, with good progress. His parents join a Facebook group for parents of children with Down syndrome.

The Diagnosis

Clinical Diagnosis: Down syndrome, mosaicism.
Genetic Diagnosis: 47,XY,+21/46,XY.

Further Reading

Bull, M.J., Trotter, T., Santoro, S.L., Christensen, C., Grout, R.W., Council on Genetics, et al., 2022. Health supervision for children and adolescents with Down syndrome. Pediatrics 149 (5) e2022057010. PMID: 35490285.

Bunt, C.W., Bunt, S.K., 2014. Role of the family physician in the care of children with Down syndrome. Am. Fam. Physician 90, 851–858. PMID 25591185.

Whooten, R., Schmitt, J., Schwartz, A., 2018. Endocrine manifestations of Down syndrome. Curr. Opin. Endocrinol. Diabetes Obes. 25, 61–66. PMID 29135488.

Case 11.13.3

Neurologic and Muscular Cases: Seizures in an Infant

Cristina Trandafir[1,4], Tara Newcomb[2] and Lynne M. Kerr[3]

[1]*Division of Pediatric Neurology, Department of Pediatrics, University of Utah Health, Salt Lake City, UT, United States;* [2]*Division of Pediatric Neurology, University of Utah Health, Ashland, NE, United States;* [3]*Division of Pediatric Neurology, Department of Pediatrics, University of Utah Health, Shriners Hospital for Children, Salt Lake City, UT, United States;* [4]*Division of Pediatric Neurology, Department of Pediatrics, Baylor College of Medicine, Texas Children's Hospital, Houston, TX, United States*

Abstract

A 3-month-old male with hypotonia and developmental regression develops multifocal seizures refractory to treatment. He is found to have a pathogenic variant in the *KCNT1* gene and as a consequence, he receives personalized care.

The Case

Samuel is seen at 3 months of age by you, his pediatrician, after a recent diagnosis of seizures. He was born at term with uneventful prenatal and postnatal history. He had normal development during his first 2 months of life. He was admitted to the hospital 3 weeks prior to this visit for episodes of eye deviation with tonic posturing of his extremities, lasting 1–2 min, with no significant cyanosis or respiratory distress. These episodes were captured on EEG and confirmed to be seizures. The EEG showed multifocal epileptiform discharges from both hemispheres. His brain MRI was normal. He was started on levetiracetam with several days of seizure freedom. When seizures recurred, phenobarbital was added, with minimal improvement in seizure frequency. Samuel now has 6–15 seizures per day. Although he breastfeeds well, he is not smiling anymore, has poor tracking, and has developed significant head lag. On exam, he is hypotonic with head lag and poor visual tracking. There is no family history of seizures or developmental delays.

The Issues

- **How concerned are you by his seizures?** The increase in seizure frequency with very poor to no response to antiepileptic drugs (AEDs) is very concerning. You discuss this concern with the parents.
- **Why has he had regression in development since onset of seizures?** You advise the parents that loss of developmental milestones might be the result of uncontrolled seizures, part of a genetic syndrome (which includes seizures and development delay), or might be a side effect of the AEDs. His development will need to be monitored.
- **Is the normal brain MRI reassuring?** Although the brain MRI ruled out a structural etiology for the seizures, a genetic or metabolic condition should be considered. You check the results of Samuel's newborn screening, and they were normal. You advise the parents that a genetic evaluation seems appropriate. You also consider additional

Genomics in the Clinic. https://doi.org/10.1016/B978-0-12-816478-5.00082-0

investigations, including complete blood count, comprehensive metabolic panel, ammonia, lactic acid, thyroid functions, plasma amino acids, urine organic acids, and acylcarnitine profile to be done prior to the genetic evaluation.

Seizures in infancy vary from benign infantile epilepsy to early infantile epileptic encephalopathy (EIEE). Benign infantile seizures tend to respond well to antiseizure medications and do not affect development. Lack of response to AEDs with worsening seizures over a short period of time is suggestive of a more severe epileptic condition.

The Referral

- **Genetic Evaluation:** A full pregnancy history, medical history, and a family history are obtained by the geneticist and genetic counselor. There is no consanguinity and there is no family history of seizures or other neurological abnormalities. Samuel has two healthy siblings. A detailed physical evaluation is performed by the pediatric geneticist. Samuel does not have dysmorphic features; he is mildly encephalopathic, with poor visual tracking and diffuse hypotonia. He does have developmental regression. His newborn screening was negative and the metabolic screening sent by his pediatrician was also negative for abnormalities.
- **Genetic Counseling:** The geneticist and genetic counselor discuss with Samuel's parents potential genetic causes of epilepsy. There are hundreds of genes that are known to cause epilepsy when mutated. The inheritance may be autosomal recessive, autosomal dominant, or X-linked (Sections 2.7.1, 2.7.2). Given the negative family history of seizures and other neurological abnormalities, a de novo mutation (Section 2.4.1) is possible.
- **Genetic Testing:** Samuel does not have any dysmorphic features or other symptoms except for seizures and developmental regression. A multigene epilepsy panel (Section 6.2.3) is recommended. The results will be available in about 2 weeks. In the meantime, Samuel is admitted to a children's hospital due to uncontrolled, nearly continuous seizures, with difficulties in feeding and increased somnolence.
- **Results:** The epilepsy panel results show a pathogenic variant in the *KCNT1* gene. Autosomal dominant (Section 2.7.1) mutations in *KCNT1* are associated with malignant migrating partial seizures of infancy (MMPSI) and with nocturnal frontal lobe epilepsy. The variant identified in Samuel is known to cause activation of the potassium channel and is considered a gain-of-function mutation that confers enhanced or new activity on coded protein. Samuel's phenotype is consistent with MMPSI, which is an early infantile epileptic encephalopathy (EIEE14). The seizures are known to be resistant to conventional AEDs and result in profound developmental delays and intellectual impairment. There are multiple case reports showing that quinidine, which is an old antiarrhythmic drug that inhibits the *KCNT1* channel (counteracting the gain of function induced by the pathogenic variant), results in improvement in seizures in some but not all of the patients with gain-of-function mutations in the *KCNT1* gene. Several other reports on patients with *KCNT1* pathogenic variants have shown improvement in seizures with ketogenic diet and benzodiazepines.
- **Management and Recommendations:** The genetic results are discussed with the parents. Both parents undergo targeted variant testing and are negative for the pathogenic change found in Samuel. This is a rare de novo mutation in a dominant gene, and his offspring will have a 50% risk of inheriting the mutation (Section 2.7.1). Clonazepam, a conventional antiepileptic benzodiazepine, quinidine, and the ketogenic diet are started during the hospital admission, with a decrease in seizure frequency. There are several reports on patients with *KCNT1* pathogenic variants showing improvement in seizures with these therapeutic interventions. A gastrostomy tube placement is scheduled. Because cardiac arrhythmia and pulmonary hemorrhage were previously reported in a couple of patients with *KCNT1* variants, Samuel will be monitored for these conditions.
- **Caveats:** Just because someone does not have a family history of seizures, it does not exclude a genetic etiology for patient's epilepsy. A patient with normal brain structure and drug-resistant epilepsy should always have a genetic evaluation.
- **Outcome:** Upon discharge, Samuel is referred to a pediatric epilepsy clinic. Over the next year, quinidine is discontinued due to lack of significant improvement in seizures. The ketogenic diet is optimized and Samuel continues on several antiepileptic medications, with stable seizure control and occasional worsening in seizure frequency during febrile illnesses. He continues to show global developmental delays and receives outpatient therapies.

The Diagnosis

Clinical Diagnosis: Early infantile epileptic encephalopathy type 14 (malignant migrating partial seizures of infancy).
Genetic Diagnosis: *KCNT1* gene, heterozygous pathogenic variant.

Further Reading

Barcia, G., Fleming, M.R., Deligniere, A., Gazula, V.R., Brown, M.R., Langouet, M., et al., 2012. De novo gain-of-function KCNT1 channel mutations cause malignant migrating partial seizures of infancy. Nat. Genet. 44, 1255–1259. PMID 23086397.

Bearden, D., Strong, A., Ehnot, J., Digiovine, M., Dlugos, D., Goldberg, E.M., 2014. Targeted treatment of migrating partial seizures of infancy with quinidine. Ann. Neurol. 76, 457–461. PMID 25042079.

Yoshitomi, S., Takahashi, Y., Yamaguchi, T., Oboshi, T., Horino, A., Ikeda, H., et al., 2019. Quinidine therapy and therapeutic drug monitoring in four patients with KCNT1 mutations. Epileptic Disord. 21, 48–54. PMID 30782581.

Case 11.13.4

Neurologic and Muscular Disease: Clubfeet and Lack of Facial Expression

Tara Newcomb[1], Russell J. Butterfield[2] and Lynne M. Kerr[2]

[1]*Division of Pediatric Neurology, University of Utah Health, School of Medicine, Ashland, NE, United States;* [2]*Division of Pediatric Neurology, Department of Pediatrics, University of Utah Health, Shriners Hospital for Children, Salt Lake City, UT, United States*

Abstract

A 12-month-old male born with clubfeet has no facial expression and almost constant drooling and is found to have congenital myotonic muscular dystrophy. His mother was not previously known to have myotonic muscular dystrophy, and the phenomena of trinucleotide repeat expansion and anticipation are discussed in the context of the family history.

The Case

Jonathan comes to see you for his 12-month-old well-child exam. He is new to your practice. On reviewing his history, he was born breech at 37 weeks gestation by Cesarean section and had clubfeet that were successfully treated with casting and bracing. He spent a few weeks in the NICU for problems breathing and feeding and was noted to have micrognathia and a high arched palate. His developmental milestones are delayed, but he is making progress, now able to sit, get to sit, and roll. Language skills are delayed for age. His mother worries that he has no facial expressions. On exam, you note almost constant drooling, decreased reflexes, a myopathic facies, and possibly decreased muscle bulk, although strength is normal within the limits of the exam.

The Issues

- **Could he have something genetic?** Being born with clubfeet and having developmental delays particularly in motor skills are two major findings. You suspect an underlying condition that could be genetic.
- **How concerned are you about the developmental delay and myopathic face**? The lack of expression due to weak facial muscles is fairly striking, and this plus his decreased reflexes, low muscle bulk, and history of being breech presentation and having bilateral clubfeet make you suspicious for an underlying problem with muscle, such as myopathy or muscular dystrophy. His motor delays are consistent with this.
- **Is there any helpful family history?** On further probing you find out that his mother has had cramps in her muscles for years and when you check her and Jonathan's 10-year-old sister who is also present at the appointment, they both have grip and eye opening myotonia; they are unable to open their hands from a fist or their eyes quickly when asked. You refer the family for a genetic evaluation.

Genomics in the Clinic. https://doi.org/10.1016/B978-0-12-816478-5.00083-2

The Referral

- **Genetic Evaluation:** A full pregnancy history, a medical history, and a family history and medical examination on Jonathan and both mother and father are performed. Jonathan's medical history is significant for breech delivery, club-feet, and motor delay, and his exam is significant for micrognathia and a high arched palate. Grip myotonia in his mother is again noted but not in his father. Jonathan's sister's history is significant for an individualized education program for occupational therapy for fine motor problems, but she is otherwise healthy.
- **Genetic Counseling:** A probable diagnosis of myotonic dystrophy (DM1, or myotonic muscular dystrophy) is discussed with the family. Myotonic dystrophy, actually myotonic dystrophy type 1 (type 2 will not be discussed further and is due to a defect in a different gene), is an autosomal dominant (Section 2.7.1) multisystem disorder that affects the brain, skeletal muscles, eyes, heart, gastrointestinal tract, and endocrine systems. Myotonic dystrophy type 1 can present as a severe, early childhood or congenital form, as seen in Jonathan (Fig. 11.13.4.1), or more typically with onset in adolescence or young adulthood, as suspected in Jonathan's sister and mother. DM1 is slowly progressive over time. It is the most common muscular dystrophy with severity depending on the size of the trinucleotide repeat expansion (triplet repeats—Section 2.4.1) that causes it. The CTG trinucleotide repeat occurs in a noncoding region of the *DMPK* gene. Affected individuals have >50 CTG repeats. Normal is 5−37 CTG repeats and 38−50 repeats are considered a premutation, at risk for expansion in the next generation. Severity is roughly correlated with the size of the expansion.
- **Results:** Testing is sent on Jonathan for confirmation of myotonic dystrophy, assessing the CTG trinucleotide repeat expansion (Section 6.4.1) in the *DMPK* gene. A positive result is found on Jonathan's sample. There is a CTG expansion in the *DMPK* gene with 1150 repeats. Due to the presence of grip myotonia in his mother and Jonathan's sister, *DMPK* repeat testing is performed on both of them. The results in his mother are 150 repeats, and in Jonathan's sister are 500 repeats. *DMPK* repeat testing is also performed in the maternal grandmother, who is found to have a normal number, and in the maternal grandfather, who has 61 repeats.

  In this family, the phenomenon of anticipation is observed. Anticipation (Sections 2.4.1, 6.4.1) in a genetic sense means that the condition appears at an earlier age and more severely than in previous generations. This is due to an expansion in the number of CTG repeats through germline transmission. In myotonic dystrophy type 1, anticipation can be seen in both genders but is more dramatic when inherited from a female. Women with the condition even if asymptomatic are more likely to have infants with congenital myotonic dystrophy. Anticipation occurs in neurologic disorders such as myotonic dystrophy type 1 and Huntington disease (Section 11.13.12) that are caused by a trinucleotide repeat expansion.
- **Management and Recommendations:** Jonathan will be referred to a multidisciplinary neuromuscular clinic for management including neurology, physical medicine, physical therapy, pulmonary, and more. Associated problems in Jonathan will include intellectual disability (he is too young to test at this age), swallowing and speech difficulties, and weakness in his distal muscles such as feet and hands. He will need to be screened over time for cataracts, diabetes, and heart arrhythmias, the latter of which can be fatal if not treated.

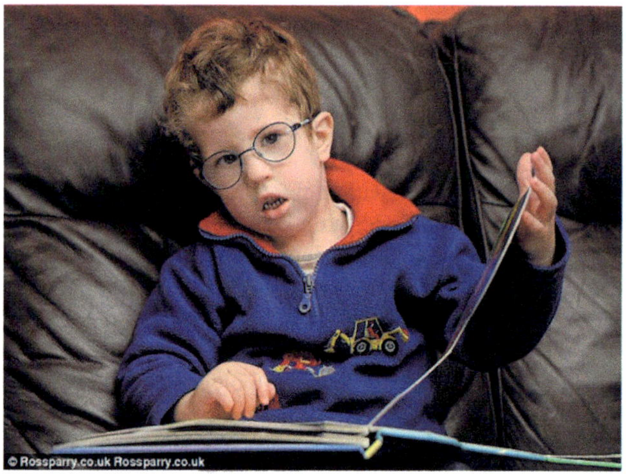

FIGURE 11.13.4.1 Young boy with myotonic dystrophy type 1. *Reprinted with permission from SWNS Media Group.*

- **Implications for Other Family Members:** Myotonic dystrophy is autosomal dominant (Section 2.7.1), so if Jonathan's mother goes on to have other children, each subsequent child will have a 50:50 chance of having myotonic dystrophy. Because of the phenomenon of anticipation, children born with the disorder may be milder such as in Jonathan's sister, or more severe like Jonathan. Prenatal testing for myotonic dystrophy is available. Jonathan's sister also has a 50:50 chance of having a child with myotonic dystrophy and her children may have more repeats than she does with more severe phenotypes due to anticipation.

  Both Jonathan's mother and sister also will need to be followed by neuromuscular specialists as well as cardiology for the risk of arrhythmias. They will potentially need pacemakers in the future. Cataracts will need to be monitored for in both mother and sister. The mother and sister will need to be monitored for cataracts.
- **Differential Diagnosis:** The differential diagnosis includes central nervous system disorders, as well as many less common muscle and nerve diseases, such as distal muscular dystrophies and Dejerine–Sottas disease, a peripheral neuropathy characterized by onset occurring in infancy.
- **Caveats:** An important consideration for individuals with myotonic dystrophy is safety perioperatively. Although myotonic dystrophy does not increase the risk for malignant hyperthermia above the general population risk, succinylcholine should be avoided. Cardiac rhythm, airway and breathing, and gastric motility need to be closely monitored. Sedatives and analgesics may have prolonged effects and should be used sparingly or avoided.
- **Outcome:** Jonathan is followed regularly in a neuromuscular clinic and starts having speech therapy to help with swallowing and facial muscles, physical therapy to help with gross motor skills, occupational therapy to help with fine motor and hand skills. His development improves, but he is still behind his peers. His mother is found to have an arrhythmia and a pacemaker is placed. Thus far, his sister remains healthy although continues to have problems with fine motor control.

The Diagnosis

Clinical Diagnosis: Patient—myotonic dystrophy type 1, congenital form. Mother and Sister—myotonic dystrophy type 1, adult form.

Genetic Diagnosis: Patient, Mother, and Sister—*DMPK* gene, heterozygous full mutation triplet repeats.

Further Reading

Bird, T.D., September 17, 1999. Myotonic dystrophy type 1. Adam, M.P., Mirzaa, G.M., Pagon, R.A., Wallace, S.E., Bean L.J.H., Gripp K.W., Amemiya, A., (editors). In GeneReviews. University of Washington, Seattle (WA), 1993-2023 [updated March 25, 2021]. PMID 20301344.

Ferschl, M., Moxley, R., Day, J.W., Gropper, M., 2016. Practical Suggestions for the Anesthetic Management of a Myotonic Dystrophy Patient. Myotonic.org, San Francisco, CA. Available at: https://www.myotonic.org/sites/default/files/MDF_LongForm_AnesGuidelines_01C.pdf.

Myotonic dystrophy foundation, 2016. Anesthetic Management of Patients with Mytotonic Dystrophy—Risks and Recommendations. Myotonic dystrophy foundation.org, San Francisco, CA. Available at: https://www.myotonic.org/sites/default/files/MDF%20Anesthesia%20Guidelines%20FNL%202016%202%202%20.pdf.

Case 11.13.5

Neurologic and Muscular Cases: Abnormal Gait in a Toddler

Tara Newcomb[1], Russell J. Butterfield[2] and Lynne M. Kerr[2]

[1]*Division of Pediatric Neurology, University of Utah Health, School of Medicine, Ashland, NE, United States;* [2]*Division of Pediatric Neurology, Department of Pediatrics, University of Utah Health, Shriners Hospital for Children, Salt Lake City, UT, United States*

Abstract

A 2-year-old boy with unusual gait is found to have Duchenne muscular dystrophy like his maternal uncle.

The Case

When Andy's mother brings him in at 2 years of age for his well-child check she tells you, his pediatrician, that he walks funny. He sits down on the floor to play with a toy but when he wants to get up, his mother has to help him. When he is on his feet he can walk but awkwardly, and when you try to get him to run he does so with more of a fast walk with a waddle that is quite obvious. When his mother takes his pants off to get him ready for his immunizations she notes that his legs look really strong and shows you his calves which are indeed quite large.

The Issues

- **Are you concerned about his presentation?** Certainly by 2 years of age, both walking and running gait should be normal and rising from a sitting position should be able done independently. Other than the motor issues, he also seems to have calf hypertrophy. You wonder about a neurologic problem.
- **Is there any relevant family history?** Andy's mother tells you that her mother's brother died of muscular dystrophy but her mother was checked and was told she could not be a carrier because apparently a "muscle test" showed a normal value. She calls her mother for more information and tells you that her uncle had Duchenne muscular dystrophy (DMD) and her mom had a normal creatine kinase level when she was about 10 years of age and was told she was not a carrier.
- **What is the next step?** You are pretty sure you know the correct diagnosis but have your nurse draw a creatine kinase level from this toddler and ask the mother to stick around for results. Andy's creatine kinase comes back and it is markedly elevated at almost 26,000 U/L (normal up to 300 U/L in male children). Putting this result with his clinical presentation, you tell Andy's mother that he most likely has DMD. You refer for genetic evaluation and pediatric neurology evaluation.

Genomics in the Clinic. https://doi.org/10.1016/B978-0-12-816478-5.00084-4

The Referral

- **Genetic Evaluation:** DMD causes progressive muscle degeneration and weakness, first apparent in the lower extremities but then becoming apparent in the upper extremities as well. Although it is almost certain that, with this very typical clinical scenario and elevated creatine kinase level of 26,000 U/L, this toddler has DMD, it is important to confirm this by genetic testing. Genetic results will define whether the boy might be eligible for medication based on his specific mutation and allow family planning (Section 4.4.3) based on carrier status (Section 4.1). A detailed family history does not reveal any additional affected family members although, of significance, Andy's mother is the oldest of three sisters and no brothers.
- **Genetic Counseling:** DMD is inherited in an X-linked recessive (Section 2.7.2) pattern. Due to the family history of the mother's uncle who died of DMD, Andy most likely has an inherited form, although in about one in three cases, DMD arises from a de novo mutation. In males, who only have one X chromosome, one mutated X chromosome is sufficient to cause the condition. In females, one abnormal gene is not enough to cause DMD, although a small percentage of carriers may have muscle weakness and cramping. Males cannot pass the gene to their sons, but all of their daughters will be carriers.
- **Genetic Testing:** Mutations (Section 2.4.1) in *DMD*, the gene encoding dystrophin, are found in about 98% of cases. If Andy is not found to have a mutation, he will need a muscle biopsy to confirm DMD and to consider other muscular dystrophies such as limb-girdle muscular dystrophy. Pathogenic variants are usually deletions (65%), point mutations (26%), or duplications (7%) (Section 2.4.1). Mutations in the dystrophin gene also cause Becker muscular dystrophy which is a milder form of the disease and less common than DMD. Andy's test results show a deletion in exons 45–52 which disrupts the reading frame, one of the more common variants associated with DMD.
- **Management and Recommendations:** Andy is also seen by pediatric neurology for ongoing management. Boys with DMD are commonly treated with oral corticosteroids such as prednisone or deflazacort. Treatment with corticosteroids slows progression, prolongs walking, and improves life expectancy by several years. Deflazacort has fewer side effects than prednisone but is very expensive and may not be covered by insurance until after prednisone is trialed. Common side effects include weight gain and behavior changes and some families choose to not continue these medications because of these problems.

  Dystrophin, the affected protein in DMD, is also present in the heart and affected boys develop a cardiomyopathy in their early teens. Female carriers also are at risk for cardiomyopathy and should be followed periodically by cardiology. Respiratory muscles are also affected and pulmonary function should be monitored regularly. Boys with DMD often need respiratory support in their later teens, first at night, then gradually during the day. Survival into the 30s is possible with proactive cardiac and pulmonary care and some individuals with DMD live into their 40s. New guidelines emphasize the increased life expectancy in boys with DMD who have received proactive care and have shifted the focus of care to improving quality of life. We are hopeful that genetic therapy (Section 7.4) will soon be available for boys with DMD.
- **Implications for Other Family Members:** In subsequent pregnancies for Andy's mother, any males have a 50% chance of having DMD and females have a 50% chance of being carriers. Prenatal diagnostic testing (Section 4.4.2) is available if the variant in the affected family member is known.

  Andy does have a 6-year-old brother who is healthy and developing well, and an 11-year-old sister who may be a carrier. Although some healthcare professionals would test Andy's sister to find out her carrier status, it is controversial about whether girls under 18 years of age should be tested (Section 10.1.2). The American Academy of Pediatrics suggests waiting until the child reaches 18 years of age and allows her to make the decision about whether to determine her carrier status, as it is unclear whether the knowledge of her carrier status during childhood could cause social and/or psychological harm.
- **Differential Diagnosis:** Although proximal weakness and large calves are compelling for DMD, there are a few other possible conditions in the differential diagnosis. These include spinal muscular atrophy, congenital myopathies, other muscular dystrophies (especially limb girdle), and polymyositis.
- **Caveats:** Until genetic testing became routine, testing for DMD was largely driven by the creatine kinase level, which is typically high along with liver enzyme levels (AST, ALT), due to leaky muscle membranes, followed by muscle biopsy. While markedly elevated in affected boys, creatine kinase may be normal in carriers. It also is not a good disease marker: boys with creatine kinases of 25,000 U/L are not necessarily more severe than boys with lower creatine kinase levels.
- **Outcome:** Andy's parents choose to start prednisone when the time comes (approximately 4 or 5 years of age). His specific mutation (deletion of exons 45–52) is eligible for treatment with an FDA-approved therapy targeting skipping

of exon 53 (Section 7.4). This therapy, which is expected to slow progression of his weakness, requires weekly IV infusions and close follow-up in the neuromuscular clinic, and his parents elect to start treatment. His gait improves initially, but he continues to be less coordinated than his peers. The family understands that this will change over time. Andy's mother has a normal cardiac evaluation and plans for follow-up exam in 5 years. She discusses her diagnosis with her sisters who have not yet started having children and encourages them to be seen for genetic counseling and testing.

The Diagnosis

Clinical Diagnosis: Patient—Duchenne muscular dystrophy. Mother—obligate carrier.
Genetic Diagnosis: Patient—*DMD* gene, heterozygous pathogenic deletion.

Further Reading

Aartsma-Rus, A., Ginjaar, I.B., Bushby, K., 2016. The importance of genetic diagnosis for Duchenne muscular dystrophy. J. Med. Genet. 53, 145−151. PMID 26754139.

Birnkrant, D.J., Bushby, K., Bann, C.M., Apkon, S.D., Blackwell, A., Brumbaugh, D., et al., 2018a. Diagnosis and management of Duchenne muscular dystrophy, part 1: diagnosis, and neuromuscular, rehabilitation, endocrine, and gastrointestinal and nutritional management. Lancet Neurol. 17, 251−267. PMID 29395989.

Birnkrant, D.J., Bushby, K., Bann, C.M., Alman, B.A., Apkon, S.D., Blackwell, A., Olson, A.K., et al., 2018b. Diagnosis and management of Duchenne muscular dystrophy, part 2: respiratory, cardiac, bone health, and orthopaedic management. Lancet Neurol. 17, 347−361. PMID 29395990.

Birnkrant, D.J., Bushby, K., Bann, C.M., Apkon, S.D., Blackwell, A., Colvin, M.K., et al., 2018c. Diagnosis and management of Duchenne muscular dystrophy, part 3: primary care, emergency management, psychosocial care, and transitions of care across the lifespan. Lancet Neurol. 17, 445−455. PMID 29398641.

Grimm, T., Kress, W., Meng, G., Muller, C.R., 2012. Risk assessment and genetic counseling in families with Duchenne muscular dystrophy. Acta Myol. 31, 179−183. PMID 23620649.

Kerr, L.M., Butterfield, R., 2013. Duchenne and Becker Muscular Dystrophy. Department of Pediatrics University of Utah, Salt Lake City, Utah. Available at: https://www.medicalhomeportal.org/diagnoses-and-conditions/duchenne-and-becker-muscular-dystrophies.

Case 11.13.6

Neurologic and Muscular Cases: Abnormal Movements and Posturing

Giulietta Maria Riboldi and Steven J. Frucht

Department of Neurology, The Marlene and Paolo Fresco Institute for Parkinson's and Movement Disorders, New York University Grossman School of Medicine, New York, NY, United States

Abstract

A 13-year-old boy, with a history of obsessive–compulsive disorder and anxiety, presents with hand shakiness and posturing of the hands upon writing. He is diagnosed with myoclonus–dystonia syndrome due to a pathogenic variant in the epsilon sarcoglycan gene. There is a paternal family history of alcoholism, and tremors in his father, both due to the same dominant condition.

The Case

Johnny is a 13-year-old boy coming to see you, his pediatrician, for the onset of shakiness of his hands when completing tasks or holding objects. His father has had jerkiness of his hands since his 20s but he has never been bothered by it. For this reason, when Johnny first manifested his symptoms his parents were not too concerned. However, since the symptoms became more severe over time they decided to ask you about them. You recall that since third grade, Johnny has manifested an obsessive–compulsive disorder and a lot of anxiety. You had not recommended any medications, but previously referred him to behavioral psychology for counseling. Johnny's development has been normal, and he does well in school. You examine Johnny and notice that his movements are quite irregular and jerky. His exam otherwise appears normal.

The Issues

- **What workup should be done to assess a diagnosis for Johnny's symptoms?** Because he is relatively young, you recommend a brain MRI and an EEG to rule out seizures. You also recommend basic metabolic screening. All the assessments are normal. You discuss with Johnny's parents that even though his father has had some of these symptoms, it is important to rule out an underlying neurologic disorder. You recommend a referral to pediatric neurology.
- **What is the most likely diagnosis?** The pediatric neurologist is able to see Johnny in 2 weeks with his parents. In addition to the irregular and jerky movements, the neurologist notes that Johnny also presents with abnormal posturing of his right arm and hand when he is writing. The neurologist suspects that Johnny may have myoclonus–dystonia syndrome and discusses with him and his parents about the possible causes and approaches for this disorder. The workup thus far has been appropriate, but the neurologist also recommends screening for the measurement of copper in urine and plasma and of ceruloplasmin to screen for Wilson disease, which come back normal.

Genomics in the Clinic. https://doi.org/10.1016/B978-0-12-816478-5.00085-6

- **What are the causes of Johnny's condition?** The myoclonus–dystonia syndromes are rare heterogeneous conditions which, in the majority of the cases, are due to a genetic mutation or pathogenic variant. These conditions are frequently misdiagnosed with forms of tremor. Behavioral and psychiatric symptoms may be present too. The neurologist explains to Johnny and his parents that a specific pathogenic variant can be identified in a large number of these patients, especially when there is a positive family history.
- **Are there other symptoms that Johnny may manifest?** Johnny's parents are very much concerned about the possible progression of the disease should the genetic mutation be found. The neurologist notes Johnny's behavioral history and tells Johnny and his parents that these traits have been frequently associated with myoclonus–dystonia syndromes. Depression and alcohol abuse have been reported as well. In very rare cases, epilepsy has been reported. However, Johnny already has been assessed for possible seizure and his EEG is normal. The phenotypic presentation of the disease can be very variable between affected subjects. Usually the dystonia remains stable, while myoclonus can progress with age. However, a few cases have been reported in which the symptoms improved over time.
- **Could Johnny's father's symptoms be related?** Johnny's father asks whether his symptom of jerkiness of his hands may be a symptom of the same disease. The neurologist explains that the disease can present with variable degrees of severity in different members of the same family. Therefore, it is possible that the father's symptoms are a manifestation of the same condition. However, only a positive genetic test can confirm this. Johnny's parents indicate that they would like to proceed with the genetic test. Johnny expresses his consent to move in this direction as well. Therefore, the neurologist refers them to genetics.

The Referral

- **Genetic Evaluation:** The geneticist and genetic counselor start the visit by collecting Johnny's clinical and family history.

*An accurate developmental and clinical history can indeed reveal details that can help in the differential diagnosis. Family history (*Section 3.2*) as well can be informative of specific clinical conditions and it is crucial to guide the counseling in the most accurate way.*

Johnny was born at term after an uneventful pregnancy. He had normal development and he always has been an active and bright child. Johnny has one younger sister who is 5 years old and healthy. Johnny's father is a 38-year-old man, and, except for the jerkiness of his hands, does not have any relevant clinical history. In the last few years he started to drink an increased amount of alcohol. He explains that the last years have been rough at work and he noticed that alcohol was calming him down and was improving his hand jerking significantly. In his family there is a history of alcohol abuse in his older brother and in one of his maternal aunts. Johnny's mother is 37 years old and she is healthy. There is no relevant clinical history in her family. Both Johnny's paternal and maternal family are from Ireland. There is no history of consanguinity (Section 3.2.4).

- **Genetic Counseling:** The genetic counselor explains that a pathogenic variant in the gene sarcoglycan epsilon (*SGCE*) is the most common cause of myoclonus–dystonia and can be identified in 30%–40% of cases with a positive family history. Forms of myoclonus–dystonia syndrome associated with mutations in the *SGCE* gene are referred to as DYT-SGCE or dystonia type 11 (DYT11). Other genes that have been associated with myoclonus–dystonia, such as *KCTD17*, *ADCY5*, *CACNA1B*, *RELN*, and *GNB1*, are much rarer.

  The disease is transmitted in an autosomal dominant fashion (Section 2.7.1). Therefore it is enough to have one mutated copy of the gene to manifest the disease. Mutations of *SGCE* have incomplete penetrance with variable expression (Section 2.8). This means that not everyone with a mutated gene will manifest the disease and also that the disease can manifest with different degrees of severity and combinations of symptoms among affected individuals. Males and females are equally affected by this disorder. However, it has been observed that when the mutation is inherited from the mother it manifests much less frequently in the offspring. This is caused by maternal imprinting (Sections 2.7.5, 6.4.3), where *SGCE* is silenced by methylation of its promoter during female gametogenesis. Thus, when passed to the progeny it causes disease only in the 5% of cases in which maternal imprinting is lost.

*In very rare cases, the disease has presented in subjects with a maternal uniparental disomy of chromosome 7, that is, subjects who inherited both chromosomes 7 (where the SGCE gene is located) from their mother and none from the father. These subjects also have Russell–Silver syndrome (*Section 11.7.4*), a growth disorder, due to imprinting of chromosome 7.*

Based on these observations, given that Johnny is symptomatic and that his father is presenting with symptoms, as well as some of his family members, most likely Johnny inherited the disease from the father.

The genetic counselor also mentions that in addition to Johnny's symptoms, alcohol abuse in the family also is consistent with myoclonus–dystonia syndrome because alcohol is known to greatly improve myoclonus in this setting. However, this feature is not exclusive to myoclonus–dystonia syndrome associated with *SGCE* mutations.

- **Genetic Testing:** Given Johnny's clinical history (onset in the second decade of myoclonus and dystonia associated with obsessive–compulsive disorder and anxiety) and the possible family history (jerkiness of the hands in his father), the yield of identifying a pathogenic variant of the *SGCE* gene is high. Since mutations in the other genes associated with myoclonus–dystonia syndromes are very rare compared to those in *SGCE*, the geneticist decides to start the genetic assessment with a targeted gene analysis (Section 6.2.1) for *SGCE.* Thus far, point mutations as well as deletions of *SGCE* have been reported in subjects with myoclonus–dystonia. Therefore both sequencing of the gene (Section 6.2.2) as well as testing for deletion/duplications (Sections 2.4.1, 6.1.3) should be performed. Johnny undergoes the test which is positive for a pathogenic variant in *SGCE*, thus confirming the clinical diagnosis.

  Since specific treatments for this condition or disease modifying therapies are not available, genetic testing is not necessary in Johnny's father. Johnny's father nonetheless also decides to undergo the test to confirm his diagnosis. He is able to get site-specific targeted testing (Section 6.2.1), and he is found to have the same mutation identified in Johnny.
- **Management and Recommendations:** There is no targeted gene therapy (Section 7.4) treatment for this condition. There are, however, a number of symptomatic therapies that can be attempted in order to address the symptoms. In particular, for the treatment of myoclonus, antiepileptic medications with GABAergic mechanisms have been used, such as benzodiazepine (clonazepam), topiramate, and valproic acid. A clinical trial using zonisamide has shown this medication to be effective in both myoclonus and dystonia. Given the beneficial effect of alcohol, sodium oxybate, a sodium salt of γ-hydroxybutyrate, can be used as well in refractory cases. For dystonia, levodopa should be attempted first in the pediatric population, followed by anticholinergic medications (such as trihexyphenidyl) or botulinum toxin injections especially in case of focal dystonia. Over the years, surgical approaches with deep brain stimulation have been shown to be an option addressing both myoclonus and dystonia particularly in treating dystonia, mostly targeting the globus pallidus internus.
- **Implications for Other Family Members:** If Johnny's mutation had been a de novo variant (meaning that originated in him and not inherited from his parents), the risk for his sister to develop the disease would have been the same as the general population. Instead, because one of their parents has the mutation that was passed to Johnny, she has a risk of 50% of having inherited the same mutation. However, because she is currently asymptomatic, she is not of reproductive age, and no preventive treatments are available, there is no need for her to be tested at this time (Sections 10.1.2, 11.18.1).

  The risk of developing symptoms from having a mutation in the *SGCE* gene will depend on whether the mutation was inherited from the father, in which case there will be almost full penetrance (Section 2.8), or from the mother, in which case only 5% of her offspring will manifest the symptoms. If Johnny's father inherited the mutation from one of his parents, the risk for his own siblings of having the mutation would be 50%. Usually the disease presents in the first or second decade, a much younger age than any of Johnny's father's siblings, so they are less likely to have the condition. However, disease-onset has been reported to be as late as 80 years of age.
- **Differential Diagnosis:** The differential diagnosis for the myoclonus–dystonia syndromes should include conditions that can present with isolated myoclonus, isolated dystonia, or a combination of the two, given the variable presentation of the disease. For myoclonus, myoclonic epileptic syndromes can be considered (such as Unverricht–Lundborg disease, Lafora disease, or sialidosis, among the others). However, in these cases seizures are usually prominent in addition to neurological manifestations. Other forms of dystonia, such as the ones related to mutations of *ANO3* and *GNAL*, or levodopa-responsive dystonia, associated with mutation of *GCH1* or more rarely of tyrosine hydroxylate (TH), can also mimic the myoclonic dystonia disorders.

  In some cases, the initial presentation of myoclonus can be misdiagnosed as essential tremor, especially when the patients are not assessed by experts in the field and when dystonia is mild. Hereditary benign chorea associated with the mutation of *NKX2-1* is another possible differential diagnosis for this condition.
- **Caveats:** DYT-SGCE (DYT11) myoclonus–dystonia is much less common when there are no other affected family members, in which case de novo mutations in other genes are suspected in unrelated probands.

- **Outcome:** Johnny is started on a low dose of clonazepam with significant improvements of his myoclonus. However, since he still presents with writer's cramp (focal dystonia), botulinum toxin injection is administered with resolution of this symptom. The injection has to be repeated every 3 months. After he starts working with a therapist, the obsessive–compulsive symptoms and anxiety improve as well. Johnny's father is started on clonazepam with great benefit to his myoclonus. He also is able to cut back on his intake of alcohol, and eventually notifies his family members about the condition.

The Diagnosis

Clinical Diagnosis: Patient and Father—dystonia type 11, myoclonus–dystonia syndrome.
Genetic Diagnosis: Patient and Father—*SGCE* gene, heterozygous pathogenic variant.

Further Reading

Fearon, C., Peall, K.J., Vidailhet, M., Fasano, A., 2020. Medical management of myoclonus-dystonia and implications for underlying pathophysiology. Park. Relat. Disord. 77, 48–56. PMID: 32622300.

Jones, H.F., Morales-Briceno, H., Barwick, K., Lewis, J., Sanchis-Juan, A., Raymond, F.L., et al., 2019. Myoclonus-dystonia caused by GNB1 mutation responsive to deep brain stimulation. Mov. Disord. 34, 1079–1080. PMID 31034681.

Marce-Grau, A., Correa, M., Vanegas, M.I., Munoz-Ruiz, T., Ferrer-Aparicio, S., Baide, H., et al., 2019. Childhood onset progressive myoclonic dystonia due to a de novo KCTD17 splicing mutation. Park. Relat. Disord. 61, 7–9. PMID 30642807.

Menozzi, E., Balint, B., Latorre, A., Valente, E.M., Rothwell, J.C., Bhatia, K.P., 2019. Twenty years on: myoclonus-dystonia and ε-sarcoglycan—neurodevelopment, channel, and signaling dysfunction. Mov. Disord. 34, 1588–1601. PMID: 31449710.

Nardocci, N., 2011. Myoclonus-dystonia syndrome. Handb. Clin. Neurol. 100, 563–575. PMID 21496608.

Case 11.13.7

Neurologic and Muscular Cases: Dystonia Family History

Giulietta Maria Riboldi and Steven J. Frucht

Department of Neurology, The Marlene and Paolo Fresco Institute for Parkinson's and Movement Disorders, New York University Grossman School of Medicine, New York, NY, United States

Abstract

A 24-year-old woman, with writer's cramp and a family history of dystonia type 1 due to pathogenic variant in the *TOR1A* gene, is found to have the same condition.

The Case

A 24-year-old woman, Rachel, comes to see you, her general practitioner, because she has been worried about her 8-year-old niece, her older brother's daughter. The niece was recently diagnosed with generalized dystonia due to a pathogenic variant of the torsin 1A (*TOR1A*) gene that is known to cause dystonia type 1 (DYT1, or DYT-TOR1A). She has been told by her brother that this is a genetic disorder that usually manifests in children, but in some cases can have a later onset. Rachel is planning to get married soon, and thus she is concerned also about family planning and risk for her own children. Therefore, she wants to discuss with you the implications of this genetic diagnosis for her and what the next steps should be.

The Issues

- **Could Rachel have some subtle symptoms of an underlying neurologic condition?** Rachel has been healthy all her life. She is not taking chronic medications and she has never been hospitalized. She is very athletic and active and just started her internship in a law firm. She only reports that in the last few years, during college, she started experiencing some difficulties writing. She thinks that this was related to the fact that she was writing a lot for her assignments, but now she is concerned that this may be a symptom of dystonia type 1 (DYT1 or DYT-TOR1A).
- **How is DYT1 (DYT-TOR1A) dystonia transmitted?** Rachel is worried that because this is a genetic disorder, she definitely would have the same condition. You discuss with Rachel that, from what you know, even if the condition is in the family, she would not necessarily have inherited the causative variant. You also suggest that she meet with a genetic counselor to further discuss any concerns and questions related to her niece's diagnosis and preconception family planning.
- **Should Rachel undergo a workup?** Because of her family history and the possible writer's cramp, you refer Rachel for an evaluation with a movement disorder specialist. Rachel is seen by a neurologist one month later. The neurologist describes that dystonia is a form of abnormal involuntary movement, mostly characterized by abnormal posturing of one or more body parts, such as the hand or fingers in the case of writer cramp. DYT-TOR1A (DYT1) is one of the

Genomics in the Clinic. https://doi.org/10.1016/B978-0-12-816478-5.00086-8

most common genetic causes of early-onset generalized dystonia. Affected children usually start manifesting the symptoms during the first decade. In the majority of the cases the disease initially presents with a pure dystonia involving one foot or leg, while the upper limbs are more rarely affected in children at the onset. The symptoms have a tendency to spread over months or years to other body parts, usually sparing the neck and the head. Cases of generalized dystonia are the more severe ones. In rare cases DYT1 can present in adults. However, onset after the age of 26 years is very rare. Later onset forms remain isolated to a single body region in approximately 20% of the cases and more often symptoms start in the upper limbs.

- **Does that mean that Rachel is affected?** Rachel says that no one else in her family ever presented with this disorder before. Therefore, she cannot understand how it is possible that there may be others with this variant in the family. The neurologist explains to Rachel that there can be significantly variable expression of the disease within the same family. Indeed, it is common in a family of children with generalized dystonia to identify relatives with other manifestations of dystonia, usually having writer's cramps or isolated lower limb dystonia. The neurologist does feel that Rachel likely is affected with this condition and agrees to refer her for a genetic evaluation.

It is helpful to bring a copy of familial genetic testing results to a genetic evaluation. In that way, the genetic counseling can be the most accurate. In addition, at times it can be helpful to enclose a copy of the results for the lab doing the testing on other family members.

The Referral

- **Genetic Evaluation:** Rachel is seen in genetics for counseling 2 weeks later. She explains her story and her concerns and questions and shows her niece's testing result. The genetic counselor collects a detailed family history in order to further discuss Rachel's specific situation and her risks for this condition. Rachel is one of five siblings. She has two sisters and two brothers. Her parents are alive and are in their 60s. Her mother has diabetes mellitus and her father has high blood pressure. Both her parents are of Ashkenazi Jewish ancestry. Only one of her siblings, the father of the girl affected with DYT1, is married and has children. He has one daughter and one son. All of the other siblings are younger than Rachel and less than 18 years of age. There is no history of cancer in the family. One of her maternal uncles has been "limping" on his left leg since he was young, but he has never been diagnosed with any specific disorder. The genetic counselor tries to collect information also about Rachel's sister-in-law's family history because this may add information about the heritability of her niece's condition, but Rachel does not know much about that side of the family. She only recalls that her sister-in-law is from Ireland and that she is an only child. As far as she knows no one has a similar condition on that side of the family.
- **Genetic Counseling:** DYT-TOR1A (DYT1) dystonia is an autosomal dominant disorder (Section 2.7.1). This means that it is sufficient to have the pathogenic variant in only one copy of the causal gene, *TOR1A*, in order to have the condition. Because this particular variant has reduced penetrance (Section 2.8), not everyone who has the variant will manifest the disease. In particular, for this specific disorder only 30% of those with the variant will present with dystonia throughout their life, while 70% will not. In the majority of the cases it is inherited from one of the parents. Only in rare cases will patients have a de novo pathogenic variant (Section 2.7.1 and 7.1.3).

  The genetic counselor discusses with Rachel that there is a possibility for her to have a pathogenic variant in the *TOR1A* gene. Indeed, for her niece to be affected, as reported above, either Rachel's brother or her sister-in-law most likely has the pathogenic variant of *TOR1A*. This variant is much more frequent among Ashkenazi Jews (estimated carrier frequency: 1:1000–1:3000) compared to other ancestry because of a founder effect. For this reason, there is a higher chance that her brother has the variant, since her sister-in-law is Irish Catholic. If Rachel's brother is the one with the variant, then Rachel herself is at risk of having the same variant in the *TOR1A* gene. If one of their parents has passed the variant to Rachel and her brother, her other siblings have a 50% chance of harboring the same variant. Moreover, her maternal uncle's symptoms (undiagnosed "limping") could also be a manifestation of the same disease. Rachel's ancestry, a positive family history (niece and possibly uncle), and onset of writer's cramp ("difficulties writing") before the age of 26 years put her at a higher risk for being affected.
- **Genetic Testing:** Rachel wants to know if she needs to be tested for the condition. The genetic counselor explains that this is her personal choice. It is not required for her to be tested at this time because she does not have a disabling symptom of DYT-TOR1A (DYT1), and in her situation the genetic test may not change treatment considerations for her current symptom (while it could be important for more severe cases as explained below). The counselor also discusses that she can wait to see the results of the test in her brother and sister-in-law, who already planned to be tested, since her risk may be reduced, although it would not be eliminated, if her brother were negative. On

the other hand, Rachel could decide on just undergoing the genetic test because her genetic background (Ashkenazi Jewish ancestry) already puts her at higher risk of having a variant, and she has concerns about the future of her new family. The genetic counselor further discusses with Rachel the importance of having a good insurance coverage and of thinking about the long-term plan before performing a genetic analysis that can be definitive for a condition (Section 10.2).

In the great majority of the cases, the pathogenic variant (Section 7.1.3) of *TOR1A* causing DYT-TOR1A (DYT1) is represented by a deletion of three base pairs in the *TOR1A* gene (c.907_909delGAG). Only very few cases of DYT1 have been related to other variants of the gene identified by sequencing and gene-targeted deletion/duplication analysis. Therefore, most of the time a targeted test (Section 6.2.1) for the common deletion is performed once DYT1 is suspected. Testing for a *TOR1A* variant can be performed also in the context of multigene panels (Section 6.2.3) or exome sequencing (Section 6.3.1), particularly when suspecting other possible causes for the symptoms. Rachel decides to perform the site-specific, targeted genetic test for *TOR1A* for the pathogenic variant found in her niece.

- **Results:** The analysis comes back positive for the common deletion c.907_909delGAG in the *TOR1A* gene. Rachel is scared that she will develop the same symptoms as her niece, who is now presenting with involvement of her four limbs and trunk. The genetic counselor explains to Rachel that although she has the variant, she has only a 30% possibility of developing symptoms, and the disease can manifest very differently even in subjects with the same variant. There is not a clear genotype–phenotype correlation (Section 2.8). Because she has writer's cramp, as confirmed after she saw the movement disorder specialist, it is likely that she already has signs of the disease. The spreading of the symptoms (causing generalized dystonia or only spreading to another body part) occurs in 60%–70% of cases overall. However, for subjects with onset of symptoms in the arm, the risk of spreading to other body parts is 50%. The genetic counselor also explains to Rachel that DYT-TOR1A is not associated with other symptoms and it is not known to cause a reduced life expectancy.
- **Management and Recommendations:** Although there are no treatments targeting the pathogenic mechanisms of the disease, a number of symptomatic approaches should be attempted in such patients. Medications commonly used in dystonia, such as anticholinergics, benzodiazepines, as well as baclofen, can be effective alone or in combination. In cases of focal dystonia, such as writer's cramp or isolated dystonia of the foot, botulinum toxin injection represents a valid option. In addition, for children with early-onset generalized dystonia, having a pathogenic variant of the *TOR1A* gene is a predictor of positive outcome for treatment with Deep Brain Stimulation (DBS). Although invasive, in cases of severe generalized dystonia this has been proven to be a very effective and a long-lasting treatment option.
- **Implications for Other Family Members:** Although their parents have not been tested, Rachel's other siblings have a risk of 50% to have the same *TOR1A* variant, since both she and most likely her brother have this variant. However, there is no urgency for them to be tested (Section 10.1). Indeed, they are all asymptomatic and they are not planning to have a family soon. Even if they have the variant, knowing it in advance would not make a difference in their management since there are no preventive treatments and since they still have only a 30% probability of manifesting the disease even if they have the pathogenic variant. However, testing may be justified regarding concerns for long term or career planning. Regarding her older brother, who is 30 years old, despite his possible affected status, he is now at a low risk of developing symptoms, since those begin very rarely after the age of 26.

  Rachel's future children will have a risk of 50% of inheriting the variant. If so, they still will have only a 30% risk of developing the symptoms, which can manifest as generalized early-onset dystonia, or focal or multifocal dystonia. Should they inherit the variant, there is no way to predict if they will manifest the disease and the extent of the symptoms. Because a rare autosomal recessive (Section 2.7.1) presentation of arthrogryposis, developmental delay, strabismus, and tremor has been described in children with homozygous variants of *TOR1A*, it would be worth considering screening Rachel's husband for the variant as well. In case Rachel decides not to perform prenatal testing (Section 4.4.2) of the variant for her future pregnancies, or use other reproductive options, her children do not need to be tested if not symptomatic, since preventive treatments do not exist. Should they become symptomatic they should be tested to confirm the diagnosis. This will be a very important piece of information, especially in the case of generalized dystonia, since it will have implications on treatment options and outcomes.
- **Differential Diagnosis:** Writer's cramp presenting in young adults or in the older population is usually idiopathic. More rarely it can be the initial sign of an underlying condition. Additional assessments should be considered in case of progression of the symptoms, co-occurrence with other neurological manifestations, or in case of a positive family history. Generalized dystonia in children is instead more commonly associated with genetic (such as DYT-TOR1A or DYT1, DYT-THAP1 or DYT6, DYT-KMT2B or DYT28) or metabolic conditions. In children who present with dystonia, it is always important to attempt levodopa, in the eventuality of levodopa-responsive dystonia (DYT-GCH1 or DYT5a, DYT-TH or DYT5b).

- **Caveats:** Because of the possibility of the autosomal recessive condition associated with the *TOR1A* variant, and its variable presentation even in members of the same family, it is worthwhile for the individual's partner to consider being tested as well.
- **Outcome:** Rachel's niece undergoes deep brain stimulation with an almost complete resolution of her symptoms. After the surgery, it takes a few months to start seeing some benefits, as is usually observed in these cases. Rachel starts treatment with botulinum toxin injection for her focal dystonia (writer's cramp) every 3 months with benefit. Her brother (the father of the affected niece) decides to have genetic testing and he is found to have the same variant as his sister and presumably his daughter who then undergoes testing and is positive for the variant. However, given his current age, there is a high likelihood he will remain asymptomatic.

The Diagnosis

Clinical Diagnosis: Patient and Niece—dystonia type 1. Brother—dystonia type 1, asymptomatic.
Genetic Diagnosis: Patient, Brother, and Niece—*TOR1A* gene, heterozygous pathogenic deletion.

Further Reading

Bragg, D.C., Slater, D.J., Breakefield, X.O., 2004. TorsinA and early-onset torsion dystonia. Adv. Neurol. 94, 87–93. PMID 14509659.

Cif, L., Vasques, X., Gonzalez, V., Ravel, P., Biolsi, B., Collod-Beroud, G., et al., 2010. Long-term follow-up of DYT1 dystonia patients treated by deep brain stimulation: an open-label study. Mov. Disord. 25, 289–299. PMID 20063427.

Kabakci, K., Hedrich, K., Leung, J.C., Mitterer, M., Vieregge, P., Lencer, R., et al., 2004. Mutations in DYT1: extension of the phenotypic and mutational spectrum. Neurology 62, 395–400. PMID 14872019.

Kariminejad, A., Dahl-Halvarsson, M., Ravenscroft, G., Afroozan, F., Keshavarz, E., Goullee, H., et al., 2017. TOR1A variants cause a severe arthrogryposis with developmental delay, strabismus and tremor. Brain 140, 2851–2859. PMID 29053766.

Case 11.13.8

Neurologic and Muscular Cases: Slowness and Loss of Facial Expression in a Young Adult

Giulietta Maria Riboldi and Steven J. Frucht

Department of Neurology, The Marlene and Paolo Fresco Institute for Parkinson's and Movement Disorders, New York University Grossman School of Medicine, New York, NY, United States

Abstract

A 32-year-old man, whose father has late onset Parkinson disease, develops inward turning foot, slowness, and progressive loss of facial expression and is found to have pathogenic variants in the *PRKN* gene as a cause for his autosomal recessive Parkinson disease.

The Case

Andrew is a 32-year-old man who comes to see you, his general practitioner. In the last year, Andrew started to develop an inward turning of his right foot when walking, but not when running or at rest. Recently, he has developed some slowness and loss of his facial expression. He says that he is taking no medications or drugs. He has no rigidity on exam. You wonder about Parkinson disease, although he is very young. Andrew reports that his father has Parkinson disease.

The Issues

- **What should be included in Andrew's initial evaluation?** Early-onset parkinsonism has signs and symptoms that are characteristic of Parkinson disease, such as the rigidity or the bradykinesia before 40 years of age. It usually is associated with other phenotypic manifestations that can present with specific alteration of the brain MRI, such as brain iron deposition or basal ganglia calcification. Therefore, in subjects with early-onset parkinsonism, it is worth performing an imaging study as part of the initial assessment. Also, previous history of exposure to medications that can induce parkinsonism (such as dopamine receptor blockers) needs to be ascertained. Andrew has never been on such medications. You recommend a brain MRI, which comes back normal.
- **What are the next steps in Andrew's assessments?** Because of his young age and his family history, as well as no clear etiology, you refer him to neurology. The neurologist diagnoses him with Parkinson disease. Andrew has a lot of questions regarding the cause of his disease. The neurologist discusses with him that the pathogenic mechanisms of Parkinson disease are not fully elucidated. However, in recent years a great number of genetic mutations have been associated with Parkinson disease, both as causal variants or genetic risk factors. Because of the early onset of his

Genomics in the Clinic. https://doi.org/10.1016/B978-0-12-816478-5.00087-X

symptoms the neurologist raises the issue of genetic testing. Andrew is very interested to know more about genetic testing and seems determined to pursue it.

- **What is the likelihood of discovering a causal genetic mutation or pathogenic variant?** The neurologist explains to Andrew that typically Parkinson disease is a multifactorial condition (Section 2.7.3) and overall only 2%–3% of all sporadic cases are associated with a specific Mendelian (Section 2.7.1) genetic cause. However, in early-onset Parkinson disease cases with an age of onset less than 40 years, the percentage of pathogenic variants (Sections 2.4.1, 7.1.3) increases up to 13%, and it is even higher in patients with a positive family history. Among the genes associated with Parkinson disease, a number of them also are associated with early age of onset, such as *PRKN* and PTEN-induced putative kinase 1 (*PINK1*). *PRKN* is the most common gene involved in this category of patients, followed by *PINK1*, while mutations in *DJ1* that codes for an oxidative stress sensor are far rarer. Pathogenic variants in these three genes are present in about 14% of early-onset cases. This also means, however, that in the great majority of cases, even among early-onset Parkinson disease where a genetic cause is expected, it is not possible to identify a causal mutation. Therefore, negative results are common even after extensive genetic testing. The neurologist discusses this with Andrew in order to manage his expectation of finding a definitive answer as the cause of his disease.
- **Are there any clinical features that can raise the suspicion for one specific pathogenic variant?** The neurologist explains to Andrew that so far, studies assessing large populations of patients with these pathogenic variants highlight some recurrent features in affected subjects, which, however, are not highly specific. For example, an initial presentation with dystonia, usually of the foot, as well as the presence of hyperreflexia are common in patients with variants of the *PRKN* gene. Patients with *PRKN* and *PINK1* variants usually do not present with cognitive impairment, while impulse control disorders are more frequent in subjects with *PRKN* mutations. Patients with triplication of the alpha-synuclein (*SNCA*) gene can present with an earlier age of onset as well but usually they manifest a more severe phenotype with prevalent cognitive impairment. Due to the high degree of overlap of the phenotypes associated with these variants, it is not possible to predict a specific variant based only on the clinical manifestations (Section 2.8). The neurologist sends the genetic testing and refers him to genetics to discuss the implications of the results when they are found to be positive.

The Referral

- **Genetic Evaluation:** The geneticist and genetic counselor meet with Andrew to discuss the results of his genetic testing. Before discussing the details of the genetic analysis, they collect Andrew's family history. Andrew has one brother, who is 5 years older than him and in good health. His mother is 62 years old and has hypertension. His father is 65 years old and he was diagnosed with Parkinson disease at the age of 57 years. No one else in the family has been diagnosed with Parkinson disease or any other movement disorder. Both Andrew's paternal and maternal families are originally from England, but they come from two different areas of the United Kingdom. There is no history of consanguinity in the family.
- **Results:** Andrew's genetic test identifies a homozygous (Section 2.7.1) variant in the *PRKN* gene previously identified in patients with Parkinson disease, but not seen in controls, thus considered to be pathogenic.
- **Genetic Counseling:** Mutations of the *PRKN* gene have been associated with autosomal recessive inheritance (Section 2.7.1), either the same, homozygous, pathogenic variant, or two different, compound heterozygous, variants. Most likely he inherited one variant from his father and one variant from his mother. As an alternative, he could have inherited one variant from one of his parents and then acquired a de novo variant on the other allele at conception. Regarding the clinical presentation of this specific variant, there are no clear evidences for a genotype-phenotype correlation. Intragenic deletions and insertions (Sections 2.4.1, 7.1) as well as single-nucleotide variants such as nonsense, missense, and splice site variants (Sections 2.4.1, 7.1) in the *PRKN* gene have been reported.

  Andrew wants to know if his pathogenic variants or mutations are the same that caused his father's Parkinson disease. His father is likely a heterozygous carrier of the variant that he then passed to Andrew. Because *PRKN* variants are autosomal recessive, the father's Parkinson disease does not appear to have the same genetic cause as Andrew's. If Andrew's father had two *PRKN* variants, he would have been expected to present with an early age of onset Parkinson disease. There is still debate in the literature regarding the risk of developing signs of Parkinson disease in carriers of only a single detectable pathogenic variant in the *PRKN* gene (pseudodominant condition) (Section 2.8). The only way to address Andrew's question would be for his father to undergo genetic testing.
- **Management and Recommendations:** The identification of specific variants can be informative about the risk of developing some features of the disease and their clinical management. Indeed, patients with Parkinson disease and variants of *PRKN*, other than an early age of onset, usually present with a slowly progressive disease with a good

response to treatments (levodopa or dopamine agonist). Motor fluctuations and dyskinesia can develop early after starting the medications but, fortunately, these patients usually have a good response to deep brain stimulation (DBS). Sleep benefit or improvement of symptoms after sleeping is common in patients with variants in *PRKN* and *PINK1* as well. Thus, for Andrew, knowledge about his specific pathogenic variants can make him a suitable candidate for some gene-targeted clinical trials, if available.

For patients with pathogenic variants in PRKN, PINK1, and DJ1, early-onset dyskinesia and fluctuations of the motor symptoms are frequent once treatment with levodopa has been started, even with low doses of levodopa.

- **Differential Diagnosis:** There are other conditions that can present with parkinsonism but are associated with other phenotypical manifestations. This is the case for dopamine-responsive dystonia due to mutations of the GTP cyclohydrolase 1 (*GCH1*) gene. Mutations of tyrosine hydroxylase (*TH*) or sepiapterin reductase (*SPR*) also can be levodopa-responsive dystonia but are usually associated with an onset in childhood and more severe phenotypes. Wilson disease should be considered in patients presenting with early-onset Parkinson disease, as its diagnosis will allow specific treatments able to improve the neurological symptoms and prevent severe complications. Early-onset parkinsonism can be found also in forms of neurodegeneration with brain iron accumulation or in case of basal ganglia calcification. There are also other genetic risk factors or modifying genes (Section 2.8) that increase the risk of developing early-onset Parkinson disease. One example would be mutations in the beta-glucocerebrosidase gene (*GBA*), which have been associated with a younger age of onset of Parkinson disease as well.
- **Implications for Other Family Members:** Andrew's brother has the same risk that Andrew had of having inherited a pathogenic variant of *PRKN* from each of their parents. Assuming their parents carry one mutated copy of the gene each, Andrew's brother has 25% chance of having a *PRKN*-related Parkinson disease. Even in the occurrence that he did not inherit any *PRKN* variant, his brother still has up to a 2% probability of developing Parkinson disease after the age of 60, which is the risk in the general population.

  Andrew's offspring will inherit one mutated copy of the gene from him. Therefore, they will be carriers for this variant. They will manifest *PRKN*-related Parkinson disease only if they also have inherited a second variant from their mother or in the case that they have de novo pathogenic variant.
- **Caveats:** There is still debate whether carriers of a single pathogenic variant can have an increased risk of developing Parkinson disease. The mechanism still needs further elucidation as to whether there may be an undetectable second variant in the same gene or other modifying genetic variants.
- **Outcome:** Andrew is started on a dopamine agonist with improvement of his motor symptoms as well as of the foot dystonia. However, after about 1 year on the medication he starts to develop moderate dyskinesia. He is switched to a low dose of levodopa multiple times during the day. This helps control the fluctuations of his symptoms while limiting dyskinesia. He is able to continue working and doing most of the activities he used to be able to do before. After about 10 years he still retains a good response to the levodopa, which is still beneficial for his motor symptoms. Unfortunately over time, his dyskinesia and motor fluctuations become more difficult to manage. For this reason, he undergoes DBS of the bilateral globus pallidus internus. The procedure is successful, allowing a satisfactory control of his motor symptoms and a significant reduction of the daily dose of levodopa with consequent improvement of his dyskinesia.

The Diagnosis

Clinical Diagnosis: Patient—Parkinson disease, early onset. Father—Parkinson disease.

Genetic Diagnosis: Patient—*PRKN* gene, homozygous pathogenic variants. Father—*PRKN* gene, heterozygous pathogenic variant.

Further Reading

Brüggemann, N., Klein, C., April 17, 2001. Parkin type of early-onset Parkinson disease. Adam, M.P., Mirzaa, G.M., Pagon, R.A., Wallace, S.E., Bean L.J.H., Gripp K.W., Amemiya, A., (editors). In GeneReviews. University of Washington, Seattle (WA), 1993–2023 [updated April 23, 2020]. PMID 20301651.

Domingo, A., Klein, C., 2018. Genetics of Parkinson disease. Handb. Clin. Neurol. 147, 211–227. PMID 29325612.

Kasten, M., Hartmann, C., Hampf, J., Schaake, S., Westenberger, A., Vollstedt, E.J., et al., 2018. Genotype-phenotype relations for the Parkinson disease genes Parkin, PINK1, DJ1: MDSGene systematic review. Mov. Disord. 33, 730–741. PMID 29644727.

Lubbe, S.J., Bustos, B.I., Hu, J., Krainc, D., Joseph, T., Hehir, J., et al., 2021. Assessing the relationship between monoallelic PRKN mutations and Parkinson's risk. Hum. Mol. Genet. 30, 78-86. PMID: 33448283.

Riboldi, G.M., Frattini, E., Monfrini, E., Frucht, S.J., Di Fonzo, A., 2022. A practical approach to early-onset Parkinsonism. J. Parkinsons Dis. 12, 1-26. PMID: 34569973.

Yu, E., Rudakou, U., Krohn, L., Mufti, K., Ruskey, J.A., Asayesh, F., et al., 2021. Analysis of heterozygous PRKN variants and copy-number variations in Parkinson's disease. Mov. Disord. 36, 178–187. PMID 32970363.

Zhu, W., Huang, X., Yoon, E., Bandres-Ciga, S., Blauwendraat, C., Billingsley, K.J., et al., 2022. Heterozygous PRKN mutations are common but do not increase the risk of Parkinson's disease. Brain 145, 2077-2091. PMID: 35640906.

Case 11.13.9

Neurologic and Muscular Cases: Progressive Gait Impairment and Genitourinary Dysfunction

Giulietta Maria Riboldi[1] and Steven J. Frucht[1]

[1]*Department of Neurology, The Marlene and Paolo Fresco Institute for Parkinson's and Movement Disorders, New York University Grossman School of Medicine, New York, NY, United States*

Abstract

A 38-year-old male presents with worsening gait impairment, impotence, and sphincter control problems and is found to have X-linked adrenoleukodystrophy.

The Case

Tony is a 38-year-old gentleman from North Carolina. In the last 2 years he started to develop a progressive gait impairment. Moreover, he has been having difficulties controlling his sphincters and has manifested impotence. He comes to see you, his family physician, for an assessment. You notice that he appears tanned and ask if he has recently been out in the sun. He says that he has noticed that his skin has become darker since last summer. When you examine him, you note some spasticity with increased reflexes, and some cerebellar dysfunction on neurological exam.

The Issues

- **What are the correct assessments for Tony's findings?** Because of the abnormal neurologic exam, you order a brain and spinal cord MRI which shows white matter abnormalities. You refer Tony to a general neurologist for further workup and management.
- **What are the potential causes of white matter abnormalities in adults?** Secondary causes, such as toxic or cardiovascular disorders, inflammatory diseases, such as multiple sclerosis or neuromyelitis optica, infections, or neoplastic conditions, need to be ruled out first when considering a patient with white matter abnormalities. Tony has no risk factors for any of these secondary causes and the white matter abnormalities as well as his clinical history are not suggestive for an inflammatory disease, such as multiple sclerosis, or neoplastic lesions. Once those are excluded, genetic causes of leukodystrophies should be considered.
- **Which assessments should be performed in the context of leukodystrophies?** A clear genotype−phenotype correlation is not always present in the setting of leukodystrophies. In most cases, these disorders manifest with a combination of motor symptoms (such as cerebellar signs or spasticity), cognitive decline, psychiatric abnormalities, and

Genomics in the Clinic. https://doi.org/10.1016/B978-0-12-816478-5.00088-1

sometimes sphincter impairment. Specific clinical features and the pattern of demyelination of the central nervous system as documented by the imaging studies can help distinguishing between the different subtypes of leukodystrophies. In some cases, when one specific type of leukodystrophy is suspected clinically, biochemical assays can be performed to confirm the diagnosis even without the need for a genetic test.

- **Which specific leukodystrophy is suspected in Tony's case?** The combination of symptoms presented by Tony as well the appearance of his brain and spinal cord MRI are suggestive for the X-linked adrenoleukodystrophy (X-ALD). This is a form of leukodystrophy that is transmitted through pathogenic variants or mutations in a gene on the X chromosome. This disorder can manifest with different phenotypic features, even within the same family. Because of the complexity of the different presentations of leukodystrophies and the treatment implications related to the identification of a possible diagnosis, the neurologist refers Tony to meet with a genetic counselor before sending any test.

The Referral

- **Genetic Evaluation:** The genetic counselor meets with Tony and explains that the leukodystrophies are a heterogeneous group of inherited disorders characterized by abnormalities of the white matter (either hypomyelination or demyelination of the white matter). Those are inherited diseases caused by different genetic variants that can manifest in childhood or adult age. The combination of the clinical presentation, imaging studies of the central nervous system, biochemical assays, and genetic tests are fundamental for establishing a diagnosis of these conditions. X-linked adrenoleukodystrophy can affect multiple members in the same family, mostly males. A three-generation pedigree (Section 3.2.1) is obtained. In Tony's family his sister's son, who is now 9 years old, started developing difficulties focusing at school and some behavioral changes over the last 2 years. His sister's family lives in Kansas. Tony's mother presented with a gait impairment in her late 50s. She died of a heart attack. In both his maternal and paternal side of the family there are apparently no other symptomatic members. Both sides of the family are originally from England and there is no known consanguinity in the family.
- **Genetic Counseling:** X-linked adrenoleukodystrophy is a rare X-linked (Section 2.7.2) genetic disorder with a prevalence of 1:20,000 to 1:50,000 individuals. It is caused by a pathogenic variant in the *ABCD1* gene, which encodes for the peroxisomal membrane protein ATP-binding cassette subfamily D member 1 (ALDP). Variants of this gene cause increased levels and impaired balance of very long-chain fatty acids (VLCFAs with greater than 22 carbons, C22). Being an X-linked disorder, X-linked adrenoleukodystrophy is expressed in all the males, who carry the inherited pathogenic variant in the *ABCD1* gene, although the condition presents with a spectrum of different manifestations. In early childhood it can manifest with hypotonia. When presenting in the younger age group it initially causes mostly behavioral abnormalities, followed by progressive motor and cognitive impairment. Adrenal insufficiency can be present as well. The adult form usually manifests as adrenomyeloneuropathy (AMN), with lower limb weakness and spasticity, sphincter abnormalities, and adrenal insufficiency. Brain involvement is present in half of the cases and is progressive in only a small percentage of them. Female carriers may manifest a milder phenotype, usually characterized by spasticity of the lower limbs. Cases of homozygous or compound heterozygous, pathogenic variants (Section 2.7.2) in females are very rare and are associated with earlier onset and more severe presentations. In 95% of cases, affected carriers have inherited the pathogenic variant of the *ABCD1* gene from one of their parents while the rest of the cases present with a de novo variant.
- **Genetic Testing:** When X-linked adrenoleukodystrophy is suspected, different tests can be performed to confirm the diagnosis. In affected males, in the majority of the cases, increased levels of VLCFAs are sufficient to reach a diagnosis. This test is recommended for Tony. This assay is not always accurate in female carriers, thus genetic testing is required. The majority of the variants can be identified by gene sequencing (Section 6.2.2), although 3% of the pathogenic variants are represented by larger deletions or duplications (Section 2.4.1) within the *ABCD1* gene. In those cases where the clinical presentation is not clearly suggestive for a specific type of leukodystrophies, genetic panels (Section 6.2.3) should be considered. Next-generation sequencing approaches (such as exome or genome sequencing—Sections 6.3.1 and 6.3.2) may be considered as well to broaden the spectrum of the differential diagnosis.
- **Management and Recommendations:** X-linked adrenoleukodystrophy is a complex and progressive disorder. In particular, the involvement of the central nervous system as well as of the adrenal gland (that can cause symptoms of Addison disease with adrenocortical insufficiency, sometimes representing the only clinical symptom in affected males) can be severely debilitating and even life-threatening. Therefore, establishing a diagnosis for this disorder can guide appropriate screening for possible complications. Indeed, in subjects with an established diagnosis of X-linked adrenoleukodystrophy, regular brain MRIs as well as monitoring adrenal function are recommended. The diagnosis and the follow-up screening may also guide treatment of this condition. For patients that present with neurological

involvement on brain MRI, hematopoietic stem cell transplantation can be considered. Early diagnosis can lead to early treatment, before the neurological symptoms become too severe. This is crucial in order to try to stabilize brain lesions and prevent serious side effects of the treatment as observed in more advanced stages of the disease. Confirming the diagnosis for Tony may also allow him to be enrolled in ongoing and future clinical trials.

For all of these reasons, the test for X-linked adrenoleukodystrophy has been recently added to the panel of newborn screening tests (Section 5.4.2) in a growing number of states in the United States.

- **Results:** Tony's biochemical test confirms the presence of increased VLCFAs (concentration of C26:0, ratio of C24:0 to C22:0, ratio of C26:0 to C22:0 are measured), supporting the clinical diagnosis of X-linked adrenoleukodystrophy. Genetic testing is performed as well through targeted single gene testing (Section 6.2.1). This shows a known, pathogenic variant in the *ABCD1* gene.
- **Implications for Other Family Members:** Tony's mother would be the obligate carrier (Section 2.7.2) because Tony received only the Y-chromosome from his father. His mother did present with some milder symptoms, as expected in female carriers. Tony's nephew, who presented with behavioral issues, may be affected with X-linked adrenoleukodystrophy as well. Should this be the case, Tony's sister would be an obligate carrier of the same variant. Being female, she may remain asymptomatic all her life. Tony's nephew was not diagnosed at birth since he was born in Kansas where the test for X-linked adrenoleukodystrophy is not yet part of the newborn screening program.

  Tony also has another sister, who is 45 years old and asymptomatic. However, she still has a 50% chance of having inherited the variant from her mother. Tony has two daughters and one son. He is wondering whether they may have inherited the variant from him. The genetic counselor explains to Tony that his daughters have a 100% risk of being a carrier for the variant, while his son definitely did not inherit the variant from him since he inherited only the Y sex chromosome from his father.

  In Tony's family, each of his maternal aunts and uncles may have a 50% chance of having inherited the variant if it was transmitted by Tony's maternal grandmother. If, however, Tony's maternal grandfather was the one who passed the variant to his children, only Tony's aunts would be at risk of having inherited the variant on one of their X chromosomes, but the risk would be 100%.
- **Caveats:** It is important to keep in mind that *ABCD1* has a number of pseudogenes (Sections 2.3.4 and 6.2.3) on other chromosomes that may confound the genetic testing results. Genetic screening for adrenoleukodystrophy is variably included in newborn screening across different states in the United States.
- **Differential Diagnosis:** With the occurrence of white matter abnormalities and neurological manifestations in adult patients, once secondary causes (such as vascular, inflammatory, neoplastic, and toxic conditions) have been ruled out based on the clinical history, brain imaging studies, and analysis of the cerebrospinal fluid, genetic testing for leukodystrophies should be considered. Among this group of disorders, some clinical features can help in the differential diagnosis. These include, for example, xanthomas in cerebrotendinous xanthomatosis, absent puberty in *Pol-III*-related disorders, ovarian failure in progressive leukodystrophy with ovarian failure, leukoencephalopathy with vanishing white matter, parkinsonism and/or motoneuron disease in the adult polyglucosan body disease, palatal myoclonus in Alexander disease, retinitis pigmentosa, hearing loss and ichthyosis in Refsum disease, as well as migraine and psychiatric symptoms in Cerebral Autosomal Dominant Arteriopathy with Subcortical Infarcts and Leukoencephalopathy (CADASIL).
- **Outcome:** Tony is assessed by an endocrinologist and he is diagnosed with adrenal insufficiency which requires appropriate treatment. Because of the slow but progressive worsening of his neurological symptoms he also is referred for an evaluation for possible hemopoietic stem cell transplantation, in order to reduce the speed of progression of the neurological manifestations of the disease. Indeed, a prompt diagnosis while his symptoms are still mild makes him an ideal candidate for this treatment, which does not seem to be as effective when applied in more advanced stages of the disease, and more severe side effects have been reported as well.

The Diagnosis

Clinical Diagnosis: X-linked adrenoleukodystrophy, adrenomyeloneuropathy
Genetic Diagnosis: *ABCD1* gene, hemizygous pathogenic variant.

Further Reading

Engelen, M., Kemp, S., Poll-The, B.T., 2014. X-linked adrenoleukodystrophy: pathogenesis and treatment. Curr. Neurol. Neurosci. Rep. 14, 486. PMID 25115486.

Kohler, W., Curiel, J., Vanderver, A., 2018. Adulthood leukodystrophies. Nat. Rev. Neurol. 14, 94–105. PMID 29302065.

Mallack, E.J., Turk, B., Yan, H., Eichler, F.S., 2019. The landscape of hematopoietic stem cell transplant and gene therapy for X-linked adrenoleukodystrophy. Curr. Treat. Options Neurol. 21, 61. PMID 31768791.

Wiens, K., Berry, S.A., Choi, H., Gaviglio, A., Gupta, A., Hietala, A., et al., 2019. A report on state-wide implementation of newborn screening for X-linked adrenoleukodystrophy. Am. J. Med. Genet. 179, 1205–1213. PMID 31074578.

Case 11.13.10

Neurologic and Muscular Cases: Progressive Ataxia

Giulietta Maria Riboldi and Steven J. Frucht

Department of Neurology, The Marlene and Paolo Fresco Institute for Parkinson's and Movement Disorders, New York University Grossman School of Medicine, New York, NY, United States

Abstract

A 40-year-old female with a family history of amyotrophic lateral sclerosis develops progressive ataxia. She is diagnosed with a spinocerebellar ataxia due to an increased number of CAG trinucleotide repeats in the *ATXN3* gene.

The Case

Brenda is a 40-year-old woman who is seen by you, her general practitioner, for progressive balance impairment. She has noticed that she loses her balance when walking and when going upstairs, and she frequently feels dizzy. Her husband has noticed that her speech has become somewhat difficult to understand. You are aware that her father had amyotrophic lateral sclerosis (ALS). You note that she has some slurring of her speech, which she has not had in the past. On exam she has dysmetria on finger to nose test and is unable to walk heel to toe without losing her balance. You feel that she has cerebellar ataxia.

The Issues

- **What is the appropriate initial workup?** You order a metabolic panel which includes a complete metabolic profile, vitamin B12, folate, vitamin E, and celiac panel, which are all unrevealing. You also order a brain MRI, which shows a shrinking of the cerebellum.
- **What are the causes of cerebellar ataxia?** You discuss with Brenda that cerebellar ataxia can be caused by a number of metabolic, infectious, and autoimmune conditions as well as genetic disorders. Since the metabolic screening was negative, you refer her to neurology. The neurologist confirms the diagnosis of progressive spinocerebellar ataxia (SCA).
- **What is the next step?** Brenda wants to discuss what the required next steps are in order to identify the cause of her symptoms. You are not sure if her family history is relevant, but you question a genetic cause. You tell her that the next step will be genetic testing to look for known gene mutations associated with her symptoms. You refer Brenda to genetics for further evaluation.

Genomics in the Clinic. https://doi.org/10.1016/B978-0-12-816478-5.00089-3

The Referral

- **Genetic Evaluation:** The genetic counselor meets with Brenda who is joined at the visit by her husband. A detailed three-generation pedigree (Sections 2.7.1 and 3.2) is obtained. Brenda and her husband have two sons who are 15 and 17 years old and healthy. Brenda says that in her family no one else has presented with symptoms that exactly resembles hers. However, her father was diagnosed with ALS at 58 years of age. He died in his 70s. Her paternal grandmother had some balance issues since her 60s but she was eventually diagnosed with Parkinson disease (Section 11.13.8). Since her relatives were diagnosed with very specific conditions that do not match with her diagnosis, Brenda is convinced that no one else in the family has had the same condition she has. Other than the neurologic symptoms, Brenda reports normal vision and no other impairments.
- **Genetic Counseling:** The genetic counselor explains to Brenda that certain genetic conditions can present with variable phenotypes across affected members even in the same family. Thus, it is possible that her father and her paternal grandmother were actually affected by the same condition that Brenda has. This would suggest an autosomal dominant inheritance (Section 2.7.1) since multiple generations appear affected by the disorder. The genetic counselor explains that genetic tests represent a clinical tool to try to better diagnose different clinical conditions. The genetic counselor also explains that subjects with a positive family history have a higher chance of having a positive genetic test.

  The cerebellar ataxias are heterogeneous disorders that can manifest in both adult and pediatric populations. The mode of inheritance and associated features can help make a specific diagnosis. Cerebellar ataxias can present in an autosomal dominant or autosomal recessive fashion (Section 2.7.1). The autosomal dominant SCAs present more frequently with an adult age of onset, while the recessive forms usually start to manifest in the pediatric age. Specific clinical features that can help discriminate among the different forms are, for example, loss of vision and retinal changes in SCA7, seizures in SCA10, psychiatric symptoms and chorea in SCA17, prominent cognitive impairment in SCA30, or spastic paraparesis, ptosis, diabetes, and ophthalmoparesis in SCA28. However, these are not always highly specific and can be absent in some subjects. Differentiating among these forms is not always easy. More than 30 forms of autosomal dominant SCAs based on clinical presentation or genetic diagnoses have been described so far. A family history of SCA, ALS, and parkinsonism in the same family, as seen in Brenda's family, is highly suggestive for SCA type 3 (SCA3), also known as Machado–Joseph disease.
- **Genetic Testing:** Genetic panels (Section 6.2.3) are usually required to identify the pathogenic variants or mutations for the ataxias. The autosomal dominant SCAs frequently are caused by a trinucleotide repeat expansion (Section 6.4.1). These expansions can be found both in the exons, or coding regions of the genome, and in the introns or noncoding regions of the genome (Sections 2.3.3 and 2.3.4). Therefore it is very important to specifically look for all possible mutations and triplet expansions that could escape detection using standard sequencing methods. Thus, a gene panel for the autosomal dominant SCAs including both screening of known trinucleotide expansions as well as sequencing coding and noncoding regions of those genes in which canonical mutations have been reported is recommended.
- **Results:** The genetic testing shows that Brenda has a 70 cytosine–adenine–guanine (CAG) trinucleotide expansion (Sections 2.4.1 and 6.4.1) in the ataxin 3 (*ATXN3*) gene, confirming the diagnosis of SCA3. Brenda wants to know if this is informative about the severity of her disease. The genetic counselor explains that unaffected subjects also have an expansion of the same CAG trinucleotide repeats in that region of the genome, but usually only 12–44 repeats. More than 60 CAG repeats will cause disease in 100% of cases (full penetrance—Section 2.8). There is a gray zone between expansions of 45 and 59 repeats in which the penetrance of the disease is reduced. This means that subjects with intermediate expansions may or may not manifest the disease and sometimes present with a milder phenotype. SCA3 is not the only disorder which presents with CAG expansions. Since CAG encodes for the amino acid glutamine, all of these disorders are referred to as polyglutamine diseases (e.g., dentatorubral–pallidoluysian atrophy, spinal bulbar muscular atrophy, and Huntington disease—Section 11.13.12).

  An inverse correlation with the number of repeats and the age of onset has been suggested, as well as a direct correlation between the extent of the expansion and the severity/type of clinical manifestations. However, this is not always the case, thus it is usually difficult to determine prognosis only based on the number of CAG expansions. It is possible that the reason why Brenda's father and paternal grandmother may have manifested the disease at an older age is because they had a lower number of CAG repeats. It is known that this expansion can cause genetic instability which leads to increased numbers of the CAG repeats between one generation and the next. Since there may be a correlation between the number of expansions and the age of onset, it is not infrequent to see an earlier age of onset in the offspring of affected subjects. This phenomenon is called anticipation (Sections 2.4.1 and 6.4.1) and it is usually more pronounced when inherited through the father. This would explain why Brenda's father and paternal grandmother had a similar age of onset of their symptoms, while she, who inherited the mutation from her father, has manifested the condition earlier.

- **Management and Recommendations:** Unfortunately no specific treatments are available for SCA3 and for the other degenerative forms of cerebellar ataxia. Instead, Brenda is referred to physical therapy as well and speech therapy to work on her balance and her speech impairments and to try to maximize safety and fall prevention. She also starts working with an occupational therapist to address her difficulties with manual dexterity. Major complications of the disease that should be cautiously monitored since they can be life-threatening are progressive cachexia due to severe dysphagia, as well as aspiration pneumonia.
- **Implications for Other Family Members:** Brenda wants to know if her sons should be tested for the pathogenic variant. Since SCA3 is an autosomal dominant condition, her sons each had a 50% chance of inheriting the allele of *ATXN3* which carries the pathogenic expansion. Since no cure or preventive treatments are available to date, genetic testing in minors is generally not advised for this condition (Sections 10.1.2 and 11.18.1). After the age of 18 years they can decide whether they would like to be tested to know about their mutation status, which can be informative for career and family planning. In that case, site-specific targeted testing (Section 6.2.1) for the *ATXN3* expansion would be recommended because the mutation in the family now is known.
- **Differential Diagnosis:** The differential diagnosis of adult onset cerebellar ataxia should include secondary systemic ataxia, such as autoimmune/paraneoplastic conditions, infective cerebellar diseases, prion disease, and metabolic disorders such as deficiency of vitamins B12 or E. These are treatable conditions, thus should be suspected and ruled out first. Clinical history, brain MRI, and additional tests (such as a lumbar puncture) can help diagnose these conditions. In terms of hereditary degenerative cerebellar ataxia in subjects with onset in their 30s, different SCAs can be considered, based on concomitant symptoms as well as frequency of the different forms in different geographic areas. The most common form of the SCAs worldwide is SCA3, followed by SCA2 and SCA6. There are also some forms of autosomal recessive (Section 2.7.1) degenerative ataxia that can be considered in adult cases with no clear family history of a dominant inheritance. These include adult onset Friedreich ataxia (due to a GAA repeat expansion in the frataxin gene) that is associated with reduced deep tendon reflexes (in most cases) and positive Babinski sign as well as cardiomyopathy and diabetes, cerebrotendinous xanthomatosis, fragile X–associated tremor/ataxia syndrome (mostly in males) (Section 11.13.11), spastic paraplegia type 7 (SPG7), and the autosomal recessive spastic ataxia of Charlevoix–Saguenay (ARSACS) to be suspected particularly in subjects from the region of Charlevoix–Saguenay in Quebec (Canada) presenting with ataxia and prominent spasticity.
- **Caveats:** Pathogenic expansion of triplets, as in SCA3, goes undetected in even extensive next-generation sequencing assessments such as exome sequencing (Section 6.3.1). Therefore, when suspected, specific tests to assess for gene expansions should be requested.
- **Outcome:** Brenda's symptoms continue to slowly progress over the course of the years. In particular her balance becomes more problematic and she ends up initially requiring the support of a walker, and after about 10 years, a wheelchair. She also develops swallowing difficulties, for which she continues to work with the speech therapist in order to minimize the risk of aspiration that could lead to respiratory complications, and in certain cases also to death in terminal patients. Life expectancy in these patients is usually of 10–30 years after the onset of the symptoms.

The Diagnosis

Clinical Diagnosis: Spinocerebellar ataxia type 3, Machado–Joseph disease.
Genetic Diagnosis: *ATXN3* gene, heterozygous full mutation triplet repeats.

Further Reading

de Silva, R.N., Vallortigara, J., Greenfield, J., Hunt, B., Giunti, P., Hadjivassiliou, M., 2019. Diagnosis and management of progressive ataxia in adults. Practical Neurol. 19, 196–207. PMID 31048364.

Klockgether, T., Mariotti, C., Paulson, H.L., 2019. Spinocerebellar ataxia. Nat. Rev. Dis. Prim. 5, 24. PMID 30975995.

Case 11.13.11

Neurologic and Muscular Cases: Intention Tremor and Difficulty with Balance

Randi J. Hagerman
MIND Institute, Endowed Chair in Fragile X Research, University of California, Sacramento, CA, United States

Abstract
A 67-year-old man with hand tremors, balance problems and increasing muscle weakness, and fatigue is found to be a premutation carrier (55–200 CGG repeats) in the *FMR1* gene and he has the fragile X–associated tremor/ataxia syndrome.

The Case

You evaluate a left hand tremor in a 67-year-old man, Tom, whom you had last seen 2 years ago when he presented with an intention tremor of his dominant right hand. This tremor now interferes with his handwriting and with the use of the computer mouse, and in the past few months it has caused him to spill food or liquid. Over the last year he also has noted balance problems. He feels unstable on a daily basis and he has fallen three times in the last 6 months. He also has felt muscle stiffness for about 6 months, and his wife has said that his gait seems to have a shuffling aspect to it. He feels that his muscles are weak and his knees give out on occasion when he walks, so he rests frequently. He has experienced chronic fatigue, and he notices that he has to take a nap at least once a day for at least 2 hours in order to function. He also has had bladder incontinence on occasion over the last year.

His past medical history includes hypertension that began at age 40 years, for which he has been treated with a beta blocker. Two years ago he was diagnosed with sleep apnea and continuous positive airway pressure (CPAP) was recommended, but he did not tolerate the CPAP mask. He also has tinnitus in the left ear that began 10 years ago, and hearing is decreased on the left side but he refuses to wear a hearing aid. He is feeling depressed because of these neurological problems. He has a long history of anxiety that began in college, and he occasionally experiences a panic attack when he is overwhelmed by stimuli such as in crowds at the mall. He continues to have erectile dysfunction, which he has had for the last 6 years. He was previously a very athletic individual, both swimming and running at least three times a week, which helped with his anxiety symptoms. However, over the last year he has not exercised regularly because he feels tired and weak.

You note on physical examination that Tom has normal vital signs. He appears to have a masked facies. He occasionally slurs his speech. His arms show some rigidity. His gait is unsteady. You recommend an MRI of the brain and referral to a neurologist who specializes in movement disorders.

Genomics in the Clinic. https://doi.org/10.1016/B978-0-12-816478-5.00090-X

The Issues

- **Are there any clues in the family history?** Tom and his wife have two daughters. One daughter has a son with normal development, and the other daughter has been unable to get pregnant at age 30 years and is undergoing a workup for premature ovarian failure. The patient's father died of Alzheimer disease at age 80 years. His mother, who has a long history of anxiety and depression, also has had tremor and ataxia since age 75 and she is still alive without dementia. The family history is negative for Parkinson's disease.
- **What does the neurologist find on exam?** On examination Tom's blood pressure is 144/72, heart rate is 58, height is 1.96 m, and weight is 72 kg. His cranial nerve testing demonstrates difficulties with smell. He cannot smell rubber on the reflex hammer and he thinks the coffee smell is citrus. His pupils are equal and reactive to light but he has nystagmus with extreme lateral gaze. His muscle strength is normal, but he has somewhat of a masked facies. His hearing is approximately 25% of normal on the left and normal on the right. His gag reflex is intact, his tongue strength is normal, and the rest of his cranial nerves are normal, but his voice is weak. His right and left upper extremities show 2+ rigidity with contralateral movement of his arms. He has rebound of 6 cm with pressure on extended arms. His deep tendon reflexes are 1+ in the right and left upper extremities, absent at the knees bilaterally, and absent at the ankles. Babinski reflexes are negative. Vibration sense is absent in the right and left great toes but normal in the upper extremities. Position sense and pinprick sensation are normal. His finger to nose touching demonstrates a kinetic tremor on the right side greater than the left, a positional tremor but no tremor at rest. His primitive reflexes show a positive snout reflex for greater than 4 taps but an absent jaw jerk, a 1+ palmomental reflex bilaterally, and an absent grasp reflex. He is unsteady initially with tandem walking, and his heel-to-shin movements suggest a subtle ataxia with the right leg more than with the left leg. His pull test is positive.
- **What does the MRI show?** His MRI demonstrates significant white matter disease bilaterally in the insula, and subtle suggestion of white matter disease in the splenium of the corpus callosum and in the middle cerebellar peduncles. His ventricles are mildly dilated and there is mild atrophy of the cerebellum. There is some evidence of increased perivascular spaces perhaps related to his hypertension.
- **What is the likely diagnosis?** The neurologist considers the fragile X—associated tremor/ataxia syndrome (FXTAS), found in premutation carriers for fragile X syndrome (Section 5.2). This is because of the family history of the daughter with possible premature ovarian failure, and with the middle cerebellar peduncle sign on MRI.

FXTAS is a neurodegenerative disorder caused by elevated FMR1 mRNA levels that are two to six times normal in premutation carriers. Symptoms begin in the 60s with tremor or ataxia, then later neuropathy pain in the legs, and then cognitive decline. White matter hyperintensities occur in the central nervous system, particularly in the middle cerebellar peduncle, the insula, and the splenium of the corpus callosum. Brain atrophy is seen, and FXTAS inclusions occur in the nuclei of neurons and astrocytes throughout the CNS and in the peripheral nervous system. FXTAS occurs in approximately 40% of males and in 16% of females with the premutation who are in their 60s.

- **What is the next step?** Tom has stage 3 FXTAS because his tremor is severe enough to interfere with activities of daily living. The neurologist orders *FMR1* DNA testing to quantify the number of CGG repeats (Section 6.4.1) on the *FMR1* gene to see if a premutation is present. He refers Tom for a genetic evaluation. He also recommends Tom have cognitive testing.

There are six stages of FXTAS related to motor symptoms: Stage 1 is without clinical tremor, stage 2 is with subtle tremor, stage 3 is with significant tremor or ataxia interfering with activities of daily living, stage 4 is use of a cane or walker, stage 5 is use of a wheelchair on a daily basis, and stage 6 is bedridden.

The Referral

- **Genetic Evaluation:** Tom meets with the geneticist and genetic counselor who review the *FMR1* DNA testing. The CGG triplet repeat number (Section 6.4.1) is 90. This is in the middle range for premutation and puts him at high risk for FXTAS. Based on the diagnostic criteria, he has Definite FXTAS because he has tremor and ataxia combined with MRI findings of the middle cerebral peduncle sign, in addition to typical findings of white matter disease in the splenium and in the insula, along with atrophy and dilated ventricles. He also has neuropathy symptoms with no reflexes in lower extremities and absent vibration sense and, when asked, reports that he sometimes has pain in his feet at night.

For fragile X gene mutations, a premutation consists of 55–200 CGG repeats, and a full mutation has more than 200 CGG repeats, and the full mutation causes fragile X syndrome. From 45 to 54 CGG repeats is the intermediate range, from which expansion should not occur. Below 45 is in the normal range.

- **Genetic Counseling:** They discuss that fragile X mutations and FXTAS (Sections 5.2, 11.12.3, and 11.17.3) and accompanying conditions are always inherited and never spontaneous. Tom received the premutation from his mother. In maternal meiosis there is possibility of expansion of the premutation number, and it is likely that her premutation is smaller than his. In addition, he would have passed the premutation to all of his daughters, with minor changes that stay in the premutation range; however, his daughters are at risk to have children with the full mutation (Section 2.4.1).

Males always pass their single X chromosome to all of their daughters and to none of their sons, who inherit the Y chromosome. Thus if this patient had any sons, none of them would be at risk for having inherited a premutation. There is much less expansion of the CGG repeats during male meiosis, and there can be reversion to a smaller number of CGG repeats in their daughters but all will have a premutation.

- **Implications for Other Family Members:** Further evaluation of his family is suggested, and they are all seen. His mother's testing demonstrates a CGG repeat number of 78. On examination she also has classical features of FXTAS, with both tremor and ataxia; however, her MRI does not demonstrate white matter disease in the middle cerebral peduncle. She does have white matter disease in the splenium, as well as significant atrophy of the cerebrum and cerebellum, and this is consistent with FXTAS in women. The mother's history of long-term psychiatric problems is also consistent with the fragile X–associated neuropsychiatric disorders (FXANDs), but she does not have a history of early menopause so no fragile X–associated primary ovarian insufficiency (FXPOI).

Only 16% of females with FXTAS have the middle cerebral peduncle sign. The problems of depression, anxiety, and/or chronic fatigue are present in approximately 50% of premutation carriers.

His daughters are each found to have a premutation on one allele: 72 CGG repeats and 98 CGG repeats, respectively, with the other allele having normal values of 33 repeats. His daughter with the 72 CGG repeats, at 30 years of age, is undergoing a workup for ovarian failure. She is subsequently diagnosed as having FXPOI. Genetic counseling for her includes that she is at a 50% risk to pass on a premutation or a full mutation with each pregnancy. Her sister with the 98 CGG repeats will pass on the full mutation in 50% of her pregnancies because the eggs having the X chromosome with a premutation of such a high number of CGG repeats will almost always demonstrate expansion to the full mutation when passed to offspring. Her reproductive options include prenatal genetic testing by CVS or amniocentesis, in vitro fertilization, egg donation, or adoption (Sections 4.4.3 and 4.5). Both daughters should be offered testing for AGG interruptions to assess stability and likelihood of expansion (Section 11.12.3).

FXPOI is the most common inherited cause of ovarian insufficiency or early menopause before 40 years of age, occurring in about 20% of women with a premutation. Findings include oligomenorrhea or amenorrhea along with hypogonadotropic hypogonadism. The estrogen deficiency can lead to decreased bone density and cardiovascular disease.

- **Management and Recommendations:** The movement disorder neurologist follows Tom for FXTAS, and he has noted some Parkinsonian symptoms including masked facies, stiff and shuffling gait, and partial loss of sense of smell. The neurologist recommends Sinemet, and it proves helpful for his tremor and stiffness.

Approximately 60% of those with FXTAS *can have Parkinsonian symptoms.*

Tom's cognitive testing on the WAIS-IV demonstrates verbal comprehension at 105, perceptual reasoning index 98, working memory index 95, perceptual speed index 89, and full-scale IQ score of 97, and all of these scores are in the normal range. His Mini Mental State Examination is 24/30 and his Behavioral Dyscontrol Scale 2, a test of frontal dysfunction, is 19/27, so he demonstrates executive function deficits that are very common in those diagnosed with FXTAS. On the Structured Clinical Interview for DSM Disorders (SCID, a psychiatric standardized assessment), he meets criteria for Major Depression and currently this is moderate, but he is not suicidal. He has lost interest in activities, and he is not social. He also meets criteria for generalized anxiety disorder. Thus, his overall IQ is normal but he has mild executive function and memory deficits, which are common when individuals present with tremor and ataxia. Both counseling and a selective serotonin reuptake inhibitor antidepressant, specifically sertraline, are recommended for his significant anxiety and depression, in addition to exercise and antioxidants. His psychiatric problems proceeded his FXTAS by many years, and they are consistent with FXANDs.

The family is also referred to their nearest family support center (Section 3.5) listed on the National Fragile X Foundation (NFXF) Website at www.fragileX.org and also to the nearest Fragile X Clinical Research Center (FXCRC) also on the NFXF web site (Section 7.4.4). Fragile X Clinical Research Centers offer the latest in interventions for those with premutation disorders in addition to those with fragile X syndrome and support to the families.

- **Outcome:** Tom returns 6 months later and his depression has resolved with counseling, sertraline, and exercise, and his cognitive testing has also improved. His tremor and ataxia have persisted but Sinemet has helped his tremor and he is currently being followed by the movement disorder neurologist and a physical therapist who is teaching him balance exercises which he carries out on a daily basis. His mother is also being followed by a movement disorder neurologist. Both of his daughters are pursuing pregnancy.

The Diagnosis

Clinical Diagnosis—Patient and Mother - fragile X–associated tremor/ataxia syndrome and fragile X–associated neuropsychiatric disorders. Daughters - premutation carriers; one with fragile X–associated primary ovarian insufficiency.
Genetic Diagnosis—Patient - *FMR1* gene, hemizygous premutation triplet repeats. Mother and Daughters - *FMR1* gene, heterozygous premutation triplet repeats.

Further Reading

Fink, D.A., Nelson, L.M., Pyeritz, R., Johnson, J., Sherman, S.L., Cohen, Y., et al., 2018. Fragile X associated primary ovarian insufficiency (FXPOI): case report and literature review. Front. Genet. 9, 529. PMID 30542367.

Hagerman, R.J., Hagerman, P., 2016. Fragile X-associated tremor/ataxia syndrome - features, mechanisms and management. Nat. Rev. Neurol. 12, 403–412. PMID 27340021.

Hagerman, R.J., Protic, D., Rajaratnam, A., Salcedo-Arellano, M.J., Aydin, E.Y., Schneider, A., 2018. Fragile X-associated neuropsychiatric disorders (FXAND). Front. Psychiatr. 9, 564. PMID 30483160.

Hall, D.A., Robertson, E., Shelton, A.L., Losh, M.C., Mila, M., Moreno, E.G., et al., 2016. Update on the clinical, radiographic, and neurobehavioral manifestations in FXTAS and FMR1 premutation carriers. Cerebellum 15, 578–586. PMID 27287737.

Case 11.13.12

Neurologic and Muscular Cases: Chorea

Giulietta Maria Riboldi and Steven J. Frucht

Department of Neurology, The Marlene and Paolo Fresco Institute for Parkinson's and Movement Disorders, New York University Grossman School of Medicine

Abstract

A 54-year-old female has developed choreiform movements, depression, and increasing irritability. She is diagnosed with Huntington disease, as her genetic testing shows an increased number of trinucleotide repeats in the huntingtin gene.

The Case

A 54-year-old woman, Ana Maria, is referred to see you, a general neurologist, for an onset of involuntary movements that you diagnose as chorea. The movements apparently started a few years ago, but they were initially very subtle. Recently they became more noticeable and her children started to become concerned. She also has a history of depression that has become more severe over the last few years. Her family also has noticed that her behavior has changed recently as she has become more irritable. She comes from Venezuela, but she has not seen the rest of her family in a long time.

The Issues

- **Is chorea always a genetic disease?** Chorea is characterized by involuntary, hyperkinetic, unpredictable movements, which flow from one side of the body to the other. A number of patients present with secondary forms of chorea, usually associated with systemic or metabolic disorders. These conditions can be medically managed in the majority of the cases. Therefore, secondary forms of chorea should be considered as part of the first line of assessments when patients present with these symptoms. You order bloodwork to test for the conditions that can be associated with adult onset chorea. All the tests are unrevealing. You also order a brain MRI, which shows mild shrinking of the caudate nucleus bilaterally.
- **What are the most common genetic forms of chorea?** Ana Maria is interested in learning more about other possible causes of her chorea. Huntington disease (HD) is the most common genetic form of adult onset chorea. Once secondary causes are ruled out, HD should always be considered and discussed with the patient before hypothesizing other more rare and uncommon forms of chorea.
- **Is a positive family history always mandatory for suspecting Huntington disease?** Ana Maria does not report any family history of chorea or other neurological or psychiatric conditions. Although the majority of patients with HD present a positive family history, individuals affected with this disorder can report no affected relatives because of a number of reasons: (1) the disease manifests mainly in adults so that the proband's relatives may have died before presenting with symptoms of HD; (2) in some cultures, a positive family history of HD is hidden by the proband because of the perceived shame or stigma associated with the disease for the entire family; (3) parents are not the biological

Genomics in the Clinic. https://doi.org/10.1016/B978-0-12-816478-5.00091-1

parents as in nonpaternity; and (4) a parent of the affected subject has a trinucleotide expansion in the huntingtin (*HTT*) gene still in the range of a reduced penetrance (explained below).

- **What are the implications of a genetic test for Huntington disease?** HD is a severe degenerative condition. The most prominent manifestation is a progressive generalized chorea which will involve the limbs, trunk, and the oromandibular region. There are other more subtle neurological manifestations that can be picked up at the clinical evaluation, such as a motor impersistence (inability to sustain certain motor activities such as hand grip or tongue protrusion), a latency of voluntary ocular saccades, and impaired speech. Moreover, the disease will cause a progressive cognitive deterioration, as well as behavioral and psychiatric symptoms (depression, anxiety, irritability, psychosis, hallucinations, obsessive–compulsive behavior, impulsiveness, and increased suicidal risk). The triad of chorea, cognitive impairment, and psychiatric disorder is typical for HD. Because of these manifestations, a definitive diagnosis of HD, especially when the patients are still only mildly symptomatic, can represent a real psychological, emotional, and financial burden not only for the patient but for the entire family as well. Indeed, the disease is transmitted in an autosomal dominant (Section 2.7.1) fashion, and therefore, relatives of the affected individuals are at risk for the disease.

  You consider referring Ana Maria for genetic testing for HD. Because of the severe implications associated with this diagnosis, it is preferable to refer these patients to specialized HD clinics. These centers offer well-established and multidisciplinary approaches for the diagnosis and the care for this severe neurodegenerative disorder. Clinical expertise, psychiatric approaches, and supportive care are offered to the patients and their families. When this option is not available, proper genetic counseling for patients and their family as well as psychological support are required.

The Referral

- **Genetic Evaluation:** After you discuss about the possibility of the genetic diagnosis with Ana Maria, she asks for a referral to genetics. The geneticist and genetic counselor evaluate the patient. A detailed clinical and family history (Sections 2.7.1 and 3.2.1) is collected with as much detail as possible, in order to identify existing affected individuals in the family. Ana Maria has three children in their 30s. She does not report any family history of chorea or other neurological or psychiatric conditions. Questions about other conditions in the family are collected as well, to identify possible elements that may support the diagnosis or suggest alternative causes. With the help of other family members, the genetic counselor is able to discover that the patient's father, who had been estranged from the family, was admitted later in life to a psychiatric hospital and died soon after. Also, one of his brothers used to be referred by the rest of the family as "strange" and he died in his 70s after choking on food because of progressive swallowing difficulties. A psychiatric history is also present in one paternal cousin, but the family lost track of her. Both sides of the family are Hispanic and they live in a small town in Venezuela, which is well known to have numerous clusters of families affected with HD.
- **Genetic Counseling:** Because of the classical clinical presentation (chorea, psychiatric impairment, and cognitive decline), a significant family history, and the geographic origin of patient's family, Ana Maria is at a high risk of testing positive for HD. The genetic counselor explains that a positive genetic test for HD can have implications for insurability and long-term plans for any patient. Ana Maria and her family are encouraged to make sure they are well prepared for these issues if they arise. The geneticist with the counselor then follow up with counseling about the disease, exploring with the patient genetic testing, interpretation of possible results, as well as possible implications for the patient and her family.

  HD is caused by the abnormal expansion of the number of cytosine–adenine–guanine (CAG) trinucleotide repeats (Sections 2.4.1 and 6.4.1) in the huntingtin (*HTT*) gene on chromosome 4, which results in an aberrant long polyglutamine tract in the protein. This is an autosomal dominant condition (Section 2.7.1); thus a pathogenic number of repeats on one allele is sufficient to cause the disease. Subjects carrying an abnormally high number of repeats are at risk for transmitting larger expansions to their progeny. Indeed, an increased number of CAG repeats in the *HTT* gene can cause instability or further expansion of the number of repeats during gametogenesis. This can occur especially during spermatogenesis, thus causing the phenomenon of anticipation particularly through paternal transmission. With anticipation, the offspring will have an even larger expansion and manifest the disease at an earlier age compared to their affected parent. Indeed, an inverse correlation between number of triplet repeats and age of onset of the disease has been reported.

  There is a phenotypic spectrum of manifestations in HD. Other than the classical form, there can be a mild presentation mainly characterized by attenuated motor manifestations. This is more frequent in patients with a late onset of the disease, typically with small expansions. On the contrary, subjects younger than 18 years of age can present with a juvenile form of HD. These patients usually have large trinucleotide repeat expansions in the *HTT* gene and manifest with atypical features mainly characterized by parkinsonism (rigidity, bradykinesia, hypomimia, resting tremor), seizures, myoclonus, and psychiatric symptoms.

- **Genetic Testing:** The specific test to assess the number of CAG repeats in the *HTT* gene is nearly 100% sensitive and specific (Section 6.4.1). Test results always will present the number of expansions of both alleles of *HTT*. Normal alleles of *HTT* have less than 27 CAG repeats. If there are more than 40 repeats, this will cause HD symptoms in 100% of the cases (full penetrance—Section 2.8). The presence of 36–39 CAG repeats is considered a gray zone with a reduced penetrance of the disease. Subjects with 27–35 CAG repeats will not manifest the disease, but they are at high risk of expansion of the CAG repeat during the formation of their gametes, causing the disease in the offspring. Cases of HD with an adult onset usually have 36–55 CAG repeats on one of the *HTT* alleles, while juvenile onset is usually associated with more than 60 CAG repeats on an *HTT* allele. Disease-modifying clinical trials (Section 7.4.5) are ongoing; therefore a confirmatory genetic test in a symptomatic subject can provide a possibility of participating in available or future clinical trials. Because of the lack of disease-modifying or preventive approaches at present, testing of asymptomatic minors is not recommended (Sections 10.1.2 and 11.18.1). Trinucleotide repeat testing in *HTT* is recommended for Ana Maria. After two sessions of psychological counseling, with full agreement, she is tested and found to have 48 CAG repeats in the *HTT* gene.
- **Management and Recommendations:** The management of patients with HD usually requires a multidisciplinary approach, combining movement disorder experts, psychiatrist, psychological support, as well as rehabilitation interventions. For the symptomatic treatment of chorea, a dopamine blocking agent (such as a neuroleptic) or dopamine depleters (such as tetrabenazine and its analogues) should be used when the symptoms are prominent, affect daily activities and balance, and increase the risk of falling. Physical therapy can be helpful as well. The treatment of the motor symptoms will have to be coordinated with the management of the sometimes severe psychiatric manifestations. Indeed, the medications used for these two categories of symptoms of the disease can in some cases be synergic and in other cases need to be used with caution because of possible side effects. As the disease progresses, dysphagia can become prominent, exposing these patients to the risk of malnutrition and aspiration pneumonia. These eventualities should always be monitored and can benefit from speech and swallow therapy.
- **Implications for Other Family Members:** Ana Maria's children had a 50% chance of inheriting the mutation. Offspring of subjects with an intermediate allele (i.e., 27–35 repeats) are at high risk of inheriting a pathogenic allele (more than 36 triplet repeats) because of the meiotic instability of the expansion. Asymptomatic adult family members with a positive family history of HD (predictive testing) may consider testing for family, career, or retirement planning purposes, but likewise need expertise and support before making the decision.
- **Differential Diagnosis:** The differential diagnosis of adult onset chorea includes: systemic and metabolic conditions, systemic lupus erythematosus (associated with calcification at brain MRI), antiphospholipid syndrome, other autoimmune disorders including autoimmune and paraneoplastic encephalitis (where specific autoantibodies can be detected on serum or liquor), polycythemia vera, hyperthyroidism, cerebrovascular lesions, hyperglycemia, electrolyte imbalance, infectious disease (such as HIV), and certain medications. Other genetic forms of chorea, which, however, are much rarer than HD, include mutation of the *C9ROF72* gene; HD-like 2 due to CTG/CAG trinucleotide repeat expansion in the junctophilin 3 (*JPH3*) gene, which has been reported in African families; prion disease, caused by mutations in the *PRNP* gene; familial basal ganglia calcifications in which mutations of *SLC20A2*, *PDGFRB*, *PDGFB*, and *XPR1* have been identified; some forms of neurodegeneration of brain iron accumulations, such as neuroferritinopathy or aceruloplasminemia, characterized by the specific pattern of iron deposition at brain MRI, as well as impairment of the levels of ferritin, ceruloplasmin, and iron storage; spinocerebellar ataxia 17 (SCA17); chorea-acanthocytosis, characterized by prominent lingual-buccal dyskinesia, often triggered by eating, associated with acanthocytosis on blood smears, and that are caused by the mutations of the vacuolar protein sorting 13 homolog A (*VPS13A*) gene or are X-linked, as in the so-called McLeold syndrome due to the mutation of the *XK* gene, in which chorea is associated with myopathy and cardiac involvement.
- **Caveats:** Currently the repeat expansion cannot be detected by clinical sequence–based multigene panels (Section 6.2.3), or exome sequencing (Section 6.3.1), but they can be captured by whole genome sequencing (Section 6.3.2).
- **Outcome:** Ana Maria continues to be followed by psychology and also psychiatry. She is started on tetrabenazine, and her choreiform movements improve. However, the disease continues to progress. She develops severe cognitive impairment and dysphagia and over about 10 years she becomes dependent in all the activities of her daily living. Two of her children decide they want to undergo testing, while one of them decides to wait. All of them continue to follow up with regular psychological counseling.

The Diagnosis

Clinical Diagnosis: Huntington disease, adult onset.
Genetic Diagnosis: *HTT* gene, heterozygous full mutation triplet repeats.

Further Reading

Bean, L., Bayrak-Toydemir, P., 2014. American College of Medical Genetics and Genomics Standards and Guidelines for Clinical Genetics Laboratories, 2014 edition: technical standards and guidelines for Huntington disease. Genet. Med. 16, e2. PMID 25356969.

Ghosh, R., Tabrizi, S.J., 2018. Huntington disease. Handb. Clin. Neurol. 147, 255–278. PMID 29325616.

Section 11.14

Psychiatric Cases

Case 11.14.1

Psychiatric Cases: Early-Onset Dementia

Giulietta Maria Riboldi and Steven J. Frucht

Department of Neurology, The Marlene and Paolo Fresco Institute for Parkinson's and Movement Disorders, New York University Grossman School of Medicine, New York, NY, United States

Abstract

A 45-year-old woman has developed difficulty with memory and cognitive skills and is found to have early dementia due to a pathogenic variant in the presenilin 1 gene, *PSEN1*, causing early-onset Alzheimer disease.

The Case

Samantha is a 45-year-old, high school history teacher who comes to see you, a primary care physician, for an initial evaluation for possible cognitive decline. Over the last year Samantha started complaining of difficulties teaching her usual classes, although she has been doing this work for 15 years. Over the last year she had to change some of the curriculum she was teaching because of a reassessment of the programs at her school. While she was studying to prepare the programs for the new classes, she realized that she was struggling to remember details and put events in order while she was explaining them to the class. This really surprised her since she easily mastered her job in the past. She is very concerned that she is developing a form of dementia because a number of her family members presented with cognitive impairments early in life.

The Issues

- **Could Samantha have dementia?** You discuss with Samantha that dementia is defined as a deterioration of the cognitive functions with an impact on patient's daily life. Usually dementia arises later in life. The diagnosis of dementia is based on clinical presentation, ruling out organic causes, and neuropsychological assessments that explore the affected cognitive domains. You recommend a workup for metabolic causes, vitamin B12 deficiency, and thyroid function. Extensive bloodwork testing is unrevealing.
- **What is the next step in the workup?** Other tests, such as brain imaging studies (like brain MRI, brain CT scan, or PET studies) or the analysis of the cerebrospinal fluid, may be important for the differential diagnosis and to rule other causes (e.g., infection, structural lesions, leukodystrophies – Section 11.13.9, vascular lesions). If the diagnosis of dementia is confirmed, ruling out treatable conditions (e.g., depression, cerebrovascular disorders) is crucial. Samantha's neuropsychological tests rule out depression and are supportive of a significant cognitive impairment consistent with dementia and suggestive for Alzheimer disease (AD). Her brain MRI shows hippocampal atrophy. You refer her to neurology.
- **What is early-onset dementia?** The neurologist notes no additional findings such as seizures, myoclonus, tremors, or parkinsonism, that can be indicative of specific forms of dementia. A spinal tap is negative. After explaining her clinical history, Samantha tells the neurologist that she is really worried she is developing AD. The neurologist discusses that

Genomics in the Clinic. https://doi.org/10.1016/B978-0-12-816478-5.00092-3

dementia can be associated with neurodegenerative processes (such as AD or frontotemporal dementia), or secondary causes (such as depression, vascular disorders, endocrine abnormalities, infections, and other systemic disorders), where the latest have been ruled out by her previous assessments. Degenerative forms of dementia more frequently develop later in life. However, there are cases defined as early-onset dementia which occur when the onset of the symptoms is before the age of 65 years. These cases are more frequently due to genetic mutations compared to forms of late onset dementia. Early-onset dementia represents about 2% of all cases of AD, which is the most common form of dementia.

- **Could Samantha's case be considered a form of familial dementia?** Samantha says that she always thought that genetic disorders occur only when other members of the family are affected with the same symptoms. The neurologist explains that in the case of dementia, familial forms are more likely when there are more than three family members affected by similar manifestations of the disease. However, in rarer cases a genetic mutation can present in one individual for the first time (de novo mutation) even without a positive family history. Given her young age, Samantha is referred for genetic evaluation.

The Referral

- **Genetic Evaluation:** The genetic counselor meets with Samantha and her husband to discuss possible genetic causes of dementia, their implications for her and her family, and about genetic testing options. A genetic diagnosis of dementia can be challenging because of the information it can provide not only of her prognosis, but also for the possible impact on family members. Samantha and her husband, who has established power of attorney for her since her recent diagnosis of dementia, express understanding and are determined to move forward with the discussion. The genetic counselor collects a detailed family history (Section 3.2) from Samantha with the help of her husband. Samantha and her husband have two children in their early 20s who are both healthy. Samantha lost her parents when she was 11 years old after they died in a car accident. At that time both her father and her mother were in their early 40s. They had no sign of dementia up to that point. However, one on her maternal uncles was diagnosed with dementia at the age of 50, and he died about 10 years later. Her maternal grandmother started to present some cognitive complains in her late 50s and she died at age 67 years after a severe infection that led to a prolonged hospitalization. Because of their premature death, it is not possible to rule out that one of her parents, most likely her mother, may have died before presenting with symptoms of dementia.
- **Genetic Counseling:** Dementia can be caused by pathogenic variants (Section 7.1.3) in a number of genes. In the majority of the cases, the clinical presentation and the associated symptoms such as parkinsonism, brain stem involvement, myoclonus, seizures, and motor neuron disease support the diagnosis of different gene variants. In addition, assessing the affected cognitive domains as well as any abnormalities on imaging studies can help identify specific diagnosis. Samantha's neuropsychological tests and imaging studies are suggestive for AD. At present, the three most common genes where mutations have been linked to AD include the amyloid precursor protein (*APP*), presenilin 1 (*PSEN1*), and presenilin 2 (*PSEN2*). However, variants in these explain only a small portion of the cases of early-onset AD. Moreover, specific polymorphisms of the apolipoprotein E (*APOE,* specifically the *APOE4* allele) represent the major genetic risk factor for AD. Specific phenotypes can be associated with mutations in some of these genes. The genetic counselor also discusses the importance of considering life insurance and a long-term disability plan before proceeding with any genetic testing, although health insurance should remain unaffected (Section 10.4).
- **Genetic Testing:** Although some specific phenotypic features may be more frequently associated with one specific variant, in the majority of cases it is not possible to identify unequivocal genotype–phenotype correlations. Therefore, in order to identify the causative variant it is usually preferable in these cases to rely on gene panels (Section 6.2.3) that include the targeted genes (e.g., *APP*, *PSEN1*, and *PSEN2*). In some cases, exome sequencing (Section 6.3.1) can be performed, should the clinician and the patient also want to explore rare genetic mutations associated with other forms of early-onset dementia. The genetic counselor, Samantha, and her husband have a discussion about possibly identifying genetic mutations unrelated to the presenting condition (incidental and secondary findings) with exome sequencing and the chance of finding a variant of unknown significance (Sections 7.1 and 7.2). Samantha and her husband agree to test the most frequent gene variants associated with her condition. A gene panel is sent.
- **Results:** Samantha's panel results show a heterozygous pathogenic variant (Section 2.3.4) in the *PSEN1* gene. Identification of pathogenic variants of the *PSEN1* gene in a subject presenting with early-onset dementia implies important prognostic information. Forms of dementia associated with this gene, which explain 70% of cases of early-onset AD, can manifest with some atypical features (e.g., ataxia, seizure, myoclonus, as well as parkinsonism) compared to classical AD. Thus, identification of this mutation will prevent unnecessary further assessments to look for other causes that

may be prompted by the atypical presentation of the disease. Unfortunately, life expectancy is shortened about 7 years in patients presenting with these mutations compared to that for other genetic forms of AD.

- **Management and Recommendations:** While there is no cure for AD, the symptoms can be mitigated by the use of a cholinesterase inhibitor as well as an *N*-methyl-D-aspartate (NMDA) receptor antagonist, modulating the neurotransmitters in the brain. Samantha is started on donepezil (anticholinesterase inhibitor) with some benefit to her symptoms. No further specific treatments are currently available for the management of the cognitive impairment in AD. Instead, it is very important to keep patients cognitively engaged. Anxiety, depression, as well as sleep problems can be present as well. It is important to treat these additional symptoms since they can all further worsen the cognitive complaints. As AD progresses, patients become increasingly dependent and need constant surveillance. Spatial disorientation, even in familial places, is common in patients with AD, also in the early stages of the disorder. Moreover, sudden or drastic changes of environment, surrounding people and places can cause severe disorientation and anxiety in these patients.
- **Implications for Other Family Members:** Mutations of *PSEN1* are autosomal dominant (Section 2.7.1). This means that each of Samantha's children has a 50% risk of having inherited the mutation and if so, will also present with early-onset dementia because the mutation has an almost full penetrance by the age of 60 years. Unless Samantha's mutation is de novo, Samantha inherited it from one of her parents, most likely her mother, who died at a young age before manifesting the symptoms and had relatives with dementia. Should this be the case, Samantha's siblings also have a 50% chance of having the same mutation and therefore of manifesting the disease, and, likewise, for their children. Since no preventive treatments are available, genetic testing in minors is not generally recommended (Sections 10.1.2 and 11.18.1), while adults can decide whether they would like to be tested for the mutation after adequate genetic counseling. If any of Samantha's family members are interested in testing or manifest symptoms of dementia, site-specific gene analysis (Section 6.2.1) for the *PSEN1* gene mutation can be recommended instead of the gene panel. Similar recommendations for financial planning, long-term disability insurance, and life insurance would be made for family members as above (Section 10.4).
- **Differential Diagnosis:** Specific conditions that can present with early dementia include CADASIL (Cerebral Autosomal Dominant Arteriopathy with Subcortical Infarcts and Leukoencephalopathy—Section 11.4.7), frontotemporal dementia, leukodystrophy, Lewy body dementia as well as Parkinson disease (Section 11.13.8), prion diseases, familial brain calcifications, cerebral amyloid angiopathy, and some genetic metabolic disorders.
- **Caveats:** Mutations in the described genes explain only a small portion of cases of early-onset AD. Further testing including exome sequencing should be considered in order to capture the most recently reported mutations (Section 6.3.1).
- **Outcome:** Samantha is able to complete the teaching year and the following year but finds it increasingly difficult and eventually she has to resign. Her husband continues to attentively take care of her as her disease progresses. Her children elect not to be tested since preventive and mutation-specific therapeutic approaches are not available yet.

The Diagnosis

Clinical Diagnosis: Alzheimer disease, early-onset.
Genetic Diagnosis: *PSEN1* gene, heterozygous pathogenic variant.

Further Reading

Kuruppu, D.K., Matthews, B.R., 2013. Young-onset dementia. Semin. Neurol. 33, 365–385. PMID 24234358.

Rossor, M.N., Fox, N.C., Mummery, C.J., Schott, J.M., Warren, J.D., 2010. The diagnosis of young-onset dementia. Lancet Neurol. 9, 793–806. PMID 20650401.

Scheltens, P., Blennow, K., Breteler, M.M., de Strooper, B., Frisoni, G.B., Salloway, S., et al., 2016. Alzheimer's disease. Lancet 388, 505–517. PMID: 26921134.

Cases 11.14.2

Psychiatric Cases: Psychosis and Learning Disabilities

Francis J. McMahon
Human Genetics Branch, Intramural Research Program, National Institute of Mental Health, National Institutes of Health, Bethesda, MD, United States

Abstract
A young man with psychosis and learning disabilities is found to have a chromosome 22q11.2 deletion.

The Case

Jamal, a 22-year-old man with a history of learning disabilities is seen by you, a family physician, through the Employee Assistance Program at the warehouse where he has worked since finishing vocational school 3 years earlier. He complains that coworkers are "watching" him and that he can hear them discussing him "behind my back," even when he is home with his parents, where he maintains a basement apartment. He reports problems with reading in school. He states that he recognized that he was somehow "different" his whole life, and always felt like he was a "bad person." He reports that he gets along well with his parents, but senses that they are "ashamed" of him and blame themselves for his problems. Physical examination reveals a moderately obese young man with unusual facies, and a systolic heart murmur easily detected by auscultation.

The Issues

- **What are you most concerned about?** You are concerned about his mental health and request a psychiatric consultation. Mental status exam reveals a guarded, withdrawn young man with restricted affect who speaks slowly and uses frequent neologisms. He admits to hearing voices arguing about his performance at work and seems suspicious of the examiner. He denies depressive or manic symptoms. The psychiatrist diagnoses psychosis and recommends treatment with a second-generation antipsychotic.
- **What is the next step?** Because of his psychosis, as well as his physical exam findings of dysmorphic facies and heart murmur, you decide to refer Jamal for a genetic evaluation.

The Referral

- **Genetic Evaluation:** The history is reviewed and physical examination performed. Medically, he has been healthy other than recurrent otitis media when younger. There is no family history of mental illness. His exam shows normal

Genomics in the Clinic. https://doi.org/10.1016/B978-0-12-816478-5.00093-5

height and head circumference, obesity, mildly dysmorphic features with prominent nasal root and maxillary hypoplasia, normal intact palate and normal extremities.

- **Genetic Counseling:** The geneticist and genetic counselor tell Jamal that they think he could have the 22q11 deletion syndrome, also known as velocardiofacial syndrome or DiGeorge syndrome (Section 11.9.2), based on his history of learning disabilities, as well as his facial features and recent history of psychosis. This is a microdeletion disorder that is typically de novo at conception, but in some situations can be inherited.

*The widespread availability of reliable copy number variation (CNV) testing (*Section 6.1.3*) coupled with the nontrivial incidence of pathogenic CNVs among people with a broad range of psychiatric diagnoses pose new questions about when and whether CNV testing should be performed in patients diagnosed with a mental illness. From the standpoint of clinical utility, current evidence does not indicate a role for CNV testing in establishing a psychiatric diagnosis or selecting appropriate treatment. On the other hand, more than 10% of psychiatric patients with comorbid intellectual disability carry pathogenic CNVs.*

- **Management and Recommendations:** In light of the psychiatric and physical findings, a chromosomal microarray test (Section 6.1.3) is ordered. If the condition is confirmed as a newly diagnosed adult with 22q11.2 deletion syndrome, he would need 22q11.2 deletion syndrome relevant laboratory tests (Section 11.9.2) and follow-up as needed that would include family planning and prenatal counseling; neurology, ophthalmology, audiology, dental, and orthopedic assessment; abdominal ultrasound, and cardiac evaluation for the heart murmur. In addition, nutritional consultation for weight loss is recommended. In all cases, careful attention to and treatment of the comorbid medical conditions associated with 22q11.2 deletion syndrome that can mimic psychiatric illness or medication side effects, including hypocalcemia, hypothyroidism, and seizures, can help in optimizing treatment.
- **Results:** The microarray reveals a 3-Mb heterozygous interstitial deletion on 22q11.2, as suspected. When told the diagnosis, Jamal is grateful that a cause for his troubles has finally been identified and requests a session with his parents present. In that session, his parents express relief to finally know his diagnosis. The parents agree to undergo testing to determine if Jamal inherited the condition from a parent as the presentation of 22q11.2 deletion syndrome can be variable and mild. When they are told neither of them carry the deletion and Jamal is the first in the family to have this condition, they are relieved as they had always blamed themselves for their son's difficulties.

This illustrates the "personal utility" of genetic testing, even when the results do not alter the psychiatric diagnosis.

- **Outcome:** Membership in a local 22q11 deletion syndrome support group (Section 3.5) is recommended. Jamal responds well to treatment, makes new friends and meets a new girlfriend at the support group, and is able to continue his work at the warehouse. He is aware that he is at a 50% risk of passing on the deletion to offspring. He will seek prenatal counseling for future family planning.

The Diagnosis

Clinical Diagnosis: 22q11.2 deletion syndrome.
Genetic Diagnosis: Chromosome 22q11.2 microdeletion.

Further Reading

Fung, W.L., Butcher, N.J., Costain, G., Andrade, D.M., Boot, E., Chow, E.W., et al., 2015. Practical guidelines for managing adults with 22q11.2 deletion syndrome. Genet. Med. 17, 599–609. PMID 25569435.

Schneider, M., Debbane, M., Bassett, A.S., Chow, E.W., Fung, W.L., Van Den Bree, M., et al., 2014. Psychiatric disorders from childhood to adulthood in 22q11.2 deletion syndrome: results from the International Consortium on Brain and Behavior in 22q11.2 Deletion. Am. J. Psychiatr. 171, 627–639. PMID 24577245.

Wolfe, K., Strydom, A., Morrogh, D., Carter, J., Cutajar, P., Eyeoyibo, M., et al., 2016. Chromosomal microarray testing in adults with intellectual disability presenting with comorbid psychiatric disorders. Eur. J. Hum. Genet. 25, 66–72. PMID 27650969.

Case 11.14.3

Psychiatric Cases: Bipolar Disorder and Stevens–Johnson Syndrome

Francis J. McMahon

Human Genetics Branch, Intramural Research Program, National Institute of Mental Health, National Institutes of Health, Bethesda, MD, United States

Abstract

A 56-year-old woman with bipolar disorder develops Stevens–Johnson syndrome following treatment with carbamazepine and is found to be positive for an HLA haplotype associated with a higher risk for the syndrome on pharmacogenomic testing.

The Case

A 56-year-old Taiwanese woman, Daiyu, is seen in the emergency department of the local hospital after a suicide attempt. Her bipolar illness had worsened in recent years, with several episodes of mania and depression in the previous 6 months. Daiyu's illness had been managed with a combination of lithium, valproic acid, risperidone, and lamotrigine. On mental status exam, she is agitated, irritable, and feels hopeless, with paranoid ideation, suicidal thoughts, and flight of ideas. Daiyu is admitted to inpatient psychiatric care, for her rapid-cycling bipolar disorder. She fails to improve despite adjunctive treatment with haloperidol. On the third day of her hospital stay she is started on carbamazepine, used occasionally as an adjunctive treatment for bipolar disorder. This is followed by gradual stabilization of mood, decreased paranoia, and cessation of suicidal thoughts. On the morning of discharge, she complains of pruritus, which is treated with diphenhydramine. She returns to the emergency department later the same day with widespread erythema and fever that rapidly progresses to Stevens–Johnson syndrome.

The Issues

- **What is the immediate management?** All medications are discontinued and Daiyu is admitted to intensive care, where she receives IV fluids and supportive care. Psychiatric consultation is sought to address further management.
- **Could there be an underlying genetic cause?** In light of the onset of severe cutaneous symptoms shortly after starting treatment with carbamazepine, this appears to be a drug reaction. The psychiatrist considers pharmacogenomic testing.
- **Has this association been reported previously?** Serious adverse events are rare with drugs commonly used in psychiatry but have occurred in people taking some less commonly used drugs. The anticonvulsant carbamazepine can provoke severe adverse cutaneous reactions, such as Stevens–Johnson syndrome.

Genomics in the Clinic. https://doi.org/10.1016/B978-0-12-816478-5.00094-7

The Referral

- **Genetic Evaluation:** Daiyu has no additional medical history. Her family history is significant for mood disorders in both parents and a brother who died by suicide at age 32. This is consistent with the high heritability of psychiatric disorders (Section 5.7.2).

*Large family studies conducted in the latter half of the 20th century established the strong familiality of psychiatric disorders, at least when diagnosed by reliable operational criteria. They demonstrated that about 25% of the first-degree relatives of probands with bipolar disorder live with a mood disorder themselves, while about 10%–15% of first-degree relatives of probands with schizophrenia develop schizophrenia. Segregation patterns, however, are largely not consistent with a single major gene disorder (*Section 2.7.3*).*

- **Genetic Counseling:** It is known that immunologic markers have been associated with drug reactions, as found on genome analysis, in certain individuals. In particular, high-risk HLA haplotypes occur in diverse populations (Section 5.7.4). Alleles in two different HLA genes have been identified that confer substantial risk for adverse cutaneous reactions in carriers exposed to carbamazepine or the related drug, oxcarbazepine.
- **Management and Recommendations:** The psychiatrist orders an HLA panel for HLA-B sequencing (Section 6.2.3). She recommends avoiding carbamazepine in favor of increased valproic acid and risperidone.
- **Results:** The HLA panel reveals that Daiyu is heterozygous for the HLA-B*1502 haplotype that predisposes to Stevens–Johnson syndrome after exposure to carbamazepine. This haplotype is seen more commonly in persons of Asian ancestry.

This case underscores the importance of HLA screening in patients being considered for treatment with carbamazepine, particularly of Asian background. Some evidence also suggests these alleles also increase risk in patients exposed to lamotrigine, which is more commonly used by psychiatrists today. However, these alleles do not explain all cases of drug-related acute cutaneous reactions.

- **Outcome:** Daiyu gradually improves on the new medication regimen and is discharged home when cutaneous symptoms are largely resolved.

The Diagnosis

Clinical Diagnosis: Stevens–Johnson syndrome predisposition.
Genetic Diagnosis: HLA-B*1502 heterozygous haplotype.

Further Reading

Chen, C.B., Hsiao, Y.H., Wu, T., Hsih, M.S., Tassaneeyakul, W., Jorns, T.P., Sukasem, C., Hsu, C.N., Su, S.C., Chang, W.C., Hui, R.C., Chu, C.Y., Chen, Y.J., Wu, C.Y., Hsu, C.K., Chiu, T.M., Sun, P.L., Lee, H.E., Yang, C.Y., Kao, P.H., Yang, C.H., Ho, H.C., Lin, J.Y., Chang, Y.C., Chen, M.J., Lu, C.W., Ng, C.Y., Kuo, K.L., Lin, C.Y., Yang, C.S., Chen, D.P., Chang, P.Y., Wu, T.L., Lin, Y.J., Weng, Y.C., Kuo, T.T., Hung, S.I., Chung, W.H., Taiwan Severe Cutaneous Adverse Reaction, 2017. Risk and association of HLA with oxcarbazepine-induced cutaneous adverse reactions in Asians. Neurology 88, 78–86. PMID 27913699.

Gershon, E.S., Delisi, L.E., Hamovit, J., Nurnberger Jr., J.I., Maxwell, M.E., Schreiber, J., Dauphinais, D., Dingman 2nd, C.W., Guroff, J.J., 1988. A controlled family study of chronic psychoses. Schizophrenia and schizoaffective disorder. Arch. Gen. Psychiatr. 45, 328–336. PMID 3355320.

Li, X., Yu, K., Mei, S., Huo, J., Wang, J., Zhu, Y., Zhao, Z., 2015. HLA-B*1502 increases the risk of phenytoin or lamotrigine induced Stevens-Johnson syndrome/toxic epidermal necrolysis: evidence from a meta-analysis of nine case-control studies. Drug Res. 65, 107–111. PMID 24871931.

Tsuang, M.T., Taylor, L., Faraone, S.V., 2004. An overview of the genetics of psychotic mood disorders. J. Psychiatr. Res. 38, 3–15. PMID 14690766.

Section 11.15

Pharmacogenomic Cases

Case 11.15.1

Pharmacogenomic Cases: Warfarin for Atrial Fibrillation

Stuart A. Scott
Department of Pathology, Stanford University, Stanford, CA, United States; Clinical Genomics Laboratory, Stanford Medicine, Palo Alto, CA, United States

Abstract
A 65-year-old man, who is on anticoagulant therapy for atrial fibrillation with an elevated internationalized normal ratio, is found to be a poor metabolizer of warfarin by pharmacogenetic testing.

The Case

A 65-year-old man named Tao schedules an appointment with you, his primary care physician. Tao has new onset atrial fibrillation and very recently was put on warfarin. He was prescribed 5 mg, but his internationalized normal ratio (INR) measurements have been elevated. Tao is nervous about his warfarin dose because his sister previously had a major bleed while on warfarin, and his children told him that they heard that warfarin bleeding "is genetic." He looked it up on the internet and found out that there are genetic tests available, but he does not really understand "how it works." Tao would like you to pursue this test for him, even if he has to pay for it himself.

The Issues

- **How does warfarin work?** Warfarin is an anticoagulant that inhibits clot formation. It blocks the enzyme epoxide reductase leading to depletion of the reduced form of vitamin K that serves as a cofactor of gamma carboxylation of vitamin K−dependent clotting factors. As a result, warfarin inhibits the synthesis of biologically active forms of the clotting factors II, VII, IX, and X, as well as the regulatory factors protein C, protein S, and protein Z.
- **What does a pharmacogenetic test tell us about warfarin?** The primary enzyme involved in warfarin metabolism is CYP2C9 and the target enzyme of warfarin is VKORC1. Both enzymes are known to have variable levels of activity, which is due in part to common genetic variants in the genes that encode them. Patients who carry variant alleles in *CYP2C9* and/or *VKORC1* have lower therapeutic warfarin dose requirements compared to patients that do not carry variant alleles for both the initial and the maintenance doses. Warfarin sensitivity dosing recommendations based on pharmacogenetic test results are denoted in the US Food and Drug Administration (FDA) drug label and from the Clinical Pharmacogenetics Implementation Consortium (CPIC) (Section 5.6.2). It is estimated that the currently identified clinical and genetic variables together account for approximately 45%−55% of the variation in warfarin dosing.

Genomics in the Clinic. https://doi.org/10.1016/B978-0-12-816478-5.00095-9

- **How accurate are these test results?** Warfarin pharmacogenetic testing performed in a Clinical Laboratory Improvement Amendments (CLIA)-certified, clinical genetic testing laboratory can be considered very accurate; however, the specific variant alleles included in pharmacogenetic testing panels may vary between laboratories (Section 6.2.3). The most commonly tested variant alleles are *CYP2C9*2*, *CYP2C9*3*, and *VKORC1* c.-1639G > A. The frequencies of variant *CYP2C9* and *VKORC1* alleles differ across ethnicities (Section 3.2.3), and it is important to recognize that some populations harbor variant alleles that may not be included in typical clinical panels (e.g., *CYP2C9*5*, **6*, **8* among African Americans).

  In addition, rare heterozygous *VKORC1* coding variants are a dominant (Sections 2.3.4, 2.7.1) cause of warfarin resistance. These rare variants make VKORC1 less susceptible to warfarin inhibition and generally result in doses in excess of 80 mg/week to maintain appropriate anticoagulation. Homozygosity (Sections 2.3.4, 2.71) for these rare variants results in combined deficiency of vitamin K–dependent clotting factors type 2 (https://www.omim.org/, OMIM #607473), which is responsive to oral administration of vitamin K. Of note, these warfarin resistance variants are *not* typically included on warfarin pharmacogenetic testing panels (Section 6.2.3). In addition, several drugs have been shown to be either inducers or inhibitors of CYP2C9, which can also influence warfarin dosing. You decide to refer Tao for pharmacogenetic testing.

Although the clinical validity of CYP2C9 *and* VKORC1 *(as well as some other genes with smaller effects) has been well established with respect to warfarin dose requirements, the clinical utility of pharmacogenetic-guided warfarin dosing has yet to be unequivocally determined.*

The Referral

- **Genetic Evaluation:** Tao is originally from China. He is not a smoker and does not have liver disease. His height is 165 cm and weight is 63.5 kg. His INR measurements have been unstable since his initial warfarin dose of 5 mg/day. As reported, his family history is notable for a sister who had several major bleeds while on warfarin; however, no pharmacogenetic testing has been performed for him or other family members. In addition to pharmacogenetic variants, warfarin dose requirements are influenced by multiple variables, including patient age, diet, gender, body size, comorbidities, and concomitant medications. The geneticist agrees that pharmacogenetic testing is indicated.
- **Genetic Counseling:** The *CYP2C9* and *VKORC1* pharmacogenetic variants are inherited as autosomal codominant traits (Section 2.7.1). They are not acquired somatic mutations (Section 2.4.2). Although specific alleles may have differing effects on enzymatic activity (e.g., *CYP2C9* variants), they are fully penetrant (Section 2.8). There are numerous *CYP2C9* variants with different frequencies in various ethnic and racial groups; however, *CYP2C9*2* (p.R144C) and *CYP2C9*3* (p.I359L) are the most common in the general population.

For current information on variant CYP2C9 *alleles, refer to the Pharmacogene Variation (PharmVar) Consortium:* https://www.pharmvar.org/gene/CYP2C9.

  Like other CYP450 enzymes, CYP2C9 participates in the metabolism of several clinically important drugs and xenobiotics. In addition to warfarin, CYP2C9 also metabolizes tolbutamide, phenytoin, losartan, and antiinflammatory drugs such as ibuprofen. Thus, carriers of variant *CYP2C9* alleles may also have altered metabolism of other clinically important medications. Tao does not recall being prescribed any of the additional drugs metabolized by CYP2C9 when he is asked.

For additional pharmacogenetic information on CYP2C9, *refer to the Pharmacogenomics Knowledge Base* (pharmgkb.org) *and CPIC* (cpicpgx.org).

- **Genetic Testing:** Tao consents to a warfarin sensitivity pharmacogenetic test and a blood sample is sent to the affiliated clinical genomics laboratory. A few days later you receive the report, which explains that Tao is a CYP2C9 "poor metabolizer" with a genotype of **2/*3* and has a *VKORC1* c.-1639G>A "high sensitivity" genotype of A/A.
- **Management and Recommendations:** According to the FDA warfarin product insert and the warfarin CPIC guideline (Section 5.6.2), a lower daily warfarin dose is expected to result in sufficient anticoagulation for Tao based on his genotyping results.

  To facilitate physician interpretation of clinical *CYP2C9* and *VKORC1* genotyping (Section 7.1), the FDA updated the "Dosage and Administration" section of the warfarin label to include a table containing stable maintenance doses observed in large patient cohorts with different combinations of *CYP2C9* and *VKORC1* genotype results (Table 11.15.1.1).

TABLE 11.15.1.1 **Range of expected therapeutic warfarin doses based on CYP2C9 and VKORC1 genotypes from the product insert.**

VKORC1	CYP2C9					
c.-1639G>A	*1/*1	*1/*2	*1/*3	*2/*2	*2/*3	*3/*3
G/G	5–7 mg	5–7 mg	3–4 mg	3–4 mg	3–4 mg	0.5–2 mg
G/A	5–7 mg	3–4 mg	3–4 mg	3–4 mg	0.5–2 mg	0.5–2 mg
A/A	3–4 mg	3–4 mg	0.5–2 mg	0.5–2 mg	0.5–2 mg	0.5–2 mg

The label recommends healthcare providers to "Consider these ranges in choosing the initial dose." Moreover, the label also states that patients with variant CYP2C9 genotypes may require a prolonged time (more than 2–4 weeks) to achieve an optimal INR effect for a given dosage regimen.

- **Caveats:** Warfarin pharmacogenetic variants are only one variable implicated in interindividual warfarin dosing requirements. As such, not all patients who carry warfarin sensitivity alleles will require lower warfarin dose requirements. In addition, and as stated above, the frequencies of variant *CYP2C9* and *VKORC1* alleles differ between ethnicities (Section 3.2.3), so a negative genotyping test result will not necessarily rule out the presence of a clinically significant variant.
- **Outcome:** Tao's INR measurements have been stable since his daily warfarin dose was reduced to 1.5 mg. You recommend that his sister also be tested to establish if she too carries any warfarin sensitive *CYP2C9* and/or *VKORC1* variant alleles.

The Diagnosis

Clinical Diagnosis: CYP2C9, poor metabolizer. VKORC1, high sensitivity.
Genetic Diagnosis: *CYP2C9* gene, compound heterozygous variants. *VKORC1* gene, homozygous variants.

Further Reading

Dean, L., 2012. Warfarin therapy and VKORC1 and CYP genotype. In: Pratt, V.M., Mcleod, H.L., Rubinstein, W.S., Scott, S.A., Dean, L.C., Kattman, B.L., Malheiro, A.J. (Eds.), 2012. Medical Genetics Summaries. Bethesda (MD). PMID 28520347.

Johnson, J.A., Caudle, K.E., Gong, L., Whirl-Carrillo, M., Stein, C.M., Scott, S.A., et al., 2017. Clinical pharmacogenetics implementation consortium (CPIC) guideline for pharmacogenetics-guided warfarin dosing: 2017 update. Clin. Pharmacol. Ther. 102, 397–404. PMID 28198005.

Case 11.15.2

Pharmacogenomic Cases: Clopidogrel for Coronary Artery Disease

Stuart A. Scott

Department of Pathology, Stanford University, Stanford, CA, United States; Clinical Genomics Laboratory, Stanford Medicine, Palo Alto, CA, United States

Abstract

A 56-year-old woman, who has had a second myocardial infarction while on antiplatelet therapy, is found to be a CYP2C19 poor metabolizer of clopidogrel based on pharmacogenetic test results and is switched to a different medication.

The Case

Your patient, Diane, had an acute myocardial infarction 1 year ago (in a different city from where she lives now), and was managed with acute coronary angioplasty and a drug eluting stent. She was placed on dual antiplatelet therapy with clopidogrel and aspirin; however, 6 months ago she again presented with an acute myocardial infarction following minimal exertion. Diane was admitted to your local hospital where coronary angiography revealed a stent thrombosis, which was treated with clot extraction and balloon angioplasty. She is now 56 years old and during her first annual wellness visit with you, on further history, she says that she used to be a heavy smoker prior to her first myocardial infarction. She reveals to you that she has had intermittent chest pain ever since her first myocardial infarction. She is very concerned about further complications related to her stent. She carefully read the clopidogrel package insert and noticed the black box warning stating how "poor metabolizers" can be prescribed a different medicine because clopidogrel may not work for them.

The Issues

- **How does clopidogrel work?** Clopidogrel is a second generation thienopyridine that undergoes hepatic biotransformation to an active metabolite, which binds irreversibly to the P2Y12 receptor and inhibits ADP-mediated platelet activation and aggregation. However, only a minor amount of the prodrug (about 15%) is available for transformation to the active metabolite, which is mediated by the cytochrome P450 system, in particular CYP2C19. Wide interindividual variability in ex vivo platelet aggregation is common among patients treated with clopidogrel, which is influenced by age, comedications, diabetes, disease activity, renal failure, and pharmacogenetic variation in the *CYP2C19* gene.
- **What does a pharmacogenetic test tell us about clopidogrel?** Importantly, patients who carry *CYP2C19* loss-of-function alleles (Section 2.4.1) (e.g., *CYP2C19**2*, **3*) have different phenotypes that confer a decreased ability to

Genomics in the Clinic. https://doi.org/10.1016/B978-0-12-816478-5.00096-0

metabolize (Section 5.6.2) a specific medication or class of medications, including clopidogrel. Thus, patients who carry these alleles and who are treated with clopidogrel for high-risk indications (e.g., acute coronary syndrome/percutaneous coronary intervention, ACS/PCI) are at an increased risk for adverse cardiovascular events (e.g., cardiovascular death, myocardial infarction, stroke) and stent thrombosis. Clinical pharmacogenetic tests are currently available for the *CYP2C19* gene, as well as other genes involved in drug response variability.

The FDA clopidogrel product insert includes a black box warning, which states that CYP2C19 poor metabolizers treated with clopidogrel have higher event rates than patients with normal CYP2C19. In addition, antiplatelet therapy recommendations based on *CYP2C19* pharmacogenetic test results are available from the Clinical Pharmacogenetics Implementation Consortium (CPIC).

- **How accurate are these test results?** *CYP2C19* pharmacogenetic testing performed in a Clinical Laboratory Improvement Amendments (CLIA)-certified clinical genetic testing laboratory can be considered very accurate; however, the specific variant alleles included in pharmacogenetic testing panels (Section 6.2.3) may vary between laboratories. The most commonly tested metabolism variant alleles are *CYP2C19*2*, **3*, and **17* (Section 5.6.2). The frequencies of variant *CYP2C19* alleles differ across ethnicities (Section 3.2.3), and it is important to recognize that some populations harbor rare variant alleles that may not be included in typical clinical panels (Section 6.2.3). You recommend referral to genetics to discuss pharmacogenetic testing.

  Although the clinical validity of CYP2C19 has been well established with respect to clopidogrel pharmacokinetics, pharmacodynamics, and clinical outcomes among high-risk patients (e.g., ACS/PCI), the clinical utility of pharmacogenetic-guided clopidogrel dosing has yet to be unequivocally determined.

*Of note, patients treated with clopidogrel who carry two increased activity CYP2C19*17 alleles are classified as CYP2C19 ultrarapid metabolizers, Who have an elevated bleeding risk compared to normal metabolizers.*

The Referral

- **Genetic Evaluation:** Diane is seen for a genetic evaluation. She is African American, and her height is 150 cm and weight is 63.5 kg. In addition to *CYP2C19* variants, clopidogrel response is influenced by multiple variables, including patient age, body size, comorbidities, and concomitant medications. Given her history and the stent thrombosis 6 months after initiating dual antiplatelet therapy, the geneticist suspects a potential drug metabolism issue with clopidogrel.
- **Genetic Counseling:** The *CYP2C19* pharmacogenetic variants are inherited as autosomal codominant traits (Section 2.7.1). They are not somatic mutations (Section 2.4.2). There are over 35 known *CYP2C19* variants with different frequencies in various ethnic and racial groups; however, **2* (c.681G>A) and **3* (c.636G>A) are the most commonly tested loss-of-function alleles (Section 7.1.3). In addition, laboratories often test for the increased activity **17* (c.-806C>T) allele.

For current information on variant CYP2C19 alleles, refer to the Pharmacogene Variation (PharmVar) Consortium: https://www.pharmvar.org/gene/CYP2C19.

Like other CYP450 enzymes, CYP2C19 participates in the metabolism of several clinically important drugs and xenobiotics. In addition to clopidogrel, CYP2C19 also is involved in the metabolism of voriconazole, antidepressants (tricyclic antidepressants; selective serotonin reuptake inhibitors (SSRIs)), and proton pump inhibitors. Thus, carriers of variant *CYP2C19* alleles may also have atypical metabolism of other clinically important medications. A *CYP2C19*-targeted variant (Section 6.2.1) pharmacogenetic test is recommended.

- **Genetic Testing:** Diane consents to a *CYP2C19* pharmacogenetic test and a saliva sample is sent to your affiliated clinical genetic testing laboratory.

*For additional pharmacogenetic information on CYP2C19, refer to the Pharmacogenomics Knowledge Base (*pharmgkb.org*) and CPIC (*cpicpgx.org*).*

- **Results:** A few days later you receive the report, which explains that Diane is a CYP2C19 "poor metabolizer" with a genotype result of **2/*2*. This result indicates that Diane is not metabolizing clopidogrel into an active metabolite as efficiently as patients with normal CYP2C19.

  Reported meta-analyses suggest that clopidogrel-treated acute coronary syndrome/percutaneous coronary intervention (ACS/PCI) patients who are *CYP2C19*2* carriers have an increased risk compared to patients with a normal (i.e., *1/*1) genotype for both major adverse cardiovascular events (hazard ratio [HR] 1.55, 95% confidence interval [CI] 1.11–2.17 for heterozygotes; HR 1.76, 95% CI 1.24–2.50 for homozygotes) and stent thrombosis (HR 2.67, 95% CI 1.69–4.22 for heterozygotes; HR 3.97, 95% CI 1.75–9.02 for homozygotes). The significance of *CYP2C19*2* in stent thrombosis risk among these ACS/PCI patients has been the most notable, with reported odds ratios for

*CYP2C19**2* carriers ranging from 1.75 to 4.68. Thus, Diane had a higher risk for both a second myocardial infarction and a stent thrombosis.

- **Management and Recommendations:** Both the FDA black box warning and CPIC recommendations indicate that alternative antiplatelet medications should be considered among patients who have undergone percutaneous coronary intervention and who are CYP2C19 poor metabolizers (i.e., *CYP2C19**2/*2). This should be discussed with cardiology.
- **Caveats:** CYP2C19 is only one variable implicated in clopidogrel response. As such, not all patients who are CYP2C19 poor metabolizers will have adverse events. In addition, and as stated above, the frequencies of variant *CYP2C19* alleles differ between ethnicities (Section 3.2.3), so a negative genotyping test result will not necessarily rule out the presence of a clinically significant variant.
- **Outcome:** After reviewing the *CYP2C19* pharmacogenetic test results with you, Diane's new cardiologist switches her antiplatelet agent from clopidogrel to ticagrelor. She does well, and 5 years later still has not had another myocardial infarction.

The Diagnosis

Clinical Diagnosis: CYP2C19, poor metabolizer.
Genetic Diagnosis: *CYP2C19* gene, homozygous variants.

Further Reading

Dean, L., 2012. Clopidogrel therapy and CYP2C19 genotype. In: Pratt, V., Mcleod, H., Rubinstein, W., Dean, L., Kattman, B., Malheiro, A. (Eds.), Medical Genetics Summaries. Bethesda, MD. PMID 28520346.

Mega, J.L., Simon, T., Collet, J.P., Anderson, J.L., Antman, E.M., Bliden, K., et al., 2010. Reduced-function CYP2C19 genotype and risk of adverse clinical outcomes among patients treated with clopidogrel predominantly for PCI: a meta-analysis. JAMA 304, 1821–1830. PMID 20978260.

Scott, S.A., Sangkuhl, K., Stein, C.M., Hulot, J.S., Mega, J.L., Roden, D.M., et al., 2013. Clinical pharmacogenetics implementation consortium guidelines for CYP2C19 genotype and clopidogrel therapy: 2013 update. Clin. Pharmacol. Ther. 94, 317–323. PMID 23698643.

Case 11.15.3

Pharmacogenomic Cases: Thiopurine for Acute Lymphoblastic Leukemia

Stuart A. Scott

Department of Pathology, Stanford University, Stanford, CA, United States; Clinical Genomics Laboratory, Stanford Medicine, Palo Alto, CA, United States

Abstract

An 11-year-old boy has intolerance to 6-mercaptopurine for his acute lymphoblastic leukemia and based on pharmacogenetic testing is better treated on a much lower dose.

The Case

Alex is 11 years old and 1 year ago presented to you, a pediatric oncologist, with persistent fever and weakness with lethargy. He is the only child of Ying and Jun, both originally from the Fujian province in southeast China. His bone marrow examination and immunophenotyping were consistent with acute lymphoblastic leukemia (ALL), and cytogenetic analysis identified trisomy 21. He was treated as per standard risk protocol for B-cell ALL, and bone marrow examination postinduction confirmed remission. However, he is now showing signs of intolerance to 6-mercaptopurine as he has had several episodes of oral ulcers and severe neutropenia. As part of Alex's care team, pharmacogenetic testing is now being considered prior to changing his maintenance treatment.

The Issues

- **How do the thiopurines work?** Thiopurines (i.e., azathioprine, mercaptopurine, and thioguanine) are used for the treatment of childhood ALL, autoimmune diseases, inflammatory bowel diseases, lupus, and transplantation. Azathioprine and mercaptopurine are used for nonmalignant immunologic disorders, mercaptopurine for lymphoid malignancies, and thioguanine for myeloid leukemias. All three agents are inactive precursors that are metabolized to active thioguanine nucleotides (TGNs), which are inactivated by thiopurine methyltransferase (TPMT) coded for by the *TPMT* gene. The TGNs are analogs of the nucleic acid guanine and are incorporated into RNA and DNA, ultimately inhibiting several metabolic pathways and inducing apoptosis. However, about 10% of the general population have variants that cause intermediate levels of TPMT activity, and less than 1% have low or undetectable activity, which results in significantly increased risks for TGN toxicity and life-threatening myelosuppression.

  In addition, variant alleles in another gene, *NUDT15,* which strongly influences thiopurine tolerance in patients with ALL and those with inflammatory bowel diseases, have been identified. As a nucleoside diphosphatase, NUDT15 catalyzes the conversion of cytotoxic thioguanine triphosphate (TGTP) metabolites to the less toxic thioguanine

Genomics in the Clinic. https://doi.org/10.1016/B978-0-12-816478-5.00097-2

monophosphate. Consequently, mutations leading to defects in NUDT15-mediated degradation of TGTP result in more TGTP available for incorporation into DNA, and subsequent DNA damage and cell death.

- **What does a pharmacogenetic test tell us about thiopurine toxicity?** Importantly, TPMT activity is inherited as an autosomal codominant trait, in which both alleles are equally expressed (Section 2.7.1). Patients, who having no mutations (i.e., homozygous for normal alleles **1/*1*), have lower levels of TGN metabolites and a low risk of myelosuppression. However, with conventional doses of thiopurines, a high proportion of patients who are heterozygous for loss-of-function *TPMT* alleles (Sections 2.4.1 and 2.7.1), thus carrying a single variant allele (e.g., **1/*2*, **1/*3A*, **1/*3B*, **1/*3C*, **1/*4*), show moderate to severe myelosuppression, and patients with two loss-of-function *TPMT* alleles (e.g., **3A/*3A*, **2/*3A*, **3A/*3C*, **3C/*4*, **2/*3C*, **3A/*4*) universally experience severe myelosuppression (Section 7.1). Clinical pharmacogenetic tests are currently available for the *TPMT* gene, as well as other genes involved in drug response variability.

  Likewise, patients who are heterozygous for loss-of-function (Section 7.1.3) *NUDT15* alleles (e.g., **1/*2, *1/*3*) are at significantly increased risk for severe myelosuppression, and patients who are homozygous or compound heterozygous (Section 2.7.1) for loss-of-function *NUDT15* alleles (e.g., **2/*2, *2/*3, *3/*3*) have very low tolerance for standard thiopurine doses and greatly increased risk for leukopenia, neutropenia, and myelosuppression.

  The FDA azathioprine and mercaptopurine labels state that patients with severe myelosuppression should be evaluated for *TPMT* or *NUDT15* deficiency, and that patients with homozygous *TPMT* or *NUDT15* deficiency require substantial dose reductions. In addition, thiopurine dosing recommendations based on *TPMT* and *NUDT15* pharmacogenetic test results are available from the Clinical Pharmacogenetics Implementation Consortium (CPIC) (Section 5.6.2).

- **How accurate are these test results?** *TPMT* and *NUDT15* pharmacogenetic testing performed in a Clinical Laboratory Improvement Amendments (CLIA)-certified clinical genetic testing laboratory can be considered very accurate; however, the specific variant alleles included in pharmacogenetic testing panels may vary between laboratories (Section 6.2.3). The most commonly tested *TPMT* variant alleles are **2*, **3A*, **3B*, and **3C*, and the most commonly tested *NUDT15* variant alleles are **2* and **3*. Inherited TPMT deficiency is the primary genetic cause of thiopurine intolerance in Europeans and Africans, whereas risk alleles in NUDT15 are more common among Asians and Hispanics. The frequencies of variant *TPMT* and *NUDT15* alleles also differ across ethnicities (Section 3.2.3), and it is important to recognize that some populations harbor rarer variant alleles that may not be included in typical clinical panels. You are aware of all of this and recommend a referral for Alex to be seen in genetics to discuss testing for these alleles.

Most clinical TPMT *and* NUDT15 *tests interrogate specific variants by targeted genotyping (*Section 6.2.1*). As such, the normal* *1 *allele is assigned in the absence of a detected variant. Therefore, if a patient harbors a rare or previously uncharacterized allele, it may be undetected by the clinical genotyping test and reported as* *1.

The Referral

- **Genetic Evaluation**: Alex is seen by the geneticist for his recent intolerance to 6-mercaptopurine and to consider pursuing a *TPMT/NUDT15* pharmacogenetic test from the CLIA-certified Molecular Pathology laboratory at your academic medical center. A family history is taken, and his ethnicity (Asian American) is discussed in the context of the medications.
- **Genetic Counseling**: The *TPMT* and *NUDT15* pharmacogenetic variants are inherited as germline autosomal codominant traits (Section 2.7.1). They are not acquired somatic mutations (Section 2.4.2). There currently are over 40 known *TPMT* variant star (*) alleles and 20 known *NUDT15* variant star (*) alleles with different frequencies across ethnic and racial groups. However, the function and clinical significance of all *TPMT* and *NUDT15* variant alleles is not always defined. Therefore, clinical pharmacogenetic testing panels typically restrict their tests to only include variant alleles with known functional and clinical effects.
- **Genetic Testing**: In addition to targeted *TPMT* and *NUDT15* gene tests (Section 6.2.1), some clinical genetic testing laboratories may offer larger oncology pharmacogenetic panels (Section 6.2.3) that include additional genes associated with other chemotherapies and medications. Ying and Jun consent to Alex having a *TPMT/NUDT15*-targeted pharmacogenetic test and a saliva sample is sent to the clinical laboratory.

Complementary clinical TPMT *laboratory tests are available that measure thiopurine metabolites in erythrocytes: TGNs for patients treated with mercaptopurine, azathioprine, and thioguanine, and* methylmercaptopurine nucleotides *(MeMPNs or MeTIMP) for those treated with mercaptopurine or azathioprine. However, therapeutic drug monitoring tests that distinguish* NUDT15 *metabolizer phenotypes in erythrocytes are not currently available clinically.*

- **Results**: You receive the report, which details that Alex is a TPMT "normal metabolizer" with a genotype result of **1/*1*; however, his *NUDT15* genotype is **1/*2*. This result indicates that Alex has a NUDT15 "intermediate metabolizer" phenotype due to the heterozygous **2* no-function allele that he inherited from one of his parents.

For current information on variant TPMT *alleles, refer to the* TPMT *Nomenclature Committee:* https://www.hmv.liu.se/tpmtalleles?l=en. *For current information on variant* NUDT15 *alleles, refer to the Pharmacogene Variation (PharmVar) Consortium:* https://www.pharmvar.org/gene/NUDT15. *For additional pharmacogenetic information on* TPMT *and* NUDT15, *refer to the Pharmacogenomics Knowledge Base* (pharmgkb.org) *and CPIC* (cpicpgx.org).

- **Management and Recommendations**: CPIC recommendations for NUDT15 intermediate metabolizers are very specific: Start with reduced starting doses (30%–80% of normal dose) if normal starting dose is $\geq$75 mg/m^2/day or $\geq$1.5 mg/kg/day (e.g., start at 22.5–60 mg/m^2/day or 0.45–1.2 mg/kg/day) and adjust doses of mercaptopurine based on degree of myelosuppression and disease-specific guidelines. Allow 2–4 weeks to reach steady state after each dose adjustment. If myelosuppression occurs, and depending on other therapy, emphasis should be on reducing mercaptopurine over other agents. If normal starting dose is already <75 mg/m^2/day or <1.5 mg/kg/day, dose reduction may not be recommended.
- **Caveats**: Poor tolerance to thiopurines could also be related to other clinical factors independent of germline pharmacogenetic variants. Thus, clinicians must always continue to evaluate markers of disease progression and/or of myelosuppression to adjust thiopurine doses up or down from the genotype-directed doses. In addition, as stated above the frequencies of variant *TPMT* and *NUDT15* alleles differ between ethnicities, so a negative genotyping test result (i.e., **1/*1*) does not rule out the presence of a rare clinically significant *TPMT* or *NUDT15* variant.
- **Outcome**: Alex's mercaptopurine intolerance is consistent with his *NUDT15 *1/*2* genotype result, as NUDT15 intermediate metabolizers have increased risks for thiopurine-related leukopenia, neutropenia, and myelosuppression. As such, you and the rest of Alex's care team proceed with a 50% dose reduction of mercaptopurine and continued monitoring for myelosuppression. He has a much better reaction and is able to finish his course of treatment without incident.

The Diagnosis

Clinical Diagnosis: NUDT15, intermediate metabolizer. TPMT, normal metabolizer.
Genetic Diagnosis: *NUDT15* gene, heterozygous variant.

Further Reading

Dean, L., 2012. Thioguanine therapy and TPMT genotype. In: Pratt, V.M., Mcleod, H.L., Rubinstein, W.S., Scott, S.A., Dean, L.C., Kattman, B.L., Malheiro, A.J. (Eds.), Medical Genetics Summaries. Bethesda (MD). PMID 28520351.

Moriyama, T., Nishii, R., Perez-Andreu, V., Yang, W., Klussmann, F.A., Zhao, X., et al., 2016. NUDT15 polymorphisms alter thiopurine metabolism and hematopoietic toxicity. Nat. Genet. 48, 367–373. PMID 26878724.

Relling, M.V., Schwab, M., Whirl-Carrillo, M., Suarez-Kurtz, G., Pui, C.H., Stein, C.M., et al., 2019. Clinical Pharmacogenetics Implementation Consortium guideline for thiopurine dosing based on TPMT and NUDT15 genotypes: 2018 Update. Clin. Pharmacol. Ther. 105, 1095–1105. PMID 30447069.

Section 11.16

Cancer Cases

Case 11.16.1

Cancer Cases: Breast Cancer Gene Mutation Found Incidentally

Audrey L. Fan
Cancer Genetics and Genomics, Stanford Healthcare, Palo Alto, CA, United States

Abstract
A 32-year-old woman is found to have a *BRCA2* pathogenic variant as a secondary finding on exome sequencing.

The Case

A 32-year-old woman named Janet schedules a same-day appointment with you, her primary care physician. At the appointment, she tells you that she had recently been notified that she has a "BRCA mutation" and is worried about her risk for cancer. Janet shows you a letter and test results that were sent to her. The letter explains that she had participated in a community health research initiative and contributed a blood sample for research analysis. The research had included exome sequencing (Section 6.3.1) and, during the consent process, she had agreed to be notified if the research study identified "medically actionable" results (Section 7.2). The study had attempted to call her by phone, but she had moved out of state, and they had not been able to reach her by phone. They had subsequently sent her this letter and genetic testing report by certified mail.

The Issues

- **Why is this important information?** You tell her that both women and men who carry a germline *BRCA1* or *BRCA2* pathogenic variant or mutation (Section 7.1.3) have an elevated lifetime risk of developing certain cancers (Section 8.1.2). Since many of these cancers (e.g., breast, ovarian, prostate) can be either prevented or detected at an early, more treatable stage, carriers of a *BRCA1* or *BRCA2* pathogenic variant should follow a screening regimen.
- **Is the patient reporting her genetic test result accurately?** Many patients may report that they had a genetic test but are confused about the type of testing, or the significance of results. For example, genomic profiling of a patient's tumor, which may detect somatic variants (Section 6.4.4), does not always carry direct implications for the patient's family members. Fortunately, Janet brought a copy of her test report and supporting documentation.
- **Should you and the patient be concerned about the result's validity?** Janet says that she has **no** family history of breast or ovarian cancer and she wonders if the test result is incorrect. She asks you, "Maybe they got my result wrong? Or maybe these results don't apply to me?" When patients report that they had genetic testing with an outside provider, it is important to review a copy of the results. It is appropriate to recommend that the test results be reviewed by a medical professional prior to any follow-up clinical actions being taken (Section 9.1.7). You refer her for a genetic evaluation and instruct Janet to bring her results to the genetic appointment.

Genomics in the Clinic. https://doi.org/10.1016/B978-0-12-816478-5.00098-4

The Referral

- **Genetic Evaluation**: The cancer genetic counselor takes a targeted family history (Section 3.2) in order to discuss these results in the context of this information and her family history. It is important that the methodology for the release of results be reviewed. Some research studies may perform genetic analysis and return results without confirmatory testing by a CLIA-certified lab. Promptly ascertaining CLIA confirmation for these results is recommended before basing clinical actions on these results. Prior to Janet's appointment, the genetic counselor had reviewed the test results and confirmed that the research study had ordered clinical genetic testing through a CLIA-certified laboratory.
- **Genetic Counseling**: The genetic counselor explains that the *BRCA2* variant is associated with a condition called "Hereditary Breast and Ovarian Cancer syndrome" (see also Section 11.16.2). On review of the database ClinVar, multiple labs have classified the identified *BRCA2* variant in Janet as "pathogenic" (Section 7.1.3). Thus she is at increased lifetime risk for both breast and ovarian cancer, as well as several others (e.g., pancreas, melanoma). She should be referred to a high-risk cancer screening program.

  The absence of a family history could be because she has a de novo variant, or there is variable expressivity or incomplete penetrance (Section 2.8). It is possible for several women in a family to carry the same *BRCA2* variant without having the same personal history of cancer. This variant is associated with younger ages of onset of breast and ovarian cancer (Section 8.1.2), but not all female BRCA variant carriers will be diagnosed with cancer. It is possible that Janet has female relatives who were at increased risk to develop these cancers, but passed away from non-cancer-related causes prior to developing them. It is also possible that Janet inherited this variant from her father, and that her closest relatives who are carriers are all male. Since the most significant BRCA-related cancer risks are breast and ovarian cancer, most men who carry a BRCA variant will not develop cancer. In fact, when the genetic counselor collects and analyzes Janet's family history, she points out that Janet's father did not have any sisters, and that her paternal grandmother was an only child and died at a young age. It is possible that Janet has no family history of breast or ovarian cancer because there have been no (or few) female BRCA variant carriers in her family.

  Since a BRCA variant is associated with adult-onset cancers in men and women, it is important that family members—especially all first-degree relatives (e.g., siblings or parents, who each have a 50% chance to carry this variant)—should be notified. The genetic counselor can provide assistance as how to speak with and help family members (Section 7.4.3).
- **Management and Recommendations**: Janet reports that she heard that women who have a BRCA mutation will have a radical surgical procedure to remove their healthy breasts (a bilateral prophylactic mastectomy). The genetic counselor confirms that this is one risk-reduction option available to her, but most women in her situation do not choose to do that immediately. It may not feel like it, but she does not have to make this type of decision quickly. For most newly identified female BRCA variant carriers, an immediate referral to a breast surgeon to discuss a prophylactic mastectomy is **not** indicated. Unless the patient requests otherwise, the most important first step is to implement and educate the patient about the appropriate breast screening protocol. Since she does not have any symptoms of breast cancer, her first step will be a breast screening protocol (Section 8.3.1).

  Initiate Screening Protocol: Guidelines to review appropriate risk-reducing and preventative options are published and updated regularly by the National Comprehensive Cancer Network (NCCN). For a 32-year-old asymptomatic BRCA mutation carrier, the following referrals/recommendations should be made:
  - Starting from age 30, an annual breast mammogram as well as an annual breast MRI
  - If patient has completed childbearing, consideration of hysterectomy with bilateral salpingo-oophorectomy to reduce ovarian cancer risk
  - Referral to a genetic counselor to review results and facilitate testing for at-risk family members.

  Janet is referred to a breast surgeon to discuss the above recommendations.
- **Implications for Other Family Members**: Janet is concerned that her 8-year-old daughter or 11-year-old son may develop cancer too. She would like to have her two children tested. You acknowledge and explain that each child has a 50% chance to also carry the pathogenic variant, but share that, currently, testing children for adult-onset conditions is not recommended (Sections 10.1.2, 11.17.1, 11.18.1). The Committee on Bioethics of the American Academy of Pediatrics advises against genetic testing of children for conditions that have their onset in adulthood: "Genetic testing for adult-onset conditions generally should be deferred until adulthood or until an adolescent interested in testing has developed mature decision-making capacities." The genetic counselor also explains that the cancers associated with a *BRCA* pathogenic variant are adult-onset cancers, and there are currently no risk-reducing medical interventions that would be recommended for children.

Additionally, the genetic counselor points out that her daughter may not want to know her own carrier status until she is at least in her 20s (the youngest age at which it may be appropriate to consider certain cancer screenings). Often patients in their late teens or early 20s seek BRCA testing because they want to be proactive. However, the genetic counselor says, "since some people feel empowered and proactive because they are taking steps to reduce their risk, I ask them to imagine how it feels to know that they are at high risk, but waiting 5–7 years before their first cancer screening. More than a few women decide to postpone testing." See also Section 11.18.1 for additional discussion.

Janet's two brothers may be also at an increased risk to develop early-onset prostate cancer or even breast cancer. Each of her brothers has a 50% chance also to be a carrier. The genetic counselor says that she would be happy to see other family members to help coordinate single site testing (Section 6.2.1) or refer them to a local cancer genetic counselor.

- **Caveats**: It is possible that Janet will never develop any cancers; this variant gives her an increased risk for developing it. There are risk models (Sections 8.2, 8.3.1) that can help predict this increased risk over the genetic population.
- **Outcome**: Janet meets with the breast surgeon and decides to be followed by annual mammograms and MRIs. She also meets with a gynecologic surgeon because she is interested in a prophylactic hysterectomy with bilateral salpingo-oophorectomy since she plans to have no further pregnancies (Section 11.16.2). Her brothers are referred for genetic counseling and one of them is found to have the *BRCA2* variant.

The Diagnosis

Clinical Diagnosis: Hereditary breast and ovarian cancer syndrome.
Genetic Diagnosis: *BRCA2* gene, heterozygous pathogenic variant—incidental finding.

Further Reading

Caga-Anan, E.C., Smith, L., Sharp, R.R., Lantos, J.D., 2012. Testing children for adult-onset genetic diseases. Pediatrics 129, 163–167. PMID 22144703.

NCCN Clinical Practice Guidelines in Oncology (NCCN Guidelines®) for Detection, Prevention, and Risk Reduction: https://www.nccn.org/guidelines/category_2.

Case 11.16.2

Cancer Cases: Ovarian Cancer Family History

Rebekah A. Moore[1,*], Takae M. Brewer[1], Joanne Ngeow[2,3] and Charis Eng[4,5]

[1]*Genomic Medicine Institute, Cleveland Clinic, Cleveland, OH, United States;* [2]*Cancer Genetics Service, Division of Medical Oncology, National Cancer Centre Singapore, Singapore;* [3]*Lee Kong Chian School of Medicine, Nanyang Technological University, Singapore;* [4]*Genomic Medicine Institute, Center for Personalized Genetic Healthcare, Cleveland Clinic Lerner Research Institute, and Taussig Cancer Institute, Cleveland, OH, United States;* [5]*Department of Genetics and Genome Sciences, Germline High Risk Focus Group, Case Comprehensive Cancer Center, Case Western Reserve University School of Medicine, Cleveland, OH, United States*

Abstract

A 36-year-old woman with family history of ovarian cancer and a known *BRIP1* pathogenic variant pursues site-specific, targeted gene testing and is found to be negative.

The Case

Saundra, a 36-year-old female, comes to see you, her gynecologist, for an annual pelvic exam. She expresses no health concerns or changes to her medical history. She had previously reported a family history of ovarian cancer diagnosed in her mother last year when she was age 62 years, ovarian cancer diagnosed in her maternal grandmother in her late 60s, and basal cell carcinoma diagnosed in her father in his 50s and 60s. You had previously discussed the family history of ovarian cancer (Section 8.1.2) with Saundra, who reported that her mother would be pursuing genetic testing for hereditary cancer syndromes as part of her cancer workup and treatment plan in her home state across the country. When reviewing her family history at today's visit, Saundra reports that her mother did undergo genetic testing last year, which was positive. Saundra does not have a copy of the report with her today, but believes her mother tested positive for a pathogenic variant in the *BRCA1* gene. She says she would like to pursue genetic testing for the familial mutation.

The Issues

- **Does Saundra need to provide a copy of her mother's genetic test results before pursuing genetic testing?** Yes, a copy of Saundra's mother's genetic test report is needed to help coordinate genetic testing for Saundra to confirm the

[*] Author RM was employed by Cleveland Clinic at the time of writing the manuscript, but is currently employed by AstraZeneca Pharmaceuticals.

Genomics in the Clinic. https://doi.org/10.1016/B978-0-12-816478-5.00099-6

pathogenic variant in the family and which laboratory performed the testing. Often patients assume the gene is *BRCA1* because they hear about this gene in the media.

- **What other topics should be discussed?** As Saundra is probably aware based on her mother's recent diagnosis of ovarian cancer, screening options for ovarian cancer are limited. You review with Saundra that while she has a history of normal pap smears, this is only reassuring for her cervical cancer risk. If Saundra also carries the familial mutation, then recommended management will be risk-reducing salpingo-oophorectomy (RRSO) (Section 8.3.1). You refer her for cancer genetic counseling (Section 8.2).

The Referral

- **Genetic Evaluation**: Saundra is able to obtain a copy of her mother's genetic test results prior to her genetic counseling appointment. A three-generation family history (Sections 2.7, 3.2.1) is obtained during the visit, and no additional risk factors are noted on her father's side. Her medical history is negative. When reviewing the mother's results during the appointment, the genetic counselor points out that the patient's mother tested positive for a *BRIP1* mutation and was negative for variants in *BRCA1*, *BRCA2*, and other genes included on the multigene panel (Sections 6.2.3, 8.2.1).
- **Genetic Counseling**: The genetic counselor discusses that Saundra still has a 50% chance to carry the familial variant (Section 2.7.1), but *BRIP1* mutations are associated with different cancer risks than *BRCA1/2* mutations. Pathogenic variants in *BRIP1* are associated with a 5.8% lifetime risk of developing ovarian cancer by age 80. Since Saundra's mother pursued a multigene panel (Section 6.2.3) and only carries a pathogenic variant in *BRIP1* and Saundra's paternal family history is not concerning for a hereditary cancer syndrome, Saundra can undergo site-specific, targeted gene testing (Section 6.2.1) for her mother's pathogenic *BRIP1* variant and does not need to pursue a multigene panel herself. The genetic counselor reviews the Genetic Information Nondiscrimination Act of 2008 (GINA) with the patient before coordinating testing. There is further discussion about insurance discrimination and GINA in Section 10.2.
- **Genetic Testing**: Saundra elects to pursue site-specific testing for the familial *BRIP1* pathogenic variant.
- **Management and Recommendations**: Due to the increased risk of ovarian cancer and lack of effective ovarian cancer screening options, National Comprehensive Cancer Network guidelines currently recommend RRSO between ages 40–45 years. Currently, there is insufficient evidence to determine if there is also an increased risk of breast cancer, and there are no increased breast screening recommendations based on *BRIP1* mutation status.
- **Implications for Other Family Members**: All first-degree relatives of Saundra's mother are at a 50% risk for carrying the same *BRIP1* mutation. Genetic counseling should be obtained so that they can understand their risks and options for testing. If Saundra tests positive, her three children will also be at 50% risk, but testing children for adult-onset conditions is not currently recommended until they reach age 18 years or at the age at which increased surveillance would be recommended (Sections 10.1.2, 11.16.1).
- **Outcome**: The test does not identify the familial *BRIP1* mutation, and Saundra is recommended to follow general population screening guidelines. Her sister and maternal aunt have the same test and are found to be negative.

The Diagnosis

Clinical Diagnosis: Family history of ovarian cancer with *BRIP1* pathogenic variant.
Genetic Diagnosis: *BRIP1* gene, pathogenic variant negative.

Further Reading

Ramus, S.J., Song, H., Dicks, E., Tyrer, J.P., Rosenthal, A.N., Intermaggio, M.P., et al., 2015. Germline mutations in the BRIP1, BARD1, PALB2, and NBN genes in women with ovarian cancer. J. Natl. Cancer Inst. 107, djv214. PMID 26315354.

National Comprehensive Cancer Network, 2022. NCCN Clinical Guidelines in Oncology (NCCN Guidelines): Genetic/Familial High-Risk Assessment: Breast, Ovarian, and Pancreatic. Version 2.2022-March 9, 2022. [Online]. https://www.nccn.org/professionals/physician_gls/pdf/genetics_bop.pdf.

Case 11.16.3

Cancer Cases: Family History of Lynch Syndrome

Takae M. Brewer[1], Rebekah Moore[1,*], Joanne Ngeow[2,3] and Charis Eng[4,5]

[1]*Genomic Medicine Institute, Cleveland Clinic, Cleveland, OH, United States;* [2]*Cancer Genetics Service, Division of Medical Oncology, National Cancer Centre Singapore, Singapore;* [3]*Lee Kong Chian School of Medicine, Nanyang Technological University, Singapore;* [4]*Genomic Medicine Institute, Center for Personalized Genetic Healthcare, Cleveland Clinic Lerner Research Institute, and Taussig Cancer Institute, Cleveland, OH, United States;* [5]*Department of Genetics and Genome Sciences, Germline High Risk Focus Group, Case Comprehensive Cancer Center, Case Western Reserve University School of Medicine, Cleveland, OH, United States*

Abstract

A 25-year-old female is found to have Lynch syndrome due to a familial pathogenic variant in *MSH2*, a mismatch repair gene.

The Case

Debbie, a 25-year-old woman, comes to your clinic to establish care with you as a primary care physician. She has recently moved from another state due to her husband's new job. She has been healthy without any comorbidities. Debbie is currently enrolled in graduate school and would like to start a family once she receives her master's degree. While providing her family history, Debbie reports multiple relatives with cancer, including her mother with uterine cancer diagnosed at age 48, a maternal uncle with colon cancer diagnosed in his early 40s, and maternal grandmother with colon cancer diagnosed in her mid-40s. Debbie tells you her mother had genetic testing done which came back "positive for Lynch syndrome." Debbie expresses her concerns about her own risk of developing cancer in the light of her family history.

The Issues

- **Is Debbie's mother having Lynch syndrome important?** Yes. Lynch syndrome (LS) is inherited in an autosomal dominant fashion (Section 2.7.1). If this result is confirmed, Debbie has a 50% chance of carrying the same LS-associated gene mutation as her mother. If Debbie is found to be positive for the familial pathogenic variant or mutation, enhanced surveillance for LS-associated cancers would be recommended.

[*] Author RM was employed by Cleveland Clinic at the time of writing the manuscript, but is currently employed by AstraZeneca Pharmaceuticals.

Genomics in the Clinic. https://doi.org/10.1016/B978-0-12-816478-5.00100-X

- **How is her family history of cancer important?** Debbie's family history is highly suggestive of LS even without knowledge of the mutation-positive status of her mother, based on the Amsterdam II criteria (Table 11.16.3.1), which identifies individuals at-risk for LS based on family history. Debbie's mother developed uterine cancer at age 48, which is one of the LS-associated tumors. Furthermore, Debbie has a maternal grandmother and an uncle who were diagnosed with colorectal cancer (CRC) (Section 8.1.3) in their 40s.
- **What other pertinent questions should you ask Debbie?** It is important to confirm her mother's positive genetic testing results and clarify what mutation has been identified so that the genetic specialist can review all the pertinent information. Debbie should be instructed to obtain a copy of her mother's genetic testing results and bring it to the visit with a cancer genetic specialist, where Debbie's concerns about her cancer risk will be addressed.

The Referral

- **Genetic Evaluation:** Prior to an appointment with a genetic counselor, Debbie obtains a copy of her mother's genetic testing results, which confirms that Debbie's mother has a pathogenic mutation (Section 7.1.3) in *MSH2*. The genetic counselor obtains a three-generation pedigree (Sections 2.7, 3.2.1) from Debbie, which reveals no cancers on her father's side. Based on this information, the genetic counselor recommends that Debbie undergo *MSH2* site-specific testing (Section 6.2.1), which tests for the same mutation detected in Debbie's mother.
- **Genetic Counseling:** The genetic counselor informs Debbie that LS (Section 8.1.3) is associated with an increased risk of multiple types of cancer in both men and women. The cancer risk, however, may vary, depending on which gene is involved. LS is caused by germline mutations (Section 2.4.1) in any of the LS-associated DNA mismatch repair genes (*MLH1*, *MSH2*, *MSH6*, *PMS2*) and the *EPCAM* gene. In general, the CRC risk ranges from 8.7% to 61% by the age of 80 (general population risk is 4.2%). The risk of developing endometrial cancer by age 80 is approximately 13%–57% (general population risk is 3.1%). The risks of developing other LS-associated tumors are also elevated compared to the general population.

  Debbie mentions her desire to start a family soon. While most LS is caused by a mutation in one allele (monoallelic), constitutional mismatch repair deficiency syndrome (CMMRD) is caused by a biallelic mutation in a LS-associated gene (a mutation in the same gene on each allele, one inherited from each parent). CMMRD is a rare, autosomal recessive (Section 2.7.1) disorder which is associated with childhood leukemias, lymphomas, brain tumors, intestinal polyposis, and colorectal adenocarcinomas. If Debbie is positive, and if her husband also carries an *MSH2* mutation, each of their future children would have a 50% chance of inheriting a monoallelic mutation (thus having LS), a 25% chance of a biallelic mutation inherited from both parent (thus having CMMRD), and a 25% chance of not inheriting either *MSH2* mutation (thus, being at the general population risk to develop cancer).
- **Results:** Debbie consents to the testing, which comes back positive for the same pathogenic variant (Section 7.1.3) in *MSH2* detected in her mother, establishing her diagnosis of LS. Thus, Debbie should follow surveillance and preventative recommendations based on her mutation-positive status in *MSH2* (Section 8.3.3). Genetic counseling and consideration of genetic testing is recommended for Debbie's husband in order to clarify their future children's risk for CMMRD.
- **Management and Recommendations:** Individuals with LS are predisposed to multiple types of cancer, and screening strategies for this high-risk group are often complex (Sections 8.1.3, 8.3.3). Patients with LS like Debbie should be

TABLE 11.16.3.1 Amsterdam II criteria.

At least three relatives with Lynch-associated cancers and the following criteria must be met:
1. One should be a first-degree relative of the other two
2. At least two successive generations should be affected
3. At least one colorectal cancer should be diagnosed before age 50
4. Familial adenomatous polyposis excluded
5. Tumors should be verified by pathological examination

Adopted from Vasen, H. F., Watson, P., Mecklin, J. P. & Lynch, H. T. (1999). New clinical criteria for hereditary nonpolyposis colorectal cancer (HNPCC, Lynch syndrome) proposed by the International Collaborative group on HNPCC. *Gastroenterology, 116*, 1453–1456. PMID 10348829.

referred to appropriate specialists including a gastroenterologist as well as a gynecologist, a urologist, and a neurologist, who are familiar with screening guidelines and managements for LS. Risk-reducing salpingo-oophrectomy is recommended for women who have completed childbearing and should be individualized based on multiple factors including the type of mutations found.

- **Implications for Other Family Members:** It is important that all at-risk relatives be informed of the recommendation to be tested for LS. Since the mutation was inherited from Debbie's mother, Debbie's siblings, maternal aunts, and maternal uncles are also at risk of carrying the familial *MSH2* mutation (Section 2.7.1) and testing for this specific mutation is strongly recommended for them to determine if they also have an increased risk for LS-associated cancers. Testing Debbie's maternal aunts and uncles will also clarify which of her maternal cousins are at risk for the familial *MSH2* mutation.
- **Caveats:** If both paternal and maternal sides of the family have significant family history of cancer, then a multigene panel (Section 6.2.3) may be a better choice than site-specific testing (Section 6.2.1). While site-specific testing is appropriate for confirming the familial mutation, which may be inherited from one parent, a multigene panel may detect other pathogenic mutations which may have come from the other parent.
- **Outcome:** Debbie meets with high-risk specialists including a gynecologist and a gastroenterologist. She undergoes a colonoscopy during the following month, which is normal. Debbie's husband requests his primary care physician to place a referral to genetic counseling. He is scheduled to see a genetic counselor for his own testing.

The Diagnosis

Clinical Diagnosis: Family history of Lynch syndrome with uterine cancer and *MSH2* pathogenic variant.
Genetic Diagnosis: *MSH2* gene, heterozygous pathogenic variant.

Further Reading

Bonadona, V., Bonaiti, B., Olschwang, S., Grandjouan, S., Huiart, L., Longy, M., Huiart, L., Longy, M., Guimbaud, R., Buecher, B., Bignon, Y.J., Caron, O., Colas, C., Noguès, C., Lejeune-Dumoulin, S., Olivier-Faivre, L., Polycarpe-Osaer, F., Nguyen, T.D., Desseigne, F., Saurin, J.C., Berthet, P., Leroux, D., Duffour, J., Manouvrier, S., Frébourg, T., Sobol, H., Lasset, C., Bonaïti-Pellié, C.; French Cancer Genetics Network, 2011. Cancer risks associated with germline mutations in MLH1, MSH2, and MSH6 genes in Lynch syndrome. JAMA 305, 2304–2310. PMID: 21642682.

Moller, P., Seppala, T., Bernstein, I., Holinski-Feder, E., Sala, P., Evans, D.G., Lindblom, A., Macrae, F., Blanco, I., Sijmons, R., Jeffries, J., Vasen, H., Burn, J., Nakken, S., Hovig, E., Rødland, E.A., Tharmaratnam, K., de Vos Tot Nederveen Cappel, W.H., Hill, J., Wijnen, J., Jenkins, M., Green, K., Lalloo, F., Sunde, L., Mints, M., Bertario, L., Pineda, M., Navarro, M., Morak, M., Renkonen-Sinisalo, L., Frayling, I.M., Plazzer, J.P., Pylvanainen, K., Genuardi, M., Mecklin, J.P., Möslein, G., Sampson, J.R., Capella, G.; Mallorca Group, 2018. Cancer risk and survival in path_MMR carriers by gene and gender up to 75 years of age: a report from the Prospective Lynch Syndrome Database. Gut 67 (7), 1306–1316. PMID: 28754778.

NCCN, National Comprehensive Cancer Network, 2022. NCCN Clinical Practice Guidelines in Oncology (NCCN Guidelines). Genetic/Familial High-Risk Assessment: Colorectal. Version 1.2022 - June 8, 2022., 1.2022. https://www.nccn.org/professionals/physician_gls/pdf/genetics_colon.pdf.

Rana, H., Syngal, S., 2017. Biallelic mismatch repair deficiency: management and prevention of a devastating manifestation of the Lynch Syndrome. Gastroenterology 152 (6), 1254–1257. PMID: 28363489.

Case 11.16.4

Cancer Cases: Prostate Cancer

Takae M. Brewer[1], **Rebekah Moore**[1,*], **Joanne Ngeow**[2,3] **and Charis Eng**[4,5]

[1]Genomic Medicine Institute, Cleveland Clinic, Cleveland, OH, United States; [2]Cancer Genetics Service, Division of Medical Oncology, National Cancer Centre Singapore, Singapore; [3]Lee Kong Chian School of Medicine, Nanyang Technological University, Singapore; [4]Genomic Medicine Institute, Center for Personalized Genetic Healthcare, Cleveland Clinic Lerner Research Institute, and Taussig Cancer Institute, Cleveland, OH, United States; [5]Department of Genetics and Genome Sciences, Germline High Risk Focus Group, Case Comprehensive Cancer Center, Case Western Reserve University School of Medicine, Cleveland, OH, United States

Abstract

A 54-year-old male with metastatic prostate cancer is found to have a *BRCA1* pathogenic variant.

The Case

Brian, a 54-year-old man, comes to see you, his primary care physician, for a routine annual physical exam. About 5 months ago, he was diagnosed with high-grade prostate cancer (Gleason score of 9 with a higher risk for aggressive nature of the tumor), which was treated with radiation therapy and androgen deprivation therapy (ADT). His prostate-specific antigen (PSA) had been barely detectable since he completed the initial treatment. Unfortunately, his PSA started to increase about a month ago, and the subsequent radiographic workup revealed metastatic lesions in his lumbar spine. He currently continues ADT and is receiving zoledronic acid, a therapeutic agent specific for bone metastases. He reports occasional hot flashes and mild fatigue but is otherwise doing well. You ask about family history. Brian reports his mother had breast cancer at age of 42 and eventually died from it. He has four full brothers and two full sisters. His oldest brother has a personal history of prostate cancer, but is currently doing well. He is of Black/African American ancestry.

The Issues

- **Is his personal history of prostate cancer significant?** The National Cancer Comprehensive Network (NCCN) recommends patients who develop high-grade (Gleason score greater than 7) or metastatic prostate cancer at any age should be referred to genetic counseling and testing. There are several genes, which predispose mutation-positive individuals (Section 2.4.1) to developing prostate cancer. Approximately 12% of metastatic prostate cancer diagnoses are estimated to be due to variants in prostate cancer predisposition genes compared to 4.6% of localized prostate cancer.

* Author RM was employed by Cleveland Clinic at the time of writing the manuscript, but is currently employed by AstraZeneca Pharmaceuticals.

Genomics in the Clinic. https://doi.org/10.1016/B978-0-12-816478-5.00101-1

- **Is family history of prostate cancer important?** Yes. Strong familial clustering and association of prostate cancer risk exists. Among patients with family history of prostate cancer, the highest association has been observed when first-degree relatives are affected and if diagnosed at younger than 65 years of age. Brian has multiple risk factors including his Black/African American ancestry, young age, and having a first-degree relative (brother) with prostate cancer. However, any patient with high-grade or metastatic prostate cancer should be referred to genetics regardless of family history (Section 8.4.3).
- **How should you educate Brian before referring him to genetics?** Brian should be made aware that the incidence of metastatic prostate cancer is increased in individuals with certain hereditary cancer syndromes. He will benefit from genetic testing because identifying pathogenic germline mutations (Section 7.1.3) may have therapeutic implications for his metastatic prostate cancer. Positive results also may impact his close blood relatives, especially his children and full siblings. You refer him for cancer genetic counseling and send the information about his family history ahead of time.

The Referral

- **Genetic Evaluation:** A three-generation pedigree (Sections 2.7, 3.2.1) is obtained. In addition to the individuals with cancer as relayed, Brian has two healthy sons and a daughter who are in their late 20s. Brian is concerned that his sons might be at a higher risk of developing prostate cancer in the future. The genetic counselor tells Brian that he is eligible for genetic testing because of his personal history of metastatic prostate cancer.
- **Genetic Counseling:** The genetic counselor explains to Brian that there are several genes, which predispose mutation-positive individuals to prostate cancer, including *BRCA1* and *BRCA2* (Section 8.1.2). If any mutations in these genes are identified, the information should be shared with his medical oncologist. Since multiple genes exist which are known to increase the risk for prostate cancer, multigene panel testing (Section 6.2.3) is often appropriate unless there is a known pathogenic mutation in the family. Among many candidate susceptibility genes, a typical prostate cancer multigene panel includes: *BRCA1*, *BRCA2*, DNA mismatch repair (MMR) genes (*MLH1*, *MSH2*, *MSH6*, *PMS2*), *EPCAM*, *HOXB13*, *TP53*, *ATM*, *CHEK2*, and *NBN*.

  The genetic counselor explains to Brian that these cancer genes are inherited in an autosomal dominant pattern (Section 2.7.1). Thus, if Brian is found to carry a pathogenic mutation (Section 7.1.3), each of his children (regardless of gender) has a 50% chance of having inherited the same mutation. The genetic counselor informs Brian that he should share genetic testing information with his children and close relatives especially if his results come back positive. The genetic counselor also reviews the Genetic Information Nondiscrimination Act of 2008 (GINA) with the patient before coordinating testing. There is further discussion about insurance discrimination and GINA in Section 10.2.
- **Genetic testing:** Brian expresses his understanding and decides to pursue genetic testing. He consents for a multigene cancer panel and a peripheral blood specimen is sent for analysis.
- **Results:** The multigene panel results come back positive for a pathogenic mutation in *BRCA1*. The genetic counselor tells Brian that this information will be shared with his medical oncologist for potential use of certain chemotherapy regimens and/or targeted therapy such as PARP inhibitors (a group of pharmacological inhibitors of the enzyme poly ADP ribose polymerase), if indicated. Brian remembers that the genetic counselor mentioned his children being potentially at risk of having the same mutation and asks for advice how to communicate with them. The genetic counselor writes up a family-share letter for his at-risk family members (Section 7.4.3), including his sons, daughter, and siblings to encourage them to be referred to the genetics for evaluation.
- **Management and Recommendations:** Brian should share his genetic test results with his oncologist and cancer treatment team to determine if additional chemotherapy or systemic treatments, including PARP inhibitors, are an appropriate options. Prostate cancer screening is important for men who test positive for a pathogenic variant in one of the prostate cancer predisposition genes. For *BRCA1* mutation-positive individuals, NCCN recommends prostate cancer screening be *considered* starting at age 45. NCCN guidelines recommend prostate cancer screening for *BRCA2* mutation-positive individuals (Section 8.3.1) starting at age 45. For this group, the Philadelphia consensus panel 2017 recommends a baseline PSA at age 40, or 10 years earlier than the youngest prostate cancer diagnosed in the family. For mutation-negative men, prostate cancer screening may be still indicated especially for those with African American ancestry and with a family history of prostate cancer. The Advanced Prostate Cancer Consensus Conference (APCCC) 2017 by the European Association of Urology reached a strong consensus (92%) that prophylactic prostatectomy is not to be performed in the presence of a germline *BRCA1*, *BRCA2*, or *ATM* mutation.

Some expert groups recommend expanding PSA screening beyond BRCA1/2 mutation-positive individuals. The Philadelphia consensus panel 2017 recommends, as outlined by NCCN Prostate Early Detection guideline, that HOXB13-positive individuals should get a baseline PSA at age 40 years or 10 years earlier than the youngest prostate cancer diagnosis in family.

For women with a *BRCA2* mutation (Section 8.3.1), breast cancer screening starts at age 25 with annual breast magnetic resonance imaging (MRI) with addition of annual mammograms at age 30. Discussion of risk-reducing mastectomy should be individualized. Furthermore, risk-reducing bilateral salpingo-oophorectomy (RRSO) is typically recommended for *BRCA2* mutation-positive women for ovarian cancer prevention. Hysterectomy may be performed at the time of RRSO, which may be more important for *BRCA1* mutation-positive individuals. In the case where hysterectomy is performed along with RRSO, low-dose estrogen therapy, if indicated, should be safe. A high-risk gynecologic oncology specialist should be involved to discuss further the risks, benefits, and timing of RRSO, which is typically after women complete childbearing.

- **Implications for Other Family Members:** Since Brian is found to carry a pathogenic mutation, his sons and daughter also should be referred for genetic counseling for a risk assessment and discussion of appropriate genetic testing options. If his sons test negative, this means they have informative negative results, which may decrease their risk of developing prostate cancer. They should discuss risks and benefits of prostate cancer screening with their primary care physicians.
- **Caveats:** It should be noted that if no pathogenic variants had been identified on Brian's genetic testing, his sons still have an increased risk of developing prostate cancer compared to the general population because of their African American ancestry and family history of a first-degree relative with prostate cancer.
- **Outcome:** Brian's children contact Brian's genetic counselor. They state they have referrals to genetics from their primary care physicians and would like to pursue genetic testing. After meeting with the genetic counselor for pretest counseling, a site-specific, targeted genetic test (Section 6.2.1) in *BRCA1* (the same mutation detected in Brian) is ordered. One son's test is negative, while the tests for the other son as well as his daughter are positive for the pathogenic variant in *BRCA1*. Brian's children make their primary care physicians aware of their genetic testing results as well as their father's diagnosis of prostate cancer. Brian's son with the informative negative result for the *BRCA1* mutation may still have a higher risk for prostate cancer compared to the general population. Brian's son, who tested mutation positive, should be encouraged to participate in a prostate cancer screening program when age-appropriate. Brian's daughter is scheduled to meet a high-risk breast specialist and gynecologist to discuss increased breast and ovarian cancer surveillance (Sections 8.3.1, 11.16.1 and 11.16.2), and possible prophylactic surgery.

The Diagnosis

Clinical Diagnosis: Patient—hereditary breast and ovarian cancer syndrome with prostate cancer.
Genetic Diagnosis: Patient, Daughter, and a Son—*BRCA1* gene, heterozygous pathogenic variant.

Further Reading

Gillessen, S., Attard, G., Beer, T., Beltran, H., Bjartell, A., Bossi, A., Blanchard, P., Briganti, A., Bristow, R.G., Bulbul, M., Caffo, O., Castellano, D., Castro, E., Cheng, H.H., Chi, K.N., Chowdhury, S., Clarke, C.S., Clarke, N., Daugaard, G., De Santis, M., Duran, I., Eeles, R., Efstathiou, E., Ekeke, O.N., Evans, C.P., Fanti, S., Feng, F.Y., Fonteyne, V., Fossati, N., Frydenberg, M., George, D., Gleave, M., Gravis, G., Halabi, S., Heinrich, D., Herrmann, K., Higano, C., Hofman, M.S., Horvath, L.G., Hussain, M., Jereczek-Fossa, B.A., Jones, R., Kanesvaran, R., Kellokumpu-Lehtinen, P.-L., Khauli, R.B., Klotz, L., Kramer, G., Leibowitz, R., Logothetis, C., Mahal, B., Maluf, F., Mateo, J., Matheson, D., Mehra, N., Merseburger, A., Morgans, A.K., Morris, M.J, Mrabti, H., Mukherji, D., Murphy, D.G., Murthy, V., Nguyen, P.L., Oh, W.K., Ost, P., O'Sullivan, J.M., Padhani, A.R., Pezaro, C.J., Poon, D.M.C., Pritchard, C.C., Rabah, D.M., Rathkopf, D., Reiter, R.E., Rubin, M.A., Ryan, C.J., Saad, F., Sade, J.P., Sartor, O., Scher, H.I., Sharifi, N., Skoneczna, I., Soule, H., Spratt, D.E., Srinivas, S., Sternberg, C.N., Steuber, T., Suzuki, H., Sydes, M.R., Taplin, M.-E., Tilki, D., Turkeri, L., Turco, F., Uemura, H., Uemura, H., Urun, Y., Vale, C.L., van Oort, I., Vapiwala, N., Walz, J., Yamoah, K., Ye, D., Yu, E.Y., Zapatero, A., Zilli, T., Omlin, A.; Advanced Prostate Cancer Consensus Conference APCCC 2017, 2018, 2019. Management of patients with advanced prostate cancer: the report of the advanced prostate cancer consensus conference APCCC 2017. Eur. Urol. 73 (2), 178–211. PMID 32001144.

Kicinski, M., Vangronsveld, J., Nawrot, T., 2011. An epidemiological reappraisal of the familial aggregation of prostate cancer: a meta-analysis. PLoS One 6 (10). PMID 22073129.

Pritchard, C.C., Mateo, J., Walsh, M.F., De Sarkar, N., Abida, W., Beltran, H., Garofalo, A., Gulati, R., Carreira, S., Eeles, R., Elemento, O., Rubin, M.A., Robinson, D., Lonigro, R., Hussain, M., Chinnaiyan, A., Vinson, J., Filipenko, J., Garraway, L., Taplin, M.E., Al Dubayan, S., Han, G.C., Beightol, M., Morrissey, C., Nghiem, B., Cheng, H.H., Montgomery, B., Walsh, T., Casadei, S., Berger, M., Zhang, L., Zehir, A., Vijai, J., Scher, H.I.,

Sawyers, C., Schultz, N., Kantoff, P.W., Solit, D., Robson, M., Van Allen, E.M., Offit, K., de Bono, J., Nelson, P.S., 2016. Inherited DNA-repair gene mutations in men with metastatic prostate cancer. N. Eng. J. Med. 375 (5), 443–453. PMID 27433846.

National Comprehensive Cancer Network, 2022b. NCCN Clinical Practice Guidelines in Oncology (NCCN Guidelines): Prostate Adenocarcinoma. Version 1.2023-September 16, 2022. [Online]. https://www.nccn.org/professionals/physician_gls/pdf/prostate.pdf.

National Comprehensive Cancer Network, 2022a. NCCN Clinical Practice Guidelines in Oncology (NCCN Guidelines): Genetic/Familial High-Risk Assessment: Breast, Ovarian, and Pancreatic. Version 1.2023- September 7, 2022. [Online]. https://www.nccn.org/professionals/physician_gls/pdf/genetics_bop.pdf.

Case 11.16.5

Cancer Cases: Pancreatic Cancer

Takae M. Brewer[1], Rebekah Moore[1,*], Joanne Ngeow[2,3] and Charis Eng[4,5]

[1]*Genomic Medicine Institute, Cleveland Clinic, Cleveland, OH, United States;* [2]*Cancer Genetics Service, Division of Medical Oncology, National Cancer Centre Singapore, Singapore;* [3]*Lee Kong Chian School of Medicine, Nanyang Technological University, Singapore;* [4]*Genomic Medicine Institute, Center for Personalized Genetic Healthcare, Cleveland Clinic Lerner Research Institute, and Taussig Cancer Institute, Cleveland, OH, United States;* [5]*Department of Genetics and Genome Sciences, Germline High Risk Focus Group, Case Comprehensive Cancer Center, Case Western Reserve University School of Medicine, Cleveland, OH, United States*

Abstract

A 55-year-old man with pancreatic cancer, but no known risk factors and family history of cancers, is found to have a pathogenic variant in *BRCA2*.

The Case

Jordan, a 55-year-old man, comes to see you, his primary care physician, for a follow-up visit and to request a medication refill. He had been healthy except for mild hypertension until he was diagnosed with pancreatic cancer about 12 weeks ago. His pancreatic cancer fortunately was considered operable and he underwent a successful Whipple procedure immediately after his diagnosis. Jordan tells you his recovery from surgery was uneventful. He is now receiving adjuvant chemotherapy. He is tolerating the treatment well except for occasional nausea, which is controlled with antiemetics as needed. Jordan tells you he has never smoked and wonders what could have caused his pancreatic cancer. He has two adult children and expresses his desire to stay alive to watch his grandchildren grow up.

The Issues

- **Is his personal history of pancreatic cancer important?** Yes. Approximately 10% of pancreatic cancer is estimated to be heritable. Thus, all patients with a diagnosis of pancreatic cancer are candidates for genetic testing regardless of family history and should be referred to a cancer genetic specialist to discuss this further. Known genes which are associated with an increased risk of pancreatic cancer when mutations are present include *BRCA1*, *BRCA2*, *CDKN2A*, *APC*, Lynch syndrome–associated genes (Sections 8.1.3, 11.16.3), and *STK11*.
- **What other pertinent questions should you ask Jordan?** In addition to inquiring about a family history of pancreatic cancer or other cancers, you should also ask about other possible risk factors for pancreatic cancer. Smoking is a known risk factor for pancreatic cancer, but Jordan is a nonsmoker. Other risk factors for pancreatic cancer include being overweight, diabetes, chronic pancreatitis, and exposure to certain workplace chemicals (dry cleaning and metals

[*] Author RM was employed by Cleveland Clinic at the time of writing the manuscript, but is currently employed by AstraZeneca Pharmaceuticals.

Genomics in the Clinic. https://doi.org/10.1016/B978-0-12-816478-5.00102-3

specifically). It is still recommended that Jordan be referred to a cancer genetic specialist based on his personal history of pancreatic cancer alone, but these risk factors may provide a potential explanation for Jordan's diagnosis if hereditary pancreatic cancer syndromes are ruled out.

- **How should you educate Jordan before referring him to genetics?** Jordan should be made aware that the incidence of pancreatic cancer is increased in certain hereditary cancer syndromes. As mentioned above, any patient with pancreatic cancer meets the criteria for a genetic referral. The genetic specialist will provide comprehensive evaluation and provide genetic counseling to discuss risks and benefits of genetic testing. Positive genetic test results may impact treatment options for the affected individuals (Section 7.4.2) as well as management for close blood relatives (Section 7.4.3). You refer him for cancer genetic counseling.

The Referral

- **Genetic Evaluation:** A three-generation pedigree is obtained (Sections 2.7, 3.2.1). No other family members have a history of any kind of cancer. The genetic counselor tells Jordan that he is eligible for genetic testing based on his personal history of pancreatic cancer.
- **Genetic Counseling:** The genetic counselor explains to Jordan that there are multiple genes which, when mutated, predispose individuals to pancreatic cancer, including *BRCA1* and *BRCA2* (Section 8.1.2). If a pathogenic mutation (Section 7.1.3) is identified, the information should be shared with his medical oncologist. Jordan may benefit from targeted therapy such as PARP inhibitors (a group of pharmacological inhibitors of the enzyme poly ADP ribose polymerase), if he is found to have a germline *BRCA1/2* mutation in the setting of metastatic disease.

  Since multiple genes exist which are known to increase the risk of pancreatic cancer, multigene panel testing (Section 6.2.3) is often appropriate unless there is a known pathogenic mutation in the family. Among many candidate predisposition genes, a typical pancreatic cancer multigene panel includes *BRCA1*, *BRCA2*, *PALB2*, *STK11*, *CDKN2A* (*p16*), Lynch syndrome–associated DNA mismatch repair (MMR) genes (*MLH1*, *MSH2*, *MSH6*, *PMS2*), *EPCAM*, *TP53*, and *ATM*. For those who have history of chronic pancreatitis, which increases the risk of pancreatic cancer, genes such as *CFTR* and *PRSS1* may be also considered.

  The genetic counselor explains to Jordan that most of these cancer genes are inherited in an autosomal dominant pattern (Section 2.7.1), with rare exceptions such as the *CFTR* gene which is inherited in an autosomal recessive pattern (Section 2.7.1). Thus, if Jordan is found to carry a pathogenic mutation, each of his children has most likely a 50% chance of having inherited the same mutation. The genetic counselor informs Jordan that he should share genetic testing information with his children especially if his results come back positive. The genetic counselor also reviews the Genetic Information Nondiscrimination Act of 2008 (GINA) with the patient before coordinating testing. There is further discussion about insurance discrimination and GINA in Section 10.2.
- **Genetic Testing:** Jordan decides to pursue genetic testing. The genetic counselor recommends a multigene cancer panel (Section 6.2.3) which includes pancreatic cancer predisposition genes, for which Jordan consents. A peripheral blood specimen is drawn and sent for analysis in a clinical lab.
- **Results:** Jordan's genetic testing results come back positive for a pathogenic variant in *BRCA2*. The genetic counselor tells Jordan this information will be shared with his medical oncologist since this may influence his treatment options in the future. Jordan expresses his concern about his children who may carry the same *BRCA2* mutation. Jordan states he has a son and a daughter who are both in their late 20s. The genetic counselor writes up a family-share letter (Section 7.4.3) for his son and daughter to encourage them to obtain referrals to a genetic service for evaluation.
- **Management and Recommendations:** The goals for screening high-risk individuals are to detect and treat early-stage pancreatic cancer as well as precancerous lesions. However, there are currently no consensus guidelines regarding screening strategies for the high-risk population. The international Cancer of the Pancreas Screening (CAPS) consortium recommends screening for high-risk individuals. Please refer to the most recent consensus summary by CAPS consortium for details if needed.

  Screening for high-risk individuals should be performed at experienced medical centers which have a multidisciplinary team familiar with surveillance strategies for this unique population. When screening is appropriate, the American College of Gastroenterology guidelines recommend endoscopic ultrasound and/or magnetic resonance imaging be performed annually, starting at 50 years of age, or 10 years younger than the earliest age of pancreatic cancer diagnosis in the affected family. Patients who are not candidates for surgical interventions should not be screened.

  Prophylactic surgery, such as pancreatectomy, is not recommended for asymptomatic individuals without evidence of suspicious lesions or pancreatic abnormalities. There are no biomarker or enzyme tests which can be reliably utilized for screening purposes.

- **Implications for Other Family Members:** Now that Jordan has been found to carry a pancreatic cancer predisposition gene, each of his children has a 50% chance of carrying this same mutation (with some exceptions as discussed below), as do other first-degree relatives. Testing family members can help clarify their risks. Accumulating evidence shows there is an association between pancreatic cancer risk and family history of the disease. Jordan's son and daughter come to see the genetic counselor together. After pretest counseling with the genetic counselor, they decide to pursue a site-specific, targeted mutation test (Section 6.2.1) in *BRCA2* (the same mutation detected in Jordan). His son's results come back negative while his daughter tests positive for the pathogenic mutation in *BRCA2*.
- **Outcome:** Jordan is alive 5 years later. Jordan's children make their primary care physicians aware of their genetic testing results and have a discussion about the risk of and potential screening for pancreatic cancer. Jordan's daughter is positive for the *BRCA2* mutation, giving her a genetic diagnosis of hereditary breast and ovarian cancer (HBOC) syndrome (Sections 8.1.2, 8.3.1, 11.16.1, 11.16.2). Jordan's daughter is referred to see high-risk specialists including a gynecologist and a breast specialist to further discuss management for her HBOC, further screening and preventative recommendations. HBOC increases her risk of not only pancreatic cancer but also breast and ovarian cancers (Section 8.3.1). She should be referred to other specialists as needed. Her partner should also consider genetic counseling and genetic testing based on the possibility of Fanconi anemia, an autosomal recessive condition causeed by biallelic pathogenic variants in *BRCA2*, for their children. As Jordan's son does not carry the familial *BRCA2* mutation, he does not need to pursue additional cancer screening outside of general population recommendations.

The Diagnosis

Clinical Diagnosis: Patient—hereditary breast and ovarian cancer syndrome with pancreatic cancer.
Genetic Diagnosis: Patient and Daughter—*BRCA2* gene, heterozygous pathogenic variant.

Further Reading

Decker, G.A., Batheja, M.J., Collins, J.M., Silva, A.C., Mekeel, K.L., Moss, A.A., et al., 2010. Risk factors for pancreatic adenocarcinoma and prospects for screening. Gastroenterol. Hepatol. (N Y) 6, 246–254. PMID: 20567579.

Goggins, M., Overbeek, K.A., Brand, R., Syngal, S., Del Chiaro, M., Bartsch, D.K., Bassi, C., Carrato, A., Farrell, J., Fishman, E.K., Fockens, P., Gress, T.M., van Hooft, J.E., Hruban, R.H., Kastrinos, F., Klein, A., Lennon, A.M., Lucas, A,. Park, W., Rustgi, A., Simeone, D., Stoffel, E., Vasen, H.F.A., Cahen, D.L., Canto, M.I., Bruno, M.; International Cancer of the Pancreas Screening (CAPS) consortium, 2020. Management of patients with increased risk for familial pancreatic cancer: updated recommendations from the International Cancer of the Pancreas Screening (CAPS) Consortium. Gut 7–17. PMID: 31672839.

National Comprehensive Cancer Network, 2022a. NCCN Clinical Practice Guidelines in Oncology (NCCN Guidelines): Genetic/Familial High-Risk Assessment: Breast, Ovarian, and Pancreatic. Version 1.2023- September 7, 2022. [Online]. https://www.nccn.org/professionals/physician_gls/pdf/genetics_bop.pdf.

National Comprehensive Cancer Network, 2022b. NCCN Clinical Practice Guidelines in Oncology (NCCN Guidelines): Pancreatic Adenocarcinoma. Version 1.2022-February 24, 2022. [Online]. https://www.nccn.org/professionals/physician_gls/pdf/pancreatic.pdf.

Rawla, P., Sunkara, T., Gaduputi, V., 2019. Epidemiology of pancreatic cancer: global trends, etiology and risk factors. World J. Oncol. 10, 10–27. PMID: 30834048.

Syngal, S., Brand, R., Church, J., Giardiello, F., Hampel, H., Burt, R., American College of Gastroenterology, 2015. ACG clinical guideline: genetic testing and management of hereditary gastrointestinal cancer syndromes. Am. J. Gastroenterol. 223–262. PMID: 25645574.

Tersmette, A., Petersen, G.M., Offerhaus, G.J., Falatko, F.C., Brune, K.A., Goggins, M., Rozenblum, E., Wilentz, R.E., Yeo, C.J., Cameron, J.L., Kern, S.E., Hruban, R.H., 2001. Increased risk of incident pancreatic cancer among first-degree relatives of patients with familial pancreatic cancer. Clin. Cancer Res. 7, 738–744. PMID: 11297271.

Case 11.16.6

Cancer Cases: Adrenocortical Carcinoma and Leukemia

Shao-Tzu Li[1], Takae M. Brewer[2], Rebekah Moore[2,*], Charis Eng[3,4] and Joanne Ngeow[1,5]

[1]Cancer Genetics Service, Division of Medical Oncology, National Cancer Centre Singapore, Singapore, Singapore; [2]Genomic Medicine Institute, Cleveland Clinic, Cleveland, OH, United States; [3]Genomic Medicine Institute, Center for Personalized Genetic Healthcare, Cleveland Clinic Lerner Research Institute, and, Taussig Cancer Institute, Cleveland, OH, United States; [4]Department of Genetics and Genome Sciences, Germline High Risk Focus Group, Case Comprehensive Cancer Center, Case Western Reserve University School of Medicine, Cleveland, OH, United States; [5]Lee Kong Chian School of Medicine, Nanyang Technological University Singapore, Singapore

Abstract

A 12-year-old female with adrenocortical carcinoma, acute lymphoblastic leukemia, and strong family history of cancers diagnosed at a young age has Li–Fraumeni syndrome, and genetic testing reveals a *TP53* germline pathogenic variant.

The Case

Lucy is a 12-year-old female, who has achieved all her developmental milestones. She was referred to you, her hematologist, because of her recent diagnosis of acute lymphoblastic leukemia (ALL), which will require a bone marrow transplant. On medical and family history, you note that she was diagnosed with adrenocortical carcinoma at age 4 years for which she underwent surgery and chemotherapy. Her latest abdominal magnetic resonance imaging (MRI) indicates no signs of recurrence or residual tumor. You identify a strong family history of cancer: her father was diagnosed with sarcoma at age 18, two of her paternal aunts were diagnosed with breast cancer at age 31 and age 33, and her paternal grandfather, who was a nonsmoker, died from lung cancer at age 48.

The Issues

- **Does her family history of cancer suggest a leukemia-predisposing condition?** Yes. Her family history is significant for young cancer diagnoses, and Lucy herself has had two primary cancer diagnoses. Genetic testing, in the setting of genetic counseling, is recommended for Lucy and her family (Section 8.1.5).
- **Why is it important to do genetic testing for a leukemia-predisposing condition?** It is critical to identify a leukemia-predisposing condition as it can influence the therapeutic decision-making and donor selection when considering allogeneic hematopoietic stem cell transplantation (allo-HSCT) (Section 8.3.5). Many of the leukemia-

[*] Author RM was employed by Cleveland Clinic at the time of writing the manuscript, but is currently employed by AstraZeneca Pharmaceuticals.

Genomics in the Clinic. https://doi.org/10.1016/B978-0-12-816478-5.00103-5

predisposing conditions are predicted to have poor treatment outcomes because of increased occurrence of treatment-related toxicities and high risk of relapse or refractory disease. You refer Lucy for genetic evaluation.

The Referral

- **Genetic Evaluation:** Full medical and family histories with three-generation pedigree (Sections 2.7, 3.2.1) are obtained by the cancer genetic counselor, and a physical examination is performed by the clinical cancer geneticist. Her physical examination is normal. The family history of young-onset breast cancers, sarcoma, and leukemia fulfills the clinical criteria for classic Li–Fraumeni syndrome.
- **Genetic Counseling:** As an adolescent, Lucy should be involved in the conversations. The geneticist and genetic counselor discuss with Lucy and her parents that her personal and family histories of cancer are suggestive of an underlying genetic condition called Li–Fraumeni syndrome. Li–Fraumeni syndrome is an autosomal dominant (Section 2.7.1) cancer predisposition syndrome caused by germline pathogenic variants in the *TP53* gene. It is characterized by soft tissue and bone sarcomas, early-onset breast cancers, brain tumors, adrenocortical carcinomas, and leukemias. Other less commonly associated cancers are lymphomas, gastrointestinal malignancies, malignant phyllodes tumors, Wilms tumors, melanoma, and lung cancers. Up to 40% of children with low-hypodiploid ALL, defined as leukemia cells containing 32–39 chromosomes, carry a germline pathogenic variant (Section 7.1.3) in *TP53.* As hypodiploid ALL is associated with poor prognosis, allo-HSCT is often recommended. A diagnosis of low-hypodiploid ALL in a child, in addition to suggestive family history, should trigger a referral to genetic specialists and consideration of testing for Li–Fraumeni syndrome.
- **Genetic Testing:** The process of genetic testing is explained to both Lucy and her parents. Either full sequencing of the *TP53* gene alone (Section 6.2.2) or on a multigene panel testing (Section 6.2.3) using next-generation sequencing which includes *TP53* and other leukemia predisposition genes can be considered on performing genetic testing. However, the selection of genes that are associated with adult-onset conditions is discouraged in testing children such as Lucy (Sections 10.1.2, 11.18.1). Genetic testing for adult-onset conditions in children should not be undertaken without medical and psychosocial justification and a discussion with parents. A large panel may reveal incidental findings, which are not immediately actionable, but could potentially cause psychosocial distress. During the consultation, the treating clinician and genetic counselor explore Lucy's understanding and feelings surrounding genetic testing. Lucy and her parents agree to purse single-gene testing for *TP*53. Additionally, in view of her diagnosis of leukemia, cultured skin fibroblasts are a preferred specimen type for genetic testing in search of germline mutations to avoid confusion with somatic mutations (Section 6.4.4) due to her leukemia.

As genetic testing using cultured skin fibroblasts can take up to 12 weeks, buccal swabs, saliva, or DNA extracted directly from a skin biopsy without culture can be considered in urgent cases. However, as these samples can be contaminated by peripheral blood, somatic pathogenic variants can be present and be inaccurately reported as germline pathogenic variants. Thus, the confirmation test using cultured skin fibroblasts will be needed for final interpretation. Skin fibroblast culture is the only option for individuals, who have undergone allogeneic bone marrow transplant.

- **Results:** Lucy undergoes genetic testing on skin fibroblasts, and a pathogenic variant (Section 7.1.3) in *TP53* is found, consistent with the genetic diagnosis of Li–Fraumeni syndrome.
- **Management and Recommendations:** Li–Fraumeni syndrome can cause multiple types of aggressive cancers at young ages and subsequent screening strategies and management are often complex. Current National Comprehensive Cancer Network (NCCN) guidelines provide management guidelines for individuals with Li-Fraumeni syndrome. Please refer to the most recent NCCN guidelines as these are updated annually. Lucy should have:
  - Comprehensive physical examination every 6–12 months and prompt assessment by medical professionals for any medical concerns.
  - Annual whole-body MRI with organ targeting protocols for detection of adrenocortical carcinoma, brain tumor, tissue sarcoma, and osteosarcoma as well as other types of cancer.
  - Colonoscopy and upper endoscopy every 2–5 years starting at age 25 or 5 years before the diagnosis of the youngest family member (if diagnosis is younger than 30 years).
  - Annual dermatologic evaluation starting at age 18.
  - For women, breast screening starting at age 20 years with annual MRI. At age 30, annual mammograms can be added to alternate with the MRI. Risk reduction mastectomy should be discussed and individualized.

  It also is recommended that Lucy avoid known carcinogens and minimize unnecessary exposures to diagnostic and therapeutic radiation.

Genetic counseling for children diagnosed with a hereditary cancer syndrome should be an ongoing process. A revisit to genetic counseling is recommended for Lucy in her mid-to-late teenage years and again at the time of family planning to review the age-specific risks and management, information sharing with at-risk relatives (Section 7.4.3), and evaluation for additional psychosocial support.

- **Caveats:** The surveillance should be tailored based on the patient's own clinical situation and preferences. The comprehensive risk management protocol for individuals with *TP53* pathogenic variants can be found in the modified Toronto protocol as well as the NCCN guidelines. Negative genetic results cannot always rule out the possibility of leukemia-predisposing conditions (Section 7.1.4).
- **Implications for Other Family Members:** As Li–Fraumeni syndrome is an autosomal dominant genetic condition (Section 2.7.1), the likelihood of Lucy's first-degree relatives of having the familial *TP53* variant is 50%. Referral for genetic counseling to consider predictive genetic testing is recommended for her relatives to identify who is at risk of cancer and to start appropriate surveillance and management early. The presentation and age of onset can vary for different individuals even in the same family. Therefore, it is important for Lucy's parents, siblings, and the rest of her family members to consider predictive testing to know their carrier status and be aware of their risks. Based on Lucy's family history of cancer, it is likely that the *TP53* mutation was inherited from her paternal side of the family.
- **Outcome:** One of Lucy's siblings who is an HLA-matched donor tests negative for the familial pathogenic variant in *TP53* gene and is subsequently selected to be her donor for allo-HSCT. Lucy undergoes this successfully and is currently in remission. Lucy's father has single site mutation testing (Section 6.2.1) and is positive for the same variant. The paternal side of relatives is recommended to undergo genetic counseling and predictive genetic testing to understand their carrier status. For those who test positive for the familial mutation, routine Li–Fraumeni syndrome surveillance is recommended.

The Diagnosis

Clinical Diagnosis: Patient and Father—Li–Fraumeni syndrome.
Genetic Diagnosis: Patient and Father—*TP53* gene, heterozygous pathogenic variant.

Further Reading

Botkin, J.R., 2016. Ethical issues in pediatric genetic testing and screening. Curr. Opin. Pediatr. 28, 700–704. PMID: 27606958.

Chompret, A., Abel, A., Stoppa-Lyonnet, D., Brugiéres, L., Pagés, S., Feunteun, J., et al., 2001. Sensitivity and predictive value of criteria for p53 germline mutation screening. J. Med. Genet. 38, 43–47. PMID: 11332399.

Druker, H., Zelley, K., McGee, R.B., Scollon, S.R., Kohlmann, W.K., Schneider, K.A., et al., 2017. Genetic counselor recommendations for cancer predisposition evaluation and surveillance in the pediatric oncology patient. Clin. Cancer Res. 23, e91–e97. PMID: 28674117.

Holmfeldt, L., Wei, L., Diaz-Flores, E., Walsh, M., Zhang, J., Ding, L., et al., 2013. The genomic landscape of hypodiploid acute lymphoblastic leukemia. Nat. Genet. 45, 242–252. PMID: 23334668.

Kratz, C.P., Achatz, M.I., Brugières, L., Frebourg, T., Garber, J.E., Greer, M.C., et al., 2017. Cancer screening recommendations for individuals with Li-Fraumeni syndrome. Clin. Cancer Res. 23, e38–e45. PMID: 28572266.

Li, F.P., Fraumeni Jr., J.F., Mulvihill, J.J., Blattner, W.A., Dreyfus, M.G., Tucker, M.A., et al., 1988. A cancer family syndrome in twenty-four kindreds. Cancer Res. 48, 5358–5362. PMID: 3409256.

National Comprehensive Cancer Network, 2022. NCCN Clinical Guidelines in Oncology (NCCN Guidelines): Genetic/Familial High-Risk Assessment: Breast, Ovarian, and Pancreatic. Version 1.2023- September 7, 2022. [Online]. https://www.nccn.org/professionals/physician_gls/pdf/genetics_bop.pdf.

Case 11.16.7

Cancer Cases: Macrocytic Anemia and Idiopathic Pulmonary Fibrosis

Shao-Tzu Li[1], Takae M. Brewer[2], Rebekah Moore[2,*], Charis Eng[3,4] and Joanne Ngeow[1,5]

[1]*Cancer Genetics Service, Division of Medical Oncology, National Cancer Centre Singapore, Singapore, Singapore;* [2]*Genomic Medicine Institute, Cleveland Clinic, Cleveland, OH, United States;* [3]*Genomic Medicine Institute, Center for Personalized Genetic Healthcare, Cleveland Clinic Lerner Research Institute, and Taussig Cancer Institute, Cleveland, OH, United States;* [4]*Department of Genetics and Genome Sciences, Germline High Risk Focus Group, Case Comprehensive Cancer Center, Case Western Reserve University School of Medicine, Cleveland, OH, United States;* [5]*Lee Kong Chian School of Medicine, Nanyang Technological University, Singapore*

Abstract

A 50-year-old male with chronic macrocytic anemia, pulmonary fibrosis, and family history of cytopenia has inherited a bone marrow failure syndrome. Clinical testing reveals short telomere syndrome, and molecular testing reveals a heterozygous *TERT* germline, pathogenic variant.

The Case

Lim, a 50-year-old male, comes to see you, his primary care physician, for a routine follow-up appointment for his chronic macrocytic anemia, which was incidentally found at age 35. Over the years, his low leukocyte and platelet counts continue to worsen. His recent bone marrow exam shows hypocellular bone marrow and signs of myelodysplastic syndrome (MDS). You note that you referred him to a pulmonologist for a chronic cough, which was persistent for 1 year, and he has been diagnosed with idiopathic pulmonary fibrosis. He reports that his 25-year-old daughter also has anemia and asks you if his daughter's anemia could possibly be inherited from him.

The Issues

- **Does anything about Lim's presentation seem unusual?** Although anemia is not rare, he has had a macrocytic anemia, which is less common than microcytic. He also has developed MDS as well as idiopathic pulmonary fibrosis.

Often a history of three seemingly unrelated phenomena can indicate an underlying genetic condition.

- **Should Lim be referred to genetics based on the suspicion of hereditary disease?** Yes. Although cigarette smoking, viral infections, and a personal history of gastroesophageal reflux disease are the most common environmental causes for idiopathic pulmonary fibrosis, a small percentage of those affected can be attributed to an underlying germline predisposition, which often presents with other clinical features such as bone marrow failure and leukemia.

[*] Author RM was employed by Cleveland Clinic at the time of writing the manuscript, but is currently employed by AstraZeneca Pharmaceuticals.

Genomics in the Clinic. https://doi.org/10.1016/B978-0-12-816478-5.00104-7

Some hereditary diseases present with specific physical features and/or clinical manifestations. A physical exam on Lim is otherwise normal, and you note no short stature, dysmorphic features, abnormal head circumference, or abnormal skin or hair pigmentation. The absence of clinical features does not exclude the presence of a germline predisposition to an underlying genetic condition. In view of his clinical diagnosis of MDS and pulmonary fibrosis, as well as family history of anemia, you make a genetic referral (Section 8.1.5).

- **How do you respond to Lim's question about his daughter's anemia possibly being inherited from him?** There are many potential causes of anemia in his daughter's age group. Therefore, a full anemia workup should be performed on her. The possibility of inherited forms of bone marrow failure should be considered when there is family history of cytopenia, aplastic anemia, MDS, and/or leukemia.

The Referral

- **Genetic Evaluation**: A full medical and family history with a three-generation pedigree is obtained. A physical examination is performed by the geneticist. When asked specifically, Lim reports having hypopigmentation and early graying hair before age 30, as well as fragile and curling nails which were missed by his referring clinician. His head circumference is 53.5 cm, slightly below average, and his height is 160 cm, which is short and below average. On family history, his daughter was born prematurely at 34 weeks gestation and has had anemia since age 25. He says that most of his siblings are short. One of his maternal aunts died at age 36 from endometrial cancer. The rest of his family history is unremarkable, and there is no consanguinity. Inherited bone marrow failure syndrome (IBMFS) is suspected in view of his personal and family history.
- **Genetic Testing**: As the clinical spectrum of IBMFS is very broad, and classical symptoms may be subtle or absent, both the telomere length test and the chromosome breakage test usually are used as the first line of analysis for making a clinical diagnosis. The presence of abnormally short telomeres increases the suspicion for a short telomere syndrome (STS) which also known as telomere biology disorders. A positive result on the chromosome breakage test would indicate the presence of Fanconi anemia and other chromosome instability syndromes. The telomere length test and the chromosome breakage test are performed on Lim. His chromosome breakage test is normal but his telomere length using peripheral blood shows less than the first percentile for age-matched normal controls.
- **Genetic Counseling**: The geneticist and genetic counselor discuss with Lim that an underlying autosomal dominant (Section 2.7.1) genetic condition, STS, is suspected, given the extremely short telomere length, family history of cytopenia in two generations, as well as the personal history of both MDS and idiopathic pulmonary fibrosis. STS is characterized by extremely short telomere length and affects the regenerative capacity of involved cells/organs in multiple organ systems by the premature loss of progenitor stem cells. Shortened telomeres are commonly associated with bone marrow failure, including cytopenia, aplastic anemia, MDS, and leukemia. Additionally, STS can include a history of skin rash, abnormal skin pigmentation, fine thin dysmorphic nails, poor nail or hair growth, early gray hair or early hair loss, and dental findings such as increased dental caries, aggressive periodontitis, enamel hypoplasia, or oral leukoplakia. Idiopathic pulmonary fibrosis is the most common pulmonary manifestation seen in these patients (70%). Microcephaly, short stature, developmental delay, liver disease, osteoporosis, immunodeficiency, and increased risk of cancer are also associated with STS. Molecular testing is recommended to confirm this and to share with potentially affected family members.
- **Results**: A multigene panel test (Section 6.2.3) for IBMFS is performed on skin fibroblasts and a pathogenic variant (Section 7.1.3) in the gene *TERT* is found in Lim's germline DNA. This genetic finding confirms his diagnosis of a *TERT*-related STS and is consistent with his personal history of MDS and idiopathic pulmonary fibrosis.
- **Management and Recommendations**: The following surveillance strategies, medical management, and lifestyle modifications are discussed with Lim based on the positive genetic testing result:
  - Follow up with a hematologist, and undergo regular testing for complete blood count (CBC) with differentials and bone marrow aspirate and biopsy at the discretion of the treating hematologist.
  - Referral to an otolaryngologist, and a dermatologist for annual surveillance of increased risk of oral cancer, head and neck cancer and skin cancers. For affected females in his family, they should be referred to a gynecologist to screen for gynecologic cancers.
  - Undergo annual pulmonary function tests starting either at diagnosis or at an age when the individual can appropriately perform the test (typically age is 8 years old).
  - Maintain good daily oral hygiene and routine dental hygiene visits every 6 months. The individual's dentist should be made aware of the increased risk of head and neck squamous cell cancers and perform a thorough examination at each visit.

- Timing of treatment—In many mild cases where the rate of progression is slow, patients may not require treatment for many years. Treatment decisions are individualized to each patient and discussed with clinicians who are experienced in treating this condition.
- Medications—Anecdotal reports of using steroid drugs (known as androgens) in patients with bone marrow failure have been shown to improve blood cell counts.
- Stem cell transplant/bone marrow transplant—Treatment for severe bone marrow failure disorders due to telomere shortening is bone marrow/stem cell transplantation, in which a patient's bone marrow or peripheral blood stem cells are replaced with those from a healthy donor. Patients with STS are prone to side effects from certain drugs used in standard transplants and should undergo a modified transplant regiment in a center with experience in this particular condition.
- Lung and liver disease—Lung or liver transplantation may also be appropriate for some patients whose pulmonary fibrosis or liver cirrhosis is the primary complications.
- For those who are candidates for hematopoietic stem cell transplantation, transfusion of red cells or platelets should be minimized if not necessary. All blood products should be leukoreduced and irradiated.
- It is prudent to minimize exposures to therapeutic radiation since data on radiation side effects are limited.
- The combination of androgens and granulocyte colony-stimulating factor (G-CSF) should be avoided because of possible splenic peliosis and rupture.
- The patient should be encouraged to avoid toxic factors (respiratory, hepatic, medullary), especially tobacco smoke, alcohol, environmental toxins, and cytotoxic drugs.
- Individuals who are on androgen therapy are recommended to have liver function tests every 3 months, and cholesterol, triglycerides, and liver ultrasounds every 6 months.

- **Caveats**: In view of Lim's diagnosis of MDS with signs of malignant transformation, cultured skin fibroblasts are a preferred sample type for genetic testing when searching for germline mutations. Peripheral blood may not be useful because somatic pathogenic variants (from circulating malignant cells) could possibly being mistaken for germline pathogenic variants (Section 6.4.4).
- **Differential Diagnosis**: Fanconi anemia and (Section 11.9.5) other chromosomal breakage syndromes are included in the differential diagnosis. Additionally, a normal or negative result on the chromosome breakage test or telomere length test does not exclude the possibility of other bone marrow failure syndromes or hereditary hematological malignancies, due to the genetic and clinical heterogeneity (Section 2.8) of the IBMFSs. Other laboratory evaluations such as the exocrine pancreatic test or the erythrocyte adenosine deaminase assay can be performed to query for the other syndromes, such as Shwachman–Diamond and Diamond–Blackfan anemia.
- **Implications for Other Family Members**: In view of the autosomal dominant inheritance pattern (Section 2.7.1) of *TERT*-related STS, Lim's first-degree relatives have a 50% chance of having the familial *TERT* pathogenic variant (Section 7.1.3), which would predispose them to telomere length shortening and increased risk of bone marrow failure, idiopathic pulmonary fibrosis, cancers, and other short telomere-related diseases. The clinical presentations and age of onset can vary for different individuals even in the same family. Therefore, genetic counseling and discussion of predictive single variant genetic testing (Section 6.2.1) is recommended for Lim's daughter, his siblings, and other relatives. They will better understand their risks and start appropriate surveillance and management if testing is done and is positive.
- **Outcome**: Lim dies a year later from complications of lung infection. His daughter and his siblings are referred for genetic counseling to understand their carrier status of the familial *TERT* pathogenic variant. They are to be referred for appropriate surveillance if testing is positive.

The Diagnosis

Clinical Diagnosis: Patient—*TERT*-related short telomere syndrome.
Genetic Diagnosis: Patient—*TERT* gene, heterozygous pathogenic variant.

Further Reading

Giri, N., Pitel, P.A., Green, D., Alter, B.P., 2007. Splenic peliosis and rupture in patients with dyskeratosis congenita on androgens and granulocyte colony-stimulating factor. Br. J. Haematol. 138, 815–817. PMID: 17760812.

Kam, M.L.W., Ngeow, J.Y.Y., 2021. Telomere biology disorders. Npj Genomic Medicine 6, 36. https://doi.org/10.1038/s41525-021-00198-5.

Savage, S.A., Cook, E.F. (Eds.), 2015. Dyskeratosis Congenita and Telomere Biology Disorders: Diagnosis and Management Guidelines. https://teamtelomere.org/wp-content/uploads/2018/07/DC-TBD- Diagnosis-And-Management-Guidelines.pdf. (Accessed 30 March 2019).
Shimamura, A., 2009. Clinical approach to marrow failure. Hematol. Am. Soc. Hematol. Educ. Progr. 329–337. PMID: 20008218.
Wu, Z.-H., 2016. Phenotypes and genotypes of the chromosomal instability syndromes. Transl. Pediatr. 5, 79–83. PMID: 27186524.

Section 11.17

Direct-to-Consumer Cases

Case 11.17.1

Direct-to-Consumer Cases: Breast Cancer Family History

Angela E. Scheuerle
Department of Pediatrics, Division of Genetics and Metabolism, Department of Pathology, Division of Genetic Diagnostics, McDermott Center for Human Growth and Development, University of Texas Southwestern Medical Center, Dallas, TX, United States

Abstract
A 35-year-old father with a paternal family history of breast cancer has tested "negative" for breast cancer on direct-to-consumer testing. Formal diagnostic testing found a pathogenic variant in the *PTEN* gene. He is interested in testing on his healthy 6-year-old daughter.

The Case

Louise is a healthy 6-year-old female with normal growth and development who is new to your practice. Family history shows that the paternal grandmother died due to breast cancer at age 55 years and the paternal aunt has just been diagnosed with breast cancer at 33 years. You have concerns about Louise's breast cancer risk. Louise's mother, Olivia, says that she is not concerned because breast cancer cannot be inherited through the father and the family all did "one of those saliva DNA tests from the drug store" that said both the father and children tested normal for breast cancer.

The Issues

- **Could Louise have inherited a breast cancer risk from her paternal line?** You, her pediatrician, know that breast cancer can be inherited in an autosomal dominant manner (Section 2.7.1) from either side of the family. The sex of the parent is not determinant. You are unsure about Louise's specific risk, whether to do other testing, or when tumor surveillance should start. What implications might this testing have for Louise's father?
- **Is the direct-to-consumer testing adequate to answer the question about breast cancer risk?** What does a "normal" direct-to-consumer (DTC) test result mean in this context (Section 9.3)? The mother says that the DNA test looked at "the important breast cancer genes." You ask the mother if the paternal grandmother or paternal aunt had any genetic testing after the cancer diagnosis, or whether either woman had other cancers or health problems. She doesn't know.
- **Are there other concerns about the "DNA tests from the drug store"?** You know that medical devices, medications, and testing sold over the counter without a doctor's supervision carry some risks and that DNA testing has privacy and other concerns (Section 9.4). It is unlikely that the family thought about all the benefits and risks of testing before doing it. You decide to refer for a genetic evaluation.

Referral for medical genetic assessment will be helpful to address the inherited cancer question and to determine what, if any, other testing is appropriate for Louise and any other family members. In this case, the referral could be just for the

Genomics in the Clinic. https://doi.org/10.1016/B978-0-12-816478-5.00105-9

parents, with further genetic evaluation of Louise if indicated. A copy of the DTC testing results should be brought to the appointment or sent ahead of time.

The Referral

- **Genetic Evaluation**: Louise's parents, Olivia, age 32, and Roderick, age 35, are referred to genetics for discussion of the family history of breast cancer. They bring the DTC testing result with them. More detailed family history reveals that Roderick's mother and maternal aunt both died from metastatic breast cancer in their 50s. Roderick's mother was diagnosed sometime in her late 40s and she also had thyroid problems. Roderick's sister, age 33, recently had a breast mass found by routine mammography and is undergoing chemotherapy. Roderick considers himself healthy. He has some hearing loss that he attributes to his work and recently had some skin tags removed. Olivia's family history is unremarkable.

  The DTC testing result reports "normal" *BRCA1* and *BRCA2* genes (Section 8.1.2). The disclaimer notes that this testing is only for selected variants, will not find deletions or duplications (Section 2.4.1), and does not evaluate for triplet repeat expansion diseases (Section 2.4.1).
- **Genetic Counseling**: A detailed three-generation pedigree is obtained (Sections 2.7, 3.2). Additional information is that Roderick's cousin, his maternal aunt's son, has autism. In addition to Louise, Olivia and Roderick have a healthy and normally developing 3-year-old boy. They have had no difficulty conceiving and Olivia's pregnancies have been uneventful.

  The strong family history of breast cancer in Roderick's family is very concerning. Given the thyroid abnormalities in his mother, his cousin's autism, and Roderick's own hearing loss and history of skin tags, this suggests a possible diagnosis of Cowden syndrome in the family. Cowden syndrome is an autosomal dominant cancer syndrome caused by pathogenic variants in the *PTEN* gene (Section 8.1.3). It has a high penetrance but widely variable expressivity (Section 2.8).
- **Results**: Sequencing of *PTEN* in Roderick is done and he is found to have a pathogenic variant. Individuals with a *PTEN* pathogenic variant *PTEN* hamartoma tumor syndrome are at an increased risk for thyroid, kidney, central nervous system, and colorectal cancers. In females, there is a very high risk for breast and endometrial cancers. Roderick is counseled that, even though he is male, he does have some rudimentary breast tissue and certainly is at risk for breast cancer.
- **Management and Recommendations**: Roderick is referred for appropriate tumor surveillance protocols including regular thyroid ultrasound, colonoscopy to assess for polyps, and dermatology exams, and at age 40 years additional screening such as renal ultrasound and imaging should be done. Females in their 30s are recommended to have regular mammograms, endometrial biopsies, and breast MRIs. Tumor surveillance recommendations in children are being developed. Roderick is advised to share the genetic test result with his sister and cousin so that they can alert their own doctors and be tested. Evaluation of Louise and her brother by a clinical geneticist is recommended and testing them should be strongly considered no later than age 10 (when thyroid tumor surveillance would start if they are affected) after appropriate counseling.
- **Caveats**:
  - Limitations of DTC testing: DTC genetic testing was not helpful in the assessment of genetic cancer risk in this family. This was complicated by the parents' misunderstanding of the genetics. There are two major concerns. The first is that the DTC genetic testing company (1) did not test *PTEN* at all; (2) tested *PTEN* but did not find the pathogenic variant; or (3) tested *PTEN*, found the pathogenic variant (Section 7.1.3), but did not report it. There are various technical, interpretive, and regulatory reasons for these failures (Section 9.1), but this highlights the difference between testing as provided by DTC companies and that done under medical, particularly geneticist, direction.

    The second point is that the false reassurance of "normal" *BRCA1* and *BRCA2* testing and the misunderstanding of cancer inheritance are a result of poor pretest counseling (Section 8.2). The larger DTC companies have detailed information on their websites and may provide a phone number for counseling, but the burden is entirely on the consumer. Genetic information is complex. A majority of the population has a poor understanding of basic genetic science. DTC genetic testing simplifies genetic testing to the detriment of many who see this as an inexpensive way to get genetic information. There was no benefit to this family from the DTC testing. In fact, it was falsely reassuring. The appropriate genetic referral was driven by good medical history taking at the pediatrician's office.
  - Broad genetic testing of healthy minors and asymptomatic populations: Studies are under way to evaluate the usefulness and ethical implications of genome testing in neonates and healthy children. Such testing has positive potential in the area of personalized medicine, but that is overbalanced by cost, legal ramifications, and ethical

concerns (Sections 9.2, 11.16.1, and 11.18.1). The Hastings Center reviewed use of broad genome testing as a screening tool in children and concluded that it is not appropriate at this time.

Genomic testing as part of health screening is of limited usefulness in asymptomatic populations. How will the results be stored so they are both retrievable to the future adult but not released inappropriately? There are potential discriminatory or insurance misuses and the possibility of generating unnecessary distress or false reassurance. Genetic science in this area is very new and identification of a variant would require health resources even if that variant is reported as benign. These are the difficulties with genomic screening in certified diagnostic laboratories (Section 6.3.2). DTC genetic testing faces all of these concerns as well as technical and ethical concerns regarding reporting of background medical information by the consumer, sample collection and handling, violation of privacy, and long-term control of the sample and/or data derived from it.

- **Outcome**: After appropriate counseling about predictive genetic testing, Louise undergoes targeted genetic testing for her father's *PTEN* pathogenic variant and is found not to have inherited the father's variant. Her 3-year-old brother is noted to have macrocephaly and behavior problems, though his development is normal. *PTEN* variant testing is recommended because of the association with macrocephaly and autism. In this case, testing is not considered "predictive" because the child is showing a concerning phenotype. The brother is diagnosed with the same *PTEN* pathogenic variant found in his father.

The Diagnosis

Clinical Diagnosis: Patient and Son—Cowden syndrome. False-negative reassurance by direct-to-consumer testing.
Genetic Diagnosis: Patient and Son—*PTEN* gene, heterozygous pathogenic variant.

Further Reading

Johnston, J., Lantos, J.D., Goldenberg, A., Chen, F., Parens, E., Koenig, B.A., et al., 2018. Sequencing newborns: a call for nuanced use of genomic technologies. Hastings Cent. Rep. 48 (Suppl. 2), S2−S6. PMID 30133723.

Ross, L.F., Saal, H.M., David, K.L., Anderson, R.R., American Academy of Pediatrics, & American College of Medical Genetics & Genomics, 2013. Technical report: ethical and policy issues in genetic testing and screening of children. Genet. Med. 15, 234−245. PMID 23429433.

Wasson, K., Cook, E.D., Helzlsouer, K., 2006. Direct-to-consumer online genetic testing and the four principles: an analysis of the ethical issues. Ethics Med. 22, 83−91. PMID 17115513.

Yehia, L. Eng, C., November 29, 2001. PTEN hamartoma tumor syndrome. Adam, M. P., Ardinger, H.H., Pagon, R.A., Wallace, S.E., Bean, L.J.H., Gripp, K.W., Mirzaa, G.M. Amemiya, A., (editors). In GeneReviews. University of Washington, Seattle (WA), 1993−2022 [updated February 19, 2021]. PMID 20301661.

Case 11.17.2

Direct to Consumer Cases: Predisposition of a Multifactorial Condition by Direct-to-Consumer Testing

Angela E. Scheuerle

Department of Pediatrics, Division of Genetics and Metabolism, Department of Pathology, Division of Genetic Diagnostics, McDermott Center for Human Growth and Development, University of Texas Southwestern Medical Center, Dallas, TX, United States

Abstract

A 9-year-old girl undergoes direct-to-consumer testing and is said to be at high risk for celiac disease by having the high-risk HLADQ2 haplotype.

The Case

History and physical exam in your pediatric office confirm that Tessa, a 9-year-old girl, is healthy, developing normally, and is well within normal limits for all her growth parameters. She has no history of chronic or recurrent gastrointestinal problems, no vomiting, abdominal pain, or recurrent diarrhea or constipation. At the end of the visit her mother asks to be referred to a dietician because the "cheek swab DNA test" done on Tessa showed that she should be on a gluten-free diet.

The Issues

- **Can DNA analysis be used for nutritional guidance?** Tessa is a healthy child with normal growth and development and no gastrointestinal complaints. Her nutritional status is not an obvious medical concern. There is no medical reason to restrict her diet. Because there are no concerns for making a dietary referral, you advise the mother to continue with a regular diet without the gluten restriction.
- **Are there other concerns about the "DNA tests from the drug store"?** You know that medical information on the internet can be misinterpreted, outdated, or just wrong. Some of it can be exploitative. On the other hand, there seems to be genetic information for many things these days and celiac testing is antibody based. You are unsure what genetic information was identified on the direct-to-consumer testing (Section 9.1.5) and decide to refer her to genetics.

Genomics in the Clinic. https://doi.org/10.1016/B978-0-12-816478-5.00106-0

Referral for medical genetic assessment will be helpful to address the questions about genetic predisposition to autoimmune disease. In this case, the referral could be just for the parents, with further genetic evaluation of Tessa if indicated. A copy of the direct-to-consumer testing results should be brought to the appointment or sent ahead of time.

The Referral

- **Genetic Evaluation**: Tessa is a healthy, growing child. There is no indication of a chronic health disease or genetic condition. Her newborn screening testing was normal. The family has a normal, culturally appropriate diet. No one else in the family is on a medically restricted diet of any kind.

  True celiac disease is a multifactorial disorder (Section 2.7.3) with specific diagnostic criteria. It has been associated with human leukocyte antigen (HLA) haplotype (Section 9.1.3) DQ2 or DQ8. The direct-to-consumer genetic testing evaluates two specific nucleotides, one each of *HLA-DQ2* and *HLA-DQ8*. It does not sequence either gene. Testing on Tessa showed the "high-risk" *HLA-DQ2* haplotype.
- **Genetic Counseling**: A detailed three-generation pedigree is obtained (Sections 2.7, 3.2). There is no history of growth or developmental problems. There is no one with chronic gastrointestinal complaints, or any problems gaining weight. There are no autoimmune diagnoses, joint pains, or anemia.

  Tessa's family is counseled that, firstly, it is important that any variant found by direct-to-consumer testing should be confirmed in a certified clinical diagnostic lab before any medical decision-making. The parents are given the option of confirming the direct-to-consumer test result at this time or waiting until clinically indicated. It should be noted that the presence of the variant *HLA-DQ2* haplotype is only one component of a formal celiac disease diagnosis and most people with the variant never develop a gastrointestinal or autoimmune disease.
- **Management and Recommendations**: Because Tessa is asymptomatic, the appropriate course would be not to gluten-restrict her diet. If at some time in the future Tessa does develop chronic gastrointestinal symptoms, weight loss, or other concerning features, she should seek formal evaluation from a pediatric gastroenterologist before making any dietary changes.
- **Caveats**:
  - Limitations of direct-to-consumer testing: Direct-to-consumer genetic testing is not held to the same technical and reporting standards as medical diagnostic testing done in Clinical Laboratory Improvement Amendment (CLIA)-certified laboratories (Section 9.3). A study has recently shown that one direct-to-consumer company reported a 40% false-positive rate for pathogenic cancer variants. Direct-to-consumer genetic testing should never be considered diagnostic testing and should not be used for medical decision-making. Any reported finding of concern should be reevaluated in the context of the patient, with formal diagnostic lab retesting as appropriate.

    In this case there was likely no benefit to the direct-to-consumer testing (Section 9.4). It raises the possibility of an unnecessary dietary restriction on the child that would have social and financial impacts even if she remains on a nutritionally balanced diet. An inappropriate diagnosis could have been documented in her medical or school record.
  - Direct-to-consumer testing for lifestyle prediction: Genetic risk assessments for Mendelian conditions (Section 2.7) can be quite straightforward. In medical conditions with a recognized genetic predisposition mixed with environmental components, such as type 2 diabetes, estimating risk of pathology is much less precise. Testing that purports to predict a child's athletic prowess, or an adult's likelihood of male pattern baldness, is no better than predictions that can be made based on a thorough history and physical exam. Claims that a test can provide specific nutrition recommendations, including for a gluten-free diet, likely have no basis in fact and may be setting up the consumer for exploitation.
- **Outcome**: The parents decide not to pursue further testing. Two years later, Tessa continues to be a healthy, growing child without any gastrointestinal concerns. While the mother continues to watch Tessa's gluten intake at home, she realizes that Tessa eats unrestricted meals at school and other places without apparent ill effects. No other testing is indicated.

The Diagnosis

Clinical Diagnosis: Concern for gluten intolerance/celiac disease by direct-to-consumer testing.
Genetic Diagnosis: HLA-DQ2 heterozygous haplotype, not confirmed.

Further Reading

ACMG Board of Directors, 2016. Direct-to-consumer genetic testing: a revised position statement of the American College of Medical Genetics and Genomics. Genet. Med. 18, 207–208. PMID 26681314.

Gardner, C.D., Trepanowski, J.F., Del Gobbo, L.C., Hauser, M.E., Rigdon, J., Ioannidis, J.P.A., et al., 2018. Effect of low-fat vs low-carbohydrate diet on 12-month weight loss in overweight adults and the association with genotype pattern or insulin secretion: the DIETFITS Randomized Clinical Trial. JAMA 319, 667–679. PMID 29466592.

Johnston, J., Lantos, J.D., Goldenberg, A., Chen, F., Parens, E., Koenig, B.A., et al., 2018. Sequencing newborns: a call for nuanced use of genomic technologies. Hastings Cent. Rep. 48 (Suppl. 2), S2–S6. PMID 30133723.

Tandy-Connor, S., Guiltinan, J., Krempely, K., Laduca, H., Reineke, P., Gutierrez, S., et al., 2018. False-positive results released by direct-to-consumer genetic tests highlight the importance of clinical confirmation testing for appropriate patient care. Genet. Med. 20, 1515–1521. PMID 29565420.

Taylor, A.K., Lebwohl, B., Snyder, C.L., Green, P.H.R., July 3, 2008. Celiac disease. Adam, M.P., Ardinger, H.H., Pagon, R.A., Wallace, S.E., Bean, L.J.H., Stephens, K., Amemiya, A., (editors). GeneReviews. University of Washington, Seattle (WA), 1993–2020. [updated January 31, 2019]. PMID 20301720.

Case 11.17.3

Direct-to-Consumer Cases: Identification of Birth Family History by Direct-to-Consumer Testing

Angela E. Scheuerle

Department of Pediatrics, Division of Genetics and Metabolism, Department of Pathology, Division of Genetic Diagnostics, McDermott Center for Human Growth and Development, University of Texas Southwestern Medical Center, Dallas, TX, United States

Abstract

A 35-year-old woman who was adopted was able to find relatives through direct-to-consumer testing, learns about a syndrome involving "X" in the family, and undergoes testing to find that she is a fragile X premutation carrier.

The Case

Kendra, age 35, has information for you, her internal medicine physician, about the saliva DNA direct-to-consumer (DTC) testing she and her family did. She is very happy with it. She had been adopted and has been able to locate some birth relatives through the testing company. Kendra has already made contact with a maternal half-sister, who was also adopted out. Through this half-sister, Kendra has learned that someone in the family has intellectual disability and "syndrome X". Kendra wants to know more about this and whether it might impact her health or that of her own three children. She says that there was no information about a "syndrome X" on the saliva testing. Could it be an abnormal chromosome test?

The Issues

- **What is the intellectual disability syndrome?** "Syndrome X" certainly could be a sex chromosome problem (Section 2.2). In a male it might mean Klinefelter syndrome (47,XXY) (Section 11.12.5). In a female it might be Turner syndrome (45,X) (Section 11.7.5) or 47,XXX. It might also be fragile X syndrome (Sections 5.2, 11.12.3, and 11.13.11), and you think that is the more likely diagnosis because of the family history of intellectual disability. Before knowing whether and how much to be concerned about Kendra and her children, you would have to know more about what the family is calling "syndrome X".
- **Is the direct-to-consumer testing adequate to answer the question about chromosome problems or fragile X syndrome?** What does a "normal" test result mean in this context? You know that some types of DNA testing will find duplications and deletions of chromosomes (Sections 2.2 and 2.4), but usually that is done with a karyotype (Section 6.1.1) or chromosome microarray (Section 6.1.3). You think that fragile X syndrome is a common single gene disorder and it seems to you that it should have been picked up if it was abnormal.

Genomics in the Clinic. https://doi.org/10.1016/B978-0-12-816478-5.00107-2

- **Are there other concerns about how direct-to-consumer DNA testing links family members?** Discovering ancestry and finding "lost" family members seems like a nice use of DTC testing, particularly for people who were adopted and consent to be contacted. It is a positive thing that Kendra's half-sister was willing to be contacted. Of course, she had submitted a DNA sample, too. You wonder if there would be a way to find their mother through some sort of DNA database, and you refer Kendra to Genetics.

Referral for medical genetic assessment will be helpful to address the intellectual disability syndrome question and to determine what, if any, other testing is appropriate for Kendra and any other family members. In this case, the referral is just for Kendra, with further genetic evaluation of her children if indicated. A copy of the direct-to-consumer testing results (Section 9.3) *should be brought to the appointment or sent ahead of time.*

The Referral

- **Genetic Evaluation:** Kendra is referred to Genetics for discussion of the family history of intellectual disability and "syndrome X". She brings the DTC testing result with her. More detailed history reveals that Kendra does not have any health concerns. She completed college and works as a bank manager. She is married with three healthy biological children. There were no problems conceiving and her pregnancies were uncomplicated. The biologic relatives she recently contacted told her about this "syndrome X″ that causes intellectual disability.

  The DTC testing report makes no mention of any chromosome abnormality or any condition that could be interpreted as "syndrome X". The disclaimer notes that this testing will not find deletions or duplications and does not evaluate for triplet repeat expansion diseases (Section 2.4.1).
- **Genetic Counseling:** A detailed three-generation pedigree is obtained (Section 3.2). The person with intellectual disability on Kendra's side is her half-sister's 6-year-old son. He carries a diagnosis of autism (Sections 5.2 and 11.12.4). In addition, the maternal half-sister has infertility with ovarian failure at age 27.

  The combination of early ovarian failure in Kendra's half-sister combined with autism in that woman's son raises the possibility of fragile X syndrome. This is an X-linked condition (Section 2.7.2) caused by a CGG triplet repeat expansion (Sections 2.4.1, 6.4.1) in the gene *FMR1*. Kendra herself is healthy at 35 without any fertility problems, but she could be a fragile X syndrome premutation carrier.
- **Results:** Kendra is tested and found to have one allele of 27 repeats (normal) and one with 62 repeats. The latter allele is a premutation, not pathogenic itself, but it could expand further to pathogenicity in subsequent generations.
- **Management and Recommendations:** Fragile X syndrome is caused by expansion of a CGG base pair triple repeat. Normal is less than 45 CGG repeats, definitively abnormal is greater than 200 CGG repeats, and repeat sizes between these two extremes are unstable and can expand over subsequent generations. CGG repeat sizes in the 80–90 range in women can cause premature ovarian failure, which probably could explain this finding in Kendra's half-sister. Because Kendra is a premutation carrier, she is also at risk of developing premature ovarian failure and/or fragile X-associated tremor/ataxia syndrome (Sections 5.2, 11.12.3, 11.13.11). As long as Kendra's children are developmentally normal, there is no indication for testing them. Testing should be done if any show learning or unusual behavior problems. Each should seek genetic counseling prior to childbearing.
- **Caveats:**
  - Limitations of DTC testing: Next-generation sequencing and mutation scanning, both used in DTC genetic testing, do not evaluate triplet repeat sizes (Section 6.4.1); triplet repeat size is invisible to the testing technology. The larger DTC genetic testing companies disclose that testing does not evaluate the triple repeat diseases. The consumer must be cognizant of how that applies in each case. In a family with a known triplet repeat expansion disease, such as myotonic dystrophy (Section 11.13.4) or Huntington chorea (Section 11.13.12), a "normal" DTC genetic testing result can be falsely reassuring.
  - Privacy implications of DTC testing: It is now estimated that enough people of Northern European origin have submitted samples to DTC genetic testing services that European individuals or their descendants residing in the United States may be identifiable to at least the third cousin level. This fact was used by law enforcement in 2018 to detect a serial murder, which made headlines, and the same technique has been used to identify other criminals. DNA has the advantage of being unique in all individuals other than identical multiple births, so with proper techniques it is unlikely to misidentify a suspect. There are concerns about technical errors and about the potential for a database to be used for other purposes.

On the other hand, DNA testing can identify biologic (or nonbiologic) relationships that are not expected (Section 9.4). As early as 2010 there were questions raised about loss of anonymity for sperm donors. The same risks have not been raised for egg donors, though the risk might be present. Additionally, women who gave a child into a closed adoption may be put at risk if the anonymity is violated. Misattributed parentage, either of the mother or the father in this era of gamete and embryo donation, would be easily identified by DTC genetic testing. There is also the possibility that DTC testing could be used by a man as paternity testing without informing the mother or the child.

- **Outcome:** Adopted children may know little to nothing about their family medical history, which can be a disadvantage. In this case, the family was well served by identification of Kendra's half-sister. Kendra was able to learn about fragile X syndrome and have testing. She now knows she has an expanded repeat into the premutation range. This is unlikely to impact her health, but she may have passed the expanded repeat to her children. While fragile X screening is common today for pregnant women, knowing the specific risk will be helpful to her three children in the future.

The Diagnosis

Clinical Diagnosis: Fragile X syndrome, premutation carrier.
Genetic Diagnosis: *FMR1* gene, heterozygous premutation triple repeats.

Further Reading

Erlich, Y., Shor, T., Pe'er, I., Carmi, S., 2018. Identity inference of genomic data using long-range familial searches. Science 362, 690–694. PMID 30309907.

Hunter, J.E., Berry-Kravis, E., Hipp, H., Todd, P.K., June 16, 1998. FMR1 disorders. Adam, M.P., Ardinger, H.H., Pagon, R.A., Wallace, S.E., Bean, L.J.H., Stephens, K., Amemiya, A., (editors). In GeneReviews. University of Washington, Seattle (WA), 1993–2022 [updated November 21, 2019]. PMID 20301558.

Lehmann-Haupt, R., 2010. Are Sperm Donors Really Anonymous Anymore? Slate [Online]. Available from: https://medium.com/open-source-family/are-sperm-donors-really-anonymous-anymore-dbf71e65fcbf.

Moray, N., Pink, K.E., Borry, P., Larmuseau, M.H., 2017. Paternity testing under the cloak of recreational genetics. Eur. J. Hum. Genet. 25, 768–770. PMID 28272533.

Ross, L.F., Saal, H.M., David, K.L., Anderson, R.R., American Academy of Pediatrics, & American College of Medical Genetics & Genomics, 2013. Technical report: ethical and policy issues in genetic testing and screening of children. Genet. Med. 15, 234–245. PMID 23429433.

Case 11.17.4

Direct-to-Consumer Cases: Metabolic and Neurologic Disease Risk by Direct-to-Consumer Testing

Giulietta Maria Riboldi and Steven J. Frucht

Department of Neurology, The Marlene and Paolo Fresco Institute for Parkinson's and Movement Disorders, New York University Grossman School of Medicine, New York, NY, United States

Abstract

A 45-year-old man is found to have a heterozygous variant for Gaucher disease on a direct-to-consumer test, which puts him at increased risk for developing Parkinson disease.

The Case

Isaac is a 45-year-old healthy man. He performed a genetic ancestry test through a commercial company and the results reported a heterozygous pathogenic variant of the *GBA* gene (c.1226A>G; N409S [previous nomenclature N370S]), that is one of the most frequent among Ashkenazi Jews. The report mentions Parkinson disease. He comes to see you, his general practitioner, for his annual well visit and wants to discuss these test results, which he has brought with him. You have been seeing him for some time and you know he never had any medical history relevant to a neurologic disorder thus far.

The Issues

- **What is the meaning of the genetic finding?** You review the result and explain to Isaac that the *GBA* gene codes for an enzyme protein called glucocerebrosidase. You also explain to Isaac that Gaucher disease (Section 11.9.4), an autosomal recessive (Section 2.7.1) condition, is associated with homozygous or compound heterozygous pathogenic variants or mutations of this gene (Section 7.1.3). Because Isaac only has a variant on a single copy of this gene, and thus he can be referred to as heterozygous or a carrier, he should not be affected with Gaucher disease.
- **What is Gaucher disease?** You discuss that the protein encoded by the *GBA* gene is an enzyme that has an important function in our cells particularly in organelles called lysosomes. When mutations are present, the enzyme does not function normally. Gaucher disease is a lysosomal storage disorder that can manifest with different degrees of severity (Section 11.9.4). The different phenotypes of the disorders are referred as Gaucher disease types 1, 2, and 3 based on the severity of the symptoms and the presence of neurological manifestations (types 2 and 3) or only the classical systemic symptoms of hepatosplenomegaly, thrombocytopenia, and osteopenia (type 1).

Genomics in the Clinic. https://doi.org/10.1016/B978-0-12-816478-5.00108-4

- **Is this diagnosis relevant to Isaac's family history?** Isaac's mother is 68 years old now and she is affected with Gaucher disease. He says that his maternal grandfather had Parkinson disease. Isaac asks if he or his mother will develop Parkinson disease. You recommend referral to genetics to discuss the findings of the testing, and the ramifications for Isaac and his family.

The Referral

- **Genetic Evaluation:** The genetic counselor meets with Isaac to review the meaning and implications of his recent genetic test results. A detailed family history is obtained. Isaac has two children, both in their late teens. He also has six siblings (four sisters and two brothers) ranging in age from 25 to 52 years. One of his brother's daughters was recently diagnosed with Gaucher disease. The rest of his nieces and nephews are healthy. Isaac's mother was diagnosed with Gaucher disease type 1 at a young age, and she is now doing well, thanks to regular intravenous infusion of the recombinant glucocerebrosidase enzyme (Section 11.9.2). Isaac's maternal grandfather was diagnosed with Parkinson disease in his late 50s and he died at 72 years of age. Toward the end of his 60s he started to develop a progressive cognitive impairment. Both Isaac's paternal and maternal sides of the family are Ashkenazi Jewish. There is no known history of consanguinity in the family.
- **Genetic Counseling:** The geneticist with the genetic counselor explains to Isaac that mutations of the *GBA* gene are associated with Gaucher disease, when in a homozygous or compound heterozygous (Sections 2.3.4 and 2.7.1) state. A large number of mutations in this gene have been associated with Gaucher disease, and up to about 8% of Ashkenazi Jews carry a single *GBA* gene mutation. *GBA* mutations are defined as mild or severe based on whether they are causative of Gaucher disease type 1 (only systemic symptoms) or Gaucher disease types 2 and 3 (also neurological involvement).

  Isaac is concerned about the risk for his family members of having the same mutation. The genetic counselor explains that the risk for Isaac's offspring to have inherited his single mutation is 50%. If their mother is also a carrier, they could have Gaucher disease as well. In addition, the counselor mentions that it should be his decision whether to disclose the results of his genetic test with his siblings and the rest of the family (Section 3.4 and 7.4.3). Should he decide to do so, the genetic counselor can help explain these findings, either by a letter to be shared with Isaac's family or by meeting with them in person, should they desire.
- **Results:** Isaac's pathogenic variant in the *GBA* gene, the c.1226A>G mutation, or protein change p.Asn409Ser or N409S, is known to be more common among those of Ashkenazi Jewish ancestry compared to other populations because of a founder effect (Sections 2.7.1 and 3.2). Moreover, this specific mutation is associated only with Gaucher disease type 1, either in a homozygous state or in a compound heterozygous state. Indeed, the N409S mutation is considered to be protective against the neuropathic forms of Gaucher disease even when in combination with other mutations of the same gene.

  In recent years it has also been observed that patients with Gaucher disease and their relatives, who may be carriers of one mutation, have an increased incidence of Parkinson disease. This observation led to the discovery that carrying even a single mutation in the *GBA* gene increases the risk of developing Parkinson disease. Further studies have shown that in patients with Parkinson disease, mutations in the *GBA* gene are more common than in subjects without Parkinson disease. Therefore, mutations in this gene can predispose to the manifestations of Parkinson disease and thus represent a risk factor for this progressive neurologic condition. However, only about 9%–10% of individuals who carry a mutation in the *GBA* gene will develop Parkinson disease. Thus, the majority of the carriers will never develop Parkinson disease. This phenomenon is called reduced penetrance (Section 2.8), although the frequency of penetrance seems to increase with age. In the general population over the age of 60 years, only 1%–2% will develop Parkinson disease. This risk is increased five-fold in a carrier of one *GBA* mutation and up to 10–20 times in patients with two mutations, thus in those with Gaucher disease. Unfortunately, no biomarkers are available to predict who will develop Parkinson disease among the carriers of this mutation. Based on the studies reported in large populations of subjects with Parkinson disease and mutations of the *GBA* gene, some common traits have been identified in this population, including an earlier age of onset, an increased incidence of cognitive impairments, and a higher incidence of non-motor symptoms. However, none of these features is highly specific for this mutation.

  Isaac's pathogenic variant is one of the most frequently identified in patients with Parkinson disease who carry a mutation in the *GBA* gene, especially in Ashkenazi Jews. This pathogenic variant is commonly associated with Gaucher disease type 1 in individuals whether or not they carry a severe mutation of the *GBA* gene on the other allele. When Parkinson disease is associated with Gaucher disease types 2 and 3, whether in a homozygous or heterozygous state, the Parkinson phenotype can be more aggressive in terms of earlier age of onset and severity of motor and non-motor symptoms.

- **Management and Recommendations**: Some of the commercially available genetic direct-to-consumer tests, such as the one Isaac did, actually look only for the most frequent pathogenic variants of the *GBA* gene (Section 9.1). Therefore, the geneticist says, Isaac should consider doing a more comprehensive gene sequencing (Section 6.2.2) test of the *GBA* gene in a clinical CLIA-approved laboratory. This would be informative for Isaac should another mutation be identified because this may indicate a diagnosis of Gaucher disease. If this is the case, specific assessment and treatments for Gaucher disease, such as enzyme replacement therapy or gene therapy, should be considered and he should be referred to a Gaucher clinic.

  Regarding Parkinson disease, the genetic counselor explains to Isaac that his mutation is only a risk factor for Parkinson disease. There is no biomarker that can help predict whether he will develop the disease or not. Since currently Isaac has no symptoms or complaints suggestive for Parkinson disease and since there are no current approaches available to prevent or delay the onset of Parkinson disease, no specific assessment or interventional approaches should be put in place at the moment for this condition.
- **Caveats:** In our genome there is a pseudogene of *GBA* (Section 2.3.4) with 98% homology to the original gene. The pseudogene probably arose from a duplication of the *GBA* gene. Therefore, methods other than actual sequencing of the gene can have misleading results because of the difficulty in discriminating between variants localized on the gene and on the pseudogene.
- **Implications for Other Family Members:** Since Isaac's mother has Gaucher disease, she carries a mutation (Section 2.4) of the *GBA* gene on both alleles and will have passed one mutation to all of Isaac's siblings, thus making them all at least carriers for Gaucher disease. Indeed, should Isaac's father also be a carrier of a *GBA* gene mutation, they had 50% chance of inheriting two mutations in the *GBA* gene, and therefore to also have Gaucher disease. Whether his siblings are carriers of one or two mutations of this gene, they will all have an increased risk of developing Parkinson disease compared to the rest of the population, similarly to Isaac. One of Isaac's nieces is already affected with Gaucher disease. This means that his brother (the affected girl's father) is a carrier of a *GBA* mutation (that most likely he inherited from his mother) and that his wife is a carrier as well. Alternatively, the niece could have a de novo mutation on the allele inherited from her mother, but this is much less likely.

  The only way to confirm whether or not the siblings are carriers of mutations in the *GBA* gene is to perform a genetic test. This should be carried out in a CLIA-approved laboratory. Indeed, a number of different mutations have been identified as being pathogenic in the *GBA* gene. In order to identify these mutations, it is necessary to perform full sequencing of the gene (Section 6.2.2).

  Based on what is known about the familial mutation, Isaac's mother has an increased risk of developing Parkinson disease compared to the rest of the population. However, it is not definitive that she will develop the disease. Regarding his maternal grandfather, who had a diagnosis of Parkinson disease, it would have been possible to discern whether he was a carrier of the mutation in the *GBA* gene by gene sequencing (Section 6.2.2). Indeed, there are no highly specific features that can allow to definitively distinguish between carriers and noncarriers of mutations of the *GBA* gene in patients with Parkinson disease only based on their clinical presentation.
- **Outcome:** Isaac undergoes full sequencing (Section 6.2.2) of the *GBA* gene. He is not found to have another mutation, and thus he is a carrier for Gaucher disease and not affected. He is aware of his risk for developing Parkinson disease, but does not change his lifestyle in any major way. He discusses his results with his children, who choose not to undergo testing at this time. He shares the information about his genetic testing results with the rest of his family, and all of his siblings undergo testing.

The Diagnosis

Clinical Diagnosis: Patient - Gaucher disease carrier with increased risk for Parkinson disease. Mother - Gaucher disease.

Genetic Diagnosis: Patient - *GBA* gene, heterozygous pathogenic variant.

Further Reading

Blandini, F., Cilia, R., Cerri, S., Pezzoli, G., Schapira, A.H.V., Mullin, S., et al., 2019. Glucocerebrosidase mutations and synucleinopathies: toward a model of precision medicine. Mov. Disord. 34, 9–21. PMID 30589955.

Rana, H.Q., Balwani, M., Bier, L., Alcalay, R.N., 2013. Age-specific Parkinson disease risk in GBA mutation carriers: information for genetic counseling. Genet. Med. 15, 146–149. PMID 22935721.

Riboldi, G.M., Di Fonzo, A.B., 2019. GBA, Gaucher disease, and Parkinson disease: from genetic to clinic to new therapeutic approaches. Cells 8, 364. PMID 31010158.

Case 11.17.5

Direct-to-Consumer Cases: Risk for Late-Onset Alzheimer Disease by Direct-to-Consumer Testing

Giulietta Maria Riboldi and Steven J. Frucht

Department of Neurology, The Marlene and Paolo Fresco Institute for Parkinson's and Movement Disorders, New York University Grossman School of Medicine, New York, NY, United States

Abstract

A 68-year-old woman is found to be a homozygous carrier of the allele *APOE4* from a commercially available genetic screening test, putting her at higher risk for developing late-onset Alzheimer disease.

The Case

Dorothy is a 68-year-old African American woman who comes to see you, her family doctor, for her annual physical evaluation. Recently it was her birthday and her children gave her a present of a commercially available kit for ancestry and genetic discovery. Through the test she was able to map her ancestry, mostly from North Africa. However, the results also reported that she is a homozygous (Sections 2.3.4 and 2.7.1) carrier for apolipoprotein E4 *(APOE4)*, which, the report says, puts her at a higher risk for developing Alzheimer disease. She is very surprised about the results, because she has no family history of dementia, but is also very scared and would like to discuss the meaning of these results with you.

The Issues

- **Does Dorothy have dementia?** You discuss with Dorothy that dementia is mostly a clinical diagnosis. Late-onset dementia is defined as onset of a cognitive impairment after the age of 60−65 years affecting daily activities. Alzheimer disease is the most common form of late-onset dementia and it is characterized by a predominant impairment of memory because of the selective loss of neurons in those area of the brain responsible for memory. Dorothy has no cognitive complaints at this time. She is able to attend to her daily activities, take care of her house and of the bills, as well as function well at her work as a teacher. Since there are no signs suggestive of dementia you believe that she does not need to undergo further testing.
- **How accurate are these types of tests?** Although you do not know a lot about direct-to-consumer testing (Section 9.1), you are aware that the testing can be partial or incomplete and should be confirmed in a Clinical Laboratory Improvement Amendments (CLIA)-approved laboratory. You tell Dorothy that perhaps she should be seen in clinical genetics, familiar with interpreting this type of testing, to discuss and confirm the results.

Genomics in the Clinic. https://doi.org/10.1016/B978-0-12-816478-5.00109-6

- **What is the next step in the workup?** Since Dorothy does not complain of any impairment in her cognitive functions and there are no worrisome signs on her history, no further assessments, such as neuropsychological tests or imaging studies, i.e., brain MRI or positron emission tomography (PET) scan, are recommended. However, Dorothy is very much concerned about the implications of the genetic findings for herself and her family. Therefore, you suggest that she pursue a genetic evaluation.

The Referral

- **Genetic Evaluation:** Dorothy meets with the genetic counselor shortly thereafter. She explains about the recent findings of the *APOE* genotypes on a commercially available genetic screening. The genetic counselor explains to Dorothy that in order to provide appropriate counseling she needs to know more about the clinical history of her family (Section 3.2) to assess the risk of hereditability of dementia. Dorothy explains that she is the third of four siblings. She has two older brothers (who are 5 and 6 years older than her) and one younger sister (who is 2 years younger). Her siblings are in good health, except for her oldest brother who had a myocardial infarction when he was 62 years old, but luckily he recovered well from it. Their parents both lived into their early 80s. Dorothy has two children, one daughter and one son, who are 38 and 36 years, respectively. In her family, only one of her aunts, on her father's side, developed dementia later in her life, as far as she knows. She is not aware of any further specific clinical conditions running in her family. Dorothy's normal health history is also reviewed.
- **Genetic Counseling:** The genetic counselor explains that Alzheimer dementia is the most frequent neurodegenerative condition in adults. In the majority of the cases (95%), it starts after the age of 65 years. The causative mechanisms of Alzheimer disease are not clearly understood yet. Possibly a combination of environmental and genetic factors contributes to the development of this disease. Although no genetic variants that are directly causative have been identified, a number of genomic susceptibility loci have been described. These are specific variants in one's genome that can predispose to the disease. The major genetic risk factor for late-onset Alzheimer disease is a specific polymorphism of the apolipoprotein E gene (*APOE*). This protein has three variants (Section 2.4.1) (E2, E3, E4). These variants can affect the functionality of the protein and thus be associated with pathological conditions. In particular, compared to the rest of the population, the E4 variant increases the risk of developing Alzheimer disease by 3–4 times when present on only one of the two alleles (heterozygous) (Section 2.3.4) and 9–15 times when both alleles are in the E4 form (homozygous) (as in the case of Dorothy). Heterozygotes have a 10%–20% chance of developing Alzheimer disease by age 75 years, whereas homozygotes subjects have a 25%–35% risk by the same age. However, the fact that this is considered a genetic risk factor means that carrying this variant is not sufficient to develop Alzheimer disease. Other factors are probably involved that we do not know about at this time, but there is a lot of ongoing research to discover them.

  On the other hand, a few gene variants have been described to cause early-onset Alzheimer disease, meaning subjects with disease onset before the age of 65. These are variants in the genes of the amyloid precursor protein (*APP*), presenilin-1 and presenilin-2 (*PSEN1*, *PSEN2*) (Section 11.14.1). These are not relevant for Dorothy at her current age.
- **Genetic Testing:** Direct-to-consumer testing is not always considered scientific or definitive. In order to confirm the results of the commercial kit, the genetic counselor recommends repeating the test for the *APOE* variant through a CLIA-approved genetic laboratory. Since Dorothy has no symptoms and there is no evidence of early-onset dementia in the family, there is no need at this time for her to be tested for the other genes, such as *PSEN1*, *PSEN2*, and *APP*, associated with early-onset Alzheimer disease, or other genes associated with other forms of dementia, such as frontotemporal dementia. The genetic testing confirms the homozygous *APOE4* allele.
- **Management and Recommendations:** There are no preventative treatments for late-onset dementia. Therefore, even in the presence of a risk factor, there are no specific recommendations except for pursuing a healthy lifestyle that would prevent other possible aggravating factors, such as cardiovascular risks.
- **Implications for Other Family Members:** Being a carrier of specific *APOE* alleles can increase the risk for Alzheimer disease. Since Dorothy is homozygous for *APOE4*, her children are carriers for the same allele. This will increase their risk to develop Alzheimer disease. If Dorothy inherited both alleles from each parent, her siblings as well are at risk of being either a carrier or homozygous for the same *APOE* allele, thus possibly carrying an increased risk of developing Alzheimer disease themselves.
- **Differential Diagnosis:** Late-onset dementia can be caused by a number of other conditions. In particular, secondary causes, such as chronic vascular diseases, vitamin B12 deficiencies, and thyroid dysfunctions, should be ruled out because these are treatable. In terms of neurodegenerative conditions, frontotemporal dementia should be considered as well. This usually manifests earlier in life, with an average onset in the sixth decade, and with a very distinctive cognitive profile. Neuropsychological tests and imaging studies can help discriminate between these two conditions.

- **Caveats:** The literature considers *APOE* genotyping not fully specific and to have little role in asymptomatic or predictive testing for Alzheimer disease at this time, since it does not affect management, morbidity, or mortality. In addition, there could be potential adverse psychosocial effects, depending on the age of the individual when doing the testing. Furthermore, although genetic testing results are not supposed to affect health insurance, there are no current regulations about life insurance and long-term care (Section 10.4). The most common allele of *APOE* is the E3 form, which is not associated with Alzheimer disease. The E2 form, instead, seems to be associated with a decreased risk of Alzheimer disease, while it contributes to type III hyperlipoproteinemia, a familial form of hyperlipidemia (Section 11.4.6) which can predispose to cardiovascular conditions. Through research studies, many other genes, in addition to *APOE,* have been found to increase the risk for Alzheimer disease. Although a combination of variants in these genes, or polygenic risk scores (Section 5.7.5), have been proposed for predictive risk, they have shown no clinical utility to date.
- **Outcome:** Dorothy develops Alzheimer disease in her mid-70s. Her sister, who is 2 years younger than Dorothy, manifests Alzheimer disease when she turns 72. Their two brothers, now in their 80s, never developed dementia. Recently over the holidays, Dorothy's siblings were each given an ancestry and genetic discovery kit by their children that includes testing for the *APOE* alleles. One of her brothers is homozygous for *APOE4* as well, while her sister is just heterozygous for the same allele. This is consistent with this variant being a risk factor and not a causative genetic variant for Alzheimer disease.

The Diagnosis

Clinical Diagnosis: Patient and Sister - dementia, late-onset Alzheimer disease. Brother - asymptomatic.
Genetic Diagnosis: Patient and Brother - *APOE* gene, homozygous variants. Sister - *APOE* gene, heterozygous variant.

Further Reading

Andrews, S.J., Fulton-Howard, B., Goate, A., 2020. Interpretation of risk loci from genome-wide association studies of Alzheimer's disease. Lancet Neurol. 19, 326–335. PMID 31986256.

Bird, T.D., October 23, 1998. Alzheimer disease overview. Adam, M.P., Mirzaa, G.M., Pagon, R.A., Wallace, S.E., Bean, L.J.H., Gripp, K., Amemiya, A., (editors). In GeneReviews. University of Washington, Seattle (WA), 1993–2023 [updated December 20, 2018]. PMID 20301340.

Yamazaki, Y., Zhao, N., Caulfield, T.R., Liu, C.C., Bu, G., 2019. Apolipoprotein E and Alzheimer disease: pathobiology and targeting strategies. Nat. Rev. Neurol. 15, 501–518. PMID 31367008.

Case 11.17.6

Direct-to-Consumer Cases: Wellness Report By Direct-to-Consumer Testing

Angela E. Scheuerle

Department of Pediatrics, Division of Genetics and Metabolism, Department of Pathology, Division of Genetic Diagnostics, McDermott Center for Human Growth and Development, University of Texas Southwestern Medical Center, Dallas, TX, United States

Abstract

A 35-year-old male has purchased dietary supplements based on a direct-to-consumer DNA test for wellness.

The Case

Jahn is a 35-year-old male who has been a patient of yours for about 12 years. He has always been generally healthy and is on no chronic medications. His appointment is a routine annual evaluation. Reviewing his medication list, you notice addition of about a dozen dietary supplements. Some of them are listed only by trade name and their ingredients are not detailed. Jahn explains that he did a cheek swab DNA test at the state fair to look at his nutritional status. When the doctor called with the result, she suggested ways for Jahn to improve his diet and, with her advice, he subscribed to this supplement program. His insurance did not cover the test or the supplements, and he wants to know if you have any suggestions for generic substitutes.

The Issues

- **Did he have a legitimate test?** You wonder about any medical test that might be done at a community event. The marketing of it as a "genetic test" makes it sound cutting edge and technologically sophisticated, but the supplied test report is not medically interpretable. The document Jahn was given does not report genetic results in a standardized manner. It is a list of genes and proteins that were found to have variants. With each variant there is a recommendation for supplementation with a particular vitamin, mineral, or herb.
- **Is there a medical indication for the supplements?** Jahn is a healthy patient whom you have known for many years. You know that he eats a standard, varied diet without problems. He reports no symptoms of illness or compromise and all objective assessments have been well within normal parameters. It is most likely that he is receiving adequate nutrition through his normal diet and does not need the supplements.
- **Is this an appropriate use of genetic testing?** The medical literature suggests that genetic testing used in this way has not been shown to have clinical utility (Section 9.3). There are genes that can be tested reliably to ascertain increased risk for lifestyle-related entities such as lipid biology, alcohol metabolism, and enzyme deficiencies related to digestive abilities (e.g., lactose intolerance, fructose intolerance, caffeine sensitivity), but definitive nutritional supplement results such as Jahn was given are likely not to be of benefit. The US Food and Drug Administration definition of Health Fraud

Genomics in the Clinic. https://doi.org/10.1016/B978-0-12-816478-5.00110-2

Scams is: "products that claim to prevent, treat, or cure diseases or other health conditions, but are not proven safe and effective for those uses" (FDA). It notes that such scams can delay appropriate diagnosis, waste money, and, in some cases, cause serious injury or death. Jahn had a "test" and then was sold expensive supplements from the same source.

The Referral

- **Genetic Evaluation:** It is difficult for practitioners to evaluate the credibility and scientific reliability of newer genetic companies to assess whether there is clinical utility for the tests offered (Section 9.1.6). A genetic referral is likely not necessary, but could be explored for interpretation if the report is complex. Jahn, however, does not have a personal clinical or family-based indication for genetic services. Without a specific reason for a genetic referral, the visit may add to the time and cost burden on Jahn. Further, genetics clinics may not be open to referrals for counseling of DTC testing.
- **Recommendations:** Referral to a dietician can be considered. Professional review of Jahn's diet may reassure him that the supplements are not necessary. Also, a dietician can evaluate the supplement ingredients to confirm that they are not potentially harmful. As always, general lifestyle recommendations, such as exercise, sun protection, moderation of alcohol, not smoking, regular sleep habits, and stress reducing activities, should be made.
- **Caveats:** Jahn could be counseled to call his insurance and alert them to the possibility of insurance fraud (Section 9.2). If he shared other personal financial information with the testing company, he should take precautions. He could also check with the Department of Health and Human Services (DHHS), to further investigate the legitimacy of the company. There are a number of new companies marketing this type of testing to the public and generating a wellness report that may suggest recommendations without clinical validity. Increased scientific support and oversight is needed.
- **Outcome:** Jahn meets with a dietician. None of the supplements are medically indicated, but several seem harmless and he opts to continue them.

The Diagnosis

Clinical Diagnosis: None. Dietary deficiencies only by direct-to-consumer testing.
Genetic Diagnosis: None. Asymptomatic individual.

Further Reading

Floris, M., Cano, A., Porru, L., Addis, R., Cambedda, A., Idda, M.L., Steri, M., Ventura, C., & Maioli, M., 2020. Direct-to-consumer nutrigenetics testing: an overview. Nutrients 12, 566. PMID 32098227.

U.S. Food and Drug Administration. Health Fraud Scams [Online]. FDA.gov. Available: https://www.fda.gov/consumers/health-fraud-scams. Accessed January 16, 2022.

U.S. Department of Health and Human Services. Fraud Alert: Genetic Testing Scam [Online]. U.S. Department of Health and Human Services. Available: https://oig.hhs.gov/fraud/consumer-alerts/alerts/geneticscam.asp. Accessed January 16, 2022.

Section 11.18

Ethical Issues Cases

Case 11.18.1

Ethical Issues Cases: Predictive Testing in an Adolescent with Positive Family History

Lynn Wein Bush[1], Amy Kritzer[2] and Cheryl D. Lew[3]

[1]*Division Genetics and Genomics, Department of Pediatrics, Boston Children's Hospital, Pediatrics, Harvard Medical School and Center for Bioethics, Harvard Medical School, Boston, MA, United States;* [2]*Division of Genetics and Genomics, Department of Pediatrics, Boston Children's Hospital, Harvard Medical School, Boston, MA, United States;* [3]*Department of Pediatrics (Emerita), Department of Pediatrics, and Emeritus Faculty, Center for Pediatric Bioethics, Children's Hospital Los Angeles, Keck School of Medicine at University Southern California, Los Angeles, CA, United States*

Abstract

An anxious mother with family history of breast cancer requests predictive BRCA testing for her 13- and 16-year-old daughters.

The Case

Ms. Schwartz accompanies her healthy 13-year-old daughter, Alona, for an annual physical exam with you, their local pediatrician. When you are updating Alona's recent history, your patient reports she recently began menstruation, and her mother interjects: "It's time for genetic testing." Ms. Schwartz requests you "do extra lab work today so Alona can have the BRCA (breast cancer susceptibility gene) test given our family history." She adds that Alona's sister, Stephanie, who just turned 16 years old, is scheduled for her yearly check-up with you next week and "needs BRCA testing too." Ms. Schwartz explains she initially held off asking you "to test my girls for BRCA," but as years have passed describes increasingly feeling "very, very anxious they have it" and expresses wanting to know "now" so they can be as proactive as possible or "relieved if some miracle happens and they don't have it." You emphatically respond to Ms. Schwartz that Alona and Stephanie *must* wait for testing, preferably at a minimum after 18 years of age, ideally later.

The Issues

- **Did the informed consent-assent process present factual information to enhance patient decision-making?** Even the earliest stage of the informed consent-assent process (Section 10.1) requires adequate information and counseling for good medical decision-making. No family history has been obtained, and any pathology or genetic testing results of the mother have not been disclosed.

There have been modifications in professional guidelines and opinions regarding predictive testing in minors. Currently, the joint policy statement from the American Academy of Pediatrics and the American College of Medical Genetics is

Genomics in the Clinic. https://doi.org/10.1016/B978-0-12-816478-5.00111-4

ethically more nuanced, permitting such testing for first time while noting "generally" not recommended before reaching adulthood, allowing for acknowledgment of significant parental anxiety as a reason for testing as well as weighing preference of older child/adolescent and some family circumstances. Counseling is required, however.

[Of note: Section 11.16.1, *"Breast Cancer Gene Mutation Found Incidentally," does not recognize the evolution of this broader scope for consideration of parental or child preference. Moreover, that case differs from this regarding younger age of children (8–11 years old vs. 13–16 years old), and some may argue those children should be presented with the opportunity to assent to testing as an adolescent if they want. Currently, when families with a known pathogenic variant (Section 7.1.3) in the BRCA gene request such testing (be they the parent or child), clinicians typically recommend counseling with potential testing starting at age 18 or early 20s.]*

- **Is there good ethical reasoning on the clinician's side to reduce harms and maximize beneficence** (Section 10.1.2)**?** You, the pediatrician, and the mother are talking about testing for the BRCA genes, and not considering possibility of other contributory gene variants. Moreover, this was relayed as a presumed adult-onset variant in a child or adolescent not recognizing the possibility for an earlier onset that will need much earlier monitoring. There is more to breast cancer screening than BRCA testing, with many genes to consider. Therefore negative BRCA results do not necessarily mean no breast cancer risk. In families with a strong history of cancer a genetic consult is essential including a good three-generation pedigree.
- **Can predictive testing be done in adolescence in a way to reduce the ethical and psychosocial challenges associated with variants of uncertain significance (VUSs)?** If there is a known genetic diagnosis in the family to narrow the predictive test to a targeted test, then testing can be considered in an "unaffected" or presymptomatic individual. If not, a geneticist or counselor would stress that the most important and informative person to test is the individual with cancer. Most genetic professionals would not offer panel testing to an adolescent because the more genes tested, there is an increased risk for VUSs (Sections 7.1.5 and 10.1.4). Targeted testing (Section 6.2.1) would be a different situation because it reduces the risk for detecting VUSs.

Ordering the appropriate test is yet another reason for you, the pediatrician, to seek a genetic referral for this family.

- **Is there an "Open Future" Issue?** In this scenario, you, the pediatrician say parents are not allowed to choose predictive testing for their child because of a child's "*right* to an open future," meaning a right to have one's future options kept open until one is capable of making one's own decisions. This would preclude the opportunity for minors to decide for themselves as an adult whether they want testing or not. However, it is not an absolute "right" and counseling ought to weigh what is "in the best interest of the child" with significant other factors such as family values and circumstances as well as adolescent older-patient preferences.
- **Was there engagement of the older child-adolescent?** You, the pediatrician, did not facilitate conversation and engagement with Alona to see if she has opinions regarding genetic testing. There is similar concern whether her sister Stephanie is aware that her mother seeks predictive testing for her.

The assent process is documented in professional guidelines of the American Academy of Pediatrics and American College of Genetics and Genomics, emphasizing active discussion and engagement of the older child/teen and preferably assent in the clinical setting (required in research).

Because the decision-making process with genome sequencing is more complex than most clinicians realize, it is critical that transparency and engagement with parent, child, and clinician be encouraged, recognizing that it is not always feasible (Section 10.1.2).

- **Are there always psychological factors with predictive genetic testing in minors?** Psychological distress regarding the return of genomic information, especially during the adolescent period of identity formation, has long been argued as a given for all involved, yielding strong recommendations not to permit sequencing for children. However, recent reviews of the literature found serious adverse psychological outcome to be uncommon and to date, empirical research demonstrating distress has largely not supported long-term negative psychological impact, though research is needed with more nuanced assessment measures to better identify varying levels of distress and who may be at most risk (Section 10.2).
- **Will waiting until the children are older reduce uncertainty?** In this scenario, you, the pediatrician, respond that Alona and Stephanie can find out *definitively* when or if they are tested as an adult. However, since new variants are reported on an almost-daily basis and risk assessment has many contributory factors, variant interpretation may still not be definitive in a decade. Thus they may also need other types of surveillance.

The Referral

- **Genetic Evaluation:** A geneticist or genetic counselor can assess for the child's understanding of testing and ability to assent (Section 10.1.2). A referral is made to a geneticist and/or genetic counselor who specializes in cancer genetic counseling. Further information is gathered to determine if testing should be offered.

Relevant family history includes: types of cancer seen in the family (for example, breast, uterine, ovarian, colon), age of onset of cancer in the family, relationship between the person seeking testing and the individual with cancer, and presence of a known pathogenic variant in the family.

Family history reveals that the maternal grandmother died at age 42 from metastatic breast cancer and a maternal great aunt had a history of ovarian cancer at the age of 45. There is a known pathogenic *BRCA1* c.185delAG variant that was previously identified in the grandmother, and the mother, Ms. Schwartz, is also a known carrier. Thus, hereditary breast and ovarian cancer syndrome (Sections 8.1.2, 8.3.1, 11.16.1, 11.16.4, and 11.16.5) is present in this family.

- **Genetic Counseling:** A careful family history will evaluate for the presence of an autosomal dominant condition in the family (Sections 2.7.1 and 3.2.1). Not all genetic variants are disease causing. It is important that a clinician who is experienced in the interpretation of genetic testing review the results (Section 7.1) prior to considering targeted variant analysis (Section 6.2.1) for family members. When evaluating anyone with a family history of cancer it is important to evaluate for other associated features of familial cancer syndromes such as mucocutaneous pigmentation or hamartomas in Peutz–Jeghers syndrome, or macrocephaly, hamartomas, or history of autism in Cowden syndrome (Section 11.17.1). In autosomal dominant conditions there is a 50% risk of passing the variant onto the next generation (Section 2.7.1). This risk is the same for each and every offspring of that individual, irrespective of the number of affected or unaffected individuals. In many breast cancer syndromes there are also risks for males in the family.
- **Management and Recommendations:** As pathogenic BRCA variants are associated with only adult-onset cancers, there are no risk-reducing or surveillance recommendations for individuals under the age of 18. Finding out that one is a genetic disease carrier may potentially cause complex psychological and psychosocial issues (Chapter 10). Diagnosis in an asymptomatic individual can affect insurability in the future for the girls as there are currently no legal protections for life, long-term care, or disability insurance (Section 10.4). If there is no known genetic diagnosis, then consider surveillance starting at the age 10 years younger than the first breast cancer diagnosis in the family.
- **Differential Diagnoses:** While *BRCA1* and *BRCA2* gene mutations cause the most common inherited forms of breast cancer, breast cancer can be a part of other cancer syndromes. A complete physical and family history is essential when considering the differential.
- **Caveats:** Many families misinterpret the 50% risk to all offspring. It is important that they understand the 50% risk applies to each child of the genetic disease carrier regardless of how many children test negative or positive. Also, carrying the variant merely increases the risk to develop breast cancer and the other related cancers, but does not indicate certainty of the cancer.

  Although there are currently protections against discrimination if you are a genetic disease carrier for employment and health insurance, there are no protections for life or long-term care insurance. Finding out this information as teenagers could impact their ability to get insurance as adults (Section 10.4).

  While BRCA genetic testing is "generally not recommended" under 18 years of age, it is permitted, particularly with a family history of breast cancer and heightened anxiety about being affected (see above and Section 10.1.2). Referral to genetics should be made for families seeking testing.
- **Outcome:** Alona and Stephanie get targeted testing (Section 6.2.1) a few months after voluntarily requesting genetic testing for breast cancer. They each assent following a few separate counseling sessions with the genetic counselor and clinical geneticist, who finds them still preoccupied, like their mother. Both clearly articulate their wish to learn their status, feeling they will have relief if they do not test positive and can have better life planning and decision-making if they test positive for breast cancer. The family pays out-of-pocket, as their insurance will not provide coverage. The results show that Stephanie is a carrier of the known familial variant, but that Alona is not. Based on the known ages of cancer in the family, surveillance is recommended to start at the standard age of 25 years for Stephanie. The family expresses relief that "at least they now know," although the family acknowledges that they still have "a lot to process."

The Diagnosis

Clinical Diagnosis: Family history of hereditary breast and ovarian cancer syndrome with *BRCA1* gene pathogenic variant.

Genetic Diagnosis: Sister and Mother—*BRCA1* gene, heterozygous pathogenic variant.

Further Reading

Bush, L., 2014. In the best interest of the child: psychological and ethical reflections on traditions, contexts, and perspectives in pediatric clinical genomics. Am. J. Bioeth. 14, 16–18. PMID 24592832.

Bush, L.W., Bartoshesky, L.E., David, K.L., Wilfond, B., Williams, J.L., Holm, I.A., et al., 2018. Pediatric clinical exome/genome sequencing and the engagement process: encouraging active conversation with the older child and adolescent: points to consider-a statement of the American College of Medical Genetics and Genomics (ACMG). Genet. Med. 20, 692–694. PMID 29565417.

Fenwick, A., Plantinga, M., Dheensa, S., Lucassen, A., 2017. Predictive genetic testing of children for adult-onset conditions: negotiating requests with parents. J. Genet. Couns. 26, 244–250. PMID 27680566.

Katz, A.L., Webb, S.A., Committee on Bioethics, 2016. Informed consent in decision-making in pediatric practice. Pediatrics 138, e20161485. PMID 27456510.

Ross, L.F., Saal, H.M., David, K.L., Anderson, R.R., American Academy of Pediatrics, & American College of Medical Genetics and Genomics, 2013. Technical report: ethical and policy issues in genetic testing and screening of children. Genet. Med. 15, 234–245. PMID 23429433.

Case 11.18.2

Ethical Issues Cases: Preconception Sex Selection

Lynn Wein Bush[1], Amy Kritzer[2] and Cheryl D. Lew[3]
[1]*Division Genetics and Genomics, Department of Pediatrics, Boston Children's Hospital, Pediatrics, Harvard Medical School and Center for Bioethics, Harvard Medical School, Boston, MA, United States;* [2]*Division of Genetics and Genomics, Department of Pediatrics, Boston Children's Hospital, Pediatrics, Harvard Medical School, Boston, MA, United States;* [3]*Department of Pediatrics (Emerita), Department of Pediatrics, and Emeritus Faculty, Center for Pediatric Bioethics, Children's Hospital Los Angeles, Keck School of Medicine at University Southern California, Los Angeles, CA, United States*

Abstract
A 28-year-old woman requests preconception gender selection based on the X-linked condition in her family, ornithine transcarbamylase deficiency.

The Case

You are a general obstetrician-gynecologist in a midsize practice with a full waiting room. A 28-year-old patient, Joan, enters your office saying she wants to become pregnant "but can't have a boy." In the brief office visit, you explore what this means to her while cognizant of completing her scheduled gynecological exam without running too far overtime. Upon further questioning, Joan states "Many boys in my family had medical problems, including early death and serious immune disorders, and others had weird behavior and intellectual disabilities." You realize that she is asking for sex selection for a future pregnancy because of a likely X-linked condition in the family.

The Issues

- **Are there ethical distinctions of nonmedical sex selection?** Many individuals and countries would argue that there is, at most, limited ethical justification for sex selection when based on parental gender-trait preference versus to avoid serious X-linked disease based on family history (Section 2.7.2). The greatest concern is that nonmedical sex selection reinforces discrimination and stereotypes (e.g., favoring males). Some clinicians make exceptions to their practice when there is a family with multiple children of one sex and the mother requests a child of the opposite sex. And some fertility clinics advertise this service welcoming sex selection as a matter of parents-to-be autonomy to select trait of their choice. The family should be aware during the consent process that even if sex selection is performed, there is the potential that later in life the child's gender identity may be discordant with their chromosomes, as possible with any individual.

Genomics in the Clinic. https://doi.org/10.1016/B978-0-12-816478-5.00112-6

- **Is it true that X-linked disorders result in medical problems that primarily affect boys?** You, the obstetrician, immediately think of fragile X syndrome, which can have potential impact on females too (Sections 11.12.3 and 11.13.11). Also, you have little knowledge of X-linked severe immunodeficiency and other metabolic disorders.

While traditionally X-linked disorders were thought to only affect males, we now know for many conditions, females may be affected as well (Section 2.7.2).

- **Are there some harms versus benefits even if a condition is X-linked?** Concerns exist that many conditions have a spectrum of phenotype, severity, and/or onset that might impact females too. There are also disability rights issues to take into consideration.
- **Are there professional responsibilities to ensure requisitioning appropriate testing?** You, the obstetrician, suggest a clinical test that sorts and can enrich for male or female sperm. You realize that there are details on family history that need to be clarified and this is not an optimal test for this circumstance with a familial history of ill-defined disorder. Part of professional responsibility includes an informed consent process with sufficient and correct information. Thus, you refer your patient for a genetic evaluation to help determine risk within her family.

The Referral

- **Genetic Evaluation:** Joan is referred for preconceptual genetic counseling. A comprehensive three-generation pedigree (Sections 2.7 and 3.2) is taken on both the patient and her partner to assess for potential genetic conditions from her partner's family as well. When multiple male members of a family have a similar medical history or developmental disability it raises concern for a possible X-linked condition (Section 2.7.2). Additional questions include whether any members of the family have other concerning features such as premature ovarian failure or tremor, which are common features in fragile X premutation carriers (Section 11.13.11), or milder presentations of learning disabilities or autism among female members of the family, also raising concern for an X-linked condition.

  Additional relevant family history includes:
  - Do males exhibit autism or severe intellectual disability, but females have mild learning disabilities?
  - Are there any physical features shared by the individuals such as dysmorphic or unusual facial features (Section 5.1), poor growth, short stature, immunodeficiency, musculoskeletal abnormalities, or birth defects?
  - Have there been infant deaths in the family?
  - Is there a history of recurrent miscarriages in the family which might raise concern for a genetic condition such as a balanced chromosomal translocation (Section 2.2.1)?
  - Is there a history of severe, recurrent illness in the family in children?
- **Genetic Counseling:** In the absence of a known diagnosis, Joan is counseled that she has a 50% chance of having a boy and each of her sons would have a 50% chance of having the X-linked condition if she is a carrier. She is also counseled that there may be risks for female children as well. Further information is requested on her family to try and clarify the condition.

  It is also possible that the mode of inheritance within the family is not X-linked. Autosomal dominant conditions (Section 2.7.1) can have reduced penetrance or variable expressivity (Section 2.8). It is possible that members of a family also carry the genetic mutation, but the mutation is asymptomatic or mildly symptomatic. This can occur in both single gene disorders as well as chromosomal deletion or duplication syndromes (Section 2.2.1). In taking a history, the genetic counselor also evaluates for the presence of offspring of both sexes as some pedigrees can appear to have X-linked inheritance if there are only a few children or if most children are of one gender by chance alone.

  If no diagnosis is known, carrier panel(s) (Sections 4.4.1 and 6.2.4) could be offered but are not always comprehensive. Without a known genetic diagnosis it can be difficult to define risk. Even with negative genetic carrier testing there is residual risk as multigene panels (Section 6.2.3) generally contain only a subset of genes. Comprehensive carrier panel screening is offered to maximize the information carried by the patient.

  Several weeks later, based on information gathered from the family, it is determined that Joan has a family history of ornithine transcarbamylase (OTC) deficiency (Section 11.6.4), an X-linked inborn error of metabolism, and that both her nephew through her sister and her maternal aunt's son have been affected. Because of this, her sister is an obligate carrier. Joan has a 50% chance of also being a carrier. If she is a carrier of an *OTC* gene mutation, she has a 50% chance of passing that genetic variant onto any child (Section 2.7.2).
- **Genetic Testing:** As Joan is not yet pregnant she is an excellent candidate for targeted variant testing (Section 6.2.1). Testing is discussed including the need to obtain insurance approval beforehand. If the genetic variant in the family is known, targeted testing can be offered to Joan. If not, she, or ideally an affected individual, will need full sequencing of the gene (Section 6.2.2), encoding the enzyme. She finds out that the mutation in her nephew has been identified.

Joan is aware that because the lack of the urea cycle enzyme results in excessive accumulation of nitrogen, there is accumulation of ammonia (hyperammonemia) that results in vomiting, refusal to eat, progressive lethargy, and coma. She is very concerned that she could pass this onto any son. Insurance approval is obtained, and Joan undergoes the targeted testing. She is found to be a carrier.

- **Management and Recommendations:** In the case of OTC deficiency, referral to a maternal–fetal medicine specialist familiar with metabolic conditions should be made, ideally prior to embarking upon a pregnancy. For carriers of an *OTC* gene mutation, further medical evaluations can be offered. While generally more severe in males, woman can be symptomatic to varying degrees including being as severely affected as males. Mothers of children with OTC deficiency are at risk of medical complications as well, particularly in the postpartum period.

  Further discussion is conducted about the following: options for preconception genetic testing, use of in vitro fertilization with preimplantation genetic testing (Section 4.5.1), egg donation (Section 4.5.2), prenatal testing via chorionic villous sampling, or amniocentesis (Section 4.4.2). If a woman is carrying a baby with known or possible OTC deficiency, a specific delivery plan should be created to ensure the health of the mother and baby. The baby should be delivered at a tertiary care facility with a level III neonatal intensive care unit (NICU) and specialists in metabolism.
- **Differential Diagnoses:** OTC deficiency is one of many inborn errors of metabolism that can present with serious health complications for the baby. As it is X-linked, woman can be symptomatic as well. This presents a unique situation where mothers may find themselves as patients as well. While there are many X-linked conditions that cause autism and intellectual disability, there are many autosomal dominant and recessive conditions as well. Determining the exact genetic diagnosis allows for the most accurate risk assessment for the family. Fragile X syndrome remains the most common inherited form of intellectual disability and autism and is offered on many carrier panels.

  There are many other X-linked conditions that can present within a family that can have serious medical complications, such as hemophilia, a disorder of abnormal clotting (Section 11.9.1); Fabry disease, a rare inherited lysosomal storage disorder of fat metabolism resulting from markedly deficient activity of the lysosomal enzyme, α-galactosidase A; severe combined immunodeficiency, a rare syndrome characterized by little or no immune response due to genetic defect affecting T cells and at least one other type of immune cell; and hypohidrotic ectodermal dysplasia, a disorder characterized by partial or complete absence of sweat glands and often underdevelopment/absence of mucous glands in respiratory/gastrointestinal tracts as well as abnormal hair, nails, and teeth (Section 11.11.3). Referral to genetics is the most appropriate next step in families with a strong family history suggesting a possible genetic diagnosis.
- **Caveats:** Understanding the potential risk for females in X-linked conditions is essential in accurate risk counseling and assessment. In vitro fertilization with or without preimplantation genetic testing is not covered by all insurance companies. It is important for families to understand the benefits provided by their health plan.
- **Outcome:** Joan is overwhelmed to discover that females who are carriers of only one *OTC* mutation can be affected too. When she finds out her insurance will not cover preimplantation genetic testing and in vitro fertilization (IVF), she decides not to undergo preconception sex selection. Rather, she becomes pregnant and is awaiting results from a chorionic villous sampling (Section 4.4.2) to see if the developing fetus is affected.

The Diagnosis

Clinical Diagnosis: Family history of ornithine transcarbamylase deficiency with *OTC* gene pathogenic variant.
Genetic Diagnosis: *OTC* gene, heterozygous pathogenic variant.

Further Reading

ACOG Committee on Ethics, 2009. ACOG committee opinion no. 439: informed consent. Obstet. Gynecol. 114, 401–408. PMID 19623005.

Faden, R.R., Beauchamp, T.L., 1986. A History and Theory of Informed Consent. Oxford University Press, New York.

Kalfoglou, A.L., Kammersell, M., Philpott, S., Dahl, E., 2013. Ethical arguments for and against sperm sorting for non-medical sex selection: a review. Reprod. Biomed. Online 26, 231–239. PMID 23337421.

Lichter-Konecki, U., Caldovic, L., Morizono, H., Simpson, K., August 29, 2013. Ornithine transcarbamylase deficiency. Adam, M.P., Mirzaa, G.M., Pagon, R.A., Wallace, S.E., Bean, L.J.H., Gripp, K., Amemiya, A., (editors). In GeneReviews. University of Washington, Seattle (WA), 1993–2023 [Updated May 26, 2022]. PMID 24006547.

Chapter 12

Communicating Genetics and Genomics

Isha Gupta[1], Antonie D. Kline[2] and Ethylin Wang Jabs[3,*]

[1]*Igentify, New York, NY, United States;* [2]*Harvey Institute for Human Genetics, Greater Baltimore Medical Center, Baltimore, MD, United States;* [3]*Department of Genetics and Genomics Sciences, Department of Pediatrics, Department of Cell, Development, and Regenerative Biology, Icahn School of Medicine at Mount Sinai, New York, NY, United States*

Abstract

Explaining medical concepts and ensuring that patients understand why they are being seen by a doctor are integral parts of clinical care. The complexity of these explanations will vary with levels of patients' medical literacy, but basic explanations encompass the core elements of each concept. The very essentials of genetic concepts may not be obvious or known to most patients. Discussions about genetics can be challenging because patients may not know what a gene is or how it affects their bodies. Thus, it becomes necessary to teach patients basic genetics before explaining genetic conditions or genetic testing. These concepts become especially important after the initial evaluation, when results will often determine the next steps in care. Examples of how to explain these concepts to patients are presented.

It is increasingly likely that a primary care provider, whether in pediatrics, internal medicine, or obstetrics/gynecology, or other specialties, such as neurology or psychiatry, may be ordering and explaining genetic testing and results. These may precede a referral to genetics, or even replace a genetic visit under certain circumstances. In any case, the explanations below can be used as a foundation to explain basic genetic concepts, specific tests, and return of results by practitioners. For the purpose of standardization, these descriptions, which can be given to patients, will be for an English-speaking patient or parent/guardian with a high school education. The diagrams at the end of the chapter may be useful to help explain visually these concepts.

12.1 Overview of Genetic Discussion with Families

Tailoring explanations of genetic concepts and results to patients at various levels of literacy as well as from different backgrounds can be challenging. Patients may also be getting information about genetics via third-party sources, such as friends and family, who have experienced genetic testing, or from the internet. After an initial genetic appointment, patients will often do their own research so the clarity of the explanation of a result becomes crucial. For these reasons, genetic counselors and clinical geneticists are helpful and are trained in interpreting and expressing genetic concepts for patients as well as helping them process this information. Also helpful is the summary letter from a genetic consult, not only for patients to receive a reminder of what was discussed, but also for referring physicians because the concepts may be new to them as well.

In clinical practice, patients may arrive with previous genetic testing results ordered by primary providers, specialists, or other geneticists. Depending on how long ago testing was done or how much the patient understood at the time, an

[*] Author EWJ was employed by Icahn School of Medicine at Mount Sinai at the time of writing the manuscript, but is currently employed by Mayo Clinic

Genomics in the Clinic. https://doi.org/10.1016/B978-0-12-816478-5.00113-8

explanation of basic genetics generally is necessary for patients to understand what has already been done and what can be done moving forward. Genetic testing for patients may be tiered, starting with basic screening tests and progressing to more comprehensive or more specific testing, depending on the clinical indication.

Currently, genetic testing requires informed consent, meaning that the patient or their legal guardian needs to have a good understanding of the recommended tests and choose to move forward with the testing before the clinician can proceed with ordering them. Clear and concise explanations regarding the basics of genetics and the specific test are very important. There are many different types of results possible with each test and there may be various potential follow-up tests recommended. All of this should be understood as best as possible, along with agreement to do the testing, and these both would comprise informed consent by the patient or parent/guardian. See Section 10.1 for a more detailed explanation.

12.2 Counseling about Genetic Concepts

Below are various types of common genetic concepts in clinical genetics and how they might be explained at a basic level to a patient or a family. These explanations can be further simplified or elevated, depending on the patient's understanding and can be useful either prior to ordering genetic testing or once the results have returned.

12.2.1 Cell, Chromosome, DNA, Gene, Variant

Our bodies are made up of trillions of cells. Cells are the building blocks of our bodies and contain our genetic information, which is packaged into chromosomes (Figs. 12.1 and 12.2).

Chromosomes contain more than 20,000 genes that include the information to make us who we are and instruct our organs and bodies to work properly. These genes are made of DNA (deoxyribonucleic acid) (Fig. 12.1) and are the instructions for making the proteins that do the work for us to function. DNA is made up of four different chemicals designated by the letters A, C, G, and T. Each combination of these letters forms the spelling or the code of a gene, which in turn codes for a specific protein. Each protein that is made carries out the function coded by the particular gene.

There are typically 46 chromosomes in each cell. Chromosomes exist in pairs, so every cell has 23 pairs of chromosomes. We get half of our chromosomes from our mother and half from our father. As chromosomes exist in pairs, genes also exist in pairs with one copy of each gene from the mother and the other copy from the father. The chromosome pairs are numbered 1 to 22. The 23rd pair determines our sex. Generally, one X chromosome and one Y chromosome are present in cells of males and two X chromosomes are present in cells of females.

It is important that we have exactly 46 chromosomes for our bodies to develop as expected (Sections 2.1 and 2.2). If we have extra or missing genetic material, this can affect how our bodies are made or function. The sequence or order of the letters of each gene (like the spelling of a word), is vital and a change (like a misspelling of a word) can affect the type, production, location or function of the protein and, in turn, how the body is made and how it works. These changes or mutations in genetic information also are known as variants. If a variant or mutation is known to affect proteins in a way that will change how the body functions, it is considered pathogenic (Fig. 12.3).

12.2.2 Modes of Inheritance

Dominant and Recessive

There are two major ways in which mutations, or variants, in genes act to cause genetic conditions. A dominant pathogenic variant is a single change in one gene that leads to some change in the body. If a parent carries a dominant variant, which causes a symptom or condition in that parent, there is a 50% chance (like a flip of the coin—heads or tails) to pass this variant to each future child (Fig. 12.4). In this way, children can directly inherit a condition from an affected parent. Sometimes a member of the family is the first person with a dominant condition because a new (de novo, not previously present in a parent), dominant variant can occur at conception. Dominant conditions can present variably, in that a mildly affected parent can have a more severely affected child, and vice versa.

Recessive pathogenic variants require both copies of the gene to carry a variant to lead to a change in the body. If only one copy of the gene has a recessive variant, the other unaffected copy will be able to work well enough for the body to function as expected. Both parents have to carry a variant in the same gene in order for there to be a chance of having an affected child.

When both parents, each carry only one copy of a variant in the same gene there is a 25% chance for an affected pregnancy (Fig. 12.5). Those parents would be referred to as carriers because they would be unaffected by the condition. A parent who is affected will pass one variant to any future child, who would then be a carrier, unless the other parent is a carrier or affected as well. If one parent is affected and the other is a carrier, there is a 50% chance of an affected pregnancy. If both parents are affected with the same recessive condition, every pregnancy would be affected, as every copy of that gene passed on would have a variant (Section 2.7.1). If each parent is a carrier for different recessive conditions, each with one variant on different genes, the future children could be carriers and not be affected by any of the conditions. In general, recessive conditions tend to be similar in clinical symptoms and severity among affected individuals.

X-linked

Genes on the X chromosome function differently in males who have a single X compared to females with two X chromosomes. A pathogenic variant on the X chromosome will act as dominant in a male, since there is no other copy present, and that male will have the condition or symptom associated with the variant. A variant on one X chromosome in a female can act in a dominant or recessive manner, as discussed above. A dominant variant on the X chromosome in a female would cause the condition or symptom (Fig. 12.6). With a recessive variant on one of the X chromosomes, the other copy of the gene normally works well enough for there to have no effect, so that female should be an unaffected carrier. However, males with one X chromosome with a "recessive" variant do not have another normal copy to compensate and would be affected, so for X-linked conditions, males are more severely affected.

Usually one of the X chromosomes in each cell in every female is randomly shut off by a process called X inactivation. However, it is possible in some of these cases for the X inactivation not to be random. Sometimes, the X chromosome without a variant can be the one inactivated in most cells, resulting in most of the X chromosomes with the recessive variant to stay active. Thus, some females carrying a variant on the X chromosome could exhibit some features of a condition because of this skewed X inactivation (Section 2.7.2).

Mitochondrial

In addition to all of the genetic material on the chromosomes in every cell's nucleus (center of the cell), there is a small ring of DNA within all mitochondria (in the cell, but outside the nucleus), the energy producers of all cells, called mitochondrial DNA. This codes for the proteins used by the mitochondria for energy production. Mitochondrial DNA in humans is passed to children only through mothers via the egg because the mitochondria from sperm are lost at conception. Thus, variants in genes of the mitochondrial DNA are inherited only through the mother (maternally inherited) (Fig. 12.7). In addition, because each cell has hundreds to thousands of mitochondria, the amount of mitochondria with the inherited variant may vary, known as heteroplasmy. The higher the number of mitochondria with the variant determines whether the condition or symptom will present more severely in the child (Sections 2.7.4 and 11.6.5).

Somatic

Somatic cells are all the cells of the body except the sperm or egg cells. Changes in genes can happen after conception in some of the somatic cells in the body and would, in turn, never be inherited from a parent. This can occur due to many factors, including the environment or other external agents. These changes are called somatic variants and can happen any time after conception through adulthood. If somatic variants occur in an important gene that functions in a specific cell, they can lead to a condition related to that cell type. An example of this is when cancer can develop because of somatic variants within cells involved in the type of cancer. Somatic variants can be found in a breast cancer tumor, usually in a gene, which then can no longer prevent breast cancer. See also Section 12.4.3 below for a discussion of mosaicism (Fig. 12.8).

12.3 Counseling about Genetic Tests

Below are explanations for the most common general genetic tests ordered, and how to describe them at a basic level to a patient and family.

12.3.1 Karyotype

A karyotype is a picture of all of the chromosomes from one cell of an individual and it is organized into 23 pairs. The purpose of ordering a karyotype is to see how the chromosomes look, including the total number of chromosomes, if there are any major extra or missing pieces, or if their arrangement is abnormal (Sections 2.2 and 7.1). This is done usually on cells taken from a blood sample or from a prenatal specimen of fetal cells, following chorionic villus sampling (CVS) or amniocentesis (Section 4.4.2). The cells are grown in a lab, and the chromosomes are isolated, stained, and seen under a microscope. With a computer, chromosomes are photographed and organized by size, shape, and staining pattern in order to study them (Fig. 12.2). This is the most basic genetic test and generally has a turnaround time of about 1–2 weeks. For this reason, this may be performed before more in-depth testing (Section 6.1.1).

We expect to see 46 chromosomes in every cell. A variation seen in a karyotype does not always cause health problems. For example, translocations are changes in which pieces of, or sometimes entire, chromosomes are attached to or exchanged with another chromosome (Section 2.2.1). If a translocation does not involve a break or interruption in a gene or a loss/gain of DNA sequence, then it may not cause a problem if the total amount of material has remained the same. Missing or extra pieces of chromosomes usually have effects on how the body functions. Karyotypes may also be done to determine a person's sex chromosomes by seeing if they have two X chromosomes, or one X and one Y, or some other variation in number or loss/gain of pieces of X and/or Y chromosomes.

12.3.2 Fluorescence In Situ Hybridization

Fluorescence in situ hybridization, known as FISH, "fishes" to see if a specific chromosome or piece of a chromosome is lost, gained, or rearranged to a different location (Section 6.1.2). This is one of the fastest genetic tests and is done when a specific chromosomal condition is suspected, or to screen for commonly found extra chromosomes such as three chromosomes 21 in Down syndrome, or missing sex chromosomes on a prenatal specimen (Section 4.4.2). A fluorescent probe specific for the part of the chromosome being "fished" is added to the cells and will glow when it finds and attaches to that specific part of the chromosome. Each probe should attach to two chromosomes (or one for X and for Y if in a male) or each copy. If the probe does not glow as two spots or in the correct location, then this test can determine whether a chromosome or a piece of it is missing, extra, or in an abnormal position. This test takes a short amount of time because it does not involve growing cells (culturing) in the lab or looking at the DNA sequence. It is very specific and is often accompanied by performing a karyotype or other higher resolution tests that can give more detailed results.

12.3.3 Microarray

A microarray (Section 6.1.3) is a test that looks at chromosomes at a higher resolution than a karyotype. It uses a computer to see if there are very small extra or missing pieces of the chromosomes that may not be detected by a karyotype using a microscope. If the microarray detects a tiny deletion or duplication, that area can be analyzed further to see if there are genes in that region that are relevant to the symptom or condition of the patient. It often cannot check for a balanced rearrangement of chromosome pieces, which can be detected by a karyotype, so that a karyotype and microarray can be complementary tests and both may be necessary. Many findings on a microarray may not be important and might just be a benign change inherited from a parent who does not have the condition. Sometimes parental studies are recommended to check for this.

Stretches of DNA sequences that are the same on both copies of a chromosome are called areas of homozygosity and can be detected with a microarray (Sections 2.7.1, 6.1.3, 7.1.1 and 12.4.2). A patient with parents from common ancestry may inherit the identical stretch of chromosomes from each of them. With microarray testing, the patient is informed that results may reveal relatedness between parents.

12.3.4 Single Gene Analysis

This test reads the DNA of one gene, to see if there is a change that could be causing a symptom or a genetic condition (Fig. 12.3). Reading the DNA of this gene is known as sequencing, meaning the test is looking at the sequence of chemicals (represented by letters A, T, C, G) to see if they are in the expected order, with nothing added or removed or changed (Fig. 12.3). Generally, a test to sequence only one gene is ordered when a change in that gene is usually related to the patient's symptoms or disorder and most likely to give a positive result (Section 6.2.2).

If a relative has already had genetic testing and a family member is tested to see if the same change or variant has been inherited, it is called targeted variant testing, which is different from full gene sequencing (Section 6.2.1). Targeted variant testing is exactly as it sounds—the exact area in the gene where the variant was found in the relative is targeted for testing by the laboratory. This testing only finds out whether or not the family member has the same variant. It does not look for other variants in the rest of the gene. Targeted variant analysis can only have a positive or a negative result, while single gene sequencing can have a positive, negative, or variant of uncertain significance result (see below) (Sections 7.1 and 12.4).

12.3.5 Panel Testing

Many genes can be associated with the same symptom(s) or condition. A gene panel test sequences several genes (Fig. 12.3), each of which with a variant that could contribute to the presenting symptom or condition (Section 6.2.3). It does not always test every variant or gene known to cause the symptom or condition, but tests the genes believed to be the most likely to give a positive result. Generally, a panel is ordered when variants in several different genes can cause the same symptom or when it is uncertain exactly which disorder a patient could have causing that symptom. Panel tests also can be used to screen for common conditions prevalent in certain ancestral groups as well as for prenatal screening of treatable or severe conditions to facilitate family planning (Section 12.3.8). Multiple variants in several genes might be found in one panel. Again, there can be a positive, negative, or variant of uncertain significance result (see below) (Sections 7.1 and 12.4).

12.3.6 Exome and Genome Sequencing

These tests are generally the most extensive and usually in the final tier of tests that are offered in clinical genetics (Section 6.3). Our genetic information in total is called the genome, which is our entire DNA composed of more than 3 billion chemicals or, A, C, G, T letters, organized into sequences that code for more than 20,000 genes. The parts of genes that actually code for the proteins that do the work (exons) to make the body develop and function are referred to as the exome, which is only a small part of the genome, comprising about 1%. Exome sequencing reads the chemical sequence of all of the genes in the coding regions, looking for variants that seem related to the patient's symptoms or condition. Since this test sequences so much information, it generally takes a long time to perform and interpret—as long as a few months. Likewise, genome sequencing looks at the entire genome, including not only the exome but also introns (the DNA sequence in between coding regions of genes that can help regulate their function), and some labs include the mitochondrial genome sequence as part of the test (Section 12.2.2). This test takes even longer than exome sequencing. The lab that performs these tests looks at the variants (pathogenic, benign, and also those of uncertain significance) found during the sequencing and assesses the relevance of each one to the patient's symptoms or condition (Section 12.4.1). Often many variants are found, and the report can be very long (Section 7.1.1).

12.3.7 Metabolic Testing

Screening for metabolic disorders (inborn errors of metabolism) (Section 5.4) is often recommended in both pediatric and adult patients. Metabolic pathways in the body are carried out at every step by special proteins called enzymes. These pathways break down food and other substances into useful chemicals or metabolites that can then be used by the cells. Variants in genes responsible for making these enzymes can cause them to work not properly or at all. If one step of the pathway is blocked, the chemicals made up to that point of the pathway will start to accumulate, and those that should have been made after that point will be very low or completely absent. Metabolic tests look for levels of these chemicals in the blood, urine, or spinal fluid. Because most metabolic tests are screening tests, any positive result found will likely be followed by more specific and accurate diagnostic testing. Some of these tests are DNA based while others look at the responsible enzyme itself. This testing can be very important because some of these conditions are treatable, since many enzyme pathways can be diverted to make an alternate, unharmful product by using things like vitamins or other protein factors. Many of these tests are used for newborn screening. Most metabolic disorders are recessive, although a few are X-linked (Section 12.2.2).

12.3.8 Carrier Screening

Carrier screening is usually done to find out if an unaffected individual is carrying a recessive variant in a gene associated with a symptom or condition (Section 12.2.2, Fig. 12.5). If present in one copy in the parent, it means that the variant could be passed on to any future children 50% of the time. In this case, the individual's partner can be checked to see if he or she also is carrying a recessive variant in the same gene. If so, each future child would have a 25% chance to inherit both copies of the gene with the variants and thus have the condition or symptom. This type of testing tends to be done preconceptually or prenatally and is usually performed by panel testing (Section 12.3.5) or single gene sequencing of the DNA (Section 12.3.4).

12.3.9 Noninvasive Prenatal Screening

Noninvasive prenatal screening is done during pregnancy on the mother's blood. It can detect pieces of DNA present circulating in the blood from both the mother and the developing baby. It looks at large fragments of DNA and compares them to the expected ratio of DNA from the mother and baby. If the amount is abnormal, it could indicate an imbalance of genetic material carried by the baby. Because it is a screening test, if abnormal, prenatal diagnostic testing (Section 4.4.1) is often suggested to confirm the result. If the result is normal, there could still be a genetic condition that is not detected by this test.

12.4 Counseling about Results

Below are descriptions of the most common results from the above testing, and how they can be explained at a basic level to a patient or family.

12.4.1 General Results

There are three major types of results from a genetic test (Sections 7.1.3, 7.1.4 and 7.1.5). Variants or changes that are found have five different classifications: "pathogenic," "likely pathogenic," "variant of uncertain significance," "likely benign," and "benign." Pathogenic and likely pathogenic results are generally treated as positive, and benign and likely benign results are treated as negative. Positive results show that the genetic change found is known to cause or likely cause symptom(s) or condition. Negative results show that no genetic change was found or expected to have any significant effect on the functions of the body. It does not mean necessarily that a genetic cause has been completely ruled out for the patient's condition, just that one was not found. At this time there was nothing found in the medical literature or databases known to be informative or relevant to the condition.

There is also a variant of uncertain (unknown) significance known as a VUS, for short. This means that the genetic change found may not have been seen either before by any laboratory or found in a large group of unaffected and unrelated individuals. It is uncertain whether that change in the gene is responsible for a disruption in the protein function in the body to cause the symptom(s) or condition. When a VUS is found, the next step can be to test parents, if available, for the specific VUS (or targeted variant testing) to see if they carry it or not. If a parent has the variant and does not have the same symptom(s) as the patient, then the variant is unlikely to be causing the symptoms or condition. Thus, the VUS may be benign or likely benign, but this may not be a definitive answer since in some cases the parent may actually have a milder form of the condition that cannot be clinically diagnosed.

An inconclusive result means that the test that was performed did not yield a satisfactory or definitive result. This does not imply anything about whether there are any variants present, but it means that the clinical laboratory could not determine any result relevant to the original reason for the test. In the prenatal setting, it could mean that the initial sample did not yield enough cells to determine a definitive result, and additional testing may be indicated. In a cancer genetic test, it could mean that the sample showed no variant that could explain the cancer. In other situations, it could imply that the cells did not grow sufficiently to produce any result, and a repeat test may or may not be indicated.

Exome and genome sequencing will become more informative because more and more is being discovered about various genes and variants. For someone who already had his or her exome or genome sequenced, it can be reanalyzed later, when new medical information becomes available. VUSs may be changed to being either pathogenic or benign, over time.

12.4.2 Areas of Homozygosity and Uniparental Disomy

There are thousands of normal differences between everyone's DNA composed of chemicals or letters of a sequence, without affecting the function of the gene itself. Since everyone gets two sets of genetic information, one from each parent, each chromosome of a pair is slightly different from the other and each gene of a pair also is slightly different. We all carry changes, or variants, in our genes and many of them are unique to our families and us. Benign variation in DNA sequences cause differences in features like hair color or height. If both parents are from the same family or a small population with less genetic diversity, it is possible that they will both pass the same or a similar set of genetic information to their children. Within this common genetic information can be gene changes (variants, mutations).

Regions that have the same exact DNA sequence between two chromosomes are called areas of homozygosity, also called a loss of heterozygosity. Areas of homozygosity can be detected on a microarray (Sections 2.7.1, 6.1.3 and 7.1.1), since it looks at the chromosomes in close detail. Areas of homozygosity can be searched to see if there is one or more genes that could be related to the patient's symptoms or condition. The letters within the gene can then be sequenced to search for variants that are significant (Fig. 12.3). Typically, people will carry only one copy of a variant and would not manifest any symptoms or a condition, especially in a recessive condition (Section 12.2.2). If a variant occurs in an area of homozygosity, there are then two copies of the same gene containing the identical variant, and it is possible that it could be causing the symptoms (Section 2.7.1).

At times, areas of homozygosity can result when a chromosome is duplicated. Having two copies of a chromosome, or part of a chromosome, from only one parent is called uniparental disomy (Section 2.7.5). There are known parts of chromosomes which, when inherited only from one parent, can cause specific disorders (Sections 11.7.4 and 11.13.1). At times, uniparental disomy can produce different symptoms or conditions, depending on whether that area is from the mother or the father. There are other areas in which being inherited only from one parent does not cause a condition.

12.4.3 Mosaicism

A genetic test can show two different results and suggest something called mosaicism (Sections 2.2.2 and 2.4.2). In art, a mosaic is a work with pieces of different colors creating one image. Mosaicism in genetics implies something similar, with not every cell containing the same genetic information. This can happen early in development, in which a cell that divides has had an abnormal change, and all the cells that come from that cell have that same change (Fig. 12.8). The change may have less of an effect on the body than if present in every cell. An example would be having an extra chromosome, such as in Down syndrome, in some but not all of the cells in the body. Mosaicism also can be present only in certain parts of the body, or in specific organs, and it can be difficult to find exactly where the cells with the change are. We can also have something called gonadal mosaicism, in which sperms or eggs have a genetic variant not present in the rest of our body. This can lead to a parent passing on a dominant condition (Section 12.2.2), even though the parent has no signs of this condition.

12.4.4 Translocations

A translocation can be seen on a karyotype (Sections 2.2.1 and 12.3.1). This means that a part of one chromosome was broken off and attached to another one. A translocation can be inherited from a parent or happen after the sperm and egg first combined genetic information to form the first cell of the body at conception.

Balanced Translocation

A balanced translocation seen on the karyotype means that chromosomes, or pieces of chromosomes, are rearranged, but there is the same total amount of genetic information, with nothing extra or missing (Section 2.2.1). This can be important in the prenatal setting when, in some cases, a chromosome containing a balanced translocation from the parent is passed to a child resulting in an imbalance or unbalanced translocation. Sometimes a balanced translocation is found by chance and is not related to the symptom or condition for which the karyotype test was ordered in the first place. If unsure of the clinical significance of a balanced translocation, testing can be done to see if any of the initial or reattachment breaks in the chromosomes disrupt any genes from functioning. A microarray test may help to determine this (Section 12.3.3).

Unbalanced Translocation

An unbalanced translocation seen on the karyotype means that chromosomes, or pieces of chromosomes, are rearranged, and there is missing or extra genetic information (Section 2.2.1). Sometimes one of the parents could carry a translocation involving the same chromosomes that actually is balanced. Their sperm or egg, which formed the cell of the future baby, contained only part of the involved chromosomes, and since the other parent gave the expected amount of chromosomal material, there is too much or too little. We always recommend testing each parent to see if either carries the balanced form of the unbalanced translocation found in the patient. If the parent does not have the symptoms and has the same imbalance, then the translocation is unlikely to have caused the condition. If the parent has similar symptoms and has the same imbalance, then the translocation is more likely to be causing the condition.

12.4.5 Deletions and Duplications

Monosomy

Monosomy indicates that only a single chromosome instead of the normal pair of chromosomes is present in every cell. On karyotype analysis, 45 chromosomes are found rather than the normal amount of 46 chromosomes. It is important for us to have 46 chromosomes in every cell, to create the proteins needed for our bodies. Missing a whole chromosome usually affects the body quite seriously. A single X chromosome, whether present in every cell or mosaic (Section 12.4.3), leads to Turner syndrome, generally the only monosomy condition able to survive after birth (Section 11.7.5).

Trisomy

Trisomy indicates that three chromosomes instead of the normal pair of chromosomes, one chromosome from each parent, is present in every cell. On karyotype analysis, 47 chromosomes are found rather than the normal amount of 46 chromosomes. It is important for our bodies to have 46 chromosomes to create the normal amount of proteins. Having an extra chromosome gives the body extra information to process and can affect how the body functions.

Major Deletion

The karyotype or microarray shows a deletion of part of a chromosome. Missing many genes in this region affects several systems of the body. This may explain many of the presenting symptoms. Unaffected parents do not have the deletion.

Major Duplication

The karyotype or microarray shows a duplication of part of a chromosome. Having too much of many genes in this region affects several systems of the body. This may explain many of the presenting symptoms. Unaffected parents do not have the duplication. Duplications can be less severe than deletions.

Microdeletion

The microarray shows a small deletion in one chromosome. This area is known to contain certain gene(s) whose functions can be impacted by this loss. We may recommend testing to determine if either of the parents carries this same deletion. If a parent has the same small deletion and does not have the same presenting symptoms or condition, then it is unlikely that this small deletion caused them. If neither parent has the same small deletion, it is a new change in the patient and more likely to be related to the symptoms. If a similar small deletion has been reported in the medical literature in other unrelated patients with similar symptoms (Sections 11.2.4, 11.4.5 and 11.13.5), it is most likely to be the cause. If there are no reported cases with similar symptoms, then it is uncertain if the deletion is the cause.

Microduplication

The microarray shows a small duplication in one chromosome. This area is known to contain certain gene(s) whose functions can be impacted by this extra information. We may recommend testing if either of the parents carries this same duplication. If a parent has the same small duplication and does not have the same presenting symptoms or condition, then it is unlikely that this small duplication caused them. If neither parent has the same small duplication, it is a new change in the patient and more likely to be important. If a similar small duplication has been reported in the medical literature in other unrelated patients with similar symptoms, it is most likely to be the cause. If there are no reported cases with similar symptoms, then it is uncertain if the duplication is the cause.

12.4.6 Triplet Repeats

Some genes contain repeated sequences of the chemicals (represented by letters A, T, C, and G) that make up the DNA. Often there are three chemicals repeated over and over, called triple or triplet repeats (Section 2.4.1). Certain conditions are caused by having too many triplet repeats in those genes (Sections 11.12.3, 11.13.10, 11.13.11 and 11.13.12). Genetic testing can count the number of triplet repeats present. In many cases, if a parent carries more than the typical number of repeats in a specific gene, that number can increase or expand when the gene is passed on to their children. These parents are called carriers and the increased number of repeats in that gene is called a premutation. If the number of repeats increases to more than a certain number in the egg or sperm before conception, the child will have a "full expansion" of the triplet repeat and thus have the condition. Many of the conditions caused by this phenomenon have neurologic involvement.

12.5 Concluding Remarks

The entire counseling session can be complex in content and emotionally charged. Delivering important information to the patient and its implications to the entire family must be done accurately and carefully by explaining the benefits as well as the limitations of testing and interpretation of the results. Often parents blame themselves for their children having inherited conditions they have, or changes that they carry. We try to reassure families, particularly in the case of a de novo (new) change, that it is not the parents' fault and they are not responsible for their child having a condition. These changes happen by chance, and sometimes are not compatible with life. Most importantly, genetic results from the patient may lead to contacting other family members to inform them of the results, and in turn, they may decide to have testing as well, known as cascade testing. If counseling is done well, genetic counselors and geneticists have a positive and impactful relationship with grateful families, who are guided to appropriate and expert management, support groups, clinical trials, and other relevant research programs (Section 7.4).

12.6 Visual Aids to Explain Genetic Concepts to Families

The diagrams below can be used to illustrate basic genetic concepts to families. Compendia of additional aids are available through MedlinePlus of National Library of Medicine; National Institute of Health, National Human Genome Research Institute (https://www.genome.gov/health/For-Health-Professionals); Greenwood Genetic Center Visual Aids (https://www.ggc.org/publications).

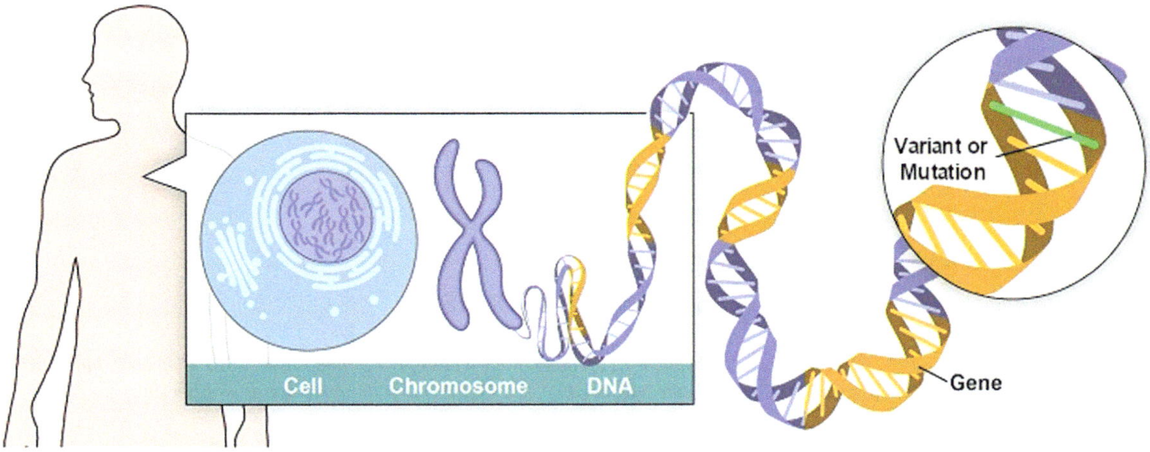

FIGURE 12.1 Cell, chromosome, DNA, gene, variant, or mutation. *Source: National Institute of Allergy and Infectious Diseases (NIAID). Courtesy: National Institute of Allergy and Infectious Diseases. Modified from original.*

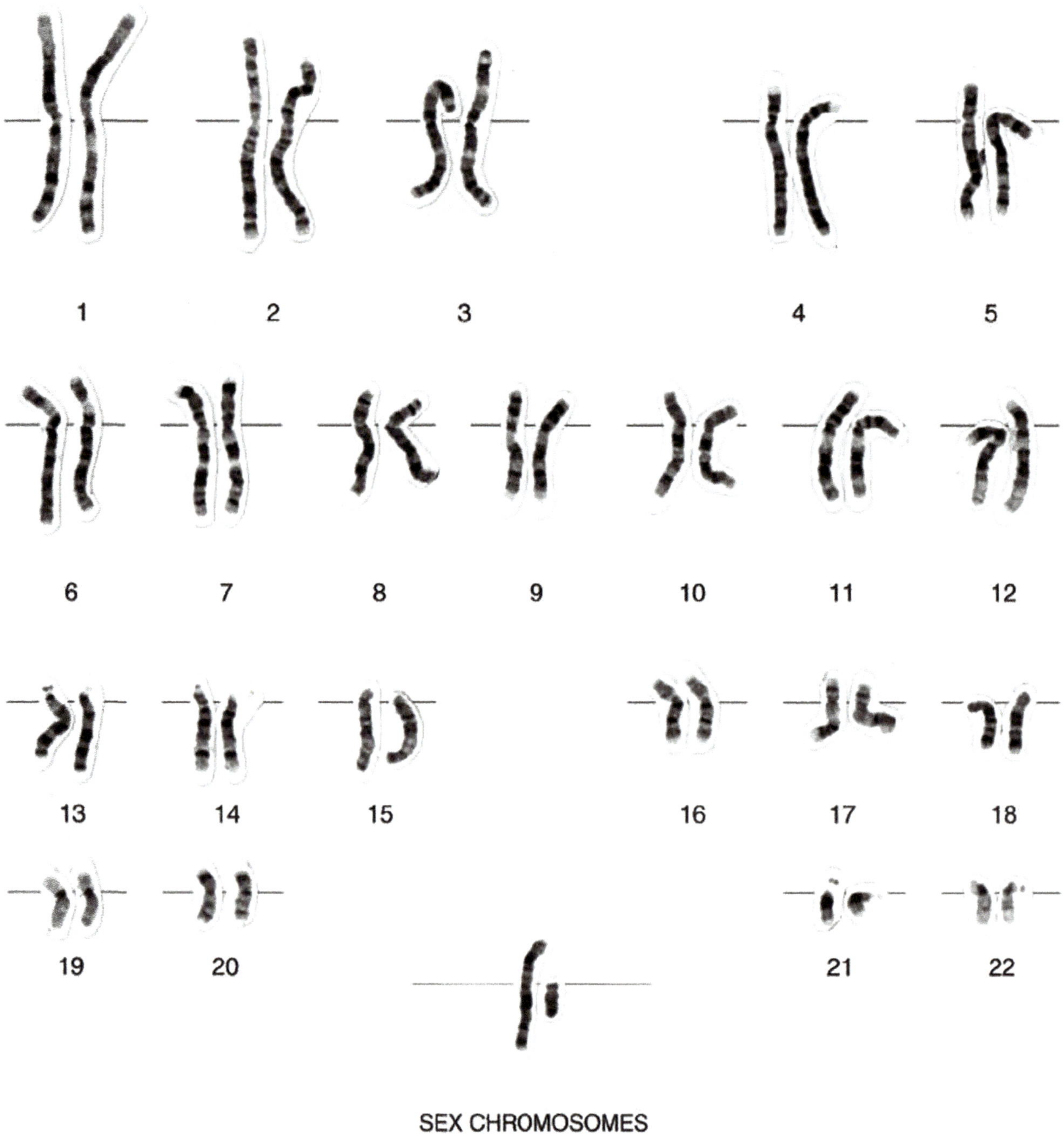

FIGURE 12.2 Chromosomes. *Source: Reprinted from Nussbaum, R.L., McInnes, R.R., Willard, H.F., Hamosh, A., 2016. Chapter 2, Introduction to the human genome. In: Thompson & Thompson Genetics in Medicine, eighth ed., p. 15, Fig. 2–11, with permission from Elsevier.*

	Substitution	Deletion	Insertion
Common Sequence	TGGGTTCTC	TGG~~GTT~~CTC	TGGGTTCTC
Variant Sequence	TGGGTTATC	TGGCTC	TGGGTTAAACTC

FIGURE 12.3 Types of variants.

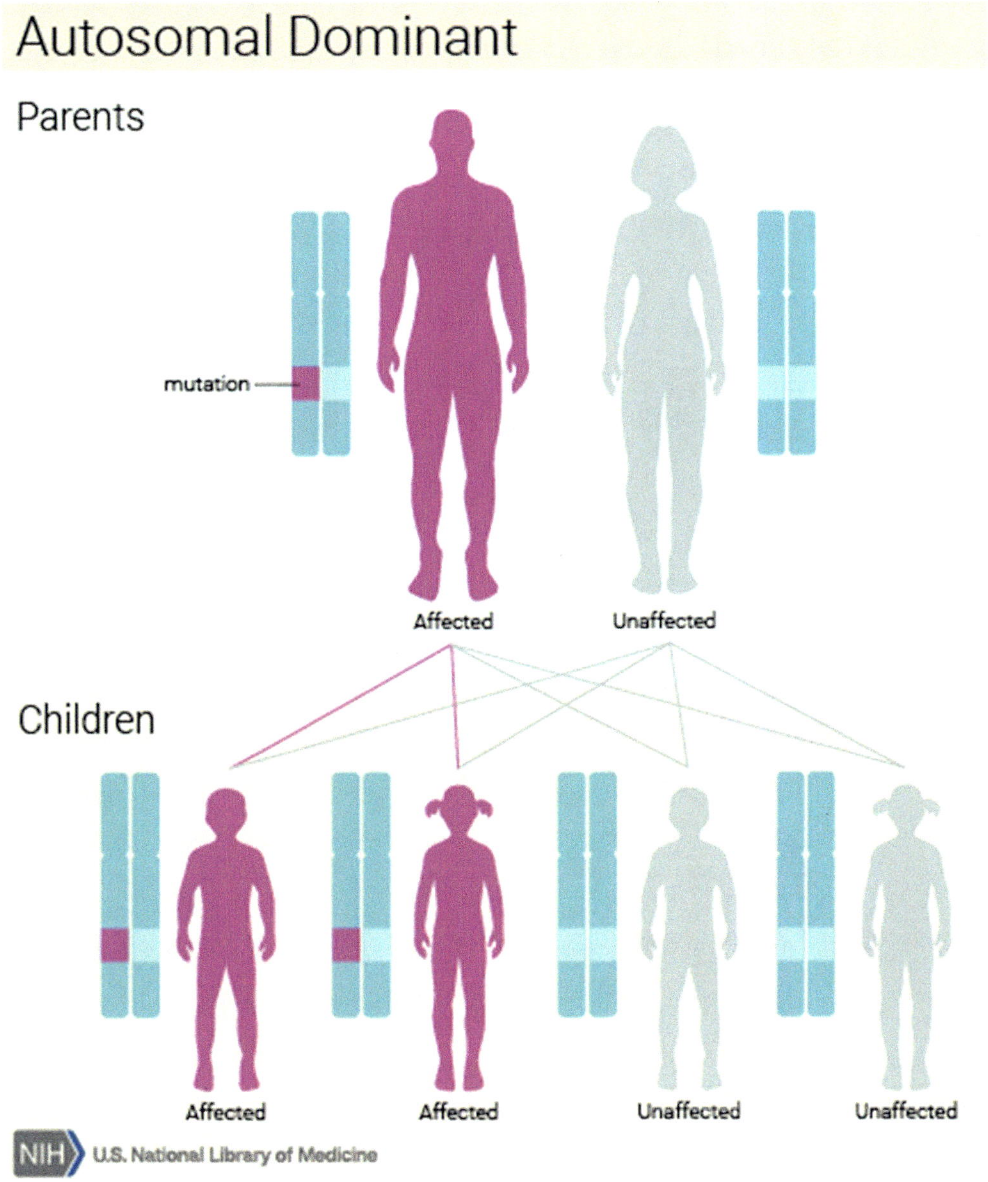

FIGURE 12.4 Autosomal dominant inheritance pattern. *Source: MedlinePlus, National Library of Medicine.MedlinePlus brings together authoritative health information from the National Library of Medicine (NLM), the National Institutes of Health (NIH), and other government agencies and health-related organizations.*

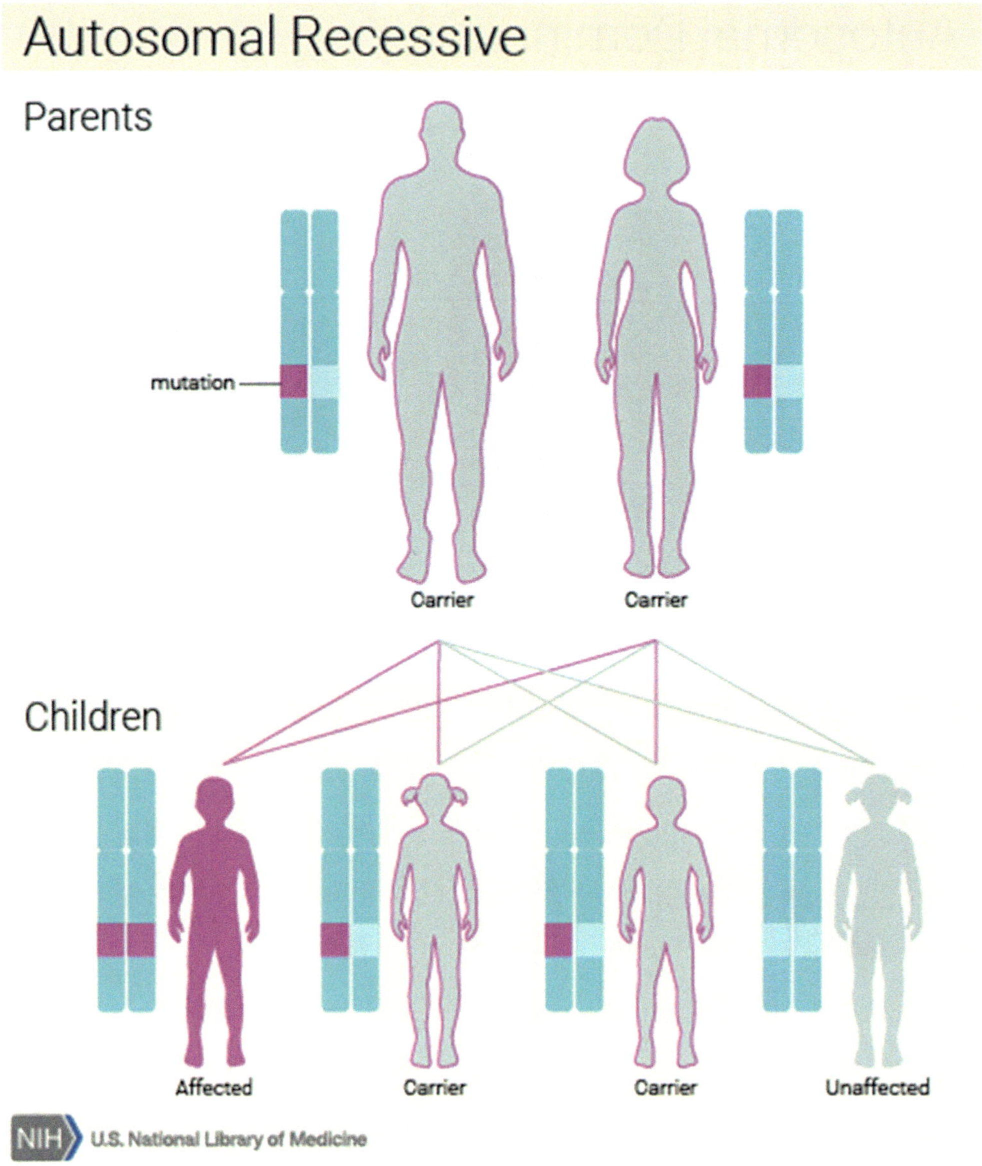

FIGURE 12.5 Autosomal recessive inheritance pattern. *Source: MedlinePlus, National Library of Medicine. MedlinePlus brings together authoritative health information from the National Library of Medicine (NLM), the National Institutes of Health (NIH), and other government agencies and health-related organizations.*

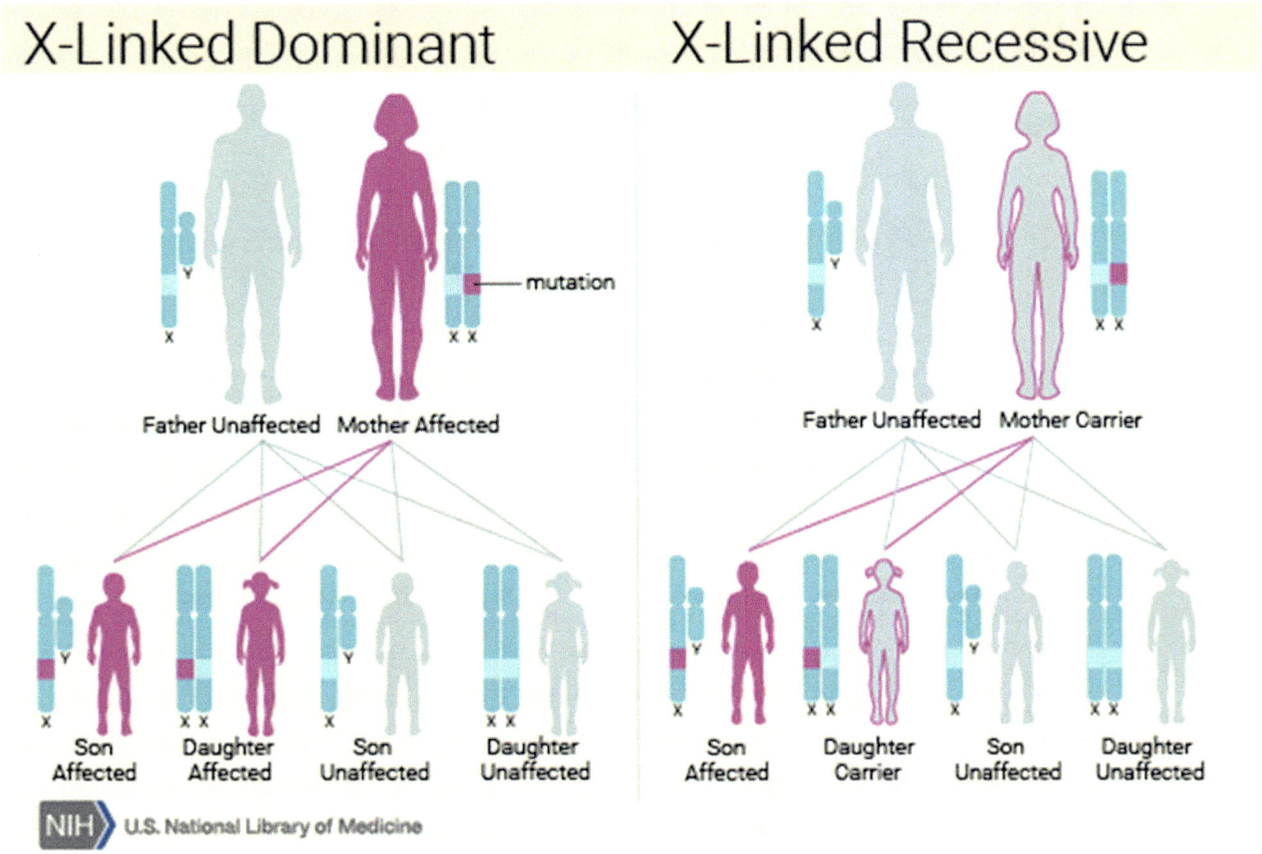

FIGURE 12.6 X-linked dominant and recessive inheritance patterns. *Source: MedlinePlus, National Library of Medicine. Composite adapted from original images. MedlinePlus brings together authoritative health information from the National Library of Medicine (NLM), the National Institutes of Health (NIH), and other government agencies and health-related organizations.*

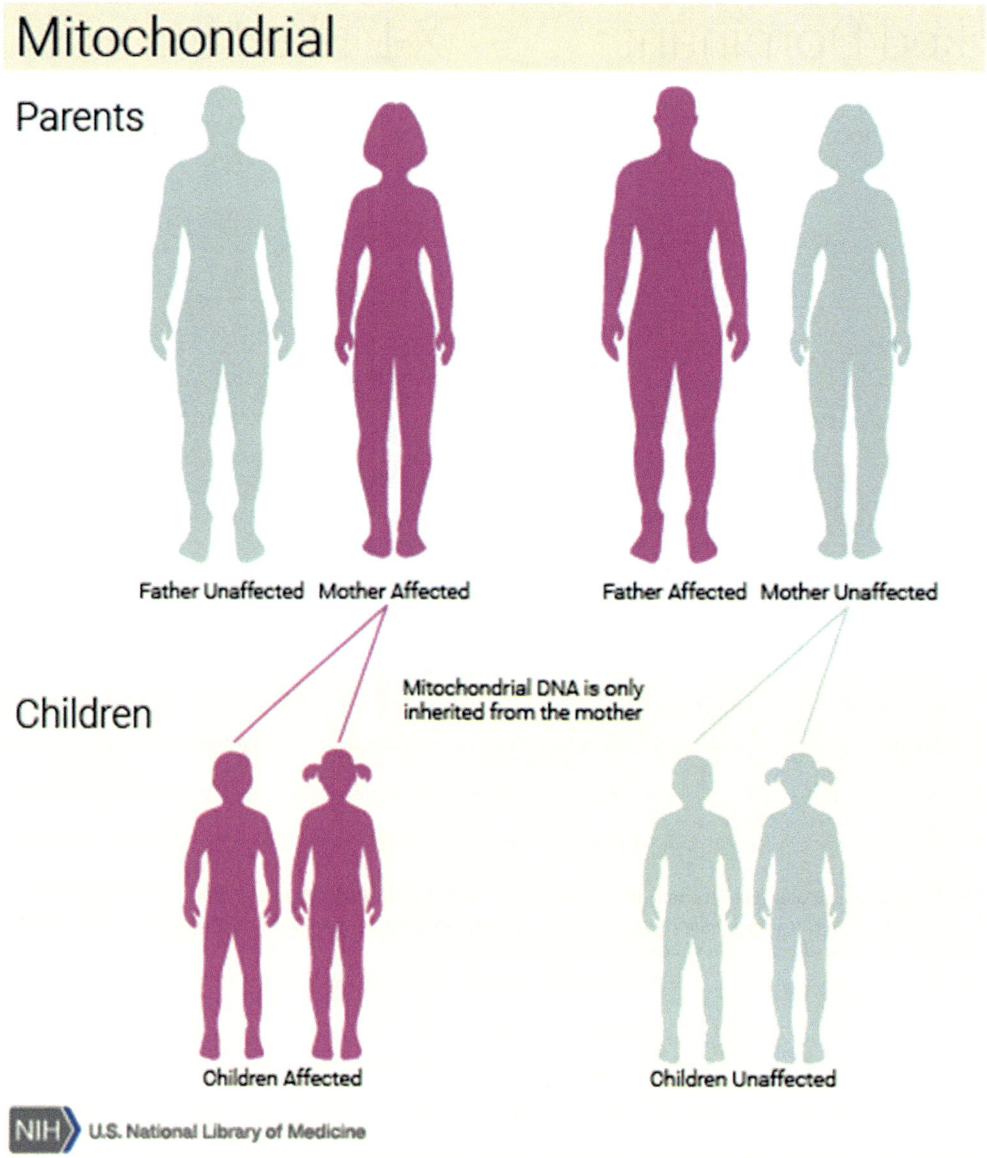

FIGURE 12.7 Mitochondrial inheritance pattern. *Source: MedlinePlus, National Library of Medicine. MedlinePlus brings together authoritative health information from the National Library of Medicine (NLM), the National Institutes of Health (NIH), and other government agencies and health-related organizations.*

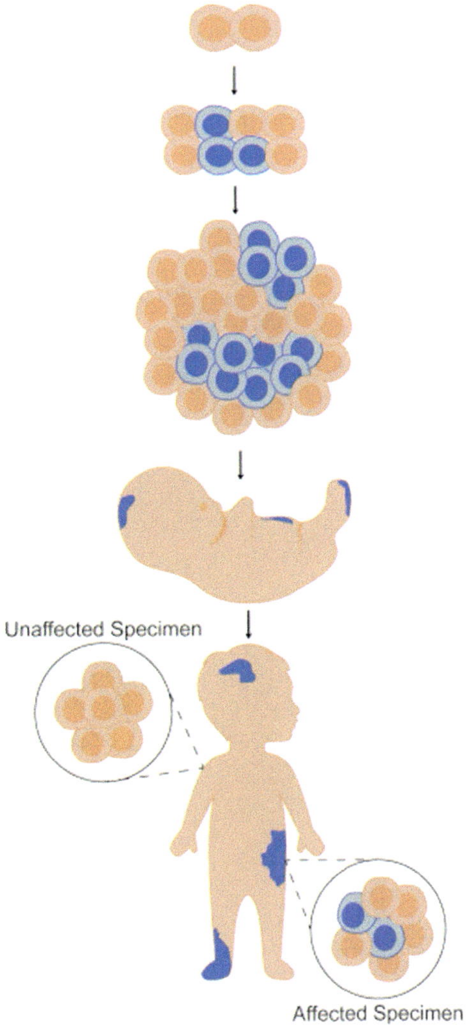

FIGURE 12.8 Mosaicism. *Source: McNulty, S.N., Evenson, M.J., Corliss, M.M., et al. 2019. Diagnostic utility of next-generation sequencing for disorders of somatic mosaicism: a five-year cumulative cohort. Am. J. Hum. Genet. 105, 734–746.*

Index

'*Note:* Page numbers followed by "f" indicate figures and "t" indicate tables.'

9780128164785